Ideas and Practices, 1300–1850

Volume 1

Cotton Mather's *Curiosa Americana*, Scientific Letters to the Royal Society

Ideas and Practices, 1300–1850

Ideas and Practices, 1300–1850, focuses on the era in which long-familiar ways of thinking about politics, religion, society and the natural world transformed fundamentally. The series offers a venue for scholars to publish ambitious works in intellectual history which explain – or question – that transformation, sometimes characterised as 'the crisis of the European mind'. Its coverage stretches from the medieval period into the nineteenth century, so that different perspectives and explanations can be brought to bear on this pivotal moment in the history of Western intellectual life.

The series encourages submissions of monographs and coherent essay collections which explore political thought, political economy, theology and religious belief, natural philosophy, scholarship and literature in Europe and across the Mediterranean and Atlantic worlds. It will also publish scholarly editions of important texts not readily available.

We are happy to consider both conventional and Open Access models of book publishing. Further information about the series and guidance on submitting a proposal can be found on the Durham University IMEMS Press website.

Other books in the series may be viewed on the Boydell & Brewer website.

Cotton Mather's *Curiosa Americana*, Scientific Letters to the Royal Society

A Critical Edition

Edited with an Introduction by
Reiner Smolinski, Kenneth P. Minkema, and
Andrew J. Juchno

Durham University
IMEMS PRESS

First published 2025

A Durham University IMEMS Press publication
in association with
The Boydell Press
an imprint of Boydell & Brewer Ltd
PO Box 9, Woodbridge, Suffolk IP12 3DF, UK
and of Boydell & Brewer Inc.
668 Mt Hope Avenue, Rochester, NY 14620–2731, USA
website: www.boydellandbrewer.com
and with the
Institute of Medieval and Early Modern Studies
University of Durham

Any permissions requests, including translation requests, should be directed to
Boydell & Brewer at the address above

Our Authorised Representative for product safety in the EU is Easy
Access System Europe – Mustamäe tee 50, 10621 Tallinn,
Estonia, *gpsr.requests@easproject.com*

ISBN 978-1-914967-07-8

A CIP catalogue record for this book is available
from the British Library

Printed and bound in Great Britain by
TJ Books, Padstow, Cornwall

Contents

"Curiosa Americana"

Illustrations

The editors, contributors and publisher are grateful to all the institutions and persons listed for permission to reproduce the materials in which they hold copyright. Every effort has been made to trace the copyright holders; apologies are offered for any omission, and the publisher will be pleased to add any necessary acknowledgement in subsequent editions.

Abbreviations

A	*Angelographia* (1696)
AAS	American Antiquarian Society (Worcester, Mass).
AB	*Angel of Bethesda* (1722)
"AB"	"Angel of Bethesda" (Holograph Manuscript. 1693–1724)
ABD	*Anchor Bible Dictionary*
ABJ	*Angel of Bethesda* (1972)
AC	*Armour of Christianity* (1704)
ACPQ	*American Catholic Philosophical Quarterly*
ADB	*Allgemeine Deutsche Biographie*
AHMD	*American Heritage Medical Dictionary*
AHPB	*Annals of the History and Philosophy of Biology*
AL	*American Literature*
AMR	*American Marriage Records Before 1699*
AMSI	*An Account of the Method and Success of Inoculating the Small-Pox* (1722)
ANF	*Ante-Nicene Fathers.* 10 vols
ARCS	*Annals of the Royal College of Surgeons of England*
ARH	*Agricola. Or, The Religious Husbandman* (1727)
AS	*Annals of Science*
B	*Bonifacius* (1710)
"BA"	"Biblia Americana." Holograph Manuscript, 1693–1728 (MHS)
BA	*Biblia Americana.*
BE	*Boston Ephemeris* (1683)
BHM	*Bulletin of the History of Medicine*
BJCP	*British Journal of Clinical Pharmacology*
BJHS	*British Journal of the History of Science*
BM	British Museum (London, UK)
BL	British Library (London, UK)

BO	*Boanerges* (1727)
"BO"	"Boanerges, or Work of the Day" (c. 1716)
BS	*Brontologia Sacra* (1695)
BU	*Blessed Unions* (1692)
C	*Columbanus* (1722)
CA	"Curiosa Americana" Holograph letters at MHS, RSL, BM
CC	*A Christian at his Calling* (1701)
CCF	*Compassions Called for* (1711)
CE	*Catholic Encyclopedia*
CGR	*Christodolus. A Good Reward* (1725)
CH	*Church History*
CJCJ	*The Correspondence of John Cotton Junior* (McIntyre & Travers 2009)
CM	*Clio Medica. Acta Academiae Internationalis Historiae Medicinae*
CMHS	*Collections of the Massachusetts Historical Society*
CNH	*Collections of the New Hampshire Historical Society*
CP	*Christian Philosopher* (1721)
CPI	*Christianus per Ignem* (1702)
CPS	*Christian Philosopher* (Solberg, 1994)
CR	*Circulation Research*
D 1-2	*Diary of Cotton Mather* (1911–12)
D 3	*Diary of Cotton Mather for the Year 1712* (1964)
DI	*Diluvium Ignis* (1726, 1730)
DL	*Decennium Luctuosum* (1699)
EAL	*Early American Literature*
EC	*Eighteenth Century*
EID	*Emerging Infectious Diseases*
EMJ	*Erciyes Medical Journal*
Ephemerides	*Miscellanea Curiosa sive Ephemeridum*
ERIP	*Essay for the Recording of Illustrious Providences* (1684), by Increase Mather
F.R.S.	Fellow of the Royal Society
HB	*Human Biology*
HH	*Hor-Hagidgad* (1727)

HJ	*The Historical Journal*
HM	*Hatzar-Maveth* (1726)
HTR	*Harvard Theological Review*
HULB	*Harvard University Library Bulletin*
JAF	*Journal of American Folklore*
JAMA	*Journal of the American Medical Association*
JEE	*Journal of Ethnobiology and Ethnomedicine*
JH	*Journal of Heredity*
JHB	*Journal of the History of Biology*
JHD	*Journal of the History of Dentistry*
JHI	*Journal of the History of Ideas*
JHM	*Journal of Herbal Medicine*
JHMAS	*Journal of the History of Medicine and Allied Sciences*
JINA	*Journal of Integrative Nephrology and Andrology*
JMB	*Journal of Medical Biography*
JRSM	*Journal of the Royal Society Medicine*
JSH	*Journal of Social History*
KJV	King James Version (1611)
LCL	Loeb Classical Library
LCM	*Life of … Cotton Mather* (1729)
LDE	*Life and Death of Mr. John Eliot* (1691)
LDRM	*Life and Death of … Richard Mather* (1670)
LJS	*Journal des Sçavans*
LM	*Literature and Medicine*
LMDM	*Letter, About a Good Management under the Distemper of the Measles* (1713)
LT	*The Life and Times of Cotton Mather* (Silverman, 1984)
LXX	Septuaginta
MAM	*Manuductio ad Ministerium* (1726)
MCA	*Magnalia Christi Americana* (1702)
MEFA	*Man Eating the Food of Angels* (1710)
MH	*Medical History*
MHR	*Massachusetts Historical Review*
MHS	Massachusetts Historical Society (Boston, Mass.)
MIH	*Modern Intellectual History*

Miller-Keane	*Miller-Keane Encyclopedia & Dictionary of Medicine*
MM	*Massachusetts Magazine: or Monthly Museum*
MP	*Mather Papers* (1868)
MPL	*Late Memorable Providences Relating to Witchcrafts and Possessions* (1691)
MPW	*Memorable Providences, Relating to Witchcrafts and Possessions* (1689)
MS	*Mediterranean Studies*
MVHR	*Mississippi Valley Historical Review*
NC	*The Negro Christianized* (1706)
NEJM	*New England Journal of Medicine*
NEQ	*New England Quarterly*
NN	*Northeastern Naturalist*
NO	*New Offer to the Lovers of Religion and Learning* (c. 1714)
NPNFi	*Nicene and Post-Nicene Fathers* (First Series). 14 vols
NPNFii	*Nicene and Post-Nicene Fathers* (Second Series). 14 vols
NRRS	*Notes and Records of the Royal Society of London*
OP	*Ornamental Piety* (1707)
OT	*Observable Things* (1699)
P	*Parentator* (1724)
PA	*Paterna* (1976)
PAAS	*Proceeding of the American Antiquarian Society*
PAMA	*Pennsylvania History: A Journal of Mid-Atlantic Studies*
PAPS	*Proceedings of the American Philosophical Society*
Pauly	*Paulys Real-Encyclopedie der Classischen Altertumswissenschaft*. 49 vols
PCSM	*Publications of the Colonial Society of Massachusetts*
PG	*Patrologia Graeca*
PL	*Patrologia Latina*
PM	*Psychological Medicine*
PP	*Pietas in Patriam* (1697)
P&P	*Past & Present*
PMHS	*Proceedings of the Massachusetts Historical Society*
"PrT"	"Problema Theologicum" (c. 1703; 1994)
PS	*Psalterium Americanum* (1717)

PT	*Philosophical Transaction of the Royal Society*
OED	*Oxford English Dictionary*
RES	*Review of English Studies*
RQ	*Renaissance Quarterly*
RSL	Royal Society of London
RSS	Royal Society of Surgeons (London)
RWSV	*The Right Way to Shake off a Viper* (1711, 1720)
SAIT	*Some Account Of what is said of Inoculating or Transplanting* (1721)
SC	*Science in Context*
SCM	*The Stone Cut out of the Mountain* (1716)
SLCM	*Selected Letters of Cotton Mather* (Silverman, 1971)
SP	*Science in Progress*
SPCK	*Society for Promoting Christian Knowledge*
SSM	*Seasonable Thoughts on Mortality* (1712)
TAGU	*Transactions American Geophysical Union*
TC	*The Thankful Christian* (1717)
TCSM	*Transactions of the Colonial Society of Massachusetts*
TD	*Terribilia Dei* (1697)
TDP	*Things for a Distress'd People to Think upon* (1699)
TDR	*Thoughts for the Day of Rain* (1712)
TFOV	*Treacle Fetch'd Out of a Viper* (1707)
TIC	*Thirty Important Cases* (1699)
TOJBU	*Targums of Onkelos and Jonathan Ben Uzziel* (Etheridge)
TP	"Triparadisus." *The Threefold Paradise of Cotton Mather* (1995)
TRR	*Triumphs of Reformed Religion* (1691)
TT	*Triumphs over Troubles* (1701)
V	*Victorina* (1717)
VGT	*Voice of God in a Tempest* (1723)
VH	*A Voice from Heaven* (1719)
VT	*Vanishing Things* (1718)
VUL	Vulgata
WA	*Work upon the Ark* (1689)
WASR	*The World Alarm'd. A Surprizing Relation* (1721)

WFD	*Warnings from the Dead* (1693)
WFS	*Will of the Father submitted to* (1713)
WIW	*Wonders of the Invisible World* (1693)
WJM	*Western Journal of Medicine*
WM	*Winter-Meditations* (1693)
WMQ	*William and Mary Quarterly*
WP	*Winter Piety* (1712)
WWG	*Wonderful Works of God Commemorated* (1690)

Acknowledgements

We wish to thank the Royal Society of London, the British Library, the Wellcome Library, the Massachusetts Historical Society, and the American Antiquarian Society, especially Brianne Barrett, for permission to edit and publish Cotton Mather's "Curiosa Americana" letters. Most of all, we gratefully acknowledge Dr. David Lupher, Emeritus Professor of Greek, Latin, and Ancient Mediterranean Studies, University of Puget Sound, for translating Mather's Latin correspondence and for providing helpful annotations. Thank you, David.

Published with assistance from the Anneliese John Daehne und Horst Pfeiffer Smolinski Memorial Fund in memory of Otto Dietrich of Querfurt (KZ Sachsenhausen).

Curiosa Americana: General Introduction

> Your excellent *Essay towards The Natural History of the Earth*, ha's obliged and even commanded, the true Friends of *Religion*, and *Philosophy*, to serve you with as many Communications as they can, that may be subservient unto your Noble Intention. But the *Letters* wherewith you have honoured me, have laid *me* under your more particular Commands, to supply you with such *Subterraneous Curiosities*, as may have been in these Parts of *America* mett withal. I do with much Alacrity apply myself immediately to obey your Commands, in one Remarkable Instance, wherein I apprehend myself best able to do it; reserving for another Opportunity (if perhaps, I find myself a Master of) any further Treasures, worthy to be laid before a Person of so clear a Judgment, & so reaching a Sagacity. And that I go *cheerfully* about this Point of my Obedience, you will conclude it, when you see me begin with assuming the Air of telling you, That you must look upon all the *American Curiosities* which are sent you.[1]

So began Cotton Mather's "Curiosa Americana" in 1712, a rich collection of epistolary essays on early modern science and the American natural world. One of only eight colonial fellows of the Royal Society of London, Mather obliged Dr. John Woodward's request over the span of many years. As secretary of the Royal Society, geologist, and member of the Royal College of Physicians, Woodward (1665–1728) particularly welcomed Mather's far-ranging Curiosa on the flora and fauna of the New World – including medicinal plants, passenger pigeons, rattlesnakes, fossils, and moose – as well as remarkable phenomena like monstrous births, Parhelia, and hurricanes. They also included Mather's commentary on pressing cultural issues such as matters of colonial identity, race, and slavery, which Mather invariably

1 In "Cotton Mather to John Woodward, November 17, 1712" (EL/M2/21), Royal Society of London, London, UK. A transcript of the letter is found in the Frederick Lewis Gay manuscript, pp. 1–27, and in MS 3339, fols 99ff, Sloane Papers, British Museum, London, UK. Curiosa letters will be referenced as CA followed by their series and item numbers. A complete list of the letters and their archival locations is given in Appendix 1.

blended with natural philosophic observations to create an invaluable record of both rarified intellectual life and the mundane world.

The Reverend Cotton Mather, D.D. F.R.S (1663–1728) is well known in the annals of American history. Grandson of Richard Mather (1596–1669) and John Cotton (1585–1652) – two of the most prominent spiritual founders of the Massachusetts Bay Colony – and son of Increase Mather (1639–1723), influential Puritan clergyman in Boston, president of Harvard College, and the colony's principal delegate to renegotiate Massachusetts' Charter privileges during the Glorious Revolution of 1688–89, Cotton Mather was heir to the religious and political aspirations of Puritan New England at the time of Boston's transformation from a City upon the Hill to the commercial center of the British Empire in North America.[2] He was educated at Harvard, pastored Boston's Second Church (Congregational) for more than forty years, received an honorary Doctor of Divinity degree from the University of Glasgow, became an elected member of the Royal Society, corresponded with important intellectual leaders on three continents, and published more than 450 works on all aspects of theology, history, social reform, natural science, and medicine. If Mather's voluminous output and intellectual achievements mark him as America's first polymath, he remains a controversial representative of Puritan hegemony in American history: His reputation is still overshadowed by the tragedy of Salem witchcraft (1692–93),[3] a truly minor event when compared to the excesses of European witchcraft in the sixteenth and seventeenth centuries.[4] Nonetheless, Mather's indirect involvement in the Salem trials cannot be denied, for his most controversial book on the subject, *The Wonders of the Invisible World* (1693), ambidextrously criticized the spectral evidence admitted at the Court of Oyer and Terminer as proof of guilt but defended the judges for condemning nineteen of the accused to death by hanging.

On a personal level, Mather was a celebrated preacher, delightful conversationalist, and loving father of fifteen children (of whom only two survived him).[5] Many of his most significant works are still reprinted, read (and reread) today, and several that remained in manuscript at the time of his death in 1728 are now being published for the first time. His contributions to the dominant academic disciplines of his time are now much better understood than they were less than half a century ago. This new appreciation is fostered not only by the new evenhanded biographies published since the

[2] See M. Peterson (2011) and (2019), esp. ch. 3–4.

[3] E.B. Holifield (2010).

[4] P.T. Leeson and J.W. Russ (2017).

[5] K. Silverman (1984), R. Kennedy (2015).

1970s but also by a number of modern scholarly editions of some previously published and unpublished works.[6] They include *Selected Letters* (1971), *The Angel of Bethesda* (1972), *Paterna* (1976), *Magnalia Christi Americana* (1977), *The Christian Philosopher* (1994), "Problema Theologicum" (1994), *The Threefold Paradise* (1995), *Biblia Americana* (2010–26),[7] and his as yet unpublished "Curiosa Americana." It has long been recognized that Mather's interest in natural philosophy was considerably more than a mere avocation;[8] while preparing for the ministry as a student at Harvard, he was equally devoted to the study of the natural sciences, particularly medicine, which he planned to pursue as an alternative calling if his speech impediment were to incapacitate him as a public preacher in the ministry. This dual focus on divinity and the sciences is most apparent in his medical handbook *The Angel of Bethesda* (the first book on diseases and their cures written by a colonial American of European descent) and in his *Christian Philosopher*, a hands-on celebration of the new discoveries in astronomy, Newtonian Science, biology, botany, mineralogy, zoology, and anthropology – all for the glory of God, but also to see "Good Done."

It comes as no surprise that historians who examined Mather's writings on natural philosophy have primarily turned to *The Christian Philosopher* (London, 1721), a manual of the best knowledge on the natural sciences whose publication was sponsored by the Royal Society. It is a richly rewarding work in which he harmonized the dominant scientific theories of the early Enlightenment with his theological purview. More to the point, Mather's grasp of the scientific currents, especially in medicine, is particularly evident in his "Curiosa Americana." They are largely on par with those published during his lifetime in the early issues of the *Philosophical Transactions* of the Royal Society (1665–), the German *Miscellanea Curiosa Medico-Physica Academiae Naturae Curiosorum* (1670–), and even the French *Journal des Sçavans* (1665–). It is therefore astounding that his Curiosa letters have received sparse scholarly attention. This neglect is unfortunate but can be easily explained: The manuscripts of the Curiosa essays are dispersed in the archives on both sides of the Atlantic – the Royal Society of London, the British Library, the Massachusetts Historical Society, and the American Antiquarian Society –

6 R. Middlekauff (1971), D. Levin (1978), K. Silverman (1984), and R. Kennedy (2015).

7 K. Silverman (1971), G.W. Jones (1972), K.B. Murdock (1977), W.U. Solberg (1994), J. Mares (1994), R. Smolinski (1995), R. Smolinski and J. Stievermann (2010–26).

8 See especially, O.T. Beall, Jr. and R.H. Shryock (1954, 1979); G.W. Jones, ed. *ABJ* (1972); and P.A. Watson (1991).

and have never been collected as a whole, let alone edited and annotated, or made available in a scholarly edition. Mather's "Curiosa Americana" essays – some of which he read as a young man to the short-lived Boston Philosophical Society (1683–87)[9] – are as wide-ranging in scope as they are novel in their focus on things uniquely American. They shed new light on the American flora and fauna, Native American medicines, astronomical observations, the new epidemiology practiced in Boston, and much more. As well, they attest to the flourishing of scientific pursuits among colonists.

Mather's Curiosa and their variegated contents place him and his network of colonial contributors at the center of knowledge production in the British Atlantic during the early Enlightenment.[10] That position was not without its perils and pitfalls. In the eyes of the Royal Society, Mather and his cohort were at once valuable witnesses to the American natural world, with intimate knowledge of an exotic land, but their English colleagues viewed them as second-rate intellectuals, chained by a parochial mind, and far-removed from the lecture halls of Oxford and Cambridge. What is more, from the pen of a faithful remnant of the not-so-distant Puritan past, these letters fused the religious and the reasonable to an extent that few texts in early modernity had or would have thereafter. In crafting the "Curiosa Americana," then, Mather himself joined the specimens of plants, fossils, and Indian medicines as yet another oddity produced by and exported from the American continent.[11]

This volume has two goals, one particular and the other general. First, this volume introduces readers to the breadth of written sources – theological, philosophical, and scientific – and communication networks that propelled the production of knowledge in the eighteenth century. There was no shortage of information and synthetic commentary that those on the shores of New England could supply to the London intelligentsia. Second, this volume places Mather and his network squarely in the context of trends in early modern intellectual life. In presenting for the first time the entirety of the "Curiosa Americana," transcribed and edited from the original manuscripts, and fully annotated, we show the ways in which ideas and practices converged in dialogue across the Atlantic.

To that end, this introduction to the world of the "Curiosa Americana" proceeds in three parts. First, it examines the life of the principal author Cotton Mather and the network of people he relied on to gather information. Second, it dissects the form of the Curiosa letters and provides a stylistic analysis.

9 See O.T. Beall, Jr. (1961).

10 C. Winterer (2012), P. Findlen (2019).

11 M.P. Winship (1994, 1996), S. Parrish (2006), W.W. Woodward (2010).

Finally, it offers a historiographical analysis of the Curiosa, showing that they have been underused despite their prescient observations.

A Life in God, A Life in Science

The Reverend Cotton Mather is a frequent subject of historical study – and for good reason. A colonial polymath, Mather involved himself fully in the intellectual world of the early modern Atlantic, contributing to far-ranging academic disciplines. He was just as comfortable commenting on the trajectory of comets and the heliocentric revolution of planets as he was philosophizing on their Maker.

For all of the many directions in which Mather found himself pulled, his involvement in natural philosophy extended beyond passing interest to become a central feature of his life. At Harvard College, where he studied to become a minister like his father, Mather committed equal attention to pursuing natural sciences as he did to divinity. "For my Declamations," he tells us in his autobiography, "I ordinarily took some Article of Natural Philosophy, for my Subject, by which Contrivance I did kill Two Birds with One Stone. Hundreds of Books I read over; & I kept a Diary of my Studies" (*PA* [1976], p. 8).[12] Burdened with a speech impediment, he had planned to take up a career in medicine should public preaching become an impossibility.[13] Even after securing a ministerial post in Boston's Second Congregational Church, Mather authored a handbook that was the first treatise on diseases and their cures written by an American of European descent, to which he gave the apt title *The Angel of Bethesda* (1972). From his schoolboy days onward, the goal for Mather was always to maintain a life in God and a life in science.

As was true of many early moderns interested in both the temporal and the eternal, Mather saw these areas of examination as self-reinforcing. That perspective and his passion for the subject matter naturally transferred themselves to his authorship of the "Curiosa Americana." In July of 1711, the year after he received an honorary Doctor of Divinity degree from the University of Glasgow, Mather first recorded in his diary,

> There is one good Interest, which I have never yet served, and yet I am capable of doing some small Service for it. The Improvement of Knowledge

12 He jotted down extracts and quotations in his "Quotidiana" and emerging manuscript of "Biblia Americana." See for example (*D* 1:545, 548; 2:162, 178, 669).

13 This was not an uncommon trajectory. See P.A. Watson (1991). Mather also composed one of the earliest essays on how to overcome speech impediments, "Ephphatha, or Some Advice to Stammerers," *ABJ* (1972), pp. 226–32.

> in the Works of Nature, is a Thing whereby God, and His Christ is glorified. I may make a valuable Collection of many Curiosities, which this Countrey has afforded; and present it unto the Royal Society. May the glorious Lord assist me, in this Performance. (*D* 2:85–86)[14]

For Mather, then, the natural world glorified God, and nature in turn could help him to know the Creator and the laws that governed His Creation.[15]

In the "Curiosa Americana," Mather's knowledge of natural philosophy certainly shone through. Substantively and topically, the texts paralleled in content material printed in contemporary scientific journals. Especially within the Royal Society's own *Philosophical Transactions*, there was a good deal of overlap between Mather's relations and the findings of less overtly pious members. For instance, Mather's November 27, 1712 letter to Richard Waller "On the Rattlesnake" (CA 1.11) foreshadowed remarkably "An account of the rattlesnake" that Paul Dudley (1675–1751), Chief Justice of Massachusetts's Supreme Court of Judicature, published with the Society some twelve years later.[16] There was, however, one significant difference between Mather and his peers in the early Enlightenment scientific community. Propelled by his self-avowed purpose to glorify God, Mather consistently referenced Scripture and biblical precedent alongside standard natural philosophical observations. That effort required a good deal of assistance, for although he could readily supply theological commentary, there was only so much physical material from the American world that Mather could gather on his own. To achieve a marriage of God and science, Mather often relied upon the help of friends, family, and neighbors to bring him information worth sending across the Atlantic to the Royal Society.

To furnish the Society with appropriate matter for the *Philosophical Transactions* (*PT*) "by Annual Communications to increase their Treasures" (*D* 3:94, 2:247, 265–66), Mather enlisted the help of numerous individuals. His information network included his uncle, the Rev. Samuel Mather (1650–1728), pastor-physician of Windsor, Connecticut, on whom Cotton could rely for "notable Things that have occurr'd unto him, in his Practice. His more singular Observations on Diseases and Remedies; his botanic Discoveries, and whatever else may be entertaining and serviceable to the World" (*D* 2:739–40). Other contributors were missionaries to New England's indigenous peoples and Native practitioners who provided knowledge about "their Peculiar

[14] Also see his diary entry for November 20, 1712 (*D* 3:93–94).

[15] Mather fully participated in the "doxological" consideration of nature. See D.B. Hindmarsh (2018).

[16] Paul Dudley, "IV. An Account of the Rattlesnake." *PT* 32.376 (April 30, 1723): 292–95.

Methods in Curing Diseases" as well as their "Peculiar Medicines they have among them" (*D* 3:110); unnamed "Physicians in this Countrey" furnished him with "knowledge of the Botanicks of the Countrey" and "of rare cures or cases occurring to them" (*D* 2:208); shipwrecked sailors shared tales of the effects of near starvation (*D* 2:71); Samuel Penhallow (1665–1726), friend of the Mathers and justice of the superior court of New Hampshire, sent him details about the discovery of petrified wood and the unusual life and death of "two very sympathizing Sisters at Hampton" (*D* 2:175–76, 291); sworn testimonials from local Bostonians about the appearance of a well-attested ghost (*D* 2:383); his nephew Thomas Walter of Roxbury (1696–1725) scouted for a "Collection of *Plants*, peculiarly American" (*D* 2:371). Mather observed cases of lithotomy and lithopedia, by attending a post-mortem autopsy (*D* 2:396, CA 1.4; 1.5, CA 3.8); he knew about the personal observations on the devastating blizzard of 1717 first hand (*D* 2:440, 455); a ship captain gave him an account of a submarine eruption of a volcano in the Atlantic (*D* 2:626; CA 6.1); a Boston gardener showed him the effects of accidental plant hybridization (CA 3.12); enslaved Africans in Boston shared their experience of smallpox inoculation as practiced in Africa (CA 3.10, 6.2, 6.14, 7.2); his young friend and correspondent John Winthrop (1681–1747) of Connecticut helped him with a collection of marine fossils as well as specimens of a water dove – like that in Noah's Ark (CA 4.1, 5.9, 7.4; *MP* 435, 436, 447); Thomas Robie (1689–1729), a young Harvard graduate, supplied Mather with astronomical observations on an eclipse (CA 4.2) and an account of alkaline salt produced from rotting wood (CA 8.7). Last but not least, the list of helpers includes unnamed ministers, physicians, and friends and correspondents of the Mathers, who provided many curiosa on comets, earthquakes, eclipses, Parhelia, an aurora borealis, rattlesnakes, passenger pigeons, tapeworms and the like, all of which became embedded in Mather's eight series of "Curiosa Americana" letters to the Royal Society (1712–24).

One of the more remarkable features of that information network was that it included ample people whom modern historians would term colonial others. Granted, Mather expressed what now read as troubling racial opinions for most of his life, and even conveyed some of those prejudices to the Royal Society. For instance, in a 1716 communication (CA 3.2) "A Monstrous Calf," Mather informed John Woodward that "this Year a *Cow* in our Vicinity, brought forth a *Calf*, which had so much of an Humane Visage, as to make the attentive Spectators apprehensive, that the poor Animal had been imprægnated by a Beastly Negro."[17] Of course, enough white settlers were convicted of bestiality that Mather's suspicions can only be chalked up to the

[17] On bestiality in New England, see J.M. Murrin (8–43).

racially discriminatory attitudes that were all too pervasive in his day. At the same time that he held such denigratory views, however, Mather relied on enslaved African people in Boston – one of whom he had received as a gift from his congregation[18] – to be educated about African smallpox inoculation practices, which became the subject of no less than six Curiosa letters.[19] From Indigenous peoples and their Protestant missionaries, too, Mather learned of new means to cure the diseases that plagued the New English coast.[20] Common folk like farmers, sailors, and housewives joined in to provide accounts of the peculiarities of quotidian life. This extensive network turned Mather into something of a clearing house for colonial oddities.

To secure the help of friends and relatives, Mather relied primarily on goodwill. His intellectual courtship of Penhallow best illustrates the means by which Mather converted personal relationships into usable scientific knowledge. For instance, in September of 1712, Mather wrote to Penhallow to "request one more Favour," a simple enough request that required little work on Penhallow's end.[21] Recalling an account of the discovery of a fossilized tree stump he had heard from another New Hampshire resident, Mather requested that Penhallow verify some details about a deeply buried "*Limb of a Tree*, with Evident Marks Left upon it, of the *Axe*, which it had been hew'd withal." Such relics of New England's prehistoric inhabitants Mather deemed as important as the mysterious hieroglyphics carved into a boulder on the banks of the Taunton River.[22] To entice Penhallow's response and alluding to his project for the Royal Society, Mather concluded that with Penhallow's help, "I should be able to answer some Obligations that ly upon me, from the other side of the Water; and should with your leave, do myself the Honour, of using your Name in the Communication" (*D* 2:175–76). Although Mather would write on the ostensible vestiges of New England's indigenous peoples some eleven years later,[23] he never cited Penhallow's name.

18 *D* (1:579).

19 See, for instance, CA 3.10, CA 6.1, CA 6.2, CA 6.3, CA 6.4, and CA 7.1. See also T.H. Brown (2247–49).

20 See CA 1.2, CA 1.11, CA 3.11, CA 3.12.

21 CA 7.7 "*Subterranea*" and "Cotton Mather to Samuel Penhallow" (19*d*. 7*m*. [September] 1712), in *D* (2:175–76).

22 CA 1.12 "Thunder & Lightning," letter to Richard Waller (November 28, 1712). An extract of this letter, along with Mather's drawing of the hieroglyphs, was published in the *PT* 29.339 (January 1, 1714): 69–70, and appendix, fig. 8.

23 CA 7.7 "*Subterranea*," Mather's letter to James Jurin (June 8, 1723).

Until this point, we have considered only the New English side of this knowledge exchange, orchestrated by Mather in collaboration with his colonial correspondents. Yet it goes without saying that Mather pitched his material to the community of natural philosophers in a particular time. The Royal Society, founded in 1660 by Christopher Wren and Robert Boyle among others, was a royally chartered society that began at Gresham College, London, with the fitting motto, "Nullius in verba" – *Take nobody's word for it*. By the time that Mather engaged with the illustrious organization, his primary points of contact were its secretaries John Woodward (1665–28), Richard Waller (c. 1650–1715), John Chamberlayne (1666–1723), James Jurin (1684–1750), and James Petiver (c. 1663–1718) – all distinguished natural scientists, historians, or physicians in their own right. Mather was, by all accounts, eager to join their ranks as a Fellow of the Royal Society, which was no small feat for a mere provincial, an "American," as Mather professed himself to be, at a time when that term still conjured up images of "savage Indians." That fact alone made Mather something of an eccentric, for there had been only seven others before him, including John Winthrop (1605/6 to 1676), colonial governor of Connecticut and original member of the Royal Society (January 1 1661/62), and William Byrd II of Westover (1674–1744), a wealthy Virginia planter and slaveholder, who was elected to the Council of the Society in 1696.[24]

The process of Mather's initiation into the Royal Society was no less curious. In July of 1713, after receiving the first packet of Curiosa letters, Waller proposed to the Society that Mather be considered a candidate for membership.[25] Four days later, Mather's candidacy was approved, to his great delight. In the following year, however, Mather again received word from the Society, this time alerting him to the fact that membership required attendance of an initiation ceremony which Mather had missed by dint of geography. Confusion continued for a number of years, for though he decided to append the official F.R.S. to his name on some publications he was not listed in published

24 On colonial membership in the Royal Society, see R.P. Stearns, "Colonial Fellows" (1951) and F.R. Freemon (1985).

25 Professor William R. Manierre II has argued that historians have followed George Lyman Kittredge's conclusion ("Election," pp. 84–85) that Mather himself did *not* volunteer his election to the Royal Society (*D* 3:94, *n.* 10). However, Manierre believes that Mather's diary entry for November 27, 1712, may suggest otherwise: "I am related unto many Societies," Mather recorded in his diary. "I will See whether a Modest Intimation of such a Tendency, may not introduce me as a Member of the *Royal Society*" (*D* 3:93–94). Once Mather had submitted the first packet of thirteen essays, Manierre argues, his election was all but assured. They were published in summary fashion in *PT* 29.339 (1714): 62–71. Cf. Appendices 2 and 3.

lists of Society members. Although some Royal Society members seemingly assumed Mather's full membership, others challenged his right to the honorific title. It was not until 1723, a full ten years after Mather's original nomination, that he was listed a fully elected member of the Society.[26] Thus he was precariously positioned, and the letters reflected that tumult and the anxiety it bore.

The Letters and Their Arrangement

By dint of his ministerial profession, Cotton Mather had a way with words. To sustain the frequency of letters such that their volume would redress any existing prejudices against him as a colonial correspondent on the periphery of the British Empire, Mather adopted a formulaic approach to authorship. The characteristic structure, style, and tone of virtually all of Mather's Curiosa epistles evince the following pattern: After introductory remarks (serving as a rhetorical hook), Mather established his credentials – he was not just a well-published New England preacher but a credible authority on the subject of natural science. He surveyed the standard experts on the topic at hand, thereby providing a train of authors from the ancients to the moderns and critiquing where they missed their mark. These introductory remarks to establish his own legitimacy were then followed by the centerpiece, his American specimens, of which he supplied detailed scientific descriptions along Baconian lines recommended in the pages of Thomas Sprat's *History of the Royal-Society of London, For the Improving Natural Knowledge* (1667). Mather's concluding remarks often end on one of his self-deprecatory witticisms to elicit approval and to encourage continued correspondence. Significantly, his ornate diction – reminiscent of his baroque style in his celebrated ecclesiastical history *Magnalia Christi Americana* (1702) – is scaled back in his scientific letters and always clear and to the point. Yet his obsequious tone (sounding his place in the great fathom-chain of being in British society) tends to be distasteful to our modern sensibilities. It may even come across as insincere. For while conventionally humbling himself in the face of his correspondents' superior station, his letters bespeak anything but his intellectual inferiority. Or as he protested in his letter (18*d.* 8*m.* [Oct.] 1715) to Sir William Ashurst (1647–1720), wealthy banker, MP, and sometime Lord Mayor of London, Mather's peers in England seemed to believe "that a poor *American* must never be allow'd capable of doing any thing worth any ones regarding; or to

[26] See G.L. Kittredge, "Election" (1912), and James Jurin's letter to Cotton Mather (December 16, 1723), in Appendix 9.

have ever look'd on a Book" (*D* 2:331).[27] Provincial colonial or no, Mather was certain he could play his part in the transatlantic Republic of Letters.

In each of the eight series of Curiosa letters, Mather presented the subject matter as foreign to the Royal Society members, a uniquely American phenomenon about which he and his informants had intimate knowledge. In an exposé on animal hybridization, for instance, Mather conveyed, "We have a little Animal called a *Musquash*, pretty frequently lodging about the Brooks that water our Countrey" (CA 3.1). That possessive refrain – *our* country – symbolized much, for Mather was wont to label each account as a particularly "American Story" (CA 2.7). From that closeness to the natural subject, Mather established his ultimate authority. Aided by wit and displays of learning, he could comfortably assure Royal Society members that "This Relation may be relied upon. Tis here quæstioned by no body" (CA 8.2). We can turn now to sample the letters and their contents.

Curiosa Americana Series 1 (November 1712)

Cotton Mather was anything but unproductive: In the three years alone before he sent off his First Series of letters to the Royal Society, he composed and preached three to five sermons every week and wrote and published more than thirty-four sermons, pamphlets, and tracts, including his book *Bonifacius* (1710), a duodecimo-sized volume of more than two hundred pages on "doing good" – the social responsibilities of each member of the commonwealth, of magistrates, ministers, physicians, lawyers, merchants and artisans, the rich and poor, parents and children, masters and servants. It is by far his most popular book, and it was reprinted more than any of his other works. In fact, half a century after Mather's death, Benjamin Franklin acknowledged that *Bonifacius* manifestly inspired Franklin's own social activism and philanthropy.[28]

[27] Even a century after Mather's death such prejudices persisted. Perhaps still smarting from the UK's loss of the War of 1812 to the young United States, Sydney Smith (1771–1845), an Anglican clergyman and cofounder of *The Edinburgh Review* 33 (January 1820), Art III, pp. 69–80, rises to his stinging crescendo, "In the four quarters of the globe, who reads an American book? or goes to an American play? or looks at an American picture or statue? What does the world yet owe to American physicians or surgeons? What new substances have their chemists discovered? or what old ones have they analyzed? What new constellations have been discovered by the telescopes of Americans?" (79–80). See also R.E. Spiller (1929).

[28] "Letters from Benjamin Franklin to Samuel Mather, 12 May, 1784." In Miscellaneous Bound Manuscripts, MHS, <https://www.masshist.org/database/533>.

Mather was also busy ministering to his congregation of about 1,500 souls,[29] the largest in English America; every week he visited with three to four families in his parish and left with them sermons and pamphlets on appropriate subjects he had published;[30] he went on preaching tours along the New England seaboard, visited "praying Indian" settlements as commissioner of the Society for the Propagation of the Gospel (SPG), tried to complete his commentary "Biblia Americana" (his most voluminous work, exceeding three million words), and corresponded with members of the English aristocracy, scientists, ministers, and missionaries as far away as India. And he worried about the War of the Spanish Succession (1701–14), the decline of the Protestant cause in Europe, and the signs of the Second Coming on the horizon. It is all the more astounding that he found time to compose his first and longest set of "Curiosa Americana" essays.

The topical diversity of Mather's set of thirteen letters (First Series), composed during the week of November 17–19, 1712, is telling. They reveal that like few others in the English colonies he was preoccupied with how America's flora and fauna differed from that of the Old World. His epistolary essays can be grouped into distinct classes of subjects. Five essays (CA 1.2, 1.5, 1.6, 1.7, and 1.11) cater to John Woodward's interest in medicine and botany (Woodward was a member of the Royal College of Physicians) and focus on Native American cures, medicinal herbs unique to the New World, and their application (in terms of preparation and dosage) to specific ailments. In each case Mather established his expertise on specific diseases, their treatment, and his extensive familiarity with the iatrochemical and iatromechanical merits of the botanical remedies found in the medical cabinets of his European peers, which he complemented with his Native American medicines. Thus, in CA 1.2, for example, he emphasizes the alexipharmic virtues of "*Asarum*" in the cure of dysentery; the anti-inflammatory potency of what Mather calls "*Anterisypelas*" (i.e., anti-erysipelas) if taken inward as a "chymical Oyl" for the cure of certain eczemas; the healing qualities of "*Partridge-berry*" when taken as an aromatic tea in the cure of edema; "*The Bleeding Root*" for the "Infallible" cure of "*Jaundice*"; a medicinal decoction of what some Native Americans call "*Taututtipaug*," a ginger-like root, for the cure of "that filthy Disease (the *Lues Venerea*)," better known as venereal disease; and the juice of the roots of

[29] As we learn from a letter of the Rev. Increase Mather (December 16, 1698) to William Stoughton, Lt. Governor of the Massachusetts Province, the Mathers preached "to 1500. souls" who "use ordinarily to attend in our Congregation" (*Diary of Samuel Sewall* 1:493).

[30] *D* 1:114–15, 538–39; *PA* (1976), pp. 42, 85.

"*Boar-thistle*," when compounded with "*Divels-bitt*," a patent medicine for the cure of "Scrophulous Ulcers" popularly known as "*Kings Evil*."

Mather was also greatly interested in the emerging debates about human anatomy. American cases of malformed and abortive fetuses ("monstrous births") and mysterious "stone babies" all too often shocked those who performed autopsies (CA 1.5, 2.3). Case studies of this topic were perennially in vogue in the medical literature of the day.[31] To Mather and his peers, they were no longer prodigies contrary to nature or signs of God's wrath but clear examples of medical pathology, marking a scientific shift in the movement away from supernatural marvels to natural cases of human embryology in the early modern period. On a more esoteric level, Mather added in letter CA 1.6 telling examples of remedies revealed in dreams (a branch of *Ars divinatoria*), including the case of an anonymous gentlewoman (Mather's first wife) who dreamt of a freshly cut piece of sheep's wool to be placed on her cancerous breast; several cases of miraculous recoveries from mortal wounds (CA 1.7), including victims of Indian raids shot through with bullets or clinging to dear life with hatchets buried in their brains; finally (CA 1.11, 8.2), the medicinal virtues of a rattlesnake's gall bladder, whose spirituous juice, appropriately dubbed "*Trochisci Connecticotiani*," serves as a proper antidote against snake venom especially when mixed with "*Indian* Meal." Indeed, Mather's Native American medicine chest opened up a new world of wonders.

New medicines fetched from the New World, birth defects, and "stone babies" were not the only issues Mather touted in the First Series of his Curiosa letters: two letters (CA 1.3, 1.11) deal with indigenous birds and reptiles; three (CA 1.8, 1.9, 1.12) with astronomical, meteorological, and geological observations; one (CA 1.4) with the function and formative power of the imagination, a hotly debated subject in its time; one (CA 1.10) with supernatural proof of the post-mortem activities of the human soul; and one (CA 1.13) with the longevity of the patriarchs, the incredible fertility of the antediluvians, and their diet – topics dear to Newtonian physico-theologians (like Thomas Burnet, William Whiston, and Sir Isaac himself) who tried to uphold biblical verity. To be sure, Mather's conjecture about the birth rates and longevity of the ancients was not necessarily of interest to the Royal Society; there were other, more tangible, reasons to be concerned with the fertility rate of the populace ancient and modern. Underscoring his argument with first-hand knowledge of a mother in New England who gave birth to "no less than Six and twenty Children, by Two Husbands," and another who

[31] See also CA 3.1 and 3.2. "Joshua Moodey to Increase Mather" (25d. 8m. 1683), in *MP* (1868), pp. 361–62. Representative accounts are legion in the *Philosophical Transactions* and the German *Ephemerides*.

"has brought forth no fewer than *Thirty nine* Children," Mather maintained that "if such a *Polytokie*, were attended with *Longævitie*, like that of the First Ages, Mankind would not require near *four hundred Years*" as it does now "to *Double itself*." Mather's unusual illustration and his pre-statistical calculations would help medical arithmeticians, especially in the Royal Society's College of Physicians, to establish tables for births, deaths, and diseases to measure the vital statistics of the population as well as the rate of success of certain medical procedures.[32] In fact, none less than Dr. James Jurin (Mather's correspondent during the smallpox epidemic in Boston, 1721–22) did as much in his *Account of the Success of Inoculating the Small-Pox in Great-Britain*, to which he added comparative tables of the "Miscarriages in that Practice, and the Mortality of the Natural Small-Pox" (1724).[33] He published his statistics annually. Mather's wide reading across the disciplines, then, enabled him to gauge the scientific interests and trends emerging among his target audience as he deployed his American evidence in support of his essays. If anything, they bespoke the multifarious fields of interest of the first American polymath who was clearly at home in the rising disciplines of his time.

Mather's interdisciplinary reading is also apparent in CA 1.3 and 1.11. Here, he celebrates the many varieties of birds in North America, including wild turkeys superior, he reckoned, in size and weight to those in England. The American turkeys weigh between forty to sixty pounds but their savor may not meet the gustatorial expectations of the more refined English palates. Better yet, the mighty "*Pilhannaw*" – as the Indians call it – is an eagle "four times as big as a *Goshawk*," whose head is "as big as a Childs of a Year old." Mather's may be one of the first descriptions of the American bald eagle before the species established its pride of place in one of Carl Linnaeus's many editions of his *Systema Naturae* (1735–59). Curious, too, is Mather's Baconian description of "our *Humming-Bird*," their habitat, plumage, feeding and nesting habits. But most fascinating is his marvelous account of the vast flocks of Passenger Pigeons, which he calls "*Season-birds*" of which "Thousands of Millions" appear annually before they migrate to unknown regions "in some Remote and upper Quarters of the *Heaven*, within our *Atmosphaere*." They likely winter on some "*Christalline*" pellucid planetoid "between the *Earth*

32 See especially A.A. Rusnock's *Vital Accounts: Quantifying Health and Population in Eighteenth-Century England and France* (2002).

33 See esp. pp. 9–11, 17–32. Perhaps more important, in his "Advertisement" published in the *Philosophical Transactions*, Jurin called upon "Physicians, Surgeons, Apothecaries and others" involved in the practice of inoculation to send to him the names, ages, numbers of the vaccinated, and their mortality rate, in *PT* 32.370 (March 31, 1722).

and the *Moon*" (CA 1.3), as some knowing philosopher had conjectured.[34] Closely related to his interest in ornithology are some of the first accounts of the American rattlesnake (1.11). They are also the subject of three subsequent Curiosa letters (CA 3.5, 7.3, 8.3). In the best Baconian manner Mather could muster, he traced the significance of serpents among the ancients all the way to "the *Prince of Serpents*," the Mexican "*Teutlacocauhqui*." Mather's account outlines their American habitats, their method of hunting for prey, the medicinal qualities of their venom and gallbladder, as well as their mythical significance. The sheer scientific, historical, and mythical detail of Mather's Ophidia amounts to a *rigorosum*.

Astronomical, meteorological, and chronometrical observations are the subject of three of Mather's letters (CA 1.8, 1.9, 1.12). Again following his established pattern, Mather studded his cosmological annotations with references to a who's who of astronomers ancient and modern (CA 1.8). He proudly pointed out that virtually all of the famous astronomers were also men of the cloth, equally at home in divinity as they were in mathematics and the observations of the stars. Thus we learn that Native Americans have no other means to measure the passage of time but by "*Sleeps*," "*Moons*," and "*Winters*," yet surprisingly call the constellation "*Ursa Major*" "*Paukunawa*," the local Native term for the Great Bear. Perhaps more significant is that Mather's astronomical observations confirmed the disputed theory of Archdeacon Joshua Childrey (1623–70), who had claimed in his *Britannia Baconica* (1661) that in the month of February a barely visible light appears in the Pleiades best seen in clear nights and now known as Childrey's zodiacal light, though frequently attributed to the Italian-French astronomer Giovanni Domenico Cassini (1625–1712), who reported it in 1683. Yet its causes remained subject to further studies. Lest Mather's astronomical acumen was not deemed sufficient, he added two mathematical tables that more accurately translated the dates of the Julian Calendar's lunar cycles into the more precise solar cycles of the Gregorian Calendar, which was introduced in 1582. The new calendar corrected the inaccuracies of the Julian Calendar by taking out several days but was not adopted in the British Empire until 1752.

Closely related in subject matter and level of sophistication is Mather's CA 1.9 on the origin of rainbows and Parhelia, or mock suns. His insights are, as usual, embedded in scientific accounts of these celestial phenomena from biblical times to Mather's own. Again, he demonstrated his meteorological acumen in these matters, not the least in his hand-drawn depictions

[34] Charles Morton, *An Essay toward the Probable Solution of this Question: Whence come the Stork and the Turtledove, the Crane, and the Swallow, when they know and Observe the Appointed Time of their Coming* (1703).

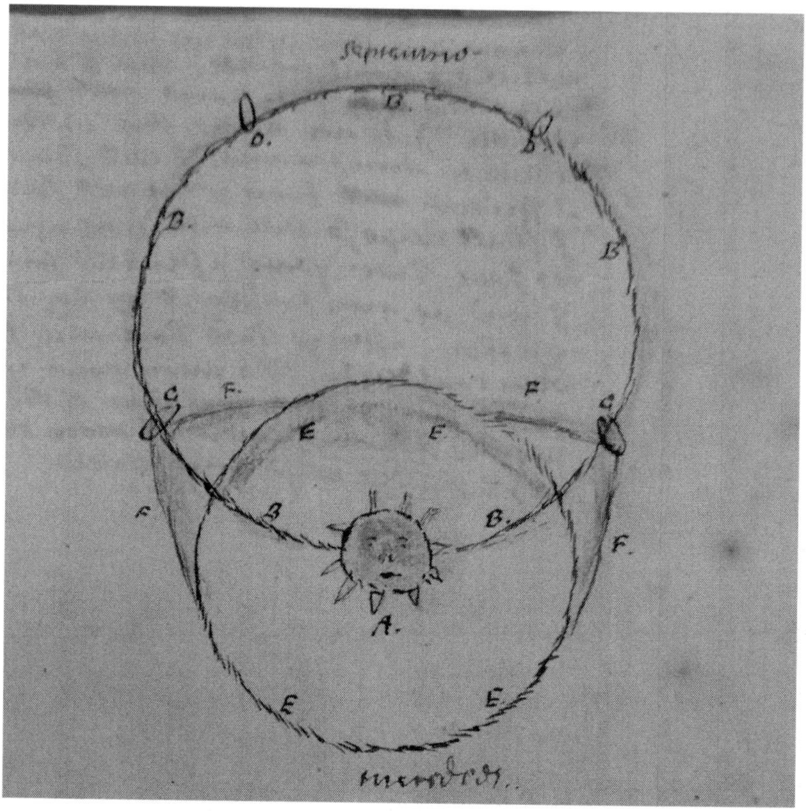

Figure 1: Cotton Mather's diagram of Parhelia. Courtesy of the Royal Society of London.

and descriptions of mock suns that both Increase and Cotton Mather had observed from their Boston perch in the 1680s. All the while he insisted in the best manner of John Spencer's *Discourse concerning Prodigies* (1663) that such mysterious events were not supernatural omens threatening danger and destruction (as was commonly believed) but phenomena quite natural, predictable, yet wonderful.

Meteorological and geological observations, including thunder, lightning, and earthquakes, are the topics of CA 1.12. Melted pewter dishes and iron braces, chimneys and steeples catching fire, and grazing cattle struck dead on the commons were subject of great interest, for they seemingly showed God's wrath – decades before Franklin's kite experiment and lightning rod demystified God's thunderbolts as mere electrical discharges.[35] Less common,

[35] See also Mather's account "*Relating Remarkables done by Thunder*," *MCA* (1702), 6.3.14–20–all linking meteorological phenomena to the voice of God, including

perhaps, were cases of a ship's compass permanently changing polarity when the vessel was struck by lightning at sea. So, too, hurricanes and their terrestrial cousin, tornadoes, provided much food for thought of their destructive force not just in the West Indies but in New England as well. The causes of earthquakes were of scientific interest as were their extent and power, occasionally making rivers flow backwards. Oddly enough, runic inscriptions found on a huge boulder (Dighton Rock) on the shores of the Taunton River raised speculations about the original inhabitants of the New World. Mather's hand-drawn copy of the inscriptions, though torn from the extant holograph manuscript sent to London, was reproduced on a separate page.[36] More than a decade after the inscription appeared in the *Philosophical Transactions*, the Rev. John Villa remembered Mather's letter and the mysterious inscription, which he showed to M. Veyssière De la Croze (a historian and orientalist at the court of the Prussian King Frederick William I in Berlin), who conjectured it might be of Japanese origin and offered to interpret them if a new drawing of all seven lines were provided him (Appendix 11).

Lest we dismiss Mather's discourse on the imagination as unexamined trivia, modern readers need to be reminded just how much European physicians from the Greek Hippocrates (c. 460–c. 370 BCE) to the French physician Benjamin Bablot (1754–1802) wrestled with the formative power of the imagination and its psychosomatic impact on the mind and body, especially in pregnant women (CA 1.4, 1.5, 3.7).[37] True, if Descartes has his say, the imagination as an immaterial faculty cannot impact the physical body directly. However, if the emotions are sufficiently agitated, so mechanical physicians argued, the humors, spirits, and vapors may upset the balance of the humoral fluids and lastingly affect and exacerbate weaknesses leading to physical malformations of embryos and mental disease in pregnant mothers. In fact, works like *De Viribus Imaginationis Tractatus* (1608), by Flemish professor of medicine Thomas Fieno (1567–1631); *The Anatomy of Melancholy* (1621–51), by Anglican scholar Robert Burton (1577–1640); *De La Recherche De la Vérité. Où l'on Traite de la Nature de l'Esprit de l'home* (1674–75), by the French Oratorian philosopher Nicolas Malebranche; and *The Force of the Mother's Imagination upon her Foetus in Utero* (1730), by the English physician Daniel

the occurrence of Mather's house being struck by lightning while he delivered his sermon *Brontologia Sacra* (1695) to his congregation.

[36] *PT* 29.339 (June 30, 1714), sep. pag. fig. 8.

[37] See also CA 6.5, item X. One of the numerous examples of this type published in the *Philosophical Transactions* is "An Account of a monstrous Foetus, resembling a hooded Monkey: Communicated by Mr. Willam Gregory of Rochester" (dated April 30, 1733), *PT* 41.461 (December 31, 1741): 764–67.

Turner (1667–1740)[38] were of great interest to Mather, who sought to explain how the "*Potentia imaginativa,*" especially when violent emotions such as anger, fear, revulsion, or unfulfilled cravings, exerted harmful vibrations on the developing fetus in the mother's womb. Likewise, Mather's Curiosa also included curious cases of sympathetic suffering, such as "puling" husbands who underwent the same physical pangs as their pregnant wives during gestation; of conditions of uncontrollable revulsion at certain medicines, foods, or actions in both men and women; and physical cures through sympathetic touch (massages), like that of the Irish "Gentleman-Stroker" Valentine Greatrakes (1629–82). The *Potentia imaginativa* as a transformative force was as pervasive a medical concept in Mather's time as the ancient chestnut told by Hippocrates and attested by St. Jerome, of a white mother who gave birth to a "Black Infant" because she had been frightened by the picture of an African.[39] Indeed, the imagination is a devilish thing!

Mather's CA 1.10 and 1.13 round out his theological convictions with scientific evidence. The relation of Joseph Beacon of Boston (CA 1.10) is particularly apropos. It involves the case of Beacon's brother who was murdered in faraway London and whose apparition appeared to Joseph in a wakeful dream in Boston on the same night giving descriptions that exactly matched the account of the murder arriving by mail some weeks later. To Mather and his peers, the sworn testimonies of the people involved proved beyond doubt that the "Foolish Philosopher of *Malmsbury*" (Thomas Hobbes) was absolutely off the mark when he dismissed such stories as "meer *Fancy*." Mather liked Beacon's story well enough to recycle it in at least four separate publications. For if the "*Plastic Power of Imagination*" can indeed impact the organs of the body as well as the spirit of persons at a distance, Mather speculated, then such apparitions may provide evidence of the reality of the invisible

[38] See Turner's earlier *De Morbis Cutaneis* (1714), pp. 102–28; and James Augustus Blondel's response *The Power of the Mother's Imagination over the Foetus Examin'd* (1729). Also see P.K. Wilson (1991).

[39] With few exceptions, the classical and medieval philosophers, anatomists, and physicians from Hippocrates, Aristotle, Galen, Avicenna, Aquinas, to the moderns from Pomponazzi, Rhodiginus, Aldrovandi, Ficino, Paré, Lemnius, Fieno, Liceti, Lycostenes, Fernel, Boerhaave, Digby, Van Helmont, Malebranche, Nieuwentyt – all the way to physicians like Bablot in the late 18th century – asserted the power of the imagination and its influence on the fetus. Perhaps the last reputable physician to do so before this hoary theory collapsed under the weight of its own collected dust was Benjamin Bablot (1754–1802), in his *Dissertation sur le Pouvoir de L'Imagination des Femmes Enceintes* (1788), pp. 199–234 (Table). See L.S. King's *The Philosophy of Medicine* (1978), M.-H. Huet's *Monstrous Imagination* (2000).

world and the existence of the soul when separated from the body. "[B]ut this not directly relating to Natural Philosophy," the editor of the *Philosophical Transaction* objected, "I omit the Particulars, tho' the Relation seems to be well attested."[40] No, such ghostly stories were proper subjects for the likes of Joseph Glanvill's many iterations of *Saducismus Triumphatus* (1681–1726), not for the scientific journal of the Royal Society. Be that as it may, to Mather and his peers they were perennially reassuring in the philosophical battle with English Deists and their fellow mechanical materialists who were shaking the pillars of biblical verity.

Also close to his heart, Mather addressed with mathematical rigor the miraculous multiplication of mankind and the longevity of the ancient patriarchs (CA 1.13). With the rise of philological and historical criticism of the Bible in the early Enlightenment (1650–1730), such marvelous stories no longer seemed to withstand the test of truth, as Deists tried their best to render threadbare the miraculous accounts in the Book of Books. If it was still hard to believe that one human pair expelled from Eden, or one family of eight survivors in Noah's Ark, could populate the earth with millions and millions of offspring in just a few centuries, then mathematical multiplication schemes along with well-attested examples from Europe, Asia, and America might stop the mouths of skeptics. To that effect, Mather proudly told the stories of certain mothers in New England who gave birth to "no less than six and twenty Children, by Two Husbands"; another gave birth to "no fewer than *Thirty nine Children*" (Kyrie eleison!) – then it was all true and the authors of the Bible did not embellish their accounts. "If such a *Polytokie*, were attended with a *Longaevitie*, like that of the First Ages," Mather posited, trying to dispel the doubts of neoterics, "Mankind would not require near *four hundred Years*, as you tell us, it now does, to *Double itself*." The fecundity of healthy New Englanders, or the vast age of numerous Puritan settlers, seemed to leave little room for doubt.

The first set of thirteen missives to the Royal Society contain an intriguing array of topics spanning multiple disciplines and responding to hot issues debated in the scholarship of the early Enlightenment. As Mather's eminent bibliographer George Lyman Kittredge recognized more than a century ago, "There is scarcely an item in these letters that cannot be paralleled in the Philosophical Transactions, or in the [German] Ephemerides of what we now style the Leopoldina."[41] To be sure, Mather's "Curiosa Americana" essays

40 *PT* 29.339 (1714), p. 67. See Appendix 6 as well as Waller's suggestions for revisions in Appendix 5.

41 Kittredge, "Scientific Communications" (1916), p. 19.

seem sometimes quaint, outright credulous, even offensive to our modern sensibilities; yet they amount to considerably more than a mere historical or antiquarian curiosity about the transference of European thought to the New World or its first flowering of science in Puritan New England. By the standards of his time, they reveal Cotton Mather's sound knowledge in several scientific fields that clearly lie outside the traditional domain of a well-trained preacher–scholar. His scientific essays bespeak his omnivorous reading, amassment of knowledge, and synthesis of the best Europe could offer in the natural sciences, meteorology, medicine, and the burgeoning fields of embryology, epidemiology, and botany – all fields that only a polymath like Mather could muster and enrich with his own American insights.

Curiosa Americana Series 2 (December 1713–June 1714)

On October 12, 1713, Mather received letters from London, addressed to him by Richard Waller, secretary of the Royal Society. The missives contained news that deeply gratified the pastor–scientist: the First Series of Curiosa had been received and read before the assembled members of the Society, who were "greatly satisfied therewithal," and it was their intention to make their author a Fellow (*D* 2:245, 246). Waller informed him, "S^{r} I tooke the same Oportunity [at that meeting] of Proposing you to be a Member of ye Society which was recieved to the satisfaction of the Members present."[42] The news overjoyed Mather, and he readily thanked heaven for this "marvellous Favour," which would place him "above the contempt of envious Men" (*D* 2:246). His exuberance, however, does not tell the whole story, for all along he had pushed the envelope and taken matters in his own hands: "And tho' I cannot presume so much upon my own merits as to dream of being thought worthy to be admitted a member of that more than illustrious Body but rather to belong to the order of the *fratres Ignorantiae*," Mather confessed his heart's desire in his cover letter to Waller, "yet I will here venture to subscribe the Promise which your Candidates make at their admission."[43] Mather's prodding Waller in the desired direction speaks volumes.[44] And yet, as good as his

42 "Richard Waller to Cotton Mather" (July 22, 1713); see Appendix 3.

43 "Cotton Mather to Richard Waller" (December 1, 1712), Appendix 2; (RSL, *Letter-Book,* XV, 44–45), in R.P. Stearns, *Science* (1970), pp. 406–07.

44 In his undated response to his nomination (c. December 1, 1713), Mather thanked Waller with a profusely self-deprecatory opening paragraph that belies his humility: "An Honour too great for me to have hoped for, the Letters werewith you have lately favoured me, have allow'd me some hope of my arriving to. The Academy of the

word, he immediately sprang into action to live up to his promise of supplying the Society with new curiosities. In his diary for October 13, he pondered "what Subjects in Philosophy may peculiarly need cultivation" in order to provide "such Studies as may be for general Benefit, and have hitherto been but little prosecuted" (*D* 2:247). So he began to scout for worthwhile Americana for his next set of letters.

Mather assembled his second set of Curiosa during the period from December 1713 to June 1714. The ten missives that make up this offering were written during a short period of momentous events for Great Britain and its colonies. In April 1713, a decade of fighting – the War of Spanish Succession in Europe, Queen Anne's War in the Americas – ended with the Peace of Utrecht (1713–14). This settlement all but secured Great Britain's dominance of North America, for with it the French ceded Newfoundland, Acadia, Hudson's Bay, and St. Kitt's, while in the Mediterranean, Spain handed over the ownership of Minorca and Gibraltar. Then in August 1714, Queen Anne, the last of the Stuart line, died without an heir. As her closest Protestant relative, Georg Ludwig (1660–1727), Elector of Hanover, was crowned king of Great Britain and Ireland, thus securing the empire for the "Protestant Interest" and ending fears that the British throne would be assumed by a Roman Catholic. This shift in monarchy attracted the nationalistic loyalties of previously ambivalent Dissenters, such as those in Mather's New England. The year 1714 would in many ways mark the beginning of English control over the North Atlantic basin and was a harbinger of its realization of imperial goals. Knowledge itself became a marker of those goals.[45]

Nearly every item in the Second Series, with the exception of the first one, "A Woollen Snow" (CA 2.2), and possibly the second, "Surprising Influences of the Moon" (CA 2.2), was composed in June 1714. However, Mather worked for some months before making final copies. In November 1713 he had begun "collecting some Curiosities, to transmit unto the Secretary of the Royal Society by which I hope to make some valuable Accession unto their Treasures." Scientific discoveries were to be kept in their proper pietistic and devotional context. In presenting his findings, Mather resolved, "I will use my most exquisite Contrivance, That Testimonies to the Glories of my Saviour may accompany them" (*D* 2:265–66).

Besides continuing the Curiosa, which would put him in contact with some of the top English scientists and naturalists of the age – Newton, Boyle, Halley –

Nascosti, at *Milain*; that of the *Innominati*, in *Parma*; and that *Incogniti*, in *Venice*, would by their Titles have been more fitt for so obscure a Person, than a room among your Honourable *Virtuosi*" CA 2.1 A Woollen Snow.

45 See J. Delbourgo and N. Dew, eds (2007); and B. Bailyn and P.L. Denault (2009).

our Boston virtuoso had other projects that were more explicitly religious in nature. Chief among them was his "Biblia Americana" (1693–1728), an immense commentary on the Bible, an extract of which Mather had sent to the Royal Society as a foretaste of sorts CA 1.1 (1712). While he had started that massive work as early as 1693, in 1714 he decided it was "high time" to bring it forth to the world. He published a broadside entitled *A New Offer, to the Lovers of Religion and Learning* early in the year, in order to gain subscribers (*D* 2:283). Copies of that printed sheet, as well as supplementary descriptions and solicitations, would accompany Curiosa sent to England over the following years.

Sometimes, Mather would include collected "specimens" in his reports. Sadly, they no longer accompany the sent letters. Even so, the first item in the Second Series (CA 2.1), "A Woollen Snow," included a "*Snow-ball*," but one which, strangely, had not melted in the course of crossing the Atlantic. This example of an "*American Snow*," Mather explained, brought biblical descriptions out of the metaphorical and into the literal. The reader may have heard or read of rains of blood, of gold, or of corn, Mather observed, but recently a town in Connecticut had a "Snow of Wool" – something, he added, "I do not remember, that I have met any where withal." And his specimen, accompanying the account, was a piece of that wool that fell from the sky, likely acquired from a local, which suggests the broad network of knowledge-gatherers Mather had at his disposal. He did not attempt to explain how the wool came to precipitate, but hoped this rare type of snow, like the snows of old used to chill wine, will "*Refresh the Souls of my Masters*."

"The Surprising Influences of the Moon" (CA 2.2) begins by recognizing the scientific contributions of the Royal Society's president, Sir Isaac Newton, particularly his demonstrations of the effects that the moon has on "*sublunary* Bodies." Here, Mather was reflecting Copernican cosmology and traditional beliefs in judicial astrology but also the more recent general obeisance accorded to Newton and to the new science by early evangelicals as a way of connecting spirituality to the natural world. Figures such as the English artist and astronomer John Russell (1745–1806) embarked on detailed studies of the moon not only to increase scholarly knowledge but as an exercise in piety. For his part, Mather collected accounts that, he thought, spoke to the influence of the waxing and waning of the moon on different sorts of wood, and on prevailing winds.

From these general considerations of natural occurrences, Mather moved, in CA 2.4 and 2.5, to accounts of animals, namely, the moose and the passenger pigeon (*Ectopistes migratorius*).[46] Responding to inquiries from abroad about

[46] Members of the Royal Society asked Richard Waller (July 22, 1713) to request more specific information about the "Mouse Deer" (Appendix 3, PS). Paul Dudley,

the types of American moose, he described the northern and the southern varieties. In regard to pigeons, Mather, in his opening of this letter, indicated that the addressee, John Woodward, had asked either him specifically or the fellowship more generally for more information on this topic, for Mather proposed "to answer your Desire of a treat with a further Mess of our *Pigeons*" (CA 2.5). First, these were to be distinguished from quails, which the ancient Israelites ate while wandering in the wilderness. But "a better Dish than *that*," Mather wrote with a figurative wink of the eye, "you would have in our *Pigeons*." They are delicious, inexpensive, and can be prepared in several ways. And they are easy to catch, whether by shooting them out of the sky, or by trapping them. Here, seemingly, is an inexhaustible resource for all. The enormous flocks of Passenger Pigeons that were reported defy credibility: one "the best Part of a Mile Square" in the sky, another taking several hours to pass, blocking out the sun. It turns out, tragically, that this species was not as inexhaustible as Mather and his contemporaries assumed, because it is now extinct from overhunting.

Many of the installments of the Second Series describe peculiar workings of the human body. This theme starts with CA 2.3, "Monsters," specifically, "monstrous" or deformed births. As attested in early modern literature from the most learned treatises to popular almanacs, Europeans had a fascination with this subject because they sought to establish a psychosomatic relation between a pregnant mother's imagination and her uterine fetus.[47] Here Mather, hiding titillating content in Latin, associated joined twins or other malformed infants with God's judgments on whoever used contraceptive herbals or who was lascivious. Curiosa 2.7 also considers twins, ancient and contemporary, who were not monstrous but who had a strong and evident "sympathy" for each other – a "very strange Agreement" between them – in their moods, likes, and dislikes. Mather relates the story of the Moulton sisters of Hampton, New Hampshire, who were so sympathetically united that, unable to live apart since birth, they died within hours of each other. From Dunstable, Massachusetts, came the account in CA 2.8 of "Uncommon Ideots" – severely mentally

"A Description of the Moose-Deer in America," *PT* 31.368 (August 31, 1721): 165–68. N.B.: the first published piece in *Philosophical Transactions* on pigeons doesn't show up until the 20th century. There was interest in other seemingly exotic birds. See James Douglas, "The Natural History and Description of the Phœnicopterus or Flamingo," *PT* 29.350 (December 31, 1716): 523–51.

47 F.e., Ambroise Paré, *De Monstris et Prodigiis* (1573); Fortunio Liceti, *De Monstrorum Natura* (1616); Ulisse Aldrovandi, *Monstrorum Historia* (1642); William Derham, "Monstrous Births" (*PT* 26.320 [April 1, 1709]:310); Daniel Turner, *De Morbis Cutaneis* (1714), pp. 102–28; Timothy Sheldrake, "A Monstrous Child" (*PT* 41.456 [1740]: 341–43) See also L. Daston & K. Park (2001).

impaired individuals – who, suffering in youth from epileptic-like seizures, lost the use of their intellect and bodily functions to the point that they were "wholly deprived of almost every thing in the World," excepting the most basic urge to eat.

A final set of Curiosa in the Second Series treats the ability of the human body to endure deprivations and damage – and, in some cases, to cure itself by not-entirely known means. The sixth letter in this series gives us a "starving story" of a young man who was marooned for twenty-six days on a small island off the northeastern tip of Long Island (CA 2.6). Finally rescued, he initially seemed to be on the road to recovery, but then fell into "Languishments," became covered with lice, and died in a few weeks. Such tales of fasting were popular at the time and Mather himself went on to include similar stories in his published works.[48]

If the human body could be denied sustenance to a point, it could also emit some strange things, and CA 2.9 "An Unusual Discharge" and 2.10 "An Ear Strangely Molested," provide striking examples. As a physico-theologian in his own right, Mather reflected the doxological impulse to praise the wisdom of the Creator and the wonders of nature, such as fashioning human bodies so that they could eject foreign objects that are "noxious."[49] As he does with so many Curiosa, Mather collects testimonies from a gallery of classical and contemporary authors, this time to testify to the "Sagacity of Nature." We have accounts of people, Mather asserted, swallowing all sorts of objects – pins, stones, bullets, even knives – only to have the body expel them through an abscess or some other way. To bring it home, Mather gives us the story of a woman from Concord who had an "ear of Rye" suppurate out of her shoulder, which had accidentally lodged in her respiratory system after she had had a fall.

In "An Unusual Discharge" (CA 2.9), Mather had noted the marvel of the "Ear-drum," so it was fitting that the final Curiosa in this series described "An Ear Strangely Molested" (CA 2.10). Mather may have chosen to place this account last, because this part of human anatomy seems to have been a special focus of Woodward's "Inquisitive Contemplations." Earwax, in particular, is remarkable for the way it protects the auditory canal from foreign objects. To the annals about ears Mather added the story of a Newtown, Massachusetts, man in whose auditory canal a fly deposited its eggs. Intense pain ensued. Repeated efforts to flush out the contents proved incomplete and agonizing. Finally, lamp oil, and other such "mollifying things," were poured into his

[48] For an earlier example, "*Relating to wonderful Sea Deliverances*," see his *MCA* (1702), 6.1.3–9.

[49] See D.B Hindmarsh (2018).

ear, which brought out twenty or more "Lusty maggots," whereupon the man became "easy." In closing, Mather could not resist a witticism, assuring Woodward that he would not have treated him to a letter with "such things as these crawling in it" if he were sure the information was not acceptable. And so with the typical closing, "Your faithful & hearty Servant," Mather prepared to send off his second "decade" of Curiosa.

Curiosa Americana Series 3 (July–September 1716)

Having sent to the Royal Society his huge manuscript of *The Christian Philosopher* in 1715, Mather resumed his Curiosa correspondence in the summer of 1716. The first eleven missives of this Third Series Mather authored successively from July 2–13 1716, and sent across the Atlantic in the hands of Samuel Woodward, secretary of the Province of Massachusetts Bay. Addressed to Dr. John Woodward, their titles are: CA 3.1 "Monstrous Impregnations"; 3.2 "A Monstrous Calf"; 3.3 "The Nidification of Pigeons"; 3.4 "A Triton"; 3.5 "A Serpent Securely Handled"; 3.6 "A Strange Mischief to the Eyes"; 3.7 "Strength of Imagination"; 3.8 "The Stone Mistaken"; 3.9 "Surprising Influences of the Moon" (a repetition of the previously sent 2.2); 3.10 "Curiosities of the Small Pox"; and CA 3.11 "The *Fagiana*." Along with these eleven letters, Mather included a fresh copy of the Second Series, for he was not sure if they reached their intended destination or were lost altogether. Later that year in September, Mather sent two more Curiosa letters in the hands of Judge Edward Loyd, this time addressed to James Petiver (September 24, 1716): CA 3.12 "*Curiosa Botanica*" and 3.13 "A Singular Lime-stone."

Mather's first two letters took their cue from installments in the first two series by treating the subject of "monsters," which was then a capacious label for physical anomalies in humans and animals. "Concerning the proper meaning of the word 'monster,'" Mather eyed the learned "hair-splitters" among his readers who employed the term loosely to designate "whatever things exceed the common order of things of their kind" and whatever unusual creature the common folk liked to gawk at on market days and in country fairs. However, true monsters were "only those living beings that have a construction of their limbs foully and markedly different from that of the perfect structure of their progenitors." Indeed, such a monster had been discovered in New England: In CA 3.1 "A Monstrous Impregnation" he drew Woodward's attention to the reported interbreeding of a cat and a muskrat in Roxbury, Massachusetts. Some of the litter was muskrat afore and cat aft, some with paws like that of a muskrat, others like those of a cat. In that same vein, CA 3.2 "A Monstrous Calf" broached the suspected miscegenation between a man and a bovine. To demonstrate to Woodward that the cow "had so much of a human visage" that bestiality was a highly plausible explanation for its appearance, Mather provided a drawing that has since been lost. Too unscientific, naïve, and downright gullible for the Society's revered journal? Think again. See "A Relation of two Monstrous Pigs, with the resemblance of Humane Faces," published in the Society's *Philosophical Transactions* by Dr. Edward Tyson (1651–1708), an English physician, anatomist, member of the College of Physicians.[50] Belief in the possibility of interspecies generation was not uncommon at the time. So, too, Mather discussed related questions about biological conceptions of plants, insects, and mammals long before modern genetics were at the cusp of shifting disciplines in the emerging fields of entomology, embryology, and biology. The hallowed interpretations of nature and her causation gradually gave way to Baconian models of empiricism in biology, medicine, and anatomy, which were held up to newly established standards of normalcy by which to examine evidence.

Closely related are the debates about equivocal generation, preformationism, and epigenesis – all precursors to modern genetics.[51] Read closely, several of Mather's Curiosa letters implicitly speculate about the theoretical rudiments of heredity. Did vermin like worms or insects like flies and bees spontaneously generate from the slime of muddy water, rotting vegetation, even from carcasses of animals – as the ancients had prated about abiogenesis? Or did such "simple" entities reproduce sexually and emerge from eggs, as the Italian biologist Francisco Redi (1626–97) seemed to have irrevocably proved in his *Experimenta circa Generationem Insectorum* (1671)? Furthermore, did organisms like the microscopically small animalcules in the blood, semen, or seeds of plants come about through pangenesis (as the followers of Darwin would argue in the nineteenth century)? Or were these animalcules fully pre-formed miniscule entities endowed with heart, nerves, and limbs (homunculi), and embedded in the spermatozoa or ova, though virtually invisible, and passed on during sexual intercourse, while their growth in the uterus was merely an augmentation of the tiny pre-existing beings (preformationism)? The French philosopher Nicolas Malebranche (1638–1715), the Dutch microscopist Jan Swammerdam (1637–1680), and his countryman, the physician Nicolaas Hartsoeker (1656–1725) and their prominent disciples certainly thought so. And viewed under a microscope, his homunculi in the spermatozoa looked something like this:

[50] *PT* 21.259 (December 31, 1699): 431–34.

[51] See J. Needham (1959), P.J. Bowler (1971), S.A. Roe (1981), and C. Pinto-Correia (1997).

Or, perhaps, embryos and plants progressively developed from the undifferentiated mass of an egg cell or spore after fertilization (epigenesis)? Finally, did organisms begin in the egg rather than the sperm as William Harvey (1578–1657), the discoverer of blood circulation, famously posited in his *De Generatione Animalium* (1651), and then develop after fertilization (ovism vs spermism)? Every one of these questions had its proponents and detractors who vigorously championed their favorite scheme. Mather's knowledge of medicine allowed him to keep abreast of the European debate. Among many other sources, he welcomed the rational harmony of cause and effect offered by the Aberdeen minister and mystic George Garden (1649–1733), who strained to reconcile the different theories in his perceptive "A Discourse concerning the Modern Theory of Generation."[52] To Mather, who favored preformationism, this theory made good sense; it helped to explain why children more often than not looked like their parents or even grandparents – boys like their father, girls like their mother. Alas, he soon discovered that preformationism was complicated by well-known cases of hybridity, such as in mules, the offspring of a donkey and a mare, the reported hybrid issue of a musquash and a cat in the wilderness, or the cross-pollination of Indian corn, gourds, and squashes. If preformation was really feasible, then all progeny

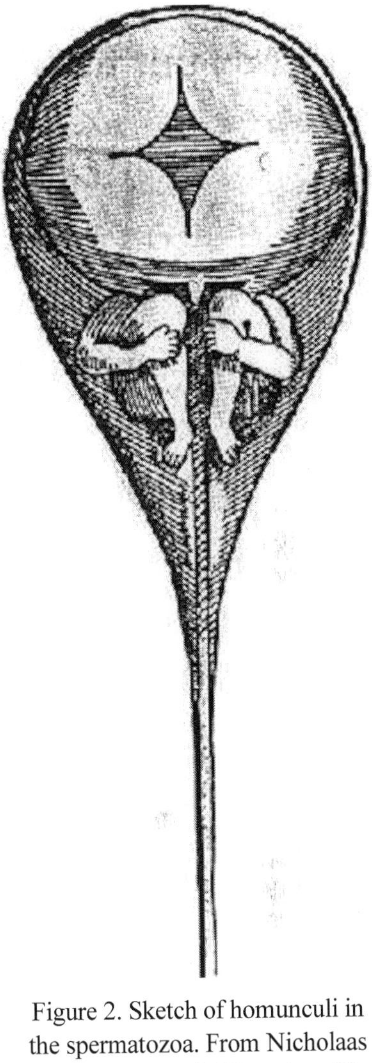

Figure 2. Sketch of homunculi in the spermatozoa. From Nicholaas Hartsoeker, *Essay de Dioptrique* (Paris, 1694), p. 230. Courtesy of Bibliothèque interuniversitaire de santé, Paris.

[52] *PT* 17.192 (February 1691): 474–83.

should either look like their paternal or maternal progenitors – not share traits with both. Something did not quite jibe as yet. "But oh! the palpable Darkness under which we are languishing!" Mather groaned in his *Biblia Americana*, when he perused the volumes of *Anthropologia Nova* (1707), by James Drake, M.D. (1667–1707), fellow of the Royal Society's College of Physicians:

> While I was pleasing myself with Speculations, which I thought were growing towards a fair *Hypothesis*, about the *Generation* carried on in the *Animal World*, I found myself compelled by Dr. *Drake*, to Destroy all again, or at least suspend it, until a, *Melius Inquirendum*, should be served upon it. He minds me, That my *Hypothesis* [preformationism] won't account fairly & fully for *Mix'd Generations*. For besides the Vast Variety of *Mungril Dogs*, Animals of Remoter Kinds have had their prolific Mixtures. The Mixture of a *Bull* and a *Mare*, produces a *Gimar*.[53]

Extraordinary for their emerging discussions of early genetics, Mather's CA 3.1 "Monstrous Impregnations" and CA 3.2 "A Monstrous Calf" would have comfortably fit on the pages of the *Philosophical Transactions* and the German *Ephemerides*, which also featured relations of hybrids and monsters the world over. Again we see how Mather is at the forefront of the proto-genetic debates as he tries to harmonize conflicting theories with his all-pervasive teleology as a means to combat an increasingly mechanistic and despiritualized universe.

In his next letter, "Nidification of Pigeons" (CA 3.3), Mather returned to a familiar subject – "my New English *Pigeons*." Adding to the observations made in CA 2.5 "The Pigeons," Mather focused this ornithological examination on the nest-building (nidification) designs of pigeons. Remarkably instinctual to Mather were the distances at which the birds placed their twigs – not too close so as to restrict air flow yet not too far apart such that eggs could fall through – the intricate nests bespoke for Mather nothing less than evidence of divine teleology. He exhorted Woodward, "You will give me Leave, to add, that the Management of our Birds cannot but compel us into the Sense of an Intelligent Being, who has imprinted on these little Animals, a Disposition to Actions of so agreeable a Tendency" (CA 3.3).

If his insights into natural causation were ahead of their time, his next Curiosa letter was like a plunge into cold water. Mather's discerning mind was on full display in his account of tritons (mermaids) and the recent sighting of one

[53] CA 3.1 "Monstrous Impregnations." See also his commentary on Gen. 6:4 (*BA* 1:598–99); a Gimar (jumart): "offspring of a bull and a mare or she-ass" (*OED Online*, accessed May 22, 2022). According to the *OED*, "Gimar" first appeared in the English language in 1690, in John Locke's *Essay Concerning Humane Understanding* (1690).

in Connecticut (CA 3.4). Aware of the subject's controversial nature, Mather narrated the process by which he came to believe in the nearly unbelievable. As Mather told Woodward, he approached with heavy skepticism the tales of tritons told by Pliny, Theodore Gaza, and Julius Scaliger. And though in reviewing the literature Mather found several accounts that might have contained fragments of truth, his disbelief remained firm. It was not until sources close to him provided a first-hand account that Mather was willing to change his mind: "But now at last, my Credulity is entirely conquered; and I am compelled now to beleeve the Existence of a *Triton*; For such an one, ha's just now been exhibited in my own Countrey; and the Attestations to it, are such that it would be a Fault in me at all to quæstion it." This morphology of conversion, so to speak, is an invaluable record of early Enlightenment modes of proof – of doubting the verity of a theory until positive proof affirms the proposition. And yet, the moderns did not hesitate to resort to mythical proofs just as much as the ancients did when they needed to shore up their own argument.

As if Mather wanted to strike a balance, the letters in the middle of the packet are relatively brief and unencumbered of fantastical material. In CA 3.5, "A Serpent securely handled," Mather returned to the American rattlesnake, an object of fascination for many English readers and would-be colonists. Quelling the prevailing fear of the snake's poison, Mather recounted a potion that Indigenous peoples developed and successfully used against snakebites. In CA 3.6, "A Strange Mischief to the Eyes," Mather passed on the tale of mariners who found themselves temporarily blinded by the vapors excreted from peas rotting in their ship's hold. In CA 3.7, "Strength of the Imagination," he continued a theme from earlier series, writing on the *Potentia Imaginativa* – a psychosomatic link between a pregnant mother's imagination and her fetus in the uterus leading to malformations of the unborn, a topic widely debated among European physicians and anatomists until the end of the eighteenth century. In the absence of any other explanation, the *Potentia Imaginativa* remained the most salient cause of deviations in human anatomy. In CA 3.8, "The Stone mistaken," Mather recounted that a deceased neighbor whose physicians performed a lithotomy to remove bladder stones in fact had a tumorous bladder. The medical causation of bladder and kidney stones was of great interest to him, for both his paternal grandfather and his father died from this painful affliction, and Cotton Mather himself was afraid that his sedentary lifestyle was conducive to the same disease. His chapter on the subject "Flagellum. The STONE," included in his medical handbook *The Angel of Bethesda*, testifies to the horrific agony caused by this condition.[54]

[54] *ABJ* (1972), pp. 82–92.

Apparently wanting for new material, Mather then resubmitted CA 2.2, which he feared had been lost in transit. This apprehension turned out to be more than justified: the uncertainties of transatlantic deliveries as well as the incumbent payment for postage exacted from the recipient were true liabilities. These factors seemed logical explanations why several of his Curiosa letters vanished inexplicably.

Mather's tenth letter (CA 3.10 "Curiosities of the Small-Pox") of this series was highly significant, for in it he began to lay the groundwork for advocating smallpox inoculation. Observing the periodic spread of that epidemic disease in Boston, he noted that smallpox ravished the population every twelve years. He recounted, too, that there was some initial success in stopping the onslaught by applying the methods advanced by the English physician Thomas Sydenham. In the most memorable portion of the letter, however, Mather then wrote on the inoculation technique he learned from his African slave Onesimus: "I had from a Servant of my own, an Account of its being practised in *Africa*. Enquiring of my Negro-Man *Onesimus*, who is a pretty Intelligent Fellow, whether he ever had the *Small-Pox*; he answered, both, *Yes*, and, *No*; and then told me that he had undergone an Operation, which had given him something of the *Small-Pox*, & would forever præserve him from it… ." In essence, Onesimus' narrative confirmed the report of Dr. Emanuel Timonius (1669–1720), a Greek physician in Ottoman Turkey, on attaining immunity through the practice of inoculation. Mather had read this account in the *Philosophical Transactions* and studied it very carefully.[55] Nudging Dr. John Woodward, a member of the Royal Society's College of Physicians, Mather wrote, "How does it come to pass, That no more is done to bring this Operation, unto Experiment & into Fashion, in *England*? when there are so many Thousands of People, that would give many Thousands of Pounds, to have the Danger and Horror of this frightful Disease well over with them. I beseech you, Syr, to move it, and save more Lives, than Dr. *Sydenham*" (CA 3.10). Mather promised himself he would be on the lookout for signs of the next smallpox epidemic and share this new procedure with Boston's medical establishment. It was prescient, indeed.

To this packet of ten letters Mather appended what he described as "a Sort of *Postscript*"– a brief account of a medicinal plant known as Fagiana (CA 3.11). He included a physical specimen of the plant that James Petiver (c. 1663–1718) catalogued in 1717 as an object worthy of study.[56] As a matter of fact, the notable English biologist, pharmacist, and herbalist incorporated the

[55] *PT* 29.339 (June 30, 1714): 72–82.

[56] On Petiver, see K.A. James (2004) and R.P. Stearns, "James Petiver" (1952).

description of Mather's Fagiana in his catalogue *Petiveriana III, seu Naturae Collectanea* (1717) and immortalized Mather's name in this herbal:

> *Fagiana* or *Cancer* Rape, *C.[ottonus] M.[atherus] Orobanche* ramosæ *N. Angliæ,* sub *Fagis* nascens. *Nob.* A Pultiss of this Plant laid to the Part, and a *Tea* of it inwardly, does Wonders in *Cancerous* and *Scrofulous* Cases, as also in *Malignant Fevers*; it strengthens and invigorates after *Fatigues* and Weariness. *Grows* only under *Beech Trees* in *shady Places* from *Midsummer* till *Autumn.*[57]

Just over a month after meeting his Curiosa obligations for the year, Mather addressed two additional letters to Petiver. The first, CA 3.12 "*Curiosa Botanica,*" reports the case of accidental hybridization of Indian corn, gourds, and squashes, hinting that these plants may reproduce sexually. This letter has been identified as the first such recorded discovery in modern times. Again, Mather's Curiosa stands out as a major contribution to botany.[58]

His next letter treated the more mundane subject of shipwrecks (CA 3.13 "A Singular Lime-Stone"). In lieu of bones, Mather included a specimen of Spanish Pieces of Eight, heavily encrusted with limestone and retrieved from the bounty of a sunken Spanish galleon in the West Indies, provided by his friend, church member, later governor of Massachusetts, and treasure seeker Sir William Phips (1651–95). Clearly, Mather's Curiosa spanned the spectrum of what he deemed interesting to his correspondents at the Royal Society.

Curiosa Americana Series 4 (July 1717–January 1717/18)

The Fourth Series of Curiosa, sent in early 1718, is Mather's smallest assemblage, consisting of four letters written over the space of nearly a year and a half. These months were busy and trying for Mather – a mixture of sorrow and bitterness. In December 1716, his eldest daughter Katherin died of consumption. As a last-ditch effort, he desperately pleaded with his friend John Winthrop in New Haven for "a Little Quantity of *Culver's Root,*" which grows in "your vicinity at Lebanon" and is "Famous for the cure of Consumption." Katherin had been languishing for more than six months, and the physicians

57 *Petiveriana* III (1717), p. 12 (sep. pag.), item # 294, in *Jacobi Petiveri Opera* (1764), vol. 2.

58 See C. Zirkle (1934, 1935, 1969); R.P. Stearns, *Science*, pp. 416–17; B.E. Goulet et al. (2017).

had all but given up on her.[59] Less than two months later, she was interred on Copp's Hill Burying Ground a little more than two stone's throws from Mather's North Church. In *Victorina* (1717), his published eulogistic sermon on her, Mather openly confessed his grief: "The *Death* of such an One, and Especially if there has been in such an One a *Constellation* of every thing that could Endear a Daughter, cannot but be a *Killing* Thing unto a Parent."[60] June 1717 saw the start of a long, wearying, and bankrupting process of settling the entangled debts of the Nathan Howell estate, which Mather had unwisely agreed to manage for his new wife, an ordeal that would severely test his marriage to Lydia Lee George (c. 1670–1734). In November, Lydia's daughter Katherine and her two children moved into Mather's house pushing out Mather's own, while Cotton was heartbroken when he learned of accusations against his wayward son Creasy (1699–1723) for fathering a child out of wedlock (*D* 2:484). At some point in this period, too, Mather was "Indisposed in my Health," as he states in his final installment. Small wonder that Mather's Curiosa during this period were few and sporadic.

"Fossils" (CA 4.1), the first item, dated July 25, 1717, was accompanied by a box of "Twenty or Thirty" American paleontological samples, possibly marine in nature.[61] Five years earlier, Mather had promised to supply Woodward with "such *Subterraneous Curiosities*, as may have been in these Parts of *America* mett withal" (CA 1.1) but was then unable to live up to his promise. When, years later, he received Woodward's gift of *Naturalis Historia Telluris Illustrata* (1714), ostensibly as a reminder, Mather realized the urgency of the matter and reiterated his pledge to "make a Collection of our Fossils" and apologized for being "overwhelmed with Employment" (Appendix 12). Meanwhile, Mather reached out to John Winthrop, the son of Connecticut's governor, asking him to "make as full a Collection as may be of the Fossils; (the Names written on each little Bundle:)," adding as an enticement that he would send them to Woodward "in *Your Name*."[62] Winthrop was happy to comply, and Mather was finally able to send to Woodward (July 25, 1717) "a Box, which contains between Twenty and Thirty of such Things as you have

59 "Cotton Mather to John Winthrop" (15^{d}. VIII^{m}. [September], 1716), in *MP* (1868), p. 420; *SLCM*, pp. 220, 224; and *D* 2:372–74, 388–92.

60 *Victorina* (1717), "Introduction," p. i.

61 See Appendices 8 and 12.

62 "Cotton Mather to John Winthrop" (15^{d}. 8^{m}. [October] 1716), in *MP* (1868), p. 419. See CA 4.1. For a subsequent request for marine fossils (which never arrived), see Mather's letter to John Winthrop (9^{d}. III^{m}. [May] 1720), in *MP* (1868), p. 438.

asked for" (CA 4.1).[63] Why this back-and-forth about a few buried pebbles? Fossilized specimens of seashells and marine animals were of great interest to Woodward, who was expecting to include images of these paleontological samples in the third edition of *An Essay towards a Natural History of the Earth* (1723). Woodward was all the more eager to receive fossils from America because of a dispute he had with Elias Rudolph Camerarius, Jr. (1673–1734), a German colleague at the University of Tübingen, who sparred with him in *Dissertationes Taurinenses Epistolicae, Physico-Medicae* (1712) over the question whether seashells found deep in the earth and even on top of mountains were of marine origin and thus proof of Noah's flood, or whether they were sports of nature (lusus naturae) and had thus naturally grown in the places where they were found.[64] Woodward had responded to Camerarius' assault in kind in *Naturalis Historia Telluris Illustrata & Aucta Una cum Ejusdem Defensione; Praesertim contra Nuperas Objectiones D. El. Camerarii* (1714). But Woodward wanted more evidence to bolster his refutation in the next edition. By sending a copy of his repudiation of Camerarius to Mather, Woodward emphasized the urgency of the matter.[65] Mather's CA 4.1, then, was a promissory note of sorts, acknowledging that he was still trying to procure additional geological specimens for Woodward.

The second letter, "An Eclipse Observ'd" (CA 4.2), was addressed to none other than Edmond Halley (1656–1742), the great English astronomer and physicist.[66] Again, this is something of a placeholder, for even as Mather was gathering more samples for Woodward, so too had he vowed to send the famed astronomer "a Manuscript Journal of a late Voyage round o^r^ Globe" (a vow he was unable to fulfill). In the meantime, he provided a set of observations of a solar eclipse gathered by Thomas Robie (1689–1729), a promising young astronomer in his own right. Robie published a yearly almanack and in the one for 1717 he laid out the dates for two lunar and two solar eclipses. "The fourth and last will be a Great *Eclipse* of the *Sun*," Robie informed his readers, "on the 23 of *September* about Noon; Visible to us if it be Fair," beginning 12:29

63 In his article "Election" (p. 83, *n.* 2), G.L. Kittredge traces the network of correspondence between Mather, Winthrop, and Woodward on the latter's request for American fossils. See also *SLCM*, pp. 305–06.

64 The geophysical concept of tectonic plate shifts was not developed until the early 20th century.

65 Also see Appendix 8.

66 Mather had tried to reach out to Halley years before, when he requested John Woodward to ask Halley to "enrich his Transactions more plentifully with Communications of what we call *natural Philosophy*, and also of *Medical Curiosities*, as well as of mathematical Rarities" (Appendix 12).

p.m. and ending 3:11 p.m.[67] Mather appended Robie's calculations, taken at Cambridge, to his letter, dated October 1, 1717.

Starting at the end of February and extending through the end of March 1717, the east coast of North America, from Virginia to Maine, experienced the "Great Snow" (CA 4.3). In as sustained a nor'easter as any on record, anywhere from five to as much as fifteen feet of snow fell. The frigid cold meant that the frozen, compacted snow would recede very slowly. From Philadelphia came reports in late March that the Delaware River was "full of Ice again, and we have as Winter-like Weather as any we have had, and a great Snow upon the ground."[68] Up north, in Piscataqua, "the Snow is said to be five foot deep on a level up the country; we Travel here altogether with Snow Shoes, Horses not being able to pass."[69] Conditions in Boston were no better. "February ended with Snow," the *Boston News-Letter* reported, "and March begins with it, the Snow so deep that there is no Travelling."[70] Mail delivery was delayed everywhere; the sessions of the Massachusetts Superior Court of Judicature for March were postponed "by reason of the Difficulty and Danger of the Road, occasion'd by the vast body of Snow still remaining"; and Governor Shute proclaimed April 4 as a day of public fasting and prayer – though not that many could get out of their houses to attend church.[71]

From his study on Ship Street, Mather passed the time as best as he could. "As mighty a Snow," he told his diary during the last week of March, "as perhaps has been known in the Memory of Man, is at this Time lying on the Ground; and as mighty a Thaw must be look'd for." Perhaps trying to find some humor, he mused, "Would it not be a seasonable, may it not be a serviceable Action, to entertain the People of God, at the Lecture, with a Discourse on Psal. cxlvii. 18. *He sendeth forth His Word, and melteth them*" (*D* 2:439).[72] The following week he wrote, "Such Storms and Heaps of Snow" were still to be seen "as were hardly ever known in the Depth of Winter." Worship services were cancelled, commerce was halted. Many people were "thrown into Straits and Wants"; looking to "Do Good," Mather opened his house to a widow and

67 T. Robie (1716). On Robie's scientific thought, see F.G. Kilgor (1939) and Stearns, *Science* (1970), pp. 426–34.

68 *Boston News-Letter* (April 8, 1717).

69 *Boston News-Letter* (March 11, 1717). See also "John Winthrop to Cotton Mather" (September 12, 1717), *MP* (1868), pp. 681–83.

70 *Boston News-Letter* (March 4, 1717).

71 *Boston News-Letter* (March 18, April 1, 1717).

72 He did in fact follow through with this plan and preached his sermon on April 28, 1717 (*D* 2:507).

her son who were "exposed unto Difficulties" (*D* 2:440). By May 24 he had resolved, "I would make some useful Collections of the rare Occurrences, which were in the Storms and Snows, that we had in the Conclusion of our Winter. I may make a good Use of them. And communicate them [to] the Royal Society" (*D* 2:455). It was not until early December that he had composed CA 4.3 "An Horrid Snow," which not only dwelt on the sheer volume of snow, but on how livestock were buried for weeks, most dying in place, buried in drifts, yet some sheep managed to survive by eating the wool off the backs of other snowbound sheep.[73]

Dated on the first day of the new year (new style), Mather wrote up the information for his final contribution to this series, "A Prodigious Worm" (CA 4.4). In the times before refrigeration and proper preservation of meats, humans widely suffered from intestinal parasitic worms. Mather surveyed accounts from both classical and contemporary physicians with claims of the lengths these tapeworms could achieve, anywhere from twenty-seven to an astounding 800 feet. He closes with an "Instance" from "my Neighbourhood," including a letter from a local physician who described a tapeworm approaching 128 feet in length, vomited up by an enslaved West Indian man after administrating "*Mercurials*."[74] Mather closed by wishing Woodward, to whom this letter is addressed, "Deliverance from all other Annoyers as well as from those we have been discoursing of," a wish he may well have harbored for himself during this time.

Curiosa Americana Series 5 (February 1720)

The new decade proved to be momentous for Cotton Mather. For one thing, his Trinitarian credo had been shaken to the core by his erstwhile friend and correspondent William Whiston, who claimed that St. Athanasius of Alexandria (c. 293–373 CE) had perpetrated a forgery when he interpolated the "Johannine Comma" in the Greek New Testament (1 John 5:7–8) to bolster his case against his Arian opponents. Whiston (1667–1752), the successor to Sir Isaac Newton's Lucasian chair of mathematics at Cambridge, had dared to

[73] The blizzard of 1717 was so unforgettable in New England's collective memory that Henry David Thoreau, more than one hundred years later, would measure a winter's snowfall by the standards of the Great Snow of 1717, in *Walden; or, Life in the Woods* (1854), pp. 129, 139, 273, 284–85.

[74] See also Edward Tyson's "Lumbricus Latus, or a Discourse read before the Royal Society, of the Joynted Worm," *PT* 13.136 (April 10, 1683): 113–44 and Table 146. And his "*Lumbricus Teres*, or some Anatomical Observations on the Round Worm bred in human Bodies," *PT* 13.147 (May 10, 1683): 154–59.

publish this claim in *Athanasius Convicted of Forgery* (London, 1712); worse, in his subsequent *Essay Towards Restoring the True Text of the Old Testament* (1722), Whiston claimed that the prophecies about the Messiah in the Hebrew Bible did not align with the wording extolled by the Apostles in the New Testament, because the Masoretic texts had been deliberately altered or passages removed to prevent Jews in the diaspora from converting to Christianity. To fix this sacrilege, Whiston proposed to take the Apostles' passages and wording about Jesus of Nazareth in the New Testament and restore them to the Old from which they had been altered or removed. Shocked by Whiston's betrayal, Mather immediately set to work and dashed off his "Goliathus Detruncatus" (1713), his "*American Letter to the Learned Mr. Whiston*" (*D* 2:230, 247), to curb any further inroads of Arianism. However, Mather's refutation was never published because his Calvinist friend Dr. John Edwards (1637–1716) at Cambridge (UK), who was to see the manuscript through the press, had died and the manuscript was lost. Furthermore, if Whiston's frontal assault on Christ as the Second Person in the Trinity was not enough, Mather's hopes to unite all Congregationalists, Presbyterians, Independents, and General Baptists by subscription to the Trinity were dashed at the Salters Hall Conference (1719) when many of the assembled representatives refused to sign. He had hoped to promote the ties among the United Brethren on the basis of his "Maxims of Piety," the Creed reduced to no more than three dogmas, and thereby produce lasting harmony among all the major Protestant denominations in the English Empire. All turned out to be a pipedream.

These disappointments notwithstanding, Mather went to work on his many other projects, including a new set of Curiosa letters sometime in 1720. He sent the bundle of twelve essays (Fifth Series) to his former student and Harvard graduate Henry Newman (1670–1743), secretary of the Society for Promoting Christian Knowledge (SPCK) in London, to be communicated to John Chamberlayne (c. 1668–1723), secretary of the Royal Society, having heard that John Woodward had died.[75] The new package included the following essays:

> 1. A strong Digester;[76] 2. The Quadrature of the Circle; 3. The Whale; 4. Prodigies; 5. Odd Effects of the Thunder; 6. A large Description of the

75 See Mather's Letter to Henry Newman, Feb. 17, 1720, MS draft, AAS (*SLCM*, pp. 301–02); and Newman's reply, September 17, 1720, MS (AAS).

76 Parts of this lost epistolary essay may well have been embedded in "The Main Wheel scoured and oiled. Or, Help for, The Stomach depraved. [and, Vomiting.], in *ABJ* (1972), pp. 195–201, as well as in *CP* (1721), 206–07, 266–68.

Mississipi, by one who sojourned several years in that Countrey;[77] 7. A Rocquing Stone; 8. A Sea-Lion; 9. A Water-Dove; 10. The Locusts; 11. The *Aurora Borealis*; 12. A Journal of a Voyage around the World; [13. A Relation of a New Burning Island].[78]

Of these thirteen letters, eight have perished, while substantial parts of the remaining five were excerpted in Mather's other venues or published separately. This loss is unfortunate, for even the draft copies that Mather generally preserved are missing as well. Significantly, John Chamberlayne acknowledged their safe arrival, claiming he had forwarded them to Sir Isaac Newton and to Edmond Halley, and responded to Mather (August 31, 1720) with profuse praise:

> I thank you for your Noble Entertainment with which so many of my [Chamberlayne's] Friends were Regaled, before I could gett a Snap for myself, who hungred & thirsted for it, that I had not the pleasure thereof till very lately, and indeed too late to communicate the same pleasure to your Illustrious Brethren, the Gentlemen of the R.S. who have always a long Recess at this time of the year. As soon as I had perused your learned Letters, I Immediately communicated them unto Sr *Isaac Newton*, our Excellent Præsident; And he has referred them to the most Learned Mathematician and Great Traveller, Dr Halley, whose Report I expect with no

77 This may very well be a larger version of "A Short Account of a Trading Voyage performed by Joseph Kellug an English Man of New England in Company with Six French Men from Canada to Missasippi in the year 1710 in two Cannoos made of Birch Bark, with some general remarks made by Said Kellug," by Joseph Kellogg (1691–c. 1756). Paul Dudley (1675–1751) of Roxbury, wealthy son of Massachusetts' Governor Joseph Dudley (1647–1720) and justice of colonial Massachusetts' Superior Court (1718–45), appears to have made Kellogg's acquaintance in the Bay and received from him the story of Kellogg's voyage down the Mississippi River when Kellogg was a young man of about nineteen. Dudley took it down in his own hand and communicated Kellogg's "A Short Account of a Trading Voyage" to John Chamberlayne (March 15, 1720/21); it was subsequently read to the RS (May 11, 1721) but remained unpublished. The manuscript of Kellogg's account, as mediated by Dudley, is still extant in the Royal Society's Journal-Book XIII, pp. 93–94; as is Dudley's letter to Chamberlayne, which is preserved in the Society's Guard-Book (MS. D-1, no. 73). See R.P. Stearns' "Joseph Kellogg's Observations" (1936).

78 See Mather's appended list in our CA 7.0. Promissory Letter to James Jurin (May 21, 1723) and G.L. Kittredge, "Scientific," pp. 45 (43–44 *n.*). It appears that item 13. A Relation of a New Burning Island was *not* included in the 1720 shipment to Chamberlayne but separately published as a public address to him in June 1721 (*D* 2:626–27); see Kittredge's "Scientific" (pp. 45, 46–47) and "Lost Works "(pp. 440–41, 463 *n.* 2).

> small Impatience. But, as I beleeve, you expect to be informed of their Destiny, with respect to their coming to my hands, I was resolved not to slip opportunity of telling you, They are safe arrived.[79]

There is yet more to come. In another epistle to John Winthrop (12^{d} $1.^{m}$ [March] 1722), Mather expressed his frustration that even though Chamberlayne informed Mather that his "XII Letters to him [Chamberlayne] have been published … yett I have never to this Hour seen them: nor have they that have published I know not how many other things of mine, ever given the sight of them; (or else they have miscarried.) I have done, I have done, with Expectations from This World."[80] Curiously, none of the documents were ever published in the *Philosophical Transactions* in spite of Chamberlayne's asseverations to the contrary. Lost, too, were Mather's holograph manuscripts at the Royal Society. All that survives of the original thirteen documents are portions, repurposed (as was his habit) in his other publications or in different venues. Our reconstruction of his documents in the Fifth Series are largely based on these.

Although nothing of the first item, CA 5.1 "A strong Digester," appears to be extant,[81] much to our benefit, we get a glimpse of what was contained in Mather's lost CA 5.2 "The Quadrature of the Circle." Surprisingly, it comes from an unexpected source: An odd passage in *The Abuses and Scandals Of some late Pamphlets In Favour of Inoculation of the Small Pox* (1722), by William Douglass, M.D. (c. 1691–1752), provides further evidence of the Fifth Series' safe arrival in London. How Douglass, Mather's old nemesis in the matter of Boston's smallpox inoculation crisis (1721–22), could have learned of Mather's essay is not hard to determine – either Douglass had been apprised of the matter by Chamberlayne himself; or by Alexander Stuart, M.D. F.R.S (1673–1742), Douglass' correspondent at the Royal Society's College of Physicians and collaborator on Douglass' *Inoculation of the Small Pox* (1722) and *Abuses and Scandals* (1722);[82] or by Paul Dudley, F.R.S. (1675–1751), wealthy son of Governor Joseph Dudley and justice of

79 John Chamberlayne as quoted in Mather's Letter to John Winthrop, 26*d.* X*m.* [December] 1720, in *MP* (1868), p. 444.

80 "Cotton Mather to John Winthrop" (12^{d}. 1^{m}. [March] 1722), *MP* (1868), p. 455.

81 Since Mather was in the habit of repurposing useful information, some minor portions of it may have been digested in his discussion of the stomach, in "The Main Wheel scoured and oiled. Or Help for, the Stomach depraved. [and, Vomiting.]," in *ABJ* (1972), pp. 195–201, as well as in *CP* (1721), pp. 206–7, 266–68.

82 See "William Douglass' Letter to James Stuart" (September 25, 1721), Royal Society (Records, EL/D2/2 Early Letters).

colonial Massachusetts' Superior Court (1718–45), who was also Mather's competitor of sorts in vying for attention at the Royal Society. For instance, Dudley warned John Chamberlayne (April 14, 1722) that Mather had already published his *World Alarm'd* in Boston (1721), i.e., CA 5.13 "*The World Alarm'd,*" even though Mather had submitted it for publication to the Royal Society. Dudley included a copy of Mather's printed work.[83]

The animosity between William Douglass and Cotton Mather is well known. In his *Abuses and Scandals*, Douglass singled out for ridicule Mather's "The Quadrature of the Circle" and his attempt to devise a system of establishing longitudes for marine navigation at sea. Douglass mocked,

> Tho' unacquainted with the first Principles of the *Mathematicks*, [Mather] sends Home a Solution of Two the most abstruse Problems or *Desiderata* in Mathematicks, *viz.* the SQUARING OF THE CIRCLE, which is only to be expressed by *Approximation*, or an *infinite Æquation*, whose Nature is known, but whose Value cannot be reduc'd to any determined Lines or Numbers; the other is the LONGITUDE at Sea, the Discovery of which for many Years has been endeavoured by the best *Mathematicians*, encouraged by large *Præmiums*; and is generally thought will continue a Secret, till some Comet or other Cælestial Body appear within our *System*, and perform his Revolutions in a N. and S. Line, or some great *Circle* not much declining from a *Meridian*, as at present we find the *Latitude*, by the Suns Revolution in a Circle not much declining from the *Æquinoctial*, or E. & W. Line, or till an *Automaton* (Clock-work) can be contrived, which shall for some considerable Time move *æqually* or *æquably*, without being affected by the Weather or external Motion; or till we find some Contrivance, notwithstanding the Motion of the Ship, to take the *Altitudes* of the *fixed Stars*, and the *Eclipses of Jupiter's Moons*. A Man guilty of such *Absurdities*, is no good Voucher for an Experiment of Consequences.[84]

No doubt, Douglass' stinging ridicule did not miss its target and Mather must have been chagrined with puzzlement at how his antagonist had found out about the contents of the missing documents. Be that as it may, Douglass' invective did serve (in hindsight) an unintended purpose. His summary provides a glimpse of what Mather's lost Curiosa letter contained: The Quadrature of a Circle and Determining the Longitude at Sea.

Whether or not Mather was too possessed of himself in trying his hand on a two-thousand-year-old mathematical problem of squaring the circle – i.e., squaring the area of a circle exactly with that of a square with no other

[83] Stearns, *Science* (1970), pp. 462–63, *n.* 160.

[84] Douglass, *Abuses and Scandals* (1722), pp. 6–7.

Figure 3. Das ist Ein Gebenedeyter Spagyrischer Wegweiser, From Herbrandt Jamsthaler, *Viatorium Spagyricum* (Frankfurt am Mayn, 1625), p. 272. Courtesy, Wellcome Library, Medical Heritage Library.

instrument than ruler and compass – is a matter of debate.[85] That over the centuries hundreds of experts and would-be mathematicians tried to accomplish the feat – and failed – is perhaps as irrelevant as the fact that the French Society of Science of 1775, in its prestigious *Le Journal des Sçavans* (January 1, 1775), foreswore to accept any more essay submissions that attempted the task. So why not Mather? Perhaps he wanted to offer (belatedly) his own mite to the Hobbes–Wallis Controversy (1669–78) over the quadrature of the circle, a captivating topic that retained its currency in the volumes of the *Philosophical Transactions*.[86] If anything it demonstrates that he was never able to

85 In his letter to his friend Jeremiah Dummer (January 19, 1719), Mather claims that he had "never heard of any premium for squaring the circle" to be given to the person who solves this problem in geometry (*SLCM*, p. 272).

86 See A. Bird (1996); and D.M. Jesseph (2000).

resist a good challenge. For him all sciences were interrelated and serviceable, or to put it differently, natural philosophy – including geometry, mathematics, and astronomy as taught at Harvard[87] – were helpmeets to theology, the queen of science.

By the way, did anyone ever succeed in solving the quadrature (squaring) of the circle? Well, even freshmen of algebra now know that this feat was not accomplished until 1882, when the German mathematician Carl Louis Ferdinand von Lindemann (1852–1939) established for all times that squaring the circle was an *insoluble* problem of algebra, because π = 3,14159 ¥ is a transcendental constant in Euclidean geometry in which the ratio of a circle's circumference to its diameter equals π. Thus the area of a circle – no matter its size – can never be identical with that of a square without allowing for arithmetic approximation.

What about Mather's second element in his letter to Chamberlayne – discovering a simplified way to establish a ship's longitude at sea? And why would he want to bother in the first place? The Longitude Act of 1714 offered a monetary reward of £20,000 to anyone who devised a simple method of determining the longitude of a ship's location at sea within half a degree. This time, Mather may have been in it for the money.

It is well known that before telegraphs, radios, and satellites, the navigation of ships – especially when intercontinental seafaring became vital for trade in the sixteenth and seventeenth centuries – depended on a navigator's ability precisely to calculate the whereabouts of his vessel at sea within a half a degree of the globe's prime meridian east or west of a north–south axis. Doing so required accurate observations of the night sky that, in the absence of complete astronomical tables, depended on a fixed timescale at the port of departure and arrival, a ship's unwavering chronometer, a telescope, and knowledge of the position of fixed stars and lunar distance at sea. No doubt, determining a vessel's position relative to its port of embarkation and desired point of entry was fraught with surprising, often deadly, inaccuracies. Did anyone win the award during Mather's lifetime? No. Not even the English carpenter turned clockmaker John Harrison (1693–1776), who designed the most accurate and reliable maritime chronometer over a period of half a century when he published his findings in *The Principles of Mr. Harrison's Time-Keeper* (1767). He received only a portion of the grand prize.

Mather's third item in his Fifth Series of Curiosa letters is less concerned with the conundrums of geometry and chronometrical navigation than with a subject in some respects much closer to his theological purview: cetology.

[87] S.E. Morison (1936) 2:208–22; T. Hornberger (1945); R.P. Stearns (1970).

CA 5.3 "The Whale" again evinces Mather's characteristic approach: Beginning with a Plinyism not unlike that of a fable, Mather opened his account with a humdinger specimen of a whale "six hundred Foot long, and three hundred and sixty broad."[88] Such a colossus, he insisted, is pure legend, even if some of the Church Fathers indulged in it as well. Much more trustworthy were the examples of beached whales described by Gesner, Brerewood, Rondeletius, Hakewill, Hartenius, Willughby, Ray, and others, whose physical descriptions brought these mammalian giants down to their proper size, emphasized their utility as suppliers of costly oil, and recounted the nursing habits of their young.

How does the biblical story of Jonah and the whale hold up against such tangible evidence? What sort of "great fish" was it that swallowed him up in the Mediterranean, carried him unharmed in its belly, traveled around Africa's Cape of Good Hope up the shallows of the Tigris River, and vomited him out on dry land near Nineveh three days later (Jonah 1–2:10)? Surely, the biblical story comes in for another Plinyism – unless probable evidence to the contrary could be mustered to uphold the historical sense and certainty of this seeming miracle. Mather, like many of his contemporaries, wrestled with the likelihood of this venerable story as they subjected it to the test of reality and found it wanting. While we might imagine some of the old tars among the whalers of his congregation snickering in disbelief, Mather tried to rescue Jonah and the whale from mere Christological allegory. Jonah's "great fish" was a species of toothless whale shark, the "*Canis Carcharias*," or dogfish, whose throat was wide enough to gulp down anyone coming near it – as certain French travelers confirmed when they dissected a beached whale near Valencia (Spain) and found two men in its belly. From a Baconian and theological standpoint, then, Mather resolved the dispute by subjecting the biblical narrative to the test of reality. He saved it from dismissal as typological allegory by supplying contemporaneous evidence in support of its probability. The letter and the spirit of the Bible remained inviolate.

Of the ten remaining letters in the Fifth Series, only four survive in one form or another: CA 5.9 "A Water-Dove"; 5.10 The Locusts; 5.11 "*A Voice from Heaven*"; and 5.13 "*The World Alarm'd*. A Surprizing Relation, Of a New Burning Island." Their content and location are taken up in sequence.

CA 5.9 "A Water-Dove" is particularly noteworthy. As an unknown (perhaps extinct) species of American waterfowl, it served Mather's dual purpose as a specimen of scientific and theological interest. The account originated with his young friend John Winthrop (1681–1747) of Connecticut and in

[88] Pliny the Elder (23–79 CE), *Natural History* (9.2.4).

scientific manner describes the bird's migratory habits, size, coloring, mating calls, Indian name, and buoyancy in water.[89] It is this latter aspect that renders this water-dove particularly significant to Mather, for its scientific interest intersects with his theological imperative to square the biblical account of the dove's threefold emission from Noah's Ark with the test of reality. Would Noah send out a bird to scout the terrain covered by the floodwaters if the fowl could not swim as well as fly?

In the Age of Newton, all people of reason sought to solve problems by exercising their mind. This rationalist approach was not suspended when discerning minds read the Bible. This truism is also applicable to CA 5.10 "The Locusts." Like its previous letter in the Fifth Series, "The Locusts" served Mather's intent to subject time-honored miracles in the Bible to the test of reality. As the venerable story goes, when the Israelites in the Sinai desert were again lusting after the fleshpots of Egypt, the Lord quenched their desire by bringing forth "quailes from the sea, and let them fall by the campe" all around the people as far and wide as "a dayes journey … and as it were two cubits *high* upon the face of the earth" (Numb. 11:31). Mather had second thoughts about this hyperbolic story. How are interpreters to keep in line with probability if unimaginable numbers of quails piled up in heaps two cubits high all around? "The *Quails* must have been unavoidably choaked & stiffled, if they had been heaped so deep one upon another," Mather averred. "[I]t cannot be conceived without *Heaps* of *Miracles*," he cagily punned. The wise Creator of the universe would not be so wasteful as to resort to miracles when the common laws of nature sufficed. Surely, the translators must have made a mistake when they translated שְׂלָו ["*Selav*"] as quails! "But the thing will be found less Wonderful, & very Natural, if it be understood of *Locusts*, who come in thick & vast Clouds, which darken the Sky" (*BA* 2:869), Mather speculated as he cast about for a solution more in line with probability. He found his answer in *Appendix Secunda ad Historiam Aethiopiam Iobi Ludolfi continens Dissertationem de Locustis* (1694),[90] by the German orientalist and grammarian of the Ethiopic language Hiob Ludolphus (1624–1704), who supplied Mather with a linguistic retrenchment that cut the alleged miracle down to size: the Hebrew noun שְׂלָו ["*Selav*"][91] should *not* be rendered "*coturnicibus*" (quails) as all the translators have done but "*locustis*" (locusts). This

89 See "Cotton Mather to John Winthrop" (30d IXm. [November], 1719), "Cotton Mather to John Winthrop" (11d. XIm [January] 1719/20), and "Cotton Mather to John Winthrop" (14d. Im. [March] 1720–21), respectively in *MP* (1868), pp. 435, 436 *n.*, and 447).

90 Pars 2: *De Locustis*, cap. 1, p. 21, §§1–5.

91 Exod. 16:13; Numb. 11:31, 32, 33; *BA* 2:868 *n.*

simple solution went hand-in-hand with the well-known "*Acridophages*, or *Locust-Eaters*" in the region who largely lived on dried, smoked, and salted locusts throughout the year (*BA* 2:185). The certainty of the Bible was preserved once again. To be sure, Mather was not given to indulging in such Plinyisms as locusts "Three Foot long, & some Four Cubits," whose thorny legs, when dried, were employed "instead of *Sawes*," and teeth similar to those of lions – as some of Mather's hyperbolical brethren were inclined to do in their pious endeavor to instill faith by embellishing miracles (*BA* 2:184).

If the quails turned locusts (CA 5.10) bespeak the hazards of linguists to unearth meaningful equivalents for the arcane vocabulary of a dead language, then Mather's *The Aurora Borealis* (CA 5.11) was much more the stuff acceptable to the members of the Royal Society. In fact, the *Philosophical Transactions* contains numerous contributions on the subject. To be sure, Mather demonstrates (as usual) that he had done his due diligence by citing the standard astronomers great and small who held forth on this celestial phenomenon – Childrey, Cassini, Gassendi, Hooke, Hartsoeker, Spencer, Bayle, Halley. In his day, natural philosophers were still uncertain about the precise causes of the aurora (and its meaning): Was it a comet that crisscrossed the night sky and illuminated the heavens like a flaming fire? Was it the finger of God whose divine prodigies presaged wars, death, and destruction as they did in olden times? Certainly, the scientist in Mather aligned himself with those who insisted on natural explanations. He rejected occult causes, as did John Spencer and Pierre Bayle, even as Mather insisted that Cartesian mechanism (by which the material world is governed) fell short of comprehending God's almighty power, which is not limited by the laws of His great machine. Or, to put it another way, while paying lip service to the Deist vocabulary of the day, Mather, the pious physician of the soul, was supremely unwilling to relegate the Savior of mankind to the sideline of his creation. He tried all he could to harmonize the letter of Scriptures with scientific knowledge – even as he shied away from resorting to miracles.

CA 5.13 "*The World Alarm'd*" is the last item in the Fifth Series – though Mather also published this epistolary essay separately, a year after he had mailed his package of letters to John Chamberlayne. The topic of a marine volcano erupting without warning and raising a new island from the depths of the Atlantic Ocean was of particular interest to Mather; it underscored his millenarian speculation that the end-time prophecies (Matthew 24) had all been fulfilled, that the predicted Conflagration at Christ's Second Coming (2 Peter 3) was on the horizon, that this world would end with a fiery bang, *not* a whimper. For Mather and his fellow millenarians, this was the kind of palpable evidence that matched natural science with biblical prophecies –

the Messiah's return in clouds of fire was imminent, and Mother Nature herself was preparing for the event. In customary fashion, Mather spiked his narrative with authoritative sources of Greek and Roman philosophers, early Church Fathers and fellow millenarians, and (most important of all) modern travel accounts from around the globe – all confirming the multitude of fiery volcanoes and their likely explosions in Asia, Africa, Europe, and the New World. With this backdrop in place, Mather readied his readers for the high point of his discourse, the eyewitness report of New England's own Capt. John Robertson, who had observed the submarine eruption near Terceira in the Portuguese Azores the year before. The honest man provided Mather with a detailed scientific description of the event – date, time, location, witnesses, boiling seawater, and pyro-plastic effects on flora and fauna. What more could the Royal Society have wished for? Mather's lengthy account ends on a hortatory note and reminder that time is short. Whether or not this closure appealed to Chamberlayne is anyone's guess, not however Mather's flattery of associating the secretary of the Royal Society with the celebrated French natural scientist Nicolas-Claude Fabri de Peiresc, one of the greatest polymaths of early modern Europe, who commanded knowledge of sixteen languages.

Curiosa Americana Series 6 (post-1 December 1721)

The Sixth Series of Cotton Mather's Curiosa letters to the Royal Society includes some of the most important documents in the history of epidemiology as practiced in New England; they reveal his keen insights into the emerging smallpox inoculation practices just as much as they demonstrate his relentless determination to save people's lives even against their vehement opposition. To be sure, Mather was not a practicing physician in the modern sense but a perceptive student of new developments in all of the natural sciences of the day, especially Newtonian physics and early modern medicine. More important yet, he was able to discern etiological links between newly discovered microscopically small animalcules (bacteria), their multiplication in the blood, and the spread of contagious diseases such as measles and smallpox, which frequently reached epidemic proportions. He was at home in the medical theories of Hippocrates, Galen, Avicenna, Sydenham, Malpighi, Borelli, Leeuwenhoek, Hartsoeker, van Helmont, Andry, and Marten, as much as he was in the iatrochemical and iatromechanical theories of his European peers.[92]

[92] See *especially* O.T. Beall, Jr. and R.H. Shryock, *Cotton Mather: First Significant Figure in American Medicine* (1954, 1979) and G.W. Jones, *ABJ* (1972), "Introduction," pp. xi–xl.

Unfortunately, establishing the exact number and contents of some items in the Sixth Series is fraught with some difficulty. At least seven of the twelve letters in Mather's itemized list have perished (CA 6–12): either because the package of letters never arrived at its intended destination or, more likely, because the individual letters were split up after delivery and scattered among the various members of the Royal Society tasked with evaluating their content for potential publication in the *Philosophical Transactions*. Of the five surviving documents (CA 1–5), only one was published in full and the remaining ones were repurposed in different venues.

According to his diary entry, Mather shipped his bundle of letters to London sometime after December 1 1721, having worked on several items at least since November 24.[93] It is highly likely that the set was sent to his old acquaintance Henry Newman, secretary of the SPCK, who had also conveyed Mather's Fifth Series to the Royal Society – with almost the same devastating results.

Not receiving any response from London and fearing the loss of yet another batch of letters, Mather inquired about their status in his subsequent letter (CA 7.0) to James Jurin (May 21, 1723) by appending a list of the Curiosa Mather sent to the Royal Society from 1713 to 1721. The added list includes unaccounted-for items that might help Jurin, Mather hoped, track down the missing pieces. The loose-leaf list appertaining to the Sixth Series includes the following items:

> 1. *An Account of the Method & Success of Inoculating the Small-Pox* ("To Mr. [Jeremiah] *Dummer*"). 2. *Variolæ triumphatæ*; or, The Small-Pox Inoculated. 3. A Further Account, of the Small-Pox Inoculated. 4. *Curiosa Variolarum*. 5. *Nishmath-Chajim*; The probable Seat & general Cure of all Diseases. 6. The Seventh Son examined, with a Touch upon the Kings Evil. 7. *Lacus Mirabilis*. 8. An Horrible Tempest. 9. *Ostreophagi*; or, a Matchless Oister-hill. 10. *Absinthium Sempervivum*. 11. Ambergrise discovered & determined. 12. *Nævi Materni*. 13. *Melissologia*; with a New Method of Bee-hunting. ["All these to Dr. *Woodward*"].[94]

Before turning to these items, we need to understand the backstory to Mather's writings on smallpox inoculation. Unbeknownst to the medical establishment in England and her colonies in North America, this immunological method of transplanting smallpox had long been practiced as a folk remedy among Greek villagers, generally undertaken by an elderly female practitioner

93 *D* 2:660, 661–62; Kittredge, "Lost" (pp. 461–77).

94 See CA 7.0 "Cotton Mather to James Jurin" (May 21, 1723), *n.* 12.

(Mather's "inoculatrix") in Ottoman Turkey, as well as in parts of Africa.[95] Meanwhile in New England, always looking out for ways to render himself useful to his community, Mather was well-advised to keep track of the various epidemics that had periodically swept through Boston since the Bay Colony's establishment in 1630. For instance, in November of 1713, "a *Month* which *devoured* my Family," he lost his wife and three of his youngest children along with their servant maid to a devastating measles epidemic.[96] Three years later, in his earlier letter to John Woodward (July 12, 1716), our CA 3.10 "Curiosities of the Small Pox," Mather calculated that the smallpox and endemic measles contagion seemed to recur in full force every twelve years, abating thereafter until a new generation of young Bostonians had grown up who had not yet attained immunity. Better yet, while pouring over the extract of his own Curiosa letters (First Series) published in the *Philosophical Transactions* in 1714,[97] he also found an article by Em(m)anuel Timonius (1669–c. 1720), a Greek-Ottoman physician, whose report on inoculation as practiced in Greece followed Mather's summary as the next item in the same issue. Mather must have been delighted to discover Timonius' "An Account, or History, of Procuring the Smallpox by Incision, or Inoculation; as it has for some time been practiced at Constantinople,"[98] for it confirmed similar accounts of the practice in Africa Mather had received from several enslaved Africans in Boston, who showed him the scars from the incision as undertaken in their home countries.[99] Ingeniously, Mather's quickly connected the two similar accounts in 1714 but waited till he had additional confirmation before taking action.

95 Lady Wortley Mongue (1689–1762), spouse of the English ambassador in Constantinople (Istanbul), was one of the first western Europeans to have inoculation performed on her three sons (1717) under the supervision of the Scottish surgeon Charles Maitland (c. 1668–1748) while in Greece. See T. Heffernan et al. eds (2012), K. Wisecup (2013), and H. Esfandiary (2022). R.P. Stearns (1950) notes that historians disagree about the time and means of introducing smallpox inoculation in England.

96 For the depredations of this epidemic on his family, see *D* 2:248–66 (qt. at p. 266). When Boston's medical establishment declined to make their services freely available to the public at large, Mather composed and published his pamphlet "A Letter, About a Good Management under the Distemper of the Measles, at this time Spreading in the Country. Here Published for the Benefit of the Poor, and such as may want the help of Able Physicians" (1713). See also D.M. Morens (2015) and P.E. Kopperman and J. Abrams (2019).

97 Appendix 6, and *PT* 29.339 (June 30, 1714): 62–71.

98 *PT* 29.339 (June 30, 1714): 72–82.

99 CA 3.10 "Curiosities of the Small-Pox" (July 12, 1716).

His opportunity came when the next smallpox epidemic began to take its toll after the arrival of the *Sea-Horse*, a British man-of-war that carried some infected sailors on board from the West Indies (May 12, 1721).[100] "The grievous Calamity of the *Small-Pox* has now entered the Town," Mather recorded in his diary (May 26, 1721). "The Practice of conveying and suffering the *Small-Pox* by *Inoculation*, has never been used in *America*, nor indeed in our Nation. But how many Lives might be saved by it, if it were practised? I will procure a Consult of our Physicians, and lay the matter before them" (*D* 2:620–21). While he was ruminating about how best to mobilize Boston's physicians, he began composing a "Little Treatise on the *Small-Pox* … exhibiting the best Medicines and Methods, which the world has yet had for the managing of it; and finally, adding the new Discovery, to prevent it in the way of Inoculation. It is possible, that this Essay may save the *Lives*, yea, and the *Souls* of many People" (*D* 2:627–28). And so he addressed *A Letter to the Physicians of Boston* (June 23, 1721) and circulated it among the medical establishment to discuss the best way to proceed in the spreading epidemic (*D* 2:628). All but one physician refused to get involved.[101] The ensuing pamphlet warfare between Mather's supporting clergymen and the handful of anti-inoculators as represented in William Douglass, M.D. (c. 1691–1752), Boston's sole physician with an M.D. degree, is well known. It will also emerge in the Curiosa letters included in the present edition.

It is fortunate that Mather decided to send the first item of his Sixth Series (September 7, 1721) to his old acquaintance Jeremiah Dummer (1681–1739), Harvard graduate (1699), well-heeled colonial agent for Connecticut and Massachusetts in London. Dummer published it as an open letter addressed to the distinguished Sir Hans Sloane (1660–1753), president of the College of Physicians of the Royal Society. Mather's *An Account of the Method and Success of Inoculating the Small-Pox, in Boston in New-England* (1722) was certainly most apropos, given the devastating smallpox epidemic then taking

[100] A Report of the Record Commissioners of the City of Boston, Containing the Boston Records from 1700 to 1728 (1883), p. 154.

[101] On the date and remnants of Mather's *Letter*, see Kittredge, "Lost" (p. 428, *n.* 1). Much the same happened when Mather tried to unite Boston's physicians behind combating the measles epidemic in 1713 – he was forced to go it alone: "That which I have proposed unto my Physicians," Mather recorded in his diary (November 4, 1713), "I have now done myself, upon their Declining (tho' highly Approving) of the Action." For the benefit of the poor who could not afford a physician, he published and distributed at his own expense *A Letter, About a Good Management under the Distemper of the Measles* (1713), yet expecting that "doubtless my Action may expose me to some Invectives" (*D* 2:272).

its toll in Boston as well as in London, with dozens dying on a daily basis.[102] So, too, it made good sense to make it public in London as a pamphlet to reach the largest number of physicians in England rather than wait for the Royal Society to circulate it among its experts to have it vetted prior to publishing it in the *Philosophical Transactions*. Precious time would be lost. Dummer's introduction also gave Mather the cover of anonymity to ward off the fury of the Boston mob, which culminated in an attempt to assassinate Mather in the wee hours of November 15, 1721, for purportedly spreading the smallpox (*D* 2:657–58).

What is more, Dr. James Jurin F.R.S. (1684–1750), who served as secretary of the Royal Society (1721–27), might not have looked upon Mather's long, chatty letter with favor – though several smaller pieces on smallpox were published in the *Transactions* before Jurin took over as secretary in 1721. Later ones in the *Transactions* probably responded to Jurin's call for reports on the success or failure of inoculation performed in the United Kingdom.[103] At the same time, Jurin knew that the Society was ambivalent about certain medical topics, and (at any rate) he would have needed the approval of the majority of his physician colleagues to publish Mather's *Account*. A case in point, Jurin declined the essay on "the History of ye Epidemick Colick," submitted by Dr. John Huxham (1692–1768), because "many Gentlemen of ye Society object against printing much of Physick or Chirurgery in them [*Philosophical Transactions*], as not being so properly ye business of ye Royal Society." Apparently alluding to the Danish *Acta Medica et Philosophica* or to the German *Miscellanea Curiosa Medico-Physica Academiae*, Jurin regretted that "It were indeed much to be wish'd, that we had something printed under ye Title of Acta Medica or some such like, where this sort of Papers might properly come in & be constantly publish'd, for I am under a necessity of suppressing many valuable Papers of this kind."[104] Indeed, Mather knew enough about the Royal Society's slow vetting process to take matters in his own hands. There was no time to waste.

Mather's *An Account of the Method and Success of Inoculating the Small-Pox* (CA 6.1), perhaps uncharacteristic for most of his other Curiosa letters

[102] See James Postlethwayt et al., *A Collection of the Yearly Bills of Mortality, From 1657 to 1758 inclusive* (1759). J.L. Carrell (2003), G. Williams (2010), and C. Silva (2011).

[103] Jurin, "Advertisement," in *PT* 32.370 (March 31, 1722). This call for submission on inoculation was repeated as an appendix to his *An Account of the Success of Inoculating the Small Pox in Great Britain, for the Year 1724* (1725).

[104] "James Jurin to John Huxham" (August 3, 1727), in A. Rusnock, editor. *Correspondence*, pp. 364–65.

sent in previous years, opens with the testimony of his "*Garamantee* Servant" Onesimus, whom Mather had interrogated about smallpox inoculation as practiced in Africa.[105] Onesimus' version was independently validated by several other Africans in Boston who showed Mather the scar tissue on their arms as evidence of the inoculation knife cutting their skin. To Mather their simple description delivered in broken English had the ring of truth – and it confirmed the reports of Dr. Emanuel Timonius as published in the *Philosophical Transactions* in 1714, as well as that by the Greek physician Jacobus Pylarinus (1659–1718), Venetian consul in Smyrna, and published in the *Transactions* in 1716.[106] Next in Mather's *Account* follows a detailed description of the inoculation process: how the variolous matter is extracted from the pustules of a healthy survivor of the pox, how it is applied to arm or leg of a healthy person, how the patient's immune system responds during incubation and maturation – running fever, pustules developing on the skin – and how the physician's bloodletting and dietary measures mitigate a patient's discomfort. "But to what Purpose is all this *Jargon?*" Mather interrupted his own report. "And of what Significancy are most of our *Speculations?* EXPERIENCE! EXPERIENCE! 'tis to thee that the Matter must be referr'd after all: a few *Empericks* here, are worth all our *Dogmatists*." The would-be preacher-physician Cotton Mather had turned medical empiricist!

Perhaps more important, *An Account* also offered an etiological explication (right or wrong) of how inoculation is more effective and less dangerous than the common way of contracting the variola virus through inhalation of the miasma encountered in the stifling chambers of the sick and dying. Since the contagion entered the body from the arms or legs of an inoculated person, Mather theorized, the pollutant enters the body "only by the *Out Works* of the *Citadel*, and at considerable Distance from it." The patient's "*Vital Powers*" are thus "kept so clear from his Assaults, that they can manage the *Combat* bravely." Conversely, if the invasion occurs through inhalation of the lungs, the organ closest to the heart, the animalcules in the blood are instantly pumped throughout the body. As he theorized why inoculation is a safe procedure, Mather gave full credit to his sources and to Zabdiel Boylston, the one Boston physician who courageously inoculated and saved the lives of nearly three hundred anxious Bostonians.

[105] See CA 3.10.

[106] Pylarinus, "Nova & tuta Variolas excitandi per Transplantationem Methodus, nuper inventa & in usum tract: Per Jacobum Pylarinum, Venetum, M.D. & Reipublica Venetae apud Smyrnenses nuper Consulem," *PT* 29.347 (March 31, 1716): 393–99.

The original manuscript of Mather's CA 6.2 "*Variolae triumphatae*; or, The Small-Pox Inoculated," the second item on his list, has perished, probably a victim of the process of circulation among the members of the Society College of Physicians. The epistolary essay was addressed to John Woodward and probably composed in November 1721. However, as was his habit, Mather frequently repurposed previously used material in separate publications. This is the case of our CA 6.2 *Variolae triumphatae*. The Small-Pox Encountred. It was integrated in the manuscript of his medical handbook "The Angel of Bethesda" (MS 112–47), extant at the AAS, and given a new title as a chapter heading. The manuscript is embellished with pious exhortations in the opening paragraphs, an approach perfectly consistent with the purpose of his handbook. "*Variolae triumphatae*" ("Variola triumphed over") is a fitting designation for a chapter heading announcing how the variola plague (i.e., the smallpox) can be vanquished.

True to his acquired knowledge of medicine, Mather rejected the "*Old Notion*" that the contagion of the smallpox lies dormant in every human as a "*Venom Connate*" – similar to the protoplast's original sin – and was passed on through "the *Maternal Blood*" until awakened by some external agent. Mather agreed with his Danish theologian-physician Thomas Bartholinus (1616–80) that the smallpox was a "*New Distemper*," simply because neither the Bible nor the ancient physicians from Hippocrates onward mention anything about this disease. No, the disease was introduced to Europe through trade with the Orient if not "on the Wings of *Arabian Locusts*," which conquered the Levant and North Africa and carried the contagion in their saddlebags.

Variola examined under the microscope revealed "*Millions* of *Bodies*" swarming in the pustules, as the animalcules do their devastating work.[107] Worse, physicians did not agree on how best to combat the contagion, let alone distinguish between distinct and confluent pox, for to many one pustule resembled another, like one egg does another. Even the renowned Thomas Sydenham (1624–89) – though he "saved more Lives than the Wars of *Ireland* had destroy'd" (CA 3.10) – could only offer palliative purges. For Mather, inoculation (coupled with prayers) was the only remedy for this scourge. Lest his readers continued unconvinced, Mather included the regimen of three other physicians who offered advice on the disease.

The third item in this series is "A Further Account, of the Method and Success of the Small-Pox Inoculated" (CA 6.3). Originally, it consisted of a much longer and more detailed essay. The last segment of this document is the only

[107] See also Mather's "*Conjecturalies*. Or, Some Touches upon, A *New Theory* of many *Diseases*," in *ABJ* (1972), pp. 43–48.

portion still extant. It was published in the *Philosophical Transactions* (March 31, 1722) as "The Way of Proceeding in the Small Pox inoculated in New England. Communicated by Henry Newman, Esq. of the Middle Temple." As mentioned above, Newman was Mather's amanuensis in London; he is *not* the author, merely the messenger who delivered Mather's "Further Account" to the Royal Society. Its shortness, however, does not diminish its value because its hands-on trial in Boston had proven successful beyond measure. It consists of thirteen step-by-step instructions or aphorisms for physicians how to perform the inoculation procedure and how to gauge the vitals and symptoms of the vaccinated patient. As a manual for would-be inoculators it is invaluable.

Curiosa Variolarum (CA 6.4) is addressed to John Woodward (March 10, 1721/22). It is primarily valuable for some of the earliest statistics on the epidemic in Boston. Mather reported that of the more than 5,000 Bostonians who contracted the variola virus, "near nine hundred have died." However, "almost three hundred" who had been inoculated – of all ages, sexes, races, and in all seasons – have survived. Only five or six died after being inoculated, primarily because they had already been infected prior to vaccination or died of other diseases. Dr. James Jurin, as a leading member of the Royal College of Physician and secretary of the Society, was eager to ascertain the rates of success or failure of inoculation throughout the kingdom, and a statistical breakdown would provide the most convincing evidence. He incorporated two long paragraphs of Mather's *Curiosa Variolarum* in his public "Letter to the Learned Dr. Caleb Cotesworth, F.R.S. of the College of Physicians ... containing a Comparison between the Danger of the Natural Small Pox, and of that given by Inoculation." For good measure, Jurin included John Osborne's confirmation of Mather's account in a "Postscript."[108] Jurin deemed this document important enough to issue a reprint of the same letter as a separate pamphlet with two additional accounts the following year.[109] Jurin's pamphlet definitely stirred up the debate and garnered support from the Scottish surgeon Dr. Charles Maitland (c. 1668–1748), a member of the Royal College of Physicians with first-hand knowledge of the inoculation procedures performed in Turkey in 1718.[110] Maitland joined the public debate with his pamphlet *Mr. Maitland's*

[108] 108 *PT* 32.374 (December 31, 1722): 213–27; (Mather's data appears on p. 215; the "Postscript" appears on pp. 225–27.

[109] Jurin, A Letter To the Learned Caleb Cotesworth, M.D.... Containing, A Comparison Between the Mortality of the Natural Small Pox And that Given by Inoculation (1723).

[110] Maitland had supervised the successful inoculation of Lady Mary Wortley Montague's son in Constantinople, in 1718, and became a vocal advocate of inoculation in the Royal College of Physicians upon his return to London.

Account of Inoculating the Small Pox Vindicated, From Dr. Wagstaffe's Misrepresentations of that Practice (London, 1722) and included Mather's *Curiosa Variolarum* letter to Woodward (March 10, 1721–22) in its entirety.[111]

Although it will be described more fully in the next section, we should also mention here "The Case of the Small-pox Inoculated, further cleared" (CA 7.1) – it belongs to the Seventh Series – which Mather forwarded to James Jurin, his correspondent at the Royal Society, sometime after May 21, 1723. This long letter was Mather's last word on the inoculation controversy of 1721/22. For his English audience unfamiliar with the storm in Boston, he repeated the main arguments leveled against inoculation and invalidated them one by one by pointing at their irrational basis, especially when his opponents rejected inoculation as a "heathen" practice while obediently taking pills and potions recommended by Hippocrates, Galen, or Avicenna – heathen or no. Is inoculation a safe practice? "There can be no Answer comparable to that of CONSTANT EXPERIENCE," Mather insisted, and he stressed experience three times. Verifiable evidence clearly demonstrated that inoculation was "a *safe and sure Method*" approved by God's divine blessings. And those who prated "like *Pothecaries*" and amused their clients "with their Learned and Senseless *Jargon*" and "*Hypothesimanies*" will do so "till their poor Patients are sacrificed by Thousands to their Obstinacys," he jabbed. Besides, those who pretended that prophylactic inoculation undermined God's judgment on New England seemed to have no qualms about taking other preventive medicines in common use – even those devised by "pagans" in ancient times. In this manner, he exposed their contradictions even as he warned James Jurin of what English physicians were likely to face if they adopted inoculation to combat the smallpox. Jurin, Mather's correspondent at the Society, was an ideal intermediary to gather information about the success of smallpox inoculation. In fact, Jurin called on his fellow physicians to send in detailed information on their inoculation practices, asking them to keep a Register of their patients' names, age, location, decumbiture, treatment, symptoms, recovery, and eyewitnesses. Most of all, he was interested in the success of the procedure, "whether Good or Bad" … whether the practice "took Effect or no … whether the Patient died or recover'd," and whether after their recovery any of the inoculated people contracted smallpox in the natural way. Either way, Jurin promised to keep the public informed by publishing the results annually.[112]

111 Mather's letter appears in Maitland's pamphlet, pp. 58–61.

112 James Jurin, "Advertisement," in *PT* 32.370 (March 31, 1722). He repeated the same "Advertisement" in his updated statistics over the following years. On Jurin's interest in pooling data on the smallpox epidemic of the 1720s, see A. Rusnock (1996), pp. 22–27.

The next item in the package of the Sixth Series, "*Nishmath-Chajim*; The probable Seat and General Cure, of Diseases" (CA 6.5), takes its inspiration from the "soul" or "breath of life" (נִשְׁמַת חַיִּים) in the creation account (Gen. 2:7). It is, perhaps, Mather's most philosophical (and strangest) theory through which he participated in the ontological debate among his peers; it served as his vademecum to redress the inroads of pervasive mechanism and materialism he felt were threatening belief in the existence of the human soul. What is this strange concept all about? This *Nishmath-Chajim* is a spirit in man, Mather argues, which "may be of a *Middle Nature*, between the *Rational Soul*, and the *Corporeal Mass*." At its upper level, it is so fine that it is akin to spirit and interfaces with the soul; at the lower level, it interacts with gross matter. As "the *Medium of Communication*," the *Nishmath Chajim* functions as "the *Vital Ty* between them" as it receives "*Impressions* from *Both*." Put in another way, matter (Mather's corpuscles) is characterized by size, weight, form, and motion, and is therefore tangible and has a surface. Mind or soul, however, are incorporeal entities lacking the attributes of the body necessary to operate on one another. To Mather, then, the *Nishmath-Chajim* was akin to the missing link, so to speak, between soul and body; it bridged the supposed disjuncture between the physical body and the invisible and incorporeal soul or mind that ontologically exist (if at all) in a realm separate from the material body to which Descartes' famous dualism had seemingly consigned them.

To be sure, *Nishmath-Chajim* is an intriguing concept on other levels as well. It is tantamount to Aristotle's entelechy (plastic nature) of living organisms (*De anima*) that governs all vital functions in nature. It makes seeds grow roots, change into plants, and develop blossoms pollinated by insects to bear fruit. It directs the propagation of species just as much as it does the instinct of animals to build nests, lay and hatch eggs, produce milk and honey to feed their offspring, and to guard their brood against intruders; it protects nursing mothers from painful nibbles of her infants by delaying the teething process; it guards against foul-smelling meat or drink by instilling "an *Incurable Aversion* for them." It springs into action when the machine of the body is disrupted by disease or injury, and governs the process of healing. Platonists called it "anima mundi" or "world soul"; Paracelsian physicians might refer to it as the "archeus"; and the Cambridge Platonists, "plastick nature" or "hylozoism"; Enlightenment philosophes, "vitalism." It permeates all matter alike, organic and inorganic, but annuls the idea of pure mechanism.

On the level of doctrine, the concept of *Nishmath-Chajim* was even more useful to theologians like Mather, for it rendered intelligible how Adam's spiritual fall (a sin of the mind) in Paradise entailed the ultimate demise of his body and rendered logical the imputation of sin (hence death) on all descendants of the protoplasts. On the level of pneumatology, it is also "a Vehicle to the *Rational Soul*" that accounts for the sighting of specters and "*Apparitions* of the *Dead*," the "*Manes*" of the ancients or, closer to home, the manifestations of witchcraft in Salem. In short, the *Nishmath-Chajim* is a universal organon that explains everything everywhere all at once, in one unified whole.

Significant, too, is Mather's "The Seventh Son examined; with a Touch upon the Kings Evil" (CA 6.6). Here, he was at his best as he dismantled the superstitions of his age. His reductio-ad-absurdum approach, coupled with dialectic questions and answers, ensured that discerning readers could no longer accept without doubt that the seventh son in a family was endowed with special healing powers – even if the Irish "stroker" seemed to restore individuals who applied for succor. What about the healing powers of the "royal touch"? No, neither did King George I (nor any king or queen before or after) possess such curative virtues – even if the famous Dr. Samuel Johnson of dictionary fame was one of the more notable eighteenth-century contemporaries who as an infant was "touched" by the royal hand of Queen Anne.[113] Besides, the efficacy of the royal touch to heal scrofula (the King's Evil), now better known as tuberculosis of the lymph nodes, was as much dependent on the faith of the credulous as it was in the token of a gilded coin supplicants received as a gift for their pious endeavor.

Finally, Mather's CA 6.11 "Ambergrise discovered & determined." It was included in the packet for John Chamberlayne and was believed to be lost. It is a short piece about ambergris, a waxy substance extracted from the sperm whale's intestines near the genitals and appreciated for its musky scent, praised as an infallible aphrodisiac, and hailed for its transformative power of ill will among mankind – at least if Ishmael can be taken at his word in Herman Melville's *Moby-Dick* (1851).[114] CA 6.11 was published in the *Philosophical Transactions* under the title "Ambergris found in Whales. Communicated by Dr. Boylston of Boston in New-England."[115] To be sure, Dr. Zabdiel Boylston, who "communicated" the piece to the Royal Society, was *not* the author, merely an amanuensis, who presented it to the Society. We believe it was written by Cotton Mather – though based on first-hand accounts of whalers in Mather's congregation; its rhetorical style is within the scope of Mather's other pieces, and the circumstances of Boylston's close collaboration in the

[113] L.C. McHenry, Jr., and R. Mac Keith (1966).

[114] Herman Melville has an intriguing chapter on Ambergris and its medicinal benefit, in *Moby-Dick*; *or, The Whale* (1851), ch. 92. Ambergris.

[115] *PT* 33.385 (December 31, 1724): 193.

smallpox epidemic in Boston may well support our contention. What is more, Boylston had made a name for himself for having inoculated nearly three hundred people in Boston (1721–22); he was well received by the Society during his visit in London in 1724 and brought within him "a 'large lump' of Ambergris" and a description of its virtues.[116] Boylston would certainly have been given credit as its author (rather than merely as its communicator) – if the piece published in the *Philosophical Transactions* had come from his pen.[117] The same mistaken attribution of authorship has mislead some historians in another piece: *Some Account of what is Said of Innoculating* [sic] *or Transplanting the Small Pox. By the Learned Dr. Emanuel Timonius, and Jacobus Pylarinus*. Published by Dr. Zabdiel Boylston (Boston, 1721). Notice, the pamphlet was "Published" *not* authored by Boylston, and was most likely a collaborative effort – Boylston supplying numbers and details and Mather writing it up but having it published under Boylston's aegis. This conclusion is confirmed by Mather's diary entry for August 4, 1721: "I will allow the persecuted Physician [Zabdiel Boylston], to publish my Communications from the *Levant*, about the *Small-Pox*, and supply him with some further Armour, to conquer the Dragon" (*D* 2:636).

At any rate, ambergris is found in different quantities in Spermaceti bull whales, and "is contained in a Cyst, or Bag, without any inlet or outlet to it.... near the Genital Parts of the Fish." It is "moist," and emits "an exceeding strong and offensive Smell." Wonder then that it was deemed a precious commodity and used as a concoction in medical potions, perfumes, even as an ingredient in food? Juvenal was right when he joked, "lucri bonus est odor ex re qualibet," or sweet is "the odor of gain, no matter its source."[118] But whether "*naturally, or accidentally produced in that Fish*," Mather leaves it "*to the Learned to determine*."

[116] RSL Journal-Book, XIII, p. 448, quoted in R.P. Stearns, *Science* (1970), p. 438. Mather's fellow competitor Paul Dudley had his piece published as "An Essay upon the History of Whales, with a particular Account of the Ambergris found in the Sperma Ceti Whale," *PT* 33.387 [March–April 1725]: 256–69.

[117] G.L. Kittredge makes a good point in his comment on another Mather piece, our CA 6.3 "A Further Account, of the Method and Success of the Small-Pox Inoculated," which may have included "The Way of Proceeding in the Small Pox inoculated in New England. Communicated by Henry Newman, Esq; of the Middle Temple," *PT* 32.370 (March 31, 1722): 33–35. Kittredge points out that Newman was not the author; he was merely an amanuensis who submitted it to the RSL, "a formula especially used when a letter or essay was handed in or read by some person other than the author" ("Lost Works," p. 461). We believe the same is true for Mather's 6.11 "Ambergrise found in Whales. Communicated by Dr. Boylston of Boston in New-England."

[118] Juvenal, *Satire* (8.204–05).

It is unfortunate that Mather's five remaining Curiosa letters in the Sixth Series are no longer extant. Their topics are as diverse in this batch as they are in any others that he sent to the Royal Society. As indicated above, the lost items are: 7. *Lacus Mirabilis*. 8. An Horrible Tempest. 9. *Ostreophagi*; or, a Matchless Oister-hill. 10. *Absinthium Sempervivum*. 12. *Nævi Materni*. 13. *Melissologia*; with a New Method of Bee-hunting. There is no telling the story of the "astonishing lake" or the "Horrible Tempest" (probably a hurricane), on which topic Mather preached a year later and published as *The Voice of God in a Tempest. A Sermon Preached in the Time of the Storm* (1723).[119] Next, who were the pescatarians who lived on oysters and piled up their shells into a sizable mount? Mather's account has perished. So has his letter on the healing virtues and dosages of the evergreen absinthium plant (aka. wormwood) to mitigate inflammations or help digestion. Mather's *Nævi Materni*? an account of a prominent birthmark of "a Daughter of one *Brown*, a Stiller, which grows ... very Troublesome to the Marked Damsel" (*D* 2:686) has been lost as well. The likely account would have included Mather's medical remarks on moles and birthmarks, whose excrescence are attributed to the vehement imagination of a pregnant mother marking her fetus.[120] Finally, his *Melissologia*, a discussion of honey gathering, has not survived either.[121] In all these, Mather was at the top of his form – his omnivorous consumption of curious topics in virtually all disciplines, fortified with his own American curiosities gathered far and wide, and embellished with his celebration of God's Book of Nature.

Curiosa Americana Series 7 (May–August 1723)

During the late spring of 1723, Mather sent two batches of Curiosa across the Atlantic, constituting the Seventh Series, totaling eight studies plus letters introducing himself (CA 7.0) and an aspiring young naturalist (CA 7.9)

[119] The lost letter (CA 6.8 "An Horrible Tempest)" may also be based on John Winthrop's "perfect Account of a late whirlwind among you, which I am told of" that Mather requested from his New Haven friend and correspondent, in "Cotton Mather to John Winthrop (29*d.* VI*m.* [August] 1720), *MP* (1868), p. 443.

[120] See Daniel Turner's *De Morbis Cutaneis* (1714), p. 121; and "LXXXIII. De Naevis Maternis," by the Danish physician Johannes Valentinus Willius (Johann Valentin Wille), in *Thomae Bartholini Acta Medica et Philosophica Hafniensia Ann: 1674. 1675. 1676.* (1677), pp. 152–53.

[121] The topic was of interest to the Royal Society; it published Paul Dudley's "An Account of a Method lately found out in New-England, for Discovering where the Bees Hive in the Woods, in order to get their Honey," *PT* 31.367 (August 31, 1721): 165–68.

to the Royal Society's new secretary, James Jurin. Mather wrote a further letter of introduction, dated Aug. 3 (CA 7.10), for his friend and neighbor Dr. John Perkins (1676–1740) of Boston, who made the trip to "possess himself of a considerable Interest there falling to him," but also to meet Jurin. This introduction gave Mather the opportunity to state: "Twice in the Summer now Spending, I have given you the Trouble of some *American* Packetts. In the former of them, I tendred you what the *Inoculation of the Small-Pox* ha's produced. In the latter, there were Seven Letters of, *Curiosa Americana*."

Several months earlier, on May 21, Mather had completed an eight-page review for Jurin of his communications with the Royal Society, to which he appended a list of items he had previously contributed. He had also drafted "The Case of the Small-pox Inoculated" (CA 7.1), which he then had his copyist indite, amounting to twenty pages, and given the same date. We know that these two items were dispatched in early June, because in his letter introducing Isaac Greenwood (1702–45), dated June 10 (CA 7.9), Mather mentions sending "a few days ago" (CA 7.1) while Greenwood himself carried the seven other Curiosa, 7.2 through 7.8, to Jurin.

"The Case of the Small-pox Inoculated, further cleared" (CA 7.1) recapitulates, to some extent, "*An Account of the Method and Success of Inoculating the Small-pox*" (CA 6.1), sent in the year of 1721, infamous in the Boston area for the epidemic. But this composition also expands upon the earlier one, because it discusses treatment procedures and, more substantially, reviews and answers a number of scientific and religious objections to inoculation that had arisen. What is more, Isaac Greenwood was one of the individuals who had successfully undergone inoculation in Boston, so he could corroborate Mather's account with his own in-person testimony. As an indication of the continuing interest in this topic, this item was marked as having been read before the Royal Society on December 19.

The subsequent Curiosa are all dated during the period of June 3 through 10. Two of these recount archaeological finds. The source for "A Monstrous Dragon" (CA 7.2) was likely Mather's colleague, the pastor-historian Thomas Prince (1687–1758). On February 19, Mather had written a note to Prince asking for his "best account" of several phenomena, including "Of the *Leviathan* dug up at *Virginia*" (*D* 2:686). This was thought to be the backbone of a giant snake, 150 feet in length, dug up along the banks of the James River. Perhaps through Prince's connections, one joint of the supposed skeleton was brought to Boston, measuring four and a half inches in diameter, nearly half a foot long, and weighing over two pounds. The other item, "*Subterranea*" (CA 7.7), treated readers to a couple of discoveries at Springfield, Massachusetts: the first a tree limb, with axe marks on it, unearthed seventeen feet

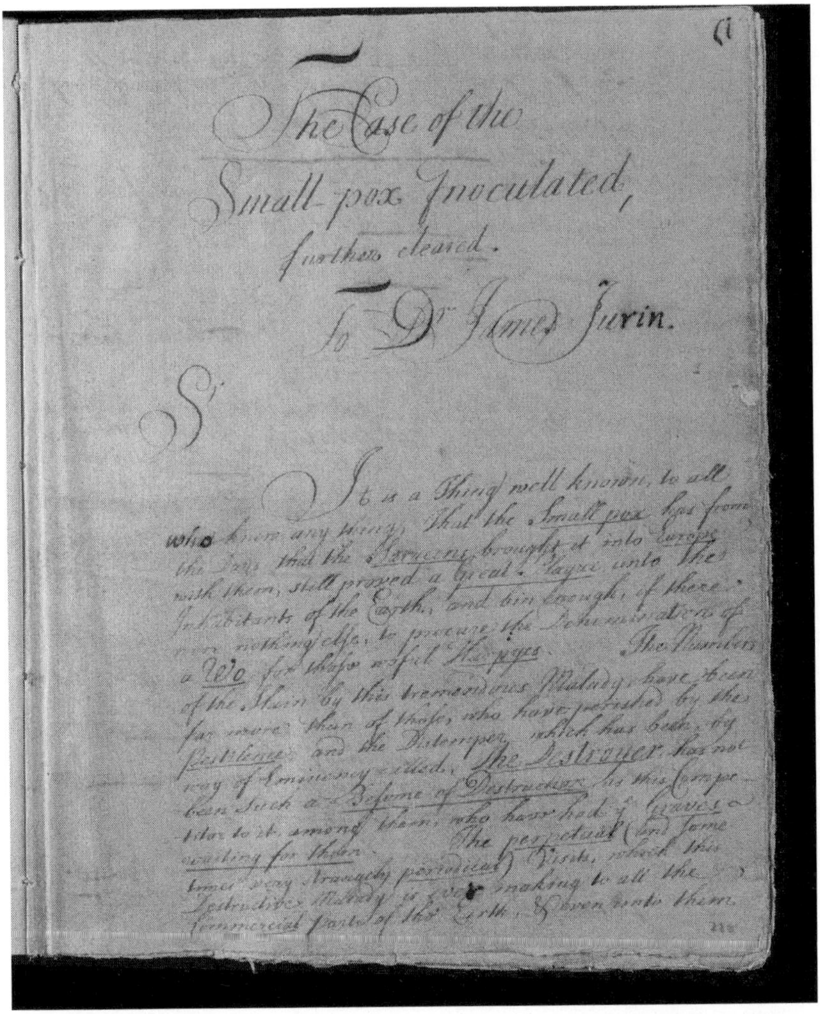
The Case of the Small-pox Inoculated, further cleared. To Dr James Jurin.

Figure 4. “The Case of the Small-pox Inoculated, further cleared. To Dr. James Jurin.” (May 1, 1723). Courtesy of the Royal Society of London.

underground; additionally, a crew digging a well found, at a depth of about twenty feet, wood “*Chips*’ and a “Parcel” of butternuts – all attestations to presence of ancient Indigenous peoples of the New World, for which John Woodward had expressed a particular interest (Appendix 8).

Likewise, this series contained a brace of accounts about animals. “*Carcharias Mirabilis*” (CA 7.4) described a large marine creature that may have been a whale shark, a topic of theological interest as well because Mather sought to demystify the account of Jonah’s survival in the belly of the Leviathan

(CA 5.3). "A New Rattlesnake" (CA 7.3) harked back to a description of the rattlesnake in the First Series (CA 1.11). This newer one, however, provided a native cure for the serpent's bite.

Speaking of snakes, Mather had credible reports of them falling from the sky. In CA 2.1, "Woollen Snow," Mather had described how, during one winter, a snow of wool descended on a southwestern Connecticut town. Now, "Mysterious Rains" (CA 7.5) discussed some other strange things that have fallen from New England skies, including barley, wheat, blood, birds, frogs, and, yes, snakes. Ever popular, reports of such inexplicable phenomena also found their way into the *Philosophical Transactions* and other venues. They addressed the vexing question of equivocal or spontaneous generation of certain species, like frogs, fish, and snakes, raining from the sky (as in the plagues visited on ancient Egypt).

"The Land Sail'd Upon" (CA 7.6) offers a mechanical invention for the consideration of the Royal Society. Joseph Morgan, Connecticut native, Presbyterian minister in New Jersey, and inveterate tinkerer, was one of many in this age of experimentation trying to design "machines" that would take advantage of different power sources. Sails power ships at sea, and windmills on land. Why not a boat on land? Morgan set about crafting one, claiming to solve various challenges that such a craft presented. Morgan and Mather corresponded, so Mather, acquiescing to Morgan's request to sponsor the project, sent the account to the Royal Society, excerpting the inventor's own description. Morgan offered to send his plans, if there was interest. For his part, Mather did not pass judgment, though neither would he "Declare how far I am in the Order of *St. Thomas*, upon it."

To wrap up this series sensationally, Mather presented "A Singular Case" of an individual who, he claimed, was "as Miserable a Patient" as Jurin would ever see. It is the story of a Boston shipwright who, on a voyage, was first so badly burnt that he was temporarily blinded and left with his windpipe exposed. After healing somewhat, he became sick with the "flux," and, incapacitated, was abandoned on an island where he lay for more than a week without food or water. He was picked up by a passing ship and left ashore again, though this time at a port, where he languished on the streets until he was put into a hospital. His leg became extremely infected, and his jawbone "perished and broke away." Amazingly, the shipwright survived and was brought back to New England – indeed, it is his own account of his suffering, in which he solicits Mather's aid, that makes up most of the final installment in the series.

Curiosa Americana Series 8 (September–December 1724)

Mather authored the Eighth and final series of the "Curiosa Americana" in the late summer and early fall of 1724. Addressed to Woodward and Jurin, their titles are CA 8.1 "Climatical Influences"; 8.2 "An *Amphisbœna*"; 8.3 "An Unusual Vomit"; 8.4 "A Tide and Storm of Uncommon Circumstances"; 8.5 "A Moveable Rock"; 8.6 "A New-found land"; 8.7 "More and Strange Effects of Thunder"; 8.8 "A Rare Discharge of Bullets"; 8.9 "Nature at Work with a Needle"; 8.10 "Uncommon Dentition." As contextual appendages, attached are two additional letters that Mather sent almost contemporaneously to his packet of ten letters. The first, CA 8.11, which Mather addressed to Jurin on October 5, advertised the completion of his new medical textbook *The Angel of Bethesda*. The second, CA 8.12, addressed to Jurin on December 15, introduced the physician Zabdiel Boylston (1679–1766), the famous inoculator during the Boston smallpox epidemic (1721–22).

Mather's first account CA 8.1 "Climatical Influences" broached the potent subject of climate and its effects on the human body. In the aftermath of the Age of Exploration in which European settlers had long-since lived on Native lands, this was something of a contentious issue. Europeans and European-Americans were all too eager to describe Indigenous peoples with any number of racist tropes as a means to create a hierarchy of societies. The source of indigenous inferiority, however, was subject to debate and those like Mather who attributed the color of their skin and leisurely habits primarily to climate – as opposed to those who speculated that Native peoples were a subclass of humans by birth – faced an immediate logical issue. If, as Mather suspected, the American climate caused Indigenous peoples to be lazy and poor parents, then white colonists could be subjected to the same degradation or "Indianization." This remained a potent question for white Americans in the eighteenth century, one that Thomas Jefferson (1743–1826) would famously take up in his widely disseminated *Notes on the State of Virginia* (1785), as did the French-American diplomat Hector St. John de Crèvecoeur (1735–1813) in his *Letters from an American Farmer* (1782).[122]

Mather next turned his sights to the myth of the Amphisbæna (CA 8.2), an oft-described two-headed serpent that found its way into the writing of Ælian, Pliny, and Galen. Mather's relation of a similar beast killed in nearby Newbury was doubly significant. For one thing, Mather's nearly verbatim copy of a *New-England Courant* article provided keen observations on the extraordinary

[122] Crèvecoeur (ch. 12) and L.A. Dugatkin (2009). For Mather's more enlightened views on skin color and climate, see his *NC* (1706), pp. 23–25; and *BA* (1:672, 674–75, 698–99); J. Stievermann, "Genealogy of Races" (2010).

snake's physical characteristics and its manner of movement, making the information useful for the Royal Society's members who retained a perennial interest in American serpents. For another, Mather's relation speaks to the standards of proof common in the Republic of Letters. Although he apologized, "I am sorry, that my Distance … made me uncapable of such an *anatomical* Proceeding upon the Animal, as would have enabled me to answer many Enquiries," Mather assured Woodward, "This Relation may be relied upon. Tis here quæstioned by no body. Afterwards I made a more exact Enquiry into the Matter, and I find it very fully Attested" (CA 8.2). Thus an ancient myth found a more reasonable instantiation on the shores of New England.

CA 8.3 "An Unusual Vomit" provided Mather the opportunity to exercise his humorous muscles. He began, "Since my Letter is not like to be read, until after you have risen from the *Table*, and perhaps have your *Stomach* well-setled with the *Liquids of the Afternoon*, I will venture to make *Strange Vomits* the Matter of your Entertainment." What followed the graphic introduction was yet another oddity from Newbury, Massachusetts – this one taken from the pen of the physician Christopher Toppan (1671–1747). Transcribing Toppan's account of a patient, Mather told of a man who belched up a common newt. According to Toppan's reasoning, the man had either ingested and gestated a newt's egg in his stomach or inadvertently swallowed a small newt right after it hatched. Mather concluded the letter in a similarly humorous tone: "Having thus contributed a little *Evet*, unto the Stores in your more curious Collections, I take Leave only to add, that the Blessings of a good & sound *Stomach*, that shall have nothing in it but what Shall agree with it …" (CA 8.4). Strange as such an account may appear nowadays, it was an ardently debated issue among physicians of Mather's time who sought to redress the pervasive problem of intestinal worms in their patients. Several articles in the German *Ephemerides* testify to similar occurrences of live lizards, spiders, and similar critters issuing from the stomachs of credulous patients.

The subsequent four letters treat a variety of natural phenomena that range from the almost unremarkable to the shockingly vivid. In CA 8.4, Mather recounted the February 1723, flooding of New England's shoreline. With winds whipping up waves and the skies darkened by clouds, the seas around Boston, Cape Cod, and Rhode Island rose to incredible heights one Lord's Day. Such pedestrian occurrences seem hardly worthy of notice for the Royal Society – unless we consider that James Jurin had specifically called on Mather to distribute a printed invitation among New England's scientists who owned accurate meteorological instruments to measure events (temperature, windspeed, baroscopic air pressure, and precipitation).[123] In fact, Jurin had published an "*Invitatio ad Observationes Meteorologicas communi consilio institutendas*" in the *Philosophical Transactions*,[124] calling on European scientists to send in data to establish a theory of meteorological formations as well as weather conditions and their impact on people and their environment. Mather's missives to Jurin, CA 8.1 "Climatical Influences," CA 8.4 "A Tide and Storm of Uncommon Circumstances," and CA 8.7 "More & Strange Effects of Thunder" – all appear to respond to Jurin's invitation. As Andrea Rusnock points out, Jurin's published call for information was blessed with some success, for meteorologists from "Uppsala, St. Petersburg, Berlin, Leyden, Naples, Luneville, Boston, and many towns in Great Britain and Ireland" sent in their observations to the Royal Society (now kept in the Meteorological Archives of the Royal Society).[125] Getting each of his international contributors to employ the same standards of measurement, however, was another matter. Jurin's formal request for an exact account of the season, date, time, barometric altitude, temperature, windspeed, and precipitation was nearly as futile as asking potential contributors to measure rainfall in decimals – no uniform metrics for liquids or instruments for temperature and air pressure being employed among his contributors.[126]

Next, alluding to CA 5.7, Mather returned to another geological oddity of Gloucester, Massachusetts, in one of his shortest letters to the Society. Seemingly self-conscious of the letter's lack of substance, Mather prefaced his relation as follows: "But tho' the thing be little above the Quality of a *Trifle*, yet because there is a little *Curiosity* in it, and some other Things as *Trivial* have been commemorated among *observable Rarities*, I will now give a Relation of another Large *Rock* in the very same Town of *Glocester* …" (CA 8.5). Then as now visitors marveled at the presence of huge boulders, vestiges of the ice age, strewn about New England's countryside. Reflecting his interest in pedology, Mather's account (CA 8.6) addressed the chemical composition of the soil in nearby Woodstock, Connecticut: he told of a type of soil that, when burned, transformed into chalk, a burgeoning topic of great interest to James Jurin and his peers, who tried to establish a taxonomy of soil samples and their fertility. Despite having finally secured a place in the ranks

[123] See Appendices 10 and 11. On the developing science of meteorology in Mather's time, see A. Wolf (1935) 1:306–24; and A. Wolf (1938) 1:274–341.

[124] *PT* 32.379 (October 31, 1723): 422–27.

[125] A. Rusnock (1996), pp. 28 and 55, *n.* 113.

[126] See Jurin's sample chart, "Invitatio ad Observationes Meteorologicas," *PT* 32.379 (October 31, 1723): 427.

of the Royal Society, Mather still feigned ignorance and intellectual humility, writing, "I have so little Skil in the *Earth*, as to be ignorant, whether I shall now exhibit any thing, that is *wholly new* to you or no. But my *Intention* will make *Atonement* for my *Ignorance*: And I will give you this Relation of something, that has appeared unto us *Americans*, what is with us esteemed, a *Curiosity*" (CA 8.6). In "More & Strange Effects of Thunder" (CA 8.7), Mather concluded this quadripartite commentary on American nature with meteorological observations of the effects of lightning. In transcribing an article from the *Boston News-Letter*, Mather passed along an exceedingly raw description of the devastating effects of lightning, four decades before Ben Franklin's famous kite experiment (1752) proved that lightning is little more than a powerful electric discharge.

The three remaining letters of the original packet of ten letters prefigure the rest of Mather's writing to the Royal Society, for in CA 8.8, 8.9, and 8.10 he treated medical marvels. In 8.8, Mather told the story of a woman who passed lead bullets through her navel after ingesting them as a treatment for irritable bowel syndrome. Again, Mather's account spoke to the standards of proof common in the Republic of Letters, for he told Jurin, "I will humbly offer you a *Matter of Fact*, which tho' not *seen* by either of us, has been so *proved* as to be incontestable" (CA 8.8). In "Nature at work with a Needle" (CA 8.9), Mather related the case of a woman who had a sewing needle removed from her arm with a magnetic plaster, and though the *Philosophical Transactions* were replete with accounts of magnetism, Mather's letter offers a rare glimpse into the inner workings of medicine in colonial America.[127] After concocting a recipe for a magnetic plaster based on the work of the Italian chemist Angelo Sala (1576-1630), Mather shared the concoction with one physician who "paid little Regard unto my *Cautions*, but used it almost at hap-hazard, & made a *Polychreston* of it, & got more than a little Money by it." It was not until Mather put the recipe into the hands of a "A more Cautious Physician than he" that the treatment found its intended purpose (CA 8.9). Mather concluded his Eighth Series of Curiosa letters (CA 8.10) with an unusual subject: he last told of a young student at his alma mater who, instead of having teeth, had a singular boney protrusion that lined his gums. Mather was interested in the

[127] See Samuel Colepress, "Extract of a letter, written by Mr. Sam. Colepress to the publisher, containing an account of some magnetical experiments; as also of an excellent liquor made of cyder-apples and mulberries," *PT* 2.27 (September 23 1667): 502–03; D.B., "An extract of a letter, written by D. B. to the publisher, concerning the present declination of the magnetick needle, and the tydes, May 23. 1668," *PT* 3.37 (July 13, 1668): 726–27; "The variations of the magnetick needle predicted for many yeares following," *PT* 3.40 (October 19, 1668): 789–90.

growth of teeth in infants and dental problems in children and adults. As good as his own experience, he included in his medical handbook a chapter titled "*Dentifrangibuluss. Or, The Anguish, and Releef, of The TOOTH-ACHE.*"[128]

The final two letters in this volume are not formal Curiosa per se, though they evidence Mather's continued interest in the production of medical knowledge. In CA 8.11, Mather advertised to Jurin the completion of his medical treatise *The Angel of Bethesda*, for which Mather hoped to interest the Royal Society to sponsor its publication, just as the Society had done for his *Christian Philosopher* (1721). Though *The Angel of Bethesda* went unpublished until the second half of the twentieth century, it achieved what Mather hoped to in his Curiosa – a fine blend of God and science to provide a holistic account of the world and humanity. In a fitting conclusion to the theme of smallpox that permeated the Curiosa, Mather introduced Zabdiel Boylston to the Royal Society (CA 8.12). Boylston had gained recognition as Mather's sole companion in the crusade to promote inoculation in Boston.

On May 20, 1725, Jurin returned his thanks to Mather for the copious letters sent the previous year. In his own words, "I have rec^{d} your large Packet of Letters to Dr Woodward & myself, wc^{h} I have at times communicated to ye Royal Society, & am commanded to return you their thanks for ye several curious relations contain'd therein." For reasons not entirely known, Mather left unaddressed what would become Jurin's last two requests of him. First, Jurin asked, "Give me leave to hope for some assistance from you with regard to my Proposal for Observations upon ye Weather" (Appendix 11). More substantially, Jurin recounted that the French historian Maturin Veyssière La Croze (1661–1739) had taken an interest in Mather's account of a hieroglyphic script engraved on a boulder on the banks of the Taunton River (CA 1.12). That "Gentleman of uncommon Learning," however, suspected that Mather had mistakenly copied the text. Thus Jurin told Mather, "M^{r} De la Croze offers his assistance towards deciphering ye whole, in case a correct copy of it can be procured. This, Sir, ye Royal Society hope for from you," adding, "I assure you it will highly oblige ye the whole Body" of the Society.[129] Perhaps the death of Mather's father (1723), his own declining health, his increased ministerial duties, or his ongoing disappointment with the Society's limited acceptance of his submission for publication in the *Philosophical Transactions* rendered him unable or disinclined to answer that call – if the letter ever reached him in

[128] *ABJ* (1972), ch. 11, pp. 61–66.

[129] "James Jurin to Cotton Mather" (May 20, 1725), Appendix 11. Nearly a decade later, Harvard's Isaac Greenwood (1702–45) supplied the Society with a fresh transcript of the mysterious hieroglyphs.

the first place. Whatever the case, this would be the last year in which Cotton Mather related American Curiosities to the Royal Society of London.

Curiosa Reception, Scholarly and Popular

Despite their intriguing variety of topics, Mather's "Curiosa Americana" letters have seen sporadic, haphazard treatment. Some tangential subjects, though, have received increasing attention. For one thing, scholarship on the history of science and the Republic of Letters has flourished, especially in the American context. Notable recent works on the former include Joyce Chaplin's *Subject Matter: Technology, the Body, and Science on the Anglo-American Frontier, 1500–1676* (2003); Susan Scott Parrish's *American Curiosity: Cultures of Natural History in the Colonial British Atlantic World* (2006); Walter W. Woodward's *Prospero's America: John Winthrop, Jr., Alchemy, and the Creation of New England Culture, 1606–1676* (2010); Cristobal Silva's *Miraculous Plagues: An Epidemiology of Early New England Narrative* (2011); and Kelley Wisecup's *Medical Encounters: Knowledge and Identity in Early American Literatures* (2013). On the latter, noteworthy publications include Theodore R. Delwiche's "Ezra Stiles and North America in the Early Modern Republic of Letters" and Caroline Winterer's "Where is America in the Republic of Letters?"[130] For another, scholarship on Cotton Mather continues to challenge Kenneth Silverman's dated characterization of Mather as America's "national gargoyle."[131] Recent works including Reiner Smolinski and Jan Stievermann's *Cotton Mather and Biblia Americana* (2010, 2011), Reiner Smolinski and Kenneth P. Minkema's *A Cotton Mather Reader* (2022), and Kristen Silva Gruesz's *Cotton Mather's Spanish Lessons* (2022) have done much to paint a fresh, human portrait of a man who was at once intensely learned and intellectually progressive while still heavily mired in the prejudices of his day. Despite the swirling scholarly interest near to the Curiosa, the collection has yet to receive its due course of treatment.

Long before there would have been such a receptive audience for the "Curiosa Americana," at least one historian believed that they merited serious scholarly attention. Almost two centuries after Mather sent his final correspondence in 1724, the eminent historian George Lyman Kittredge issued the first call for their publication. In 1916, Kittredge opined of Mather and the Curiosa:

[130] T.R. Delwiche (2023), C. Winterer (2012).

[131] K. Silverman, *LT* (1984), p. 425.

> Hence his letters to the Royal Society are among the most interesting of his many 'composures,' not only for the light they throw on the active but often misunderstood intellectual life of New England in the early eighteenth century, but for their significance in the annals of scientific theory and investigation. Some day they will be printed. Indeed, it is not very creditable to American scholarship that they were not published years ago.[132]

There have been several previous attempts to transcribe, edit, and publish the Curiosa in fruitful first, second, and third ventures at answering Kittredge's call.

Frederick Lewis Gay (1856–1916), a native of Brookline, Massachusetts, and an avid antiquarian, seems to have been the first editor interested in the Curiosa. In the tradition of nineteenth-century historical praxis, Gay was the consummate collector, bringing together a hodgepodge of literary material on New England's earliest years. In a volume that he personally transcribed and lent to George Lyman Kittredge for historical use, Gay copied out all thirteen letters of the First Series, along with CA 2.4, 7.2 through CA 7.8, and 8.1–8.10. Gay's thirty-two transcribed letters amounted to 259 manuscript pages that can still be found, in all of their neat script, at the Massachusetts Historical Society.

After Gay, Kenneth Silverman is the first modern historian to transcribe some of Mather's Curiosa and included twenty-seven partial transcripts of the seventy-three fully extant letters in his *Selected Letters of Cotton Mather*. Silverman does not seem to follow a discernible pattern in choosing which letters to excerpt.[133] In his transcriptions, Silverman tends toward brevity and thus does not allow for systematic treatment of the letters. Neither does he provide any annotations or explanations of the scientific debates that are vital to modern readers. More discernible is Andrea Rusnock's methodology in *The Correspondence of James Jurin (1684–1750)*. She fully transcribes Mather's fifteen letters to James Jurin, yet as in Silverman's case, without crucial annotations or explanations. Though her selection is brief, her transcriptions are sound.[134] This troika of transcribers, however, let the texts stand without contextual annotations making the fullness of Mather's work hard to comprehend for most readers today.

[132] G.L. Kittredge, "Scientific" (1916), p. 18.

[133] Silverman (*SLCM* [1971]) partially transcribes CA 1.1, 1.3, 1.6, 1.7, 1.8, and 1.10; in the Second, CA 2.1, 2.2, 2.3, 2.5, 2.8, and 2.9; of the Third, CA 3.2, 3.4, 3.7, 3.10, and 3.12; in the Fourth Series, CA 4.1, 4.2, and 4.3; he treats none of the missing or misplaced Fifth and Sixth Series; in Series Seven, CA 7.1; and of the Eighth Series, CA 8.1, 8.3, 8.8, 8.9, 8.11, and 8.12.

[134] A.R. Rusnock (1996): CA 7.1-7.8, 8.1, 8.3, 8.5, 8.7, 8.9, 8.11, and 8.12.

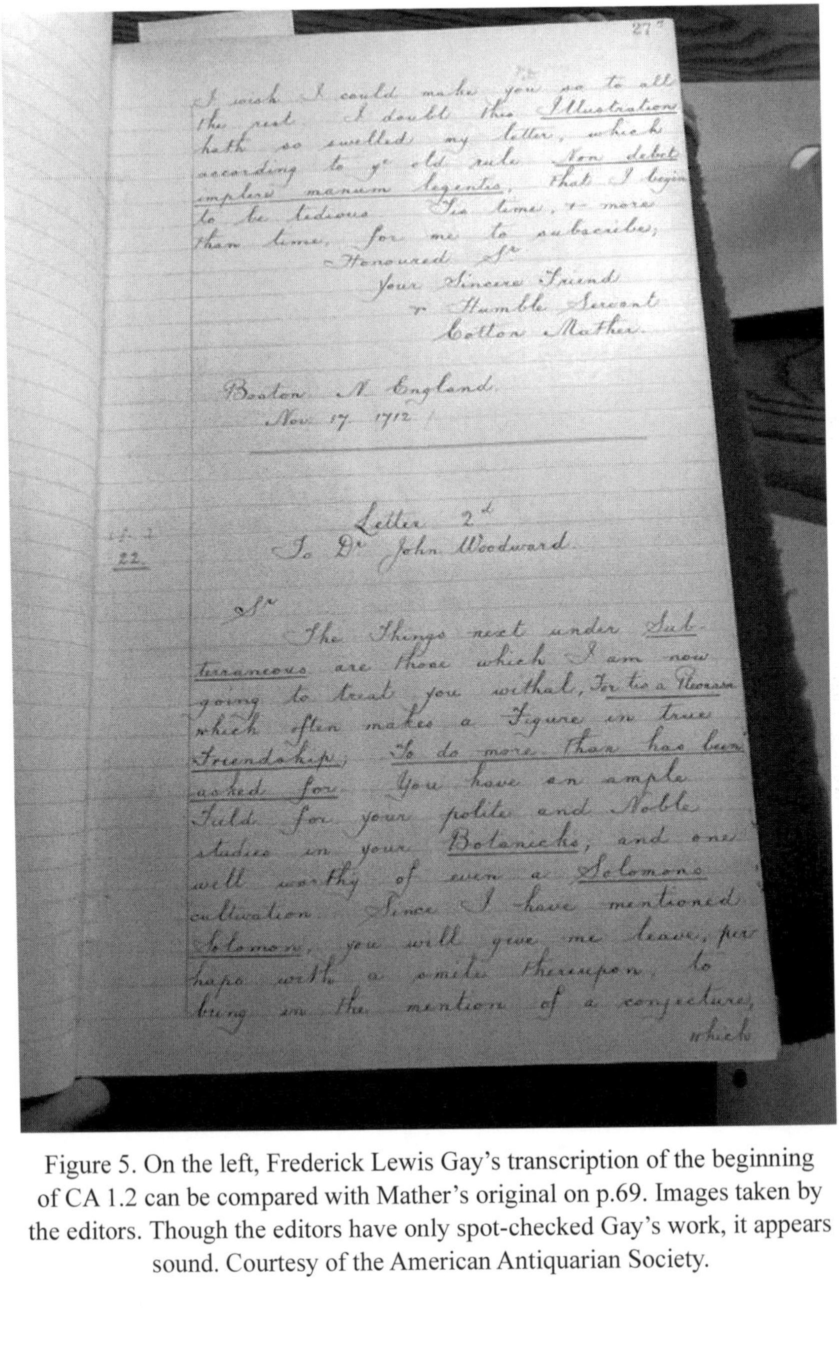
27

I wish I could make you so to all
the rest. I doubt this Illustration
hath so swelled my letter, which
according to ye old rule Non debet
implere manum legentis, that I begin
to be tedious. Tis time, & more
than time, for me to subscribe;
Honoured Sr
Your Sincere Friend
& Humble Servant
Cotton Mather.

Boston N England
Nov 17 1712

Letter 2d

1.2
22 To Dr John Woodward.

Sr
The Things next under Sub-
terraneous are those which I am now
going to treat you withal, For tis a Pleasure
which often makes a Figure in true
Friendship, To do more than has been
asked for. You have an ample
Field for your polite and Noble
studies in your Botanicks, and one
well worthy of even a Solomons
cultivation. Since I have mentioned
Solomon, you will give me leave, per-
haps with a smile thereupon, to
bring in the mention of a conjecture,
which

Figure 5. On the left, Frederick Lewis Gay's transcription of the beginning of CA 1.2 can be compared with Mather's original on p.69. Images taken by the editors. Though the editors have only spot-checked Gay's work, it appears sound. Courtesy of the American Antiquarian Society.

2 Letter 2d.

22 To Dr. John Woodward.

Sr.

The Things next unto Subterraneous
are those which I am now going to treat you withal;
For it's a pleonasm, which often makes a Figure in
true Friendship; to do more than has been asked for.
You have an ample Field for your polite and
Noble Studies in your Botanicks, and one well
worthy of even a Solomons Cultivation. Since
I have mentioned Solomon, you will give me leave
perhaps with a Smile thereupon, to bring in the
mention of a conjecture, which a learned Friend
of mine has, concerning those Writings of his,
which no Pancirol, or Salmuth, could ever yet
inform us of. Whereas we read concerning him,
He spake of Trees, from the Cedar that is in Lebanon,
Even unto the Hyssop that springeth out of the Wall;
He spake also of Beasts, & of Fowles, & of creeping things,
& of Fishes: which some think, he only spoke of
them, Studio cupidi de his rebus philosophandi So-
(and perhaps he sett up a ROYAL SOCIETY in
his Kingdome too,) and what he wrote is only in
ye Books which now make a part of or Canon;
this Gentleman conjectures, that the Books of Theo-
-phrastus and of Aristotle, on those two Subjects,
may indeed be Solomons. for, it's well-known, that
ye Last one of those two philosophers, was a no-
-torious plagiary, and burnt the Volumns out
of which he had been supplied with all his
Treasures. But we rather suppose all Solomons
Herbal to be lost, and be glad that we have
Parkinsons instead of it; rejoice that such no-
-table things have been done by the Gerhards,
and Johnstons and other Industrious and

Aside from these more collective attempts, transcriptions of a few letters have found their way to print in varying degrees of completion and annotation. A full transcript with remarkably thorough annotations of CA 1.1 is the subject of David Levin's article "Giants in the Earth" (1988).[135] A longer and more detailed version of CA 1.1 appears in the first volume of Reiner Smolinski's annotated edition of Cotton Mather's *Biblia Americana* (2010).[136] Moreover, a complete transcription of CA 2.5 with excellent notes is the subject of Arlie W. Schorger's article "Unpublished Manuscripts By Cotton Mather on the Passenger Pigeon."[137] Of a similar quality is Eric Howard Christianson's transcription and annotation of CA 4.4 in "Making Heads or Tails of It."[138] A full transcript, without annotations, of CA 8.4 is included in the Massachusetts Historical Society's *The Mather Papers*.[139] An extract of CA 1.11 appears in an early nineteenth-century assessment of Mather's scientific correspondence by none other than George Lyman Kittredge who, for reasons only known to himself, never answered his own call to edit Mather's "Curiosa Americana."[140] And CA 8.5, about a teetering rock, is partially transcribed in John J. Babson's *History of the Town of Gloucester*.[141]

In addition to efforts at transcription and annotation, several of the "Curiosa Americana" letters are regularly cited as evidence in histories of American natural philosophy and medicine. In reviewing these notices, we get a sense of the scholarly consensus on the Curiosa manuscripts. Mather's letters on smallpox, for instance, comprise much of the evidentiary bulk of Robert Tindol's 2011 article "Getting the Pox off All Their Houses." Tindol notes, "Mather discussed his work on smallpox inoculation in several brief pamphlets, and devoted somewhat more effort in his occasional reports to the Royal Society. Known as the *Curiosa Americana*, these letters have never been published collectively… ."[142] In his long-distant study of the first Curiosa series, Otho T. Beall, Jr., concludes, "Because of their variety and scope, the 'Curiosa Americana' … represent Mather as the foremost American gatherer and disseminator of the new scientific knowledge during his time."[143] Lydia Barnett

135 D. Levin (1988).

136 Cotton Mather. *Biblia Americana. America's First Bible Commentary* (1:582–600).

137 A.W. Schorger (1938).

138 E.H. Christianson (1993).

139 *MP* (1868), pp. 456–57.

140 G.L. Kittredge, "Letters" (1912), p. 174; "Scientific" (1916), p. 18.

141 J.J. Babson (1860), pp. 5–6.

142 R. Tindol (2011), p. 4.

143 O.T. Beall, Jr. (1961), p. 366.

has a less rosy view of Mather's work, writing, "The response to Mather's *Curiosa Americana* makes clear that the Royal Society did not regard Mather as their intellectual equal, but rather as their creole informant. While his British interlocutors were indeed eager for American curiosities, they refused to acknowledge Mather's authority as an interpreter of those curiosities… ."[144]

Beyond narrower academic histories, at least three of Mather's letters and their subjects have found their way into popular, more mainstream media. Locally, Mather's account of the so-called "rocking stone" of Gloucester, Massachusetts, is the subject of a 2010 article in the *Gloucester Daily Times*. It is worth comparing reports on the rock in the twenty-first and eighteenth centuries to get a sense of Mather's accuracy. Of the former, Peter K. Prybot relates his interview with a local expert: "'Some are perched on other rocks. This one just happened to land (after the glacier moved it and melted) in a way it teetered a bit,' [a local Geologist] said. 'The bottom of this one has a hollow spot, and it's balanced.' The stone rocks from east to west and vice versa."[145] Almost three hundred years earlier, Mather wrote of that same rock:

> But when a Storm arises, and the Sea beats Tempestuously upon the *Rock*, the Spectators, have a Sight that a little surprizes them; They see the *Rock*, notwithstanding the vast *Bulk & Weight* of it, move backward & forward three or four Foot, and in less than a Minute or two perform its motions with continual Repetitions.
>
> The *Rock*, whereon it stands is worn into a most regular Smoothness, as far as this Motion extends; and a *Rock* on which the Back of it rubs is also thereby worn considerably.

Mather's prescience, then, was prodigious.

Other popular retellings of Mather's "Curiosa Americana" include P.K. Prybot's article in *History Today* that focuses on Mather's description of teetering rocks (CA 8.5) and V. Scribner's on mermaids in CA 3.4, and Jill Lepore's quotation from 4.3 on the blizzard of 1717, in a 2011 *New Yorker* editorial.[146] We can see, then, that in the three-hundred years since their inception, the Curiosa remain no less curious as artifacts of a misunderstood mind that preserve, at once, accurate natural observations and early modern blends of the religious and the reasonable.

144 L. Barnett (2019), p. 228.

145 P.K. Prybot (2010).

146 V. Scribner (2018), J. Lepore (2011).

Mather's Library

The *Curiosa Americana* not only reveals another side of Cotton Mather, but it also shows him at his scholarly best – and thoroughly enjoying himself in the process. These letters and essays on matters natural, botanical, geographical, anthropological, curative, and astronomical gave him the opportunity to range through his extensive library – and his capacious memory – for classical and recent illustrations and allusions on the wide variety of subjects on which the series bore.

And an impressive library it was.[147] As a young man, Mather wrote in his diary on October 28, 1683 that he had "A Library exceeding any man's, in all this Land." His study was the center of his house, and there – with the motto above the door reminding visitors to "Be Short" – he spent many hours each day, composing on a slate "*Table-Book*" while kneeling on the floor before a chair, or reading in a "very *easy Chair*," or seated before a writing table. In 1702 he observed: "My Study, is tho' a large, yett a warm chamber, (the hangings whereof, are boxes with between two and three thousand Books in them."[148] His book collection, he said, was "the darling of my little enjoyments." That enjoyment is readily evident in the Curiosa. If he had any regrets about the anticipated losses the Great Conflagration of the Earth would cause at Christ's Second Coming, it was the loss of great libraries: "In the Midst of the *Desolations*," he vented in anguish, "there has been scarce any thing, that some whom I know, would more beg to be spared, than the *Libraries*. Their most Pathetical Cry would be, *Oh! Spare the Libraries*.[149]" A true bibliophile if there ever was one.

Mather lived at a time when it was still ostensibly possible to read everything, or nearly everything, that had been printed in Europe and America. It was only during his lifetime that the burgeoning book market put that achievement increasingly out of reach. Even so, if he could not own or read every book, he made every effort to examine any that he could. In 1706 he noted, "Seldome any new Book of Consequence finds the way from

[147] Julius Herbert Tuttle, "The Libraries of the Mathers," *PAAS* 20, pt 2 (April 1910): 269–356. See also Mather's *PA* (1976), p. 42.

[148] After Cotton Mather's death, son Samuel (1706–85) estimated the number of books in the family library at between 7000 or 8000 Volumes of the most curious and chosen Authors, and a prodigious Number of valuable Manuscripts, which had been collected by my Ancestors for five Generations" (S.G. Drake, "Introductory," *The History of King Philip's War, By the Rev, Increase Mather, D.D.* [1862], p. xxiii); Tuttle, "Libraries," p. 299; K. Silverman (*LT*, p. 263).

[149] *TP* (1995), p. 241.

beyond-Sea, to these parts of America, but I bestow the Perusal upon it" (*D* 1:548). So, too, his commonplace book "Quotidiana" saw nearly daily entries of important quotations from books Mather had read during the day. And fearing that he might not be able to track his thousands of excerpts, he "number'd every Quotation, and at the End of his Book had an *Index* of *Names, Things*, &c. so that he could presently have Recourse to the Sentences he wanted." In this way, he coached fledgling ministerial students in his manual *Manuductio ad Ministerium* (1726), you will have collected in a few years "an inexhaustible *Magazine* … from whence as a *Scribe instructed for the Kingdom of Heaven*, you may *bring out Things New And Old*, and have agreeable *Grains of Salt* for all your discourses."[150] As good as his word, Mather did the same in his encyclopedic *Biblia Americana* (1693–1728), a huge commentary on the Holy Scriptures of 4,500 folios containing more than three million words.[151] In combination with the thousands of books and stacks of manuscript sermons he owned, these "Perusals," too, provided fodder for Curiosa, as did the college library in nearby Cambridge whose rich horde of volumes in Mather's time can still be estimated in the *Catalogus Librorum Bibliothecae Collegij Harvardini* (1723).

Mather's Published Sources

It was typical for Mather, when addressing a subject of curiosity, to begin by regaling his imagined Royal Society audience with a litany of what ancient Greeks and Romans, church fathers, medieval philosophers, Reformation and post-Reformation exegetes and theologians, and, where applicable, rabbinic and even occasionally Muslim scholars had to say on the matter. After this historiographical display, he would then introduce the particular information he wished to contribute to the field. Sometimes this addition was the result of his own labor, other times it was something he had obtained at second-hand.

In collecting these instances and allusions, Mather could utilize any number of sources, going from wooden box to box in his study, opening this tome and that. He would often draw from other writers with bare acknowledgement of his indebtedness. While this practice violates modern standards, we have to understand that in Mather's time this was acceptable; such knowledge was, in

150 Samuel Mather, *Life of … Cotton Mather* (1729), p. 24; and Cotton Mather's *MAM* (1726), p. 72.

151 See R. Smolinski, "Introduction" (*BA* 1:17–30). How scholars like Mather kept track of their reading and notetaking is excellently described in A.M. Blair's *Too Much To Know* (2010), esp. ch. 2.

a sense, public domain, and other writers could crib from Mather's works the same way that he copied from them, without any offense.[152]

As a way of examining Mather's use of sources in the Curiosa, we can look at an instance of his habit of mining a work for content. Let's consider the very first epistle, "Giants" (CA 1.1). Here, Mather cites a respectable, even admirable, assortment of authors on the topic who give accounts of gigantic persons and peoples down through the ages. Yet many of the references, if they are tracked down, are taken from one book, and even from one chapter in that book, Philipp Camerarius' *Operae horarum subcisivarum*, the 1654 edition of which was in his library.

On other occasions, Mather would import passages from his own writings into a Curiosa. His "Biblia Americana" manuscript, which was growing apace in the two decades before Mather's death, was one such source. If we return to "Giants," we see that the lion's share of the composition consists of the gloss from the "Biblia" on Gen. 6:4, "There were giants on the earth in those days." In some ways, CA 1.1 is an advertisement for the "Biblia," part of Mather's tireless but ultimately unsuccessful effort to find sponsors to publish it. It was not coincidental, therefore, that we see other excerpts from and references to the "Biblia" in the Curiosa. By the same token, Mather repurposed a number of Curiosa, such as his accounts of diseases and inoculations in the Sixth Series, into separate publications. The original holograph of his treatise on the "breath of life" (CA 6.5 "*Nishmath-Chajim*)" he had sent to John Woodward in 1721 is no longer extant, but multiple versions of the lost document were reused and survive in *AB* (1722), *ABJ* (1972), and *TP* (1995) – Mather again plying his wares among his learned audience to garner a sponsor.

Contemporary Oral and Scribal Sources

It is further evident that, as a knowledge gatherer, besides relying on classical and historical accounts, Mather utilized both contemporary oral and written sources. Although he was the conduit of information, he invited others into the circle of contributors. Perhaps the most notable eye-witness statement or first-hand description in Mather's array of medically related compositions is that of Onesimus, the person of color "given" to him by his congregation, who confirmed that his Garamantes people in Africa used incision as a method to prevent or minimize the effects of smallpox, and that he himself had been immunized in this fashion. Among the several Curiosa that consider smallpox

[152] On the conventions of quotation and source handling in early modern history, see A.T. Grafton's perceptive *The Footnote A Curious History* (1997).

and inoculation, Mather cited Onesimus in CA 3.10, in CA 6.14 (though not by name), and in CA 7.1 incision is again mentioned as a common practice among Africans. Mather would advocate for preventive procedures beginning in 1716, and especially in the epidemic of 1721, but his position would bring down the ridicule and resentment of his less-enlightened contemporaries.

Other conversations on different topics also appear in the Curiosa. In CA 2.5, which gives accounts of the huge flocks of Passenger Pigeons that used to fill the North American sky, he gathered stories of multitudes of these birds caught and killed. Mather relayed (though perhaps did not so much rely on) the humorous anecdote he was told "at a Table of some Superior Gentlemen, relating some of these things: and one, whose Veracity was not to be disputed, said He had a Story that would cap all of mine." This gentleman claimed to have caught nearly 2,500 pigeons at once – merely by closing the door upon a flock that had flown into his barn. In some cases, it is difficult to tell if Mather obtained information via written communication or by a chance conversation. For instance, he does not disclose how he came by the details of the "Uncommon Ideots" of Dunstable (CA 2.8), a town approximately fifty miles from Boston. Perhaps he had spoken to a person from that place who had business in Boston; perhaps he had requested a written report from someone on the scene – Mather may even have preached there and been brought to see the individuals.

Mather also called on a number of worthies and colleagues in his acquaintance to assist in gathering information – and specimens. He was often promising to send a cache of North American fossils to John Woodward, always pleading "the Load of my various Employments" for not fulfilling that promise. But in 1717, for an installment titled "Fossils" (CA 4.1), Mather enlisted the aid of Connecticut magistrate and naturalist John Winthrop (1681–1747, Harvard class of 1700), grandson of Governor John Winthrop the Younger (1606–76) of Connecticut. Winthrop gathered a number of items – we do not know what they were – and had them delivered to Mather. Through the offices of a ship's captain going to London, "a Box, which contains between Twenty and Thirty of such Things as you [Woodward] have asked for," accompanied the Fourth Series of Curiosa. These material objects – which through the years included botanical samples, pieces of the "woollen snow" (CA 2.1) that fell on Fairfield, Connecticut, and other items – have, sadly, been lost.

Astronomical observations occasionally appear in the Curiosa, such as the one that records the effects of the moon on various species of woods (CA 2.2.). Curiosa 4.2 relates facts and figures about a solar eclipse. Mather's source for data on this phenomenon was Thomas Robie (1689–1729), Harvard class of 1708, who was a mathematics tutor at Harvard and the publisher of an almanac when he wrote up his account. He later became a physician at Salem upon marrying, in 1723, Mehitabel Sewall, the granddaughter of Boston judge and diarist Samuel Sewall, a longtime member of Boston's South Church.

Mather could rely on pastor-scholars such as himself in the Boston area and beyond interested in participating in the Republic of Letters for information that could find its way into his Curiosa, and, thereby, possibly into the *Transactions* themselves. So, for example, the source for details about the spine of "A Monstrous Dragon" (CA 7.2) uncovered in Virginia was likely Mather's colleague, the pastor-historian Thomas Prince (1687–1758). Newbury's pastor Christopher Toppan (1671–1747), Harvard class of 1691, assisted with an account of the "Amphisbaena" (CA 8.2) – a snake with heads on each end of its body – as well as with an item in Series Eight, "An Unusual Vomit" (CA 8.3). Like Mather, Toppan was interested in medicine and served as his town's physician and surgeon. In CA 4.4, which discussed tapeworms that had been removed or expelled from humans, Mather included a report that he requested from a physician "in my neighborhood," John Perkins (1676–1740), providing a description and measurement of just such a parasite – reportedly 128 feet in length – that was vomited up by an enslaved man of color. Mather later provided Perkins with a letter of introduction to James Jurin (CA 7.10).

Since Mather lived in a seaport, and had many seafaring individuals in his congregation and among his acquaintance, it should not be surprising to discover that some mariners were his informants. For example (CA 5.13), a Capt. John Robertson provided an account of the underwater volcanic eruption on December 31, 1720, near Terceira in the Portuguese Azores, in the north central Atlantic. Robertson, we know from Mather, commanded a sneau "of New England," a ship with two masts, square-rigged sails, and sharp stern. He had arrived at Terceira earlier in the month, and was hired by the town's government to scout out the "Fire out of the Sea" in the distance. The eventual eruption resulted in the creation of the Dom Joao de Castro Bank, an island about one mile in length, initially above the surface, but which within a couple of years had disappeared beneath the surface of the water.

But, as we have seen with Onesimus, not all of Mather's living sources were members of the elite. From Native Americans Mather may have learned the curative qualities of local plants (CA 1.3), how to treat a rattlesnake bite, or how Natives tell time (CA 1.8). As part of the Seventh Series of Curiosa, Mather included an account from a Boston sailor named Joseph Skeath (CA 7.8), about whom little if anything is known, aside from his narrative of the harrowing ordeal he endured on a recent voyage. First badly burned, then ill with the "flux," then suffering from an infected leg, then marooned on an island and left to die, he was picked up by a passing ship and deposited at Port Royal, where he lay in the streets until put into the hospital. But here his leg became worse, and his mouth infected, so that part of his "Jawbone perished and broke away." As he wrote, he assured Mather, "I have now *Two* Considerable pieces of it in My Hand; with Six of my Teeth." Remarkably he survived and was conveyed back to New England, though likely broken in health. His narrative was in part a plea for assistance.

Another sort of mariner entirely was the Rev. Joseph Morgan (1671–c. 1742), a Connecticut native and Yale College graduate, and in 1723 the pastor of a Presbyterian church in Freehold, New Jersey. Morgan was an author, publishing *The History of the Kingdom of Basaruah* (1715), loosely based on John Bunyan's famous allegory, *Pilgrim's Progress* (1678). He was also a tinkerer. Among his inventions was a land craft powered by sails, which, he claimed, could be controlled by the devices he had perfected, or nearly so – if the Royal Society wanted to build his prototype, he would have to be taken to London to oversee production. Out of support for a fellow man of the cloth, Mather related Morgan's description in CA 7.6 "The Land Sail'd Upon," though he was coy about whether the contraption would actually work.

So we see that the compiling of the Curiosa was not merely a solitary scholarly endeavor. It was, in reality, a social activity, involving a network of individuals from across an expanse of geography, professions, and experiences. It also involved going out into nature, into the community, and beyond, for information deemed valuable for audiences back in England and on the Continent. Communication involved cooperation. As an organizer, reformer, advocate, and educator, Cotton Mather proved this dictum in this aspect of his life as in others.

Conclusion

Where, then, does all that leave the *Curiosa Americana*? Otho T. Beall, Jr. and Richard H. Shryock, in *Cotton Mather: First Significant Figure in American Medicine* (1954, 1979), are among a small handful of historians who recognize that Mather's scientific acumen, especially in medicine, is unique in the colonies and on par with much of what was published in the *Philosophical Transactions* or even in the German *Ephemerides* with its emphasis on medical case studies. They also praise him for his open-mindedness about Native American medicines, his willingness to learn, and his understanding of new medical procedures. Mather was prepared to go against the grain even if his advocacy of innovative methods of combating epidemic diseases earned him the wrath of the community. Likewise, Raymond Phineas Stearns' detailed (though selective) review of the Curiosa letters in *Science in the British Colonies of America* (1970) goes a long way to provide a balanced assessment of

Mather's epistles. Though a clergyman first and foremost, Stearns maintains, Mather is the first American of European descent whose letters reveal "a genuine *philosophical* approach to science, with scientific *ideas* and *hypotheses* of his own, in addition to the contribution of specimens and observations of natural phenomena."[153]

And yet, since their impartial assessments more than half a century ago, much new information has come to light, material which allows us to view Mather's Curiosa Americana in the context of the "medical enlightenment" of the time. Our evaluation of his missives takes into consideration the early medical debates on bacteriology (animalcules), on the precursors of genetic abnormality and its causes, on spontaneous generation and preformationism, and on the kinetic and alleged transformative influence of a pregnant mother's imagination on the developing fetus. The editors believe that the peculiar place of these letters is best seen in light of this new framework – the developing medical theories in microbiology, bacteriology, embryology, and genetics, made possible by the development of powerful microscopes in the late seventeenth century. That this has not been recognized before is partly the result of whiggish historians who are not prepared to set aside their presentist lenses or who were unable to look at Mather's medical handbook and "Curiosa Americana" letters in their totality. Physicians among them particularly eschewed meddlesome clergymen who practiced medicine with prayers and uncouth potions in tow. And for those who broached the subject, ridiculing and brushing aside the standard practices of the past as hopelessly delusionary or downright horrific was much easier than to examine Mather's medical writings in the context of contemporaneous medical theories and practices – barbarous or no.[154] For the first time, we make *all* extant and widely dispersed "Curiosa Americana" accessible in a fully annotated edition set in the context of the scientific debates of their time and place.

[153] See R.P. Stearns, *Science* (1970), pp. 403–26, quote is on 426. *Italics* his.

[154] See for instance Oliver Wendell Holmes, Sr.'s *Medical Essays 1842–1882* (1895), ch. 6, The Medical Profession in Massachusetts.

Chronology

1629	Settlement of Salem, Massachusetts
1630	Settlement of Boston, Massachusetts begins with the arrival of John Winthrop's fleet
1634–52	John Cotton, pastor of the First Church of Boston
1636	Harvard College founded, Cambridge, Massachusetts
1636–69	Richard Mather, teacher of the First Church of Dorchester
1639	Increase Mather, son of Richard Mather, born in Dorchester
1662/1663	Cotton Mather, son of Increase Mather, grandson of Richard Mather and John Cotton, born in Boston, February 12, 1662 (Old Style)
1674–79	Cotton Mather, age 11, admitted to Harvard; baccalaureate degree, at age 15
1678–79	Church membership in his father's North Church, Boston (Second Congregational church); studies medicine as alternative vocation
1680	Mather preaches his first public sermon at Richard Mather's church, Dorchester
1681	Mather receives his M.A. at age 18. Declines a call to the church of New Haven, Connecticut
1683–87	Boston Philosophical Society; founded by Increase Mather; some of Cotton Mather's early "Curiosa Americana" letters presented to this Society
1683–85	Mather serves as pastor of the North Church (ordained in 1685) to assist his father; pastors for more than forty-two years
1684	Revocation of the Old Charter
1685–1701	Overseer at Harvard; Increase Mather appointed president of Harvard
1686	Publishes his first sermon, *The Call of the Gospel*; marries Abigail Phillips of Charlestown, Massachusetts; arrival of Royal Governor Edmund Andros

1688	Increase Mather represents Massachusetts and Plymouth in England to regain Charter; Glorious Revolution in England; accession of William and Mary
1689	April 18, Glorious Revolution in New England; Cotton Mather delivers *The Declaration of the Gentlemen, Merchants, and Inhabitants of Boston*; arrest of Edmund Andros
1690	Baptizes Sir William Phips
1691–93	Salem witchcraft trials
1692	Increase Mather returns from England with Second Charter and Sir William Phips as royal governor of New England Confederation; June 10 1692, first execution for witchcraft, Salem; Cotton Mather writes "The Return of Several Ministers Consulted," warning against the use of spectral evidence to convict the accused
1693	Begins to collect material for *Magnalia Christi Americana* (published in London, 1702); begins medical handbook *Angel of Bethesda* (published 1972); begins *Biblia Americana* (ten volumes published 2010–26); publishes *Wonders of the Invisible World,* an ambidextrous defense of the witchcraft trials; publishes broadside *Rules for the Society of Negroes*
1695	Death of Sir William Phips in London
1698	Appointed commissioner of the Society for the Propagation of the Gospel in New England for the conversion of Native Americans
1699	July 9, 1699, birth of son Increase Mather ("Creasy"); Brattle Street Church ("Manifesto Church") established; begins autobiography "Paterna" (published 1976)
1700	Publishes *An Epistle To the Christian Indians* (dual language; Algonquian and English); Robert Calef blames the Mathers and their fellow ministers for instigating the witchcraft hysteria in *More Wonders of the Invisible World*
1701	Increase Mather resigns from the Harvard presidency
1702	*Magnalia Christi Americana* published in London; death of Abigail Phillips Mather, Cotton Mather's first wife; Joseph Dudley appointed royal governor
1703	August 18, 1703, marries widow Elizabeth Clark Hubbard; resigns as one of the overseers of Harvard
1706	Publishes *The Negro Christianized*; birth of son Samuel Mather

1710	Publishes *Bonifacius* (*Essay to Do Good*); begins *The Christian Philosopher*; receives honorary D.D. from the University of Glasgow
1712	Sends thirteen "Curiosa Americana" essays (First Series) to the Royal Society
1713	Elected Fellow of the Royal Society, London; measles epidemic in Boston; death of Elizabeth (Mather's second wife) and three of their children
1714	Long Extract of his "Curiosa Americana" (First Series) published in the *Philosophical Transactions of the Royal Society* 29.339 (June 30, 1714): 62–71. Appendix 2; sends ten "Curiosa Americana" essays (Second Series) to the Royal Society
1715	Marries third wife, widow Lydia Lee George; death of Governor Joseph Dudley; sends manuscript of "The Christian Philosopher" to the Royal Society
1716	Sends eleven "Curiosa Americana" essays (Third Series) to the Royal Society
1717	Persuades Elihu Yale (wealthy London merchant) to endow a new college in New Haven (Yale College); publishes *Malachi*, a pietist call for union of all Protestant denominations on the basis of his three "Maxims of Piety"; sends four "Curiosa Americana" essay (Fourth Series) to the Royal Society
1718	Bankruptcy over Lydia's inherited estate; nearly loses his prized library (largest private library in English America)
1719	Publishes CA 5.11 *A Voice from Heaven*. The *Aurora Borealis*.
1720	Sends thirteen "Curiosa Americana" essays (Fifth Series) to the Royal Society, via Henry Newman
1721	Royal Society of London publishes his *The Christian Philosopher* (first book on Newtonian Science written by an American); CA 5.13 "The World Alarm'd: A Relation of a New Burning Island"; sends fourteen "Curiosa Americana" essays (Sixth Series) to the Royal Society Smallpox epidemic in New England; with Zabdiel Boylston, a Boston physician, introduces smallpox inoculation (vaccination) in spite of the opposition of Boston's magistrates; Mather and Boylston publish *Some Account of ... Inoculating or Transplanting the Small-Pox* November 14, assassination attempt on Cotton Mather

1722	Publishes CA 6.14 *An Account of the Method and Success of Inoculating the Small-Pox* (London); *An Account of Inoculating the Small Pox Vindicated*; "The Way of Proceeding in the Small Pox," *PT* 32.370 (March 31, 1722): 33–35
1723	August 23, 1723, death of Increase Mather; sends eight "Curiosa Americana" essays (Seventh Series) to the Royal Society
1724	Death of first son "Creasy"; publishes *Parentator*, a biography of his father; sends ten "Curiosa Americana" essays (Eighth Series) to the Royal Society; finishes medical handbook "Angel of Bethesda" (publ. 1972)
1726	Publishes *Ratio Disciplinae Fratrum Nov-Anglorum* (laws and rites of New England Congregationalism) and *Manuductio ad Ministerium* (handbook for young ministerial students)
1728	Dies February 13, 1728, the day after his sixty-fifth birthday; interred in Copp's Hill Burying Ground, Boston
1729	Samuel Mather, Cotton Mather's son, published biography *The Life of the Very Reverend and Learned Cotton Mather*

Note on the Text and Editorial Principles

In spite of their age and wear-and-tear, Mather's surviving "Curiosa Americana" epistles are in excellent condition. To be sure, several of the documents have frayed or are damaged at the bottom, in the margins, or gutter of the pages, but most have been carefully restored by archivists. Only in a few cases did the restoration contribute to the loss of text. However, this blemish did not significantly impede our transcriptions simply because in most cases the lost matter could be reconstructed either from Mather's draft collated with the sent copy or from the printed works he cites in his letters. Thankfully, such loss of matter is extremely rare. Nonetheless, where applicable, we indicate the loss of words and our restoration of the text in our footnotes and render the reconstruction in braces { }. Similarly, we have made every effort to collate Mather's draft copies with the scribal and sent copies when available. Differences between them, especially when Mather introduces corrections or revisions in the manuscripts, are noted in our transcriptions and footnotes. Furthermore, Mather was in the habit of repurposing matter in different locations, by adding or deleting passages where he deemed appropriate or necessary. If needed, we supply diplomatic transcriptions of our copy texts and Mather's variants by inserting extroverted < > or inverted > < arrows in front of and after the interpolation. See especially CA 6.5 "*Nischmath-Chajim*."

Differences in spelling, capitalization, and punctuation tend to occur between Mather's holograph drafts and authorized scribal copies he sent to the Royal Society. Of the roughly eighty Curiosa letters only about fifteen were not in Mather's hand. With few exceptions Mather's spelling is surprisingly modern; his seemingly indiscriminate use of punctuation marks – commas, semicolons, colons, and dashes – is governed by the unwritten conventions of his time, which differ from one writer to another; however, such differences cause no problems. Neither does his frequent underscoring of innumerable words, phrases, even entire paragraphs. This convention, common in the homiletic discourse of the period, serves him to draw attention to significant words or thematic patterns. Underlining also signaled to typesetters and printers when such passages were to be rendered in *italics* in the printed text; double underscoring as, for instance, in "EXPERIENCE" or "GOD" was nothing out of the ordinary either. Typesetters conventionally rendered them as "Experience" or "God." We have retained the typesetters' convention in all cases.

However, his indiscriminate use of capitalization and lowercasing of nouns, gerunds, verbs, adjectives, or adverbs is another matter. Anyone who has ever examined Mather's handwriting will notice this distinct feature in *all* of his manuscripts. All things considered, he generally does *not* distinguish between capital and lowercase forms of such alphabetic letters as *c, e, g, l, m, o, p, s, u, v, x, y,* and *z*. Thus capital- and lowercase phrases like "the **I**llustrious and **p**ious **M**aster **c**artwright" or "he **B**arely touched **Y**et **s**acrilegiously **R**emoved" (emphasis added) are legion. We have adopted the convention established in our edition of Cotton Mather's *Biblia Americana* (2010), pp. 203–09, and "regularized" words like those above as "the **I**llustrious and **P**ious Master **C**artwright" and "he **B**arely touched **y**et **S**acrilegiously Removed." Likewise, Mather's calligraphic use of both the archaic long-tailed *ſ* side-by-side with *s*, as in "Poſſession" or "Paſſion," have been regularized as "*Possession*" and "Passion." All things considered, nothing is gained by retaining Mather's convention, and no change of meaning occurs through modernization.

Finally, conventional abbreviations in the holograph such as

y^{t} = that	y^{e} = the	y^{m} = them	y^{r} = their
o^{r} = our	w^{th} = with	w^{ch} = which	$afores^{d}$ = aforesaid

have been regularized in our transcriptions. However, we have elected to retain the common enough ampersand & and *&c* along with Mather's apostrophe in such third-person singular verbs as *ha's* (has), and his conventional ligatures æ (ae), Æ (AE), œ (oe) and Œ (OE).

SERIES ONE

I.I. DESCRIBING 'BIBLIA AMERICANA' & GIANT FOSSILS.

November 17, 1712.

Letter. the 1st.

To Doctor JOHN WOODWARD.

Sir,

Your excellent *Essay towards the Natural History of the Earth,*[1] ha's obliged and even commanded, the true Friends of *Religion*, and *Philosophy*, to serve you with as many Communications as they can, that may be subservient unto your Noble Intention.[2] But the *Letters* wherewith you have honoured me, have laid *me* under your more particular Commands, to supply you with

1 John Woodward (c. 1665–1728), physician, natural historian, and secretary of the Royal Society and the Royal College of Physicians. Mather refers to Woodward's popular *An Essay toward a Natural History of the Earth and Terrestrial Bodies* (1695, 1702, 1723). He presented Mather a copy of *Naturalis Historia Telluris Illustrata & Aucta* (1714), a defense against Elias Rudolph Camerarius (1673–1734), professor of Medicine at Tübingen, who had sparred with Woodward in *Dissertationes Taurinenses Epistolicae, Physico-Medicae* (1712) on the origin of fossils and Noah's flood (Appendix 12).

2 As early as July 1711, Mather was contemplating how a "valuable Collection of many [scientific] Curiosities, which this Countrey has afforded" might be made serviceable to the Royal Society of London (*D* 2:85–86). The present epistle is the first of the "Curiosa" Letters (First Series, November 17–November 29, 1712) Mather sent to the Royal Society, which abstracted and published them in the *PT* 29.339 (June 30, 1714) 62–71: "An Extract of several Letters from Cotton Mather, D.D." (Appendix 6). Significantly, it is the same *PT* issue in which Emanuel Timonius published his account of smallpox inoculation in Constantinople, which served Mather during the smallpox inoculation controversy in Boston, in 1721. In fact, Timonius' essay follows Mather's abstracted correspondence in the *PT* (pp. 72–82). The abstract of Mather's twelve letters (First Series) also appeared in translation in the first volume of Michel de la Roche's *Bibliothèque Angloise* tome premier, premiere partie (1717), pp. 99–106.

such *Subterraneous Curiosities*,[3] as may have been in these Parts of *America* mett withal. I do with much Alacrity apply myself immediately to obey your Commands, in one Remarkable Instance, wherein I apprehend myself best able to do it; reserving for another Opportunity (if perhaps, I find myself a Master of) any further Treasures, worthy to be laid before a Person of so clear a Judgment, & so reaching a Sagacity. And that I go *cheerfully* about this Point of my Obedience, you will conclude it, when you see me begin with assuming the Air of telling you, That you must look upon all the *American Curiosities* which are sent you, as being in some sort *Subterraneous*; inasmuch as you know, who have understood the Inhabitants of *America*, by, *The Things under the Earth*, which the Sacred Oracles tell us, are to *bow in the Name* of our Glorious JESUS.[4] But indeed one must be as ungeographical as the Dayes of *Lactantius*,[5] to admitt the Interpretation; and therefore I shall endeavour a nearer Approach unto your Satisfaction.

There ha's lately been brought into our Light from the *Subterraneous World*, in our Neighbourhood, a Rarity, which appears to me worthy of a Consideration with all Mankind, as well as with a *Woodward*, or a *Kircher*.[6] One whom I verily take to have been an Inhabitant of the *Antediluvian World*, has had, not a *Resurrection*, tis true, but yett ha's been pull'd out of the Bed, in which he had lain, doubtless ever since that wondrous Revolution, which made the wicked Children of Man to *groan under the Waters*.[7] For truly, the Men, who associate themselves with a late *French* Author, in a fond Conjecture, that such Bones may be the *Petrifying Sports* of Nature,[8] provoke me to

3 See also Mather's Letter to John Woodward (July 25 1716), CA 4.1 "Fossils" and Appendix 12.

4 Philip. 2:10.

5 Lucius Caecilius Firmianus, aka. Lactantius (c. 250–c. 325 CE), an early Christian apologist who in *Institutiones Divinae* (3.24), *ANF* (7:94–95), ridiculed the idea of the earth's poles. If there were poles, he argued, people's heads would be higher than their feet – or so the argument went. The great St. Augustine of Hippo (354–430) was of like mind in his *City of God* (16.9), in *NPNFi* (2:315–16). See also *MCA* (1.1.1–2 §5), *TP* (1995), p. 110 and *BA* (1:473–74).

6 German Jesuit polymath at the Collegio Romano, Athanasius Kircher (1602–80) published *Mundus Subterraneus* (1664–78), a voluminous study of subterraneous rivers, lakes, volcanic magma, minerals, and fossils.

7 Job 26:5. For the full version of Mather's discussion on the giant fossils, which he takes to be the bones of an antediluvian Nephilim (Gen. 6:1–4), see *BA* (1:582–99). See also *SLCM* (110–12) and D. Levin (1988).

8 I.e., *Lusus naturae*; the paleontological speculation that petrified remnants of seashells, fish, trees, and dinosaurs (a term not devised until the early 19th century) were not the relics of real bones or plants but merely the sports of fertile nature

say of them, as *Abarbanel* does upon some of his Rabbis for some strange glosses on the *Isaian* Prophecies, *Videntur omnes hi Sapientes percussi fuisse cæcitate*;[9] & to look on them as opinionative Persons, under the Power of an Humour, not worthy to be argued with. I am going to give an Account of that Rarity; but it shall be under the Advantage of Transcribing somewhat that was præpared for the Public diverse Months ago; it shall be a pure Transcript of a Copy that ha's been some while waiting for Publication.[10]

It seems not improper for me to give you first of all some Account, How and where I mett with the Discourse, which I am now to lay before you. And herein also I shall do that which perhaps may gratify the Inquisitive Genius of a Person who wishes very well unto the *Common wealth of Learning*. It is fetch'd from an *Amassment of Treasures*[11] which I could wish, did not ly too like the *Subterraneous* Ones; and that the Library in which they lodge may not be so like the horrid Cellars of *Indoustan*, where your Silver and Gold after they have been brought out of their *Subterraneous* Condition, and circulated thro' the Business of the World, again return into it, and are again buried by the covetous *Mogols*,[12] & have no more to do on the face of the Earth. My Meaning is, tho' I need not explain what I mean; I wish they had the *Publication* which they are waiting for, and that your Presses would return to print something else besides your *Politicks*, & serve to better Purposes than to vent the *Ill Humours* of your Nation. I wish it may come into the Minds of Ingenious & Opulent Men, to do Mankind the good Office of bringing forth what has been prepared for common Benefit.[13]

(*lusus naturae*) imitating itself. See *BA* (1:582 ff). On early interpretations of fossils, see P. Rossi (1984), esp. chs 1–5; P. Findlen (1990), R. Rappaport (1990, 1997), A. Mayor (2000 and 2005).

9 Isaac ben Judah Abarbanal (Abravanel), a 15th-century Portuguese Jewish philosopher and biblical expositor, author of his Hebrew commentary *Perush 'al Nevi'im ahronim,* aka. *Commentarius celeberrimi Rabbi Ishak Abarbanel* (1642): "All these wise men seem to have been struck by blindness."

10 Mather refers to his holograph manuscript "Biblia Americana," a biblical commentary of more than three million words, currently edited in a ten-volume edition. Hereafter "BA."

11 The manuscript of Mather's "BA" commentary on the Bible for which he hopes to attract subscribers.

12 I.e., Moguls, the Muslim rulers of India in the 16th and 17th centuries. See also Mather's *B* (1710) §19, p. 148.

13 See Mather's advertisement of his "BA," in *B* (1710), pp. 200–06; and his *NO* (1714).

There is an *American* Friend of yours, who tho' he never travelled out of *America*, ha's had the Honour to be Related unto one of your *Europæn* Universities;[14] and ha's been desirous to oblige a Number of the best People in *Europe*, with a Composure, which now arises to Two considerable Volumns in *Folio*, wearing the Title of, BIBLIA AMERICANA.[15] He had long since been of Your Excellent *Boyl*'s Opinion, That *you should no more measure the Wisdome of God concealed in the Bible, by the Glosses & Systems of Common Expositors, than estimate the Wisdome He ha's express'd in the Contrivance of the World, by* Magirus's, or Eustachius's *Physicks*;[16] and agreed with him in hopes, that learned Men would go on to make more admirable Explications and Discoveries, in that *wonderful Book*, than what usually occurr'd in the *Vulgar* (tho' very *useful*) *Annotations*. Having the Honour of a Correspondence, with the Incomparable *Witsius*,[17] one of the greatest Men that the last

[14] Universitas Glasguensis (Glasgow), which bestowed a D.D. on Mather in 1710 (*D* 2:40, 63–64). See *LT*, pp. 222–25.

[15] Mather's plans to compile his "BA" are first mentioned in August 1693 (*D* 1:169–71). By 1706, he had amassed material for two folios (*D* 1:563–64, 567), composed an advertisement, "An American Offer to serve the Great Interest of Learning and Religion" as "*Prodromus* and *Engaging* for that work" (*D* 1:570), and sent the introductory advertisement and essay to London. He sent yet another draft of his prodromus to John Woodward (November 17, 1712) as Mather's first "Curiosa Americana" letter for publication in *PT*. He published another advertisement, *A New Offer* (c. 1714/1715) and wrote several letters to influential friends at home and abroad to garner support (*D* 2:284, 309–10; 314–15). By the end of his life, his "BA" remained unpublished: The manuscript had swollen to nearly five thousand pages in folio and more than three million words – too large and too costly to be published at the time (*BA* 1:17–41).

[16] Robert Boyle (1627–91), an Anglo-Irish natural philosopher, chemist, physicist, and important member of the Royal Society. Mather paraphrastically quotes from Boyle's *Excellency of Theology, Compared with Natural Philosophy* (1671), part 1, sec. 1, pp. 47–48, and refers to both authors referenced in Mather's letter: Johannes Magirus (1560–96), German Lutheran philosopher, professor of physics at the Lutheran University of Marburg (Germany), and commentator on Claudius Ptolemy's *Geography* and the nine crystal spheres of Ptolemy's *Almagest*. Today, he is mostly remembered for his oft-reprinted commentary on the Peripatetics in *Physiologiae Peripateticae* (1616), which was part of John Harvard's original bequest to Harvard and there used as a textbook. Bartolomeo Eustachio (c. 1510–74) was an Italian anatomist. Mather appears to refer to Eustachio's *Opuscula Anatomica* (1564), a work with valuable plates depicting the muscles and nervous system of the human body.

[17] Hermann Witsius (1636–1708), a distinguished Dutch theologian and professor at Franeker and Utrecht, whose work on covenant theology *De Oeconomia Foederum Dei cum Hominibus* (1677) was highly prized by Mather and his Reformed

Age produced, he laid before him, the Plan of his Undertaking, and received the Encouragement of that *Great Man* to go on with what he had in hand. And the Truth is, That *Hero* (for, I know not why *Learning* also should not have its *Hero*'s,) was himself not the Least in furnishing this Author with rich Materials for him to work upon, and in verifying his own Report; *Neque profectò officio hic suo defuerunt Illustres Animæ, quæ quicquid in Philosophiâ invenerant solidi; quicquid in Antiquitate Hebræa, Græca, Romana, eruditi; quicquid in Arte Hippocratica, aut Juris Civilis prudentia, quod aliqua se cum Sacris nostris Literis affinitate commendaret; id omne ad eas illustrandas insigni dexteritate contulerunt. Atque ità usu venit, ut Theologia, suis cæteroquin opibus contenta, nunc Exoticis quoque Divitijs, velut opimis Spolijs locupletata, cultius quàm unquam in Publicum prodierit, majestatisque suæ confessionnem, æmulis etiam, atque invitis extorserit.*[18] Your Author laid aside his other Employments.[19] He preached his weekly Sermons, & performed all the other pastoral Duties, to the greatest Church in *America*; & answered in a Measure the Expectations, which the various Relations he sustained, might cause his Neighbours and more here in all these Colonies, to have upon him. He gave the Public at the same Time, above Two Hundred other Books on various *Arguments* & in various *Languages*.[20] But neither these, nor any other Obstructions could retund his Passion for the Favourite Work of his, *Biblia Americana*. These Things are not mentioned, as bespeaking any Esteem for

---

peers. Mather here comments on Witsius' oration "*Oratio de Felici an Infelici Reipublicae Literariae, hoc seculo, statu*," delivered on March 16, 1698, in Utrecht, in *Miscellaneorum Sacrorum Tomus Alter* (1700) 2:937–51.

18 The Latin citation is from Witsius' above-mentioned oration published in *Miscellaneorum Sacrorum Tomus Alter* (1700) 2:942–43. Addressing how the other arts are shedding more light on theology by God's providential design, Witsius says, "Nor, indeed, have illustrious spirits shirked their duty in this. Whatever they had found of solid worth in philosophy, whatever learning they had discovered in Hebrew, Greek, or Roman antiquity, whatever they had found in the Hippocratic art or in the knowledge of civil law which could recommend itself through some affinity to our sacred writings, all this they have applied with remarkable skill to the task of illuminating those writings. And so it has come about that theology, though in other respects content with its own resources, now additionally made wealthy with foreign riches, as with the greatest spoils of war [*opimis spoliis*], it has stepped forth into the public eye more elegantly than ever, and has extorted even from its unwilling rivals an admission of its own majesty." (David Lupher's translation).

19 Proud of his own humility, Mather likes to refer to himself in the third person to tone down his self-praise, a rhetorical device not unheard of in his time.

20 By the end of his life, Mather had published more than 450 titles (see T.J. Holmes, *Cotton Mather: A Bibliography*, 3 vols (1940)).

the Author, but rather to Abate something of your Esteem for the Work, & bring it down to the most proper *Alloy*; You must not look for too much in it; and yett in a few Years, the gracious Hand of Heaven ha's enabled him to prepare for you, much more than is expressed in the *Bill of Fare*, which I will now exhibit you.

He ha's given us the *Sacred Scriptures*, in the *Order of Time*, wherein the several & successive Occurrences may direct the Placing & Reading of them.[21] *This Alone* were itself a valuable Commentary. But then, the *Common Translation* is here also amended & refined; with a due Modesty, tis true; & only where the most Judicious Philology ha's offered much Reason for it. You have here also a Rich Collection of *Antiquities*, which the laborious Researches of learned Men in our Dayes have recovered, for a sweet Reflection of Light on the Heavenly Oracles; The *Agriculture*, the *Architecture*, the *Tacticks*, but above all, the *Idolatry* of the former Ages, you will find explaned in this Collection. The *Types* in the Bible, are here accommodated with their *Antitypes*; and those Paragraphs of it that appeared the least fruitful with Instruction, are now to be perused with a vast *Mixture* of Holy *Profit* and *Pleasure*. The Lawes of the *Israelitish* Nation, in these *Pandects of Heaven*[22] are here interpreted, & the Original and Intention thereof rescued from the Misinterpretations of some famous Writers. *Gold* is here fetch'd out from the Dunghil of the *Talmuds*, & other Jewish Writers, not only to Illustrate the *Oracles once committed unto the Distinguished Nation*, but also to demonstrate the Truth of Christianity. *Natural Philosophy* is here called in. The fairest *Hypotheses* (and yours,[23] my most honoured Doctor!) of these Grand Revolutions, the *Making*, and the *Drowning*, and the *Burning*, of the World, are offered. The *Meteors*, the *Minerals*, the *Plants*, the *Animals*, the *Diseases*, the *Astronomical* affaires, & the Powers of the *Invisible World*, mentioned in the Book of God, are here considered, with the *Best Thoughts of our Times* upon them. The *Chronology* of this Book, & of the World, is here clear'd from all its Difficulties; and the *Clock* of *Time*, sett and kept Right, in its whole Motion from the Beginning of it. The Sacred *Geography* here survey'd; *Paradise* and *Palæstine* are particularly laid out. And many Notable & Enlightning things are contributed unto this Work, from the Modern Travellers. Here is a particular History of the City *Jerusalem*, in its marvellous Vicissitudes, from the Dayes of *Melchizedek*

[21] Mather refers to his chronology, "The Old Testament, in the Order of History" (*BA* 1:213–68).

[22] A comparison to Justinian's *Digest* of Roman civil law.

[23] Woodward, *An Essay toward a Natural History of the Earth ... With an Account of the Universal Deluge* (1695). See *BA* (1:490–91).

down to ours; with an Account of its present Condition. But this is not all; For you have here, what the Author calls, *A sort of Twenty-ninth Chapter of the Acts*; or, an Elaborate and Entertaining History, of what has befallen the *Israelitish Nation*, in every Place, from the Birth of the Glorious Redeemer to this very day; And the present Condition of that Nation, the Reliques of the *Ten* as well as the *Two Tribes*, and of their ancient Sects, yett (several of them) existing also in the several Parts of the World, where they are now dispersed, *at this Time*, when their speedy Recovery from their Sad and Long Dispersion, is hoped for. More than so; You have here the *Histories* of all Ages, brought in, to show how the *Prophecies* of this Invaluable *Book*, have had their most punctual *Accomplishment*; and strongly established *Conjectures* on such as yett remain to be accomplished. With the true Doctrine of the *Chiliad*, brought in as a *Key*, to very much of the Wealth which the Church of God enjoyes in this Book of the Kingdome.[24] Syr, All these things are but a Part of your Entertainment, in the Work, which I may præsume to call, as they did the *Octapla* of *Origen* long ago[25]; *Opus Ecclesiæ*; & which yett lies like the Impotent Man at the Pool of *Bethesda*;[26] waiting till Heaven stir up the Inclinations of some Generous Minds to appear as *Benefactors*, in the Publishing of it.

One of the *Illustrations,* whereof there are Thousands, in that Work, is now to perform the Office, which this Letter begun upon. Here it is; Take it as it is; I will no longer detain you from it.

An *Illustration*,

On that Passage, in Gen. 6.4. *There were Giants on the Earth in those Dayes*.[27]

QUA{E}STION.

Concerning the Dayes before the FLOOD, the glorious Historian ha's told us, *There were Giants on the Earth, in those Dayes*.[28] Could any undoubted

24 *B* (204, 205). On Mather's millennialism, see *TP* (1995), ch. 3, esp. p. 72.

25 Origen (c. 185–c. 254) was a biblical critic and exegete of Alexandria (Egypt), whose *Hexapla* became a premier *Opus Ecclesiae*, a work of the church. Mather means Origen's famous *Hexapla*, a compilation of the Old Testament in six parallel columns: the Hebrew text, a transliterated version of the Hebrew text, and four Greek versions. And yet, it was not uncommon to refer to it as *Octapla* (eightfold edition), because Epiphanius of Salamis (*On Weights and Measures*, 19) argues that two additional translations had been discovered and added to Origin's six columns (See James Elmer Dean's edition, p. 31).

26 John 5:2–9.

27 See *BA* (1:582ff).

28 I.e., Moses, the assumed author of the Pentateuch; the quote is from Gen. 6:4.

Ruines and Remains of those GIANTS be found under the *Earth*, among the other *Subterraneous Curiosities* in *our Dayes*, it would be an Illustrious Confirmation of the *Mosaic History*, and an admirable Obturation on the Mouth of *Atheism*![29]

ANSWER.

Then lett the Inquisitive Part of Mankind, know that the *Bones* of those who were certainly some of the *Antediluvian* GIANTS, have been found under *the Earth*, in these later Ages. Below the *Strata* of *Earth*, which the *Flood* left on the Surface of it, in the other *Hæmisphære*, such enormous *Bones* have been found, as all Skill in Anatomy, must pronounce to belong unto *Humane Bodies*, and could belong to none but GIANTS, in Comparison of whom, OG, and, GOLIAH, and all the Sons of *Anak*, must be hardly so much as *Pygmies*. But, that AMERICA too, as tis but agreeable, may throw in Something to the Treasures of the BIBLIA AMERICANA, I will surprize you, with telling you, That the *Men*, who were able to *have turned the World upside down, came hither also*: [How? Lett any Man alive tell me, and he shall be to me, a greater *Apollo*, than he that could lead me to the Tomb of *Cœl{ius}*].[30] And the *Bones*, most probably of the Antediluv{ian} GIANTS, have here been mett withal.

I will not go to trouble you, with the Opinion of the *Ancients*, who suppose the *Original* of those *Giants* to be, *The Sons of God coming in unto the Daughters of Men*; and that the *Sons of God* were *Angels* or *Divels*; to which the Name of *Nephilim*, which may signify, *Fallen Ones*, may indeed agree well enough. This the LXX induced the Fathers to imagine; and they had *Philo* and *Josephus*, to countenance the *Imagination*;[31] But it is a *Vain* one. I do not think, That it was an Error in the learned *Henricus Vorstius*, to say, that the Scripture does not mention the *Original* of the *Giants*;[32] No, it

[29] For this and the following paragraphs, see *BA* (1:582–98).

[30] MS damage affects the end of this and the following line. The reference is possibly to Coel Hen, or Coelius of Ayrshire, a 5th-century CE Roman ruler of Britain (the basis for the "Old King Cole" of the nursery rhyme).

[31] Seventy Interpreters (Septuaginta or LXX) of the Bible; Philo Judaeus, 1st-century BCE Hellenistic Jewish philosopher and author of *Quaestiones et Solutiones in Genesin, I*), sec. 92, on the Bible; and Flavius Josephus, 1st-century CE Roman-Jewish historian, author of *Antiquities of the Jews* (1.3.2).

[32] Guilielmus Henricus Vorstius (Willem Hendrik van der Vorst, d. 1652), a 17th-century Dutch rabbinic scholar and translator of פרקי רבי אליעזר *Capitula R. Elieser*, aka. *Pirke Rabbi Eliezer* (1644), cap. 22, p. 50, an aggadic-midrashic chronological history, attributed to tanna Eliezer ben Hyrcanus (1st–2nd century CE).

only declares the *Time*, when God punished the World with them. And our Saviour ha's abundantly taught us, how disagreeable *Matrimonial Affairs* are to the *Angelical Nature*.[33] Nor is there any Strength in the attempts of *Drusius*, to defeat the Argument,[34] which is fetched from the Words of or Saviour, to confute the *Dream* of the *Old Men* about this matter.

Tho' we yett meddle not with the *Original* of *Men*, we may touch upon that of their *Name*. The Name / נפילים / *Nephilim*, comes from / נפל / *To Fall*. And without expounding it, of *Men Fallen from an Heavenly to an Earthly Temper*, [A Frivolous Exposition that some have given; and which would, alas, fill the World with *Giants*!] Why should not the Name intend as much as, *Irruentes, Prosternentes, Dejectores*;[35] Men who took a Vast Liberty to *Fall upon* other Men, and make them to *Fall*. If they were of such an horrid Magnitude, as the *Monster* I am anon to shew you, the Etymology would be agreeable enough unto them. I doubt, they were a Sort of Men, as unlikely as any upon *Earth*, to deserve the Character that *Bolducus*, in his Treatise, *De Ecclesia ante Legem*, ha's provided for them; who saies, They were calle{d}*Nephilim*. *Quòd ob veri Dei cultum, à Reliquis, in eo distinguerentur, quòd Deum prostrationibus adorarent, adeò ut illâ ceremoniâ eos ab improbis secerneret*;[36] Because they did use to *Fall down* prostrate upon the *Earth*, in their Adorations before the God of Heaven.

An Extremity of Incredulity, ha's led some to think, That the *Antediluvian Giants* were but *Metaphorical* Ones; That they were *Giants* for *Quality*

In his appended "Animadversiones in Pirke" (pp. 184–85), Vorstius rejects R. Tzadok's comment (*Capitula*, cap. 20, p. 50) about the genealogy of the "fallen ones" (nephilim). It is uncertain (says Vorstius), for the Bible only mentions the "time" when they consorted with the daughters of man and that God punished the world with these giants. And their supposed fertility, their dances, etc. ("stolidis Rabbinorum ineptiis annumeranda sunt") "are to be numbered among the stupid absurdities of the rabbis."

33 Matth. 22:30.

34 Johannes Drusius, aka. Johannes van den Driesche (1550–1616), a Flemish Protestant divine and author of *Ad Loca Difficiliora Pentateuchi* (1617), Loca Genes., cap. 18, pp. 34–35, who discourages the dreams of old men once again to be young and married (Matth. 22:30). See *BA* 1:584, *n.* 131.

35 "Invaders, Overthrowers, Conquerors."

36 The Latin quotation is a paraphrase from the French theologian Jacques Bolduc, or Jacobus Bolducus (d. 1646), in *De Ecclesia Ante Legem* (1626, 2nd ed., 1630), lib. 1, cap. 3, pp. 17–21, and reads, "because for the sake of worshiping the true God, they [Nephilim] might in this way be marked out from other men: that they worshiped God with such prostrations that He would through that ceremony distinguish them from the wicked."

only, & not for *Quantity*. We will not count the *Gigantomachia* of *Goropius Becanus* worth considering.[37] *Temporarius*[38] has endeavoured more elaborately to show, That the Name is of a *Scythian* Original, and means no more than, *A Man with long Hands*, which are no other than a World of Rapacious Mortals, are at this Day furnished withal; as well as he who reaches from *Versailles* to *Peru*.[39] And indeed, *Philo* the Jew, strips these *Giants* of all their Procerity, & makes these *Children of Heighth*, no more than ανθρωπους της γης, *Homines Terræ*,[40] or another *Sort of Men*, than the *Men of Heaven & of God*.

But the *Apocryphal Baruch* shall be a better Interpreter for us; who [chap. 3.26.] expressly saies of these *Giants*; they were ευμεγεθεις επισταμενοι πολεμον· *Ingenti corporis magnitudine præditi, reique bellicæ scientissimi*.[41] The *Apocrypha* several times, calls them *Giants*. And afterwards *Chrysostom* calls them, Ηρωας ευμηκεις· *very tall Hero's*; and, υψηλους *High Ones*.[42] *Theodoret* calls them, παμμεγεθους ανθρωπους· *Prægrandes Homines*.[43] And *Cyprian* tells us, They were those, *Quibus corporum Deformitas et Magnitudo data fuit, in pœnam effrænatæ Libidinis Parentum*.[44] But I have a greater Authority than all of this; Namely, That of the *Giants* themselves. The

37 Johannes Goropius Becanus, aka. Jan van Gorp (1519–73), a Dutch physician and linguist. "Gigantomachia" was the first book of his *Origines Antwerpianae* (1569), lib. 2, pp. 128–226; in which, among other things, he claimed that ancient Flemish was the language spoken in the Garden of Eden. See D.W. Baker (2019).

38 Jean du Temp, or Johannes Temporarius, a 16th-century French Protestant jurist and author of *Chronologicarum Demonstrationum* (1596), lib. 1, pp. 10–11. See also *BA* (1:584–85).

39 I.e., the Roman Catholic pope and Louis XIV of France. Mather may have learned of the well-publicized expedition of Amédée-François Frézier (1682–1773) in January 1711/12, funded by Louis XIV, to chart the coastlines of Chile and Peru and to gather other information and samples. Frézier published on his return *Relation du Voyage de la Mer du Sud aux Cotes du Chily et du Perou, Fait pendant les années 1712, 1713 & 1714* (Paris, 1716). Incidentally, he famously brought back the strawberry. See the depiction of this delicious fruit between pp. 70–71 of this edition.

40 Philo Judaeus, *De Gigantibus* (13.60): "Men born of the earth."

41 Apocryphal Baruch (LXX): these giants "were [by nature] of so great stature, and so expert in war." Mather drops the Greek diacritics deliberately. See *MAM* (§6, p. 29).

42 Archbishop of Constantinople and early Church Father John Chrysostom (c. 347–407), *Homily on Genesis* (22.12), in *The Fathers* (82:77).

43 Bishop Theodoret of Cyrus (c. 393–c. 466), *Quaestiones in Octateuchum: Genesis* (Quest. 48, p. 47, line 25) calls them "Rather large men."

44 Not Cyprian but the Patriarch Cyril of Alexandria (d. 444), in *Glaphyra in Pentateuchum* (lib. 2) [*PG* 69.56]: "To whom deformity and bulk of the body were given as a punishment for the unbridled lust of their parents."

Wretches *under the Earth* are in some Sort brought up from thence, before the *Resurrection*, when they are *to Bow the Knee unto the Name of JESUS*. The *Giants* that once *Groaned under the Waters*, are now *found under the Earth*, and their *Dead Bones* are *lively Proofs* of the *Mosaic* History.[45]

Of all those *Curiosities*, I know none that exceeds what ha's lately been found in an *American* Plantation, adjoining to *New England*. And its being found in *America* makes it yett the more curious and marvellous. For, I beseech you, How did the *Giant* find the Way hither?

The *Post-diluvian Giants* mentioned in the Sacred Scriptures, were *Puisny* Things in Comparison of ONE ABOVE SEVENTY FOOT HIGH; and yett we have here the undoubted Reliques of SUCH AN ONE.

What *Julius Capitolinus*, and others, report of the Tyrant *Maximin*, That he was *Eight Foot High*;[46] and what Nicknames *Herodian* tells us, were therefore putt upon this *Typhon*, and *Cyclops*, tis no longer worth the while to mention.[47]And as little worth a mention, is the *Swevian* in the Army of *Charlemaign*, of whom *Aventinus* tells us, That he would mow down his Enemies, and spitt good Numbers of them like so many little Birds, on his huge *Spear*, which he would so carry in his hand about the Camp.[48] We shall hardly regard, what old *Mela* tells us, of *Indians* who were so Tall, that *Elephants* were to them instead of *Horses*.[49] We shall as little regard, the *Americans* near the Streights of *Magellan*, to whom they assign a Stature sufficiently extended, and a *Neck* as long as a *Mans Arm*.[50] We shall be far from Trembling before

45 Job 26:5.

46 The Roman historian Julius Capitolinus (3rd century CE), in his notoriously unreliable *Historia Augusta: Maximini Duo* (6.8; 8.5–6; 28.8), the life of Roman Emperor Gaius Iulius Verus Maximinus Thrax (235–38 CE), argues that Maximinus was "six inches over eight feet" tall and that some nicknamed him "Cyclops, some Busiris, and others Sciron, not a few Phalaris, and many Typhon or Gyges." See also Hesiod, *Theogony* (lines 823–35).

47 Herodian, early 3rd-century CE Greek historian, author of *The History of the Empire after Marcus* (6.8.1, 7.1.1–5, 12), relates that Maximinus's nicknames were Typhon (the hundred-headed Titon) and Cyclops (Homer's one-eyed giant).

48 In his *Annales Boiorum* (1554), lib. 4, pp. 364–65, 420–21, 450, the Bavarian humanist Johannes Aventinus, aka. Johann Georg Thurmayr (1477–1534), relates the story of the giant Suevian, who served in the army of Roman Emperor Charlemagne (747–814).

49 The Roman geographer Pomponius Mela (1st century CE), in *De Chorographia Libri Tres* (3.52–53), tells the story of his tall natives of India.

50 The giant Americans near the Strait of Magellan in Terra Del Fuego (southern tip of Chile) are the Tehuelche people of Patagonia.

the *Giant* seen by *Pigafetta*, taller than other Men, from the Girdle upwards.[51] Lett us make room for *Greater Company*.[52]

We will not be too urgent in demanding a Consideration, for what *Pliny* relates; That *Lucius Flaccus*, and *Metellus*, in the *Cretian War*,[53] found the Bodies of some *Giants*, that were *Thirty Cubits long*; and the Earth opening in an Earthquake, there was another cast up, which extended unto *Forty Cubits*. We may suspect a *Cretian* Story; tho' one of their *Poets* may once in his Life, be a *True Witness*. And, I find, the Credit of *Pliny*'s Relations runs pretty low among the Learned in our Dayes; Plinys Relations if his own knowledge be true, tis an Error but he is not Valued by the Learned. I know what they mean, when they call a thing, *A Plinyism*.[54] However, we may give the more Credit unto *This*, because we have in the Grave *Plutarch*, much such another:[55] yea, a *Skeleton* amounting to *Three Score Cubits*.

But, lett us go on. The *Egyptian* Traditions of the *Gigantic Emephimi*, related by *Franciscus Patritius*, were not altogether without some Foundation.[56]

[51] The Italian traveler Antonio Francesco Pigafetta of Vicenza (c. 1498–1534) accompanied the Portuguese explorer Ferdinand Magellan (1519) and marveled in his journal *Le Voyage et Navigation* (1525) at the giant Patagonians against whom Magellan's sailors were puny. Pigafetta's full account is in *Magellan's Voyage* (1994), ch. 8, p. 46, but Mather may have gleaned his information from Thomas Molyneux's "An Essay concerning Giants," *PT* 22.261 (February 28, 1700), pp. 487–508, esp. 505.

[52] In this paragraph (as in some of the following), Mather draws on *Operae Horarum Subcisivarum Sive Meditationes Historicae Auctiores ... Centuria Prima* (1644), vol. 1, cap. 82, pp. 376–85, an oft-reprinted history by Philipp Camerarius (1537–1624), a German scholar and counselor at Nuremberg. The Mathers owned the three-volume Frankfurt 1644–50 edition. All citation references are to the 1644 edition.

[53] The Roman general Lucius Flaccus and Quintus Caecilius Metellus Creticus, Roman consul of Crete (69–c. 65 BCE), crushed the pirates who had fought the Romans during the long Cretan War (c. 106–66 BCE). According to Solinus (*De Mirabilibus Mundi*, cap. 1), floodwaters laid bare a gigantic skeleton measuring thirty-two cubits (c. 47 feet), which Flaccus and Metellus examined at length.

[54] Pliny the Elder (c. 23–79 CE) was a Roman naturalist and philosopher known for his exaggerations (Plinyisms). He does not refer to this particular fossilized monster, but mentions one forty-six cubits tall (c. 65 feet), in his *Natural History* (7.16.73–74).

[55] Plutarch (c. 46–119 CE), a Greco-Roman philosopher, historian, and biographer tells the story of the likely remains of a dinosaur 60 cubits tall (c. 85 feet; c. 27 m), in his *Vitae parallelae: Sertorius* (9.3–4); *Plutarch's Lives* (8.9).

[56] Via Camerarius' *Operae Horarum Subcisivarum* (1644), 1:382, Mather refers to the Italian philosopher Francesco Patricius (1529–97), professor at Ferrara, who

Reineccius in his *Monumenta Brandenburgica*, affirms, That in his Time, there were dug up the *Bones* of Men, whose *Legs* reached up to the *Waste* of other Men.[57] *Fazellus* as an Ey-witness affirms, That in *Italy*, there were dug up the *Bones* of Men, which were of a stupendous Magnitude. *Camerarius* in his *Operæ Succissivæ*[58] is much affected, with the huge Humane *Bones*, found at *Puteoli* in *Italy*; on which a Poem of *Pomponius Lætus* concludes thus;

Hinc bona posteritas immania corpora servat,
Et tales mundo testificatur avos.[59]

I find our P. *Martyr* [in Jud. 1.10.] quoting it from *Philostratus*; That the Carcase of a *Giant*, was found of no less than *Twelve Cubits* long; of another, no less than *Twenty two Cubits*; of another, no less than *Thirty*.[60]

One way of Judging alwayes allow'd, ha's been, *Ex Ungue Leonem*.[61] I know not why we_may not as well Judge, *Ex Dente Gigantem*.[62] The *Teeth*, which are a very durable Sort of *Bones*, found below the Surface of the Earth, may serve as well as any *Tongues*, to tell us, what *Men* were once upon it.

---

relates the tale of the Egyptian "*Emephimi*," the race of colossal giants who battled the Egyptian gods Isis and Osiris, in his *De Legendae Scribendaequae Quae Historiae Ratione, Dialogi Decem* (1570), Dial. III, p. 57.

57 Camerarius (1:81–82). Mather refers to *Origines Illustris Stirpis Brandeburgicae* (1581), p. 70, a Latin translation of a genealogical history of the Margraves of Brandenburg, by Reinerus Reineccius' (1541–95), professor of history at Frankfurt.

58 Camerarius (1:383). The Italian Renaissance historian Thomas Fazelli (1498–1579) describes the excavation of giant fossilized bones in Sicily, believed to be of human origin, in his *De Rebus Siculis Prioris Decades Duae* (1558), lib. 1, cap. 6, p. 23.

59 Quoted at second-hand from Camerarius (1:382). The Latin distich is from Pomponius Laetus and reads, "From henceforth good posterity preserves the immense bodies, and such is the testimony of our ancestors to the world." The adapted lines are from "Carmen in gigantum ossa quae Puteolis visuntur" (*Opera Omnia* [1510]), as rendered in Julius Pomponius Laetus, aka. Guilio Pomponio Leto (1425–97), the Italian humanist, classical scholar, and editor of ancient Latin manuscripts.

60 Camerarius (1:383, 384). See *Most fruitfull & learned Commentaries of Doctor Perter Martir* (1564), cap. 1, fol. 16^{V}. In *Common Places* (1583 ed.), ch. 13, "Of Giants," p. 130, Peter Martyr Vermigli (1499–1562), Italian Reformed theologian and author of a commentary on the book of Judges (1:10), quotes from *Heroicus* (7.9; 8.3–14), a fictional prose dialogue between a learned vinedresser and a Phoenician sailor at the ruins of the shrine of Protesilaus in the Troad, by Flavius Philostratus, a Greek sophist of Athens (3rd century CE).

61 "A lion from his claw."

62 "A giant from his tooth."

Austin, in his famous Treatise, *De Civitate Dei*, [L. 15. c. 4.][63] complains of those, who would not beleeve, that there had been *Literal Giants*, of an unusual and stupendous Magnitude.[64] He saies, That *most Noble Poet Virgil*, did beleeve it, when he brought in his Champion, snatching and throwing a *Stone*, where of he sais,

Vix illud lecti bis sex cervice subirent,
Qualia nunc hominum producit corpora tellus.[65]

To confute them, he carries them down, into the *Sepulchres* opened of later Dayes, where they should find, *Incredibilis magnitudinis ossa mortuorum.*[66] Yea, He affirms that he himself with several more, saw with their own Eyes, the TOOTH of a Man, which would æqual at least an *Hundred* of the *Teeth* of ordinary Men. *Vidi ipse non solus, sed aliquot mecum, in Uticensi litore Molarem hominis Dentem, tam ingentem, ut si in nostrorum Dentium modulos minutatim, concideretur, centum nobis videretur facere potuisse. Sed illum GIGANTIS alicujus fuisse crediderim.*[67]

To gett nearer home; *Cambden* in his *Britannia*,[68] when he comes as far as *Nesse-point*, in *Essex*, makes a Citation from *Ralph de Coggeshal* (who

[63] Mather erroneously lists [L. 15. c. 4.]. It should be [L. 15. c. 9.] as it appears in his source (Camerarius 1:384).

[64] Camerarius (1:384). The Christian theologian and philosopher St. Augustine of Hippo (354–430 CE) complained in his *De Civitate Dei* (15.9) [*PL* 41.448] of the incredulity of the people and invokes the Roman epic poet Publius Vergilius Maro (71–20 BCE) as proof.

[65] Camerarius (1:384). "Scarce twice six strong men of such frames as the earth now brings forth could lift it on their shoulders" (Virgil, *Aeneid* 12.899–900), in *NPNFi* (2:291). Or, rather: "Twice six picked men, men with such bodies as the earth produces nowadays, would have scarcely managed to hoist that weight onto their shoulders."

[66] "Bones of the dead of incredibly large size."

[67] Camerarius (1:384), Mather's source, directs him to St. Augustine, but Mather took the trouble to consult the text of St. Augustine's *De Civitate Dei* (15.9) from whose work he quotes. It reads, "I myself, along with some others, saw on the shore at Utica a man's molar tooth of such a size, that if it were cut down into teeth such as we have, an hundred, I fancy, could have been made out of it. But that, I believe, belonged to some giant" (*NPNFi* 2:291).

[68] Camerarius (1:383). William Camden (1551–1623), an English historian and author of *Britannia* (1586), relates the story of Ralph de Coggeshal(l) (fl. 1207–26), abbot of Coggeshall (Essex), and chronicler during the reign of King Richard I (1157–99), in *Camden's Britannia, Newly Translated into English* (1695), "Essex," col. 351. This translation was done by Edmund Gibson of Queen's College, Oxford

wrote between Four and Five hundred Years ago,) in these Terms. 'In the Time of K. *Richard*,[69] on the Sea-shore, in a Village called, *Edulfiness*, were found *Two Teeth* of a *Giant*, of such a prodigious Bigness, that *Two Hundred* of such *Teeth*, as Men ordinarily have now, might be cutt out of one of them. These I saw at *Cogshal*, and handled with great Admiration.' *Cambden* adds; Another, I know not what, *Gigantic Relique*, was found near this Place, in the beginning of Q. *Elizabeth*, by the Noble R. Candish.[70]

And here were enough to prevent any ones Quarrelling with *Fulgosus*,[71] for his Report, That the *Teeth* of *Giants*, which *Teeth* weigh'd each of them, *Three Pound*, were found at *Drepanum* in *Sicily*.

But *America* too, will come in, to shelter the Reputation of these Historians. They may *shield* themselves with the *Teeth* lately dug up, at *Albany*.[72]

Had *Johannes Cassion*'s Book, *De Gigantibus*, ever come over into *America*, I do not know, but I might have had a larger Entertainment for you.[73] But what matters it, as long as the *Giants* themselves have come over to *America*! In reading of *Zaraet* and *Acosta*, and *Laet*,[74] I find, That the *Americans* in the *Southern* Regions, have Traditions of *Giants*, who were fought and kill'd

(1669–1748), later bishop of London. He was a learned antiquarian and Anglo-Saxonist; Samuel Pepys (among others) helped Gibson in this translation.

69 King Richard I (1157–99) of England.

70 Richard Candish, aka. Cavendish (c. 1530–1601), an English courtier and author during the reign of Queen Elizabeth I of England (1558–1603).

71 The Italian Renaissance humanist, antiquarian, and doge of Genoa, Baptista Fulgosus, aka. Battista Fregoso (1452–1504), mentions the giant teeth in his *Chronique de Baptista Fulgosus* (1494), lib. 1, c. 6, which were found in the Sicilian Drepanum (modern Trapani), NW Sicily. See also Fulgosus' *de dictis factisque memorabilibus* (Milano, 1509), lib. 1, cap. 6, fol. F6^r ("De Gigantibus").

72 *Boston News Letter*, July 30, 1705. On the Claverack fossils, see D.E. Stanford (1959); C.E. Lutz (1982); P. Semonin (2000), pp. 15–40; and T. Sweet (2021), pp. 32–48.

73 Cameriarius (1:383). Mather misspells the patronymic of the French Protestant antiquarian Jean de C(h)assanion (1531–98), whose *De Gigantibus, eorumque Reliquiis* (1580), cap. 9, pp. 49–51, debates the *Gigantomachia* (1569) of the Flemish linguist Joannes Goropius Becanus (see above), arguing that Becanus' giants were not giants of physical size but merely of political power (tyrants).

74 Mather refers to *The strange and delectable Historie of ... Peru* (1581), bk 1, ch. 5, fols 6–7, by Agustín de Zárate (1514–c. 60), Spanish historian for the Spanish crown in Peru; see also *The Naturall & Morall History of the East and West Indies* (1604), bk 7, ch. 3, pp. 501–02, by the Spanish Jesuit Fray José de Acosta (c. 1539–1600); and the annotations on Hugo Grotius' *De Origine Gentium Americanarum* (1643), p. 83, by Joannes de Laet (1593–1649), Dutch geographer of Antwerp.

by a Man descended from Heaven. A Spanish Commander had the Curiosity to Dig, where they said, the *Battel* was fought; and there he found Bones enough to make a *Skeleton*, the *Teeth* whereof were Four Inches long, & Two Broad. *Acosta* relates, that another much larger was discovered at *Mexico*; for one of his *Grinders* was as big as a Mans Fist; & the other Parts of the Body in Proportion. Other Bodies were found there, Three times bigger than the ordinary Indians.

But at last we dig them up in our *Northern* Regions too.[75]

The First printed Account, the Public had of it, is from *New York*, July. 23. 1705. to this Purpose.[76]

> There is a prodigious TOOTH brought hither, supposed from the Shape of it, to be one of the great *Teeth* of a MAN. It weighs *Four Pounds* and *Three Quarters*. The Top of it, as sound and white, as a *Tooth* can be; but the *Root* is much Decay'd. Yett one of the Fangs of it, holds Half a Pint of Liquor. It was lately Dug up, a great Way under ground, in the Side of a Bank or Hill, Thirty or Forty Foot above it; at or near a Place called *Claverack*; about Thirty Miles on this Side *Albany*, {and is lookt upon here as a mighty wonder, wither the Tooth be of Man or Beast:} They also Dug up several *Bones*, which as they came to the Air, crumbled away. {They say} One of them, which is thought to be a *Thigh Bone*, is Seventeen Foot long. There is since, another *Tooth* taken up, in the same Place; which is a *Fore-Tooth*, Broad and Flatt, and is a Broad as a Mans Four Fingers, {which I have not yet seen; but the other I did see, and was brought here by a Gentleman of the Assembly:[77]} They dug up several Trees, in the same Place, of great Bigness.

Afterwards I was able to use the Language of *Austin*, *Vidi ipse non solus, sed aliquot mecum*.[78] Very many Gentlemen, and I myself also, have had the Satisfaction, of Seeing and Handling, and Weighing one of these TEETH, dug up, with other Proportionable Bones, at the Place aforesaid. All that I have

[75] See also *BA* (1:597–98) and the letter of Joseph Dudley (1647–1720), governor of Massachusetts (1702–15) to Cotton Mather (July 10, 1706), in *CMHS* (second series), vol. 2 (1814): 263–64; and *BA* (1:592–94). See also D.E. Stanford (1959) and D. Levin (1988).

[76] *Boston News Letter*, July 30, 1705, which includes a report of the Claverack discovery from New York, July 23, 1705.

[77] Peter van Bruggen, of Albany, who was an elected member of the New York General Assembly from 1705–08. See D.E. Stanford, p. 61, *n.* It is telling that Mather edited out the passages in braces { }, ostensibly to strengthen the reliability of the report.

[78] St. Augustine, *Civitate Dei* (15.9): "I myself saw [it], along with others" (*NPNFi* 2:291). See also *BA* (1:590).

ever yett conversed withal, are compelled to Beleeve, That it must be an HUMANE TOOTH. It ha's the perfect Form of the *Ey-Tooth* of a MAN.[79] It has Four Prongs or Roots, and Six Distinct Faces or Flatts, on the Top; a little worn and all perfectly smoothed with Grinding. We measured it; and as it stood upright, it was *Six Inches* High, lacking an Eighth; and Round, *Thirteen Inches*, lacking an Eighth." And (tho' Part of it be broken off, & lost,) its Weight in the Scales, is, *Two Pounds* and *Four Ounces* Troy-weight. Another of the same Figure exactly, was lately produced at *Hartford* of near a Pound Weight more than this.

These *Teeth* and *Bones*, were taken up, under the Bank of *Hudsons* River about Fifty Leagues from the Sea, many Foot below the Surface of the Earth, in a Place where the Freshet every Year does Rake and Wast the Bank. And in that Place, there is a plain *Discoloration of the Ground*, for *Seventy Five Foot long* at least, it is different from the Earth in Colour and Substance; it is Judg'd by every body that sees it, to be the Ruines and Dust of the *Body*, that bore those *Teeth* and *Bones*. The *Teeth* can certainly agree to none but an *Humane Body*; and probably t'was one *Interr'd* by the *Flood*, the New Sediment whereupon gave him the Depth where we now find him covered. The *Distance* of the *Sea*, takes away all Pretension of its being a *Whale*, or any other Animal of the *Sea*, as well as the *Figure* of the *Teeth*. Nor will the *Figure*

[79] Lest we deem Mather's toothsome story too fanciful, we need only to cast a glance at several such accounts in the *Philosophical Transactions*. For instance, "Part of 2 Letters from Mr. Thomas Molyneux concerning a Prodigious Os Frontis in the Medicine School at Leyden," *PT* 15.168 (February 23, 1685): 880–81; "An Essay concerning Giants. Occasioned by some further Remarks on the large Humane Os Frontis," by Dr. Thomas Molyneux, M.D., Fellow of the King and Queen's College of Physicians in Ireland and of the Royal Society in England, *PT* 22.281 (February 28, 1700): 487–508; "A Letter of Mr. Francis Nevil[l]e to the Right Reverend St. George Lord Bishop of Clogher, R.S.S. Giving an Account of some large Teeth lately dug up in the North of Ireland," *PT* 29.346 (December 31, 1715): 367–70; "Remarks upon the aforesaid Letter and Teeth, by Thomas Molyneux, M.D. and R.S.S.," *PT* 29.346 (December 31, 1715): 370–84. Hans Sloane, "An Account of Elephants Teeth and Bones found under Ground," *PT* 35.403 (September 1, 1728): 457–71; and Sloane's "Of Fossile Teeth and Bones of Elephants. Part the Second." *PT* 35.404 (October 31, 1728): 497–514. In Mather's day, scientists in the emerging field of paleontolgy disagreed about the origin of these fossils, whether human or animal (See R. Plot, *Natural History of Oxford-shire* [1677], esp. pp. 136–39). While they generally concurred they were of mammalian origin, some believed they were the fossilized molars of elephants; others the remains of extinct species, yet others (in an attempt to square the evidence with the Bible) of giant humans like the Nephilim – the offspring of fallen angels and the daughters of man (Gen. 6:1–4; Numb. 13:33; *ABD* 4:1072–73). The debate continued until the early nineteenth century, see C.E. Lutz (1982); R. Rappaport (1997), ch. 4; T. Sweet (2021), ch. 1; L. Barnett (2019).

of the *Teeth*, or the admeasurement of the Body, in the Ground, allow it to be the Remains of an *Elephant*.

Upon the Discovery of this horrible *Giant*, the *Indians* within an Hundred Miles of the Place, agreed in pretending a Tradition, which they said, they had among them, from Father to Son, for some hundreds of Years concerning him; and that his Name was, *Maughkompos*.[80] But there is very little in any Tradition of the Salvages to be relied upon. And such a Name also may seem to be disagreeably enough imposed, by the Indians thereabouts, who have not one *Labial* in all their Language. Might I be allow'd so much of Pleasancy, while I am writing an *Illustration* upon the Sacred Scriptures, I would say, that while the Indians in our Next Neighbourhood, have no shorter a Word for, *Good-Looking-glass-makers*, than,

Nuppahknuhtopepenawutchutchuquokenn ehcheninn ummunnonag,

[*Printer*, If thou hast done thy work right, {thou}[81] wilt find *Fifty four Letters* in the Word!] those *Alban{y}Indians*, ought in Proportion, to have putt a N{ame}upon this Champion, that would at least have made a Line in *Zoroasters* Book of, *The Similitud{e.}*[82]

Upon[83] this Occasion, it is not easy to forbear some ESSAY, to investigate the *Original* of such an *Exorbitant Appearance* among the Sons of Men.

And the *First Thoughts* which now visit the Mind of One, whom I must confess to be, shallow enough in his *Philosophy*, are such as these.

We have heard from others, and our *Microscopes* also, which *our Fathers* knew not, *have told us, what wonderful things the Great* GOD ha's done, in creating the *Seminal Part* of the World.[84] Among the *Vegetables*, the *least Part* of that which we call, *The Seed*, is really so. The *True Seed* lies in so

80 Perhaps a pun on "Maw-compos" ("endowed with an able jaw or stomach"). Or, "Maushump" or "Maushop," a giant in Wampanoag lore. Cf. A. Morris (2013), esp. 706–07; L. Barnett (2018), pp. 225–32.

81 MS damage affecting the end of this and the following lines; conjectural readings are provided in braces { }.

82 Mather alludes to *Avesta*, one of the mystical books of Zoroaster, aka. Zarathustra (c. 628–c. 551 BCE), the Persian prophet and founder of Zoroastrianism and its adoration of Ahura Mazda (Ormuzd), the creator of the world.

83 The text of *BA* (1:595) resumes here.

84 Although Hans and Zacharias Janssen seem to be the inventors of a rudimentary microscope (c. 1595), their Dutch colleague and physician Anton van Leeuwenhoeck (1632–1723) improved their invention and thereby discovered bacteria (animalcules) under the microscope. However, the English physicist Robert Hooke

little Room, that it is not visible to the Naked Eye. The *Rest* serves but as a *Lodging* for it, & for its most proper *Nourishment* and *Expansion*, at the First Opening of it. But in that *little Room*, there lies the *whole Plant*, in all the *True Parts* of it; which is afterwards evolved and extended, & filled up, with its *Adventitious Nutriment*, until it be carried as far as the Original *Stamina* are capable: And *then* the *Growth* stops.[85] But perhaps towards the *Period* of the *Growth*, so weak may be the *Stamina*, that according to the *Strength* of the *Soyl* that nourishes it, & some other Circumstances, the *Growth* may be more or less, but a *little* varied.[86] And why may not the *Seeds* of *Animals* have the same said of them, that we see and say concerning the *Seeds* of *Vegetables*?[87] Yea, The *Microscopical Inquisitions* have made it more than probable; That the *True Seeds* of *Animals* floating in their suitable Vehicle, have, lying in a *Space* much less than the Naked Eye can discern, & less than the Thousandth Part of the least visible Grain of Sand, the whole *Bodies* of the *Animals*, even to all their *Nerves* and *Fibres*; which afterward *Grow* as aforesaid, until their original *Stamina*, cannot be much further carried out.[88] And this, by the way, takes away the common Cavil against, *The Resurrection of the Dead*; That the *Bodies* of Men, have been devoured by *Others*, and become the *Parts* of them. For according to this *Hypothesis*, there can be no such thing, as a *Confusion* of *Humane Bodies*. They can, by being Devoured, lose no more than their *Adventitious Nutriment*. And it is

---

(1635–1703), father of microscopy, was the first to publish his observations in *Micrographia* (1665). See E.G. Ruestow (1983).

[85] Here, Mather joins the age-old debate about pre-formationism vs epigenesis. In this context, he ponders the question of how the male sperm of animals, when interacting with the female ovum, contains a miniscule, pre-formed version of the fully grown animal, whose stamina (seminal fluid) makes the pre-formed fetus unfold and expand until birth (*BA* 1:427, 595–99); see also *CP* (1721), Essay XXVI, pp. 127–28; George Cheyne, *Philosophical Principles of Natural Religion* (1705), §XXXIX, p. 231; Marcelli Malpighi's *Anatome Plantarum* (1675), and William Harvey's *Exercitationes De Generatione Animalium* (1651), esp. cap. 50–51, 62–63. On the historical debate, see especially J. Needham (1959), E.B. Gasking (1966), and C. Pinto-Correia (1997).

[86] A summary of Nehemiah Grew, William Derham, John Ray, Marcello Malpighi, Anton van Leeuwenhoek, and many others who wrote on the subject. See also Mather's *CP*, Essay XXVI, pp. 131–37.

[87] On this early theory of generation, see "A Discourse concerning the Modern Theory of Generation, by Dr. George Garden of Aberdeen," *PT* 16 (1686–92): 474–83.

[88] See also "Curiosa" 3.1, Mather' s Letter on "Monstrous Impregnations," to John Woodward (July 2, 1716).

Remarkable, That *the Resurrection of the Dead*, is called, *The Filling of the Dead Bodies*. [Psal. 110.6. and, Eph. 2.1.][89]

Quære. What Provision in the Case of *Amputations*? Whether a *Seminal Principle* seated in a Part of this *original Animalcule*, which is yett unknown to us, may not be furnished by our Glorious and Almighty Creator, with a *Plastic Vertue*, to *Renew* what may happened to be lost of it? *Ulterius Inquirendum, aut forsan Ignorandum*![90]

And what should now hinder, but that GOD, having at first created Numbers of *Humane Bodies* in these *True Seeds* of them, there may be some among them, whose original *Stamina*, may be much larger than others, and capable of being Drawn forth, to the most *Gigantic Dimensions*? GOD, for the Punishment of a wicked World, might order some of these *Gigantæan Stamina*, to enter the *Bodies* of the Children of Men, perhaps in their *Food*: And so ly ready to be brought forth, as the *Propagation of Mankind* proceeded. Or what should hinder, their being all at first, in the Body of the *First Man*, according to a Conjecture formerly proposed? [See our Illustration, on Gen. 3.15.][91] The *Giants* thus brought forth, by *Parents* not exceeding the common Stature, were such a Plague unto the World, that if a *Flood* had not exterminated *them*, they would in a while, have exterminated all the rest, without a *Flood*; unto *Them alone the Earth had been given, & no Stranger* or common Man, *had passed among them*. It was most agreeable, That this Plague should come upon the World, in the Way of *Generation*; (as *another*[92] ha's done since!) and that when the Carnalities of the World, were grown to a *Gigantic* Enormity, they should be chastised with a *Gigantic Posterity*. And that we now suffer no more of this Plague, as well as that it was once inflicted, is owing to that *Vigilant* and *Immediate* PROVIDENCE of GOD, which is employ'd about the *Propagation of Mankind*: and whereof, the *Proportion* of SEXES, the *Difference* of FACES, (to which I may, for a Third Instance add, what ha's rarely been mentioned, The Variety of HANDS, written by the many Hundreds of

[89] For much the same, see *BA* 1:595–96 and Robert Boyle's *Some Physico-Theological Considerations about the Possibility of the Resurrection* (1675), pp. 13–14.

[90] "To be investigated further, or perhaps it must remain unknown." See also *BA* (1:596, *n.*164), and CA 3.1 "Monstrous Impregnations" (July 2, 1716).

[91] *BA* 1:485.

[92] Smallpox. In his medical handbook *ABJ* (1972), Cap. XX, Mather argues that smallpox was unknown to the ancients and that "it was brought into *Europe* thro' *Africa;* on the Wings of those *Arabian Locusts,* which in the *Saracen* Conquests did *spread over the Face of the Earth*" (93). He provides a similar argument on venereal disease, in his "Kibroth Hattaavah, or Some Clean Thoughts on the Foul Disease," in *ABJ* (1972), Cap. XXI, pp. 116–20.

whom learn to write of the *Same Tutors*;) are Instances that call for our Contemplation and Astonishment.[93]

But I will proceed no further in the *Speculation*. That I do proceed so far may need, and I ask, a *Pardon*. I am sensible, That this *Hypothesis* may be liable to great Objections: yett it being the best I had at hand, I was willing to make an Offer of it, with the *Modesty* and *Suspicion*, (I had almost said, the *Pyrrhonism*) wherewith all such things ought to be offered. If it be thought, I am taken with the Distemper, which they call an *Hypothesimania*,[94] I have this to plead, You shall not find me *often* to fall into it. —

The Truth is, Dr. *Drake*, in his *Anthropologia Nova*, I doubt, ha's nearly destroy'd the *Hypothesis*, which I have hitherto been labouring.[95] And I doubt, I must be driven to the same Conclusion with him; 'However old and exploded, the Opinion of a *Plastick Power* be, I must however embrace it, even tho' I know not exactly wherein it lies; At least, till I meet with something more sufficient to Resolve my doubts, than hitherto I have done.'[96]

93 The preceding paragraph is extracted from *BA* 1:597.

94 Yet another Matherian neologism that has escaped the *OED*. Mather links the Greco-Latin words "hypothesis" (a stated proposition) with "mania" (obsession).

95 James Drake, M.D., F.R.S. (1667–1707), English physician, anatomist, and author of *Anthropologia Nova* (1707), vol. 1, bk 1, ch. 25, pp. 334–35. Drake ultimately rejected preformationism because it contradicted the existence of hybrid animals, such as mules, which incorporate the characteristics of two distinct species. Mather here engages the precursors of modern embryology and genetics, who in Mather's time consisted of four basic groups of theorists: those who embraced preformation, spermism, ovism, epigenesis, or a combination of the four. Preformationists argued that plants and animals, respectively, develop from microscopically small but fully articulated replicas in seeds and, respectively, homunculi in the male sperm (spermatozoon), which merely unfold and grow during the gestation process. Ovists insisted that preformed homunculi are present in the egg from which they develop, the spermatozoa and seminal fluid merely supplying nourishment; epigenesists believed that organisms are formed from an undifferentiated mass in the seed or egg and develop after fertilization. The early debate is neatly summarized in George Garden's "Discourse concerning the Modern Theory of Generation," in *PT* 17.192 (1691): 474–83. See also Mather's CA 1.5 "Monstrous Births." Questions of this nature were hotly debated in Mather's time, and several of his Curiosa letters addressed the underlying issues. See E. Gasking (1966), C. Zircle (1969), E.G Ruestow (1983), K. Park and L.J. Daston (1981, 1998), C. Pinto-Correia (1997), M. Hagner (1999), M.-H. Huet (2004), and F. Kraemer (2023). See also CA 3.1 "Monstrous Impregnations" and *BA* 1:104–05.

96 *BA* 1:598–99. The idea of the "*Plastick Power*" of nature is derived from Plato's concept of *anima mundi* (*Timaeus*), the theory that physical nature is endowed with a spiritual or plastic (formative) power (Aristotle's entelechy) that orders and shapes the size and growth of all biological life forms. This vitalist theory made

Syr, Thus talks, and thus ends, the *Illustration*, which I was to fetch out of the Manuscript BIBLIA AMERICANA for you; and you are welcome to it; I wish I could make you so to all the rest. I doubt this *Illustration* hath so swelled my Letter, which according to the old Rule, *Non debet implere manum legentis*,[97] that I begin to be Tedious. Tis time, & more than time, for me to subscribe

Boston, N. England.
Nov. 17. 1712.

Honoured Sir,
Your Sincere Friend
& Humble Servant,
Cotton Mather.

a brief comeback but was largely discredited by the end of the 17th century. In Mather's time, the English theologian and Cambridge Platonist Ralph Cudworth (1617–88), in his huge *The True Intellectual System of the Universe* (1678), and the English botanist and natural philosopher John Ray (1628–1705), in his *The Wisdom of God Manifested in the Works of the Creation* (1691), esp. pp. 31–40, were two of the principal proponents of "plastic nature." See also Mather's *CP* (1721), Essay XXVI, pp. 131–37. For a useful review of the vitalist debate, see W.B. Hunter (1950) and L.S. King (1978).

97 Adapted from Seneca (the Younger), *Epistles* (45.13), "To be investigated further, or perhaps it should be left unknown."

I.2. ON AMERICAN PLANTS & INDIAN CURES.

November 18, 1712.

Letter. 2d.

To Dr. John Woodward.

Sir,

The Things next unto Subterraneous,[1] are those which I am going to treat you withal; For tis a *Pleonasm*, which often makes a Figure in true *Friendship*; *To do more than ha's been asked for*. You have an ample Field for your Polite and Noble Studies in your *Botanicks*; and one well worthy of even a *Solomons* Cultivation. Since I have mentioned *Solomon*, you will give me Leave, perhaps with a Smile thereupon, to bring in the Mention of a Conjecture, which a learned Friend of mine ha's, concerning those Writings of his, which no *Pancirol*, or *Salmuth*, could ever yett inform us of.[2] Whereas we read concerning him, *He spake of Trees, from the Cedar that is in Lebanon, even unto the Hyssop that springeth out of the Wall; He spake also of Beasts, & of Fowles, & of Creeping things, & of Fishes:*[3] while some think, he only *spoke* of them, *Eruditè et piè de his rebus philosophatus est*,[4] (and perhaps he sett up a ROYAL SOCIETY in his Kingdome too,) and what he *wrote* is only

1 Mather refers to his previous (first) Letter to John Woodward (November 17, 1712), our CA 1.1.

2 Guido Panciroli (1523–99), Italian antiquarian, historian, jurist, law professor, and author most famously of *Rerum Memorabilium* (1590, 1660), a popular work which draws attention to the loss of ancient knowledge in the modern world. English transl., *The History of Many Memorable Things Lost* (1638). One of Panciroli's students was the German Protestant theologian Heinrich Salmuth (1522–76), who translated his teacher's work from Italian into Latin. There are numerous Latin editions, including 1599, 1602, 1660. Mather's refers to *Rerum Memorabilium* (1660), pars prior, lib. 2, tit. I, p. 39.

3 Kings 4:53, speaking of Solomon. See Samuel Bochart, *Hierozoicon* (1663), pars prior, lib. 2, cap. 50, col. 589, lines 66–71; and *BA* (3:414–15).

4 "He philosophized learnedly and piously about these things."

in the Books which now make a Part of our *Canon*; this Gentleman conjectures, that the Books of *Theophrastus* and of *Aristotle*, on those two Subjects,[5] may indeed be *Solomons*; for, tis well-known, that at least one of those two Philosophers, was a notorious *Plagiary*, and the Volumns out of which he had been supplied with all his Treasures. Lett us rather suppose all *Solomons* Herbal to be lost, and be glad that we have *Parkinsons* instead of it; rejoice that such notable things have been done by the *Gerhards* and *Johnstons* and other Industrious and Ingenious Men, in our later Ages, to retrieve the Loss.[6] I find, that the Number of *Plants*, contained in the *Pinax* of *Bauhinus,*[7] is about Six thousand; which are all that had been described by the Authors who had written before him, or that had fallen under his own Observation. Tis true, some accurate *Botanists*, tho' they admire his Performance, complain yett of some Inaccuracies, and Repetitions in it; and of his making sometimes a Different *Species* of Plants, for some accidental Varieties. But still our Incomparable *Ray,*[8] as competent a Judge of that Matter, as perhaps any Man then in the World, supposes that there may be in the World, more than Triple

5 Panciroli, *Rerum Memorabilium* (1660), pars prior, lib. 1, tit. XXII, p. 64. Theophrastus (b. 371 BCE), Greek Peripatetic philosopher and biologist, author of *Historia Plantarum* and *De Causis Plantarum*, and a student of Aristotle (384–322 BCE), Greek philosopher and purported author of *De Plantis*. Mather's charge of plagiarism is not all that preposterous given the Renaissance concept of *prisca theologia*, the belief still regnant in Mather's time, that the pristine wisdom of God was handed down via Noah to his three sons, who carried it into all corners of the world, but corrupted it in the process by adding their own admixture of paganism. In time, the Babylonians, Egyptians, Greeks, and Romans became envious of the Israelites and their Word of God (as reestablished by Moses), copied its wisdom, and claimed it as their own – so the argument goes. See H.C. Maddux (2010).

6 John Parkinson (1567–1650), English herbalist and botanist, remembered for his *Theatrum Botanicum* (1640); Johann Gerhard (1582–1637), Lutheran theologian in Chemnitz and author of *Locorum Theologicorum* (1639); Johannes Jonstonus (Johnston) (1603–75), Polish scholar, physician, and author of his oft-reprinted *Historiae Naturalis. De Insectis libri III, De Serpentibus et Draconibus libri II: Cum aeneis figuris* (1653). Mather puts Jonston's *Historiae Naturalis* to good use in his fourth letter (November 27, 1712) to Richard Waller. See our CA 1.11 (below).

7 Caspari Bauhini, aka. Gaspard Bauhin (1560–1624), a famous Swiss physician and botanist, renowned for his *Pinax Theatri Botanici Caspari Bauhini* (1623), a compendium for the classification and cataloging of more than six thousand botanical species and genera.

8 John Ray (1627–1705), English physico-theologian and author of *Historia Plantarum Species*, 4 vols. (London, 1686–1704), a botanical handbook of the first order, which went through numerous expansions and reprints. Ray's *Historia* added considerably to those listed in Bauhin's *Pinax Theatri Botanici* (1623).

that Number. He makes a more particular Supposal, That there are in the vast Continent of *America*, as great a Variety of *Species* as with you; which is what my Letter will anon proceed upon.[9] And you know, Syr, that the *Almagestum Botanicum*,[10] which was published a dozen Years ago, ha's a *Mantissa* of above a Thousand New Exotic Plants, added unto the former Exhibitions.

For my own part, they are but few Tours, and very short ones, that I have had the Liesure to take in the *Vegetable Kingdome*.[11] I honour the Study, and tho' the Delight of *Tulipists*, ha's not escaped the severe Denomination of a *Tulipomania*;[12] yett I have been far from charging with a foolish Mispence of Time, those Gentlemen who have singled out some *One Plant*, and written a whole Treatise upon it; whereof, I thought, I could have given you a pretty good Catalogue, before the *Naturæ-Curiosi* in *Germany* began the Stroke that way, which ha's now raised the Number to a little Army.[13] But I find it impossible, for one Man intensely to *study* all things; & be at the same time employ'd in a Life of much *Action* too. Indeed, the little I have done this way, has enabled me to take some Cognisance of the *Plants*, which we find growing in the Garden of the *Sacred Scriptures*. The Thoughts of Sir *Thomas Brown* on many of those *Plants* are critical and curious;[14] but such as it would be but a Troublesome Impertinency for me now to offer them unto your Consideration.

9 See his letters to the English botanist James Petiver (c. 1663–1718), our CA 3.11 "The *Fagiana*" (July 13, 1716) and CA 3.12 "*Curiosa Botanica*" (September 24, 1716).

10 Leonard Plukenet (1641–1706), English botanist, and author of *Almagestum Botanicum sive Pythographiae* (1696) and its supplement *Almagesti Botanici Mantissa* (1700) added upon the former exhibitions. These two pre-Linnean thesauri provide the Latin names of thousands of plants and their American, African, Asian, and European origins.

11 Mather compensated for his lack of expertise by including a whole chapter on medicinal plants in his medical handbook *ABJ* (1972), Chap. LVII. "A Physick-Garden," pp. 262–69.

12 The speculative trade in tulip bulbs (tulipomania) in Holland (1634–37) that almost brought down the Dutch government – similar to the infamous South Sea Bubble (1711–20), the Wall Street market crash of 1928, or the subprime mortgage collapse (2007–08) in the United States.

13 The academy of *Naturae Curiosi*, or *Academia Naturae Curiosorum* (Leopoldine Academy), founded in 1652, whose annual publications began in 1670 under the title of *Miscellanea Curiosa Medico-Physica Academiae Naturae Curiosorum, sive Ephemeridum Medico-Physicarum Germanicarum*, a scientific journal which Mather often cites in his *BA, AB, and CP* and refers to as the "*German Ephemerides*."

14 See for example *The Garden of Cyrus* (1659), by the English polymath, physician, and natural philosopher Sir Thomas Browne (1605–82), whom Mather cites numerous times throughout *BA*.

I have taken care that the BIBLIA AMERICANA I told you of, be enriched with them. However, I cannot but bestow a Touch or two, upon some of the *Plants* in our Bible, which I find some others of your best Philosophers take notice of. *Hottinger* satisfies me, that the *Shittim Wood*,[15] employ'd so much in the Materials of the *Israelitish* Tabernacle, was the *Acacia*; and indeed such a solid Wood as the black *Acacia*, was very proper for the business; which perhaps was in the Eye of them, who render it, *Incorruptible Wood*, in the more ancient Versions. The *Trees of Aloes*, & the *Wood* thereof, sometimes occurring to us, may be best answered, as I perceive some are of Opinion, by the *Wood-Aloes* of *Syria*,[16] which is called, *Aspalatha*;[17] a little Shrub, covered with Prickles, of the Wood whereof, Perfumers make Use, to give a Consistency unto their otherwise too liquid Perfumes; and which *Cassiodorus* tells us, was of so sweet a Smell, they burnt it before the Altars instead of *Frankincense*.[18] The *Husks*, desired by the Prodigal, is doubtless the *Carob-bean*; as *Bochart* ha's demonstrated:[19] but whether this were the Diet of *John Baptist*, (which we render *Locusts*,) is doubted by some; tho' indeed not by the *Germans*, who call it *Jaans Broot*, or *Johns Bread*. We have done very oddly, when we have used the Term of *Camphire*, for the *Cyprus tree*;[20] the

15 Johann Heinrich Hottinger (1620–67), Swiss Reformed theologian, rector of the University of Zürich, professor of Oriental languages at Heidelberg, and voluminous author, most notably his *Smegma Orientale* (1658). See Mather's discussion of "Shittim" or "Gopher Wood," in *BA* 1:602, 607, 608*n*; *BA* 2:348, 357, 965–67; and "An Extract of several Letters from Cotton Mather, D.D." *PT* 29.339 (January 1, 1714), p. 63 (Appendix 6). The specifications for the Israelites' wilderness tabernacle are given in Exodus 25–30.

16 Mather provides the same documentation in his commentary on Numb. 24:6, in *BA* (2:958).

17 A genus of flowering shrub-like plant in the *Fabaceae* family.

18 Magnus Aurelius Cassiodorus (c. 485–c. 585 CE), Roman statesman and historian, author of numerous works, including his *Chronica*, a history of the Roman Empire up to 519 CE. Cassiodorus comments on the sweet-scented, unctuous quality of myrrh and aloe, in his *Expositio in Cantica Canticorum* (4:14) [*PL* 070. 1079*CD*].

19 From the parable of the Prodigal Son (Luke 15:16): "And he would fain have filled his belly with the husks that the swine did eat; no man gave unto him." Samuel Bochart (1599–1667), distinguished French Protestant biblical scholar and author of *Hierozoicon Sive Bipartitum Opus De Animalibus Sacrae Scripturae* (1663, 1692, 1712). Bochart describes the fruit and the husks (pods) of the Carob tree, aka. Ceratonia siliqua ("fructus arboris ceratoniae siliquis digitalibus inclusi"; i.e, "the fruits of the carob tree, enclosed within finger-like pods"), along with its etymology, in *Hierozoicon* (pars prior, lib. 2, cap. 57, col. 708, lines 4–32).

20 Cant. 1:14, 4:13; See also Mather's comments on Gen. 6:14, 6:22; Exod. 30:24 (*BA* 1:602, 607–08; 2:414 *n*.).

Vulgar Latin itself ha's herein done better than we. The *Apples of Gold*, in the *Proverbs*, are *Oranges* unquætionably; and the *Apple-trees* in the *Canticles*, are *Orangetrees*.[21] We confound *Honey* with *Dates*, because they have the same Name in the Hebrew. *Ludolphus* has well taught us what to think of the Plant, which in our Version goes by the Name of *Mandrakes*.[22] I know not how to withold my Assent from that accomplished Botanist, who judges that the *Gopher-Wood*, in *Genesis* is the *Juniperus arbor tetragonophyllos*; frequent in the *East-Indies*, and (as *Vossius* thinks) the Same with the *Tetragonia Theophrasti*;[23] under the Shadow whereof, the Prophet *Elijah* composed himself to sleep in the Wilderness,[24] whereto the *Juniperi gravis umbra*, in *Virgil*, may be attributed.[25] But, Syr, you are better acquainted with these things, than I can pretend to be; nor is it my Intention to divert you in any thing like *Ursin* his *Arboretum Biblicum*, on invite you to converse with his *Herbarius Sacer*;[26] or attempt a Collection of Illustrations on the Sacred *Phytology*.[27]

My Ambition is to procure an Addition to your *Botanic* Treasures from our *American* Colonies. But neither the Time of the Year, nor the Shortness of the Time for the Dispatch of my Letters, will allow my Ability to hold Pace as yett with my Ambition. I can only inform you, that as diverse *Plants* which *we* have in common with *you*, have with *us* those Vertues, which have not been

21 Prov. 25:11; Cant. 2:3. See also Mather's commentary on Cant. 2:5 (*BA* 5:477) for the same argument.

22 Hiob Ludolf, aka. Ludolph, Leutholf (1624–1704), German Lutheran Orientalist and grammarian of the Ethiopian language (Amharic). Mather frequently consulted him on questions of uncertain meanings of Hebrew words. On the mandrakes (Dudaim), he specifically refers to Ludolph's *Historia Æthiopica* (1681), translated as *A New History of Ethiopia* (1682), bk 1, ch. 9, p. 50. See also *BA* (1:981, 1038–39), where Mather agrees with Ludolph that the mandrakes about which Jacob's wives were arguing were "some rare and pleasant Fruit" (like the "Indian Figg"), rather than "a stinking *Mandrake*, which could be a fruit little worth contending for" (*New History*, p. 50). The secret is out!

23 "Tetragonia" (odorous cedar, yew), in Theophrastus, *Enquiry into Plants* (3.4.2–3); Gerrit Janszoon Vos (1577–1649), or Gerardus Vossius, a Dutch classical scholar and theologian. On "Gopher-wood" (cypress), see Gen. 6.14 (*BA* 1:602 *n.*, 607–08).

24 1 Kings 19:4–5.

25 "The shade of the Juniper tree is harmful," so relates the Roman poet Publius Vergilius Maro (70–19 BCE), in his *Eclogues* (10.76).

26 Johann Heinrich Ursin (1608–67), German scholar, Lutheran theologian, and humanist, author of *Arboreti Biblici Continuatio: Sivè Historiae Plantarum Biblicae, Libri Tres. Editio Nova* (Norimbergae, 1699), lib. 2, cap. 1, pp. 74ff. See also *BA* 2:957 *n.*

27 The study of plant biology (botany), in Ursinus, *Arboreti Biblici Continuatio* (lib. 1).

observed in *yours*; (particularly, our *Asarum* is one of our most potent *Alexipharmacks*,[28] & a Sovereign Remedy for *Fluxes*,[29] even for *Bloody* ones;) thus these Regions are stock'd with *Plants* that are peculiar to them; and with these our *Indians* do Cures many times that are truly stupendous. I will use my best Endeavours, that you may e're long have a *Catalogue*, & full *Description* of them.[30] In the mean Time, I shall only give you a more general Account of some Things, which I have in more immediate View, for my own more exact Enquiry, and e're long for your further Entertainment.

We have a *Plant*, thought peculiar to these Parts of the World; which from its marvellous Efficacy in the Cure of *Erisypelatic* Inflammations, goes among us by the Name of *Anterisypelas*.[31] It growes plentifully from Weeds, and we extract a chymical Oyl from it, which being taken inwardly, does wonders at the absorbing of *Scorbutic Salts*, and the releeving of the Salt Humours, in our Bodies, which corrode and threaten the Parts they fall upon.

We have another Plant which growes plentifully in our Woods, distinguished among our People, by the Name of, *Partridge-berry*.[32] The Leaves afford a Pretty, Pleasant, Aromatic *Tea*; This, or, a Decoction being taken for some Dayes, during which no Effect will appear, anon will carry off a *Dropsy*,[33] with an Incredible Discharge of Urine, for diverse dayes together; which being over, & the Dropsy cured, the *Tea* may be continued without any Danger of any further Discharge than that of the ordinary Urine. It becomes a fashionable *Tea* among our gouty Gentlemen; and they imagine themselves to find much Advantage by it.

28 A genus of plants in the birthwort family, commonly known as wild ginger. An alexipharmic is an antidote to poison or infection.

29 "*Fluxes*" (dysentery). Mather has much to say on this condition in "*Jehoram Visited. Or, The* Bloody-Flux remedied," in his medical handbook *ABJ*, Cap. XLVI, pp. 211–14.

30 Always on the lookout for new knowledge in medicine, Mather reached out to missionaries among Native Americans in 1712, to "Set afoot, a Design of Learning from the Indians, what are their Peculiar Methods in Curing Diseases, and what Peculiar Medicines they have among them: the Knowledge whereof may befriend and oblige mankind abroad in the world" (*D* 3:110).

31 Erisypelatic inflammation is a bacterial infection of the outer layers of the skin.

32 Partridge-berry, a native of the eastern North America, is a perennial, a small, woody, trailing vine with six- to twelve-inch, slender, trailing stems that do not climb but lie prostrate on the forest floor, and that has dark green leaves, white flowers, and scarlet berries.

33 "*Dropsy*" (edema, swelling through building up of fluids in the body). See Mather's "*Moses*, or One *Drawn Out* of the Dropsy," *ABJ*, Cap. XXIII, pp. 125–29.

We have a remarkable Plant, called, *The Bleeding Root*;[34] which is an Infallible Remedy for the *Jaundice*, and requires at most but five or six Dayes to finish the Cure.

We have another *American* Plant, which is a Specific, and a wonderful one, for the Cure of *Gangrœnes*; and ha's prevented Amputations, when our most experienced Cheirurgeons had pronounced them necessary.

We have another *American* Plant, which is a certain & speedy Cure, and does Wonders, for the bite of a *Rattle Snake*; and is admirable against all Internal as well as External Poisons.[35]

We have another *American* Plant, which in a *Quinsy*,[36] or a *Sore throat*, will cure to admiration; & when Persons have been utterly speechless by the Prevailing of the Distemper, ha's brought sudden and surprising Releef unto them. For which Cause, we call it, *Throat-weed*.[37]

We have a Sort of *Cranes-bill*,[38] which differs little from yours; except perhaps in the Colour of the *Flowre*, which with us is of a Sky-blue, and the Largeness of the *Root*, which is near that of *Ginger*. Our Indians call i{t by t}he Name[39] of *Taututtipaug*;[40] and from them we learn the Manner, and strange Effect, of using it. It is an infallible Cure, and *Safely*, and *Quickly*, and *Easily* performs it, for that filthy Disease (the *Lues Venerea*,)[41] which the just Judgment of God had reserved for our later Ages, wherein so many Fools abandon themselves to the destructive Debaucheries of Unchastity. The famous *Bolognian* Physician, you remember, hoped, that the Picture of a Gallant, with his Nose Ears off, would prove that for which his Advertisement vended it, *A*

34 This unspecified plant may well be the same that Mather called *Ophiophuga*, or snakeweed (CA 3.12 "*Curiosa Botanica*"). See G.L. Kittredge, "Letters," p. 184.

35 For Mather's knowledge of how Native Americans cure snakebites, see his Letter to Richard Waller, November 27, 1712, CA 1.11 "On the Rattlesnake," and his Letter to James Jurin, June 4, 1723, CA 7.2 "A New Rattlesnake"; see also Kittredge, "Letters," pp. 183–84.

36 Acute tonsilitis.

37 See Mather's recommendations, in "*Suffocatus*. or, A *Sore Throat;* and, Quinzey"; "*Ajutoria Catarrhi*. A Catarrh: And what we Call, A *Cold*: How to Stop it!" and "The *Breast-beater*. or, A Cough quieted," in *ABJ*, Caps. XXXV, XXXVI, XXXII, pp. 167–69; 170–72; 172–76.

38 Geranium.

39 The manuscript is damaged at the top of the page.

40 See also the description of the medicinal use of the Indians' *Taututtipaug*, which he calls *Geranium Americanum*, in CA 3.12 "*Curiosa Botanica*."

41 Venereal disease.

Sovereign Antidote against the foul Disease.[42] But since this Præservative has proved so ineffectual, and the unbridled Follies of Men are so Incurable, we must think of a Cure for the Plague which their Follies bring upon them. And I have heard of none comparable to our *Taututtipaug*. There must be a Decoction or Infusion made of the *Root*. About half a Pint of the Liquor must be drunk every day; either warm or cold, as the Patient pleases. To some, it proves at first, a little *Emetic*;[43] especially if it be drunk warm. It will putt the whole Mass of Blood into a little Ferment, that will be felt even to the fingers Ends. At the same time, the *bruised Root*, is to be made a Cataplasm,[44] unto the abominable Ulcers, if the unclean Patient have any such upon him. With no other Means, the Wretches, who have been so far gone in this foul Disease, as to ly roaring with the Anguish of it, have in about a Fortnights time received a thorough Cure. Happy, if it could have convey'd unto them the good Sense, *To Sin no more*. But you will readily now beleeve, that such a Medicine must needs be of incomparable Service in all *Scorbutic* Maladies. It ha's done Wonders in them!

I will at this time trouble you with but one Story more. We have a Sort of a *Thistle*, which we call, the *Boar-thistle*. It is the shortest, and the most prickly Sort, in all the Tribe of Thistles;[45] And it ha's a large & long *Root*, which is the thing that is to bear its Part, in the Medicine that is now to be compounded. With this, we join a Root which we call, the *Cancer-root*; and add unto it a little of what you call *Divels-bitt*.[46] Of these three Roots, we make a Decoction, or Infusion. Lett a Person languishing (yea, tho' very far gone) in the *Kings Evil*,[47] take every day a little Draught of this Liquor. At the same time, lett the Patient

42 The "Bolognian Physician" is probably the pioneering surgeon of rhinoplasty Gaspare Tagliacozzi (1544–99) of Bologna, whose *De Curtorum Chirurgia per Insitionem Libri Duo* (1597) describes the surgical process of repairing the loss of noses through grafting. For Mather's discussion of venereal disease and its potential cures, see his chapter "*Kibroth Hattaavah*. Or, Some Clean Thoughts, on, The Foul Disease," in *ABJ*, Cap. XXI, pp. 116–20.

43 Vomit-causing.

44 Poultice or plaster.

45 "Boar Thistle," also known as purple-milk thistle, is a species of annual herb in the family *Asteraceae* found in the North Atlantic region.

46 American cancer-root is a parasitic plant with yellowish flowers that grows on the roots of some species of oak; scabious "Devil's-bit" is a perennial flowering plant in the honeysuckle family, with purple lobed flowers, which grows in damp ground.

47 Scrofula, a form of tuberculosis of the lymph nodes of the neck (lymphadenitis). Mather mentions the case of a youth in the neighborhood, whose "lingring and grievous Death" he anticipated (*D* 2:204).

lay the bruised Roots, on the Scrophulous Ulcers or Tumours, and every day Renew the Cataplasm. Tis astonishing to see, how soon that *Reproach of the Physicians* will fly before this Remedy. It works a swift and a strange Cure; and I don't think, that the Gentleman who ha's lately published at *Frankfort*, what is by him called, *Praxis Medicinæ infallibilis*, ha's a more *Infallible*, among all his boasted Medicines, fetched, *è concilijs ac secretis Archiatrorum nostro tempore fœlicissimorum*.[48] It will do more good, than some Cæremonies, which from *Edward* the Confessors time to ours, have imposed on the Credulity of {the} World, for the Cure of this grievous Malady.[49]

Those *Plants*, of *America*, I can assure you are not such *Poetical* ones, as the *Moly* and *Nepenthe* of *Homer*.[50] *His* you will give me Leave to count such; For that which is now called *Homers Moly*, ha's none of the Qualities he assigns to *his*. And they must have taken it, who can dream that *Homers Nepenthe* is *Opium*. You may suspect that *Mine* are no better, while I so often use the Terms, *We have*, and make no fuller Communications of what we

48 Mather's refers to *Praxis Medicinae Infallibilis* (1711), a voluminous collection of medical studies by different physicians on different diseases, edited by Michael Bernhard Valentini (1657–1729), professor of medicine and rector of the University of Giessen (Germany). The infallible *Praxis Medicinae* went through numerous reprints. Mather's Latin passage from "Prologus" (sign 1[V]) translates "From the advice and secrets of the most productive chief-doctors of our time" but appears to be a paraphrase of "Praxis haec … Archiatrorum scilicet undique Celeberrimorum, cautelis, consiliis & secretis illustratur"; i.e, "This handbook is furnished with the cautions, recommendations, and secrets of the arch-doctors most famous everywhere." See I. Veith (1978).

49 See William Beckett, F.R.S. (1684–1738), a London surgeon and antiquarian, who (like Mather before him) rejected the belief in the efficacy of the king's touch in curing scrofula, in *A Free and Impartial Enquiry into the Antiquity and Efficacy of Touching for the King's Evil* (1722). The venerable custom, apparently first introduced by the Anglo-Saxon King St. Edward the Confessor (c. 1003–66), fell out of favor during Queen Anne's reign (1702–14).

50 Mather refers to the moly (Galanthus nivalis) Helen served at dinner during Telemachus' visit to Sparta (*Odyssey* 4.219–34). Beauteous Helen (whose face, notably, launched a thousand ships) had acquired this drug in Egypt; it was very helpful in allowing her to reminisce about the Trojan War without causing undue awkwardness in her audience (including, of course, her husband Menelaus) about her deceptive role in starting the dismal war. In another incident, Hermes gave Odysseus an antidote against Circe's forgetfulness-inducing drug (*Odyssey* 10.302–06); Odysseus's companions forgot all their woes and their homes (to boot) among the Lotus-eaters, whose nepenthes (lotus) is mentioned as a drug that brings relief from anxiety or grief (*Odyssey* 9:87–97). On "*Homers Moly*," see R. Molina-Venegas and R. Verano (2024).

have. But I will do my best, that, *you shall have*, these things, as soon as our Examinations are carried on to a little more Perfection.

In the mean time, I call to mind, that *Pliny* speaks of an Herb, which *Democritus*, called by a Greek Name, [Αισχυνομηνη] that signifies, *Bashful.*[51] I think, *Theophrastus* ha's it.[52] If it be touch'd, it contracts its Leaves; and it is the Same doubtless, with what we now call, *The Sensitive Plant*. If among the *Plants*, there be a *Bashful*, I am sure, I who am writing of *Plants*, deserve to be turn'd into one; and with a *Bashful* Blush confess my Præsumption in undertaking to write any thing unto you, on a Point, wherein I foresaw I must come off with a very mean *Performance*. But I was willing to *Begin*, and *Assay*, tho' I should not yett *Perform*; inasmuch as it would lay me under Obligations to *Go on*; and by some *Delay*, I may grow the more able {to} perform what I desire; for my *Spanish* Tutors have taught me, *La tardanza es odiosa, pero da fuerças*.[53] It will give me some *Strength*, & furnish me with both Abilities & Occasions, if I live, to do something hereafter, by which I may more fully approve myself,

Boston. N. England. Syr,

Nov. 18. 1712.

Yours, glad and proud

always to serve you;

C. Mather.

[51] Or, rather Αἰσχυνομένη (Lat. *aeschynomene*), *Mimosa pudica,* or Touch-me-not. The word is attested only in Latin, in Mather's source: the *Natural History* (24.102.167) of Pliny the Elder (23–79 CE), Roman natural philosopher (who died in the eruption of Vesuvius in 79 CE). Given that Mather was dependent upon Pliny's Latin transliteration of the Greek word, he made the easy mistake of an *eta* for an *epsilon* in the third letter from the end. Pliny attributes the noun to Apollodorus of Athens (c. 180–c. 120 BCE), a follower of the Greek atomist philosopher Democritus (c. 460–c. 370 BCE), who descriptively calls it "aischynomene" because of its shrinking leaves when touched.

[52] The Greek peripatetic philosopher Theophrastus of Eresus (c. 371–c. 287) lists the *Mimosa asperata* in his *Enquiry into Plants* (4.2.11), Appendix of Unnamed Plants, but does not use the word αἰσχυνομένη.

[53] "Tardiness is hateful, but it gives strength." On Mather's "Spanish lessons," see K.S. Gruesz (2002).

I.3. ON AMERICAN BIRDS.

November 19, 1712.

Letter. 3d.

To Dr. John Woodward.

Sir,

You expected from me a Communication of my *Subterraneous* Curiosities; and I have in one Remarkable Instance endeavoured it.[1] I will go on to endeavour your Satisfaction; but in things that are so far from *Subterraneous*, that they shall be rather of a *Cœlestial* Character and Residence. In this, you will not charge my Friendship, with the Solæcism of that Actor, who when he cried, *O Terra!* Pointed up unto the *Heavens*.[2] But you will accept from your *American* Friend, every thing, that may entertain and gratify a mind, that would know what is observed in our *Hæmisphære*.

I shall not attempt an Addition to the Treasures of your *Ornithology*.[3] As we have most of your Birds, (indeed our being without *Rooks* and *Buzzards*, and *Magpies*, and *Woodcocks*, and *Cuckooes*, ha's been improved by some

1 Referring to his first letter, CA 1.1, dated November 17, 1712.

2 An inside joke among stage players based on a witty repartee in *De Vita Sophorum* (1.542), by the Athenian Lucius Flavius Philostratus (c. 170–c. 205 CE), who puns on a passage in Euripides's *Orestes* (1496). As the story goes, Polemo, president of the Olympic games at Smyrna, laughs at the exaggerated gesticulation of a certain tragic actor who "pointed to the ground as he uttered the words, 'O Zeus!' then raised his hands to heaven at the words, 'and Earth'" [Euripides, *Orestes* (1496)]. Polemo, however, disqualified the actor from the theatrical contest, laughing, "The fellow has committed a solecism with his hand" (*Lives of the Sophists*, p. 131). The anecdote also appears in the popular manual *Chironomia: Or, The Art of Manuall Rhetorique* (1644), by J.[ohn] B.[ulwer], p. 135 (sep. pag.). See also T. Stern (2013), p. 17.

3 Mather may refer to John Ray's English translation of, and supplement to, *Francisci Willughbeii, Ornitologiae Libri Tres: Aves omnes hactenus cognitae in methodum naturis suis* (Latin, 1676; English 1678). Francis Willughby (1635–72) was an English ornithologist and ichthyologist who collaborated with his friend

aiery Gentlemen, as a Peece of Banter on our behalf,) so you have, I suppose, the most of ours. Tis true, some that we have in common, with you, yett here make almost a different *Species* from yours. Particularly, our *Wild Turkies*,[4] which are bigger, and blacker, & their Flesh tougher, than your *Tame* ones, & whereof some weigh Forty, yea, some Sixty Pounds. We will not boast of our *Pilhannaw*, a mighty kind of *Eagle*, four times as big as a *Goshawk*,[5] with an Head, as big as a Childs of {a} Year old; who when he soars aloft, makes the Feathered Tribe, in his View, to hide themselves. We behold as much Beauty and Glory in our *Humming-Bird*, the least of all the Things, that are cloathed with *Feathers*. Their *Feathers* glitter with variable Colours. Their *Bills* are long, and like a Needle. With these they suck the Honey which they find on the Blossoms & Flowers in our Gardens. They build little Nests, in the Spring, made up like a Bottom of soft Matter like Silk; Here they lay their Eggs, no bigger than small Pease; whereof they hatch Three or Four at a Time. We can rarely catch them, unless by shooting them, with Pistols charged with Sand; when weighing of them, we find them with Feathers & Entrails & all, to make but Eight or Ten Grains. They are never seen in the Winter; but are some of the *Season-birds*, whereof I propose now to invite you unto an Entertainment. And so are our wild *Pigeons*,[6] whereof Thousands of Millions visit us at their appointed *Season*. The Flights have been so great, that for four or five Miles together, they have meerly darkened the Horizon. They have been, commonly sold in the Markett-place, ready pluck'd, & drawn, for two pence or three pence a dozen, enough to make a Meal for half a dozen temperate People. We take them, either with *Gun*, or with *Nett*; it is hardly credible, how many at a time. Their Numbers are of late Years much diminished; especially on this Occasion. When the Time of their Departure ha's been at hand, about

John Ray (1627–1705), English clergyman, natural philosopher, and herbalist. See also our CA 2.5 "The Pigeons," and his commentary on Jer. 8:7 (*BA* 5:879–83).

4 It's a myth! Benjamin Franklin did *not* propose a turkey as the national symbol to appear on The Great Seal of the United States – no matter *The New Yorker*'s November 24, 1962 cover story. https://declaration.fas.harvard.edu/blog/turkey. But Franklin did not like the bald eagle either. See his letter to Sarah Bache (January 26, 1784). https://founders.archives.gov/documents/Franklin/01-41-02-0327 (both downloaded October 2, 2023).

5 Perhaps Mather intends the bald eagle, giving his spelling of the Native name for it, "Pilhannaw." Goshawk (Accipiter gentilis), or true hawk, in the family Accipitridae.

6 The American Passenger Pigeon, now extinct. See *CP*, Essay XXX: *Of the Feathered*, pp. 192–93; Mather also records his observations on the "Water-Dove" in his commentary on Gen. 8:8 (*BA* 1:643–44). See also A.W. Schorger (1938) and F.T. Lewis (1944).

Michaelmas, they have in horrible Storms miss'd their Way; & thousands of Millions have perished in the Sea, where our Ships have afterward sailed thro' them lying on the Surface of the Water, for some Leagues together. Tis odd, that tho' we have such vast Numbers of them, yett in *Virginia*, a Colony a little to the Southward of us, I am told, they are somewhat of a Rarity; which will a little fortify a Conjecture, about the *Season-birds*,[7] which I am now going to tender you.

It was my learned Friend, Mr. *Charles Morton*,[8] (whose Nephew and Names-sake, is at this time, a worthy Physician in your City,) who first led me into this *Conjecture*, (I had almost said, *Hypothesis*,) which I am now laying before you, for your Judgment upon it; how far it may pass the *Ordeal* of Reason & Experiment.

The *Little Birds*,[9] which we may call *Season-birds*, do seem to have their Annual Recesses, in some Remote and upper Quarters of the *Heaven*, within our *Atmosphære*. They are not only, *Birds [OF] Heaven*, but also, *Birds [IN] Heaven*, [Jer. VIII.7.][10] from whence at their *Appointed Times*, they take their *Journey* hither. It seems, as if they found something, either in the Temper of their own *Bodies*, or in the *Alterations* of their Lodgings above, or in the *Effluvia* of the Earth, upon the new Reflections of the Sun thereupon; which invites them to change their Quarters; and so they *know* their Seasons. If these *Birds* during their Absence from us, did reside in any Part of this *Earth*, methinks, we should after so many Ages at last, have heard, *where* it is. But no Man hath yett seen any Numbers of them, out of their *Seasons*; nor have we any but *old Wives Fables* concerning the finding of any, except a *Bird*, or so, that by some *Lameness*, ha's been left behind his Fellowes. Again; The *Birds* come in a Manner so sudden, so surprising, so unmarked, as if *dropping down from above* upon us. In a Nights time, the whole Countrey is full of them; and

7 Migratory birds appear during specific seasons of the year.

8 Charles Morton (1627–98), an English nonconformist who emigrated to Massachusetts, became minister of Charlestown, and subsequently taught astronomy at Harvard. Mather here draws on Morton's anonymous *An Essay toward the Probable Solution of this Question* (1703). His namesake appears to be Richard Morton (1637–98), who was ejected from his ministry and became a physician and distinguished member of the College of Physicians in London.

9 Mather uses much the same material (below) in his commentary on Jer. 8:7 (*BA* 5:879–83). His versatility in adapting materials to ever-new purposes is highly instructive. In this biblical gloss, he embraces Morton's argument that migratory birds might winter on the earth's satellite, the moon (Morton, pp. 35–50) or, so Mather, on a planetoid or crystalline sphere between the earth and the moon.

10 See M. Dopffel (2010).

that, not in *Flocks*, but extremely *Separated*. Probably, they keep hovering at some Distance from us, till by the blowing of certain *Winds*, they perceive agreeable *Steams* calling for them; and then down they fall, tho' disadvantaged sometimes as to their designed Pirching-places; whence, now and then a Bird, has made a Descent upon a Vessel at Sea, further off, than ever any such *Land-birds*, have been known to fly. Moreover; At the *first Coming* of these Birds, they are generally of a *Texture* very different from what they have, after they have been here a while. Their *Taste* is finer; they eat short; their *Flesh* is tender; they have little or no *Blood* in them. Which argues, that they have had another Nourishment, than what this *Earth* affords. Once more; The *Flight* of these Birds, while they abide among us, for the most part, is but *short*. And I am told, those which on your Island of *Great Britain*, are found near the Sea-shore, being disturbed, never will offer *toward* the *Sea*; wherefore, tis probable, they never came Thither, from *beyond* the *Sea*. Finally; There are very odd *Phænomena*, among these *Birds*, at or near the Time of their *Departure*. They don't grow duller; their *Flights* are higher; their *Notes* are brisker; they *gather together* as having a noble design for another World. Particularly to mention the *First Bird*, in the famous Text of the *Season-birds*; for *Bochart* ha's not corrected the Translation for *that*, tho' he ha's for some of the other.[11] The *Storks* of the *Low Countrey's* when the *Time of their Departure is at hand*, they All to a Bird assemble together; there they continue chattering for diverse days; till the Last are come into this *General Rendevous*; Then in the midst of all the Din, there is a sudden *Silence* for a little while; after which, upon some *Signal*, they all Rise together, and fly in one great Flock or Cloud, fetching many great Rounds, higher & higher still, much like unto the *Spiral Ascent* of a *Goshawk*, when she towers, till at Last, their Distance makes them appear less & less, and so they utterly disappear. Every thing looks, as if they sought a comfortable Repose, by flying *Directly upward*, which is not the Way to any other Land on our Earth. If it be now remarked, what and where the *Receptacles*? We are not so vain, as {to} imagine going with the *Ganza*'s of *Gonsales*, to the *Moon*.[12] If we do but allow to the *Moon*, such a small Distance from

[11] Samuel Bochart's "De Mundis Avibus," in *Hierozoicon Animalibus* (1663), pars post. (lib. 1, cap. 1–9, esp. cols 1–59). Passenger pigeons (like many others) were migratory, seasonal birds. In Mather's observation, a leading fowl took charge of the flock and signaled departure by rising to the sky.

[12] Domingo Gonsales (i.e. Francis Godwin), *The Man in the Moon* (1638). Francis Godwin (1562–1633), Anglican bishop of Hereford, has his fictive hero undertake an imaginary voyage to the moon with the help of a contraption tied to flying geese. The story may have been inspired by Galileo Galilei's telescopic discovery of mountains on the moon, leading to speculations that people lived there as well.

the *Earth*, as the *Arabians* after *Ptolomy* have assigned it; namely, *Forty-nine* Semidiameters of the Earth; which is in English Miles, counting according to *Cluverius*, fifty five Miles to a Degree, on the Great Circle of the Earth, 154350 Miles:[13] And if we suppose the Birds to travel at least one Third of the Year, according to which they can have but Sixty One Dayes for their Journey, they must then fly every day at least 1530 Miles; which is, omitting Fractions, about 63 Miles in an Hour; and this is nigh as fast again as the swiftest *Race-horse*, who moving at the rate of Four Miles in Seven Minutes and an half, dispatches in an Hour no more than *Two and Thirty*. We cannot consent unto this Velocity, in our *Birds*; and if we did, yett it would hardly serve; inasmuch as the true Distance of the *Moon* from the *Earth*, is doubtless much greater than what was apprehended in the Dayes of *Ptolomy*. But then, it Seems no unreasonable Præsumption, to imagine, That there are nigh unto our *Terraqueous Globe*, sundry, minute, *Planetary Bodies*, which partly from the Smallness of their Bulk, and perhaps also from some other Causes, may not be visible unto *us*, without the Help of Instruments. Until we were assisted with *Telescopes*, who ever dreamt of the *Satellits*, which move about the *Superior Planets*; or, who ever thought of catching *Saturn* by his Ears? If *Glasses* were a little more improved, and Mens Attention to the Discovery of their Glasses a little further awakened; it is possible, we might soon discern some very Significant *Globules* in our supream Regions, without coming under the Scoffs, employ'd upon those who *build Castles in the Air*. Yea, the greatest Philosopher of them who died in the former Century, did affirm in private Conversation, That some *Christalline*, or *Semi-pellucid* Bodies, between the *Earth* and the *Moon*, have already been discovered.[14] Certainly, the *Negro*'s on the Coast of *Barbary*, that know not of such a Rock as the Isle of *Ascention*,[15] which tho' it be Thirty

[13] Claudius Ptolemaeus, aka. Ptolemy (c. 100–c. 170 CE), Greco-Roman astronomer and geographer, calculated the distance of the sun and moon from the earth in his astronomical handbook *Almagest* (bks. 4–5); *Cluverius* is Philipp Clüver (1580–1622), German geographer, author of *Introductionis in Universam Geographiam* (1624), and editor of Ptolemy's *Geographia.*

[14] Mather alludes to the ancient concept of the crystalline sphere in which the fixed stars are lodged, in the Ptolemaic geocentric system. See O. Pedersen (1993), esp. pp. 268–71.

[15] A volcanic island in the South Atlantic, about 1,000 miles (c. 1,600 km) from the west coast of Africa. Mather appears to have relied for information on *John Huighen van Linschoten. His Discours of Voyages into the Easte & West Indies* (1598), ch. 95, pp. 175–76, which describes the arid condition of Ascencion Island. Van Linschoten (1563–1611) was a Dutch merchant and traveler, whose famous *Itinerario* (1596) was translated into English (1598) and several other European languages. See also Mather's commentary on Numb. 21:6 (*BA* 2:923).

Miles in Compass, has no Fresh Water on it, will ignorantly think the prodigious Multitude of *Sea-fowl*, which visit them from the Rock, to be bred on their own Shore, because they know not of such a Retirement for them. And why may not this illustrate the Case before us?

But I must not make your *Breakfast* on my *Season-birds* too long; tho' *most of them* are a more *Palatable Fowl*, than what the learned *Ludolphus* ha's turned the *Israelitish Quails* into.[16] My Præsumption, That my *Season-birds* might happen to prove not unacceptable unto so great a Philosopher, & Unto one who makes the *Natural History of the Earth*, a favourite Article of his *Philosophy*, but one who at the same time loves to see the Oracles of the *Sacred Scripture* illustrated, this is my Apology for the Freedome I have used in this Exhibition of them. With a Quill taken from one of them, I will now præsume on subscribing myself,

Boston, N. England. Syr,

Nov. 19. 1712. Yours in the Bonds of the most

Inviolable Friendship,

Co: Mather.

[16] The German Orientalist and grammarian of the Ethiopic language, Hiob Ludolph, Ludolf, aka. Leutholf (1624–1704), argued that the flying creatures that piled up in the camps of the ancient Israelites for food (Exod. 16:13, Numb. 11:31) were not "*coturnicibus*" (quails), but "*locustis*" (locusts), in Ludolph's *Appendix Secunda ad Historiam Aethiopicam ... Dissertationem de Locustis* (1694), pars sec, cap. 1, §§1–5, p. 21. See also Mather's commentary on Numb. 11:31 (*BA* 2:862–70).

I.4. ON ANTIPATHIES & THE FORCE OF THE IMAGINATION.

November 20, 1712.

Letter. 4th.

To Dr. John Woodward.

Sir,

My design in this Letter, is to entertain you, with some *American Curiosities*, which may contribute (a little, it may be, some of them,) unto those Nobler Stores, which you may have at hand, when that curious Topick, *De Viribus Imaginationis*,[1] may come under your Consideration.

Unto this Head, methinks one might refer the strange *Antipathies*, which there are found, not only in the Classes of *Inferiour Animals*, (tho' that of the *Lions* being terrified with the Crowing of a *Cock*, ha's been as *Camerarius* tells us,[2] confuted by what happened in the Court of the D. of

1 *De Viribus Imaginationis Tractatus* (1608) by Thomas Fieno, aka. Fienus, Fyens, Feyens (1567–1631), a Flemish professor of medicine at the University of Louvain. This popular work went through several reprints and editions; it examines the influence of the human imagination on diseases and their healing process in individuals. A rather free-floating English translation of this work was published in London (1657). On Fienus, see Mather's Letter to John Woodward (July 9, 1716), our CA 3.7 "Strength of Imagination," and *BA* (1:973–74). L.J. Rather (1967), E. Fischer-Homberger (1979), M.-H. Huet (1993), and H. Hirai (2017) provide helpful background information.

2 Philipp Camerarius (1537–1624), son of famous humanist Joachim Camerarius (1500–1674), was a German jurist and polyhistor, first chancellor of the newly founded University of Altdorf (near Nuremberg), and author of the popular and oft-translated *Operae Horarum Succisivarum* [*sic*], *Sive Meditationes Historicae* (1591). Mather facetiously refers to the entertaining story of the lion "much afraid of the crowing and sight of cocks, either because they [cocks] be more solarie [solitary] than the lions, or for some hidden propertie." Alas, lions only fear "white cocks," the Dominican friar Albertus Magnus (1200–80) objects but, says Camerarius, the Roman Epicurean philosopher Lucretius (c. 99–c. 55 BCE) thinks that the reason for such antipathy between lions and roosters is because "out of the

Bavaria:[3]) but also in particular Children of *Men* to various Objects, which are almost unaccountably offensive to them. Looking over my *Adversaria*,[4] I find that long since I made this brief Collection of what I had observed.

Ad Carnis, *imò et ad* Panis, *esum aliqui nullo pacto possunt adigi. Ad* Rosarum *odores fragrantissimus, quidam Deliquia patiuntur: Ad* Aceti *fœtores quidam: Alii ad* Anguillæ *vel* Ranæ *conspectum: Etiam et ad* Campanarum *Sonitus nonnulli.* Scopus *aream purgans unum aliquem horroribus miris solitus est vexare. Quiamque* Cultrum *tentaverat exacuere, cuidam è Gingivis Sanguinem fecit erumpere. Quendam à viso* pomo *ad mensam apposito mortuum* Quercetanus *memorat. Sudore ac Terrore ineffabili ad* Felis *occursum quidam corripiuntur. Imò et* muris *aspectus, viros alioquin fortes, et fugæ nescios, fugare potest. A* Caseo *ità abhorrent aliquis quòd vel è comesto pane quem tetigit culter in scindendo caseo antè adhibitus,*[5] *Lipothymià misera misotyris oriatur. Ad* Capitis Vitulini *et* Brassicæ *aspectum, syncope non raró sequebatur Viro, qui in Hypocondriâ ad latus dextrum caput vitulinum, Brassicam verò ad latus sinistrum, Nævos nempè maternos, affixos habuit. De quodam narrat* Libavius, *qui ad proprii* Filii *accessum deficiebat. Ad Rationem, lector, hæc portenta revocato*![6]

---

bodies of cocks there issue certaine toies, idols, or images which flie right against the lions, so sharpe and pricking that they cannot endure the violence of it," in *The Living Librarie, or, Meditations and Observations* (1621), bk 2, ch. 2, p. 78. The Mathers owned a copy of the Frankfurt reprint of *Operae Horarum Subcisivarum* (1644), and the lion's tale appears in ch. 22, pp. 118–19 of this Latin edition. See also *CPS* (1994), p. 364 *n*.

3 During Philip Camerarius' lifetime, there were three dukes of Bavaria: Albert V the Magnanimous (1528–79); William V the Pious (1548–97), and Maximilian I the Great (1573–1623), but the second-hand story may well have originated much earlier. As the story goes, a lion, having escaped from his enclosure, got into a coop where chickens, roosters, and other birds were feeding. Simba, far from being afraid at their crowing, "fell a killing and devouring the greatest part of them" (*Living Librarie*, p. 79). *Operae Horarum Subcisivarum* (1644), pp. 118–19.

4 According to O.T. Beall (1961), p. 368 *n.* 31, Mather's *Adversaria* contained "a body of early scientific data." It may well have been his youthful collection of no-longer-extant essays he devised for debate among the short-lived Boston Philosophical Society headed by his father, Increase Mather (1683).

5 See also Mather's Letter to John Woodward (July 9 1616), CA 3.7 "Strength of the Imagination."

6 "Some can by no means be forced to eat meat, yea, and bread. Some swoon from the most delicious fragrances of roses; and some from the smell of vinegar: others from the sight of an eel or frog: yes, even at the sound of bells: a broom cleaning the floor was apt to trouble a certain person with amazing horrors; whoever had tried to sharpen his knife caused blood to burst from a certain person's gums;

Unto this Heap of *Antipathies*, I will now add an odd one, of a Gentlewoman in my Neighbourhood, who whenever she sees any Person pare their *Nails* with a *Knife*, she presently swoons away; and it ha's ever been thus with her from her Infancy. And yett she can see any one cutt their *Nails* with *Scissers*, without any such Ill Effects upon her. You will now believe me, to be in the Doctrine of *Antipathies*, no *Erastian*.[7]

From these *Antipathies*, I will ask your Leave to pass unto some other Points, wherein the *Force of Imagination*, ha's pretty odly display'd itself.

A Physician of my Acquaintance,[8] gave an Emetic unto a Carpenter in my Neighbourhood; & such a Dose, as had used formerly to work sufficiently upon him; but the Carpenters Wife seeing the Physician give this Vomit,

---

Quercetanus [Joseph Duchesne (c. 1544–1609, French physician)] mentions that a certain person dropped dead at the sight of an apple placed on a table; some are seized by unspeakable sweat and terror at the sight of a cat, nay even the very sight of a mouse can put to flight men otherwise brave and with no thought of running away; some people are so turned off by cheese that they can't even eat the bread which a knife has touched that was previously used for cutting cheese; there arises a nasty cheese-allergic fainting spell; at the sight of a calf's head or a cauliflower, it is not uncommon for a certain man to fall into a hypochondriac fit, if either a calf's head is affixed on the right side of his abdomen or a head of a cauliflower on the left – evidently his mother's marks. Libavius [Andreas Libau (c. 1540–1616) German chemist and physician] told of a certain man, who fainted at the sight of his own son. Reader, see that you reduce these portents to rational principles." (David Lupher's translation). Mather's source for such intriguing stories of antipathies is informed by the same that his father employed, in *ERIP*, ch. 4, esp. pp. 100–03. Such stories are aplenty in Sir Kenelm Digby's *Of Bodies, and of Mans Soul* (1669), ch. 38, esp. pp. 427–29; and in Digby's *Two Treatises* (1645), ch. 38, pp. 409–10. On the power of sympathetic and antipathetic healing, see Digby's *Of the Sympathetick Powder* (1669) as well as *Theatrum Sympatheticum Auctum exhibens Varios Authores, De Pulvere Sympathetico* (1662) by the Scottish medical writer Sylvester Rattray (fl. 1650–66), whose *Aditus Novus ad Occultas Sympathiae et Antipathiae causas Inveniendas* (1658) was reprinted in his collection *Theatrum Sympatheticum*.

7 See also Increase Mather's *ERIP* (1684), p. 100, for the same reference to the Swiss theologian and physician Thomas Erastus (1524–83). Erastus authored several works in his losing battle against the novel iatrochemical approach to medicine of his German–Swiss physician and alchemist Paracelsus (c. 1493–1541). In his *Disputationum De Nova Medicine Philippi Paracelsi* (1572–73) and in some of his later works, Erastus rejected Paracelsus' Neoplatonic doctrine (widely held at the time) that the power of the imagination influences the physical body and that spirit can be transmuted into matter.

8 Probably John Perkins, a practicing physician and friend of the Mathers in the North End (*D* 2:93, 609, 627).

fancied strongly, that she felt it go down her own Throat; the issue whereof was, that the Vomit wrought not on the Husband at all, but on the Wife as vehemently as if she herself had really swallowed it. Whereupon the Physician was forced to give the Carpenter another Dose of the same Physic, in the same Quantity, with care taken that his Wife should be out of the Way; & then it had on the Man its proper Operation.

On this Occasion it may be agreeable enough to add; That we do not want Exemples,[9] of Husbands who *breed* for their Wives; and while they for diverse Months together are puling[10] under the most sickly Indispositions, their Wives have the Satisfaction that they suffer not one Moments Disorder. Yea, we have known several Husbands carry the Good Humour so far, as to *Travail* for their Wives, & Cry out with exquisite Pains, & have the Gossips attending on them; while their Wives have brought forth the Offspring, with scarce a Groan upon it.

But the *Force of Imagination*,[11] you are aware, does no where produce more surprising Effects, than in the Impressions, which it enables prægnant Women to make on their unborn Infants. You are apprised of many astonishing Instances, besides those with which *Helmont* ha's obliged us;[12] and no doubt,

9 I.e. "lack examples."

10 Crying piteously or weakly.

11 See also *ABJ* (1972), Cap V, item X, p. 31. At least until the middle of the 18th century and long before modern embryology, philosophers and physicians alike believed in the "plastic power" or formative power of the imagination to shape not only the fetus of pregnant mothers but also their sympathetic and visceral response to visual stimuli. A good starting point is a review of Mather's contemporary, the distinguished French philosopher Nicolas Malebranche (1638–1715), whose influential *De La Recherche De La Verite* (1674–75) was translated into English as *Treatise Concerning the Search after Truth* (1700). Malebranche's analysis of the formative power of the Imagination, especially in pregnant women (esp. bk II, pt 1, ch. 7), illustrates the pervasiveness of this concept. In fact, another Frenchman, M. Benjamin Bablot (1754–1802), medical counselor to King Louis XVII of France, in his *Dissertation sur le Pouvoir de L'Imagination des Femmes Enceintes* (1788), pp. 194–96, provides a substantial list of philosopher-physicians, ancient and modern – all attributing a formative ("plastic") power to the imagination. See especially the alphabetical and analytic table of the numerous authors and their sources he cites in his defense of the power of the imagination (pp. 199–234). L.S. King (1978), pp. 152–81, traces the changing views on the imagination among the principal physicians in the seventeenth and eighteenth centuries, from Daniel Sennert (1572–1637) to Benjamin Bablot.

12 Jean Baptist van Helmont (1580–1644), Flemish chemist, physiologist, and physician of Brussels, renowned for his *Ortus Medicinae* (1655). The applicable chapters in this work are as follows: pars 1, esp. "Imago Fermenti Imprægnat" (69–73), "Imago Mentis (163–69), and "Ortus Imaginis Morbosæ" (339–41).

you have read some which you leave with a, — *Fides sit Penes Authorem.*[13] I will give you now as odd an one, as most that you have mett withal. But I must præpare you for it, with telling you, That our Countrey for many Years did suffer much by the Calamity of *Buggy Pease*. While our *Pease* were in the Blossom, its thought, Swarms of unknown Flies visited the Fields, & left the Seeds of Ruine upon them. The *Pease* when Ripe, would seem fair enough; but anon in the Bag, or in the Barrel, a living, winged, ugly *Bug*, generated within the *Pea*, would eat up what was most valuable there, and eat his way thro' the Skin; & leave the Disappointed Husbandman only an insipid Shell for all the Harvest of his Labour. Now accept a Copy of a Letter to me, from a worthy Minister, whose Credit was always unreproachable.

To the Reverend Dr. *Mather*.

Reverend Syr.

The Account of that Diminutive Rarity, which you desired of me, is as followeth. My Daughter-in-Law (my Sons Wife) being big with Child, in the Year 1693. had a great Desire of a Mess of *Pease*, and knew where she might attain them by sending for them; but was discouraged by a knowledge she had, that they were annoy'd by those *Flies* or *Bugs*, which of late Years have been the common Annoyance of Pease; and being æqually poised between a *Liking* of the *Pease*, and a *Loathing* of the *Flies* in them, she chose to lett them alone. But when her Child was born, which was, Jun. 16, 1693. he had near the Middle of his Forehead, a little Inclining to one side, a fair, white *Pea*; both as to Sight and Touch exactly resembling a *Natural Pea*; and it so remained till *Pease* came to be Ripe.[14] And *then*, there began to appear a small black *Speck* in it, which gradually grew bigger and bigger; till at Length, about two Months after, it appeared full as big as a

[13] "Credit this to the author" or, let the author be responsible for the reliability of the facts.

[14] Such strange phenomena seemed all too familiar to those who believed in the power of the imagination, especially that of pregnant women. In fact, the Italian Renaissance philosopher Pietro Pomponazzi (1462–1525), in his *De Naturalium Effectuum Causis, sive De Incantationibus* (1556), cap. 4, p. 54, knowingly relates, "nam in coitu quis imaginans aliquid, producit realiter foetum ei similem quod imaginatur. & mulier prægnans, ex vehementi desiderio ut pote ciceris, similitudine ciceris foetum signatum parit: Unde Ciceronum domus dicta est" ("for instance, if during intercourse some one [woman] imagines something, she develops in reality a child similar to what she imagines, and a pregnant woman, from a vehement desire to get chickpeas, will deliver a child with the mark of a chickpea: whence it was called the house of Cicero"). Pomponazzi pointed at the Roman philosopher Cicero, whose patronymic is derived from Latin *cicer*, i.e., "chickpea" (106–43 BCE), for one of his ancestors is alleged to have received his name from a prominent chickpea-like growth on his face.

Flie in a *Natural Pea*, just before it leaves it; & seemed so loose in it, that it might be taken out upon the Point of a Pin. It was thus observed by myself, & Wife, & Daughter, & others of my Family upon a Satureday a little after Noon; and the Night following it was gone out of the *Pea*, none knew how; and left the *Pea* a meer empty Hull or Husk, which gradually withered away; so that since, the Forehead of the Child is as smooth as any other, and hath no Sign of the *Pea* remaining. I was not without some Expectation, that there might have been some Return of it, the following Summer; but there was *none*. This is the brief account of it, from, Syr,

Your Friend & Servant,
Zachariah Walker.[15]

I will rather make it a Paragraph of this Letter, than write Another on Purpose; to entertain you with another Curiosity of an *Humane Body*; which can claim Relation to the former Stories, but just enough, to come with a tolerable Grace in the same Letter with them.

The *Petrifications* happening in the several Parts of an *Humane Body* have been, as you well know, Syr, the Entertainment of the Curious. Besides those too well known Torments of Mankind under *Nephritic* Dolors;[16] and the *Stones* taken out of the Joints of *Gouty* Persons; there have come *Stones* out of opened Veins in a Phlebotomy. A *Stone* ha's been bred between the *Dura* and the *Pia Mater*, of the Brain;[17] yea, the *Brain* itself ha's been found a *Stone*. A great *Stone* was found adhæring to the *Liver* in the Countess of *Nadasti*.[18] Three *Stones* were found in the Heart of the Emperour *Maximilian* II.[19] *Albosius*[20] reports, and

15 Zachariah Walker (1637–1700) attended Harvard College (1653–55) and pastored the church of Woodbury, Conn., from 1670 until his death.

16 Kidney-related disease, including inflammation.

17 Dura: "The strongest and outermost of three membranes that protect the brain, spinal cord, and nerves of the cauda equina" (*Gale Encyclopedia of Medicine*, 2008. The Gale Group, Inc.) Accessed March 10 2022, https://medical-dictionary.thefreedictionary.com/Dura. Pia mater: "The fine vascular membrane that closely envelops the brain and spinal cord under the arachnoid and the dura mater" (*The American Heritage Medical Dictionary*, 2007. Houghton Mifflin Company). Retrieved March 10 2022, https://medicaldictionary.thefreedictionary.com/pia+mater.

18 The identity of the Countess of Nádasdy, probably a member of the Hungarian nobility, remains elusive. Her story is likely told by Albosius or Sennert (below).

19 Maximilian II (1527–76), Holy Roman Emperor of the House of Habsburg (Austria).

20 Joannes Albosius, aka. Jean d'Ailleboust (1524–94), a French doctor of medicine, personal physician of French King Henry III, and author of *Portentosum Lithopaedion, sive Embryon Petrefactum Urbis Senonensis* (1582, 1587; 1683) on the

Sennertus writing *De Lithopædiâ*[21] confirms it, That a Woman retained within her with much Misery for twenty Eight Years, an unborn Infant, which was turned into a *Stone*.[22] And Histories affirm, whole Bodies of Men turned into *Stone*, in *Africa*.

The *Tongue* more particularly ha's been a frequent Seat, wherein *Stones* have been generated. The *German Ephemerides*,[23] give several Exemples of it. But that which most emboldens me to hope, that you will count my Story not altogether unworthy to be laid before you, is, That in your *Philosophical Transactions* for the Year 1672. you have thought one worthy to be inserted, of a Man with you, who had a *Stone* bred under his *Tongue*, occasioned by his Enduring much cold in a Winter-Voyage.[24]

My Story then is this. One Mr. *Collier*,[25] an honest Man of *Hull* in New England, grew sensible of a Pain in his Throat; & made use of the common Remedies, but unto little Purpose. At last the Pain grew very extreme; especially when he drank any *Beer*; nor indeed was he able to swallow, without extraordinary Difficulty; For which Cause, he subsisted entirely on liquid Substances. After he had been for some time in this Misery, a *Stone* appeared under his *Tongue*; which anon fell into his Mouth (and so into his hand,) leaving an hole behind it, near the *Root* of the Tongue. It was of a yellowish Colour, near an Inch in Length, & in shape somewhat resembling the Tongue

formation of stones in human organs and petrified fetuses "stone babies" in female wombs. His famous account is based on that of Mrs. Colombe Chatry, who suffered abdominal pain for more than 28 years before she died in 1582. An autopsy revealed a calcified fetus she had carried in her womb since her last pregnancy at the age of 40, in *Collectanea de Diuturna Graviditate* (1683), pp. 11–18.

21 Mather refers to Daniel Sennert (1572–1637), illustrious German Iatrochemical physician, professor of medicine at Wittenberg, whose treatise *De Lithopaedia*, on petrification or calcination of human organs, probably appeared in his multivolume and widely reprinted *Sennerti Operum* (1630–50). The Mathers owned a copy of Sennert's *Thirteen Books of Natural Philosophy* (1660).

22 A lithopedion, or stone baby, can develop in the abdomen, when a fetus grows outside the uterus (ectopic pregnancy), and dies. If it is not reabsorbed by the body's fluids, it can calcify and remain in the body. See also Mather's discussion of the "mola" in CA 1.5 "Monstrous Births."

23 See for instance, Christian Friedrich Garmann's "Observatio XLIII: De Tophis Sub Lingua," *Miscellanea Curiosa sive Ephemeridum Medico-Physicarum Germanicarum Academiae* (1. Decuria 3. 1672), p. 73.

24 "An Account of a Stone Cut Out from under the Tongue of a Man," *PT* 7.83 (May 20 1672): 4062–64. The account was sent in by Martin Lister, F.R.S. (1639–1712), an English naturalist, physician, and avid conchologist.

25 Unidentified.

of a Man. But there was this odd Circumstance attending the Man, more surprising than all the rest. When he had Occasion to make Water, he was in as much Pain, as if the *Stone* had been in his Bladder, or his Kidney. While the *Stone* was in his *Tongue*, when he voided his *Urine*, he was usually thrown into a Sweat with the intolerable *Pain* and *anguish* of it.

I know not, whether, it may not putt you to *some*, if my Letter swell any bigger. I began with *Antipathies*, and I know not whether you may not have one for *long Epistles*. For which Reason, I now hasten to conclude, and subscribe,

Boston, N. England. Syr,

Nov. 20. 1712. with very great Respect,

Your most Ready Servant.

C. Mather.

I.5. MONSTROUS BIRTHS.

November 21, 1712.

Letter. 5th

Doctissimo Clarissimoque Viro
D. JOHANNI WOODWARD. M.D.

Epistolam hanc aperiens, Miraris forsan, Vir Optime, styli Mutationem, et Americanam calami. Latinitatem. At quo magis *Res Monstrosa* videatur, eò magis apta, statim censebitur, decora, atque idonea. De *Monstris* Epistolam scripturus, nolui *Linguâ Maternâ* uti, ne, quum (si tuâ cum veniâ hoc aliquandò fiat) publici Juris facta foret. *Mater* aliqua legens, Infanti nondum nato, motæ ac læsæ Imaginationis, noxam inferat.

De propriâ *Monstri* quidem *Significatione*, verborum *Cumini-Sectores* qui Accurati censeri velint, magno conatu disputarunt. At dum quæcunque *Communem* rerum sui generis *Ordinem* excedunt, à multis *Monstrorum* nomine appellantur, et quæcunque miranti *monstrantur* vulgo; quo Sensu Jesuita quidam *Gallus* apud *Anglos* in concione publicâ, B. Virginem *Mariam*, summo cum Auditorum Risu, *Monstrum* ineptè nuncupavit; ea solùm *Animantia*, quæ membrorum suorum constitutionem habent perfectè progenitorum structuræ *Turpiter* et *Insigniter* dissimilem, jam solita sunt apud Doctos *Monstra* vocarı. Quod ad *Monstri* Etymologiam attinet, Antiqua M. *Tullij* sententia asserit, *Monstra*, quòd res eventuras mortalibus *Monstrent*, suum sortiri nomen; ast verò haud videtur magis concinna hæc sententia, illâ aliorum, quæ vult *Monstra* dici, quià ejusmodi sunt quæ ob Novitatem et Enormitatem cum stupore quodam, ab unoquoque unicuique *Monstrentur*. Qualia interdum nascuntur *Monstra*, Nosti tu, Vir Doctissime, qui non solùm *Fortunium Licetum* cum *Gerardi Blasii* recensione, verùm et alios quamplurimos Historiæ ac Medicinæ Scriptores, legisti. En, Jam auctum catalogum! Non tantum *Africa*, sed et ipsa *America* quidem sua *Monstra* protulit; atque ipsa *Nov-Anglia* non *Monstris* prorsus Vacua. De *Monstris* aliquot inter *Nov-Anglos* editis, *Historiolas* has, ab omni mendacio mundas ac immunes, lubet hic adducere.

Bostonæ Nov-Anglorum, D. *Gulielmi Tinctoris* Uxor, cujus tam *cerebrum* quam *uterum* simul *Monstra* aluisse notârunt Rerum Nov-Anglicarum Scriptores, *Monstrosis* quibusdam, et *Hæreticis* Opinionibus inquinata mulier, Fœtum enixa est prodigiosè *Monstrosum*. Obstetrix parturienti huic Fæminæ fuit *Saga* vetula, cujus maleficio, et (quod credebatur) veneficio, astantes omnes mulierculæ, aut crebris vomitibus, aut somnolentis Delirijs, aut liberorum convulsionibus correptorum ejulatibus, ità vexatæ fuerunt, quòd præter Diabolicam illam *Lamiam*, non nisi una in ædibus remanserit. Non prius *partum* quam *Mortem*, sensit, *Monstrum horrendum*; quo tempore, mirum in modum *tremuit* matris, quæ et ipsa *Tremulana*, cubile ac totum cubiculum. Natum hoc atque donatum Animalculum, statim summo cum silentio, nec minore celebritate, Sepulchro condebatur: At paucis posteà Diebus, nescio quid de *Re monstrosa* garrulæ inter se mulieres mussitârunt. Ad magistratum hoc pacto delatus est hujusce Rei Rumor, cujus jussu effodiebatur Cadaver, et nullo coram Testibus reperiebatur immane hoc spectaculum atque prodigium. Huic *Monstro*, dicam? an *Spectro*, nullum omninò fuit *caput*. *Faciem* habuit in pectore sitam; *Auriculas* verò simiarum similes, ex Humeris enascentis; *Os* atque *Oculos* valdè prominentes; *Nasum* sursum curvatum; *Thoracem* ac *Dorsum* Rajæ instar Clavatæ aculeatum; *Umbilicum*, *Alvum*, et *Vulvam* coxendicis locum occupantes; et *Nates* vicissim eorum; *Brachia*, *Manus*, *Femora*, et *Crura* Regularea; in utroque pede, non *Digitos*, sed *Tres* velut Avis *Ungues*; in *Tergo* duo insolita *Orificia*, in quorum utrisque comparuerunt Carnis Massæ duæ non minus insolitæ: *Frontem* prorsus nullam; at quatuor suprà oculos *Cornua*; duo vel unciæ longitudinem excedentia, dura et acuta, alia duo paulò minora. Nihil denique hoc *Monstro* magis *Monstrum* esse potuit.

Guoburnæ Nov-Anglorum conjux *Josephi Fabri* monstrosum hujusmodi patiebatur partum. Habuit Infans *Caput*, *Collum*, ac *Brachia*, naturaliter formata; *Os* verò *pectoris* nullum, nullamque *Dorsi Spinam*; *ventrem* valdè tumidum, *Femora* per-parva, *Ossibus* carentia; *Nates* prorsus nullus; *Membrum Virile* penitus osseum; nullos pro urinà aut pro stercore *Meatus*; *Pedes* extrorsum inversos; *Calces* resupinatos, et osseum similes. Habuit et dissecto Cadavere, duo magna carnis *Frusta* ventrem replentia; *Intestina* verò in pectore jacentia.

Hartfordiæ Nov-Anglorum, *Samuelis Debellij* Uxori hoc monstrum accidit. Fuerunt Infanti (qui neque mas, neque Fæmina, sed inter utrumque extitit) in manu dexterâ, *Pollex* et sex Digiti, in sinistrâ verò quinque Digiti, cum superaddito *Pollice*; Digiti pedum *sex* in utroque pede; *venter* magnitudine insignis; *Oculi* nulli; nullum superius *Labium*, nulla *Lingua, at quiddam Radici Linguæ* non absimile; *collum* nullum, *Sinciput* nullum, nullusque meatus urinarius. *Newichavonocæ* Nov-Anglorum *Gulielmi Plaistedi* marita talem lugebat Filiam. A cincturâ deorsum ad pedes usque singulis rectis fruebatur; at non fuerat *Mulier formosa superne.* Omni{a} superius fuerant vel defectiva vel dislocata; *Caput* plus solito magnum, et *Cranio* omnibusque ossibus penitus destitutum; *Facies* quæ vel adultæ mulieris æquet; *Brachium Dextrum* omninò nullum, at *Carnis Frustulum* Brachij loco, Digito non absimile; *Brachium Sinistrum* in longitudinem sanè nimiam extensum; *Nasus* nullus; in *Fronte* verò quiddam *Narium* formam obscurè referens; *Oculi* in Genis positi; *Os* nullum, sed minuta quædam *Cavitas* non aptè disposita; *Auris* una sub *Mento*, altera in *Vertice*; *Collum* curtum, et quasi nullum; Aliquid suprà *Pectus*, Reni simile excrescens; *venter* aperitus, et quasi dissectus; omnia item *Intestina* externè hinc inde affixa. Statim à partu expiravit.

Mulier quædam *Dedhamensis*, nuper post partum valdè difficilem, nixa est Infantem, cui *duo capita.* Hi verò *Infantes*, (*Prisciani* dicam cum veniâ) statim etiam à partu *expiravit. Monstris* hisce vix licet annumerare *Molas*. In eâdem verò epistolâ referre, non absurdum omninò et absonum videtur. Temporis illud spatium, per quod *Molæ* in gravidarum uteris interdum continentur, contemplatoribus Naturæ curiosis magnam admirationem aliquandò concitat. *Molam* Annorum Quinque conspexit *Savanarola*, et miratus est. *Dodonæum* fermè attonitum reddidit illa vidua, quæ *Molam* in utero, annis plus minus Quindecim gestavit: Cujusmodi *Molam* et *Duretus* non sine Demiratione, in suis commentariis memorat. *Paræus* totos per septendecim Annos, *Molam* in Utero cujusdam gestam, rei tam insolitæ Demiratione Perculsus refert; pariter ac *Florentinus* quandam, quæ per plusquam viginti Annos *Molâ* vexabatur. Nec non ad ultimum vitæ finem, Fæminam satis ingenti *Molâ* afflictam, scribit suspicitque *Alexander Benedictus*. At cunctas inter *Molas*, illa per quam uterus cujusdam vehementer extensus, peritonæo à lineâ albâ ad costas usque spurias adeò tenaciter adhærebat, quòd vix avelli posset, *Gulielmi Fabricij* opinione, palmam fortè visa est præripere. Quæ verò omnes hasce superat, Historiolam jam habeas *Nov-Anglicanam*. Matrona quædam, cui Nomen *Longa*, Anno Suæ ætatis octuagesimo, mortem cum vitâ commutavit, postquam in ejus utero *Molam* Annos per Triginta quinque gesserat. *Abdomen* illi fuit valdè tumidum, tactui durissimum; umbilico per nimiam extensionem penitus obliterato. Cadavere transversim ad peritonænum usque inciso, apparuit, ecce, *Cartilægo* quædam firma, gibbosa, formæ penè ovalis, per Regionem sese extendens umbilicalem, peritonæo quaquaversum adeò annexa, ac si peritonæum ipsum, in massum talem condensatum esset. Originem hæc substantia ducere potuit è materia quadam albâ, lentâ ac viscosâ, in hujusce *scutellæ* cavitate, à *Molâ* exudante, quàm interclusam deindè invenimus. Hæc *Scutella Cartilaginea* à providâ Naturâ, ut partem *Molæ* prominentem, et injurijs magis obnoxiam, quasi *scutum* defenderet, atque ut *Mola* huic firmiter connexa in situ propriò suspenderetur, quæ alioquin ponderosâ suâ gravitatione, inferiora compresserat, excretionemque omninò impedierat, proculdubiò instituta est. Hâc amotâ, aspectui nostro se exposuit, *Massa* monstrosa, albicans maximi peponis formam referens, sedecim libris præponderans; quæ fuit ipse quidem uterus, qui per talem diutinam extensionem, abijt à propria Crati, et naturam corneam induit. Hujus Apertioni reclusa sunt quinque distincta Involucra, sibi invicem, perindè ac cæparum Tunicæ, implicata, ad crassitiem quinque Digitorum simul sumpta; Quorum unum cultello dividentes, in tenuissimas membranas findi exploravimus. *Mola* inclusa compages erat carnis maximè induratæ, intermixtis Frustulis osseis et cartilagineis. Organa utero ancillantia, aut forsan colliquata, aut *Molâ* sat magno absorpta, prorsus defecerunt.

Sed, Manum de Tabula. Ubi nihil præter *Monstra* conspiciendum, non diu nimis exhibenda Tabula.

Vale, Vir præstantissime. Te tuosque quàm fælicissimè valere animitus cupio.

Si *Judaica* tibi valedictio arrideret, et Benedictio, optarem

מחיים מתים יתן לך חיים ארוכים טובים ומתוקנים:

Cui magis *Christianam* addam Apprecationem.
Sospitator noster, clementissimus JESUS, te Servet Incolumem.

Bostonæ Novanglorum. || Aureis Amoris ac Honoris Catenis
Nov. 21. 1712. tibi adstrictissimus;
C. Matherus.

Translation

[November 21, 1712]

Letter. 5th[1]

To the very learned and distinguished man
Mr. John Woodward, M.D.

As you open this letter, you are perhaps surprised, most excellent man, by its shift in manner of writing and by my pen's American Latinity. But the more monstrous the subject matter appears, the more it will be deemed suitable,

1 The editors gratefully acknowledge Dr. David Lupher, professor emeritus of Classics, University of Puget Sound, for translating Mather's Latin letter and for providing helpful annotations.

fitting, and proper. As I was about to write a letter about monsters, I chose not to use my mother tongue, lest when it became a matter of public judgment (if, with your permission, that may someday happen) some mother reading it might inflict on an infant not yet born the harm caused by a disturbed and wounded imagination.[2]

Concerning the proper meaning of the word "monster," it is true, the hair-splitters who wish to be considered precise dispute with great effort. But while whatever things exceed the common order of things of their given kind have this term bestowed upon them, as well as whatever things are displayed to the wondering common folk – in which sense a certain French Jesuit in a public meeting in England, to tremendous laughter from the auditors, ineptly called the Blessed Virgin Mary a "monster" – only those living beings that have a construction of their limbs foully and markedly different from that of the perfect structure of their progenitors have now been accustomed to be called "monsters" by the learned. As for the etymology of the word "monster," the ancient view of Cicero asserts that things that *reveal* {*monstrent*} things that are going to happen receive this name.[3] But in fact this view scarcely seems more fitting than that of others who are of the opinion that monsters are given that name because they are the kinds of things that are pointed at {monstrentur} with a kind of gaping wonder by all and sundry to all and sundry because of their unusual nature and irregular shape.

You, most learned man, are aware of what sorts of monsters are born from time to time, for you have not only read Fortunio Liceti in the edition of

[2] Callous as the observation may sound to our sensibilities today, such monstrous births bespoke the effects of original sin, as Mather points out in his sermon *OP* (1707): "To be born a *Sinner*, is to be born a *Monster*. Tis as Odious, as Frightful, as to be born with the Scales of the *Leviathan*; to be born with the *Belly* where the *Head* should be; with *Eyes* in the Back-part of the *Head*; with the *Hands* growing to the Mouth; with the *Feet* made for ever to go awry every Step. I Speak Significantly; Sinner, such, even such a *Monster* would be thy very *Emblem*" (p. 9). To be sure, Mather's vivid imagery (borne out in real medical cases to which he was an eyewitness) is here deployed for soteriological ends. See also his *MEFA* (1710), pp. 35–36; CA 2.3 "A Monster"; CA 3.1 "Monstrous Impregnations"; and CA 3.2 "A Monstrous Calf." For useful background, see K. Park and L.J. Daston (1981), P.K. Wilson (1992), M. Hagner (1999), M.-H. Huet (2004), A.W. Bates (2005), and J. Spinks (2009). S. Asma (2011)

[3] Mather refers to the Roman statesman, philosopher, and scholar Marcus Tullius Cicero (106–43 BCE), who in his *De Divinatione* (1.42.93) explains the connotation of the term "*monstra*" among the ancients. Such *monstra* "'make manifest' (*ostendunt*), 'portent' (*portendunt*), 'intimate' (*monstrant*), 'predict' (*praedicunt*), they are called 'manifestations,' 'portents,' 'intimations,' and 'prodigies.'"

Gerard Blaes, but also very many other writers of history and medicine.[4] Behold! The catalogue has now been expanded. Indeed, not only Africa but also even America has brought forth its monsters; and New England itself is not utterly empty of monsters. I want to introduce here these brief accounts, free and clear of all mendacity, of several monsters that have been produced among the inhabitants of New England.

In Boston in New England, the wife of Mr. William Dyer, whose brain (as noted by writers of the events of New England) nourished as many monsters as her uterus did, a woman stained by certain monstrous and heretical views, gave birth to a prodigiously monstrous fetus.[5] The midwife for this woman in her birth-giving was an old witch, through whose spells and (it is believed) poison all the women standing by were so troubled by frequent vomitings or by drowsy deliriums or by the shrieks of their children seized by convulsions, that not a single one remained in the house apart from that diabolical witch. She was aware of the death even before the birth: a horrid monster – at which time the bed of the mother (who herself was shaking) and the whole bedroom shook in marvelous fashion. When the little creature was born and set forth, it was immediately hidden away in a grave with the strictest silence and with not even a smallish gathering. But a few days later the talkative women muttered something among themselves about the monstrous business, and in this way the rumor about this matter was reported to the magistrate, at whose order the corpse was dug up and {*nullo*?} in the presence of witnesses this frightful and prodigious spectacle was brought to light. This monster – or shall I call it a spectre? – had no head at all. It had a face located in its chest; little ears

4 Fortunius Licetus, aka. Fortunio Liceti, (1577–1657), an Italian physician and natural philosopher at Bologna and Padua, and author of numerous medical and astronomical works. Mather refers to *Fortunius Licetus De Monstris. Ex Recensione Gerardi Blasii, M.D. & PP.* (1616, 1634, 1665). The richly illustrated third edition (1665), was edited and revised by the Dutch physician Gerard Blaes (1627–82). A copy of the 1665 edition was in the Mather libraries. See also *Monstrorum Historia* (1642) by the Italian natural philosopher and physician Ulisse Aldrovandi (1522–1605) of Bologna.

5 The infamous case of Mary Dyer, friend of Anne Hutchinson, executed as a Quaker (1660). On October 17, 1637, she gave birth to a deformed stillborn fetus, which was secretly buried, though later disinterred on the orders of Governor John Winthrop (1587–1649). Mather here follows very closely the account of the circumstances and nature of the birth, apparently by Winthrop, in *A Short Story of the Rise, reign, and ruin of the Antinomians, Familists & Libertines* (1644), pp. 43–45. Even the comparison to the thornback ray is derived from Winthrop's account. Mary Dyer's husband is referred to as "D. Gulielmi Tinctoris" (*tinctor* = dyer); "D." stands for "Dominus," "Mr.," not "Doctor." Accounts of such fetal malformations are plentiful in the *PT* and German *Ephemerides*.

like those of apes growing from its shoulders; mouth and eyes considerably standing out; nose curved upwards; breast and back prickled like a thornback ray; navel, stomach, and vulva occupying the place of the hips; buttocks on the other side of them; arms, hands, femur, and legs regularly formed; in each foot, not toes, but three claws like a bird's; two odd orifices in the back, in each of which appeared masses of flesh no less unusual; no forehead at all; but four horns above the eyes, two of them exceeding an inch in length, hard and sharp, the other two a little smaller. In short, nothing could be more monstrous than this monster.[6]

In Woburn in New England the wife of Joseph Carpenter endured a monstrous birth of the following sort.[7] The infant had head, neck, and arms naturally formed; but no sternum and no backbone; a stomach much bloated; very small femurs, lacking bones; no buttocks at all; male member thoroughly bony; no passage-way for urine or excrement; feet turned outwards; heels drawn back {or flattened, *resupinatos*} and similar to bones. When the body was dissected, it had two large clumps of flesh filling the stomach; but the intestines lying in the chest.

In Hartford of New England, the following monster happened to the wife of Samuel Debell.[8] The infant (which was neither male nor female but between both) had in its right hand a thumb and six fingers, but in the left five fingers with a thumb superadded; six toes in either foot; a stomach unusual in size; no eyes; no upper lip, no tongue, but something not unlike the root of a tongue; no neck; no cheek, and no urinary tract.

In Newichawannock in New England the wife of William Plaisted mourned a daughter of the following sort.[9] From the waist down to the feet

[6] Except for slight variations in wording, Mather includes much the same description of the deformed fetus in his *MCA* (1702) 7.3.11.20. It is telling that although Mather's account of Mary Dyer's miscarriage is rife with superstition, he relegates the seeming prodigy to mere natural causes: It was affirmed "by some that were Eye-witnesses ... that these were no more *Monstrous Births*, than what it is frequent for Women, labouring with *false Conceptions*, to produce." See M.P. Winship (1994), p. 97; and J. Crawford (2005).

[7] We have rendered the Latinized name "Faber" as "Carpenter," though "Joiner" would be another possible English surname. Neither name is prominent in the published accounts of early Woburn.

[8] There's a testimony of Samuel Debell (1667) in Hoadly's transcripts, Connecticut Historical Society, cited by D.D. Hall (1989), p. 275 *n.*79.

[9] "Newichawannock" was the Abenaki name adopted in 1703 after the resettlement of part of Kittery (now in Maine); the year after Mather's letter, it was renamed Berwick. In his letter to Increase Mather (25. 8^{m} [Oct.] 1683), Rev. Joshua Moodey (1633–97) of Portsmouth (Mass.) gave a detailed description of a "monstrous birth

it had the benefit of regular features. But above it was not an attractive female. Everything above was either defective or dislocated. The head was uncommonly large and utterly devoid of the cranium and all bones; a face resembling that of a grown woman; no right arm at all, but a morsel of flesh, not unlike a finger, in the place of an arm; the left arm extended in quite excessive length; no nose; in the brow something vaguely recalling the shape of nostrils; eyes placed in the cheeks; no mouth, but a certain small cavity awkwardly arranged; one ear under the chin, the other in the hairline; neck curtailed and almost nonexistent; something like a kidney growing above the chest; stomach open and as though dissected; likewise all the intestines affixed on the outside from either direction. It died immediately upon birth.

A woman of Dedham recently, after a very difficult birth, produced an infant that had two heads. But these infants (with apologies to Priscian) died immediately after birth.[10]

It's scarcely permissible to add *molae* to monsters of this sort.[11] Still, to report them in the same letter is not entirely absurd and inharmonious. The extent of time over which *molae* are occasionally transformed in the uteruses of pregnant women often stirs up great wonder for the curious contemplators of nature. Savonarola examined and marveled at a *mola* of five years' growth.[12] That famous widow who gestated a *mola* in her uterus for more than fifteen

brought forth at Newichuwenoq by the wife of one W^{m}. Plaisted" (*MP* [1868], pp. 361–62). Most likely, Mather has in mind the story of Mary Plaisted, wife of James Plaisted, who was captured during King Philip's War (1675–78), mentioned in Mather's *DL* (1699), p. 55. See also Lt. Plaisted (p. 78). The name "Plaisted" *appears* to have been a common name in this area in the second half of the 17th century and into the 18th.

10 "With apologies to Priscian." The phrase "with Priscian's permission" (*Prisciani cum venia*) is an attempted witticism. Mather is constructing a sentence here with a plural subject (*hi infantes*) – because it was a two-headed fetus – and a singular verb (*expiravit*) – because it was still just one *infans*. Technically, this is a violation of the rules of grammar. Priscian (Priscianus Caesariensis, *fl.* 500 CE) was a famous grammarian, whose name became a synecdoche for grammatical rules. Hence, Mather is composing an irregular sentence "with apologies to Priscian." Unfortunately, since English does not distinguish the number of verbs, Mather's witticism goes lost in translation.

11 This relatively rare condition is called lithopedia. A lithopedion, or stone baby, can develop in the abdomen, when a fetus grows outside the uterus (ectopic pregnancy) and dies. If it is not reabsorbed by the body's fluids it can calcify and remain in the body. See also Mather's discussion of the "mola" in CA 1.4 "On Antipathies."

12 This was *not* the famous Dominican preacher Girolamo Savonarola, but his grandfather, Giovanni Michele Savonarola (ca. 1385–1460s), who taught medicine at

years rendered Dodonaeus almost thunder-struck.[13] Also Duretus mentions a *mola* of this sort in his commentaries not without astonishment.[14] Paraeus, struck by the astonishment caused by a thing so unusual, mentions a *mola* that had gestated in the uterus of a certain woman over the course of seventeen full years.[15] Likewise Florentinus {mentions} a certain woman who was troubled by a *mola* for more than twenty years.[16] In fact, Alexander Benedictus writes about and accepts the case of a woman afflicted by a considerably large *mola* right up to the end of a long life.[17] But among all the *molae* the one which

---

Padua, then at Ferrara. His discussion of the mola is in his *Practica Major* (1547), Tractatus VI, cap. 21, rubrica 27 ("De mola matricis"), fols 276^{V}–77^{V}.

13 Rembert Dodonaeus, aka. Dodoens (1517–85), a Flemish physician and botanist. His *Herbal,* in Henry Lyte's 1578 English translation, was in the Plymouth libraries of both William Brewster and Myles Standish. Dodonaeus's discussion of the mola appears in *Medicinalium Observationum Exempla Rara* (1585), cap. 49, pp. 107–09.

14 Louis Duret (Lodovicus Duretus, 1527–86) was a French doctor, physician to Charles IX and Henri III. His discussion of molae was in his commentary (bk 1, ch. 61) of his *Opera Practica* of Jacques Houillier (ca. 1510–62): *Adversaria, Enarrationes et scholia in Jac. Hollerii opera practica* (1571). Duret's comments on the mola are included in a later printing of Houllier's *Opera Practica* (1623), pp. 519–23. Moreover, Duret mentions a similar case in his oft-reprinted *Hippocratis Coacæ Phrænotiones* (1588), Tractatus Tertius, "De Morbis Mulierum," pp. 440–41, and again in the same tract, under # 42 "Sterilibus," pp. 478–79. On the causation of *mola,* see Aristotle's *Generation of Animals* 4.7 (776*b*25–776*a*14).

15 Paraeus, Ambroise Paré (1510–90), a French surgeon, whose work on monsters has again lionized the attention of scholars in the past half century. His discussion of *molae* is in his oft-reprinted *Opera Chirurgica Ambrosii Paræi* (1582), lib. 23, cap. 29, 33–38, 41, pp. 700–09, 713; Book 24 is entitled *De Monstris et Prodigiis* (pp. 731–97) and English, French, and German translations were published in the 20th century. See Ambroise Paré, *On Monsters and Marvels* (1982), ch. 11, pp, 43–45.

16 "Florentinus" may be Taddeo Alderotti, also known as Thaddeus Florentinus (ca. 1210–95). His works were published as *Expositiones in Arduum Aphorismorum Ipocratis Volumen* (1527). According to Nancy Siraisi (1981), he did make mention of a case of retention of an aborted fetus (p. 280), referring to *Consilia* no. 60, p. 128.

17 Alexander Benedictus, aka. Alessandro Benedetti (fl. 1450–1512), author of *De Observatione in Pestilentia* (1493) and several works on human anatomy. His discussion of lithopedia (mola, petrifications in the human body) can be found in Benedetti's edition of *Omnium À Vertice ad Calcem Morborum Signa, Causae, Indicationes & Remediorum Compositions utendique Rationes, Generatim Librim XXX conscripta* (1539), lib. 24, cap. 34–38, pp. 991–96. Here he sketches the views on *molae* from Aristotle to his own time.

seemed to take the prize, according to Guglielmus Fabricius,[18] was the one by which the uterus of a certain woman was so powerfully distended, that it clung so tenaciously to the peritoneum from the linea alba all the way to the lower ribs that it could scarcely be pulled out. But here is a little history from New England for you that surpasses all of these.

A certain woman by the name of Long, in her 80th year of age, died after she had carried a *mola* in her uterus for 35 years. Her abdomen was extremely swollen, very hard to the touch; the navel had been almost completely obliterated by the excessive expansion. When the corpse was cut into cross-wise all the way to the peritoneum,[19] behold, there appeared a mass of cartilage, firm, humped, almost oval in form, extending through the umbilical area, annexed to the peritoneum in every direction in such a manner as if the peritoneum itself had condensed into such a mass. This substance may have had its origin from a certain white, tough, sticky matter exuded from the *mola* in the cavity of this platter, the *mola* which we subsequently found closed up inside it. This cartilaginous platter was doubtless set up by a provident nature in order to defend like a shield that part of the *mola* that protruded and was more subject to injuries, and so that the *mola*, firmly connected to this, would hang securely in its place; otherwise it {the *mola*} would have compressed the inner organs with its own heavy weight and utterly impeded excretion. When this was removed, a monstrous mass was exposed to our view, whitish, suggesting the shape of a very large melon, weighing sixteen pounds – this was in fact the uterus itself, which through such prolonged bloating, had left its own frame and put on a horny nature. Hidden in an opening of this were five distinct wrappings, enfolded one within another, like the skin of onions, to the thickness of five fingerbreadths when taken together; when we divided one of them with a knife we ascertained that it was split into very thin membranes. The enclosed *mola* was a composite structure of extremely hardened flesh, with little chunks of bone and cartilage mixed in. The organs ancillary to the uterus were utterly missing, either having perhaps utterly dissolved or absorbed by the mola over a sufficiently long period of time.

But enough! Where there is nothing to be seen but monsters, the tablet should not be exhibited too long.

Farewell, most distinguished man. I heartily wish you and yours to flourish most happily.

[18] Hieronymus Fabricius ab Aquapendente, aka. Girolamo Fabrizio (1533–1619), professor of anatomy and surgery at Padua, often called "the Father of Embryology." His treatises (including *De Formato Foetu,* 1600) may be found in his *Opera Omnia Anatomica et Physiologica* (1687).

[19] The serous membrane lining the abdominal cavity and covering the abdomen.

If a Jewish farewell and benediction would meet your approval, I would choose:

[20]מחיים מתים יתן לך חיים ארוכים טובים ומתוקנים:

To which I will add a more Christian prayer: May our Savior, the most merciful Jesus, keep you safe.

One bound to you by the golden chains of love and honor;

Boston, New England. C. Mather.

Nov. 21. 1712.

[20] Roughly: "He who raises the dead will give you a godly life, good and established." The sentence faintly echoes Paul's promise, "He that raised up Christ from the dead, shall also quicken your mortal bodies" (Rom. 8:11). Mather's Hebrew citation appears to be an adaptation of מְחַיֵּה חַיִּים יִתֵּן לְךָ חַיִּים אֲרוּכִּים וְטוֹבִים וּמְתוּקָּנִין (Talmud tractate Yoma 71a), in which Rabba explains that the Sages take leave by saying, "May the One Who grants life to the living grant you a long, good, and established life." Significantly, in Mather's quote the word חיים (life or the living) is replaced by מתים (the dead). I am grateful to Professor Lisa Gordis of Barnard College for identifying the source of Mather's derivation.

I.6. REMEDIES REVEALED IN DREAMS.

November 22, 1712.

Letter. 6th.

To Dr. John Woodward

Sir,

It ha's doubtless been somewhat of a Surprize unto you, to find, what *Operations of the Invisible World*, there have been, in communicating the *Knowledge of Medicine* unto us.[1] You Remember how a *Dream* taught (the Gentleman whom the Persians call *Irsenus*) *Galen* himself, which way to cure a Malady upon himself.[2] You are well-acquainted, with what Antiquity ha's related concerning the usual Method of Dreaming for a *Cure*, after the Observation of certain Ceremonies and Superstitions in the Temple of *Æsculapius*.[3] You have particularly considered the Greek Inscriptions on the

1 Collecting memorable providences, such as physical cures revealed in dreams, was a common enough endeavor in Mather's day, and his accounts (given below) are merely a fraction of what he and his father Increase recorded elsewhere as "memorable providences." Increase Mather, inspired by the famous Rev. Matthew Poole (1624–79), set forth a set of eight rules for selecting and recording credible suprarational occurrences in his *ERIP* (1684), "The Preface" (signs. A^{3r}, A^{7r}–A^{8v}). They were to provide empirical proof of the invisible world and the existence and immortality of the soul seemingly challenged by Cartesian mechanists and Hobbesian materialists of the day. To be sure, pamphlets and books on the subject of dreams and their factual revelations are legion in the early modern period. For perceptive insights into the significance of dreams in theology, medicine, and divination, see D.D. Hall (1989), esp. 86–91; C. Gantet (2010), and J. Rivière (2013).

2 Mather refers to Aelius Galenus of Pergamon (129–c. 216 CE), Greek physician and surgeon, who wrote *De dignotione ex insomniis* (*On Diagnosis from Dreams*). As a devotee of Asclepius (god of medicine), Galen firmly believed in the diagnostic power of dreams. Mather also invokes Galen's medicinal revelations in his Seventh Letter to Woodward (November 24, 1712), CA 1.7. See S.M. Oberhelman (1983) and K.V. Ergil (1997).

3 Cicero dismisses the belief that the Greek Asclepius or the Egyptian Serapis conveyed knowledge of medicine to supplicants in dreams (*De Divinatione* 2.59.123).

Marble Table found in the Isle of *Tyber*, where the Temple of *Æsculapius* formerly stood;[4] & now to be seen in the Palace of *Mapheo*.[5] Among the modern Exemples, that of *Wallace* cured of a Consumption by a few Leaves of *Red Sage* & *Bloodwort* steeped in *Small Beer*, from the Instruction of a *Genius*, ha's not escaped you.[6] And you have no doubt employ'd a particular Thought on old *Napiers* Way of Practice, and the *Responsa Raphaelis* which he entred in his Manuscripts, to be seen in the *Musæum Ashmolæanum*, where I understand they are now reposited.[7]

In my Neighbourhood, I have mett with several such Instances.

A Man who had an obstinate Pain in his Stomach, which could be releeved by none of our Physicians, dreamt that he should find Releef, by boiling a *Perch* and a *Pastnip*[8] together, and applying this odd Cataplasm,[9] to the Region of his pained Stomach. He did it, & it helped him.

A Gentlewoman,[10] whom I may do very well to keep alive in my Memory, fell into grievous Languishments; wherein a Pain of her Breast; & an excessive

---

According to Diodorus Siculus, the Egyptians revered Isis for revealing medicinal drugs to her dreaming devotees (incubation) in her temples (*Library* 1.25.2–6).

4 The Temple of Asclepius on the Isola Tiberina, in Rome, was constructed between 293 and 290 BCE, ostensibly to ward off a plague that subsequently decimated Rome. See Livy, *Ab Urbe Condita* (10.47.6–7).

5 Probably the Palazzo Maffei in Verona (Italy), a seventeenth-century Baroque palace of the Italian banker family of Marcantonio and Rolandino Maffei.

6 The story of Samuel Wallas (Wallace) of Stamford (Lincolnshire) appeared in Increase Mather's *ERIP* (1684), pp. 205–08. An elaborate version of Wallace's story appeared in William Turner's *Compleat History of the most Remarkable Providences, both of Judgment and Mercy* (1697), ch. 2, p. 9, # 7. Wallace was miraculously cured by a remedy given to him by a mysterious stranger, later judged to be an angel. Mather includes a fuller version of this story in medical handbook *ABJ* (1972): Cap. VIII "*Raphael*. Or, Notable Cures, from the Invisible World," pp. 49–50. See also *D* (2:504, 520).

7 Richard Napier (1559–1634), English cleric, astrologer, and physician, claimed to speak with the archangels Gabriel, Michael, and Raphael; his manuscripts came into the possession of the 17th-century English antiquary Elias Ashmole (1617–92) and are now in the Bodleian Library. See also O. Hadass (2018) and D.D Hall (1990).

8 Presumably, a parsnip.

9 A poultice or plaster.

10 Mather's first wife, Abigail Phillips Mather (c. 1670–1702), daughter of Colonel John Phillips (1633–1725/26). Mather was happily married to her (1686–1702); she died of breast cancer. Mather records her struggle during her last year and her conviction that the cure revealed in her dreams would be successful (*D* 1:443–45, 448–53).

Salivation, were two Circumstances, that were become as Insupportable unto her, as they were Incureable. She apprehended (in her Sleep, no doubt,) that a grave Person appearing to her, directed her, for the former Symptom, to cutt the *Warm Wool* from a *Living Sheep*, & apply it warm unto the grieved Part; for the latter Symptom, to take a Tankard of *Spring-Water*, & therein over the Fire dissolve an agreeable Quantity of *Mastick* & of *Gum-Icinglass*;[11] & now & then drink a little of the Liquor to strengthen the *Glands*. The Experiment was made, and she found much Advantage in it. However, other Distempers growing upon her, did a few Months after putt an End unto her Dayes.[12]

My Third Instance, will be so Remarkable, that I beleeve, you will not Repent the Pains you bestow in Reading of it. A young Woman of *Boston*, called, *Lydia Ingram*,[13] after other languishing Influences, fell Ill of a Feaver. Her Feaver was at Length attended with a very great Swelling of her Stomach, and Sides, and a total Suppression of *Urine* for Ten Dayes together. Several of our most able *Physicians* did what they could for her; but at Length desisted from doing any more, because they saw, her Case was altogether Hopeless. While she lay in this Condition, she dreamt that there came into the Room, a Gentleman with the Circumstances of a venerable old Age upon his Head and Face, but of a very comely Countenance; of a middle Stature, & having a light-coloured Chamlett-Coat for his Habit. This Gentleman said, That over-hearing the Groans of one in distress, as he passed by the House, he could not but come in; and now, having the Distress'd Case of the young Woman more particularly related unto him, he gave her his Advice: *Take a Glass of White-Wine, and putt into it a little Alum; and add the Powder of a burnt beef marrow-bone, as much as may ly upon the Lid of the Civet-box*[14] *now in your hand*, [which happened now to be there, and was about as broad as a Sixpence,] *Drink this off; Do it twice in three Hours; By the Fourth you shall have Releef.* So he went away. She awoke; and informed her Physicians of what passed; only had forgott the Third Ingredient. They allow'd her to take the *White-Wine* and *Allum*, if she pleased, for they were able to do no more. She took it, but without any Effect. Hereupon she fell asleep again; and the same venerable Person came again unto her. He asked, whether she had followed his Advice? She said, she had; but found no Good of it. He asked, *What*

[11] "Isinglass" is a collagen made from the dried swim bladders of fish.

[12] Mather's account is nearly identical to that in *D* (1:444).

[13] Possibly Lydia Dowse, who in 1671 married Henry Ingraham of Boston. A Lydia Ingram was admitted to Boston's First Church on April 17, 1698, in *Records of the First Church in Boston, 1630–1868*, p. 97. Mather provides a slightly shorter version of the same story in *ABJ* (1972), pp. 50–52.

[14] A small container for fragrances.

have you done? She told him. He replied; *Don't you remember the Powder I told you of, to take as much of it as might ly on the Lid of your Civet-box? Alas, you have almost fool'd your Life away. But however – Go take what I bade you. I believe it may do you Good yett. There is nothing too hard for God, and so you'l find it*! Which last Words he also Repeted, at his then going out of the Door. She then awaking, told her Physicians of the Remedy, whereof she now gave them a more perfect Account. The Remedy was immediately Allowed and Præpared; tho' the Physicians derided it at the same time, as a meer Notion, and an Improper one too. She took it, and by the Fourth Hour, she had the Releef that was promised her; & voided Five Pints of Urine at One time; And from Wednesday at One a Clock, to Friday at One a Clock, there came from her by Measure Six Gallons! The Suppression returned, & continued Three Dayes. The same venerable Person, then a Third Time returned unto her, in her Dream. He said, *How dost do now, Child*? She said, *I grow weaker*. He said, *Did you do as I advised you*? She said, *Yes*. He said, *Did you Repeat it*? She said, *No*. He said, *Why not*? She said, *I could gett no White Wine*. He look'd very much Displeased, and said; *No White Wine! There's enough in the Town; but People that want a Will seldome want an Excuse!* She said, *I am almost gone!* He replied, *You are almost gone indeed. However, take it again; All things are possible with God!* He then went unto the Door; & then turning about again, he said with a more cheerful and pleased Countenance; *Take it, and give God all the Glory; give God all the Glory*; *Farewell!* And so he went out. She accordingly followed his Advice, & found present Releef. She went on, taking the Remedy twice or thrice a day; and God bless'd it so, that she Recovered, and quickly came abroad; & ha's been since married successively to two Husbands, & been a Mother of Children, and is to this day alive, which is fourteen Years after the Occurrences that have been related.[15]

I am writing to One, who perfectly comes into the Sentiments, which our Incomparable *Boyl* expressed in those agreeable Terms. "If it were allow'd me to envy Creatures so much above us, as are the Cœlestial Spirits, I should much more envy that Angels charitable Employment, who at sett times diffused an healing Vertue thro' the troubled Waters of *Bethesda*, than that dreadful Angels fatal Employment, who in one night destroy'd above four-score thousand fighting Men."[16] And from this Disposition, Syr, you may be

15 Mather originally ended the letter here, but struck out his closing paragraph to add what follows.

16 Robert Boyle (1627–91), Anglo-Irish natural philosopher, chemist, and inventor, who endowed the distinguished Boyle Lectureship Series, presented annually since 1692. The quotation is from Boyle's *Some Considerations* (1664), Essay V, pp. 298–99. The Mathers owned a copy of the third edition (London, 1668) of this work.

under some Temptation to suspect, that they may be *good Angels*, who discover this Benignity, & thus *Do Good* unto the Children of Men. And yett, I know, you will be content, that the Matter ly in a Dubious Uncertainty; and you will allow, that an *Evil Spirit* may *Do Good* upon an *Evil Design*.[17]

It is an odd Business, that is related by *Suidas*.[18] An Athenian called *Plutarch*, and one *Domninus* a Syrian Philosopher, both consulted *Æsculapius* upon different Cases; but he ordered both of them to *eat Pork*, for their Cure. The Athenian would not obey, but in raillery asked the God, *What he would have præscribed unto a Jew in his Case*? And obliged him to præscribe something else. The Syrian, who was really a *Jew*, notwithstanding the Law of his Religion against the Eating of Swines-flesh, conformed unto the Direction, and found himself so well after it, that he eat of it all the rest of his Life; & found himself the worse for one Single Dayes abstaining from it. His Disease was, if, Syr, perhaps you remember the Story, *A Spitting of Blood*.[19]

I am not so forgetful, as to offer any of my Remarks on these things, to a Person of your Sagacity. I leave them with you, for *yours*; but among them, I hope, you will not fail to make One, on the Forward Zeal, which you find in me, to make you a Tender of every thing in our *American* World, which may give you any sort of Entertainment, & to approve myself,

Boston, N. England. Syr,

Nov. 22. 1712. One who alwayes loves & honours you.

C. Mather.

17 I.e. "*Caco-Daemons*," which "*oftentimes pretend to be good Angels*" but are really Satan's minions. It was the apparition of a "Caco-daemon," which pretended to be the spirit of Saul, says Increase Mather, in his *ERIP* (1684), pp. 200, 202, 208. See also *BA* (1:477; 3:327).

18 This story appears in *Suidae Lexicon* (1705) 1:616–17, *voce* Δομνῖνος Domninus Philosophus; Mather also refers to it in *ABJ* (1721), CAP. VIII, p. 48.

19 Probably consumption (tuberculosis).

I.7. REMARKABLE RECOVERIES FROM WOUNDS.

November 24, 1712.

Letter. 7th.

To Dr. John Woodward.

Sir.

You will allow me to think, That in your Noble Profession, *Cheirurgery* was the elder Sister to *Pharmacy*; and your *Father* shall I say? or *Idol*, whom you call, *Æsculapius* (or *Asclepius*,) does in his Name, *Ish Calaphot*, that is to say, *A Man of the Knife*, carry the Remembrance of his principal Profession; or, take *Bocharts* Etymology for it, *Ish Calabi*; or, *A Man of the Dog*,[1] you know, that he carried about a *Dog* with him, to lick the Sores of his ulcerous, or wounded Patients.

I am writing to an Accomplished and Experienced Physician, who ha's at some time or other had his Thoughts employ'd upon that Problem of *Cheirurgery*, *What Wounds are to be esteemed mortal*. When you resume that Subject into your Thoughts, you will please to think on some of my *American* Stories, and see whether they will not render the Problem almost impossible to be decided; and beyond a Decision by the *Aphorisms*, even of that Gentleman, of whom *Suidas* reports, *That it is commonly thought, what he saies, proceeds out of a Divine Mouth, & not an Humane*; and your

1 Samuel Bochart (1599–1667) derives the Hebrew word *Ish Calaphot* from "Ish Calabi" (or rather, "is-calibi"), a transliteration of the Phoenician איש־כלבי, in *Hierozoicon*, editio quarta (1712), pars prior, lib. 2, cap 55, col. 663 (line 70). He suggests that this may have been the Phoenician name for Asclepius, the Greek god of healing, who is frequently depicted in the company of his dog; hence "*ish*-calabi." Mather uses the same reference to "*Ish Calaphot*" in his letter to John Woodward (July 10, 1716), CA 3.8 "The Stone Mistaken," as well as in his "*Flagellum.* The Stone [And other Diseases of the *Kidneys* and *Bladder*]," in *ABJ*, cap. XIV, p. 86. The story about the curative power of a dog's saliva can be found in Conrad Gesner's *Historia Animalium Liber I. De Quadru Pedibus Viviparis*. Tiguri, 1551, lib. 1, p. 178, who cites both Albertus Magnus, *De Animalibus* (22.2.1, MS, page 1363) and Aelian, *De Historia Animalium* (8.9).

Galen tells you, *What he ha's delivered, has always been esteemed as the Word of a God.*[2]

You are well-apprised, That many very *Little Wounds*, have in the *Event* proved *Mortal*, which no *Judgment* would have pronounced such.

You Remember, with whom a sleight *Scratch* of a *Comb* had done *so*; and with whom a single *Hair* sportfully pluck'd off his *Breast*, was no less than *that*. You have not forgotten, That the Incomparable Anatomist *Spigelius*,[3] at the Wedding of his Daughter, gathering up the Reliques of a Broken Glass, a Fragment of it scratched one of his Fingers; and all his exquisite skill in *Anatomy* could not prevent its producing an *Empyema*,[4] that killed him. Perhaps you have heard, that Colonel *Rossiter*, cracking a Plumbstone with his Teeth, broke his Tooth, & more than so, lost his Life: That the Lord *Fairfax* cutting a *Corn* in his Foot, cutt asunder with it, the Threed of his Life: That Mr. *Fowler* a Vintner, playing with his Child, received a little Scratch of a *Pin*, which turned into a *Gangræne*, that cost him his Life. How *Frail* a thing the Life of Man! What a *Folly* to forgett the *Frailty*![5]

But then, the *Wounds* that all the World would have pronounced *Mortal*, we have seen sometimes *outliv'd*, even by the feeblest of Men! In our Wars with our *Indians*, we have seen surprising Instances.

A Man whose Name is *Jabez Musgrave*, tho' he were shott with a Bullet that went in at his *Right Ear*, & went out at his *Left Eye*; and a Brace of Bullets, that went in to his *Right Side*, a little above his Hip, & passing thro' his Body within the *Backbone*, went out of his *Left Side*, yett recovers, & lives many Years after it.[6]

2 *Suida* (*Suda*) is a 10th-century Greek Byzantine dictionary and encyclopedia. Well known in Mather's time were the aphorisms of the Greek father of medicine, Hippocrates of Kos (c. 460–c. 377 BCE), which his disciple Claudius Galen of Pergamon (2nd century CE), the Greek physician and surgeon, collected and embellished with his own insights. The Harvard catalog of 1723 lists the six-volume edition of Galen's *Opera* (1586) as well as individual volumes of Hippocrates' works. As a devotee of Aesculapius, Galen claimed that his remedies were revealed to him in dreams by Aesculapius. On the same, see Mather's Sixth Letter to Woodward (November 22 1712), CA 1.6. See also G. Majno (1975).

3 The Flemish anatomist Adrianus Spigelius, aka. Adriaan van den Spiegel (1578–1625), professor of anatomy and surgery at Padua, is best remembered for his *De Humani Corporis Fabrica Libri Decem* (1627), a work richly illustrated with anatomical tables of the human body and its organs.

4 "A collection of pus in the cavity between the lung and the membrane that surrounds it" (Johns Hopkins Medicine, www.hopkinsmedicine.org/health/conditions-and-diseases/empyema).

5 The stories of Colonel Rossiter, Lord Fairfax, and of the vintner Mr. Fowler are taken from Mather's *DL* (1699), p. 72. For the same, see *MCA* (1722) 7.74.

6 The account of Jabez Musgrave also appears in *DL* (1699), p. 73; and in *MCA* (7.74).

The Indians making an Assault upon *Deerfield*,[7] struck an Hatchet some Inches into the skull of a Boy; & even so deep, that the Boy felt a sensible Wrench used by them to get it out. There he lay a long while weltring in his Blood. At Length, we found him; we dress'd him; considerable Quantities of his Brain came out from time to time, when we opened the Wound. And yett the Lad Recovered,{and Lived many years after it}.[8]

I know not whether the Story told by *Plato* be true; That one *Herus Armenius*, (whom *Clemens* will have to be *Zoroaster*,) being slain in War, lay Ten Dayes among the Dead, and then being brought away, and on the Twelfth Day laid on the Funeral Pile, came to Life again.[9] But it is true, That one *Simon Stone* being here wounded with Shott, in *Nine* several Places, lay for *Dead* (as it was Time!) among the *Dead*. The Indians coming to strip him, attempted, with *Two* several Blowes of an Hatchet at his *Neck*, to cutt off his *Head*; which Blowes, you may be sure, added more enormous Wounds unto those *Port-holes of Death*, at which the Life of the poor Man, was already running out, as fast as it could. Being charged hard by our Forces, they left the Man, without *Scalping* him, which else they would have also done. And ours now coming to bury the *Dead*, one of the Souldiers perceived this poor Man to fetch a Gasp: whereupon an *Irish* Fellow then present, advised 'em to give him another Dab[10] with an Hatchet, & so bury him with the rest. Our *English* detesting this barbarous Advice, lifted up the wounded Man, & poured a little *fair Water* into his Mouth, at which he coughed; then they poured a little *Strong Water* after it, at which he opened his Eyes. The *Irish* fellow, was ordered now, to hale a Canooe ashore, to carry the wounded Men up the River, unto a Cheirurgeon; and as Teague[11] was foolishly pulling the Canooe ashore, with the Cock of his Gun, while he held the Muzzle in his hand, his Gun went off, & broke his Arm, whereof he remains a Creeple to this Day. But *Simon*

7 Referring to the raid on Deerfield, Mass., in September 1675, during the notorious King Philip's War (1675–78).

8 *DL* (1699), p. 73. The line in braces – defaced and illegible in the holograph manuscript – is reconstructed from p. 73. For a similar case of recovery from severe brain injuries, see *Wounds of the Brain Proved Curable* (1682), by James Yonge (1646–1721), surgeon of the Royal Navy, Plymouth (SW England).

9 *DL* (1699), pp. 70–71; the story of Er (Herus), the son of Armenius, the Pamphylian, is told in Plato's *Republic* (10.613*E*–10.621). The Greek Church Father Clemens Alexandrinus (c. 150–c. 215 CE) identifies him with Zoroaster (Zarathustra), the founder of Zoroastrianism in Persia, in *Stromata* (5.14.193.2), in *ANF* (2.469).

10 "A blow of somewhat sharp and abrupt character" (*OED*).

11 Mather's contemptuous reference to an Irishman as "Teague" bespeaks the prejudices of his time.

Stone was thoroughly cured; & as he was born with *Two Thumbs* on one hand, his Neighbours have thought him to have at least as many *Hearts* as *Thumbs*.[12]

Some other *Wounds* we have seen, which appeared sufficiently *Desperate*; and yett the Patients have survived them.

One *John Symonds*[13] being at plough, & sitting down to Rest himself; on the Ground, a little before the Oxen; the Oxen upon some Fright, leap'd over the Boy; but so that the Plough-share took hold on his Ribs, a little below his left Pap, and there tore an hole into his Breast, so large, that a Man might have putt his hand into it. His very *Heart became visible*, without a Metaphor. His *Lungs* would fly out for two or three Inches at the Orifice; & continued so to do, every time the Wound was dressed, until the Orifice grew less, & at last closed up. Until this, the Motion of the Lungs, working against the Plaisters, & Pleqeanis,[14] did so frett the Skin, as to render them very Raw. But as the Sore healed, that Part of the Lungs grew to the Flesh, & became sound therewithal. He recovered in Seven or Eight Weeks, & became a very Healthy Man.

Looking among my loose Papers,[15] I find, that when I was a Youth, I entred a Couple of *Curiosities*, relating to the Head, which I am now upon; with my *Juvenile Remarks* upon them. I did it in the *Latin Tongue*, because I was willing to præserve my poor *Latinity*, such as it was, from growing *Rusty*; and because I had some thoughts of publishing a Collection of *Curiosities* in that Language, to the learned World, which I have since laid aside. I will here transcribe those two Passages, and give you the Trouble of *Reading* them in the Language wherein I had *written* them.

The One was this.

Mors, *non tantum per immortales Horatianæ Lyræ Modulos, sed in* Mortalium *omnium communi Suffragio*, ultima Linea rerum, *audit. Quàm levi interdum Tactu hæc* Linea Terminum *inveniat, et periodum, attentè secum si quis cogitet*, Mortis *assiduam et contemplantionem et expectationem sedulamque pro* Vitâ Æternâ *præparationem in se excitatam, non potest non sentire*; Id quoque quod vivat (*ut non minus* Christianè *quàm* Ovidianè *loquar*) *munus*

12 The entire paragraph is taken – with slight changes – from Mather's *DL* (1699), pp. 70–71. For the same, see *MCA* 7.74.

13 For the same, see *MCA* 6.2.10; perhaps this is John Symonds (1674–1757) of Topsfield, Massachusetts.

14 Conjectural reading.

15 Mather's commonplace book or "Quotidiana" (*D* 1:548). Even as he gathers material for the first series of letters (*CA* 1) to the Royal Society, he reproaches himself for not spending enough time "in transferring into my *Quotidiana*, from *The Works of the Learned*, and from the *Philosophical Transactions*, and from the German *Ephemerides*, and other things[;] those Treasures which may furnish and Enrich me, for my many Services" (*D* 3:88).

Dei Optimi Maximi habere, tum *verè admirandum, tum piè agnoscendum. È contra verò, Illa ipsa quæ vocamus* Lethalia *aliquandò* Vulnera, *Mortem non semper hominibus* Letho *proximis afferre, cum gaudia, et non sine stupore conspicimus. è vulneribus deploratis aliquandò convalescentes.*

Convaluit ità Isaacus Gulielmi. *Trium, quatuorve Annorum puer, porcorum paternorum jugationi interesse voluit. At Apri cujusdam furibundi Ictibus et Morsibus expositus, eam quâ Ventriculus loq[u]itur, pectoris partem ità vulneratam ac laceratam, quòd et ipse* Ventriculus *proruperit, ad Abdomen jam dependens, habuit. Misera atque satis territa Mater, puerum abripiens, propriâ Manu Infanti supino, et, quod mirum, Rationis usum non amittenti, Ventriculum horribiliter sauciatum, per stupendum illud vulneris orificium intrusit; eâdemque Manu per horas aliquot vulneri impositâ, futuras Ventriculi protrusiones, usque dum ejus Maritus pedibus plura Milliaria incesserat Cheirurgum accersiturus, prohibuit. Cheirurgi Applicatonibus Deus aspiravit. Quamvis enim per Dies quatuordecim nihil omninò vel ederat vel biberat hic puer absque immediatâ vomitione, ad summam tame pancraticamque Sanitatem deinceps pervenit, ac præter spem omnium, nullam sensit posteà Appetitus Læsionem vel Debilitatem. Hodiè quin etiam Vir ferè Quinquagenarius, fruitur aurâ Vitali, atque in Ecclesiâ Novangelicanâ Diaconatu fungitur*:

The other was this.

Quæ Vulnera debeant verè Lethalia *vocari, inter* Cheirurgiæ *peritos orta est controversia, et Quæstio à medicis accuratissimè pensitanda. Omnia siquidem* Vulnera Lethalia *sunt* Insanabilia; An *tamen omnia* Insanabilia *dicenda* Lethalia? *Vulnera quæ* Suâpte Naturâ *(ità Philosophicè Terras Incognitas nonum exploratas habentibus balbutire licent,) mortis Januas aperiant; ideòque Illa quæ Spiritus Vitales vel dissipant omninò vel suffocant; Illa quæ subitò purpureum vitæ Liquorem à vasis majoribus et copiosè effundunt: Illa quæ totam statim Respirationem tollunt: Vulnera pro* Lethalibus *haberi solent. Venerandus ille senex* Hippocrates, *Verissimus medendi, ac ferè perpetuus Dictator.* Cujus Scripta *(sic stat Doctissimi Heurnij Sententia)* tanquam oracuta tota amplectitur Antiquitas, inter Lethaliter vulneratos *eos quibus* persectum est cerebrum, *non sinè Ratione pronunciat. At alibi illum limitat ac lenit* Aphorismum, *dum,* Αποθνησκουσι μαλιστα, fere, *vel* frequenter, *mortem cerebro vulneratis contingere affirmat.* Galennum *quidem talia vulnera curata vidisse, in aureo ejus* De Usu partium Tractatu, *legimus; at inter miracula istiusmodi curationem vir summus posuit. Catalogus imò non perbrevis Medicorum qui* cerebri vulnera *in integrum restituta observarunt, studio* Johannis Andreæ à Cruce, *ubi de cerebro agit sauciato, posteris relictus est. Annoque suprà millesimum, sexcentesimo octuagesimo secundo,* Jacobus Junius *Anglus, libellum nobis non adhuc visum edidit, ubi de* Cerebri vulnere *per eum sanato triumphos agit; et non minus quàm* sexaginta quinque *Authores, qui istiusmodi vulnera non esse semper Lethalia, plus quam centum comprobarunt observantionibus,[con?]scripsit; ut in hic vulneribus, tum Medicorum, tum ægrorum posthac non animi succumbant. Ac paulò majora! Alter ille Hippocrates, præclarissimum* Sennertum *volo, de pluribus qui post amissam* cerebri *particulam convaluerunt, narrantes bonâ cum fide Authores, recitat non contemnendos, cujus etiam è propriæ experientiæ Testimonio Historia, de sanitate periculosissimo cerebri vulneri redditâ, est fide et notatu digna.* Vulnera cerebri (*scribit à Vega*) si ad ventres perveniant, non curantur; sed alia multa vidimus sanata, quæ ad cerebri substantiam perveniebant, et amissâ cerebri ipsius portione. *At anteriorem* cerebri ventrem *vulnus* à Galeno *memoratum fertur penetrasse. Quid verò de puero quem* Zacutus Lusitanus *per tres Annos absque* ullo cerebro *vixisse asserit, sentiendum sit, per* modestiæ Verbum *planè dicam,* Nescio. *En, Historiolam Nov-Anglicanam ab hàc re non omninò alienam!*

Puella quædam, cui Nomen Abigail Eliota, *Anuum suæ ætatis quintum forsan aut sextum agens, sub adstante plaustro cum alijs puellis ludens, ad cardinem ibi ferreum cujus partes inferiores erant acuminatæ, suum ità allisit caput, quòd suprà Dextram ejus Aurem propè verticem, per cranium, et in ipsum cerebrum vulnus pertigerit; extractoque jam Instrumento ferreo* cerebri vulnerati *frustula adhærebant. Denudato deindè puellæ capite, per cranij perforati rimam apparuit, ad ovi fortè magnitudinem Tuberculum. De* scindendo *hoc Tuberculo* duplici in diversum scindebantur hamo *attoniti cheirurgi; at in cerebrum admittendi aeris metu ducti, in aliam hujusmodi operationem se potius contulerunt. È supradicto Tumore leniter per digitos compresso cerebri et sanguinis non parvas portiones expresserunt; adhibitis posteà emplastris vulnerarijs solitis et salutiferis. Sensim verò post Dies aliquot absumpta pars illa ubi terebatur cranium, à Numati coronati semisse vix tandem tegebatur. Res tamen fœliciter successit. Tenerrimæ vulneris* cuticulæ, *dicam, vel* cicatrici, *laminam quandam argenteam, quæ externas prohiberet Injurius imposuerunt.* Pro Maris Fluxu Refluxuque *dehinc assurgebat ac subsidebat puellæ* cerebrum. *Appropinquante (*Saxonicè *loquam)* Malinâ, *turgebat ejus cerebrum usque ad Laminam, per vulneris orificium, vacuo illo spaciò sic repleto.* Lidunâ *recurrente, intrà depressiores et contractiores limites se cerebrum continebat. In Maritam et Matronam, imò et duorum tandem Infantium Matrem, adolevit hæc puella; neque ullam omninò sensit Intellectus aut Memoriæ defectum; à secundi verò partu, statim Diem suum obijt.*[16]

I doubt, you are by this time almost weary of my *Latin*. I will impose no more of *that* upon you. But because, I suspect these Relations, may have

[16] See below for translation.

raised in you some Fresh Thoughts about the *Seat of the Soul*,[17] I will ask your Leave, to add One Story more, that is not wholly foreign to the Subject I have been all this while upon.

A Woman whose Name was, *Sarah Wilkinson* falling into a *Dropsy*, was extremely swelled before her Death.[18] When she drew near her End, and her Disease increased upon her, it was the Manner to lean her forwards, with her Head and Mouth downwards; and then her Mother would with a Feather tickle her Throat; whereupon she would bring up a great Quantity of Water; sometimes it may be, a Gallon; and sometimes not so much. After such an Evacuation she would find a present Ease; and in the Water that came up, there might be seen, many Long and Broad Skins, which were sometimes the Breadth of Two Fingers. At the first Appearance of these, her Mother would encourage herself, with supposing that knotts of Worms had been dissolved, & were in this Manner brought away. But notwithstanding the Evacuation aforesaid, her Dropsy brought her to her End. She retained her Senses to the Last; and spoke sensibly to the Day of her Death. After she died her Body was opened. At the first Stroke of the Knife, a vast Quantity of Water issued out. But when they had fully opened her, unto their astonishment, they found, that there were *no Bowels* at all within her![19] Only her *Heart*, which was very small; even like a Childs; & looking as if it had been parboild. And her *Spleen*, One End whereof stuck to her Back, the Other to her Ribs. There was also a very small Bitt, either of her *Liver* or her *Lungs*; but so corrupted, they knew not which of the Two it was. No other Bowels could be found; & not so much as a bitt of a Bladder. Nor had she long before her Death, any other Evacuation, but that of the Throat, which we have mentioned.

But inasmuch as *Rare Cures*, are the chief Matter of my Letter, I will take Leave to relate unto you, a very *Odd* One, of a *Dropsy*, which arrives to me, very well-attested, while I am writing this Letter; and such an one, as I durst say, the most *Mathematical Physician* among you, would find a Difficulty, in præscribing to any of his Patients, the Repetition of it.

17 René Descartes (1596–1650) argued for the soul's location in the pineal gland in the forehead. See Descartes' *The Passions of the Soul* (secs. 30–50), in *The Philosophical Writings of Descartes* (1:340–48). Mather mentions that he was reading Descartes (*D* 1:72) as a young man.

18 "Dropsy," i.e, edema, or swelling of body tissue caused by accumulation of water. See also Mather's "*Moses*. or, one *Drawn out of*, The Dropsy," *ABJ* (1972), Cap. XXIII, pp. 125–29. A short version of Sarah Wilkinson's story appears in *MCA* (6.2.10).

19 See also *ABJ* (1972), cap. V, item XV, p. 33.

Mrs. *Elizabeth Look*, of *Tisbury*, on our *Martha's Vineyard*, by taking a grievous Cold in Childbed, grew Hydropical. Her Dropsy increased upon her for Nine or Ten Years, in which time she brought forth successively Two Children. But after this, it arose to excessive Circumstances, wherein she continued for Twenty Years together. Her Body was now swollen into enormous Dimensions. When she satt down, her Belly hung out a great Way beyond her Knees, and rose up within four Inches of her Chin. She suffered grievous Pains in her Back and Loins; & was for diverse Years unable to go, or be carried out of Doors. Her Face and Limbs were in this time extremely lean; scarce any thing but skin & bone. Her Body ever so cold, that it felt like a Dead Carcase; & such as lay by her, found it necessary for them to rise in the Night, & go to the fire, that they might warm themselves. But on Oct. 30. 1708. when she was in the Sixty third Year of her Age, it happened, that she *fell down Stairs*, from the Top to the Bottom of the *Stairs*, that led up to her Chamber. By her Fall, she broke her left Wrist, & made a large Wound in her Forhead: but she received what made amends for it. She voided by the Urinary Passage, a Water as limpid as Rock-Water, which had neither the Colour, nor Scent, of Urine; and this in somewhat large Quantities, for about a dozen Dayes together. By the fifth Day, it was perceived, that her Bulk was very much abated. But in twelve or fourteen Dayes, having voided about a dozen Gallons of the Water, aforesaid, she was perfectly reduced unto her ordinary Size, & recovered her perfect Health, & continues to this day, an Hearty, Healthy, Lively Gentlewoman, and with rather a youthful Vigour, than any visible Decayes of old Age upon her. Her Ribs, which during her Illness, had been push'd out of their true Scituation, by degrees came into their former State & Shape again; and she has no Remembrance of her Malady left upon her, but only that the *Sternum*[20] which had been turned up, so continues.

Wishing you a perpetual Freedom from all that may wound or hurt you, I now take Leave. But remain,

Boston, N. England. Syr,

Nov. 24. 1712. Alwayes at your Service.

C. Mather.

[20] Breastbone.

Translation[21]

Not only in the immortal melodies of the Horatian lyre but also in the shared opinion of all mortals, death is called the end of the line.[22] If one attentively ponders on how light a touch is sometimes required for this line to find its limit and end, one cannot fail to sense that a constant contemplation and expectation of death and a diligent preparation for Life Eternal have been stirred up in oneself. On those occasions one also must truly marvel and piously acknowledge that one owes the very fact that one is alive at all to a gift of the greatest and most excellent God (to put it no less in Christian than in Ovidian[23] terms). And yet, on the other hand, there are times when we have to acknowledge that those very wounds which we call lethal do not always bring death to those who are at death's door, on those occasions when, rejoicing and not without astonishment, we sometimes see them recovering from desperate wounds.

It was in this way that Isaac Williams[24] recovered. A lad of three or four years, he wished to get involved in the trussing of his father's pigs. But on becoming exposed to the blows and bites of a furious boar, he had that part of his chest called the stomach so wounded and torn that the stomach itself burst forth and was now hanging down to his abdomen. Snatching the boy away, his wretched and quite terrified mother with her own hand thrust the horribly wounded stomach back into that astonishing opening of the wound as her child lay supine but, astonishingly, still conscious; and, with the same hand placed for some hours on the wound, she prevented further eruptions of the stomach, the whole time that her husband had gone several miles on foot in order to fetch

21 The editors express their deep gratitude to Dr. David Lupher, Professor Emeritus, University of Puget Sound, for translating the quicksand of Mather's Latin, for his expertise, and for unraveling even the most intricate mysteries of the ancients and moderns. Most of all we wish to thank him for going the extra mile with good humor and indefatigable attention to detail. What a pleasure to work with such a rare scholar! Thank you, David.

22 Mather is quoting from the last words of Horace's *Epistles* 1.16.79: "Mors ultima linea rerum est" ("Death is the end-goal of things"). The *ultima linea* was the white line marking the end of a racecourse. These words are often found on early modern funerary monuments. In the next sentence, Mather seems to interpret *linea* to mean the racecourse itself – not the mark at the end. That still makes sense, for the adjective *ultima* can refer to the final extremity of an extent of space; hence we have translated *ultima linea rerum* "the end of the line."

23 The reference is to the first poem of Ovid's *Tristia* (1.1.20), a collection often called *De Tristibus* in Mather's day. Writing from his Black Sea exile, Ovid was telling a friend to inform other concerned friends that the poet was alive, but not well, and to add "also this: that I consider the fact that I am alive a gift of a god" ("id quoque, quod vivam, munus habere dei").

24 Unidentified.

the doctor. God inspired the applications of the surgeon. For although over the course of fourteen days this boy had eaten and drunk nothing at all, and with a complete absence of immediate vomiting, he nonetheless subsequently arrived at an utter and complete state of health, and, beyond everyone's expectation, he felt no ill effects on his appetite or any weakness. Why, the fact is that today he enjoys 'the life-giving air'[25] as a man of about forty and serves as a deacon in the church of New England.

There has arisen among those skilled in surgery a dispute as to which wounds ought to be called lethal, and the question needs to be very carefully pondered by doctors. True, all lethal wounds are incurable. But should all incurable wounds be called lethal? Wounds which by their own special nature – this is how those philosophically considering unknown lands not yet explored are allowed to babble on – open the doors of death and thus utterly dissipate or suffocate the vital sprits; those which suddenly and copiously pour out the purple liquid of life from the larger vessels; those which suddenly take away one's entire respiration – these tend to be considered wounds virtually lethal. That estimable old man Hippocrates, the truest and almost perpetual dictator[26] of healing, whose writings (according to the opinion of the learned Heurnius)[27] all of antiquity embraced as oracles, assigns to the category of the fatally wounded those whose brain has been cut into – and not without reason.[28] But elsewhere he limits and softens that aphorism, when he affirms that αποθηνσκουσι μαλιστα: death usually or frequently comes to those wounded in the brain.[29] Indeed, we read that Galen saw such wounds cured, in his golden tract *On the Use of the Parts*;[30] but the great man placed a cure of this sort among miracles. But a quite extensive list of doctors who observed wounds of the brain thoroughly restored has been left to posterity by the efforts of Giovanni Andreae della Croce,[31] in the place where he deals with wounded brains. And in 1682, the Englishman

[25] The ablative phrase *aura vitali* recalls the *aurae vitales* of Lucretius (*De rerum natura* 3.405) and Vergil (*Aeneid* 1.387–88).

[26] "Perpetuus Dictator" is Mather's favorite trope to celebrate Sir Isaac Newton.

[27] Johannes Heurnius, aka. Jan van Heurne (1543–1601), a Dutch anatomist, physician, natural philosopher, and author of numerous medical works, including *Hippocratis Aphorismi, Prognostica, Prorrhetica, Coacae Praenationes, Aliaque Decem Opuscula* (1627), a popular work which went through numerous editions and reprints.

[28] Hippocrates, *Aphorisms* (6.18).

[29] Perhaps Hippocrates, *Coae praenotationes* (1621), bk 3, Treatise 2, aphorism 2, pp. 408–09, by Ludovicus Duretus. Most likely, Mather's source is Heurnius.

[30] Galen, *De usu partium corporis humani:* perhaps bk 8, ch. 10 (below).

[31] Johannes Andreae à Cruce (Giovanni Andrea Della Croce), Venetian physician and author of *Chirurgiae ... Libri Septem* (1573).

James Yonge[32] published a little book which we have not yet seen, in which he celebrates his success concerning a wound of the brain cured by him; and he lists no fewer than 65 authors who verified through direct observations that more than one hundred wounds of this sort proved non-lethal – to show that when facing these wounds the spirits of both doctors and sick people need not give up in despair. But now for somewhat greater matters![33] That second Hippocrates, I mean the most eminent Sennert,[34] notes that one should not dismiss authors writing with good faith about the many people who have become healthy again after losing a portion of their brains – indeed his own account of health restored to a very dangerous wound of the brain, derived from the testimony of his own experience, is worthy of trust and notice. 'Though if wounds penetrate the walls of the brain they are not cured,' as de Vega[35] writes, 'nonetheless we have seen many others cured which were penetrating into the substance of the brain, even when a portion of the brain itself was missing.' But Galen[36] mentions a wound

32 Jacobus Junius; i.e., James Yonge (see above). In his *Wounds of the Brain Proved Curable* (1682), pp. A10^{r-v}, Yonge lists his 65 authorities. His account of his cure of a child near Plymouth (England) with a discussion of their claims appears in more detail (pp. 63–100). Yonge's declared aim (here attributed to him by Mather), is to demonstrate "that improbable and discouraging accidents have been cured and that therefore we never ought to despair or esteem anything absolutely mortal" (p. 59). Yonge's main claim to historical fame is that he embalmed the battered remains of the great admiral Sir Cloudesley Shovell after the Scilly Naval Disaster of 1707.

33 The phrase *paulo maiora* derives from the first line of Vergil's 4th Eclogue: "Sicilides Musae, paulo maiora canamus!" ("Sicilian Muses, let us sing somewhat greater things!").

34 Daniel Sennert (1572–1637), German physician, alchemist, professor of medicine at the University of Wittenberg. Mather was drawing here upon Sennert's *Medicine Practica* (1632), vol. 1, pt 1, ch. 23, pp. 141–42. Though not directly quoting, Mather drew upon several phrases on this page. Sennert's personal account to which he alludes is the story of the cure of a carpenter (*faber lignarius*) who suffered an inch-deep head wound from an ax that fell on his head.

35 Christopher à Vega (c. 1510–c. 1573), Spanish court physician to Philip II and philosopher at the academy of Complutense. His *Opera Omnia nunc Denuo Publici Iuris Facta* (1626) is a huge commentary on the art of medicine as practiced by Galen and Hippocrates. Mather's second-hand quote appears in Vega's commentary on Hippocrates, *Opera,* lib. 5, Aphorism XVIII, p. 654, on the vulnerability of the cerebrum and of other organs, citing Galen's commentary: "If the wounds of the brain reach the stomach, they are not treated; but we have seen many others healed, which reached the substance of the brain, and lost a portion of the brain itself." For the same, see à Vega's *Commentaria in Librum Aphorismorum Hippocratis* (1563), bk 6, Aphorism 18, p. 475, Mather probably drew on à Vega at second hand from Yonge's *Wounds of the Brain,* p. 71, who cites most (but not all) of this passage.

36 Galen, *De usu partium corporis humani,* bk 8, ch. 8 (so Yonge, who quotes this passage).

which is said to have penetrated the back wall of the brain. As for what one should think about the boy Zacutus Lusitanus[37] claims lived for three years without any brain at all, well, I am going to modestly declare that I don't know. Behold, a little account from New England not entirely foreign to this matter!

A girl whose name was Abigail Eliot,[38] probably about five or six years old, while playing with other girls under a nearby wagon, banged her head against an iron prong there, the lower parts of which were sharpened. The wound reached into her skull above her right ear near the top of her head and into her brain, and when the iron instrument was withdrawn, bits of her wounded brain were clinging to it. Then, when the girl's head was shaved, there appeared through the cleft of the perforated skull a swelling of approximately the size of an egg. The astonished doctors were split by a double hook in two directions over the question of splitting this abscess. But driven by the fear of letting air into the brain, they resorted to another operation, of the following sort. They expressed considerable portions of brain and blood from the aforesaid tumor by pressing it gently with their fingers, afterward applying the usual health-giving wound-plasters. When, after some days, that part [of the plaster] was gradually worn away where the skull was being chafed, it [the hole] was scarcely covered by a half-crown piece. Nonetheless, the matter ended well. They placed upon the very tender skin or (as, I would call it, the scar) of the wound a silver plate, to prevent external injuries. Afterwards, the girl's brain would sink and rise with the ebb and flow of the sea. When the Malina[39] (to put it in Saxon terms) approached, her brain would swell though the hole of the wound all the way to the plate, that empty space thus being filled. When the Liduna returned, the brain confined itself within more suppressed and contracted limits. This girl grew up to be a married woman, a matron, and even the mother of two infants, nor did she feel any loss of intellect or memory at all. True, though, in her second birth-giving she suddenly met her end.

[37] Zacutus Lusitanus, aka. Abraham Zacuth (1575–1642), Portuguese Marrano physician in Amsterdam, author of his oft-reprinted *De Medicorum Principum Historia libri sex* (1642, 1649), a detailed description of diseases. In the present context, Mather cites from *Praxis Medica Admiranda* (1637), lib. 1, obs. 4, p. 5, "*Sine Cerebro vixit quidam tribus annis*," a fellow who lived without a brain for three years. See also *ABJ* (1972), cap. V, item XV, p. 33.

[38] Abigail Eliot. Her story is told in *MCA* (6.2.10).

[39] *Malina* and *Liduna* were rare Gallic (not Saxon) terms for incoming and outgoing tides. They were discussed in the *Disputatio Philo-Physio-Logica* (1645) by the German doctor and neo-Latin poet Andreas Rivinus (Bachmann). While there is no record of Mather's awareness of this eccentrically learned treatise, it would have appealed to him.

I.8. ASTRONOMICAL OBSERVATIONS.

November 24, 1712.

Letter. the 1st.[1]

To Richard Waller Esq.

Sir,

Your good and honourable Character is known beyond *Thyle*;[2] known in remote *America*: And it now obliges me to join you, with an excellent Friend, in the Communication of some *American Curiosities*, from One who is willing to testify his Respects to both of you,[3] by imparting to you, what he had at hand, that might not be altogether an improper, or ungrateful Entertainment for you.[4]

When I took up the Resolution of addressing both of you with something of an *American* Production, the Fleet was near Sailing, and I had little more Time left me, than to Resolve, that I would steal so much Time, from the public Employments, which find me near as much Work as one of the Three Sons of *Noah* had in their *Ark* to exercise them,[5] as to write one Letter *per*

[1] That is, the first letter Mather writes to Waller rather than Woodward.

[2] Aka. Thule, a region (says Pytheas of Massalia), which lies "six days' sail north of Britain, and is near the frozen sea" supposedly "the most northerly region in the world," as imagined by Strabo (*Geography* 1.4.2–3). And yet, Seneca (*Medea* 374–79) has the chorus image that some bold sailor might yet discover new worlds "so that Thule might not be the remotests of lands."

[3] Presumably John Woodward and Richard Waller, who served as secretary of the Royal Society (1710–14) when John Woodward left off.

[4] Probably Thomas Robie, F.R.S (1689–1729), Harvard 1711, an avid astronomer, self-taught physician, and correspondent with William Derham and Edmond Halley of the Royal Society, on meteorological observations, including lunar eclipses and an aurora borealis. See Stearns, *Science* (1970), pp. 426–42.

[5] That Noah and his sons were busy enough tending to the animals in the sizable Ark can be seen in Mather's annotations on Gen. 6 (*BA* 1:606–23).

Day, till I had passed thro' a Fortnight of a Correspondence,[6] than which there is none that I should more wish to cultivate.

Having taken my Pen in hand, the First Thing that I enquired was, *Whether I had nothing of an Astronomical, or any otherwise Mathematical Importance, in my American Collection*, that an accomplished WALLER may esteem worthy to be look'd upon? But here, at the very Beginning of my Enquiries, I found so sensible a Balk upon my Purposes, that I was upon the very Point of cntirely desisting from them.

Tis true, *Divines* have sometimes been Studious and Accurate *Astronomers*. The brave *Alsted*, was *both* of those, and every thing else. *Copernicus*[7] eternized, like the very Stars, by his most celebrated System of them, was a Churchman; and his learned Champion *Lansbergius* was, you know, a Minister.[8] You know, *Gassendus* and *Clavius*, were both of them, Doctors of Divinity.[9]

6 Mather spent several days (November 17, 1712 to November 29, 1712) composing his first set of correspondence (First Series) with the Royal Society.

7 Johann Heinrich Alsted (1588–1638), German Reformed theologian, Ramist logician, and millenarian, authored (among numerous other books on philosophy and physics) a highly respected *Encylopædia Septem Tomis Distincta* (1630), a systematic analysis of all knowledge, much prized by Mather and his contemporaries. See H. Hotson (2000). The Polish astronomer Nicolaus Copernicus of Thoruń, aka. Thorn (1473–1543), revolutionized the regnant geocentric cosmology of Ptolemy by developing his heliocentric system in his *De Revolutionibus Orbium Coelestium* (1543).

8 The Dutch Reformed theologian Johann Philip van Lansberge (1561–1632), Flemish astronomer, mathematician, geometrician, and disciple of Copernicus, tried to improve on his master's calculations with his own (incomplete) *Progymnasmatum Astronomiae Restitutae Liber I: De Motu Solis* (1628). Today, Lansberge is principally remembered for his *Triangulorum Geometriae Libri Quatuor* (1631) and his *Tabulae Motuum Coelestium Perpetuae* (1632), a calculation of planetary positions.

9 Pierre Gassendi (1592–1655) was a renowned French atomist philosopher, astronomer, mathematician, and Roman Catholic priest. Mather appears to refer to Gassendi's *Romanum Calendarium Compendiosè Expositum* (1654), a study of the Roman calendar. The German Jesuit Christopher Clavius (1538–1612) was an astronomer and mathematician who was involved in replacing the old Julian (lunar) Calendar with the Gregorian (solar) Calendar sponsored by Pope Gregory XIII in 1582. Clavius's most celebrated works is his *In Sphæram Ioannis De Sacro Bosco Commentarius* (1585), a highly successful commentary on medieval astronomy as taught by Johannes de Sacrobosco (c. 1195–c. 1256) at the University of Paris.

The Names of those English Doctors, *Wallis*, and *Wilkins*, and *Barrow*,[10] will not be forgotten as long as that Learning which is to be called *Real*,[11] ha's any Disciples in the British Islands. *Ricciolus*[12] himself, the Compiler of that Voluminous and Judicious Work, the *Almagestum novum*, was a Professor of Theology. Tis a Remark, which I remember, *Melchior Adam*[13] ha's, in the Life of his rare *Pitiscus*; *Illud mirandum, quod homo Theologus, in Mathematum Studijs, nullo nisi se magistro, eo usque progressus est, ut editis Scriptis, Disciplinæ illius gloriam, Magnis Matheseos professoribus præripuerit.*[14] But why do I remember Exemples at such a Distance, when you have an *Harris*[15] among yourselves at this Day; the Author of that matchless Work,

[10] The English clergyman, mathematician, and founding member of the Royal Society John Wallis (1616–1703) published several works on logic, mathematics, and geometry, including his *Operum Mathematicorum* (1657). One of the founders of the Royal Society, John Wilkins (1614–72), Anglican bishop of Chester, mathematician, and natural philosopher, whose anonymously published *Discovery of a World in the Moone* (1638) and *Of The Principles and Duties of Natural Religion* (1675), were of particular interest to Mather. The English theologian, mathematician, and Lucasian professor of mathematics at Cambridge, Isaac Barrow (1630–77), is credited for his contributions to the development of calculus, especially in his *Lectiones Mathematicae* (1664). Barrow was Isaac Newton's teacher at Cambridge.

[11] Mather appears to allude to the crumbling distinction between Theology as the "Queen of Science" (scientia = divine knowledge) and Natural Philosophy (scientia = empirical knowledge) as the "helpmeet" of Theology.

[12] The Jesuit priest, physicist, and astronomer Giovanni Battista Riccioli (1598–1671) was professor of theology at Bologna. His most famous work is his encyclopedic *Almagestum Novum Astronomiam Veterem Novamque Complectens* (1651), which also included a map of the moon's craters.

[13] Melchior Adam (c. 1575–1622), Silesian theologian, biographer, and literary historian, whose biography of Bartholomaeus Pitiscus appears in *Vitae Germanorum Theologorum* (1620), pp. 833–41; Mather's Latin praise appears on p. 840 and can be rendered: "It is a remarkable fact that a man who was a theologian, with no teacher in mathematical studies but himself, advanced to such a point that, upon the publication of his writings, he snatched away the glory of that discipline from the great men who professed to learning."

[14] Bartholomeo Pitiscus (1561–1613), the Silesian astronomer, mathematician, Reformed theologian, and court chaplain for Elector Frederick IV of the Palatinate at Heidelberg, best known for his *Trigonometriæ. Sive De Dimensione Triangulas* (1595, 1600) and his *Thesaurus Mathematicus* (1615). Pitiscus is remembered for coining the term "trigonometry."

[15] The Anglican clergyman and mathematician John Harris (1666–1719) is most famous for his encyclopedic *Lexicon Technicum: Or, An Universal English Dictionary of Arts and Science*, 2 vols. (1704). Mather puts him to good use in his own *CP* (1721).

the *Lexicon Technicum*, a *Doctor of Divinity*! There is a famous Divine[16] in our Nation, at this day, of whom tis wish'd, that he had confined himself unto the *Mathematicks*, for his Lectures on which, the World has been so much obliged unto him. And I can easily come into the Wish of *Tycho Brahe* on my *Pitiscus*, *Optarem plures ejusmodi concionatores reperiri, qui Geometrica gnaviter callerent; fortè plus esset in ijs circumspecti et solidi Judicij, Rixarum inanium et logomachiarum minus*.[17] The Men whose Business it is, to instruct Men in thc Way to *Heaven*, have also busied themselves in Studying the *Heavenly* Bodies and Motions; and the Church ha's not fared the Worse for their doing so. But my own Circumstances are well-known to be so far from those of *Homo unius Negotij*,[18] that no body can imagine my *Astronomy* to be such as to purchase for me One Acre in the most remote Nook of the Dispositions *Hevelius* was once making to carry on his *Selenography*.[19] And if my Acquaintance with the *Mathematicks* be no more than that of a *Well-wisher* to them, I shall easily find my Friends alwayes ready with my Apology.

Nor can it be expected, that while Colonies are yett so much in their Infancy as ours are, and have had so many *Serpents* also to crush while in

16 This Anglican divine is none other than William Whiston (1667–1752), successor of Sir Isaac Newton to the Lucasian professorship of mathematics at Cambridge. Like Newton, Whiston was an Arian (anti-Trinitarian), rejecting the dogma of the Trinity as a forgery of St. Athanasius of Alexandria, and promptly lost his academic appointment when his beliefs became public in his *An Essay on the Revelation of St. John* (1706) and in his five-volume *Primitive Christianity Revived* (1711–12), which rejected the doctrine of the Trinity as an Athanasian forgery. See also Whiston's Arian *A Collection of Ancient Monuments Relating to the Trinity and Incarnation* (1713) and his controversial *An Essay Towards Restoring of the True Text of the Old Testament* (1722). See J.H. Tuttle, "Whiston" (1912).

17 Melchior Adam quotes the Danish astronomer Tycho Brahe's (1546–1601) praise of Pitiscus, in *Vitae Germanorum Theologorum* (1620), p. 840. Brahe was unhappy with the Copernican heliocentric theory and offered his own modified geo-heliocentric theory in which the earth remained in the center of the universe but all other planets revolved around the sun. Be that as it may, Brahe's praise of Pitiscus reads, "I would wish that many of the following preachers were well-skilled in geometry; peradventure they [preachers] could be more circumspect in their discipline, and of a solid judgment, less of empty strife and contradiction in reasoning."

18 Mather's characteristic self-denigration suggests "man of one trade"; i.e., ignorance about things other than one's calling.

19 The German–Polish astronomer Johannes Hevelius (1611–87) of Danzig, renowned for his lunar topography, built a huge telescope (1641), and published his discoveries in his *Selenographia: sive Lunae Descriptio* (1647), the first detailed map of the moon. See also Mather's *WM* (1693), p. 20, *CP* (1721), pp. 38, 42, 44, 50, and *BA* (2:1070–72).

their *Cradles* as ours have had, they can be so circumstanced, as to produce many Acute Mathematicians, or allow them the Liesure for extraordinary Inventions and Performances. You are likewise enriched so notably, by the continual Accessions to your Treasures, from your own learned Men at home, that you can expect nothing from our Side the Water, that can add unto them. Nevertheless, I have this to encourage me, that if one of my *Indian* Countrey-men should present you with no better than a little String of *Wampam* [certain little *Beads* made of our *Shel-fish* & strung on a *Bracelet*, whereof some are *White*, and of those there go *Six* for a Penny; some are *Black* or *Blew*, and of them there go *three* for a Penny; the only ancient Coin among them;] it would find a kind Acceptance with you.[20]

And therefore, the first Thing that I shall mention, shall be an *Indian* Curiositie. As they have no *Division of Time* among them, except *Sleeps*, and *Moons*, and *Winters*, thus I cannot learn, that they have divided the *Stars* into Constellations. But yett it is extremely odd, That our *Indians* have called the *Stars* of our *Ursa Major*,[21] by the Name of *Paukunawaw*, that is, *The Bear*. This they did long before the *English* came into these Territories, & before they could have any Communication with any *Europæans*.[22] And all the World knowes, there is nothing of any Figure to determine the *Names* of the Constellations; and even for the *Names* we have, honest *Schiller* ha's as good Reason for his,[23] as the *Pagans* that went before us had for theirs.

You will but bestow a Smile upon it, if I tell you, that we have a Tradition among us, and you have it particularly asserted in a Book published *by* a Gentleman[24] in the Year 1674. with a Dedication to the *Royal Society*: That in

[20] For the same, see Mather's biographies of John Eliot, *TRR* (1691), p. 78; and *LDE* (1691), p. 70.

[21] The constellation "Great Bear" in the northern hemisphere. For contrast, see Mather's *WM* (1693), sign. B$^{2r.}$

[22] Mather provides the same explanation of Native timekeeping in his biography of John Eliot, *TRR* (1691), p. 81; and in *LDE* (1691), p. 73.

[23] See also *BE* (1683), "To the Reader" (n.p.). In his atlas of the stars, *Coelum Stellatum Christianum* (1627), the German astronomer and lawyer Julius Schiller (c. 1580–1627) replaced the pagan names of the celestial constellations with the names of biblical and early Christian names – names from the Old Testament for the southern hemisphere, and from the New Testament for the northern. See M. Mendillo (2022).

[24] *An Account of two Voyages to New-England* (1674), p. 53, by John Josselyn (fl. 1638–75), an English traveler, who relates that following the lunar eclipse of July 15 1668, "a Star between the horns of the Moon in the midst" were seen in the month of November 1668. The dedication "To the Right Honourable, and most Illustrious, the President & Fellows of the Royal Society" appears on p. A3$^{r.}$

the Month of *November*, 1668. there appeared a *Star*, before the Body of the *Moon*, and within the Horns of it. Tho' perhaps my valuable Dr. *Dickinson*, would count the Sleight he ha's cast upon all your *Hypotheses*, in his, *Philosophia Vetus ac Vera,*[25] a little countenanced by such a *Phænomenon.*

Perhaps, tis hardly worth mentioning, that in these *Goings down of the Sun*, (as you use to call our Parts of the World,) *we* take perpetual Notice, of a thing which your Mr. *Childrey* in his *Britannia Baconica*, ha's advertised *You* concerning. Tis, "That in *February*, and for a little before, and a little after that Month, about Six in the Evening, when the Twilight has almost deserted the Horizon, you shall see a plainly discernible Way of the Twilight striking up towards the *Pleiades* or Seven Stars, and seeming almost to touch them. It is to be observed any clear Night; but it is best *illuc nocte.*[26] There is no such Way to be observed at any other Time of the Year, nor any other Way at that time to be perceived darting up elsewhere. What the Cause of it in Nature should be, (saith Mr. *Childrey*,) I cannot yett imagine, but leave it to further Enquiry."[27]

And by informing you, Sir, that We take notice of it, I point you, to look for the Cause, higher than in our *Atmosphære.*

I shall at this time add but one thing more. You know of how great Use, the *Fictitious Epocha*, of the Famous *Julius Scaliger*, is in managing the affairs of your *Chronology*.[28] All the Rule to find the *Julian*

[25] Mather refers to the royal physician of King Charles II, Edmund Dickinson (1624–1707), whose atomist explication of the Mosaic creation account, in *Physica Vetus & Vera* (1702), furnished Mather with much useful information, in *BA* 1:357–83.

[26] "There at night."

[27] Joshua Childrey, D.D. (1623-1670), archdeacon of Salisbury, antiquarian and astrologer, is mostly remembered for his natural history *Britannica Baconica* (1661); Mather refers to pp. 183–84. The slightly changed quote is from the "Advertisements to the Curious and Ingenious." See also *CP* (1721), Essay II, pp. 16–17. Much the same reference to Childrey's discovery appears in Mather's CA 3.11 "*The Aurora Borealis*," to John Chamberlayne (February 1720), published separately as *VH* (1719), p. 2. And, "The *Evening Glade*; first taken notice of by Dr. *Childrey*," in *PT* 25.305 (1706), p. 2220. A summary of Mather's letter (along with extracts of his First Series of letters to the Royal Society) appears in "IV. An Extract of several Letters from Cotton Mather," *PT* 29.339 (January 1, 1714), pp. 62–71. The summary of Childrey appears on pp. 65–66 (Appendix 6).

[28] Mather appears to have in mind *Opus de Emendatione Temporum* (1583) or his later *Thesaurus Temporum* (1606), by the French Renaissance humanist Joseph Justus Scaliger (1540–1609), son of the famous Italian polymath Julius Caesar Scaliger (1484–1558). In both of these revolutionary works, but especially in his *Thesaurus Temporum*, the younger Scaliger demonstrates that a comparative basis between the chronologies of the Roman, Greek, Persian, Babylonian, Egyptian, and Jewish calendars has to be established to attain a meaningful alignment of the computations of time, epochs, and calendars of the ancients. See also A. Grafton (1975).

Period,[29] wherein you had instructed me, was what you know very well. To multiply the *Solar Cycle* by 4845, the *Lunar Cycle* by 4200, and the Indiction by 6916. Then Divide the Summ of Products by 7980, and the Remainder of the Division, (without having regard unto the Quotient) shall be the Year enquired. I think, I am the Master of a small *Table*, which by the Cycles of ʘ and ☾, and *Indiction*, finds the *Julian Period*, with more Certainty and Expedition than any I have yett litt upon. I will invite you to my poor *Table*; and if you have already a better, however you will accept my *Intention*, and excuse my *Ignorance*.

A Table to find the *Julian Period*.

For	Cycle ʘ	Cycle ☾	Indict.
1–	4845	4200	6916
2–	1710	420	5852
3–	6555	4620	4788
4	3420	840	3724
5–	285	5040	2660
6	5130	1260	1596
7	1995	5460	532
8	6840	1680	7448
9	3705	5880	6384
10	570	2100	5320
11	5415	6300	4256
12	2280	2520	3192
13	7125	6720	2128
14	3990	2940	1064
15	855	7140	7980
16	5700	3360	
17	2565	7560	

29 Attributed to Julius Caesar (100–44 BCE), who called for the reform of the Lunar calendars in use, fixing the year 753, the traditional date for the founding of Rome (*Ab Urbe Condita*), as the year 1 of his new Julian Calendar. This calendar was used in the Roman Empire until it was replaced in 1582 with the more accurate lunar-solar calendar of Pope Gregory XIII. However, in the Protestant United Kingdom and the British Empire, the Julian Calendar remained in official use until 1752, when eleven days were added to the Julian Calendar to attain parity with the Gregorian in continental Europe.

18	7410	3780
19	4275	7980
20	1140	
21	5985	
22	2850	
23	7695	
24	4560	
25	1425	
26	6270	
27	3135	
28	7980	

By the Cycles, find in the Three last Columns, their Respective Numbers. Their Summ compare with 7980. If the said Summ be less than 7980, it is the Year sought for; if it be greater, the Excess gives the Year sought for; of the *Julian Period.*

From *Ann. Jul. Per.* substract, *An. Dom.* rest, 4713. And *Ann. Jul. Per.* — 4713 == Ann. Dom. The first Year of the Christian Æra, is Ann. Jul. Per. 4714. whole Cycles here,

Cycl. ☉	—	10	—	570.
Cycl. ☾	—	2	—	420.
Indict.—	—	4	—	3724.
				4714.

Cycl. ☉	—	12.	—	2280.
Cycl. ☾	—	12.	—	2520
Indict. —	—	6	—	1596
Ann. Jul. Per.				6396.
Substract.				4713
Rest Ann. Dom.				1683.

Cycl. ☉	—	1.	—	4845.
Cycl. ☾	—	10.	—	2100.
Indict. —	—	8.	—	7440.
				14393.

Substract,	7980.
Ann. Jul. Per.	6413.
Substract,	4713.
Rest. Ann. Dom.	1700.

I need not say to you; you know very well, 28 x 19 x 15 = 7980. Within the Compass of which Period, no two Years, can have the same Cycles of ʘ and ☾, and Indiction.

But if the *American Rarities* wherewith I would fain have treated you, prove no better than what I have hitherto sett before you, methinks I can compare myself to nothing more considerable than the poor *Jew*, carrying about his *Osier-Basket*, (of whom *Juvenal* saies, *Quorum cophinus fœnumque supellex*;) the, *Cistifer*, derided by the Epigrammatist, as a disgrace to those that conversed with him.[30] I will therefore, tho' I descend unto *lower Subjects*, go on for five Dayes more, *& fulfil a Week* in your Service, if my *Studies* & my *Neighbours* will give me Leave, with *a Letter a Day*, to try whether I can find nothing else that may express & answer my Zeal to gratify you. Only I must aforehand pray you to pardon an Infirmity, that I find, that I do unawares and easily fall into; Which is, That while I am writing to a learned Friend, (tho' it be in the most *Autoschediastic*[31] Manner that can be) I find a Difficulty to forbear my little Excursions, into more of *Reference* and *Quotation*, than may be decent, in so free a Conversation. Indeed, I should be more ashamed of this Infirmity, if I were conscious, (which I find, I shall be, ere I {am} aware, of committing it, even in the very Moment when I am asking Pardon for it!) of its proceeding from the Humour which *Austin* reproached in *Julian*, who was ever citing of Authors, to make a vain Ostentation of his Erudition.[32] But,

[30] In his *Satire* (3.14), the Roman satirist Juvenal (c. 55–60–127 CE) lampoons a Jewish Roman, "who with his straw-lined baskets, head-downwards," slinks to the grove of Numa in the cover of the night to render his dues at the shrine of the nymph Egeria, the divine consort of King Numa Pompilius. Not unusual for Mather's linguistic playfulness, the Latin word *Cistifer* (a bearer of a ceremonial basket or casket as a religious rite), here used as an appositive, is not listed in the *OED*.

[31] I.e. improvised.

[32] Mather alludes to St. Augustine of Hippo (354–430), whose polemic *Contra Iulianum* (4.15.75) engages Julian (c. 386–c. 455), Pelagian bishop of Eclanum (Italy), in a prolific give-and-take of source citations. No matter what critics past and present have said about Mather's compulsion to pack his lucubrations with learned citations to conceal his inferiority complex, he is more than aware of this peccadillo and can poke fun at himself.

tho' I may plead the great Exemples which I find of *Reading* being made and thought much more pleasant, for being so *dulcified*; and the real Advantage, (espcially when the Subject is none of the Greatest,) of having what is *written* somewhat *Justified* at least, if not also *Embellished*, by a Style with some *Spirit* this way putt into it; I rather chuse to say, that you shall, if you please, in your Friend who now writes, lett it pass for one of those *Natural* Weaknesses, which all Men of Breeding indulge in one another. However, I will so far keep a Restraint upon it, that it shall be within the Bounds of *Mediocrity*; & be confined unto my *First Thoughts*, with Facility flowing thro' my Pen, at the Moment of their occurring, into these mean and course *Originals*, whereof I have had no time to take and keep a Copy. They are but a few of your and my *Subcesive Hours*, which these Letters are like to call for; and you will permitt me a little of the Air of a *Camerarius* in writing of them.[33]

I wish you infinitely more than a *Julian Period* of Happiness; and am, Ambitious to be allow'd the Title, of,

Boston, N. England. Syr,

Nov. 24. 1712.

Your Friend

& Servant,

Cotton Mather.

[33] Perhaps embarrassed by his stodgy preamble to his slender lines on Indian time-keeping, Mather facetiously associates himself with the German classical scholar, legal advisor, and polyhistor Philip Camerarius (1537–1624), son of the famous classical scholar Joachim Camerarius (1500–74), whose learned *Operae Horarum Subcisivarum* (1644) Mather had employed in previous letters to the Royal Society. He appears to allude to Camerarius' entertaining mixture of meditations and essays on political, civil, historical, scientific, and personal subjects in the manner of Michel de Montaigne's popular *Essais* (1588). A copy of Camerarius' volume was in the libraries of the Mather family as well as at Harvard. An English translation was published as *The Walking Librarie* (1610) as well as *The Living Library* (1625).

I.9. RAINBOWS & PARHELIA.

November 25, 1712.

Letter. 2d.

To Richard Waller Esq.

Sir.

In the Kingdome of *Meteors*, I know, you allow the RAINBOW to be of so singular a *Beauty*, as justly to challenge an universal *Wonder*. And you will not wonder at *Plato* for making *Iris* the Daughter of *Thaumas*;[1] a Name derived no doubt from / תמה / *Admiratus est*.[2] You do as the Son of *Syrach* advises, *look upon the Rainbow, & praise Him that made it; very beautiful it is in the Brightness thereof; It compasses the Heaven above with a Glorious Circle, and the Hands of the most High have bended it*.[3] You

1 In the first part of this letter, Mather appears to allude to "A Catalogue of *Simple* and *Mixt Colours*," *PT* 16.179 (February 28, 1686): 24–32, table, by the distinguished English naturalist and secretary of the Royal Society Richard Waller (d. 1715). Plato, *Theaetetus* (155d) and Hesiod (*Theogonia* 265); Thaumas (θαῦμα), signifying "wonder," wedded Electra (daughter of Ocean) and brought forth her daughter Iris, the rainbow (if we allow Hesiod's poetic license). A similar reference occurs in "The Eternal Truth of Scriptures" (bk 1, ch. 16), in *A Collection of the Works of Thomas Jackson, D.D.* (1653) 1:54, by Thomas Jackson (1579–1640), English theologian and president of Corpus Christi College. He writes, "the Ancient Poets saw more when they feigned *Iris* to be *Thaumantis Filia*, the Daughter, or (as we of this age would say) the Mother of Wonderment: the Messenger of the great God *Jupiter* and his Goddesse *Juno*."

2 From the Hebrew root תָּמַה [tâmmah], "to be amazed," "be astonished," "wondered at." Hence, Chaldean תְּמַהּ [t^{e}mahh], "miracle," "wonder" (Jer. 4:9) [Strong's ## 8539, 8540]. Thus the Latin phrase suggests, "[shall] be regarded with wonder, admired." Such linguistic derivations, still common practice in Mather's time, bespeak the belief in the primacy of Hebrew from which the languages of the Mediterranean world were thought to have been derived.

3 Mather quotes from the KJV translation of the apocryphal *Liber Ecclesiasticus* (*Wisdom of Jesus, Son of Sirach*) 43:11–12, attributed to Ben Sirach of Jerusalem, a book on Jewish ethical teachings, dated from the 3rd to the 2nd century BCE).

consider the *Sacramental* Importance of the *Bright Cloud*, (which is, you know, the Scriptural term for a *Cloud brightened* with a *Rainbow*;) and what is implied & assured by that *Faithful Witness in Heaven*. You see no need of concluding with *Aben Ezra*,[4] or of suspecting with our Dr. *Jackson*, that there was no *Rainbowe* before the *Flood*;[5] but you can list with the Army, wherein my *Calvin* is a *Standard-bearer*;[6] that can allow a *Rainbow* in the *Old World*[7] and yett now assign a *New Use* unto it; and see with the Eyes of *Danæus*,[8] that the *New Use* which it ha's in thc *Covenant* of God, no more denies the Antediluvian Existence of it.[9] *Quam si quis Boves ante Sacrificia*

4 The classical commentator on the Torah, Abraham ibn Ezra (Aben Ezra) of Tudela, Spain (c. 1089–c. 1164) takes issue with his Jewish colleague Saadiah Gaon (892–942 CE) and glosses on Gen 9:13, 14: "Behold, I now set a bow in the clouds. Saadiah Gaon's interpretation that the rainbow was previously in existence is incorrect." For, "if we would believe what the Greek scholars tell us, that the rainbow is produced by the sun's flames, then we must assume that after the flood God strengthened the sun's light. This is the way that a person who understands the sciences will interpret the light," in *Ibn Ezra's Commentary on the Pentateuch*: Genesis (*Bereshit*), p. 125.

5 As Thomas Jackson put it, "But as *Aristotles* Philosophie is imperfect, because it leades us not, either unto the First Cause, or Last End of all things; so it is fully sufficient to confute such Divines, as think there were Rainbows before the Floud. Which opinion hath no pretence of Scripture to enforce it," in *Collection of Works* (1653) 1:54. The French philosopher Pierre Gassendi (1592-1655) says as much in his *Exercitationes Paradoxicae adversus Aristoteleos* (1649), a copy of which Mather owned (Tuttle, "Libraries," p. 328) and may well have influenced his attitude when he poked fun at the Greek philosopher as the "fumivendutous *Aristotle*" (*TP*, p, 286).

6 John Calvin, *Commentary on Genesis* (on Gen. 9:13), "I do set my bow in the cloud," which reads in part: "From these words certain eminent theologians have been induced to deny, that there was any rainbow before the deluge: which is frivolous. For the words of Moses do not signify, that a bow was then formed, which did not previously exist; but that a mark was engraven upon it, which should give a sign of the divine favor towards men," in *Commentaries of The First Book of Moses* (1:299).

7 Here, Mather deleted "before the flood."

8 Lambertus Danaeus, aka. Lambert Daneau of Castes, France (c. 1530–95), French jurist and Reformed theologian, author of *Physices Christianae* (1576, 1580), which went through several editions and reprints.

9 Mather makes the same point about the antediluvian existence of the Rainbow, in his *WA* (1689), p. 47, and in his *BA* 1:635, 669–70, 688–90. See also his Mather's two sermons on the rainbow, in *TDR* (1712), esp. Preface, pp. ii–iv; and *CP*, Essay 13. *Of the* RAINBOW," pp. 54–58.

Lege Mosis prœscripta, vel Aquam ante Baptismum non fuisse colligat.[10] The *Mathematical Philosophy* of the *Rainbow* is what you are no Stranger to.[11] You have taken into your Consideration, the Essay first made by the Arch bishop of *Spalato*, to proceed mathematically, with Enquiries after the formation of it; in his Book, *De Radijs Visus et Lucis*,[12] written before the Beginning of the former Century;[13] And the Progress made by *Descartes*,[14] who took his Hints from *Antonius de Dominis*,[15] (but makes it none of his Custome to betray his Tutors,) & gave us a Theory of the *Iris*, from the Lawes of *Refraction*, which *Lucid Rayes* do suffer, in passing thro' Diaphanous Bodies. You have observed, how Mr. *Halley* comes after the French Philosopher, and showes how the *Cartesian* Problems were more easily solved than the Author himself imagined:[16] But after all, the Incomparable

10 Danaeus, *Physices Christianae*, Tract. 2, sive Dies 2, cap. 17, pp. 42[r-v]. The Latin quote translates, "It's as if one were to conclude that there were no oxen before the sacrifices commanded by Moses or that there was no water before there was baptism."

11 Mather puts this line and much of the following material to good use in his "Preface" to his *TDR* (1712), pp. ii–iv. Although Descartes and Huygens had also provided mathematical descriptions of the rainbow, Mather probably alludes to Sir Isaac Newton's observations on the refractions of light in a drop of water, in *Optics* (1704), Experiment XVI, esp. Prop. IX, Prob. IV, pp. 127–34.

12 *De Radiis Visus et Lucis* (1611), cap. 10–16, p. 43–74, by Marco Antonio de Dominis, aka. Markantun de Dominis (1560–1624), Dalmatian natural philosopher, archbishop of Split, and primate of Dalmatia and Croatia. His interest in lenses, telescopes, and the ocean tides led him to new discoveries, including the origin and causes of the rainbow.

13 Mather provides an expanded version of Spalato's, Descartes' and Halley's explanations of the rainbow in his *CP* (1721), pp. 54–55.

14 René Descartes (1596–1650), *Dioptrique, les Météores et la Géométrie* (1637). The French philosopher laid out his law of refraction, now known as Snell's law or Snell-Descartes's law, in which he demonstrated that the angular radius of a rainbow is forty-two degrees.

15 Marco Antonio de Dominis is best remembered for his treatise on the refraction of light *De Radiis Visus et Lucis* (1611) as the cause of the rainbow (see above). Sir Isaac Newton praised Dominis for arguing that the ocean tides are caused by the gravity of the moon, an idea opposed by Galileo. Dominis fled to England to escape the Inquisition but soon recanted his "errors" and returned to Roman Catholicism. Alas for the recusant, Dominis was arrested by the Inquisition and died in the Castel Sant'Angelo, in Rome. See R.E. Ockenden (1936); W.R.J. Shea (1970).

16 The English Astronomer Royal, mathematician, and physicist Edmond Halley (1656–1742) published his observations "*Some Queries concerning the Nature of*

Sir *Isaac Newton*, that *perpetual Dictator* of the learned World, finishes the Matter in his rare Treatise of *Opticks*,[17] & ha's not only shown how the *Bow* is made, but also how the Colours are formed; how the Rayes do strike the Senses with the *Colours*, in the Order which is required by their Degrees of Refrangibility, in the Progress from the *Inside* of the *Bow* to the *Outside*; the *Violet*, the *Indico*, the *Blue*, the *Green*, the *Yellow*, the *Orange*, & the *Red*. You will give me Leave, Syr, to speak in these Terms, of Sir *Isaac Newton*; tho' there be an *Hartsoecker* who violently falls upon him, and will have all the Suppositions of that great Philosopher, to be less worth than the *Occult Qualities* of the Ancients.[18] At the same time, no doubt, you have smiled at the *Spoil*, which this Observation may committ upon many pretty *Allegories*, which you find in some Ancient and Honest Writers, fetched from the *Three Colours* of the *Bowe*; and are glad, particularly, that *Basil*, ha's better Demonstrations for the Mystery of the *Trinity*.[19] These Things are so much better known to you, that I can make no Apology for the *Little Touch* I now give upon them, except it should be this; I was willing to have you think, that your *American* Friends also are not utterly unacquainted with them. And it may serve as a more decent Introduction unto my Mention of some Occurrences, which claim some Relation to the *Rainbow*, but are not so frequently exhibited. Occurrences whereof I was myself a Spectator, and One ha's already had the honour of being publickly celebrated, by

Light, and Diaphanous Bodies," *PT* 17.206 (December 1, 1693), 998–99. See also W. R. Albury (1971).

17 For the same, see Mather's *CP* (1721), p. 56. Newton's *Opticks* (1704), Prop. IX. Prob. IV, "*By the discovered Properties of Light to explain the Colours of the Rainbow*," pp. 126–34. Mather's specific reference is to p. 130. See also Newton's Prop. III, Prop. I, "*To define the refrangibility of the several sorts of homogeneal Light answering to the several Colours*" (*Optics*, pp. 91–93).

18 Nicolaas Hartsoeker (1656–1725), Dutch Mathematician, physicist, and natural philosopher, author of *Recueil de Plusieurs Pièces de Physique, où l'on fait principalement voir l'Invalidité du Système de Mr Newton et où se trouve* (1722). In his *Essay de Dioptrique* (1694), Hartsoeker also took issue with Newton's own theories on the refraction of light.

19 Basil of Caesarea, aka. Basil the Great, 4th-century Byzantine bishop and theologian, author of *On the Holy Spirit*, in which he famously posited that all three persons of the Godhead are *homoousion*, of the same being or essence, a position encoded into the Nicene Creed. Basil's allegorical association of the rainbow (Iris) with the Trinity appears in his "Letter XXXVIII. To his Brother Gregory" (*NPNFii* 8:139–40).

the learned *Senguerdius*[20] in his *Philosophia Naturalis*, which he published above Twenty Years ago.[21]

I remember, *Du Hamel's* Nicety, *Philosophos magno in Errore versari, dum putant illic formari Irident, ubi videtur*,[22] – in That th/e *Cloud* indeed stops the Rayes of the Sun like a Wall; but the *Tinged Particles* are between the Cloud & the Sun. I am now to tell you, that I have seen *Rainbowes*, in a *Sky* wherein we could see no *Cloud* at all; *Rainbowes* made at High-Noon; with *Parhelia*'s waiting on them also:[23] but usually a terrible Storm of Snow immediately succeeded.

If *Cartesius* thought such Appearances at *Rome*, not unworthy of a Record, in his Book, *De Meteoris*,[24] you will permitt my Tender of what we have had at *Boston*.

[20] The Dutch natural philosopher and professor of philosophy at the University of Leiden, Wolferdus Senguerdius, aka. Wolfgang Senkward, Wolferd Senguerd (1645–1724), is best remembered for his *Philosophia Naturalis* (1680, 1685). Senguerdius mentions that he had corresponded (March 5 1685) with Increase Mather on the phenomena of the parhelia. See Senguerdius' *Philosophia Naturalis* (2nd ed. 1685), pars tertia, cap. 6, p. 295: "… mihi literis communicatum fuit excerptum aliquod ex epistolà, ad Ipsum die 5 Martii Anno 1685, data Bostoniæ, à Crescentio Mathero, ibidem Ecclesiæ Christi Pastore Seniore, inter alia sequential continens verba; *Non-nullis ab hinc mensibus quatuor Parelia, nec non variæ Irides nobis in Bostonia apparuerunt; similia eorum, quæ Romæ*." In fact, the diagrams of the parhelia observed by Increase and Cotton Mather are surprisingly similar to the ones Senguerdius subsequently published in his *Philosophia Naturalis* (1685), pp. 292 and 296, esp. 296. See O.T. Beall (1961), p. 370.

[21] In his biography of Increase Mather, Cotton Mather (years after his own letter to Waller) relates that in 1679, his father "formed a *Philosophical Society* of Agreeable Gentlemen, who met once a Fortnight for a Conference upon Improvements in *Philosophy* and Additions to the Stores of *Natural History*; From which the Learned *Wolferdus Senguerdius* a Professor at *Leyden* had some of the Materials, wherewith his *Philosophia Naturalis* was Enriched" (see second edition of Senguerdius [1685], p. 296). Then taking credit himself, Cotton Mather related that "in part from *their* [Boston Philosophical Society] Collections also One [Cotton Mather] that had a share in that Combination, and is now a Fellow of the Royal Society in *London*, afterwards transmitted Communications thither," in *Parentator* (1724), p. 86.

[22] Jean-Baptiste du Hamel (1624–1706), French natural philosopher, physician, astronomer, and secretary of the Académie Royale des Sciences in Paris. The Latin quotation is from Du Hamel's *De Meteoris et Fossilibus* (1660), lib. 1, cap. 5, pp. 72–73: "that philosophers are stuck in a great error as long as they suppose that rainbows are formed in the spot where they appear."

[23] See Beall (1961).

[24] René Descartes, *Les Météores*, published in his *Discours de la Méthode* (1637), "Discourse Dernier," pp. 281–94. See also S. Werrett (2001).

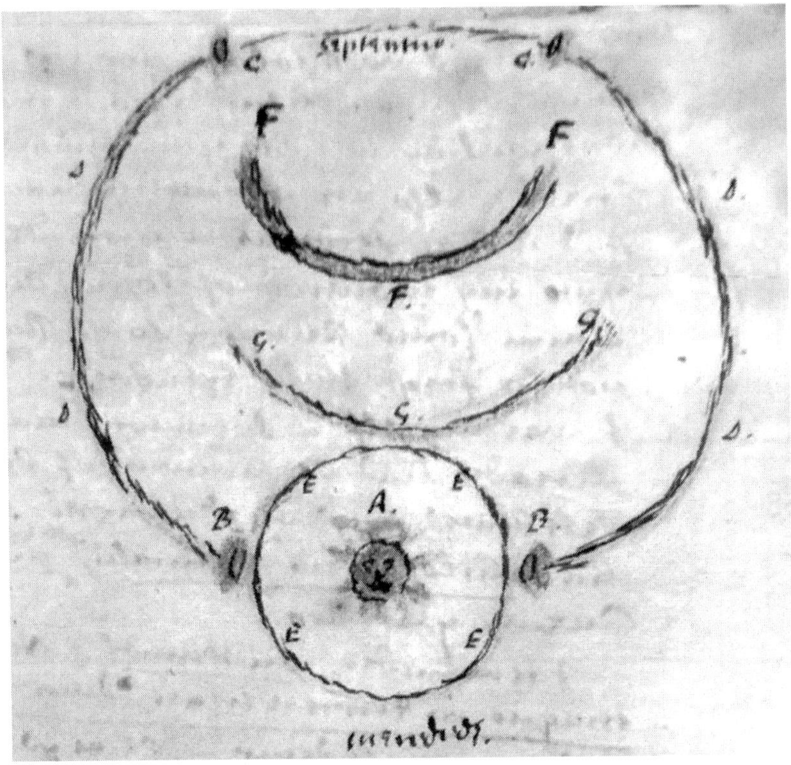

Figure 6a. Rainbows and Parhelia (Courtesy of the Royal Society of London).

On the Second of *January*, in a clear Sky, but an extremely cold Season, the Sun was from Ten a Clock, for the best Part of Three Hours, attended with Four Mock-Suns, thro' which there passed a white Circle; in the Midst whereof were Two *Rain-bowes*, according to the Repræsentation in the ensuing Schæme.

A. The Sun.
B.B. Two *Parhelij*, a little Distant from the Sun. Between which and the Sun, the Part of the Great Circle made by the Sun, was not so dense and conspicuous, as the rest of the Circle was.
C.C. Two Smaller and Fainter *Parhelij*, near our Zenith.
D.D.D.D. A Great Circle, passing from the *Sun*, thro' the Four *Parhelij*.
E.E.E.E. A Lesser Circle, having the *Sun* for the Center of it.
F.F.F. A *Rainbow* in the Midst of the Great Circle; itself consisting of more than a Semicircle.
G.G.G. Another, but a more dull, *Rainbow*; made, as it should seem, from a Reflection of the former.

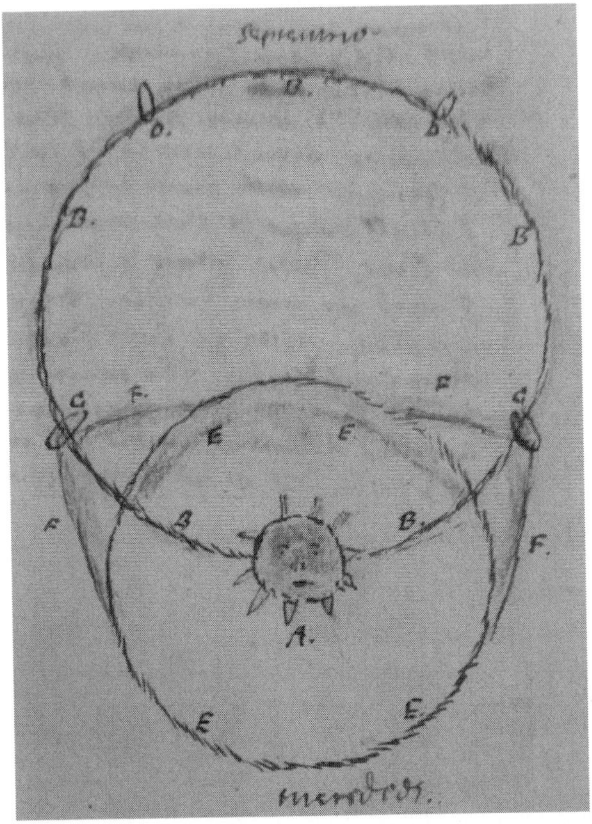

Figure 6b. Rainbows and Parhelia (Courtesy of the Royal Society of London).

About Six Weeks after this, on a day much colder than we use to have at that time of the Year, and in the Sky a little hazy, there appeared for about half an hour together, about one a Clock, Four *Sun-Dogs* attending the *Sun*, with such Circles, as you have in the ensuing Diagram.

A. The Sun.
B.B.B.B. A Great Circle; The Bottom of which cutt the Sun. The Top extended considerably to the Northward of our Zenith.
C.C. The Two Larger *Parhelia*'s.
D.D. The Two Smaller.
E.E.E.E. A Lesser Circle, which had the Sun for the Center. The upper Part of it, had the Figure, & all the Colours of the Rainbow. The lower Part, had no other Colour, than the Larger Circle; which was that of the ordinary *Halo*.
F.F.F.F. Rainbow-coloured Arches, passing from the Lesser Circle, and terminating on the Greater, a little within the Two Larger *Parhelia*'s.[25]

[25] See also Aristotle's *Meteorologia* (bk 3, chs 2–3).

I cannot say, how often we have been entertained with these Appearances. But it is a little surprizing, that they come with so much *Variety*. I do not know, but the Exhibition ha's alwayes been various, & with some notable Difference from what we saw before. I shall impose but one more upon you; which we saw, Two Years after the former. And I need no more employ any *Explaning Letters* on this, after the Two Foregoing Schemes have been considered. The main point of its difference from the first, was that of the faint inverted *Rainbow*, under the main strong one in the Middle of the Larger Circle.

Our People of a more *Prognosticating* and *Superstitious* Temper, would fain have made these Appearances to be *Ominous*, of something or other, if they could have told, *of what*? (A learned Writer, whom we consulted, said, *Changes* in the *Government*.) And you may be sure, *something did happen after them*![26] However, all the *Omen* that wiser Men could find in them, was that of, a *Storm* now a breeding; which, I think, alwayes happened.[27]

But, I hope, this Exhibition of them, will carry no such *Omen* with it. Your Goodness will *avert* it. Your *Wisdome* causes your *Face* alwayes to *Shine* upon your Friends; to look with a Serene, Benign, *Uncloudy* Aspect upon them. And, if they are emboldened from the Notice they see you take of such Things in the *Philosophical Transactions* of your more than Noble Society, to treat you with further Communications that may enlarge the Stores of your *Meteorology*, you know how to favour their Action, with a kind & candid Acceptance.[28] And therefore *this* of,

Boston. N. England, Syr,

Nov. 25. 1712.

Your *American*,

but most Affectionate,

Servant;

C. Mather.

[26] The Glorious Revolution in England (1688) and in New England (April 18 1689).

[27] Although he was not beyond reading celestial phenomena as omens of things to come, especially when he tried to arouse the attention of his parishioners for soteriological purposes, Mather had since learned that such superstitious explications were not welcomed by the Royal Society – at least not since the publication of *A Discourse concerning Prodigies* (1663, 1665), by the distinguished scholar John Spencer (1630–93), fellow of Corpus Christi, Cambridge. See also M.P. Winship (1994 and 1996).

[28] There were at least fifteen articles on parhelia and mock suns in *PT* during Mather's lifetime – all responding to Aristotle's discussion in *Meteorologia* (bk 3, chs 1–6) in one way or the other.

I.10. JOSEPH BEACON'S VISION OR DREAM.

November 26, 1712.

Letter. 3d.

To Richard Waller. Esq.[1]

Sir.

Having been informed, that you were desirous to be confirmed in the Truth of a Relation which had reached you, concerning an Apparition to One Mr. *Joseph Beacon*,[2] a Gentleman in my Neighbourhood, I address myself to discharge that Part of my Duty.

1 Mather's collection of stories from the invisible world responds to the call of Boston's ministers at the Cambridge Association (March 5 1693/94), "To Observe and Record the more Illustrious Discoveries of the Divine *Providence*, in the Government of the World," especially such memorable occurrences "well attested with Credible and Sufficient *Witnesses*" of "*Apparitions, Possessions, Enchantments,* and all Extraordinary Things, wherein the Existence and Agency of the *Invisible World*, is more sensibly demonstrated," in *TIC* (1699), pp. 71, 72. This call reiterated in substance Increase Mather's proposal first published in his *ERIP* (1684), "*The Preface*," sign. A^{7v}. Cotton Mather deploys the same Joseph Beacon story (albeit in a somewhat shorter form) in several other publications: *WIW* (1693), pp. 79–81; *MCA* 6.7.77–78; *TP* (c. 1726; 1995), pp. 115–16—all in an effort to combat the Hobbesian materialists and psycho-pannychists (soul-sleepers) of his day. In the Royal Society's abstract of Mather's "Curiosa" (Series 1712), *PT* 29.339 (June 30, 1714), p. 66, the recorder summarized the main points about Beacon, but concludes that "this [letter] is not directly related to Natural Philosophy, I omit the Particulars, tho' the Relation seems to be well attested." See also, J.P. Quincy's (1906–07) anecdotal defense of Mather's story. Mather reports another sighting of an apparition of a dead person in Boston (November 15, 1716) and hopes to employ the manifestation for pious uses. As usual, he sets out to "obtain a full Relation of the Matter; and have the Persons concerned therein, to make Oath unto it, before a Magistrate" (*D* 2:383).

2 According to Joseph Beacon's holograph letter "Fraternum Donum," written "in the year 1688," and addressed to his elder brother Nathaniel – then in London – the younger Joseph was deeply worried about the spiritual welfare of his sibling, the temptations lurking in the big city, and the kind of friends he associated with far away from his native Boston. Joseph's loving letter amounts to a premonition and

The Relation is this.

On the Second of *May*, in the Year 1687. a most Ingenious and Religious young Gentleman, Mr. *Joseph Beacon* by Name, about Five a Clock in the Morning as he lay, whether sleeping or waking, he could not say, (but judged, the Latter of them,) had a View of his Brother then at *London*, altho' he was himself now at our *Boston*, distanced from him the whole Breadth of the Great and Wide *Atlantic* Ocean. This his Brother appeared now unto him, having on him a *Bengal* Gown, which he usually wore, with a Napkin tied about his Head. His Countenance was very Pale, Ghastly, Deadly; and he had a Bloody Wound on one Side of his Forehead; *Brother*! saies the affrighted *Joseph. Brother*! Answered the Apparition. Said *Joseph, What is the Matter, Brother! How came you here*! The Apparition replied, *Brother, I have been most barbarously and inhumanely murdered by a debauch'd Fellow, to whom I never did any Wrong in my Life*. Whereupon he gave a particular Description of the Murderer; adding, *Brother, This Fellow, changing his Name, is attempting to come over unto* New England, *in* Foy *or* Wild. *I would pray you, on the arrival of either of these, to gett an Order from your Governour, to sieze the Person, whom I have now described, & then do you prosecute him for the Murder of me your Brother. I will stand by you, in the Prosecution*. And so he vanished. Mr. *Beacon* was extremely astonished at what he had seen and heard; and the People of the Family where he sojourned, observed an extraordinary Alteration, which continued many dayes upon him; in which time he also gave them an Account of this Apparition. All this While, Mr. *Beacon* had, no other Advice of any thing amiss attending his Brother, at such a Distance from him. But about the latter End of *June* following, he understood by the common Wayes of Communication, that some time in the præceding *April*, his Brother going in some haste by Night, that he might call a Coach, for a Gentlewoman, then with Child, whom he was attending upon, mett a Fellow, who had his *Doxy*[3] in his Hand, and too much *Drink* in his Head. Some Way or other, the Fellow thought himself affronted, in the Hasty Passage of this *Beacon*, and immediately with a Transporting Fury, ran in to the Fire-side of a Neighbouring Tavern; from whence he fetch'd out a Fire-fork, wherewith he grievously wounded *Beacon* on the Skull, even in that very Part, where the Apparition shewed his Wound. Of this Wound he languished until he died, on the Second of *May*, about five of the Clock in the Morning at *London*; so

is a testament to his own spirituality and mutual obligation to remind Nathaniel of his duty to God (MS Am 527, Houghton Library).

3 "A beggar's trull or wench: hence, *slang*, a mistress, paramour, prostitute; *dialect*, a wench, sweetheart" (*OED*).

that the Spectre, it seems, took the same Time, that the *Sun* takes, to pass over the Degrees of *Longitude*, into *America*. The Murderer, as we are told, was endeavouring the Escape, which the Apparition affirmed; But the Friends of the Deceased *Beacon* in *London* seized him; and pursued the Law upon him; but he found the Help of such Friends, as brought him off, without the Loss of his Life.[4]

Now for the *Evidence*, with which I can offer you this Relation; First; Mr. *Joseph Beacon* gave it me, written and signed with his own Hand, a little before his own Death; which follow'd not many Months after these things had happened. Now, he was known to be a Gentleman of such Discretion, & such Piety, and at the Time of his giving me this Relation, (the Original of which, I keep in my Possession,) he had his own Death so much in View, that his own Testimony must be very credible. But then, Secondly; The Gentlewomen of the House where Mr. *Beacon* lived, supplied me with this Additional Testimony.

> In the Beginning of *May*, 1687. Mr. *Joseph Beacon* told us, of an Apparition of his Brother *Benjamin*, as murdered; and seemed for a Week following much disordered with the Thoughts of it. And about the latter End of *June*, he had his Letters of his Brothers Death. This we attest.
>
> *Elizabeth Sharp,*
> *Elizabeth Thompson.*[5]

I know not, whether for a Story of this Nature, a fuller *Evidence*, can be with any Reason asked for. But the *Matter of Fact* being thus incontestable, we are still in the dark, about the *Manner*, wherein *Spirits* do make these *Apparitions*. There is that, which we may call, an *Internal Apparition of Spirits*. The Ancients called such things, *Phantasmata*, and, *Idola*; (not because, as a

4 In his *WIW* (1693), p. 81, Mather adds, "This History I received of Mr. *Joseph Beacon* himself; who, a little before his own Pious & Hopeful Death, which follow'd not long after, gave me the Story written and signed with his own Hand, and Attested with the Circumstances I have already mentioned." See for the same in *MCA* 6.7.78. Attestations of this type (especially when given just before a person's death) more than lived up to the standards of verification of the time. They are legion in Joseph Glanvill's ever-popular *Saducismus Triumphatus* (1681, 1689, 1700). Perhaps most famous in Mather's time was Daniel Defoe's oft-reprinted *A True Relation of the Apparition of one Mrs. Veal* (1706) as well as Defoe's more analytical *An Essay on the History and Reality of Apparitions* (1727).

5 Elizabeth Sharp (1671–1731), native of Boston, married Samuel Craft in 1693, lived in Roxbury, and had six children. Elizabeth Tomson (1654–1717) married Thomas Swift of Weymouth in September 1687, with whom she had seven children, and died in Barnstable.

Foolish Philosopher of *Malmsbury* will tell you, the things are nothing more than a meer *Fancy*;[6] but) because they thought, *Spirits* did appear nextly to the *Fancy*. They operate immediately upon that *Internal Faculty*; & exhibit themselves meerly by smiting on that, & moving of it; And so are as well to be seen when the *Eyes* are *shutt*, as when they are *open*: as well *sleeping* as *waking*. But then, there is that which we may call, an *External Apparition of Spirits*; wherein the Spirits do really present themselves unto the *Ocular View* of those to whom they make their Visits. Dr. *Burthogge*[7] thinks, The Spirits make the *Bodies*, wherein they thus appear, by the *Plastic Power of Imagination*; (whereof the Signatures made by the *Imagination* of the Mother on a *Fœtus*, are a *Prægnant* Instance!) and that what *Gravity* is in respect of elementary *Bodies*, *That* a strong *Inclination*, *Disposition*, *Passion*, or *Will* is unto *Spirits*. We seem to be left at a Loss, which of these, was *Beacons* Apparition? Yea, you are well aware, Syr, That it is not easy alwayes to determine,

6 Mather alludes to the English materialist philosopher of Malmesbury, Thomas Hobbes (1588–1679), who relegated such spiritual manifestations to the fickleness of the human fancy and decaying memory, in *Leviathan* (1651), part I, ch. 2, p. 7; ch. 12, p. 53; ch. 34, pp. 207–08. Hobbes dismisses such apparitions – whether in dreams or in the state of being awake – as nothing else but creatures of a runaway imagination. To be sure, Hobbes does not deny God's power to make incorporeal apparitions become visible, but Hobbes questions that they happen as often as men claim they do. As a true Epicurean materialist, he argues that ghosts or apparitions, which are incorporeal, cannot exist, for "the Word *Body*, in the most generall acceptation, signifieth that which filleth, or occupyeth some certain room, or imagined place; and dependeth not on the imagination, but is a reall part of that we call the *Universe*." Accordingly, "the word *Substance* and *Body*, signifies the same thing; and therefore *Substance incorporeall* are words, which when they are joined together, destroy one another, as if a man should say, an *Incorporeall Body*" (pt 3, ch. 34, p. 207). See also P.C. Almond (1994) and J. Bath and J. Newton (2006).

7 The English physician and natural philosopher Richard Burthogge (c. 1638–c. 1698) of Devon studied medicine at the University of Leiden. His ideas were shaped by Locke's sensationalist philosophy as well as by those of the Cambridge Platonists Henry More and Ralph Cudworth, but he rejected the concept of innate ideas. Mather appears to have been familiar with at least one of Burthogge's works: *Organum Vetus et Novum* (1678), *An Essay upon Reason, and the Nature of Spirits* (1694), or *Of the Soul of the World, and of Particular Souls* (1699). Burthogge argues that there is a Platonic "World Soul" which permeates and animates everything, that spontaneous generation is proof of such an anima mundi, and that the souls of humans and animals are propagated by this world soul (*Of the Soul of the World, and of Particular Souls*, esp. pp. 39–45). Burthogge's intriguing discussion of spirits and apparitions, in his *An Essay Upon Reason and the Nature of Spirits* (1694), ch. 7, sects. 1–4, pp. 161–235, is particularly relevant here. See M.W. Landes (1921). On the spontaneous generation controversy, see J. Farley (1977).

whether the *Spirits* appearing in the Shapes of *Deceased Persons*, are truly the *Spirits* of those Persons, or may not be other *Dæmons*. You may Remember perhaps, how *Chrysostom* in one of his Sermons about *Lazarus*, affirms, that oftentimes *Dæmons* would appear, pretending to be the *Souls of the Dead*.[8] And what shall we make of this; That the *Shapes* of Persons have been seen in distant Apparitions, while the Persons themselves have been yett *alive*? Of this, Mr. *Baxter* gives you a famous & a modern History.[9] I am not yett apprised of any Infallible Rules to Distinguish the Real *Apparitions* of the *Dead*, from *Diabolical Personations*. Those that are given by *Malderus*,[10] I suppose, will not satisfy you. However, the *Apparitions* are many times made, in such *Circumstances*, & on such *Occasions*, that, methinks, there is little Cause to quæstion, whether they are of the *Spirits* they pretend to be. Those of the murdered Persons, related by *Webster*, are truly Remarkable & very memorable.[11] That of Dr. *Brettons* Wife, that of *Haddock*, and that of *Avon*, related by *Glanvill*, are each of them worthy of Consideration.[12] *Beaumont*

8 The story of Lazarus being raised from the dead appears in John 11:1–44 and is expounded in St. Chrysostom's *Four Discourses of Chrysostom*. Mather alludes to "Discourse II. Concerning Lazarus – And that the Souls of those that Die a Violent Death do not become Wandering Spirits" (chs 1, 2, pp. 40–42), by the early Church Father John Chrysostom (c. 347–407), archbishop of Constantinople.

9 Mather was well familiar with *Of the Nature of Spirits* (1682) and *The Certainty of the Worlds of Spirits* (1691), both by Richard Baxter (1615–91), a leading English Puritan pastor, theologian, and controversialist. In *Certainty*, ch. 7, pp. 147–51, Baxter relates the story of Mary Goffe, who lay dying in her father's house at West Mulling, yet her spirit appeared at Rochester "about nine Miles distant," where the nurse of Goffe's infants saw Mary playing with her children. Baxter was a friend and correspondent of the Mathers and prefaced the second edition of *MPW* (1689) in Cotton Mather's *MPI* (1691) as follows: "*If any are Scandalized, that* New-England *a place of as serious* Piety *as any I can hear of under Heaven, should be troubled so much with Witches, and with Melancholios, and self-murders, as* Mr. [Cotton] Mather *tells us, I think it no wonder: where will the Devil shew most malice, but where is hated and hateth most …*" ("The Preface").

10 Johannes Van Malderen (1563–1633), Roman Catholic bishop of Antwerp, who has much to say on angelic and demonic manifestations in his *Commentaria, De Sancta Trinitate, Creatione in Genere, et Angelies* (1634), fols 139–26.

11 The English clergyman, physician, and skeptic of witchcraft John Webster (1610–82), like his Dutch colleague Balthasar Bekker in his *De Betoverde Weereld* (1691), elicited numerous attacks from pneumatologists who could not abide their Cartesian denial of demons and witches, in Webster's descriptively titled *The Displaying of supposed Witchcraft* (1677). See also M. Hunter (2017).

12 For these and the following examples Mather more than likely dipped into his father's *ERIP* (1684), ch. 7, pp. 221–22, 224–25, and 235–36. Increase Mather

ha's lately published some such Relations, as are most Unquæstionable; and as Admirable.[13] That of Sir *Charles Lee's* Lady, must alwayes cause admiration.[14] Your whole City a while ago, the Greatest City in the World, saw the Execution of several Murderers, all detected by an Apparition; and there is no Cause to think, it was not of the Person murdered. But, my honoured Friend, will excuse me, if I now press a little more of my *American* Entertainments upon him. *One Dish* is too little for a Guest, that I have such a Value for!

There have been some in the World, so unadvised, as to make *Agreements* that the first of the so Agreed, who *Died*, should after Death *appear to the Surviver*.[15] The *Compact* between *Marsilius Ficinus* and *Michael*

(p. 220) himself acknowledges as his source none less than Joseph Glanvill's *Saducismus Triumphatus*, second edition (1682).

13 *An Historical, Physiological and Theological Treatise of Spirits, Apparitions, Witchcrafts, and other Magical Practices* (1705), by John Beaumont (c. 1650–1731), an English physician, geologist, and member of the Royal Society (F.R.S.). See J. Barry (2012).

14 The apparition of Sir Charles Lee's first wife appeared to their daughter shortly before her marriage to Sir William Perkins, warning her that "by Twelve of the Clock, that Day, she [the daughter] should be with her." The nightly manifestation (or was it a premonition?) came true when physicians and surgeons were summoned to perform the usual procedure of bloodletting, but the daughter died at the appointed time and was buried next to her dead mother. The story, handed down to Beaumont by none less than William Nicholson, Lord Bishop of Gloucester (1660–72), dates to 1662 and appeared as "A Postscript," in Beaumont (pp. 398–400). In light of such revelatory nightmares, we should not be surprised that when the apparition of the young Rev. Thomas Shepard (1658–85), minister of Charlestown and Mather's friend, accosted him in a dream, "*Syr, you need not be so shie of me, for you shall quickly bee as I am, and where I am*," Mather came down with "a Fit of my *Ephialtes*, which almost killed mee. But I was immediately upon this, arrested with so violent a *cold*, as threatened mee with a mortal *Feavour*" (*D* 1:129–30).

15 Such stories were ever popular for more reasons than one. They proved (seemingly beyond doubt) that there was a world of spirits (and souls) beyond (*MPW* [1689], sec. XXXI, p. 39). They were vital to all those who set out to combat the Hobbesian materialists and modern Sadducees ("the *Soul-killing Swine* of our Days"), who denied the existence (let alone) immortality of the soul. Mather included several such stories in *TP* (1995), pp. 116–19, 121. Benjamin Franklin (1706–90) tells of a similar agreement with his old chum Charles Osborne, "that the one who happen'd first to die should, if possible, make a friendly visit to the other, and acquaint him how he found things in that separate state." Alas, Franklin confessed, Osborne "never fulfill'd his promise" when he died young in the West Indies. *Autobiography of Benjamin Franklin* (p. 27).

Mercatus:[16] and that between Major *Sydenham* and Captain *Dyke*; are Exemples that have been much discoursed on.[17] But besides the *Criminal Curiosity*, which may ly at the bottom of such *Agreements*, & the horrid *Snares* laid for the People that come into them, there is this to discourage wise People from them. The *Apparitions* which follow upon them, are usually attended with *Tragical Consequences*, to those who are addressed with them. *Lavater* and *Goulartius* have told you, what Harbingers of *Mortality*, the Apparitions of the *Dead*, use to be unto those that are visited with them.[18] They come usually to *call away*, such as they have to deal withal.

I will now go on to tell my Story.

In the latter End of the Year, 1695. a young Woman in *Boston*, whose Name was *Mehetabel Warner*, had a free Conversation with another young Woman, whose Name was, *Mary Johnson*. The Latter of these being somewhat sickly, fell into discourse with the Former, about, *The State of our Spirits after Death*.[19] The Discourse proceeded unto so much of Curiosity, that *Mary* proposed unto *Mehetabel*, an Agreement between them, That which soever of them two should *Dy first*, would come to the Surviver, and lett her know, whether there be a *State of Separate Spirits*, and whether

16 Their story appears in *Epitome Annalium Ecclesiasticorum Caesaris Baronii* (1602), p. 476 (*B-D*), of Cardinal Cesare Baroni (1538-1607), the Italian historian, and relates how the Florentine Marsilio Ficino (1433–99), the famous Roman Catholic priest, first translator into Latin of Plato's Greek manuscripts, and tutor of Lorenzo de Medici and Giovanni Mirandola, made a pact with his friend Michael Mercator that whoever of them should die first would make every effort to contact the other about the reality of the afterlife. Alas, they, too, did not live up to their pact. The story was well known and oft repeated. See Henry More's *The Immortality of the Soul* (1659), lib. II, ch. 16, sec. 6, pp. 293–94; and E. Seeman (2019).

17 For the same, see Increase Mather's *ERIP*, ch. 7, pp. 239–40, 241–43 – all from Glanvill's *Saducismus* (1681 ed.), "Relation X," pp. 219–24. See also S. Corneanu (2020).

18 In his *ERIP* (ch. 7, p. 245), Increase Mather acknowledges as his source the oft-reprinted work *De Spectris* (1570), by the Swiss Reformed theologian Ludwig Lavater (1527–86) of Zurich. An early English translation appeared as *Of Ghostes and Spirites Walking by Nyght* (1572). Simon Goulart (1543–1628) was a French Reformed theologian, humanist, translator, and author of *XL Tableaux de la Mort* (1606). It was a common belief at the time that prodigies such as ghosts, demons, comets, meteors and other unusual phenomena presaged wars, plagues, or God's divine wrath. See for instance, "An *APPENDIX* Touching Prodigies in *NEW-ENGLAND*," in Cotton Mather's *WWG* (1690), sep. pag. D^1–D^4; and M. Winship (1994).

19 No further information on Johnson or Warner has been found. Her story is also put to good use in Mather's *TP* (1995), pp. 116–17.

they *have any knowledge of what is done in this World*! When they took Leave of one another, *Mary's* last Words unto *Mehetabel* were; *Well, If I dy first, you shall hear of me*! *Mary Johnson* quickly goes to *Haverhil*, a Town Thirty Miles off *Boston*; and there she dies. It was not long, before *Mehetabel Warner* heard a Voice diverse times calling to her, Hitty, *Come here*! and with such Circumstances, that it could be concluded, the Voice came from none but the *Invisible World*. There ensued Violent and very Audible *Knocks*, from an *Invisible Hand*, upon her Mothers House, where she now resided; which continued for *Diverse Weeks*, and sometimes for the best Part of *Whole Nights* together; and the *Knocks* were heard as far off as in other Families of the Neighbourhood; and this to a considerable Distance; And they shook the very *Doors* and *Walls* of the Habitation where they were given. In the mean time, the Disturbed Neighbours gave themselves all possible Satisfaction, that here was no *Collusion*. Upon the Experiment made, of *Mehetabels* altering the Place of her Lodging, and the Noise more closely following of her; it was taken for granted, *That it came upon her account*. She had now heard of *Maries* Death; and she then told her Friends, of the Rash *Agreement* they had come into; and she verily beleeved, it was the *Departed Spirit* of that young Woman, that was now so troublesome. She resolved therefore, That the next Night, that there should be any more such *Knocking*, she would speak unto the *Knocker*, and ask, *What they would have*? The *Knocking* soon returned: *Mehetabels* Lodging was in a Chamber; Her Mother, with another of her Children, lay below. The whole House was made sensible enough of the *Knocking*, & pretty much affrighted with it; tho' they had already heard so much of it, & for so long a time. At Last, *Mehetabel* took the Courage, tho' not without a Sweating Terror upon her, to speak, and say, *In the Name of the Lord, what is it that you come for*? Immediately, a Shrill Voice, very audibly answered; Hitty, *Come hither*! Hitty, *Come hither*! Hitty, *Come hither*! Not only did *Mehetabel* hear this; but also her Mother,who is a vertuous and credible Woman; and the other Person that was with her. They heard no more after this. But not many Dayes were expired, before *Mehetabel Warner* fell sick. She lay sick, but about Four Dayes; and it could hardly be told what her Distemper was. Then she fell into some odd extatic Circumstances; and so she died. It was in the Month of *April*. 1696. when she was a little above Twenty Years of Age.

You will now *Scan* the *Evidence*, with which this Relation also arrives unto you. And Romances about *Apparitions* are so many, that it is fitt every thing of that Character should be *well-scann'd*, & will be so by Judicious Men, before it be Received. Now besides the *Evidence*, which you find the

Story in the whole Course of it, as I have told it, armed withal; I can only add, That I took the Assistence of a very Intelligent Friend, and we together nicely examined all Points of the Story, while it was yett capable of being well-attested; and we found the Demonstrations thereof to be such, that we could not be such Proficients, in the, *Ars nihil credendi*, (which you know the Humoursome *Godofridus de Valle* wrote a Book upon,) as to refuse Beleeving it.[20]

However, if there appear any thing Defective in the *Evidence* of these Relations, I will do my best, that there shall want no *Evidence*, of my desire to approve myself,

Boston. N. England. Syr,

Nov. 26. 1712. Yours most heartily.

C. Mather.

[20] Godofridus de Valle (du Val), aka. Geoffroy Vallée (1535–74), a French Calvinist turned Deist, who authored a widely discussed *De Arte nihil Credendi* (1570), *The Art of Disbelief.* Mather also dismisses the same author in his own *DL* (1699), p. 14. See also K. Kennedy (2017).

I.II. ON THE RATTLESNAKE.

November 27, 1712.

Letter. 4th.

To Richard Waller Esq.

Sir.

This Letter throwes down before you an *American Monster*,[1] that yett shall do you no other Damage, but that of engrossing so much of your Time, as may go to the Reading of it.[2] He shall not be the *Dragon*, which in *Ælian*, and *Pausanias* and *Philostratus* you find Thirty Cubits long; much less, the *Dragon* which in *Philostorgius* you find grown to forty five Cubits;[3]

1 Mather incorporates a discussion of vipers, snakes, serpents, and the medicinal use of their venom in his *CP* (1721), "Essay XXVIII. *Of* REPTILES, pp. 166–70. An earlier report on an American rattlesnake appeared as "Vipera Caudi-Sona Americana, Or the Anatomy of a Rattle-Snake," by the English physician and anatomist Edward Tyson, M.D., F.R.S. (1651–1708), in *PT* 13.144 (1683), pp. 25–58. It served Mather as one of his sources in his Letter to Richard Waller; see "Vipera Caudi-Sona," p. 26. See also Mather's CA 3.5 "A Serpent Securely Handled" (July 6, 1716). Mather's erstwhile friend Paul Dudley, F.R.S. (1675–1751), Attorney General of Massachusetts and son of Governor Joseph Dudley (1647–1720), supplied the Royal Society with yet another account, in his "*An Account of the Rattlesnake*," *PT* 32.376 (December 31, 1723) pp. 292–95.

2 See Kittredge, "Scientific Communications," pp. 11, 27.

3 Mather's primary source for his slithering monsters is Samuel Bochart, *Hierozoicon* (1712), pars posterior, lib. 3, cap. 14 ("De Draconibus"), fols 430–31, which lists a goodly number of Greco-Roman authors and their prodigious serpents cited by Mather: the Greco-Roman author and rhetorician Claudius Aelianus (c. 175–c. 235 CE), whose Phrygian dragon was "decem orgyias" long, and that of his relative around the Red Sea (according to Alexander the Great) forty cubits, in *On the Nature of Animals* (17.1); the Greek geographer Pausanias (146–176) of Lydia, whose serpents in India and Libya were up to thirty cubits long, especially the one in Corinth, in *Description of Greece* (2.28.1); the Athenian biographer and sophist Lucius Flavius Philostratus (c. 170–c. 245 CE), whose Indian dragon was thirty cubits (*Life of Apollonius*, lib. 3, caps. 6–9); the Cappadocian Philostorgius'

least of all, That of *Fifty Cubits*, mentioned by *Suetonius*, or that of Eighty five foot mentioned by *Dion*, or that of an Hundred foot long mentioned by *Strabo*, or that of an Hundred and Twenty foot long, which was killed near *Utica*, by the Army of *Regulus*; which last, had not the Skin & Jawes of it, been kept a long while at *Rome*, would have been certainly too big to have come into my Letter.[4] Tis a *Smaller Animal* with which I propose to treat you; but yet one which the *Mexicans* do with Good Reason call, *Teutlacocauhqui*, or, *The Prince of Serpents*.[5] It is a *Snake*, celebrated in the Writings of *Marcgravius*,[6] of *Nierembergius*, of *De Laet*, of *Hernandez*,[7] of

(368–c. 433) monstrous serpent forty-five cubits, or as much as fifteen fathoms long, *Ecclesiastical History* (lib. 3, cap. 11, p. 450), in *Epitome of the Ecclesiastical History of Philostorgius, compiled by Photius* (1855).

4 The Roman historian Suetonius Tranquillus (c. 69–c. 130 CE) reports in his *De Vita Caesarum: Divus Augustus* (cap. 43, sec. 3) of a serpent eighty-five cubits long, like the two-headed serpent in Etruria, eighty-five feet long, reported by Greco-Roman historian Dion Cassius (164–c. 235 CE), in his *Historiae Romanae* (50.8.4). Finally, to top it all off, Mather invokes the serpentine dragon of the incomparable Greek geographer Strabo of Pontus (c. 64 BCE–c. 24 CE), in *Geographica* (16.2.17), whose fallen leviathan in the Plain of Macra (Coele-Syria), was no less than a "plethrum" or one-hundred feet long, allegedly slain in the First Punic War, at Utica (near Carthage), by the Roman general Marcus Atilius Regulus (fl. 3rd century BCE). For a shorter version of it all, see Mather's *CP* (1721), Essay XXVIII, p. 168.

5 Mather's primary source for "Teutlacocauhqui" (Boicininga) and for most of the following author references is Joannes Jonstonus, M.D., aka. Johnston (1603–75), Polish physician and author of *Historiae Naturalis ... De Serpentibus et Draconibus Libri II* (1653), *De Serpentibus*, lib. 1, Art. 16 [*sic* XVII], punct. 1, *De Boicininga, seu Teutlacocauhqui*, p. 26.

6 Mather's second-hand reference (via Jonston) is to Margravius, aka. Georg Marcgrave, Margraf (1610–44), German naturalist, astronomer, and author of *Historia Naturalis Brasiliae* (1648), lib. 6, cap. 14, pp. 240–41. Margravius's work was bound with the work by the Dutch physician Guilielmi Pisonis, M.D., aka. Willem Piso (1611–78), *De Medicina Brasiliensi libri IV et Georgii Marcgravii de Liebstad, Historiae Rerum Naturalium Brasiliae libri* Octo … *cum Appendice de Tapuyis, et Chilensibus Joannes de Laet, Antwerpianus* (1648).

7 Joannis Eusebii Nierembergii S.J., aka. Juan Eusebio Nieremberg y Ottí (1595–1658), Spanish Jesuit and mystic, author of *Historia naturae, maxime peregrinae libris XVI distincta* (1635), lib. 12, cap. 1, pp. 268–69, which includes a full-page depiction of a rattlesnake, rattles and all. The Dutch geographer and author of *Nieuwe Wereldt ofte Beschrijvinghe van West-Indien* (1625), Johannes de Laet (1581–1649), assisted the publication of Marcgrave's and Willem Piso's *Historia Naturalis Brasiliae* (1648) in the above-mentioned works. De Laet's Dutch *Nieuwe Wereldt* was translated into several other European languages, including Latin, and

Johnston,[8] and others; but mostly under such Names as the *Brasilians* have putt upon it. I will wave their Names, and come to that, by which our Colonies have distinguished it; which (as the Dutch call it, *Raetel-Schlange*,) is that of the RATTLESNAKE.[9] It receives its Name as you know, from an Hollow Tail, of hard, thin, dry bone, very brittle, almost like to Glass, & very sonorous, which in the Motion of the Creature, makes a Noise like a *Rattle*; The Cells of which sometimes in their Number arise to Sixteen, yea, to Twenty, but ordinarily to Six or Seven; and some fancy, every Year in the Age of the *Snake* adds a New One. Of this odd *Rattle*, in our *Vipera Caudisona_Americana*, the only good Account yett given, ha's been from *Final Causes*, & is that of *Piso. Huic tam pernicioso colubro, benigna Natura Cautionis quasi gratiâ crepitaculum addidisse videtur; ut illius sonitu admonitus quilibet Homo non solum, sed et qualecunque pecus, tempestivè sibi caveat à Vicino Hoste.*[10]

Of this Creature, there are many things, which we have thought observable.[11] Particularly: we take Notice that the more Northward we go, the *Rattlesnakes* are both less Numerous, and less Venemous. We have no *Snake* so formidable; but it is affirmed, that none of them are ever seen, to the Northward of our *Merrimack* River; which is near Forty Miles to the Northward of *Boston*. However, our People think, the *Indians* in continual Hostility there,

appeared as *Novus Orbis seu Descriptionis Indiae Occidentalis Libri XVIII* (1633). His brief account of poisonous vipers in the province of Mechoacana (Mexico) appears in lib. 5, cap. 24, p. 267 and lib. 15, cap. 6, p. 555. Francisco Hernández de Toledo (1514–87), Spanish naturalist and personal physician to Philip II. His most esteemed work is *Nova Plantarum Animalium et Mineralium Mexicanorum Historia* (1650).

8 Jonstonus (Johnston), *Historiae Naturalis ... De Serpentibus et Draconibus Libri II*, p. 26. See note (above).

9 Margravius, *Historia Naturalis Brasiliae* (1648), lib. 6, cap. 14, p. 240.

10 Mather's Latin citation of Willem Piso's *De Medicina Brasiliensi Libri Quatuor* (1648), lib. 3, p. 41, is at second hand from the report of Edward Tyson's "Vipera Caudi-Sona," *PT* 13.144 (February 10, 1682/83): 53–54. Tyson dissected a rattlesnake at the repository of the Royal Society (January 1682/83) and included in his published article various depictions of the rattlesnake's interior organs, vertebrae, teeth, and rattle (fig. 1–12). See for much the same in Johnston, *Historiae Naturalis*, lib. 1, Art. XVII, punct. I, p. 26. The Latin citation from Piso reads, "Kind nature has given this very dangerous snake a little rattle as though for the purpose of giving warning, so that not only a human being but also any sort of cattle might in timely fashion save itself from the nearby enemy upon being given warning by its sound."

11 See also Mather's Letter to James Jurin (June 4, 1723), CA 7.3 "A New Rattlesnake"; and his Letter to John Woodward (September, 21 1724), CA 8.2 "*Amphisbaena*."

have pretty sufficiently supplied their Place, & done the Part of *the Serpent in the Way*![12] Another very odd thing, ha's been so frequently & constantly affirmed, by them, who have said they have been Ey-witnesses of it, that is now no longer quæstion'd among us. Tis no rare thing to see *Rattle Snakes* ly coyled at the bottom of a Tree, with their Eye fixed on something above. There we soon see a poor *Squirrel*, turning about, shrieking & crying & using all the Actions of one *Loth to go down*. However, by degrees, down he goes, & runs into the horrid Mouth of the Charming Devourer, below waiting for him, who swallowes him up alive. The Matter of Fact, is become so notorious, that our Preachers no longer scruple,[13] to employ it, as a pungent Repræsentation of the Confusion which too many under the Power of their Lusts, do run into. But the Observation, and the Curiositie, which I mainly intended for you, is of a *New Benefit* for Mankind, which has been afforded unto our Physicians by this wicked Creature.

Would you have expected, Syr, to have mett with a Rich Medicine, & a very noble Cordial, in the *Gall of a Rattle Snake*? Tis what I am going to exhibit you.[14]

The *Rattle-Snakes* have their Winter-habitations, on our Hills, in hideous Caves, and the Clefts of Inaccessible Rocks. In the Spring they come forth, & ly a Sunning themselves, but still in pretty feeble Circumstances. Our Trained Bands in some of our Countrey towns, take this time, to carry on a *War* with the *Snakes*, and make the Killing of *them*, a Part of their Discipline; Imitating, if I don't misremember, something of this Kind, which they tell us, was done of old time among the *Persians*. But some of our Physicians, lay hold on this time for a better Purpose. For, at this time, the *Cystis*, or

12 Given the several Indian uprisings practically in Mather's backyard, he shared the racist attitudes of his time.

13 Mather might well have mentioned himself. See his *BU* (1692), pp. 52–53; *HH* (1727), p. 13.

14 In his *RWSV* (1711, 1720), Mather pokes fun at calumniators who "Sambenito" the reputation of upstanding members of society, especially clergymen. These belittlers, Mather states, are like vipers whose poison aims to kill. And yet, "Our Dispensatories fetch noble Medicines from them [vipers]… . I have my self made a fine Anodyne and Febrifuge, with the *Gall of Rattle-Snakes*" (like the Indian who anointing himself with "the *Fat of Snakes*" became "Invulnerable and Impenetrable to the *Sting of Snakes*"), Mather delights to tell, yet his literal serum here is too metaphorical and ephemeral, when he recommends to "get his own Heart & Life healed of all that he can find amiss in it; by the Exercise of PIETY, which the Sibilations of Slanderers, may excite him to" ([1720], p. 38; Appendix, pp. xiii, xiv). See also his letter to James Jurin (May 21, 1723), CA 7.1 "The Case of the Small-pox Inoculated, further cleared."

Gall-bladder of the Snakes, is full of an azure & acid Spirit; a *Liquor*, which they squeeze into a Glass; but so spirituous, that if the Glass be not presently stopped, it will soon evaporate. This Liquor, we incorporate into convenient Quantities of powdered *Chalk*, or Indian Meal, and so tis præserved for our Use in Medicine.[15] We have been at a Loss for a Name to be putt upon it, but some of us have contented ourselves with that of *Trochisci Connecticotiani*,[16] because it was in our Colony of *Connecticot* that the Medicine had its first more famous Improvement.

When the Summer comes on, the Snakes have no longer this Liquor in their *Gall-bladder*. There is nothing to be now found in that Vessel, but a Black and Thick Sediment, of no known Use in Medicine. Tis thought by some, that the Spirit aforesaid, is now convey'd by certain Passages to the Head, where finding another Digestion, it becomes the *Deadly Poison*, which we are so much afraid of. How *deadly*, you may guess by this Experiment. A Traveller in this Countrey mett and kill'd a *Rattle-Snake*; but suffered the Angry *Snake*, to give a Bite before he died, unto the lower end of the Switch, with the Lashes of which he had first spoiled his Leaping. He rode on, & a fly disturbing him on one of his Temples, he rubb'd the Place, with the upper End of the Switch in his hand, unto which the Poison below had so permeated, that the Head of the poor Man swell'd immediately, and (as I remember) he died upon it. When the Head of the Snake is off, there is nothing of Poison remaining; every thing is rather useful; except the Rattle; which, it seems, is good for nothing, when the Beast can hurt no more.[17] Yea, when the *Snake* ha's been Dead a while, his *Head* also is eaten by our Dogs and Swine, without any Detriment: & sitts well enough upon their *Stomachs*. The more the Heat of the Summer comes on, the Poison of the *Snake* becomes yett more exalted; and the more *Southward*, still the more poisonous. At Cape *Fear*, one of our People sporting with a *Rattle Snake*, provoked him, & suffered him to bite the Edge of a good Broad Ax; whereupon immediately, the Colour of the Steeled Iron changed, & at the first blow he gave, when he went after this to use his Axe, the discoloured Part of the Bitten Iron, broke off without any more ado.[18] I know not whether I have now sprung a New Game, for the Gentlemen, that

[15] For much the same, see *CP* (1721), p. 169.

[16] Trochisci (troches) are powdered substances combined with sugar to form lozenges for medicinal purposes. On the Trochisci Connecticotiani, or Indian "snakeweed," found "near the River in *Connecticut*," see Nehemiah Grew's *Musaeum Regalis Societatis* (1685), pt II, sec. III, ch. 1, p. 227. For background, see Kittredge, "Scientific Letters," pp. 183–84.

[17] On the uses of the skins and bones of rattlesnakes, see Grew, *Musaeum Regalis Societatis*, pt I, sect. III, pp. 50–51.

[18] See also *CP* (1721), p. 169.

are hunting after the Liquor *Alkahest*.[19] But it is time for me to hasten unto the Properties of my *Trochees*.[20]

It is truly Alexipharmic, & Sudorific.[21] It is Anodyne to such a Degree, that I know some, who chuse to take Three or Four Grains, to compose them unto Rest, when they are tired with Travel. It is Highly deobstructive. Good in all Feavours, especially the more malignant. Women who gett cold in Childbed, & have their Lives endangered by sudden Obstructions, find this an Infallible Remedy. By Consequence, Women with Child must not dare to meddle with it. It cures Quartan Agues almost miraculously;[22] A proper Dose being taken on an Empty Stomach, twelve hours before the Fitt. The *Trochisks* may be taken in Substance; or in Water, or Wine; or almost any Way in the World. The Dose is fourteen Grains, more or less, according to the Strength of the Patient.[23]

I hope, this Letter about a *Rattle Snake*, will not prove so Ungrateful to you, as the *Viperæ Cubile*, proverbially mentioned by the Poet *Martial*.[24] If it should in any Measure {be} so, I would endeavour to abate it, with an Observation, that shall have in it more *Philology* than *Philosophy*.

Treating on a *Rattle-Snake*, it naturally carries one into an *Allegorical Meditation*.[25] Especially since the First Gospel that was ever heard in the World, &

19 A universal solvent of all substances highly sought after by Renaissance alchemists. See also Kittredge, "Scientific Letters," pp. 174–75.

20 I.e. antidotes. Native Americans, so Mather, employed "Snakeweed" to cure snake bites, see Kittredge, "Scientific Letters," p. 183. Mather also recommended it as an anti-inflammatory analgesic for measles in his *LMDM* (1713), p. 2.

21 "Alexipharmic," akin to an antidote, especially through a sudorific, which induces sweating. See also *CP* (1721), p. 169.

22 Malaria fevers that recur every four days.

23 In his funeral sermon *Victorina* (1717), p. 72, Mather relates the story of a Native American anointing himself "with the Fat of … an *Hawk*, which lives much upon *Rattle-Snakes*," believing he [the Indian] "will be fortifyed, that he will take *Rattle-Snakes*, and handle the fierce Poisoners with as much freedom, and as little Danger, as what *Alcuinus* {i.e., *Alcimus* }Writing *De Peccato Originali*, sings of. '*Mox impune manu coluber tractatur inermi*.' ['And soon the snake is handled without danger']. Mather quotes *Alcimus Avitus, De Origine Mundi, De Peccato Originali* (1545), lib. 2, p. 28; Alcimus Vitus (c. 450–519), bishop of Vienne in Gaul, poetically foretold, "And soon the harmless snake is handled without danger." See Isa. 11:8.

24 True to his philology, Mather invokes the Roman poet Martialis (c. 38–c. 104 CE), whose orator would rather smell a viper's nest than [the stench of] Bassa, in *Epigrammaton liber IV* (4.4.11).

25 See Mather's allegorical reading of the sting of the original serpent, in *TFOV* (1707) and *D* (1:580). See also the fiery flying serpents and Moses' copper serpent in the desert, in *BA* (2:922–37).

that rich Ingott of Gold, the *Third of Genesis and the Fifteenth*,[26] which being beaten out affords all the Leaves of our Bible to us, does countenance the *Allegory*; and exhibits our Great Saviour unto us, in the Glories of a mighty *Serpent-killer*.

You know very well, Syr, that when the Religious King of *Judah*, destroy'd that *Brasen* (or, I should rather say, *Coper*) *Serpent*, wherewith *Moses* had in a Sacramental Way, procured the Cure of the People, among whom the *Præsters* once in the Arabian Wilderness made a miserable Havock, he called it, *Nehushtan*;[27] and that this was in the Account commonly given of it, as much as to say, *This is no more*; There is no such Divinity in it, as pretended to. *Kimchi* sais, This Denomination is as much as if *Hezekiah* had said, *How can it come into the Heart of any Man, to think, that this Brass can do any thing for him? Tis a meer Peece of Brass. Ralbag* expounds it; *This Brass ha's no more Vertue in it, than the common Brass of their Houses*: He takes, /an/, at the End of the Word, for the Same, with, /am,/ which is, *Ipsorum*.[28] Our Dr. *Jackson* penetrates yett a little deeper into the Matter.[29] The Destruction of the *Old Serpent*, the *Divel*, was foreshadowed by the Lifting up of the *Brasen Serpent* in the Wilderness. The Design of *Moses* was to excite in the Minds of the People, a Remembrance of the *Third of Genesis & the fifteenth*, and a Faith of the Great Redeemer, who would one day come to deliver us from the Mischiefs, which the *Old Serpent* has brought upon us. It seems,

26 Gen. 3:15, commonly known as the protoevangelium: "And I will put enmity between thee and the woman, and between thy seed and her seed; it shall bruise thy head, and thou shalt bruise his heel" (KJV). On this pericope, see Mather's commentary *BA* 1:487–89.

27 Numb. 21:9, part of the account of Moses' brazen serpent on a pole (OT type of Christ on the cross) set up in the camp of the Israelites to heal those beset by fiery flying serpents. King Hezekiah (2 Kings 18:4) had the image destroyed. See *BA* 2:929–37; 3:617–18. נְחֻשְׁתָּן [nechushtan] (2 Kings 18:4) signifies "a thing of brass."

28 *BA* 3:617–18. The Sephardim Rabbi David Kimchi, aka. Radak (c. 1160–c. 1235) of Narbonne (France), esteemed biblical commentator and grammarian, and the French Talmudist and Jewish philosopher Rabbi Levi ben Gershon (1288–1344), aka. Gersonides (Ralbag), are both excerpted in *The Commentary on the Tanakh* (II Kings 18), in *The Book of Kings 2* (1998), pp. 374–75. Mather's second-hand quotes of Kimchi and Ralbag commentary are from *A Commentary upon the Two Books of Kings* (1705), p. 527 (2 Kings 18:4), by Simon Patrick (1626–1707), Anglican bishop of Ely, whose commentaries on virtually all the books of the Hebrew Scriptures served Mather as one of his principal sources throughout *BA*.

29 Mather refers to *The Humiliation of the Sonne of God by His Becoming the Son of Man* (1636), ch. XXXI, §6, pp. 361–62, by Thomas Jackson, D.D. (1579–1640), Chaplain to King Charles I of England and president of Corpus Christi College, Oxford.

Hezekiah might be moved with the greater Indignation against the Worship of the *Brasen Serpent*, into which the Idolatrous Jews were fallen, because, in truth, it was not so much a Type of our glorious *Redeemer*, as a Figure & Emblem of the *Destroyer*, from whom He does Redeem us. He expressed his Detestation of it, in the Word, *Nehushtan*; which, tho' our defective *Lexicons* have not yett made any Register of it, yett it imports no less than, *The Foul Fiend; the Old Dragon*; or, *Satan*.[30] The *Serpent* exhibited on the Pole in the Wilderness, was doubtless, as a *Slain Serpent*; or one with a *Bruised Head*. It was, q.d. *Behold him that ha's the Power of Death, even the Divel. Expect a Time, wherein the Messias will slay that Serpent, & extinguish his Power to plague & kill a miserable World*. This *Faith* procured the Healing of the *Israelites*! Indeed, when we read, Joh. III.14. *So must the Son of Man be lifted up*; it looks, as if the *Serpent* were to be a Type of Him. No, Tis an astonishing Intimation of the *Method*, wherein our Great Redeemer is to slay the *Serpent*. He was to undergo the Punishment, of a *Bruising*, something like that, which He is one day to inflict upon the *Serpent*, & leave Him eternally under it. A *Generation of Vipers* were to treat Him, as if *He* had been the *Serpent*. They were to lift Him up on a *Pole*, as a *Serpent* which they had conquered, & from which they were delivered.[31] The learned *Vitringa* observes, *Durum atque horrendum videtur, Christum comparare Diabolo, cujus Emblema Serpens est*.[32] But then, he ha's this noble Thought upon it. When our Saviour was exposed unto the Faith of His Church, on the erected Wood of the *Cross*, the *Old Serpent* also in Effect underwent a Crucifixion. We have been poisoned with *Sin* by that *Serpent*; but the Death of our Admirable Redeemer, procures in a way of Expiation to Divine Justice, the Death of our *Sin*, & so of that Old *Serpent*. Tho' the Mischief were intended unto our Lord JESUS, by His Crucifixion, yett all the Mischief redounded unto the *Old Serpent* who intended it. See Rom. VI.6. So my *Vitringa*: *Diabolus affigendus erat eidem cruci cui Ille.*

30 Both Johannes Buxtorf's *Epitome Radicum Hebraicarum et Chaldaicarum* (1607), voce: נְחֻשְׁתָּן [nechushtan], p. 520, and his fuller *Lexicon Hebraicum et Chaldaicum* (1645), p. 462, agree with Mather's proposed signification. To be sure, Christological readings were not the purpose of these lexicons.

31 The preceding pericope of Mather's Letter (MS 3^V) up to this point, also appears, with slight revisions, in his commentary on 2 Kings 18:4 (*BA* 3:617–18).

32 The citation is from *Observationum Sacrarum Libri Sex... Tomus I* (1712), 1:424, "De Supplicio Crucis Christi," lib. II, cap. XI, sec. IX, 1:424, by Campegius Vitringa (1659–1722), the Dutch Reformed professor of Oriental languages at Franeker. Mather's quote translates, "It seems harsh and shocking to compare Christ to the devil, whose emblem is the serpent."

Diaboli Vires tolli non poterant, nisi sublato peccato. Peccatum verò expiandum erat per obedientiam Christi in cruce.[33]

But I will have done. A Sort of Impatience (forever agreeable to one of my Profession) to write many Letters on the Works of *Creation*, without any Contemplations on the Glories of the wonderful *Redeemer*, to whom all His Creatures are so much endebted, ha's thrown me unawares into this Digression. But I am glad, I am Writing to One, whom, I know, such things are alwayes welcomc to; and who will alwayes bestow a candid Interpretation on whatever is address'd {to} him, from,

Syr,

Yours in all good Wishes

Boston, N. England. & Services;

Nov. 27. 1712.

C. Mather.

33 Vitringa, *Observationum*, cap. XI, sec. XX, 1:434: "The Devil had to be nailed to the same cross as He was. The powers of the Devil could not be taken away unless sin had been abolished. But sin had to be expiated through Christ's obedience on the cross."

I.12. THUNDER & LIGHTNING.

November 28, 1712.

Letter. 5th.

To Richard Waller Esq.

Sir,

You will anon conclude, that the Pen which now addresses you, is in the Hand of one belonging to the Tribe of the *Boanerges*.[1] As *Thunder* and *Earthquake*, are *both* of them implied in the Signification of that Name, this Letter shall make *both* of them your Entertainment.

I observe, that you have thought Relations of Things done in and by the *Thunder*, to be now and then Materials not unworthy of your *Philosophical Transactions*.[2] I may then inform you, That *New England* is a Countrey signalized, with Mischiefs done by *Thunder*, as much as perhaps most in the World. If things that are smitten by *Lightning*, were to be esteemed *Sacred*, this were a *Sacred Countrey*; I wish, nothing amiss gaining upon the Morals of it, might ever forfeit the Denomination. Rarely a Year passes without some Strokes from the *Thunder*, on the Persons, or Houses, or Cattel of our People; yea such *Mortal* Ones, that there have needed none of *Paulus Zacchias* his Rules,[3]

1 Mark 3:17, i.e. "Sons of Thunder." This is the title Mather gave to two other works a few decades later: (1) *BO* (1727), and an unpublished manuscript, "Boanerges, or Work of the Day" (c. 1716), a lengthy essay advocating the union among all Protestant churches. This work is unfortunately lost, but pieces thereof were likely integrated in his pietist sermon *M* (1717). See T.J. Holmes (1:84–85).

2 There are numerous accounts of thunder and lighting, and speculations on their causes and effect, in the volumes of the *PT* throughout the 17th and 18th centuries. See for instance, John Wallis' "A Relation Of an Accident by Thunder and Lightning, at Oxford," *PT* 1.13 (June 1666): 222–26; N.A. "A Relation of the Effect of a Thunder Clap on the Compass of a Ship" and "A Letter of Sir R.S," *PT* 15.157 (1684): 520–21; John Wallis' "A Letter of Dr. Wallis to Dr. Sloane," *PT* 19.231 (1697): 653–58; "Part of a Letter," *PT* 26.316, pp. 137–39.

3 The learned Italian physician, jurist, and natural philosopher Paulus Zacchias, aka. Paolo Zacchia (1584–1659), is mostly remembered for his oft-reprinted

to know whether the struck Neighbour have been Dead past Recovery; nor would any of *Jacobus Javellus* his Remedies have recovered them.[4] To enumerate the Instances of Damages done by the *Thunder* in this Land; *Houses* fir'd, (were our People not so vain, as to reckon their *Housleek*[5] a Præservative!) *Cattel* slain, *Metals* run, *Trees* torn apeeces, *Rocks* pulverized, *Bricks* vitrified, and *Ships* mortified, would be, to fill not a *Letter*, but a *Volumn*. *Zuingers* Collection of People kill'd by *Lightning*, might have a considerable Appendix and Addition to it, from these *American* Colonies.[6] The Punishment of *Burning* used sometimes among the *Jews* of old, some say was inflicted by pouring Hott Lead into the Mouths of Criminals. This Punishment was called, *Combustio Animæ*, and used in Imitation of Gods destroying Men with *Lightning*, whereby the *Inward* Parts are burnt, while the *Outward* are not hurt.[7] A, *Combustio Animæ* by *Lightning*, hath ever now and then kill'd some of our People; whereof some indeed had Marks left upon them; their *Hair* sing'd; or their *Skin* perforated with numberless little Impressions like what the smallest Shott would have made upon them; or their *Bones* dissolved & made as limber as a Grissle; but others have had no external Signs of Hurt upon them. Some

---

three-volume *Quaestiones Medico-Legales* (1621–35; 1701), in which he supplies the physical signs by which to determine whether a person struck by lightning can still be saved. See especially *Quaestiones* (1701 ed.), tom. 1, lib. 2, tit. I, Qu. XV, p. 145; lib. 5, tit. II, Qu. VIII, pp. 394, 395.

4 Giovanni Crisostomo Javelli (c. 1470–c. 1538), Italian Dominican philosopher and theologian, author of *Medicina Compendium* (1609), details how a person struck by lightning can be healed. In both instances, Mather cribs from his father's *ERIP* (1684), p. 96, where Increase Mather cites both Zacchias and Javellus.

5 The houseleek is a succulent that, among other applications, has been used to treat burned skin.

6 Theodore Zuinger, the elder (1533–88), Swiss physician and scholar, author most famously of the massive five-volume *Theatrum Vitae Humanae* (1565–1604), an encyclopedic study structured in the (cumbersome) Aristotelian–Ramist method characteristic of its time. In his fascinating *Theatrum* (1586), vol. secundi, lib. III, pp. 322, lib. VII, pp. 477, 545, Zuinger provides the rudiments of statistics listing the causes of death by drowning, suffocation, warfare, fire, lightning, impure water, tempests, shipwrecks etc. along with the sources he consulted (from the classics to his own time).

7 Mather also alludes to this punishment by "burning" in *MCA* (6.3.14) and in more detail in his commentary on Deut. 21:9 (*BA* 2:1119). Talmud Tractate Sanhedrin (52a, b) and Gemara (52b) tell the tale of a priest's adulterous daughter. Generally speaking, a condemned criminal would be buried in a dung heap up to the knees, a towel wrapped about the neck and pulled in opposite directions by the witnesses "until he opened his mouth." Next, "a wick [or boiling lead] was kindled" and shoved down his throat to "burn his entrails," in *Mishnah*, tractate Sanhedrin (7:2), p. 391.

of these have been very pious People; and so especially was the Captain of our Castle, whom the mortal Artillery of Heaven found and left Asleep upon his Bed, and confuted *Plutarchs* Opinion, That men Asleep are never smitten with *Lightning*.[8] I will add one small Curiosity more. A great Man ha's noted it, That the *Lightning* oftener falls upon Churches, than upon any other Houses, because the *Dæmons* have a peculiar Spite at Houses that are sett apart for the peculiar Service of God.[9] I know not how far *Dæmons* may by the Permission of the Glorious God, have their occasional Operations in the *Thunder*; (tho'

8 The Greco-Roman philosopher and historian Lucius Mestrius Plutarchus (c. 46–post 119 CE) in his *Symposiacs: Quaestiones Conviviales* (4.2, 4.2.4 §§664*b*, 665*e*) has Agemachus and Dorotheus claim that the growth of mushrooms is promoted by thunder and that sleeping men (like dogs) are never struck by lightning. Again, Mather relies on his father's *ERIP*, pp. 72, 76 for this ancient wisdom. See also *MCA* (6.3.14).

9 Four decades later, Benjamin Franklin knew better than that, when his kite experiment led him to theorize that lightning is merely an electrical discharge that can be conducted away from houses or church steeples by an iron lightning rod grounded in the earth. Although the Royal Society published his account of the kite eliciting electrical sparks, in "XLIV. A Letter from Mr. Franklin to Mr. Peter Collinson, F.R.S. concerning the effects of lightning," *PT* 47: 567–69 ([January 1, 1752, December 31, 1752], pp. 202–11, 289–91, 565–67), the Society was reluctant to accept his experimental conclusion, as can be seen in William Watson's "XCVI. A Letter … concerning the electrical Experiments in England upon Thunder-Clouds," *PT* 47 (December 21, 1752), pp. 567–70. Significantly, a study commissioned to determine how St. Paul's Cathedral in London could be protected from lightning was not undertaken until nearly two decades later (see James West, "XXI. Proposal of a Method of securing the Cathedral of St. Paul's from Damage by Lightning," *PT* 47 [March 9, 1769], pp. 160–69; and William Watson's "CI. Some Suggestions concerning the preventing Mischiefs, which happen to Ships and their Masts by Lightning" *PT* 52 [December 1761], pp. 629 35). Back in Boston, a tempest in a teapot brewed when the Rev. Thomas Prince (1687–1758), minister of the Old South Church and stalwart friend of Cotton Mather, seemed to implicate Ben Franklin's lightning rod as somehow causing (or contributing to) the 1755 earthquake that rocked Boston (December 5, 1755). In a post scriptum to his "Appendix" (p. 23), Prince seemed to insist that "P.S. The more *Points of Iron*" are erected round the *Earth*, to draw the *Electrical Substance* out of the *Air*; the more the *Earth* must needs be charged with it" (pp. 20, 23) – hence the increase in earthquakes – in Prince, "Appendix Concerning the Operation of God in Earthquakes by Means of Electrical Substance," in *Earthquakes the Works of God* (1755), pp. 20–23. None less than Harvard's own John Winthrop (1714–79), Hollis Professor of mathematics and natural philosophy, took exception to Prince's alleged blame-game for the moral edification of his parishioners; the ensuing debate between Winthrop's accusation and Prince's half-hearted retraction was carried out in the public forum of the *Boston Gazette* (January 26, 1756; January 28, 1756; February 23, 1756; March 1, 1756).

doubtless the *Prince of the Power of the Air*,[10] ha's as good Skill in Chymistry, as goes to the making of *Aurum Fulminans*,)[11] but I have often heard it Remark'd, That our *Temples*, and the Houses of our most valuable *Ministers*, have been much oftner *Thunderstruck*, than any other Buildings.[12] But I will supersede all further Account of our *Thunder*, with a Story, which will afford you something Rare, for your Philosophy to work upon. On July. 24. 1681. a Ship whereof one Mr. *Lad* was Master, being about an hundred Leagues off the Coast of *New England*, in Lat. 38. about 3 hours p.m. mett with a Thunderstorm, which made considerable havock aboard her.[13] One of the odd Circumstances was, the Fall of a Bituminous Matter on the Vessel, that smelt like fired Gunpowder, and continued burning in spite of all their Endeavours with much Water to extinguish it, until such time as all the Matter was consumed. But a much odder was, that when the Night came, & the Stars called for their Observation, the People found a surprising Change upon their *Compasses*. The Polarity of the *Compasses* was changed, & the *North* Point was turn'd unto the *South*, and so it continued standing all the rest of the Voyage, & for a thousand Leagues together. Yea, one of them does continue so to this day, which is above Thirty Years after; & I have it in my Eye upon my Table, at the Instant of my writing what I do about it.[14]

10 Eph. 2:2.

11 An explosive salt made of fulminic acid (HCNO).

12 Mather certainly knew what he was talking about, for on September 12 1694, during a thunderstorm, while delivering his sermon to his assembled parishioners in church, a messenger interrupted to inform him that his house had just now been struck with lightning! Did Mather break off his discourse and rush to his home? No Way! Such a miraculous confirmation from heaven was far too opportune a means for convicting his congregants to deprive them of his practical lesson of *contemptus mundi*: "*Brethren*," Mather cleared his throat, "I am just now informed, That the *Voice* of the *Glorious God* in the *Thunder*, has been very Immediately directed unto my self, by a Fall of *Thunderbolts* upon my own House, at that very instant, as far as I can judge, that I felt the powerful Impressions of Heaven upon my own Soul," in *BS* (1695), p. 4. See also Mather's *D* (1:519), *MCA* (6.3.16), and *CP* (1721), Essay XVI. *Of the Thunder and Lightning* (pp. 61–65).

13 Edward Lad was the master of the ship *Albemarl* (*ERIP*, pp. 91–93, 108–09). The story is extracted from "*A Relation of the Effect of a Thunder Clap on the Compass of a Ship*," *PT* 14.157 (Marc 20 1684): 520–21; and confirmed in an accompanying statement by Mr. Edward Randolph, in "*A Letter of Sir* R.S. *concerning the former Relation. Lond. March* 8*th.* 1684," *PT* 14.157 (March 20 1684): 521–22.

14 This was the compass of Mr. Edward Lad, captain of the *Albemarl*. In his *BS* (1695), Mather retells the same story: "I have now in my House a Marriners *Compass*, whereupon a *Thunder-clap* had this odd Effect, that the *North-point* was thereby *turned* clear about unto the *South*, and so it will veer and stand ever since unto this

You know, Syr, who ha's illustrated the Name of *Thunder*, from the *Fire* which will break forth, when Sulphur, mixed with Filings of Steel, admitts a little Water. But then, the Mysteries of the *Magnett*, require a deeper Penetration, than all your *Wards*, and *Gilberts*, and *Cabeus*,[15] and *Keplers*, and *Kirchers* have yett reach'd unto.[16] I will not attempt the Solution of what ha's befallen my *Compass*; tho' I know the common Opinion, That the Earth is a great *Magnet*, (and Mr. *Halley*'s Ingenious *Hypothesis*,[17] which by

day, tho' the thing happened above thirteen years ago." But Mather certainly knew how to turn a startling story into a tool for conversion: "I would to God that the next *Thunder-claps* would give as effectual a *turn* unto all the unconverted Souls among us!" (*BS* [1695], pp. 11–12). See also Increase Mather's *ERIP*, pp. 108–09.

15 Mather misspells it as "Cables."

16 Mather's second-hand story is again cribbed from his father's *ERIP* (1684), pp. 103, 106–07. Increase himself acknowledges as his source *Some Considerations* (1664), Essay I, p. 15, by Robert Boyle (1627–91), an Anglo-Irish natural philosopher and scientist, who in c. 1660 conducted experiments on refrigerated and heated lodestones (magnets). Boyle was one of the original founding members of the Royal Society – hence Mather's designation of him as "your." Samuel Ward (1572–1643) of Ipswich (Suffolk), England, published *Magnetis Reductorium Theologicorum Tropologicum* (1637), translated as *The Wonders of the Load-Stone* (1640). See Mather's *CP* (1721), Essay XXIV. *Of Magnetism*, p. 115; and *CPS* (1994), pp. 120–21; Boyle, for his part, acknowledges William Gilbert (1544–1603), English physician and scientist, author of *De Magnete* (1600); the Jesuit professor of theology and mathematics at Parma and Genoa, Nicolaus Cabeus, aka. Niccolo Cabeo (1586–1650), who recorded his experiments with metal filings and magnets in his *Philosophia Magnetica* (1629); so, too, *Ars Magnesia* (1631); and *Magnes sive de Arte Magnetica* (1641, 1654), by the German Jesuit polymath at the Collegio Romano Athanasius Kircher (1602–80), whom Boyle (pp. 15–16) singles out for special praise. Increase Mather himself adds to this list Seth Ward (1617–89), an English mathematician, astronomer, and bishop, who with his fellow countryman John Wilkins (1614–72), an Anglican clergyman, mathematician, and founding member of the RSL, defended the study of mathematics, chemistry, and magnetism at the Oxbridge colleges, in *Vindicia Academiarum* (1654); and Johannes Kepler (1571–1630), German astronomer and natural philosopher, who was inspired by Gilbert's concept of the earth's magnetic soul, which Kepler put to good use in his laws of planetary motion (gravitation) in *Astronomia Nova* (1609) and *Epitome Astronomiae Copernicanae* (1618) – works which subsequently informed Sir Isaac Newton's definition of the laws of gravitation.

17 Edmond Halley (1656–1742), the famous English mathematician and astronomer, undertook a scientific expedition to ascertain the magnetic fields of the earth (1701). On the subject of the earth's magnetism, Halley published, for instance, "A Theory of the Variation of the Magnetical Compass," *PT* 13.148 (June 10, 1683): 208–221; and "An Account of the Cause of the Change in the Variation of the Magnetical Needle," *PT* 17.195 (October 19, 1692): 563–78.

supposing an huge *Vacuum*, & other Globes, within the Thick Shell which we walk upon, attempts to solve the *Variation of the Compass*, & the Four Poles of the *Magnet*,) I am also aware of what others besides *Porta* write, of the *Fire* interrupting the Vertue of the *Loadstone*;[18] and of what your Honourable *Boyl* found upon Experiment, that having heated the *Loadstone*, he could change the *Polarity* of it, by the different Position in which he cooled it.[19] And some would count it enough to say, That the *Lightning* had heated my *Compass*, and it being unhung, with the *Flower-de-Luce* toward the *South*, when it cooled, this might fix the Tendency of it, that way that it ha's ever since been verging.

But from *Thunder* I obliged myself to pass unto *Earthquake*. Tis what the Countrey ha's felt several Times. The *Motions* have been sensible, and affrighting, but no *Mischiefs* have ever happened from them: nothing like what ha's been done in *Italy*, & in *Sicily*, or what you find in the old *Roman* Histories. In *January* of the Year, 1663.[20] we had six or seven violent Shakes, in the Space of Three Dayes. A Town lying on the River of *Connecticut*, ha's had them Scores of Times in a Year, for many Years together. Our Indians affirm whole Rivers to be not only stopt by *Earthquakes*, but also wholly swallowed up; as one of those which the Sacred Scripture mentions as the *Rivers of Damascus*,[21] was doubtless lost that way in After-ages. One *Josselyn* who dwelt in the Neighbourhood,[22] ha's given the Public a Narrative of it, that in the Year 1670. at a Place called *Kenebunch*, in our Province of *Main*, somewhat near the side of the River, a Peece of Clay-ground was thrown up, over the Tops of High Oaks that grew between it, and the River, and cast into the River so as to stop the Course of it; and leave an hole forty Yards Square, wherein were thousands of Clay bullets, as big as Musket-bullets, and Peeces of Clay in Shape like the barrel of a Musket. The

[18] Giambattista della Porta (1535–1615), Italian scholar of Naples, and author of *Magia Naturalis* (1558, 1589). Translated into English as *Natural Magick* (1658), Porta's applicable discussion appears in Book 7: "Of the Wonders of the Loadstone," the specific work to which Mather here alludes.

[19] Boyle, *Some Considerations* (1664), p. 15; see also his *Natural Philosophy* (1673); and D. McKie (1934).

[20] The tremors of the 1663 earthquake were felt from New York to Quebec. See S. Perley (1891), ch. 4.

[21] 2 Kings 5:12. See also Mather's *BA* (3:571–72).

[22] John Josselyn (1638–75), *An Account of Two Voyages to New-England* (1674), pp. 277–78. Josselyn (fl. 1638–c. 1675), an English traveler, spent two separate periods of time in New England: 1638–39, and again, 1663–71.

like accident, much about the same time, fell out at *Casco*, twenty Miles to the Eastward of this Place: much like what befel your *Marclay Hill*, in the Year, 1571.[23] and what after that, *Stow* tells you happened at *Hermitage* in *Dorsetshire*.[24] And the *Fish* in several Ponds were thrown up Dead on the Banks, in prodigious Quantities.

We have had our prodigious *Hail-Storms*, which have poured down vast Quantities of *Hail*, as big as *Hens-Eggs*,[25] and many of them five times as big, that have lain three or four foot thick on the Ground, unmelted in the hottest time of our sultry Summers, for three or four dayes together; and such as led us to think on the horrid *Hailstorms*, mentioned by *Jerom* and *Socrates*,[26] at both the Beginning and Conclusion of the *Gothic Troubles*, which we suppose foretold in the *First Trumpett* of the *Apocalypse*.[27] Our *Johnston* would certainly have putt in the Stones into his *Thaumatography*![28] Akin to these,

23 *The General History of Earthquakes* (1694), by R.B. (R. Burton), records that on February 17, 1571, an earthquake occurred near Kingston (Herefordshire): "the Earth began to open and an Hill called Marclay Hill with a Rock under it, made at first a mighty bellowing noise which was heard afar off, and then lifted up itself a great heighth and began to travel; bearing along with it the Trees that grew upon it, the Sheepfolds and Flocks of Sheep abiding thereon at the same time" (p. 60).

24 Mather probably refers to the oft-reprinted work *Annals, or, A Generall Chronicle of England* (1592, 1632), by John Stow (c. 1525–1605). Hermitage is a small village in Dorsetshire, southwest England. See also Increase Mather's *ERIP*, p. 325.

25 The hailstorm of June 26 1682, described in Increase Mather's *ERIP*, pp. 318–19.

26 Jerome's *Chronicle* (a Latin translation of *The Chronicle of Eusebius of Caesarea*; translated by Roger Pierce, https://tertullian.org/fathers/jerome_chronicle_03_part2.htm) reports that in the 286th Olympiad (367 CE), "a great tempest arose at Constantinople that, hail of wonderful size, it killed some men." Likewise the Byzantine Church historian Socrates Scholasticus (c. 380–c. 450) reports of the great hailstorm (with hailstones the size of "a man's hand") falling on Constantinople, in 367 CE, in his *Ecclesiastical History* (bk 4, ch. 11), in *NPNFii*, 2:100. See esp. Mather's long commentary on Rev. 9:21, in *BA* (10:538–49). Mather's source is Drue Cressener's *The Judgment of God upon the Roman Church* (1689), ch, 4, p. 46.

27 Rev. 8:7-8. See also Mather's commentary on the First Trumpet in *BA* (10:521–22).

28 Mather refers to the Polish-born Calvinist natural philosopher, physician, and polymath Joannes Jonston, aka. John Johnston (1603–75), in his popular *Thaumatographia Naturalis* (1632, 1661, 1665), Classis secunda, cap. II: De Aqua, Articulus VI, pp. 59–60 of the 1661 edition. This work went through more than forty editions. Mather calls him "Our Johnston," because, though a subject of the King of Poland, he was raised a Calvinist. Mather also makes good use of Johnston's other

I would reckon an Accident we have seen sometimes in our Winters. It ha's Rained plentifully, and at night it ha's frozen so extremely, that the Weight of the Ice of the Limbs of the Trees, ha's broken them off, yea, and splitt the very Trunks of the Trees themselves; and many Thousands of Cords of Wood, have herewith fallen off, with such Cracks as in an united Noise, have been all night long, as little surprising to our Inhabitants.

The *Hurricanes* have not visited us, to the Degree wherein the *Caribbee Islands* have had them. Yett we have had our *Whirlwinds* drive along,[29] a particular Narrow Track for diverse Miles together with such an Impetuous Fury, that nothing upon Earth ha's been able to stand before them, and if our Towns had been in their Way, they must have undergone astonishing Desolations; but the Mercy of our Good God ha's ordered another Course for them. A Thick Dark, Small Cloud ha's arose, with a Pillar of Light in it, of about Eight or ten foot Diameter, and passed along upon the Ground, in a Track scarce wider than one of our Streets, but horribly tearing down Trees, & pulling them up by the Roots, and blowing them about like Feathers in the Air; & throwing up Stones of great Weight unto a great Heighth in the Air; & breaking Fence, and Barns and Houses to Peeces, at least, those Ends of them that stood in the Way; and People that could not gett fast enough out of the Way have been bruised to Death, under the irresistible Force of the *horrible Tempest*. The Noise of the Cloud all the While so great, that[30] no other Noise of the Mischiefs done by it, could have any Notice taken of them. You will allow my mention of this; inasmuch as I think, for this also, you have in your *Philosophical Transactions*, for the Year 1669. a Relation of the like Occurrences.[31]

Having treated you with some Curiosities of *Nature*, I will finish my Letter, with some Curiosities, whereof tis doubted, whether they be *Natural* or *Artificial*.

At a Place called *Ammuskeag*,[32] a little above the hideous Falls of *Merrimack River*, there is an huge *Rock* in the midst of the River, and in the midst of the Rock, there are exactly round Pitts, a great Number of them,

works in his second Letter to Dr. John Woodward (November 18, 1712), CA 1.2, as well as in 1.11 and 3.1.

29 Tornadoes.

30 MS: "the."

31 Mather may have in mind Joshua Childrey's account of Richard Smithson's voyage to the East Indies, in "An Extract of a Letter," *PT* 40.50 (August 16 1669): 1003–09.

32 Present-day Manchester, N.H.

resembling Barrels or Hogsheads, but of different Capacities; there being some of them large enough to hold several Tuns. In these Cavities, concerning which it may be wondred, how they *should* be, and as much how they should *not* be, Artificial, the Indians of that Neighbourhood, have been wont to hide their Provisions, when they had Wars with the *Maqua's*;[33] affirming, that God on this Intention did long since cutt out those Repositories for them. If this *Rock* afford you nothing, I will carry you to one that shall a little entertain you. At our *Taunton*, by the side of a Tiding River, part in, part out, of the River, there stands a large Rock; on the perpendicular Side of which Rock next the River, there are Seven or Eight Lines, about Seven or Eight foot long, and about a foot wide, each of them; in unaccountable Characters. It is generally taken for granted; that they are *Artificial*; and there they stand *Graven, in the Rock forever*;[34] but no man as yett has been a *Zaphnath Paaneah*[35] enough to know any more *what* to make of them, than *who* it was that *graved* them. I have not yett been able to gett all the *Lines*; which, I hope, e're long I shall, when they will be at your Service. But I will here give you the two first of them. [**torn*][36]

[33] I.e., the Mohawks.

[34] See for the same reference in Mather's in *WWG* (1690), sign. A^{3v}, and his biography of John Eliot, *TRR* (1691), p. 73. Mather refers to the Dighton Rock (a forty-ton boulder), once on the banks of Taunton River and now in the Dighton Rock State Park, in Berkeley, Mass. For the various renditions of the inscriptions, see G. Mallery (1894), p. 86; and L. Barnett (2019).

[35] Gen. 41:45. Pharaoh's name given to Joseph, suggesting "one who reveals mysteries."

[36] Mather's transcript of the Dighton Rock engravings, originally appearing in the space below, was torn out from his holograph letter to Waller, most likely to enable an engraver to reproduce it on copper plates to be published with "An Extract of several Letters from Cotton Mather," *PT* 29.339 (January 1, 1714): 1, 62–72. Much the same description – along with the carving's top line – appears in Mather's "The Epistle Dedicatory" to Sir Henry Ashurst, in *WWG* (1690), sign. A^{5v}. More than a decade after the inscription appeared in the *PT*, the Rev. John Villa remembered Mather's letter and the mysterious inscription, which he showed to M. Veyssière De la Croze (a historian and orientalist at the court of the Prussian King Frederick William I in Berlin), who speculated it might be of Japanese origin. To get to the bottom of the matter, Villa reached out to James Jurin (March 13, 1725) of the Royal Society in London for help, who in turn wrote to Mather (May 20, 1725); see Appendix 11. Jurin's letter to Mather, however, remained unanswered (if it ever reached him), but the desired information was finally supplied by Harvard's Isaac Greenwood (1702–45), an elected member and active correspondent of the Royal Society, who sent to London a fresh transcript of the mysterious hieroglyphics (April 28, 1732). See especially R.P. Stearns (1970), p. 452; and E.B. Delabarre (1916 and 1917, 1928).

Figure 7. Dighton Rock Inscription. In *PT* 29, issue 339 (January 1, 1714), p. 1. Courtesy of the Royal Society of London.

It is now time for me, in more Intelligible Characters, to assure you that I am, with Impressions deeper and more lasting than upon the durable *Rock*,

Boston, N. England.
Nov. 28. 1712.

Syr,
In my Heart alwayes
disposed for your
Service;
C. Mather.

I.13. CASES OF LONGEVITY & FRUITFULNESS.

November 29, 1712.

Letter. 6th.

To Richard Waller Esq.

Sir,

When you are confuting the cavelling *Præ-Adamites*,[1] and confounding the Cavils which your Deists raise, from the Numbers of People whom guilty *Cain* was afraid of,[2] I observe you make very *Moderate Allowances*, for the quick Progression of Mankind into a Multitude. The Computation made in the, *Dissertations sur la Bible*, published lately at *Paris*, is very obvious.[3] It allowes *Cain* and *Abel* to marry in the 19th Year of the World. By the 25th tis an easy Allowance, that they might each of them have 8 Children, Males and Females. In the 50th Year, 64 Persons might be in a Direct Line descended from them; In the 74th Year they would come to 572. In the 98th to 4096. In

1 The French Marrano, theologian, biblical critic, and anthropologist Isaac La Peyrère (c. 1594–1676) aroused vehement opposition to his thesis in *Prae-adamitae* (1655), *A Theological System Upon that Presupposition, That Men were before Adam* (1655), and *Men Before Adam* (1656), in which he argued that the Mosaic creation (Gen. 1–2) actually contains two creation accounts (polygenesis): the first describing the creation of the gentiles (men before Adam); the second, the creation of Adam (and Eve), the progenitors of the Jews. See Mather's *BA* (1:99, 122, 127, 501, 825–26); *BA* (9:31). See P. Rossi (1979, 1984), R.H. Popkin (1987), and D.N. Livingstone (2008).

2 Gen. 4:13–14.

3 *Dissertationes Historiques, Chronologiques, Geographiques et Critiques sur la Bible*, 2 vols. (1711), by Louis Ellies Du Pin (1657–1719), a French ecclesiastical historian and professor at the Sorbonne, set out to refute Isaac La Peyrère's thesis that there had to be men before Adam if Cain was afraid of the people of Nod. Du Pin (vol. 1, Diss. 1, ch. 3, esp. pp. 69–84, and ch. 6) confirms the propitious fertility and multiplication of humans, and supplies Mather (*BA* 1:517, 528) with a multiplication scheme. The remarkable fertility and long life spans, says Du Pin (Diss. 1, ch. 8, pp. 135–41, and Diss. 2, ch. 5, pp. 223), was also fostered by the antediluvians' consumption of healthy fruits, a vegetarian diet, and abstinence from meat.

the 122{nd}, to 32768. But then by adding the other Children, that might be born to *Cain* and *Able* in this Time, they and their Descendents, would in 122 Years easily come to 421164 marriageable People. To these if we add, the other Children of *Adam*, & their Descendents, how surprizing a Number will presently appear unto us![4]

Thus, when you are considering the Number of People in the *Antediluvian World*, I observe, you make a liberal Abatement of the Two or Three first Centuries, (in which yett you see how plentifully the Globe just now came to be stock'd;) and you suppose a Man to have Children at Sixty, and in the next 40 Years to have 20 Children. You then with Sir *Thomas Brown*,[5] single out the *Shortest Liver* of any mentioned (except *Enoch*,) before the Flood; and from that one Stock of 700 Years, multiplying still by 20, you find the Produce to be more than One Thousand, Three hundred & forty seven Millions. Thus, *One Family*, would long before the Flood have afforded, many more than twice the Number of People, that are now living on the Earth.

And so, when you are considering the speedy & mighty Multiplication of the *Ismelites* in *Egypt*, I observe, that altho' the Word in the Hebrew Text, (which is noted by *Hacspan*; out of *Jalcut*,) intimates, they *increased like Fishes*,[6] yett you are content with *Cappellus*,[7] that *One Man* in the 30th Year of his Age begin

[4] Mather's numbers are taken from Du Pin, Diss. I, ch. 3, p. 76.

[5] Sir Thomas Browne (1605–82), renowned English physicist and essayist whose works Mather frequently cites and excerpts in his *Biblia Americana* volumes. Mather's extract is here derived from *Pseudodoxia Epidemica*, third edition (1658), bk 6, ch. 6, pp. 370–71, 373–75. See also *BA* (1:517, 528) and Sir William Petty's *An Essay concerning the Multiplication of Mankind* (1682).

[6] Theodor Hacspan, aka. Hackspan (1607–59), German professor of theology and Oriental languages at the University of Altdorf, cites *Perush Baal HaTurim al HaTorah* (2:516–17), by R. Yaakov ben Raash (c. 1275–c. 1340), on *Sefer Jalcut*, a haggadic compilation on the books of the Hebrew Bible, attributed to R. Shimon HaDarshan (13th century) of Frankfurt. Mather here refers to Hacspan's philological commentary *Notarum Philologico-Theologicarum in Varia et Difficilia Scriptorae loca Pars Prima* (1664), 1:273–74, 276. See also Mather's commentary on Exod. 1:7 (*BA* 2:117).

[7] The renowned Huguenot scholar in exile Ludovicus Cappellus, aka. Louis Cappel (1585–1658), distinguished professor of Hebrew and divinity at Saumur, calculates the fish-like multiplication of the Israelites in his *Commentarii et Notae Criticae in Vetus Testamenti* (1689), pp. 609–10, on Exod. 12:40. His slightly older brother, no less renowned, Jacques Cappellus (1570–1624), professor of theology and Hebrew at Sedan (France), offers yet another set of calculations in his *Historia Sacra et Exotica ab Adamo usque ad Augustum* (1613), p. 71. Either way, Mather tries to demonstrate that the Bible's marvels are within the scope of mathematical possibility.

to be a Father, & have in all no more than 10 Children, who also, with their Posterity should begett at the Same Age; and this One Man shall in less than 200 Years, have One Million, one hundred & eleven thousand, one hundred & ten, of his Posterity. You don't grow in your Demands as far as the Scheme of *Torniellus*, who without any Difficulty, but that of assigning 14 Children to every married Couple in 30 Years, in 210 Years, will muster, *Plusquam sexies et decies centum millia, et quadraginta septem millia hominum.*[8] Nor do you see any Necessity for *Temporarius* his Concession of Marriage at 20 Years, on which in 210 Years the Multitude will be so great, that, *Si tota Jacobi familia funditus interisset, uno tantum Josepho superstito, ex hoc uno surculo armatorum millia, non solum 600, verum etiam suprà centies centena millia, in Annis 210, sinè ullo Naturæ prodigio excitari potuerit.*[9]

But now, how prodigious would be the Increase of Mankind in these Terms, if you should insist upon such a *Fruitfulness*, as that by which many of my good Countrey Women have been signalized? It is no rare thing for an Aged Gentlewoman here to see many more than an hundred of her Offspring, before she leave the World. Indeed we can't boast, of such Fertilities, as are collected by *Caspar Schottus*, who particularly mentions one Woman of *Florence*, who had fifty two Children, but never less than Three at a Birth;[10] nor such as *Joannes Helwigius* reports in his *Observationes Physicomedicæ*,[11] where he

8 Agostino Torniello (1543–1622), Italian Roman Catholic theologian, member of the Society of St. Paul (Barnabites), esteemed for his *Annales Sacri & Profani* (1622), a popular chronology that went through numerous editions. Mather's Latin citation appears in *Annales*, cap. "Ab Orbe Condito *Annus* 2329. Tertiae Mundi Ætatis Annus 217," p. 250 (*b*), and signifies that "more than sixteen hundred and forty-seven thousand men" would have been begotten in 210 years. See also *BA* (2:211–12).

9 Joannes Temporarius, aka. Jean du Temps (c. 1535–c. 1575), French jurist, mathematician, and geographer of Lyon, whose *Chronologicarum Demonstrationum Libri Tres* (1596), lib. 2, "Canon Tertius," p. 140, is the work in question. Temporarius conjectures, "if the whole household of Jacob would have died out or perished, with Joseph alone surviving, then from this single shoot in 210 years [might come] not only 600,000 but even more than a hundred times a hundred thousand soldiers could have been called up, without any miracle of nature." See also (*BA* 2:212).

10 The German Jesuit physician and natural philosopher Caspar Schottus (1609–66) relates this story of the prolific Helionora Salviata, spouse of the Florentine citizen Batholomaeus Friscobaldus, in his popular collection of grotesquely misshapen animals and humans, *Physica Curiosae, sive Mirabilia Naturae et Artis Libri XII Comprehensa* (1662, 1667), lib. 3, cap. 29, §1, p. 470. See also *BA* (2:117–18).

11 Mather refers to the German physician of Regensburg Johannes Helwig, *Observationes physico-medicae* (1680), "Observatio CCLV. Fœcunditas mira," pp. 389-90,

brings a well-attested Relation of one Wedlock, which at thirty five times produced no less than fifty three Children. Yett we have had our Instances of what has been considerable. One *Morey* had no less than twenty three Children, by One Husband; whereof Nineteen lived unto Mens & Womens Estate.[12] One *Wharton* had no less than Twenty two. One *Bent*, at the time of my gathering these Exemples, is big with her Seven & twentieth Child. Our celebrated *Sherman*,[13] (whose Life you have in our *Magnalia Christi Americana*,) having had Six Children by a first Wife, married a Second who enriched him with Twenty more.[14] Mrs. *Phips*, who is yett Living in a very hearty old Age, ha's been the Mother of no less than Six and twenty Children, by Two Husbands; One and twenty of these were Sons;[15] and One of them was the Remarkable Sir *William Phips*, the Governour of *New-England*;[16] who together with one unto whom I have the Honour to be related,[17] had the Curiosity to satisfy themselves concerning another Mother, who ha's brought forth no fewer than *Thirty nine* Children; with the *Thirty fifth* of whom they personally discoursed, and gained those Proofs, which assured them of it, & my Assurance with theirs, is now transferred unto you.

If such a *Polytokie*,[18] were attended with a *Longævitie*, like that of the First Ages, Mankind would not require near *four hundred Years*, as you tell us, it now does, to *Double itself*. But, alas, you inform us, That but 64 out of an 100

---

who relates the story of Barbara Schmotzerin of Boenningkhaim (Württemberg), whose fertility rivaled that of her Florentine fellow *polytokos*. See also *MCA* (3.29.165).

12 See *MCA* (3.29.165).

13 Rev. John Sherman (1613–85), Puritan clergyman of Watertown, Harvard fellow, and astronomer *MCA* (3.29.162–65).

14 *MCA* (3.29.164).

15 Mrs. Mary Phips (n.d.) in *MCA* (3.29.164). See also E.W. Baker (1998), p. 10.

16 Sir William Phips (1651–95), governor of Massachusetts (1692–95), whose life Mather celebrates in his *PP* (1697) and in *MCA* (2.35–75).

17 Cotton Mather baptized William Phips (March 23 1690), who became a member of the North Church (*D* 1:148; *MCA* 2:46). The genealogical relationship between Phips and the Mathers, however, remains unclear. Both family lines are complex, complicated by second and third marriages on both sides. By the standards of Mather's time, even the most distant kinship several times removed was acknowledged as a family relationship.

18 I.e., fecundity. The *OED* identifies Cotton Mather's *MCA* (3.29.165) as the first occurrence of "polytokie" in an English imprint: "the condition of having borne many children; an instance of this. *Obsolete*."

of the Children that are born, are alive at 6 Years old; but 40 of an 100 alive at 16 Years End; but 3 out of an 100 at 60, & but 1 out of an 100 at 76.[19]

The Causes of the *Longævity* in the First Ages, and the *Abbreviation of Humane Life*, to the Term of *Threescore Years & ten*, at which it has now stood for three thousand Years, were a Noble, but obscure {Subject?} which I would humbly recommend unto our sagacious Enquiries. The Causes assigned by *Josephus*,[20] & by the Hebrew Doctors, to which *Lyra*[21] ha's added some others, are so insufficient, that they are hardly worth reciting. Indeed, one would imagine, That when a *Physician* comes to handle the Subject, we should see the *Nail hitt on the Head*. And yett the learned *Beverovicius* brings in *Peter Mexa* a Spanish Physician, assigning the most unlikely Fancy that any Mortal could have litt upon.[22] The Antediluvians, forsooth, lived upon

19 Mather employs this data repeatedly, including in *SSM* (1712), pp. 14–15; *ARH* (1727), pp. 139–40; S.J. Kunitz (1984), esp. pp. 561–64. To be sure, Mather's conjectures are not necessarily of theological interest to the Royal Society. Rather, his pre-statistical calculations would help medical mathematicians (like James Jurin) who were trying to gauge the vital statistics of the population by establishing comparative tables of births, deaths, and diseases. See *especially* A.A. Rusnock's *Vital Accounts* (2002).

20 The Jewish historian Flavius Josephus, aka. Joseph ben Matityahu (c. 37–c. 100 CE), opined that the antediluvians in Noah's time enjoyed their long lives "on account of their virtue, and the good use they made of it in astronomical and geometrical discoveries" (*Antiquities* 1.3.9).

21 Nicholas of Lyra, aka. Nicolaus Lyranus (c. 1270–c.1340), learned biblical exegete and professor at the Sorbonne. In his *Postilla Super Biblicam* (1492), vol. 1, n.p. (Gen. 5), the first printed biblical commentary, Lyra offers the same causality as Abarbanel does for the decline of nature. The patriarch's longevity was, in part, fostered by better food, greater virtue, scientific discoveries, the purity of the air, and God's providence that bestowed great fertility on his people. See also Mather's commentary on Gen. 5:3 (*BA* 1:533–35, 566–78).

22 In the present instance, Mather's principal source (*BA* 1:574–75) is Johann Heinrich Heidegger's רָאשֵׁי אָבוֹת [Rashi Aboth] *Sive de Historia Sacra Patriarcharum Exercitationes Selectae*, 2 vols. (1729 ed.), vol. 1, Exerc. XIV, §20, pp. 250–51, who cites Johannes Beverovicius and Pedro de Mexía as his sources. Johannes Beverovicius, aka. Jan van Beverwyck (1594–1647), was professor of medicine at Dordrecht. Mather, who owned at least three of the Dutch physician's medical works, here refers to his *Schat der Gesontheyt* (1636), aka. *Thesaurus Sanitatis* (2nd ed. 1638), which went through numerous editions and translations into several European languages. In this context, Mather refers to Beverovicius' *Schat der Gesontheyt* (ch. 3, p. 27), who addresses the question of the patriarchs' longevity and enlists the learned humanist Pedro de Mexía (c. 1496–c. 1552), a Spanish historian and imperial biographer, whose *Silva de Varia Lección* (1540) appeared in an English translation as *The Forest or Collection of Historyes* (1576). Mexia

Raw Flesh; [which is more than my Doctor knowes!] but we dressing our Meat, so consume the most Nutritive Part of it, that the *Vitæ Stamina* are not strengthened by our Food. *Cardan* also advanced this Notion;[23] but I think, *Scaliger* ha's with his Arguments done as bad as *beat him with the Spitt*.[24] However I apply myself to One, from whom the World justly expects nothing, but what is great, and wise, and unexceptionable.[25]

But tho' the *Period of Life*, be at such a Stand as it is, yett you know, Syr, there are *Some*, [And for the Good of Mankind, may *you* be One of the Number!] who in our Dayes go much beyond the *Stated Period*. In *Pliny* [L. 7. c. 48.] you find a Notable Assembly of *Long-Livers*, whom that Writer summons together.[26] And one *Phlegon* wrote a Book, περι Μακροβιων·[27]

---

explains the patriarchs' longevity through natural rather than supernatural causation: the absence of disease, their healthy diet, knowledge of medicinal herbs, the beneficial influence of the stars, moderate living, and proximity to the perfection of the protoplast, in *The Forest of Collection of Historyes* (1576), pt 1, ch. 1, p. 1^r–2^v.

23 The "*Vitæ Stamina*" are the "threads of life," which are controlled by the Three Fates. The Italian physician, mathematician, and astrologer Girolamo Cardano (1501–76) lectured on medicine at Milan, Pavia, Bologna, and Rome, and published more than two hundred works on medicine, physics, astrology, and mathematics. Cardano argues that since fire (cooking) separates the substances of food, our diets are less nutritious; however, if we were to eat our meat, eggs, oysters, fish, etc. raw rather than cooked, it would greatly contribute to our health and long life, in *De Subtilitate libri XXI* (1550), lib. 2, pp. 40^v–41^r. On Cardanus, see A. Grafton (1999).

24 His fellow Italian, Julius Caesar Scaliger (1484–1558), a learned philologist and physician, attacked Cardano on many issues in philology and the natural sciences. In his *Exotericarum Exercitationum Liber Quintus Decimus, de Subtilitate, ad Hieronymum Cardanum* (1557), Exercitatio XXII, Scaliger took issue with Cardano's scientific and metaphysical explanations of nature, largely because the former defended Aristotelianism against Cardano's popular *De Subtilitate* (1550), a controversial system of cosmogony, which aroused great interest at the time.

25 Mather has much to say on the "Famous Methods for the Prolongation of Life," in his medical handbook *ABJ* (1972), Appendix, pp. 14–19, providing mostly practical advice such as a moderate diet, drink, and sleep, and God's blessings. See also his "Rules of Health" for young ministers, in *MAM* (1726), §19, pp. 130–35, which include gymnastics, horse-back riding, a vegetable diet, and abstinence from "*Salt-Meats*" and from the use of tobacco.

26 See also *BA* 1:568. The Roman historian Pliny (*Natural History* 7.48.153–49.163) lists many individuals by name whose longevity rivals that of some Hebrew patriarchs.

27 The Greek historian and paradoxographer Publius Aelius Phlegon, aka. Phlegon Tralles (2nd century CE), wrote a variety of works on Roman topics. Mather refers to Phlegon's *Opuscula, De Rebus Mirabilibus, De Longaevis* [*Peri Makrobion kai thaumasion*], or *Short Works, On Marvels* and *On Long-Lived People*. His *On*

You are no Stranger to the Exemples mentioned by *Scaliger*;[28] nor to what *Vartomannus* affirms, he found among the *Arabians*.[29] M. *Pagenstecher* ha's lately given us an Essay, *Concerning the Nature and Priviledge of Old Men*, and ha's mentioned several Instances of exceeding *Old Men*, who married & had Children; among whom, that of *Platerus* his Grandfather, becoming a Father after he was an hundred Years old, is very extraordinary.[30] Old *Tom Parr* married at an hundred & twenty, & lived with his Wife Twelve Years, and expired at One hundred & fifty two.[31] *Peter Delaval* in his Travels mentions one *Jasper Dragonetti*, who at One hundred & fifteen had all his Teeth, and made no Use of Spectacles.[32] Among our *New Englanders* also, we now

Marvels describes such prodigies as hermaphrodites, fetal malformations, monstrous animals with human faces, and people of marvelous age – just the kind of stuff the Italian physician Fortunio Liceti (1577–1657) depicted in his *De Monstris. Ex recensione* Gerardus Blasii (1665), a work Mather also refers to in his Letter to John Woodward CA 3.2 "A Monstrous Calf" (July 3, 1716). See also K.E. Shannon-Henderson (2019).

28 Joseph Justus Scaliger's *Elenchus Primae Orationis Chronologia D. Davidis Parei* (1607), §§65–66, provides numerous instances of longevity to which Mather alludes.

29 The Bolognese traveler Ludovicus Vartomannus, aka. Luigi Varthema, aka. Barthema (c. 1465–1517) gives his account in his popular *Ludovicus Vartomani Novum Itinerarium* (1511), lib. 2, cap. 8, which first appeared in Italian (1510), Latin (1511), English (1576–77), and in many other European languages thereafter. Vartomannus relates that in the city of Reame (Yerim), Arabia Felix, he talked to many "that have passed the age of a hundred five and twentie yeares, and yet verie lustie and well complexioned," in (Mather's likely primary source) "The Travels of Lewis Barthema or Vertoman," in Samuel Purchas, *Hakluytus Posthumus or Purchas His Pilgrimes. The Second Part* (1625), bk 9, ch. 7, p. 1493. See also *The Travels of Ludovico Di Varthema in Egypt ... A.D. 1503–1508* (1863), p. 78.

30 There are several likely candidates to fit the bill, including the German legal scholar Alexander Arnold Pagenstecher (1659–1717), one of a long line of German, Dutch, even Mexican Pagenstechers. Mather's reference to Pagenstecher's essay and to the prolific grandfather of Platerius is uncertain. It may well be a reference to the Swiss physician Felix Plat(t)er (1536–1614), professor of Medicine at Basel and author of his oft reprinted *Praxeos Medicae Tomi Tres* (1602). An English translation appeared as *Platerus Golden Practice of Physick* (1664).

31 See *BA* (1:568). Thomas Parr (1483–1635), who is said to have lived for 152 years and 9 months, lies buried in Westminster Abbey. See *Monumenta Westmonasteriensia* (1683), §44, p. 51.

32 The Italian musicologist and traveler Pietro Della Valle (1586–1652) reports in his *Travels of Sig. Pietro della Valle, A Noble Roman, Into East-India and Arabia Deserta* (1665), "Letter 18" (August 1, 1626), p. 233, that while in Rome, he encountered the venerable Frere Gaspare Dragonetti of the Fathers of the *Della Scuole Pie* (Piarist School), who was as chipper as ever, his advanced age notwithstanding:

and then have some, that may stand in an Appendix to Phlegons *Catalogue*.[33] We cannot show, a, *Johannes De Tempore*;[34] nor such a *Persian* as *Peireskius* took to be Four Hundred Years old.[35] But we have had several that have lived above an Hundred; and some now living, are much older than the Countrey in which I am now writing; the Planting whereof began a little after the Beginning of the former Century.[36] My Neighbour *Boniface Burton*, was counted an Hundred & Twelve.[37] And while I am writing, a Woman whose Name is *Moor*, comes to visit me; who keeps the Remembrance of her Age by this Token, that she was Twelve Years of Age, when K. *Charles* I. came to the Throne.[38] One *Richard Everid* at our *Dedham*,[39] was twelve Years old, at the Time of the *Spanish* Invasion; and held out a strong Man, all this Year 1682. tho' he observed no *Rules of Health* in the

"Although he was now a hundred and fifteen years old and more" (judging by his letters of ordination in 1530 or 1531), yet F. Dragonetti was "sound and lusty, and not only sees without Spectacles, and hath his Teeth good," but continues to teach his young charges their grammar.

33 Phlegon Trallianus, *Opuscula De Rebus Mirabilibus De Longaevis* (*Book of Marvels*). See above.

34 Joannes Temporarius (see above).

35 The French philosopher Pierre Gassendi (1592–1655) relates in his *The Mirrour of True Nobility & Gentility: being the Life of the Renowned Nicolaus Claudius Fabricius Lord of Peiresk* (1657) that Lord Peiresk, aka. Nicolaus Claudius Fabricius (1580–1637), senator of Aix-en-Provençe, validates the tale that "having information from Alepo [*sic*], by the Letters of *Peter* and *John Constantine*, that there was a man in *Persia*, known to the *Capucines*, and other credible persons, who was now four hundred years old: of the verity whereof, they were undoubtedly satisfied" (*The Mirrour*, bk 5, pp. 139–40, "for 1636"). See also Mather's commentary on Gen. 5:3 (*BA* 1:534–35, *n.* 5).

36 Mather's canceled passage "~~in the Year, 1620.~~" suggests that he had in mind Plymouth Plantation (founded in 1620). Perhaps on second thoughts, he realized that the colony of Virginia was older yet by more than thirty years.

37 Boniface Burton, an early settler of Lynn, Massachusetts, died in Boston on June 13(?) 1669 – according to the merchant and mint master John Hull (1624–83), in his *Diaries of John Hull* (1857), p. 229; see also A. Lewis (1865), pp. 239, 619.

38 The Stuart King Charles I (1600–49) acceded to the throne of England in 1625. Thus Mrs. Moor was born in c. 1613, and, given the date of Mather's letter to Waller (1712), she was almost 100.

39 If Richard Everid or Everad (Everett) was twelve years old in 1588 when the Spanish Armada under King Philip II of Spain (1556–98) crossed the English Channel, Everid would have been 106 years old when he died in Dedham (Massachusetts), in 1682. F. Smith (1936), Dedham's historian mentions Richard Everett as one of the original settlers of the village, p. 14.

World. One *Richard Lenten* at our *Lancaster*,[40] arrived unto so many Years above an hundred, that he had lived in Wedlock with his Wife Sixty three Years, and yett she was thirty five Years younger than he; This Man was able to follow his Toyls at Husbandry with Vigour, but about a Month before he died. One *Clement Weaver* of *Rhode-Island,* arrived unto the Hundred & tenth Year of his Age; and had a Wife above an hundred Years old, unto whom he had been married more than fourscore Years; in all which time they never had one Word of Difference, and had this further odd Circumstance of their Friendship, that they constantly eat upon One Trencher at the Table. This Man about a Year before he died, would go to the Mill two Miles, with a Bushel of Corn upon his Back, & saw his Children to the Fifth Generation. We had an old Man, who dwelt in the Town of *Lyn.* He was brought before the Court for some Immoralities; where the Judges passing a Censure upon him, with an agreeable Piety dispensed an Admonition unto him, wherein they minded him of what the Scripture saies about, *A Sinner an hundred Years old*:[41] The profane old Wretch, as if he would boast of having lived beyond the reach of the *Curse*, replied, *Please your Honours, I am one hundred & two.* The poor Man continued many Years after this; and came into a Dotage, wherein he forgatt all affaires, and all faces, which he had been of later time acquainted withal, & became insensible of any Relation between himself & his own Children, which were alwayes about him. Nevertheless in his Time, he had a strange & strong Remembrance of many Matters & Persons & Places, which he had in his Youth conversed withal. Of these he talked continually; and sometimes having stripped himself naked, he would make towards the Seaside, & say, he was going to such a Town, or such a Man as he had in his Childhood, many Years before the first Nativity of these Colonies, a Knowledge of.

The *Hollanders* found at *Java*, a King who was One hundred & fifty Years old, and at last lost his Life in a Fight.[42] I beleeve, they know of some King or

40 This may be Richard Linton, one of the founding members of the Lancaster settlement in the Bay Colony, in the early 1640s, prominently mentioned in A.P. Marvin (1879), 1:38, 40, 41, 42ff.

41 Isa. 65:20.

42 See also *BA* (1:569). Samuel Purchas reports of the imaginary voyage of Iambulus (c. 3rd century BCE), the Greek utopian travel writer, who talks about "the inhabitants of some unknown Island, either Zeilan or Java" or "Sumatra, who lived very long lives," "namely one hundred and fifty yeeres, and for the most part without any sicknesse." If any came down with a "Fever or be sicke in his bodie, they enforce him to dye according to their Law," in *Hakluytus Posthumus or Purchase His Pilgrimes* (1625; 1905), vol. 1, ch. 8, pp. 215, 217. Whether or not this is the King of Java Mather alludes to, examples of fabulous longevity were still a commonplace in his time even as the long lives of the ancients were increasingly scrutinized and put to the test.

other, who, if they thought he would live to that Age, it would for a Reason you may guess it, give them some Uneasiness. But for a Reason that is the Reverse of *that*; such and a greater Number, of Years, and even the *Nestorean*,[43] could I heartily wish to the Excellent Friend, whom I am now addressing, if they might not be those, whereof he will say, *I have no Pleasure in them*! or if it were possible for them to be free from those Inconveniencies, whereof, *Multa senem circumvenient*.[44] But I remember, *Georgius Leontinus* at one hundred & seven, being asked, why he would be willing to live so long, answered: *Quià nihil habeo propter quod senectutem meam accusem*[45] That is the Old Age I wish you; and such an one as *Drusius* had, *cui senectus melior* (he said, he found it so,) *quàm ipsa Juventus*.[46]

But that I may *Polychronize*[47] you more agreeably and significantly, I wish you a Part in that *Holy City*, where our Great Saviour will endue His Raised People with *Immortality*, and they shall so feed on the *Tree of Life*, as to have their *Dayes like those of the Tree*,[48] and prolonged unto an *Everlasting Life*.

Boston, N. England. I am,

Nov. 29. 1712. Syr,

Ever and Entirely yours,

C. Mather.

[43] Mather appears to refer to Homer's Nestor, king of Pylos (Nestor Gerenios), one of the Argonauts who led his men into war against the Trojans in spite of his advanced age (*Iliad* 1.250 ff; 9:53 ff).

[44] Horace (*Ars Poetica* 169): "Many are the discomforts of old age."

[45] Gorgias Leontinus (483–375 BCE), the Socratic philosopher of Sicily, replied, "because I have no reason to blame my old age" (adapted from Marcus Tullius Cicero's *De Senectute* [5.13]).

[46] Mather quotes from *Ad Loca Difficiliora Pentateuchi* (1617), cap. 66, p. 84, by Joannes Drusius, aka. Jan van den Driesche (1550–1616), a renowned Dutch Orientalist at Franker and Leiden. Referring to Drusius's explication of senescence among the ancient Hebrews, Greeks, and Romans, Mather jokes that he wishes his addressee (Waller), the same that Drusius said of himself, "whose old age was better than his youth." See also *BA* (1:588, *n.* 131).

[47] Again, Cotton Mather throws the omniscient *OED* for a loop. Mather anglicized the Greek verb πολυχρονίζειν ("to prolong," Deut. 4:26 LXX). While the first mention of "polychronic" (adj.) in print occurs in 1907, its verb form "polychronize" has remained elusive. It suggests the activity of synchronizing numerous (poly) periods of time (chronos). Here (Deut. 4:26, KJV): "to prolong your life" or "lengthen your days or time on earth."

[48] Isa. 65:22.

Series Two

2.I. A WOOLLEN SNOW.[1]

December 1, 1713.

To Richard Waller, Esq.

Boston, N.E.

Sir,

You cannot but form some very strange Conjectures, concerning the Matter which upon the first Opening of this Letter, you find enclosed in it, if you allow any time for Conjecturing before you have read all the Letter over.

If I make so much Haste, as to tell you, that you have here enclosed, a *Snow-ball*, you will wonder what the *American Snow* should be made of; & you will affirm, this is the first Time that ever a Gentleman sent his Friend a Parcel of Snow wrapt up in a Letter to him; at least at such a Distance, as the Breadth of the wide Atlantic.

When you have read of Heaven *giving Snow like Wool*,[2] I know, you have taken it, as it is, for a poetical Expression; & very much approv'd and admir'd the Sacred Poetry.

But a *Snow of Wool* without a Metaphor is a thing so unusual, as to be almost Incredible;[3] And yett I must ask that such a Thing may find some Credit with you.

1 On the origin, composition, texture, and figures of snow, see Nehemiah Grew's "Some Observations Touching the Nature of Snow," *PT* 8.92 (March 25, 1673): 5193–94, 5196.

2 Ps. 147:16.

3 Mather relates the same story in his *CP* (1721), p. 60. T. Wickman (2017), p. H102, offers the following explanation for the woollen snow during "The Great Snow of 1717": "[S]ome Sheep blended in with snow landscapes, and in their winter travels the animals rubbed off some of their wool, leaving it on the surface of the snowpack." I wish to thank Diane Lee, Collections Manager of the Fairfield Museum and History Center, for drawing my attention to Wickman's article. For additional essays on the figure and shape of snow published in the *PT*, see Benjamin Langwith, "Observations on the Figures of Snow," *PT* 23.376 (January 1, 1723): 298–300.

For, Sir, I do assure you, That this *Wool* did fall from Heaven; and a considerable Quantity of it; at a Time when more common Snow was falling.

I have unhappily mislaid the large, & full, and well-asserted Account of it, with which I furnished myself, at the Time of it, which was a few Years ago: so that I cannot be so particular in my Relation, as I shall be; if I happen to recover it.

But however, my Memory sufficiently serves me; to assert so much as may afford you a tolerable Satisfaction; which is, That at a Town in one of our Colonies, called *Fairfield*, in the Depth of Winter, there fell a Snow, as at other times; But, there was a large Frozen Spott, of I know not how many Acres, which instead of the Snow, that lay in other Places, was covered with a very considerable Quantity of that Wool, whereof I now render a Specimen to your Acceptance.–[4]

The Stories that you have Read, about a Rain of *Blood*, I can guess what you think of. And what a Roman Historian relates, about a Rain of *Gold*.[5] I can guess what some others would be glad of. *Thuanus* will tell you of *Corn* rained for the Space of Two Hours, & of Two Miles together.[6] You are so well acquainted with such Relations, that it would be a perfect Rudeness in me to mind you of them. A Snow of Wool, and of what would be proper to make us a *Double Cloathing*, that we may *not be afraid of the Snow*, I do not remember, that I have any where mett withal.

4 William Derham published this and the preceding paragraphs of Mather's letter to Waller as an extract in *Philosophical Experiments and Observations of the Late Eminent Dr. Robert Hooke* (1726), p. 386. See also *CPS*, annot. p. 333.

5 Zeus sent "bloody rain drops" from heaven just before Patroclus killed Sarpedon, the Lycian leader, beloved by Zeus (Homer, *Iliad* 16.458–59; see also 11: 53–55). Roman historians not wanting, Plutarch relates the tale of showers of blood rain falling on the city in the times of King Romulus (*Romulus* 24.1). Mather's likely source is *De Nivis Usu Medico Observationes Variae* (1661), cap. IV, p. 29, by the renowned Danish physician Thomas Bartholin (1616–80), professor of anatomy at the University of Copenhagen, who alludes to "de pluvia sanguinolenta naturae" ("the bloody rain of nature"). See also R. Stothers (1979).

6 Jacques-Auguste de Thou, aka. Thuanus (1553–1617), French statesman, historiographer, and author of *Historia Sui Temporis* (1604–08). A two-volume English translation appeared in 1724–34. Mather's reference is a close paraphrase of the March 23 1548 event near Klagenfurt (Austria), relating "it rain'd Corn above two Miles about for the space of two Hours; the Inhabitants made excellent Bread of it, a sample whereof being brought to the Emperor, with some of the Grains of Wheat, which fell from the Sky, was beheld with the highest Astonishment," in *Monsieur de Thou's History of his own Time*, 2 vols. (1729–30) 1:259.

When we read of a *Faithful Messenger* compared unto, *The Cold of Snow in the Time of Harvests*,[7] I was well pleased with the *Danish* Gloss in the Margin, which explains it, concerning the *Reserved Snow*, wherewith it was usual for them to temper their *Wine*, in the time when the *Festival Heads* of *Harvest* exercised them.[8] Of this, you find many Testimonies & Intimations in Antiquity, besides those in *Athenæus*,[9] and in *Xenophon*,[10] and in *Callistratus*.[11] Both *Pliny* and *Martial*,[12] I remember complain of the Usage; tho' the following Ages imitate it, & continue it.[13]

My Letter, Sir affirms itself to be a *Faithful Messenger*; it uses alwayes all possible Fidelity in its Relations. It brings indeed another Sort of *Snow*; I shall Rejoice, if it at all *Refresh the Souls of my Masters*,[14] and if it may be allow'd the Honour of being thought to convey a Curiosity at all worthy of the notice, of that Illustrious Society; whereof, and of their Honourable Secretary, I am ambitious to be reckoned,

Sir,

a most humble

& obedient Servant.

{C. Mather.}

7 Prov. 25:13.

8 Mather's primary source is Bartholin's *De Nivis Usu Medico Observationes variae* (cap. 14, pp. 90–91), who refers to the "*glossa Danicæ*" (p. 90), relating that the ancients chilled (and refrigerated) their wine with snow.

9 Athenaeus of Naucratis (fl. 2nd–3rd century CE), *Deipnosophistae* (3.35, 3.124).

10 Xenophon of Athens (c. 430–c. 354 BCE), philosopher and historian, in his *Memorabilia* (2.1.30).

11 The Athenian orator and statesman Callistratus (d. c. 360 BCE) enjoyed imbibing snow-chilled wine, according to Athenaeus Naucratis, *Deipnosophistae* (3.125).

12 Pliny the elder (c. 23–79 CE), in *Natural History* (19.14.57), and the Roman poet Martial (c. 38–c. 104 CE), in his *Epigrammata* (9.22.708; 14.117).

13 A more detailed version of this paragraph – the substance of which is based on Bartholin's *De Nivis Usu Medico* (cap. 14, pp. 90–92)–appears in Mather's commentary on Prov. 25:13 (*BA* 5:306–08).

14 1 Cor. 16:18.

To Mr. Waller.

{Cover Letter accompanying CA (Series Two).}[15]
{To Richard Waller.}
{c. December 1, 1713.}

Sir,

An Honour too great for me to have hoped for, the Letters werewith you have lately favoured me, have allow'd me some hope of my arriving to. The Academy of the *Nascosti*, at *Milain*; that of the *Innominati*, in *Parma*; and that *Incogniti*, in *Venice*, would by their Titles have been more fitt for so obscure a Person, than a room among your Honourable *Virtuosi.*[16]

I have nothing to render me worthy of a Relation to so Illustrious a Society, as that whereto you have done me the Honour of proposing my Admission as a Member, except it be my vast Veneration for the Persons that compose it, and my firm Resolution, to have annually as long as I may live, contributed unto its Treasures, by my best (tho' mean) Communications, altho' I should never have been in this Way obliged unto it.

And now you give me a Prospect of reaching to such a Dignity, there will be some Addition to my Assiduities, as well as my Capacities, that if I be not one of the *Ardenti*, which they tell me, is a Title worn by one of the Academies at *Naples*, yett I will not be one of, the *Otiosi*, which. they say, is a Title worn by another of them.[17]

I first render my most humble & hearty Thanks unto you, for doing the Part of a Patron on my behalf, in the Recommendation you have been pleased to

15 According to G.L. Kittredge, "Election" (1912), pp. 84–87, Mather addressed this letter (December 1, 1713) to Richard Waller, Secretary of the Royal Society, who had nominated him for membership to the Society. See Richard Waller's holograph letter at the RSL (July 22, 1713), Appendix 3.

16 The Accademia dei Nascosti (i.e., the Hidden Ones") was founded in Rome, Italy; the Accademia degli Innominati (i.e. "the Nameless Ones") was founded in 1580 in the Italian city of Parma, and the Venetian Accademia degli Incogniti (Academy of the Unknown) was a learned society of freethinkers, founded in Naples (c. 1540) and Venice (1626–27). The same self-deprecating reference appears in Mather's Letter to Richard Waller (undated draft, in Kittredge, "Election," p. 86), thanking him for being instrumental in nominating and electing him to membership in the Royal Society. See also Mather's letter (March 10, 1715) to Waller (Appendix 7) and *DL* (1699), p. 7.

17 Mather's play on words suggests "Ardenti," *shining*, *ardent* or *eager*; "Otiosi," *careless* or *negligent*, *leisurely*.

give me! And I assure my self, that an affayr, which will so much strengthen my Opportunities to render myself a Master of what may be found useful & proper to be transmitted from these Parts of the World, under such a Management as yours, cannot miss of being brought unto Perfection.

I must further pray you to be my Instructor, (for, Sir, you must imagine that you have now a sort of tame Indian under your Tuition,)[18] what the Rules of my Relation will oblige me to observe, in the Point of those little pecuniary Expences, with which I am to consider the Treasurer of the Society.

Your Instruction, which my Ig{norance}[19] at so long a Distance from you, rende{rs necessary} for me, will in this, & in ever{ything} else be complied withal, by, Sir,

Your most obliged Fr{iend},

{C. Mather.}

18 An oft-repeated tongue-in-cheek self-deprecation, Mather alludes to the British view of colonials in America as no better than wild Indians, and cannibals at that. See Jonathan Swift's satire "A Modest Proposal" (1729): "I have been assured by a very *knowing American* of my acquaintance in London, that a young healthy child well nursed is at a year old a most delicious, nourishing, and wholesome food, whether stewed, roasted, baked, or boiled; and I make no doubt that it will equally serve in a fricassee or a ragout" (emphasis added).

19 The manuscript corner is torn here, making the ends of the subsequent text difficult to read. The textual reconstruction is rendered in braces.

2.2. SURPRISING INFLUENCES OF THE MOON.

1714. Sent July 11, 1716.[1]

To Richard Waller or John Woodward.

Sir,

Tho' our Incomparable Præsident, Sir *Isaac Newton*, than whom Philosophy has never had a Dictator more justly submitted to, ha's at Length obliged us with a *Theory of the Moon*,[2] which ha's performed that which all former As-t{r}onomers have thought almost impossible; Yett the History of the Influences which the *Moon* ha's upon *Sublunary* Bodies, is among the *Desiderata* of our Philosophy. With my Consent, he shall merit more than the Title of a, *Solomon Jarchi*, who gives it unto us.[3] Our most valuable Dr. *Grew*, in his *Cosmologia*,

1 Fearing the loss of his 1714 "Curiosa" letters (Second Series) during the transmission process to London, Mather included his undated CA 2.2 "Surprising Influences of the Moon" as the 9th item in his 1716 shipment (Third Series) and redated the new copy as July 11, 1716. See G.L. Kittredge, "Scientific," pp. 28–30.

2 Sir Isaac Newton (1642–1727), in his famous *Philosophiae Naturalis Principia Mathematica* (1687), esp. lib. III, Hypoth. IX, pp. 404–16, 421, 427–74, holds forth on the gravitational power and motion of the moon and the earth. He was president of the Royal Society from 1703–27. See also *A New and most Accurate Theory of the Moon's Motion; Whereby all her Irregularities may be solved, ... Written by ... Mr. Isaac Newton, And Published in Latin by Mr. David Gregory in his Excellent Astronomy* (1702), referring to David Gregory (1659–1708), F.R.S., Scottish mathematician and Savilian professor of astronomy at Oxford, who advocated Newtonian physics in his *Astronomiae Physicae & Geometricae Elementa* (1702); see lib. IV, Prop. XXIX, pp. 330–36, Prop. LXV, pp. 384–87.

3 Solomon Jarchi, aka. Shlomo Yitzchaki (1040–1105), French rabbi and commentator on the Talmud and Tanakh. The acronym by which he was known, Rashi, was sometimes expanded to "Rabbi of Israel"; alternatively, the name by which Mather refers to him, Jarchi, may have been derived in part from the Hebrew word for "moon," since Rashi was supposed to have lived in Lunel, Provençe. Rashi's classic commentary on the Chumash appeared in *R. Salomonis Jarchi, Commentarius Hebraicus in Pentateuchum Latine Versus* (1710), by Johann Friedrich Breithaupt. Rashi's gloss on Gen. 1:16 reads as follows: "They [the two great lights, sun and

ha's enumerated more than a dozen remarkable Heads of, *Effects* and *Motions* and *Changes* in the World, over which that *Satellit* of ours, ha's a sensible *Dominion*.[4] Our *Lunaticks* are not the only Instances, of the Powers which the *Moon* exerts upon our Humours. Besides the Powers of the *Moon* exercised in the Animal World, the Enumeration whereof would swell into a considerable Volumn;[5] wherein, what Monsieur *Andry* relates, confirming the Observation of *Borellus*, about the Success of Medicines for *Worms* in Humane Bodies, taken in the *Wane* of the *Moon*, would be an Article particularly wonderful;[6] What occurs in the *Vegetable World* is notable. Our Gardeners and our *Husbandmen* will multiply {the}Varieties upon us, till they make a Volumn, which neither a *Columella*, nor a *Tom Tusser* have reach'd unto.[7]

The *Georges*[8] of my Neighbourhood, enable me to add unto your Stores, when you shall direct a general Collection of these things, to be proceeded in. There are especially Two Curiosities, which they have help'd me to.

---

moon] were created equal but then, the moon was diminished because it objected by saying: 'It is impossible for two kings to rule with the same crown.' Because He diminished the moon He increased her hosts [stars] to placate her" (*Metsudah Chumash/Rashi Bereishis* 1:11).

4 Nehemiah Grew, M.D., F.R.S. (1641–1712), was an English physician, botanist, physiologist, and author of *Anatomy of Plants* (1682). In the present context, Mather refers to Grew's popular *Cosmologia Sacra: or a Discourse of the Universe* (1701), bk III, ch. I, sec. 25–29, pp. 88–89, which considers the influence of the moon on a variety of terrestrial conditions, including human health. Mather also put this work to good use in his *CP* (1721) and in *BA* (2010–).

5 See Mather's essay "Of the Moon," in *CP* (1721), Essay XI, pp. 47–52.

6 Nicolas Andry de Bois-Regard (1658–1742), French medical doctor and pioneer of orthopedics, author of *De La Génération Des Vers Dans Le Corps De L'Homme* (1700); see also Mather's *CP* (1721), Essay XXVII, p. 144. Mather's reference to Borelli's warning appears in Andry's *An Account of the Breeding of Worms in Human Bodies* (1701), ch. 10, p. 175[V], which lists *Historiarum et Observationum Medicophysicarum Centuriae IV* (1657), Cent. 1, Obs. 89–90, by Pierre Borel (c. 1620–1671), French chemist, alchemist, physician, and botanist. Borelli's work went through numerous editions and reprints. Mather also employed Andry's work in his discussion of a prodigious tapeworm extracted from the stomach of a neighbor's African servant, in Mather's letter CA 4.4 "A Prodigious Worm," to John Woodward ([December] 12, 1717). On this by-no-means rare intestinal plague, see also the account published in *The Boston News-Letter* (January 7–14, 1717).

7 Lucius Junius Moderatus Columella, 1st-century CE writer on agriculture in the Roman empire, author of *De Re Rustica*. Thomas Tusser (c. 1524–80), English poet and farmer, author of *Five Hundred Points of Good Husbandry* (1573).

8 Mather is punning on the Greek/Latin derivation of the name *George* from *Georgos/Georgius* for *farmer*, *gardener husbandman*. Here, he probably alludes to the

The One is this. If our *Chestnutt Wood*, whereof we sometimes make our Fuel, be fell'd while the *Moon* is *Waxing*, it will so sparkle in the *Fire*, that there shall be no Sitting by it in safety. If it be cutt while the *Moon* is waining, there will be no such Inconvenience.[9]

The other is this. Whatever Timber we cutt in Two Wanes of the Moon in a Year; the Wane in *August*, and the Wane in *February*; it will be forever free from Worms; No *Worms* will ever breed in it.[10]

Unto these, I will add a Third, which relates unto the *Weather*. We very much observe it in our Countrey, and govern our affairs by the Observation, That, as *the Winds are in the last Quarter of the Moon, so they generally govern in the next Three Quarters*.

{C. Mather.}

---

gardeners in his neighborhood who still maintained their garden plots behind their houses in Mather's day, even in the increasingly crowded North End.

9 Mather offers the same explanation in his *CP* (1721), pp. 51–52. See also "A Discourse concerning the most seasonable Time of Felling of Timber," *PT* 17.192 (February 1, 1691): 455–61, by none less than Robert Plot, F.R.S. (1640–96), English natural philosopher, professor of chemistry at Oxford, and secretary of the Royal Society and editor of its *PT*.

10 See *CP* (1721), p. 52.

2.3. A MONSTER.

June 21, 1714.

{To Richard Waller or Dr. John Woodward.}

{Sir,}

The Conversation which you allow me the Honour of maintaining with you, you may depend upon my Entertaining you with nothing, but what upon the strictest Inquisition, will be found so true, that I shall even challenge the Name of the late Inquisitor General in *Portugal*; whose Name, I find was, *Dom Verissimo.*[1]

What I now relate, I beheld, 15.^{d} 8.^{m} 1713. *A Woman* in my next Neighbourhood, was delivered of Two Daughters, which were so united, as to afford us a shocking Spectacle: whereof I was myself one of the Spectators. They were in all Points entirely Two lusty Children; Only from near the Top of their Breasts, to near the Bottom of their Bellies, they grew together, in such a Manner, that without a Destruction to both, it would have been impossible to have parted them. The Union was for the whole Breadth of the Chest; and their Four Paps were on their Sides, below the Arm-pitts. Their Heads lay on the Cheeks of each other: Their Arms in a mutual Embrace of their Bodies; Their Legs all stowed close, that the whole might ly in as little Room as could be.[2] It was a little surprising, that the Mother had no harder a Travail,

1 The Grand Inquisitor, Archbishop Dom Verissimo de Lencastre (1615–92), archbishop of Braga (1676–92). See A.I. López-Salazar (2013).

2 In his anonymous sermon *OP* (1707), Mather likens a born sinner to a monster in terms reminiscent of the malformed twins he had witnessed himself: "Tis as Odious, as Frightful, as to be born with the Scales of the *Leviathan*; to be born with the *Belly* where the *Head* should be; with *Eyes* in the Back-part of the Head; with the *Hands* growing to the Mouth; with the *Feet* made for ever to go awry every Step" (p. 9). Mather's theological and moral reading of monstrous bodies (human or animal) is at the cusp of shifting disciplines in the emerging field of embryology, in which such interpretations of nature and its causation gradually gave way to the Baconian model in which empirical, medical, and biological conceptions of what is normal or abnormal were developed and held up to new standards by which

in bringing such a Burthen into the World. Had these united Miserables lived, our first Thoughts cannot but in a Minute or two suggest a Variety of Cases, which might have occurred in their Circumstances, not easy to have been encountred or answered; what must the One have done, if this or that which were easy to have been supposed, had befallen the other? But the Cases were all superseded at once, by their Dying as soon as they were born.

I remember one *Bouchard* giving the Relation of a Monster, ha's this Remark.[3] *Inter Causas Monstrorum Auctores {ponunt} Iram Dei, si Viri et Mulieres, non habit{â} ratione, à Deo et Naturâ præscriptarum {legum} sibi cohabitant.*[4] And it is not amiss, the Displayes of the Divine Providence in such things, be observed, if it be done with a due Regard unto the Rules of Modesty, & of Charity; and with the Reservation, which our Saviour made upon the Case of the Man Blind from Nativity.[5]

Mihi verò de hujus modi Monstrosi partus Matre anxiò aliquando perquirenti, (cujus nomen quadam reticere, mihi est Religio,) an per Virgam sic

---

to examine evidence. Closely related to this issue are the concepts of preformationism and epigenesis. Were organisms like a fetus or seed fully pre-formed and embedded as miniscule entities (homunculi or animalculae) in the body and their growth merely an augmentation of the tiny being, or did organisms begin from an undifferentiated mass (an egg) and then develop after fertilization? Either way, the question of preformationism was complicated by hybrids, such as the sterile offspring of a donkey and a mare, or the cross-pollination of Indian corn or of gourds and squashes (*BA* 1:325, 386–87; 594–99; and CA 3.12 "*Curiosa Botanica*"); see also Mather's CA 1.5 "Monstrous Births"; CA 3.1 "Monstrous Impregnations"; and CA 3.2 "A Monstrous Calf." Questions of this nature were hotly debated in Mather's time. For background, see E. Gasking (1966), C. Zircle (1969), K. Park and L.J. Daston (1981, 1998), C. Pinto-Correia (1997), M. Hagner (1999), M.-H. Huet (2004), and F. Kraemer (2023).

3 François Bouchard (fl. late 17th century), professor of medicine, at the University of Besançon (France). Mather quotes from Bouchard's "Observatio XII(I), 'Infante Monstroso Lugduni in viam publicam die V. Martii A. MDCLXXI. exposito." *Miscellanea Curiosa sive Ephemeridum . . . Annus Tertius* (1672, 1673), 3:14–16 [17–18]. Bouchard's "Observatio XII" is thought to be the first description of Roberts Syndrome, a genetic disorder characterized by limb and facial abnormalities.

4 "Among the causes of monsters, authors place the wrath of God if men and women do not live together in accordance with the laws prescribed by God and nature." Mather's Latin quotation is highly redacted when compared to Bouchard's paragraph (p. 17, paragraph 3).

5 John 9:2–3, "And his disciples asked him, saying, 'Master, who did sin, this man, or his parents, that he was born blind?' Jesus answered, 'Neither hath this man sinned, nor his parents: but that the works of God should be made manifest in him.'"

duplicatam, culpæ alicujus pœnas lueret, statim fuit indicatum ac exploratum, quod mollior illa, et lasciva Mulier, nescio quæ medi{c}a, ac Medicamenta solitam fuisse adhibere, nè à mariti amplexibus gravida fieret, atque prolis parturiendæ ac educandæ doloribus obnoxia foret. En, Justissima Dei Judicia, ut quæ nullam omnino prolem suscipere voluisset, monstrosè gemellas, cum Summo Vitæ periculo, et non sinè Viciniæ stupore, eniteretur![6]

While I am writing of this, there occurs, what perhaps at another Time, I should not have thought worth mentioning. A Friend of mine, ha's a *Cow*, that at One birth brings him Three Calves, of three Several Colours, a Black, & a White, & a Red: In which tho' there be nothing *Monstrous*, yett because it is what is not *Common*, you will allow my having mentioned it.

But that my Epistle may not grow to a *Monstrous* Bigness, & into the Dimensions & Arguments of Two, I shall here conclude it, & subscribe –

Sir,

Your most hearty

Friend & Servant.

[C. Mather.]

6 "On a certain occasion, when I was anxiously making inquiries about the case of a mother of this sort of monstrous birth (a woman whose name it is a certain scrupulous principle of mine to withhold), in the attempt to determine if she was being punished for some crime by means of this doubled offspring, suddenly it was discovered and brought to my attention that this rather weak and lascivious woman was in the habit of applying some herbs and drugs in order not to become pregnant from the embraces of her husband and not to become subject to the sorrows of giving birth and bringing up a child. Behold, the justice of God, that the woman who had wished not to bear any offspring at all would give monstrous birth to twins, accompanied by extreme danger to her life and the astonishment of her neighbors!" (David Lupher's translation). Mather may also have been thinking of a case in his own family, for in the morning hours of March 28 1693, his wife Abigail, who had been "affrighted with an horrible *Spectre* in our Porch, which Fright caused her Bowels to turn within her; and the *Spectres* which both before and after, tormented a young Woman in our Neighbourhood, brag'd of their giving my Wife that Fright, in hopes, they said, of doing Mischief unto her *Infant* at least, if not unto the *Mother*" (*D* 1:163–64). No surprise then that when Abigail gave birth to a son (Joseph), he died of an inoperative case of obstructed bowels on April 1 1693.

2.4. CONTAINING THE DESCRIPTION OF THE *MOOSE DEERE*.[1]

June 21, 1714.[2]

To Richard Waller, Esquire.

Sir,

You remember, how long since, & by whom, a Letter began, *Si Philosopharis, benè est*.[3] I have the Satisfaction, to begin mine, with such an Assurance of your *Good Health*; which ha's been demonstrated in your particular Enquiries, after the MOOSE in our Countrey.[4]

We have Two different Sorts of *Moose*, the *Black* and the *Red*. The *Black* are called, *The Northern Moose*; The *Red*, the *Southern*.

The *Black Moose* have *Horns* that are *Flatt*; with *Flukes*; whereof some are so large, that the Indians formerly made a Sort of Trayes of them.

1 This notation is written in a different hand.

2 Composed on June 21, 1714, Mather's letter to Richard Waller was read to the assembled members of the Royal Society in October of the same year, bearing the canceled notation "Read Oct. 28. 1714" and title "Containing the Description of the *Moose Deere*" on the original at the RS [EL.M2.35]. Both the canceled notation and title are in a different hand – most likely that of Waller himself or of his assistant. This document is fuller than the MHS draft copy (MHS, MS. N-527, 327–28), which bears the canceled title [The Moose.] / A brief Account, / of the MOOSE, / J* a Creature frequent in the / New-English Colonies./

3 "If you are studying philosophy, it is well." Lucius Annaeus Seneca, *Epistulae ad Lucilium* (15.1).

4 In his letter (Boston, 30d. 1m. [March] 1714) to John Winthrop (1681–1747), Mather had apparently reached out to his friend for a description of the moose for his "Curiosa" letter to the Royal Society (*MP* ([868], p. 417). Roughly seven years after Mather had submitted his letter to the Secretary of the Royal Society, Mather's fellow New Englander Paul Dudley, F.R.S. (1675–1751) had his essay on the same topic published as "*A Description of the Moose-Deer in* America," *PT* 31.368 (January 1, 1721): 165–68.

Josselyn asserts the Dimensions of these *Horns* to have been such, that the Tips of them have been Two Fathoms or 12 foot asunder.[5]

The *Red Moose*, have large Horns; but *Round*. It runs in my Mind, I have read somewhere, That the *Horns* of these Animals are sometimes dug up in *Ireland*, tho' the whole Race have long been utterly extinguished there (*Quære*, By Murrains, or, by Hunters?),[6] & not one of them have been seen in the whole Island for diverse Ages.

We are not yett so deprived of ours; and [a][7] present of a *Buck* and a *Doe*, had been lately made unto Her Majesty,[8] by an Honourable Friend of mine, if a certain Captain would have executed his Orders, to take them on board; but both of them are lately Dead.

They are a very Stately, Courageous, Intrepid Creature: very proud in their March; Regardless of any thing that may occur in their Way.

Their Colour is a *Red*, a little darker than our *Deer*; Tho' they turn almost *Grey* towards the Spring. They then shed their *Horns*, and their Hair. When they shed their *Coat*, it is in *Flakes* or *Fleeces*, which hang together, about a Foot Square, or more.

Their *Stature* is about as big as that of an *Horse*. Indeed *Schroder* mentions One he saw killed in *Finland*, Seventeen Spans high.[9] We have them not ordinarily of such an *Stature*,[10] tho' they deserve to come into the Tribe of those, whom the Sacred Language Calls, [Job. XLI. 34.] *The Children of Heighth.*[11] And *Josselyn* affirms, That ours are sometimes Twelve foot high.[12]

5 John Josselyn (fl. 1638–75), *New-Englands Rarities Discovered in Birds, Beasts, Fishes, Serpents, and Plants of that Country* (1672), p. 19.

6 Mather inserted this question in the left margin without indicating its position in the text

7 MS damaged; text supplied from MHS draft.

8 Queen Anne died in 1714.

9 Johann Schröder, aka. John Schroder (1600–64), German physician and pharmacologist, whose *Pharmacopoeia Medico-Chymica, Sive Thesaurus Pharmacologicus*, Editio Secunda (1644), went through several editions, emendations, and translations. The work explores the medicinal uses of herbs, minerals, and animal products. Mather refers to the corrected "Editione Ultima" (1665), lib. V, Classis I, sec. II. *Alces* [i.e., elk, moose], p. 674, which identifies the height of the Finnish elk as "17. *palmos altam.*" This reference does not appear in the English translation of this work. See Johan Schroder, *The Compleat Chymical Dispensatory, in Five Books* (1669), bk 5, p. 507.

10 Mather's holograph reads "Elevation," but the word was canceled and replaced with "Stature" interpolated in the margin and written in a different hand.

11 Job 41:34 (KJV): "He beholdeth all high things: he is a king over all the children of pride."

12 Josselyn, *New-Englands Rarities Discovered,* p. 19.

Figure 8. Depictions of Unicorns. In Samuel Bochart, *Hierozoicon* (Lugduni Batavorum, 1712), ch. 27, cols 955–58. Courtesy of the National Central Library Rome.

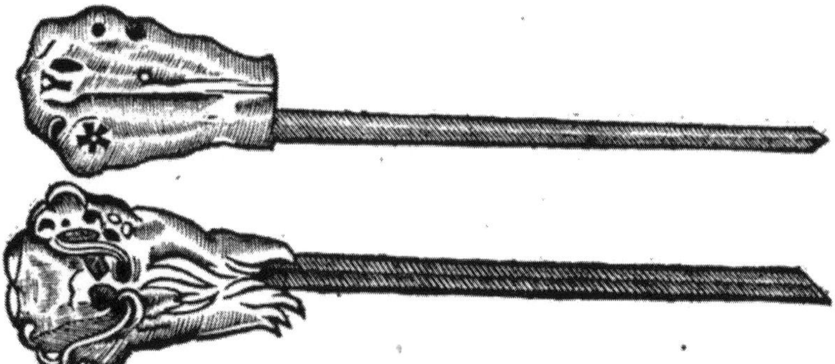

Figure 9. Unicorn Horns. In Bochart, *Hierozoicon,* pars 1, lib. 3, cap. 27, cols 956, 958. Courtesy of the National Central Library Rome.

In their Shape, they very much resemble, that which the Romantic Painters draw for an *Unicorn.*

We are now satisfied, that the *Unicorn* which pretends to support our *Arms*, is as *Fabulous* for a Beast, as our Patron St. *George* is for a *Saint*. We are well Content, that he should be chased out of our Translation of the Bible: Especially when we shall find ourselves obliged Else to read Such a Solæcism as that; [Deut. XXXIII. 17.] *The TWO HORNS of Unicorns.*[13] We are willing, that the *Reem* would be translated, as *Bochart* would have it, *The Arabian Wild-goat.*[14] But the *Doe* of our *Moose*, is really, at least sometimes, an *Unicorn*; That is to say, She has a *Single Horn* growing out on her Forehead.

Their *Horns*, are above a *Yard* in Heighth; and as much in their Distance from one another; Yea, tis no rare thing, for the Tips of the Horns to be found more than Two Yards asunder. They are for a long while of such a *Tenderness*, that if they can be catch'd, they are capable of being stewed into a Dish, the most salubrious in the World. But, as if they were *ashamed* of what befalls them in their *Annual Shedding* of them, they so hide them under Ground in their plashy Swamps, that they ly effectually *Covered* and *Concealed* there, and are not often mett withal.

They *Breed* and *Feed* much Like our *Beeves*.[15] But it is well for them, that they live in Countreyes well furnished with Vegetables, that have the heighth of *Shrubs*. As they are not sufficiently tall, to pillage the *Trees* for their nourishment; so their Necks are not long enough, without a deal of Trouble in spreading their Forefeet, for to reach the *Grass*. A *Bushy* and *Brushy* Food, is what they most live upon.

Their *Tails* are short, like a *Rabbets*, and cover nothing.

13 Deut. 33:17 (KJV): "His glory is like the firstling of his bullock, and his horns are like the horns of unicorns … " See also *BA* (1:584n; *BA* 2:337). In his letter to James Jurin (June 3, 1723), our CA 7.2 "A Monstrous Dragon," Mather dismisses all those (including painters of the coat of arms of the United Kingdom) who endow dragons with wings.

14 Samuel Bochart, "Præfatio ad Lectorem" (pp. 31–32) and *voce* "monoceros," "unicornibus," "urus," in *Hierozoicon* (1712), pars prior, lib. 3, caps. 26 and 27, cols 930–48, 948–72, traces among the ancients the rendition of the Hebrew "reem" ראם or "rem" רים (Deut. 33:17) variously associated with an elephant, wild ox, antelope, gazelle, and wild goat. An amusing depiction of the mythical single-horned and two-horned unicorn appears in ch. 27, cols 955–58 (see Figure 8).

15 Archaic term for cattle.

They are distinguished for furious Venery; And they commonly have *Three Fawns* at a time.

Their *Flesh* is a very acceptable Dish; between *Beef* and *Venison*: exceeding Fatt, Juicy, Tender and Wholesome. We find it sometimes near Six Inches thick, in Fatt.

The *Tongue* and the *Nose*, well-dress'd, is a Dish for a *Sagamore*.

And the *Flesh* of their *Fawns*, is vastly præferrible, no doubt, unto that of an *Asses Foal*, so highly esteemed among the *Romans*.

I wish I could have obtained your *Blessing*, by Regaling you, with so delicious a *Dish*. Whereas, now you must be putt off with only a mean & short *Story* of it, from One who yett hopes to be esteemed,

Boston, N. E. Sir,

Jun. 21. 1714. Your Sincere Friend,

& hearty Servant.

Cotton Mather.

2.5. THE PIGEONS.[I]

June 1714.

To Dr. John Woodward.

Sir,

Proposing to answer your Desire of a treat with a further Mess of our *Pigeons*, I must in the first Place confess, that they sometimes make me think of the *Quails*, with which the Appetites of the *Israelites* were gratified in the Arabian Deserts. The Sacred Historian mentions them as arriving in such a Number, that they were *as the Sand of the Sea*; and they were scattered not only thro' the whole Camp, but also a Dayes Journey which we may take to be Twenty Miles at least, on both sides of it: and it is added *They were, as it were two Cubits high, upon the Face of the Earth*.[2] I know, both *Jonathan*, and *Jerom* carry that Passage, as only meaning, That they flew *Two Cubits high above the Earth*;[3] A Jewish Rabbi, therefore notes upon it, that their Flight was, *Ad humani pectoris Altitudinem*,[4] that so there might be the Less

1 For Mather's discussion of American birds, see his third letter (November 19, 1712) to Dr. John Woodward, CA 1.3 "American Birds." Mather's scientific description of the now-extinct American passenger pigeon (*Extopistes migratorius*) also appears in his *CP* (1721), "Essay 30," pp. 192–95; and *CPS* (1994), pp. 204–08. See A. Schorger (1938), who also includes Mather's letter to Woodward (July 4, 1716), on the nidification of pigeons, CA 3.3. See also F.T. Lewis (1944).

2 Numb. 11:31.

3 St. Jerome (*Divina Bibliotheca Pars Prima: Liber Numerorum* 11:31) [*PL* 028. 0364B]. Jonathan ben Uzziel, in Walton's *Biblia Sacra Polyglotta* (4:260); Etheridge, *Targums Onkelos and Jonathan Ben Uzziel* (2:376).

4 In his commentary on Numb. 11:31 (*BA* 2:866), Mather identifies the "Jewish Rabbi" as "R. *Selomo*," i.e., R. Salomon (Shlomo) Jarchi (Rashi), and supplies a longer version of Rashi's gloss on the matter: "*Volabant ad Humani pectoris Altitudinem, ne quisquam in earum collectione se attollendo aut deprimendo fatigaret*": "They [quails] flew to **a height** which corresponded **with a person's heart [or breast],** so that it would not be tedious to gather them, neither to rise nor to bend" (*Metsudah Chumash/Rashi* 4:155). (The phrases in **bold** correspond to Mather's Latin extract.)

of Trouble in catching them: And *Philo* takes notice, that such a Flight was rendered; εις ευθηρον· that they might have the more *Easy Fowling* of it.[5] Nor is this Disagreeable to *Plinyes* Remark, on the *Quail*, That, *cum ad nos venit, Terrestris potius quam sublimis est.*[6] But, why may it not suffice to say, That they lay in scattered & numberless Heaps, every where, not far asunder *Two Cubits high*. And what we render *Ten Homers*, which are mentioned as the least Quantity of them gathered by any who employ'd themselves in gathering of them, a famous Interpreter, I remember, directs us with very good Reason, to read, *Heaps*, rather than *Homers*; and if I don't forgett, both *Onkelos*, and the *Arabic*, do read it so.[7] And therefore there was no need for, *À Lapide* to be at the Pains of casting up the præcise Number of the Birds then gathered by the *Israelites*, which he finds to be *Twelve Thousand Millions*.[8] On this Occasion I call to mind, *Varro* tells us, that the *Quails* visited *Italy* in the Season of them, *Immani Numero*.[9] *Pliny* and *Solin{us}* add, That such was the Number of the *Quails* then flying over the *Mediterranean*, as to endanger the Vessels, which thro' Weariness they would sometimes light upon.[10] But there was no where a greater Plenty of *Quails*, then in *Egypt*, from whence they were now brought unto the

5 Philo Judaeus (*De Vita Mosis* 1.209, line 6) translates "to be easily caught" (*On the Life of Moses* 1.37.209), in *Works*, p. 478.

6 Pliny, *Naturalis historia* (10.33.65): the quail (a small bird) "when it has come to us remains on the ground more than it soars aloft."

7 Mather's primary source (*BA* 2:866–67) is Samuel Bochart, *Hierozoicon* (1663), pars 2, lib. 1, cap. 15, col. 105, lines 58–65, and col. 106, lines 10–20. The ancient Hebrew "homer," measuring capacity (volume), amounts to c. 100 liters or c. 26 US liquid gallons (*ABD* 6:903–04). Both the Targum Onkelos and *Versio Arabica* of the Pentateuch (Numb. 11:32) are accessible in Walton's *Biblia Sacra Polyglotta* (1:593).

8 The distinguished Flemish Jesuit theologian Cornelius à Lapide, aka. Cornelis van den Steen (1568–1637), on Numb. 11:32, in *Pentateuchum Mosis Commentaria* (1626), p. 785, estimates that the meandering Israelites "collegerint duodecim millia millionum coturnicum"; i.e., "they have collected twelve thousand million quails" – indeed an unimaginable multitude rivaling the vast flocks of American Passenger pigeons.

9 The Roman historian Marcus Terentius Varro (116–27 BCE), in *De re rustica* (3.5.7), speaks of "tremendous numbers" of quails, which seasonally pass across Italy.

10 Pliny (*Naturalis historia* 10.63.5) reports that the flocks were so large they were "not without danger to seafarers." The Latin geographer and grammarian Caius Julius Solinus (fl. 3rd century CE) confirms the existence of such vast flocks in his *De Mirabilibus Mundi* (11.20).

Israelites. Many Ancient Writers tell us, They were so many in their Seasons, that the *Egyptians* not being able quickly to devour them salted them up; tho *Theocritus* tells us, There were Thirty thousand Towns in that Countrey;[11] and *Josephus* tells us, there were many more than seven hundred & fifty Myriads of People there.[12] But, while I am thus falling to a dish of the Israelitish *Quails*, *Job Ludolphus* comes in, & wholly turns my Stomach, by proving, That there were no *Quails* at all in the Story, but that all this while they were no other than Locusts, which are intended by the Term *Selan*, which we have mistranslated, *Quails*, upon the Credit of One single Jew, who is not alwayes to be relied upon.[13]

A better Dish than *that*, you would have in our *Pigeons*; A Bird which in almost every thing resembles your *Turtle-Doves*; only that it is a little Bigger. The *Numbers* of these, that visit us in their *Seasons*, are such, that I am almost afraid of giving you a true Report of them, lest you should imagine a *Palæphatus*[14] were imposing his *Incredibles* upon you. Yett it will a little answer the Intention of the Correspondence wherewith you favour me, if I do Report something of them.

I affirm to you then; That sometimes we have mighty Flocks of these *Pigeons* flying over us; thousands in a Flock; the best Part of a Mile[15] Square

11 The Greek pastoral poet Theocritus Syracusanus (c. 300–260 BCE) says much the same in his bucolic *Idyllia* (17.82–84).

12 Not far behind his predecessors is the Jewish historian Flavius Josephus (c. 37–100 CE) in estimating that Egypt received enough quails annually to feed its "seven millions five hundred thousand men, besides the inhabitants of Alexandria" (*Wars* 2.16.4), in *Complete Works*, p. 489.

13 The German orientalist at the Franconian University of Altdorf, Hiob Ludolphus, aka. Job Leutholf (1624–1704), in his *Appendix Secunda ad Historiam Aethiopicam* (1694), pars 2: *De Locustis*, cap. 1, §§1–5, asserts that the Hebrew noun שְׂלָו "*Selav*" (Exod. 16:13; Numb. 11:35) should not be rendered "*coturnicibus*" (quails) but "*locusti*" (locusts). In his commentary on Numb. 11: 31, 33, Mather came to accept Ludolph's rationale because such vast numbers of quails piling up "would have been præposterous, & but the sooner have made them stink" (*BA* 2:862–70; qtd. at 870).

14 Probably Palaephatus of Paros, extant in *Peri Apiston Historion, "On Incredible Tales,"* a work that aims to demythologize Greek legends through a skeptical Pyrrhonist interlocutor, who explains them with references to mundane occurrences.

15 The remaining manuscript pages (2r and 2v) are misfiled in the MHS holograph copy. They are included in CA 3.7 "The Strength of the Imagination," pp. 2r and 2v (MHS 323) but restored to their rightful place in our edition. G.L. Kittredge, "Scientific" (p. 16, # 5), does *not* indicate that the MS is incomplete.

occupied by a Flock; These Passing along, the Welkin in a Manner obscured & covered with them; & several Hours have run out before the Appearance of these Birds thus making the best of their Way have been over.

They have been frequently sold for Two Pence or Three Pence a dozen;[16] tho' two or three of them, Roast or boild or broil'd, may make a Meal for a Temperate Man. Yea, they are sometimes kill'd in such Plenty, that the Countrey-People feed their Hogs with them.

One of my Neighbo{urs} has killed no fewer than two & thirty dozen at one Shott.

Gentlemen have complain'd unto me, that they have litt in such Numbers on their Trees, as to break down the Limbs thereof, & spoil their Orchards.

They will sometimes roost at Night in such Numbers among our Thicketts, that our People with no other Weapons than Sticks & Poles kill Thousands of them.

One worthy Person of my Acquaintance, had a Descent of them in his Neighbourhood, in the Month of *December*, a very unusual Time of the Year; while there was yett no *Snow*, but many *Acorns* on the Ground, which, tis thought, might then draw them thither. At their Lighting on a Place of Thick Woods, the *Front* wheel'd about, the *Flanks* wheel'd inward, and *Rear* came up, [Sir, He was a *Captain*,[17] who gave me the written Relation!] and pitch'd as near to the Center, as they could find any Limb, or Twig, or Bush, to sieze upon. Yea, they satt upon one another, like Bees, till a Limb of a Tree would seem near as big as an Horse. Tis incredible to tell, how large & strong, & many Limbs were broken down, by this New Burden upon them. The breaking of them was heard at a mighty Distance. The Birds filled more than Half a Mile, about from the Center; and the Noise they made, was like the Roaring of the Sea. The Night was dark; but this Gentleman, and his Sons, with Guns, & some other less noxious Tools, laid in among them, & some they took alive with their Hands; and in the Morning found the Number of their Slain, to be one hundred & three dozen; besides, what some other People had carried away.

A few Dayes ago, I was at the Table of one superiour Gentleman, relating some of these thing{s}. And one, whose Veracity was not to be disputed, sais He had a Story that would cap all of mine. For (said he,) *I have catch'd no less*

[16] See also *CP* (1721), Essay XXX, *Of the* FEATHERED, p. 192. For nearly the same description, see Mather's third letter to Dr. John Woodward (November 19, 1712), CA 1.3 "On American Birds."

[17] Possibly the same Capt. Billinge whom Mather quotes toward the end of his missive.

than Two hundred Dozens of Pigeons, in less than two Minutes of Time & all in one Trap! The pleasant Mention of the *Whetstone* on the Occasion, obliged the Gentleman immediately to explain himself; and add; *Such a Number broke into my Barn, & but by shutting the Doors, I had them all at my Mercy*. And the Truth is; They have appear'd in such Numbers that it ha's been thought, all the Corn in the Countrey, would scarce afford them a Breakfast. But our People, enjoy as much of a Divine *Mercy* in such a Supply for our Table in the Wilderness, as if the Bird *Racham*, which the *Talmuds* tell of, were making his Appearance.[18]

Our *Indians* call these *Pigeons*, by a Name that signifies, *Wanderers*. But where they go, when they Leave us, no man alive can tell. They go not unto our English Colonies to the Southward. If you will not allow them to retire unto some Receptacles above us in our Athmosphære,[19] (which, I wonder, how you can account for your Season-birds, without supposing,) they must have a Retreat in South-West Parts of *America*, whereof we are not yett advised.

I will take Leave at this time, to add one Passage more, which my Friend, Captain *Billinge* gives me in a Letter; from whence I transcribe this Paragraph.[20]

> Once in *April*, it being the time of the Year, that those Birds were on their Nests, they came down to feed on the *Salt-Marsh*. After the Manner of *Doves*, the *Cocks* take care of the young Ones in the Nests, for One Part of the Day; & the *Hens* the other. I have often killd no less than twenty dozen at one Setting; but all generally of one Sex. The *Cocks* were alwayes by

[18] Mather alludes to the charming story in the Babylonian tractate Chullin (63a), in which R. Judah glosses that the noun רחם racham (Lev. 11:18) is "the sherakrak [vulture]." Upon which R. Johanan responds, "Why is it called raham? Because when the raham comes, mercy רחמים [rahamim] comes to the world, R. Bibi b. Abaye said, provided it perches upon something and cries 'sherak-rak.' There is a tradition that if it settles upon the ground and hisses, the Messiah will come at once, for it is said, I will hiss for them and gather them" (Soncino).

[19] In his annotations on Jer. 8:7 (*BA* 5:879–83, esp. pp. 881–82) and in CA 1.3 "American Birds," Mather speculates that certain species of migratory birds might hibernate on some invisible crystalline body in our atmosphere. See also M. Dopffel (2010).

[20] Samuel Sewall (1652–1730), Mather's long-time friend, mentions two potential candidates: "Capt. Ebenezer Billings" (*D* 3:19), apparently of Dorchester, Mass (1651–1717), who was buried on January 28, 1717/18 (*Diary* 3:162–66) and his relative Capt. [Roger] Billings, who died January 27, 1717/18 (*Diary* 3:163 *n.*). Sewall variously spells their name "Billinge" and Billings.

> far the fattest; and when we opened them, we found in their Craws, about the Quantity of half a Gill of a Substance like a Tender *Cheesecurd.* I asked some of our Indians, what those Pigeons had eaten; and why the *Hens* did not Feed on the Same? They answered, It was nothing they had eaten, but something that came naturally into their Crops, as Milk does into the Dugs of other Creatures; and that the Hens could not keep their Young alive, when first hatched; and that this nourished the young Birds, & caused them to grow faster, & fly in half the Time, that any other Birds could attain to it.[21]

All that now remains, is to wish you as many Friends, as there have been seen Fowls in the mighty Flocks I have told you of; but among them, a very particular Consideration for him, who would be,

Sir,

Heartily & Forever

at your Service.

{C. Mather.}

[21] See also *CP* (1721), p. 192.

2.6. LONG FASTING.

June 1714.

To John Woodward.

{Sir,}

Having lately sett some American Dishes before you,[1] perhaps you may be the better præpared for a starving Story, which this Letter will bring you to.

A young Man, about Eighteen Years of Age, putting ashore on a Rock, which they call *Gull-Island*, near *Long-Island*,[2] & not considering the Nature and the Danger of the Tide, lost his Canoe, which he had left unsecured. The Place contains about half an Acre; & is little other than a Rock; he also had no Provisions with him, & it was now the latter End of *January*, in One of our terrible Winters. He was confined here for the Space of Six and twenty Dayes, not having the Company of any Creature but a Dog, which he had made the Companion of his Voyage thither. The first Day of his being there, he was very Hungry; and for the Releef of his Hunger, he gathered about a Pint of Barnacles from the Rock, which was all that could be gotten.[3] With some drift Wood, which was all the fuel the Place afforded, he made a Fire, which he so managed as to keep it burning for Twenty Dayes. The Distress of his Hunger, for the first Six Dayes, after he had spent his Pint of Barnacles, was insupportable. But from thence to the End of Twenty Dayes following, he never felt any Hunger at all. His Dog remained alive with him; & yett in the greatest

[1] Referring to CA 2.5 "The Pigeons."

[2] Probably Little Gull Island rather than its much larger nearby Great Gull Island, both located at the north-easternmost tip of Long Island in the Long Island Sound. Today, the rocky Little Gull Island (c. one acre) is occupied by a lighthouse.

[3] Increase Mather tells a similar story of providential rescue, in *ERIP* (1684), pp. 67–70. See also Cotton Mather's *MCA* (1702) 6.1.3–9 and *CCF* (1711); the *Nottingham*, a galley, was wrecked on Boone Island. The narrative was obtained from Capt. John Dean (*D* 2:71). Such dramatic accounts enjoyed great popularity. See D.D. Hall (1989), pp. 92–94, 123; J. Sievers (2006); and S.J. Pitt (2012).

Extremity of his Hunger, he never had a Thought of killing him. At the End of Nineteen Dayes of entire Fasting & Famine, there fell some Snow and Rain; and he made a little Use of that; whereto some ascribe his Præservation for the last Six Dayes of his miserable Anchoritism.[4] On the twenty Seventh Day, One *Havens* passing by with a Boat, and seeing the young Man there, went & fetch'd him off, & sent him home: He was at first able to eat and walk; but he fell into Languishments, whereof in Three Weeks he died; but before he died, all the Flesh on his Bones was to appearance turned into Lice; he was all over a thick, prodigious, & astonishing Swarm of Lice; mortal Eyes never look'd on a more lamentable Object![5]

I know not whether you will allow my poor *Gull-Island* Youth to increase the Catalogue, which *Fortunius Licetus* ha's given you, when writing, *De ijs qui diu vivunt sine Alimento*;[6] or *Cyriacus Lentulus*, when writing *De Prodigiosis Inedijs*.[7] One would think that the Exemples of *Long Fasting*, which we find related by *Donatus*,[8] and by *Horstius*,[9] and by *Quercetanus*,[10] (and

4 A form of eremitic lifestyle in which Christian devotees withdraw from secular society, commit themselves to an ascetic life of prayer, and live in small walled cells or enclosures attached to monasteries or churches.

5 For similar stories of "remarkable sea deliverances," see *MCA* (1702) 6.1.3–9.

6 One of Mather's favorite and oft-quoted authors, the reference is to *De His, Qui Diu Vivunt Sine Alimento Libri Quatuor* (1612), ("Of those who live a long time without food") by Fortunio Liceti (1577–1657), an Italian physician of Bologna, who asserted that a moderate diet and episodes of fasting promoted longevity.

7 Cyriacus Lentulus, aka. Paulus Lentulus (c. 1560–1613), a Swiss medical doctor of Bern and personal physician of Queen Elizabeth I of England. Like his Italian colleague Liceti, Lentulus promoted the health benefits of fasting, in his *Historia Admiranda, De Prodigiosa Apolloniae Schreierae ... Inedia* (1604). His medical insights into fasting and starvation are based on his personal knowledge of Apollonia Schreier, a peasant villager living near Bern, who embraced long periods of fasting as an expression of her faith. See Hyder E. Rollins's (1921) discussion of such providence stories.

8 Marcellus Donatus (1538–1602), renowned Italian physician, whose popular *De Medica Historia Mirabili libri sex* (1586) went through numerous editions and reprints.

9 Gregorius Horstius, aka Gregor Horst (1578–1636), German anatomist, professor of medicine at the University of Giessen. Among his numerous medical works, Horstius published the first German edition (in Latin) of Marcellus Donatus' *Historia Medica Mirabili Libri Sex* (1613). In lib. IV, cap. XII, p. 364, and lib. VII, cap. 1, pp. 645–57, Donatus discusses some remarkable cases of fasting and starvation ("*Inediae*").

10 Josephus Quercetanus, aka. Joseph Duchesne, Du Chesne (1544–1609), French Paracelsian physician and author of, among several others, *Liber De Priscorum Philosophorum Verae Medicinae Materia* (1603).

many more Foreigners, whom to quote, were an *Ostentation* rather than a *Pertinency*,) follow'd with what we have in Dr. *Plott*,[11] and several Writers of our own, and especially the famous *Darby-shire Maid*,[12] on whom our Philosophers have entertained us with their learned Exercitations; all these might satisfy any Reasonable Man, *That some have endured Hunger for a very long while together*. I don't understand that because Mr. *Velschius* discovered a Cheat in the Abstinence of a Girl at *Ausburgh*,[13] he had Reason to say, *Nos plane persuasi sumus, inedias illas ad unam omni omninò fictas et ementitas fuisse*.[14] But this I understand; *Qui ad pauca respiciunt, citò judicant*.[15] And I have myself known, *Persons* who were *Energumens*, keep a Rigid Fast, for diverse whole Weeks together. One who had Fasted Nine Dayes, and had Liberty to eat for Three Dayes, & then Fasted for Fifteen Dayes more.[16] *Hæc ipse*

11 The English naturalist and antiquarian Robert Plott, aka. Plot (1640–96), author of the popular *Natural History of Stafford-shire* (1686), mentions two cases of "extraordinary *fasting*" of some hapless prisoners in Rome and London, in ch. 8, §§37 [misnumbered 47], 38, pp. 286–87.

12 Mather refers to Mary Taylor, of Bakewell, Derbyshire, whose story was published by John Reynolds (of Kings-Norton), *A Discourse Upon Prodigious Abstinence* (1668). Though having all the tell-tale signs of a 17th-century wonder-book, Reynolds's pamphlet provides a detailed and credible discussion of the physiological causes and effects of long-term fasting. (Mary Taylor's account appears at the end of the pamphlet.) It was reprinted the following year as *A Discourse Upon Prodigious Abstinence ... Humbly offered to the Royall Society*" (1669). See G.L. Kittredge, "Scientific Communications," p. 18. Hungry for more? See the Netflix movie *The Wonder* (2022), directed by Sebastián Lelio, based on a Screenplay by Emma Donoghue.

13 Augsburg (aka. Augustae Vindelicorum), a city in present-day Bavaria (south Germany).

14 The Italian physician Georgius Hieronymus Velschius, aka. Georg Hieronymus Welsch (1624–77), *Exercitatio De Vena Medinensi, Ad Mentem Ebusinae, Sive De Vermiculis Capillaribus Infantium* (1674), a study of worms in the human body, including Guinea Worms. The Latin passage reads, "we are completely persuaded that those fastings are all, without a single exception, fabricated and phony."

15 "[T]hose who respect few things judge quickly."

16 This is the case of Mercy Short, one of the afflicted young energumens, whose possession and episodes of fasting Mather outlines in "A Brand Pluck'd out of the Burning": "I'l forbear mentioning the marvellous Abstinence enforced, and by *Spirits* upon these People, for *Nine*, yea, for *Fifteen Dayes* together; and their eating but one meal in about Three Weeks, except perhaps now and then a bitt of *Raw and Sharp Fruit*, which would but render them the more uneasy," in W.C. Ford, "Mather-Calef Paper on Witchcraft," in *PrMHS* 47 (October 1913–June 1914), p. 249, §5. See also the case of the possession of Margret Rule, whose fast lasted from "the Tenth of *September* to the Eighteenth." Though she

miserrima vidi.[17] – The Dæmons who kept the Miserable in such an horrid Fast, seem'd to take Occasion for it, from my mentioning a well-attested Instance given by *Henricus ab Heer*, of an Energumen having pass'd thro' a Fast of such Dimensions.[18] –

Your Friend will reckon himself kept in some Sort long Fasting, & not without a Degree of Pain, if you lett the Time run on too long, before you refresh him with your Letters, which may assure him, that you esteem him, as being,

Sir,

In the Number of your

Faithful & Ready Servants.

{C. Mather.}

was tormented by extreme hunger, "her Teeth would be set," when food was provided. "Another Brand Pluckt Out of the Burning," in Robert Calef's *More Wonders of the Invisible World* (1700), Part I, p. 5, §6.

17 "These very wretched things I have seen with my own eyes," adapted from Vergil's *Aeneid* (2.1.5).

18 Henricus ab Heer(s), aka. van Heers, De Heers (1570–1636), Dutch medical doctor, private physician to the prince-archbishops of Cologne, and author of *Spadacrene. Hoc Est, Fons Spadanus; Ejus Singularia, Bibendi Modus, Medicamina Bibentibus Necessaria* (1614). Mather may have drawn on *Most True and Wonderful Narration of two Women bewitched in Yorkshire ... As it is attested under the hand of that most famous Phisitian Doctor Henry Heers* (N.P., 1658), p. 4.

2.7. A STRANGE SISTERHOOD.

June 1714.

To Richard Waller or Dr. John Woodward.

Sir,

History ha's often diverted you, with Exemples of Two Persons, having an extreme Resemblance to one another. When the Resembling Persons have been Brethren or Sisters, the Wonderment ha's been abated, tho' the Spectators have been still diverted with it. Most of all when they have been Twins, with the Abatement of the Wonderment, still the Diversion ha's rather had some Augmentation. This was the Excuse of *Toranius*,[1] when he sold unto *Mark Anthony*, the Two Boyes for *Twins*, which were not *akin*, tho' most exquisitely *Alike*, to One another; That if they had been *Twins*, their Likeness would not have been so valuable a Rarity. *Oporinus* the Printer of *Basil*,[2] and *Albertus* the Marquiss of *Brandenburgh*,[3] were as much Alike as if they had

1 According to Macrobius (*Saturnalia* 2.4.28) and Suetonius (*Lives*: *Augustus Octavian* 27.1), Toranius, aka. Thoranius Flaccus, was a slave trader in the time of Roman General Marcus Antonius (83–30 BCE). In his *Natural History* (7.10.12), Pliny the Elder, discussing human conception, relates the case of two boys, born a year apart, yet resembling one another "as though they had been twins." Their resemblance, Pliny opines, proceeded "from the imagination of the parents" or resulted from the "impressions received at the moment of conception." In the case mentioned by Mather, the resemblance of the two doppelgangers was truly uncanny, even though they were unrelated, one being "born in Asia, and the other beyond the Alps." When the cheat was discovered, Toranius defended the high purchase price for the twins, because their case was extremely rare.

2 Johannes Oporinus, aka. Herbst (1507–68), a notable printer of Basel and main figure in the printing industry of the 16th century. Among his most significant achievements is the first printed edition of the Latin *Alcoran* and of *De Humani Corporis Fabrica Libri Septem* (1543), a prized depiction of human anatomy, by the Flemish physician and anatomist Andreas Vesalius, aka. Andries van Wesel (1514–64).

3 Albert, aka. Albrecht, margrave of Brandenburg (1490–1545), cardinal, and archbishop of Madgeburg and Mainz. Though sympathetic to the Protestant cause

been Twins; But, methinks, the Historian carried on the Likeness between them in that Article, but somewhat Indifferently;[4] *As One filled all* Germany *with Wars, the Other filled the whole Christian World with Books*. Brothers that have not been Twins have sometimes had a very Brotherly Correspondence with one another. *Gaffarel* mentions two such in the City of *Provence*, who not only were very like one another, but if one were Ill, the other could not be in Order; if one were sad, the other could enjoy no Mirth; if one had his Head aking, the Head of the other could not be easy; and if one were Asleep, the other did not care to be Awake. But in Twins, the Brotherhood is wound up sometimes to a more perfect *Unison*.[5] That which may appear most surprising in the Harmony for which Twins have sometimes been celebrated, is, that both in their *Fates*, & in their *Minds*, they have been as like to one another as in their Features.

Platerus[6] relates concerning certain Twin Brothers at *Basil*, that tho' they were daily conversant with him, yett he did committ Mistakes every day, by not knowing them asunder; and their natural Inclinations also were so much alike, that what the one thought ha's at the same time come into the Mind of the other; & if one were sick the other at the same time was not well, tho' they were as far distant from one another as *Basil* is from *Campania*.[7]

---

before Martin Luther's death, the cardinal ultimately supported Johann Tetzel (1465–519) and the sale of indulgences, which sparked Luther's famous *Ninety-five Theses* (1517) and the Schmalkaldic War (1546–47) between Roman Catholic and Protestant forces, culminating a century later in the devastating Thirty Years' War (1618–49).

4 Mather may have in mind *Oratio De Ortu, Vita, et Obitu Ioannis Oporini Basiliensis* (1569), by the Silesian professor of ethics Andrea Jocisco (fl. 1550), who furnishes a biography and complete catalog of the works printed by Johannes Oporinus (Herbst), including a folio volume of "Marggrave Albrechts von Brandenburg Außschreyben" (n.p.).

5 Jacques Gaffarel (1601–81), French scholar, astrologer, and author of the controversial *Curiositez Inouyes sur la Sculpture Talismanique des Persans, Horoscope des Patriarches et Lecture des Estoiles* (1629), pp. 338–39, an oft-reprinted work translated into English as *Unheard-of Curiosities: Concerning Talismanical Sculpture of the Persians* (1650). See ch. VII. 7. *The Sixt Objection of no Force.* The story of the twins' sympathetic pain appears on p. 220. The city was actually that of Riez, in the region of Provence. Mather also employed Gaffarel's work on Renaissance wonders in *BA*.

6 Felix Platerus, aka. Platter (1536–1614), Swiss physician, remembered for classifying psychiatric diseases, and for being one of the first physicians to promote germ theory of diseases. Among his major works is *Praxeos Medica Tomi Tres* (1602), which went through several expansions.

7 I.e. the city in Switzerland and the region in south-western Italy.

Fulgosus as well as *Kornmannus* have informed us,[8] That *Medardus* and *Gerardus*,[9] who were Twin-Brothers, were on the same Day preferred unto the Episcopal Dignity, the one in the See of *Rhotomage*, the other in that of *Noviodunum*;[10] and as they lived alike, so they died alike; both expiring on one & [the] same day. Dr. *Fuller*[11] likewise informs us, concerning the Twin-Brothers, who were Sons of Mr. *Tremain*, a Devonshire Gentleman, that they were so like in their Lineaments; as to cause more merry Misunderstandings, than the Two *Sosia*'s in old *Plautus's Amphytruo*.[12] And more than this, They would feel the *like Pains*, tho' at a Distance, & without Intelligence; They æqually desired still to walk, Sitt, Sleep, Eat or Drink together. In this they differed, that the One was Captain of a Troop, the Other but a private

8 In his *De Dictis Factique Memorabilibus Collectananea* (1588 ed.), the Genoese Renaissance scholar Baptista Fulgosus, aka. Battista Fregoso (1452–1504), lib. 1, cap. 6, p. 52, has a brief paragraph on the twin brothers in his chapter "De Miraculis." Likewise, the German jurist Henricus Kornmannus, aka Heinrich Kornmann (1579–1627), in his *De Miraculis Mortuorum, Opus Novum et Admirandum in Decem Partes Distributum* (1610), pars octava, cap. 9 (n.p.), mentions Bishop Medardus, but acknowledges Fregoso as his source.

9 St. Medard, Medardus (c. 456–545), bishop of Vermandois and Noyon.

10 Merovingian Noviodunum (modern Soissons).

11 The English theologian, historian, and scholar Thomas Fuller, D.D. (1608–61), was an oft-consulted source in Mather's publications. Here, Mather relies on Fuller's *History of the Worthies of England* (1662), p. 266, who relates that the likeness of the twin brothers Nicholas and Andrew Tremaine, sons of a Devonshire gentleman, was such that they could only be distinguished "but by their several habits" and gave rise to "more mirthful mistakes than ever were acted in *Amphitruo* of *Plautus*." Alas, they were both slain as soldiers at "*New-haven* in *France*" in 1564, thus sparing them the sympathetic pain they felt for each other throughout their lives.

12 In the Latin comedy by the Roman playwright Titus Maccius Plautus (c. 254–c. 184 BCE), *Amphytruo* (*Amphitruon, Amphitryon*), the god Jupiter takes on the guise of Amphitryon to have his way with Alcmena, Amphitryon's wife, who promptly gives birth to twin boys – one from each father. Likewise, the god Mercury, who takes on the guise of Sosia, Amphitryon's slave, creates much mirth among the audience. The classic theme of mistaken identity and the confusion arising from the gods changing their guises was popular in ancient times, and gave rise to many speculations about the origin of twins, anciently believed to be the offspring of their mother's adulterous relationship with two men. For speculations on the origin of twins, see Helkiah Crooke's *Micrographia: A Description of the Body of Man* (1615), bk 5, Quest. 21, pp. 312–13: "How Twinnes, or more Infants are generated"; and Thomas Browne's *Pseudodoxia Epidemica*, second ed. (1650), bk 7, ch. 2, p. 293.

Souldier; But then, to prevent the Languishing of the Survivor, they were kill'd both together.

You are præpared now to receive my *American* Story; If you please, a *Twin-Story* to what ha's hitherto entertained you.

At *Hampton*, a Town about Fifty Miles from this Place, there were Twin-Sisters, whose Names were, *Bridget* and *Jane Moulton.*[13] The perpetual Harmony and Sympathy between the Sisters was the Observation of all the Neighbourhood.[14] They were never, *contented*, except they were together. If the one were desirous to go abroad, the other would be impatient of staying at home. If the one were merry, the other would be Airy. If the one were troubled, the other would be chagrin{ed}. When one was for Carding, the other was for Spinning. For their Dispositions and Satisfactions, there was a very strange Agreement between them. The Particulars wherein every body with Pleasure & Wonder saw how they were agreed, and would like your famous Twins of *Hippocrates*,[15] which you tell us, would, *Flere et Ridere Simul*,[16] were numberless. They lived a Virgin-Life; and in this Good Accord reached about threescore Years. Then Death after a short Sickness arrested the one of them; The other grew full of Pain, & bid her Friends not be in an hurry about her Sisters Funeral, for hers must accompany it. By dying within a few hours after her Sister, she answered their Expectations.

13 The twin sisters Bridget and Jane Moulton, daughters of John Moulton (c. 1599–1649), were both baptized on April 8 1635, in Hemsby, Norfolk (England), and died on the same day, March 19 1699, in Hampton (New Hampshire). Their genealogy – along with an extract of Mather's letter to John Woodward – appears in Joseph Dow's *History* (1893), 2:861–62. Their story also appears in the same extract of Cotton Mather's 1716 letter "To John Woodward," *CNH* 3 (1832): 122–23.

14 The story of the "*two very sympathizing Sisters of Hampton*" appears to have been conveyed to Cotton Mather, for in his letter "To Samuel Penhallow" (24*d.* 10*m.* [December] 1713), Mather (*D* 2:291) requested more detailed information about the twin sisters, who were first brought to the attention of the Mathers by their long-time friend Penhallow (1665–1726), sometime preacher to the Indians, speaker of the Assembly under Governor Samuel Shute, chief justice of the superior court of judicature in New Hampshire, and author of *The History of the Wars of New-England With the Eastern Indians* (1726).

15 Hippocrates (c. 460–c. 370 BCE), the famous Greek physician, often called the "Father of Medicine." Seeing the likeness of a pair of two brothers, Hippocrates speculates about the causes of their simultaneous sickness and recovery. For Hippocrates's generation of twins in general, see "Hippocratis De Victus Ratione Sanorum Libri Tres," lib. 1, cap. XI, p. 457, in *Operum Hippocratis Coi, et Galeni Pergamenis* (1638), tom. VI, p. 457. See V. Dasen's "Twins" (1995) and "Multiple Births" (1998).

16 "Weep and laugh together."

Mr. *John Cotton*[17] the worthy Minister of the Place, preached a Funeral Sermon, for this Occasion, on those Words; 2 Sam. I. 23. *In Their Death, they were not divided.*

The nearest unto these Congruities, would be those of a true *Friendship*; A Blessedness which I am willing to find celebrated among the Ancients, & the *Regularity* they consulted in it cried up, and this made a grand Article of an Elogy, *They died having lived Without Reproach in Friendship.* Nevertheless, I can by no means allow, that this *Astrœa* has wholly left our World,[18] & that the *Moderns & Christians* are Strangers to it, and that the Maxims of our Saviours Holy Religion would not improve unto a Degree of Strictness, & Fineness, & Solace, beyond what was ever known in Paganism.

I only wish for the Experiment, which might be made in my having oftner & nearer Opportunity of meriting the Esteem of being,

Sir,

Your true Friend,

& Servant.

{C. Mather.}

17 Rev. John Cotton (1658–1710), son of Seaborn Cotton and Dorothy Cotton, Cotton Mather's cousin and Harvard classmate, served as minister of the congregational church of Hampton (N.H.) from 1696–1710. Apparently, the sermon was never published, but Mather may have had access to the manuscript.

18 The Greek virgin goddess Astrea (star-maiden) lived on earth among men during the Golden Age. Yet at the dawning of the Iron Age and man's wickedness increasing, she ascended to the stars where she became the constellation Virgo.

2.8. UNCOMMON IDEOTS.

June 1714.

To Richard Waller or Dr. John Woodward.

{Sir,}

You find *India* to have been among the Ancients, a term for *Ignorance*; and where they knew no further, they still term'd *India*. If you *Europæan* Gentlemen still count *India* a Seat of Ignorance, we poor Inhabitants of the Western *India*, must humbly acknowledge, we give you too much Occasion. If you look upon us as generally *Ideots*,[1] or but a little superior to such, we ought not to complain of you, as having wronged us. And therefore you will not wonder at it, that you are entertain'd from hence, with a Story of some *Ideots*; but rather, that we have any body in any tolerable Manner to relate the Story.–I remember *Strabo* saies, That all the Geographers who had written of *India*, had given the World nothing but lies.[2] But you shall have none such from the Pen, which is now writing from thence unto you.

1 Mather uses the term "idiot" not in its modern pejorative sense but in the ancient Greek sense of "Idiota," suggesting "untutored" or "simple," hence simple-minded. See P. Wickham (2002) and Wickham (2007).

2 The Greek geographer, historian, and philosopher Strabo (c. 64 BCE–c. 24 CE) provides remarkable insight into the world of the ancients, in his *Geography*. He tells his reader that the common people were so simple that the way to hold their attention was to tell stories about fabulous places, people, and animals – most of all to control them through fear by mixing myths with their religious beliefs. This was the way of Homer and of most other poets and historians: Thus, "a man will lie more plausibly," Strabo argued, "if he will mix in some actual truth" – like Homer's hero Odysseus, "who told lies in the likeness of truth" (*Geographia* 1.2.9). As to the mysteries about faraway places, Strabo insists, "all who have written about India have proved themselves, for the most part, fabricators." Stories about "men without mouths," or "without noses," or "with one eye," or "with fingers turned backward" are more fetching than tales about real people (2.1.9).

I have been told, that the famous Dr. *Sydenham*,[3] having that Quæstion putt unto him, *What he took to be the greatest Cure that ever he wrought*? Replied, *The Restoring of Reason to an Ideot*![4] How much Mankind in general may want such Doctors, is not the present Subject of Remark. But I know not what our *Sydenham* could have done upon Two deplorable Objects in our Town of *Dunstable*.[5]

A Gentleman there, whose Name is Mr. S.— under paralytic Effects, is yett the Father of Two Daughters, whereof the one is now Nine, the other Three, Years of Age. They continued for several Months after their Nativity, in the same Circumstances of Sensibility, that other Infants of their Age use to have. But anon, they were taken with odd convulsive Motions, which carried a little of an Epileptical Aspect upon them. The Fitts would be short, and many of them in an Hour: But after some while the Fitts grew seldomer, & last longer; & the Screeches of the little Wretches in them would be very doleful. These Fitts anon left them, wholly deprived of almost every thing in the World, but only a little Sight, & Scent, & Hunger. Nothing in the whole Brutal World, so insensible! They move not their Limbs; you may twist them, & bend them to a Degree, that none else could bear, & they feel it not. They take notice of nothing in the World; only they seem to see & smell Victuals, at the approach of which, they will gape, & be very Restless, & make something of a Bray. They are in good Health; & eat rather more than other Children of their Age. But they lett their Excrements pass from them, without the least Regard. The Elder is ever drivelling; the Younger never has any Salivæl Discharge. They shed no Tears. They never sneeze; They have no Speech. They have no way to discover any Sentiments of their Minds. They never use their Hands to take hold of any thing. Was *Idiocy* ever seen so miserable!

I have sometimes beheld & admired the Display of the Divine Providence, in this Thing, that a Total *Deafness* and Total *Blindness*, ha's hardly ever been known to concur in One Subject. The Concurrence of Two such Incapacities in One Subject, must have made the Miserables so uncapable of being subsisted, that it seems the Compassion of Heaven forbears the Infliction of it. This Exemple is the nearest unto it, of any I have mett withal.

3 Thomas Sydenham (1624–89), a notable English physician and author of *Observationes Medicae* (1676), a standard medical textbook.

4 The works of Thomas Sydenham are filled with oft-quoted anecdotes and aphorisms.

5 A town (NW of Boston) in Middlesex County, Massachusetts.

But so much for any *Dunstable*-Ideots. I shall but bring myself into the Number, if I once imagine, that I can impose long Letters upon a Person of your superiour Circumstances, without hazard of making you weary with them. I will break off seasonably, with some Evidence of my being so far from Idiocy, that (besides my Wishes that I were no worse an One, than honest *Raymund* the Abbot, who under the Name of *Idiota*, wrote the famous, *Contemplationes de Amore Divino*;)[6] I understand my Duty, which is to subscribe,

Sir,

For many Reasons,

& with no less Respects,

Your most obedient Servant,

{C. Mather.}

6 Mather refers to *Idiotae Contemplationes de Amore Divino*, the subtitle of a book on meditations (pt 1), separately published as *Contemplationes Idiotae* (1519), by the French Father Raymundus Jordanus (fl. c. 1381), abbot of Selles-sur-Cher (southern France). His nom de plume "*Idiota*" (the simple, uneducated) kept his identity hidden for a long time. The oft-reprinted work is included in the collection *Divi Aurelii Augustini Hippon. Meditationes ... Meditationes Idiota Viri Docti De Amore Divino* (1639). See also A. Devine (1913).

2.9. AN UNUSUAL DISCHARGE.

June 1714.

To Dr. John Woodward.

{Sir,}

You doubtless remember, who & where that Gentleman was, that had the Name of *Assiddik* assign'd him, which in *Arabic* signifies, *A great Speaker of Truth*,[1] such the Reputation of his Veracity! I know not what the Merits of that *Asiatic* were on this account; but I will do my best, that the *American* who now corresponds with you, shall never deserve any other Character. And God forbid, that the Veracity of a Christian, should not be more unspotted, than that of a Mahometan. Without such a præliminary Protestation, perhaps you might have quæstioned, the Matter of Fact, with which I am now going to entertain you.

There is a notable Display of our glorious Creators Wisdome and Goodness, in the Provision made for a good Success to the Efforts of Nature, in attempts to discharge such Things as may be noxious to our Bodies.

As long ago as when *Hippocrates* wrote his Treatise, *De Alimentis*,[2] the Sagacity of Nature, in finding out Methods & Passages, for the Discharge of offensive Matters, was a Subject of Admiration.[3]

1 Abū Bakr al-Ṣiddīq (573–634), whose last name signifies "the Upright," was the companion, father-in-law, and successor to the Prophet Muhammad.

2 "On Nutriments," a work by the Greek physician Hippocrates of Cos (c. 460–c. 375 BCE), *De Alimento Liber*, Textus Hippocratis 25: Excretiones secundum naturam ventris … . *Opus Hippocratis Coi*. Tomus VI (1638), pp. 267–72.

3 Mather's primary source for this and following paragraph is *Physico-Theology: or, a Demonstration of the Being and Attributes of God, from his Works of Creation*, second ed. (1714), by William Derham F.R.S. (1657–1735), Anglican clergyman, rector of Upminster (Essex), and natural philosopher. Derham was well-known for his Boyle lectureship summarized in this work and reprinted numerous times in the eighteenth century. Mather here refers to bk 5, ch. 7, pp. 311–12, *ns.*4 and 5. Mather uses much of the same material in his *CP* (1721), Essay XXXIII, "*Of* MAN," p. 227.

It would be a long Work to transcribe, what ha's been reported, by *Sennertus*, and *Hildanus*, and *Rhodius*, and *Bartholinus*,[4] and in the *German Ephemerides*,[5] and in our *Philosophical Transactions*,[6] of those who have swallowed Knives, & Pins, & Stones, and an hundred hurtful Things, which could not gett thro' the People in the ordinary Way, but have strangely made a Way, thro' some Abscess, that no mortal could have imagined. If I don't forgett, *Valsalva* writing, *De Aure humanâ*,[7] discovers with Wonderment, certain Passages into the Region of the *Ear-drum*, which marvellously accommodate the Brain and Head, with a Way to discharge purulent & morbific Matter from them. And it runs in my Mind, that not long since casting my Eye, upon the

4 The list of distinguished physicians includes Daniel Sennert (1572–1637), a German physician and iatro-chemist, professor of medicine at the University of Wittenberg, and author of *De Chymicorum cum Aristotelicis et Galenicis Consensu ac Dissensu* (1619) and *Hypomnematum Phisicorum* (1636). Increase Mather owned two of Sennert's medical works: *Medicina Practica* (1629) and *Institutionum Medicinae Libri V* (1645). Often called the "Father of German Surgery," Guilhelmus Fabricius Hildanus, aka. Wilhelm Fabry of Hilden (1560–1634), was a prominent iatro-mechanist, who published numerous works, among which is his best-known, *Observationum et Curationum Chirurgicarum Centuriae* (1641). Ambrosius Rhodius (1577–1633), German astronomer, mathematician, and physician at the University of Wittenberg, authored numerous works, including one on optics, *De Crepusculis* (1611). *Anatome ex Omnium Veterum Recentiorumque Observationibus* (1673), by Thomas Bartholin (1616–80), a distinguished Danish physician and anatomist, went through numerous editions and reprints. See also Mather's medical handbook *ABJ* (1972) and *CP* (1721).

5 Via Derham (see below), Mather here alludes to Dr. Joseph Lanzoni's "Observatio XLIX, De Pure in coxae tumore collecto, per alvum expurgato" (the natural evacuation of a tumor from the belly of a soldier), in the German scientific journal *Miscellanea Curiosa sive Ephemerides* (1691), p. 82.

6 Mather alludes to James Yonge's Letter to Hans Sloane, "Concerning several Solid Bodies voided by Urine," *PT* 26. 323 (October 1, 1709): 420–23. This topic was certainly of great interest to Cotton Mather, for he well remembered the agony of dying from kidney stones, which his grandfather Richard Mather suffered, and of which disease Increase Mather was exceedingly afraid. Rightly so, for Samuel Sewall reported that when he called at Increase Mather's house near Cambridge Court (July 30, 1723), he heard his agonizing cries, "Pity me! Pity me!" (*Diary* 3:325–26). For Richard Mather's suffering from kidney stones, see Increase Mather's *LDRM* (1670), p. 26.

7 *De Aure Humana Tractatus* (1704), cap. 2 §14, pp. 34–59, and cap. 5 §8, pp. 106–13, by Antonio Maria Valsalva (1666–1723), Italian professor of medicine and anatomy at the University of Bologna, renowned for the art of anatomical dissection. Mather's primary source is William Derham's *Physico-Theology* (1714), bk 5, ch. 7, pp. 311–12, *n.* (4). See P. Fughelli (2019). See also *CP* (1721), Essay XXXII, p. 227, and *CPS* (1994), p. 243.

very valuable *Physico-Theology* of our learned & pious *Derham*,[8] I observed some Just Reflections of his upon the Wayes which Nature ha's to sett all things to rights.

You shall if you please, allow my Story to come into the Collection of Curiosities upon that Article.

At a Town called *Concord*, about Eighteen Miles distant from me, a young Wom{an}[9] was taken with a grievous Fitt of Coughing, which held her for an hour or two, & wher{in} she was near strangling. At last the Cough grew more Moderate; but still it was Troublesome, & she become very much Indis{po}sed. One of her Indispositions was a Tumour, which in a little While began to appear about her left Shoulder-blade. The Cheirurgeon applied his Plaister to it; and one Morning, about a Fortnight after the first Inconvenience, her Father taking off the Plaister, saw a Speck in the Skin, of such Maturity towards a Suppuration, a little below the Shoulder-blade, that he resolved upon opening it. He did so, and unto his Astonishment pulled out a large Ear of Rye; after which the Place healed fast enough, and she returned unto her former Health & Ease. It seems, that the young woman, going in too much hast, with an Ear of Rye in her Mouth upon the Action of driving some Hogs out of the Field, received a Fall, the Violence whereof drove the Ear of Rye down the *Aspera*. But then, for the safe Passage of the Ear of Rye out a little below the Shoulder-blade, I leave this to your Investigation, who have gone beyond *Mundinus* in your Study of Anatomy.[10]

My wishes for your Prosperity, & the Articles thereof, which I wish for, are so many, that —— *citius numerabis Aristas*.[11]

{C. Mather.}

8 The famous Boyle lectures (1711–12), in Derham's *Physico-Theology*. Mather here refers to bk 5, ch. 7, pp. 312, *n.* (5). See also Mather's *CP* (1721), pp. 226–27.

9 Manuscript damaged, making the ends of the next three lines difficult to read.

10 A prominent Italian physician and anatomist Mundinus, aka. Raymondo Mondino de Liucci, or de Luzzi (c. 1270–1326), was professor of surgery at the University of Bologna, who promoted dissection of human cadavers. Among his well-illustrated works on human anatomy is his *Anatomia Mundini* (1531), first published in Padua (1475–78). Mather's view of the work is similarly dismissive in *CP* (1721), p. 289: "What is the Anatomy of *Mundinus*, if compared to our *moderns?*"

11 Mather ends his epistle to John Woodward with a witty citation adapted from Ovid's *Ex Ponto* (2.7.25): "You will more quickly count the ears of a crop," alluding to the "ear of rye" the young woman swallowed.

2.10. AN EAR STRANGELY MOLESTED.

June 1714.

To Dr. John Woodward.

{Sir,}

In the *Ear* of Man, I know, you have had your Inquisitive Contemplations entertained with a Collection of Wonders. The Auditory Passage ha's given you more than a few of the wonderful Things, in that glorious Collection, which ha's invited, & even compelled your Praises of the great GOD, *who planted the Ear.*[1] You have been surprised, when you have considered how Artfully this Passage is tunnelled; But it has doubtless added unto your Surprisal, when you have considered how that Passage is marvellously secured against the Invasion of noxious Insects, by a bitter & a nauseous Excrement afforded by the Glands appointed for such a Purpose. This Ear-Wax you are satisfied, is elaborated by the *Glandulae Ceruminosæ*,[2] which {are} described by some later Anatomists; for you reject the Opinion of the Elder ones, that make it an Excrement of the Brain, & say with *Schelhammer* of it, *Nil Absurdius*.[3]

1 Psal. 94:9; Mather's more detailed discussion of the sense of hearing appears in *CP* (1721), Essay XXXII, pp. 244–53; his principal source is the popular *Physico-Theology*, sec. ed. (1714), bk 4, ch. 3, pp. 114–38, by the distinguished English physico-theologian and elected member of the Royal Society William Derham (1657–1735).

2 Mather, *CP* (1721), p. 246. The ceruminous tubular glands in the inner ear secret cerumen, or earwax. On the source and function of earwax, see also *Anthropologia Nova* (1707), vol. 2, bk 3, ch. 12, pp. 565–66, by James Drake, M.D. (1667–1707), fellow of the College of Physicians and of the Royal Society.

3 Derham, *Physico-Theology* (bk 4, ch. 3, p. 122, *n.* 12), who cites the Danish physico-theologian Caspar Berthelsen Bartholinus (1585–1629), who dismissed the old anatomists' belief that earwax was the excrement of the brain, in Bartholinus' *Institutiones Anatomicae* (1641), lib. 3, cap. 9, p. 296. There is "nothing more absurd," laughs Günther Christoph Schelhammer (1649–1716), professor of anatomy and surgery at the University of Jena (Germany), in *De Auditu Liber Unus* (1684), pars primus, cap. 2, p. 32; see also pars altera, cap. II, Ax. IX. (Scholion), p. 118.

I am not now going to insist on the medicinal Vertue of the *Earwax*; or mind you, how *Pliny* celebrates the *Sordes ex Auribus*, for curing the Bites of *Men*, which he saies, *Inter Asperrimos numerantur*;[4] [If I might have spoken a little Figuratively, I would have said, That *Stopping our Ears* may be a good Cure for those *Wicked Bites*!] and of *Scorpions* & of *Serpents*; or what a *Balsame* some others have since reported it from their own Experience. Nor am I about pressing the revival of an old Practice, which the Physicians used in the Dayes of *Hippocrates*; which was, to taste the *Ear-wax*, as well as to see the *Urine* of their Patients, that they might form their Prognosticks on their Diseases.

I am only observing this Remarka{ble}[5] Thing, that where the *Meatus Auditorius*[6] is lon{g} enough to allow an Harbour for an{y} Insects, there this *Earwax* is constantly to be mett withal. Whereas, *Birds*, whose Ears are covered with Feathers, & where the drum lies but a little Way within the Skull, have none of it.[7]

But unto this Observation I will annex a Relation of an odd Occurrence, which lately happened. At a Place called *Newtown*, at about Seven or Eight Miles distance from me; there lives one Lieutenant *Fuller*, who one day thought he felt something like a Fly, make his Way into his Ear, & prove pretty Troublesome to him there. He requested a Lad with a little Engine to endeavour the fetching of it out; but the Endeavour was ineffectual. Anon, that is, after a few Minutes more he perceived the Fly himself to fly out, of his Ear & catch'd him in his Hair. Within a little While, he felt a Pain in that Ear, which grew so insupportable, that he was in a Manner distracted with the Anguish of it. For the Releef of his Anguish a Variety of Means was used; his Friends made various Applications; but all to little Purpose. At Length, a Syringing of his Ear with a convenient Liquor, was thought upon; which upon the Trial, fetch'd away two or three well-grown Maggots. But this Operation prov'd so painful, that the poor man chose rather to dy, (as he concluded he must) than to have it proceeded with. What they next went upon, was to pour Lamp-Oyl into his Ear, & some other such mollifying things, which fetch'd

4 Derham, *Physico-Theology*, p. 122 (*n.* 13); Pliny the Elder, *Naturalis Historia* (28.8), discussing the medicinal uses of earwax ("sordes ex auribus"), "filth from the ears," argues that a human bite is considered to be "one of the most dangerous of all" and "human earwax is the best remedy for it" – provided it is taken from the sufferer's own ears.

5 The manuscript corner is torn, rendering the end of the next several lines illegible.

6 "The internal auditory canal" (Derham, *Physico-Theology*, p. 121, *n.* 10).

7 Derham, p. 121, *n.* 10 and 11; Mather's *CP* (1721), p. 246.

away so many more lusty Maggots, as made up the Number of above Twenty in all. Whereupon the Man became easy as he was before.

I was advised to think, that such an unusual Thing, might pass muster among your Curiosities. Else I would not have præsumed upon treating you with a Letter that should have only such things as these crawling in it. Tho' if my Letter might hope for the Fate of a Cheese, there would be Gentlemen who would not count it much the worse, for a few of these attending of it. However, studious to keep as clear from them as may be, I would alwayes apply myself to your Service, with the affectionate Zeal, of being esteemed,

Sir,

Your faithful & hearty Servant

{C. Mather.}

Series Three

3.I. MONSTROUS IMPREGNATIONS.

July 2, 1716.

To Dr. John Woodward[1]

Sir,

The Philosophy of *Generation*, was become one Subject of a short Contemplation with your Servant.

I thought, Among the *Vegetables*, the *least Part* of that which we call, *The Seed*, is really so. The True *Seed* lies in so *Little Room*, that it is not visible to the Naked Eye:[2] The *Rest* serves but as a *Lodging* for it, and for its most proper *Nourishment* and *Expansion* at the first Opening of it. But in that *little Room* there lies the *whole Plant*, in all the *True Parts* of it; which is afterwards evolved, and extended, and filled up with its *Adventitious Nutriment*, until it be carried as far as the original *Stamina*, are capable. And *then*, the Growth stops. But perhaps towards the *Period* of the Growth, so weak may be the *Stamina*, that according to the *Strength* of the Soyl in which the Plant is lodged, the *Growth* may be more or less, tho' but a little, varied.[3]

And then I thought, why may not the *Seeds* of *Animals*, have the like said of them? Yea, The *Microscopical Inquisitions*, have made it more than probable; That the *True Seeds* of *Animals*, floating in their *Suitable Vehicle*, have, lying in a *Space* much less than the Naked Eye can discern, the whole *Bodies* of the *Animals*, even to all their *Nerves* and *Fibres*:

1 The editors have elected to choose Mather's draft (MHS) as their copy-text rather than the British Library's Sloane MS. copy (Call # 3340, ff. 277–97b), which is in the hand of the Royal Society's copyist who misread and mistranscribed several words and phrases. Nonetheless, we have collated the MHS draft with the Sloane copy.

2 See also *BA* (1:595); *CP* (1721), Essay XXVI, "Of the Vegetables," pp. 128–29.

3 This and the following two paragraphs also appear in Mather's commentary on Gen. 6:4 (*BA* 1:595–96). See preformationism and epigenesis (below).

which afterwards *Grow* as aforesaid, until their original *Stamina* can be no further carried out.[4]

This by the Way, thought I, takes away the common Cavil against, *The Resurrection of the Dead*; That the *Bodies* of Men, have been devoured by *others*, & become the *Parts* of them. For according to this *Hypothesis*, there can be no such thing, as a *Confusion* of *Humane Bodies*;[5] They can, by being devoured, lose no more than their *Adventitious Nutriment*. And it {is} remarkable, That the *Resurrection of the Dead* , is called, *The Filling of the Dead Bodies*. [Psal. CX.6. and Eph. II.1.].[6]

But then, what Provision here, against the Mischief of *Amputations*?[7] Must we suppose, a *Seminal Principle* seated in a Part of the Original *Animalcule* which is yett unknown to us, and this furnished by our Glorious & Almighty Creator, with a *Plastick Vertue*, to *Renew* what may be lost of it?[8]

4 See also *BA* (1:595).

5 See Augustine, *City of God* (22.20).

6 The Anglo-Irish natural philosopher, chemist, and theological writer Robert Boyle (1627–91) provides the physiological underpinnings of how even the most miniscule corpuscles of the decomposed human body may serve God in the resurrection and restoration of the original body, in *Some Physico-Theological Considerations about the Possibility of the Resurrection* (1675), esp. pp. 13–39.

7 Mather tried to find a solution to this troublesome question as it pertained to his belief in the resurrection (and complete restoration) of the body at Judgment Day (*TP* pp. 123–26; *BA* 1:596). Although the Swiss naturalist Abraham Trembley, F.R.S. (1710–84) has been credited rightfully for his discovery that primitive polyps are able to rebuild lost limbs, in his *Memoires pour servir A L'Histoire d'un Genre de Polypes D'Eau Douce, a Brase en Forme de Cornes* (1744), Mather did conjecture in his medical handbook "Angel of Bethesda" (1693–1724) that such a feat of restoring lost limbs might be accomplished by the "Nishmath-Chajim" (*vis formatrix*, plastic spirit) that permeates the whole body as a medium between the immaterial soul and the material body responsible for healing wounds and restoring health of a diseased person, in "Cap. V. *Nishmath-Chajim*," *ABJ* (1972), p. 30. See also CA 6.5 "*Nishmath-Chajim*" (item 5). On Trembley's discovery, see A. Vartanian (1950).

8 *BA* (1:596). The concept of Nature's "*Plastic Power*" (*vis formatrix*, hylarchic principle) is derived from Plato's idea of *anima mundi* (*Timaeus*), the theory that Nature is endowed with a spiritual power that (by God's direction) orders and forms the size, growth, and shape of all biological life forms. This Spirit of Nature is an incorporeal substance (lacking intelligence), which permeates all matter in the universe and acts as a formative (plastic) power that through God's *fiat* mechanically shapes and moves all matter. Revived by the Cambridge Platonists to combat Cartesian mechanism, this vitalist theory made a brief comeback but was largely discredited by the end of the 17th century. See Ralph Cudworth's learned (but nearly impenetrable) discussion, in *The True Intellectual System of the Universe*

I was trying for a more mechanical Account of this Matter; without flying to *Plastic Vertue*; which if I call *unintelligible*, I must confess my *Mechanism* to be so too.[9]

The Words of the Acute *Leibnitz* made a sufficient Impression {on} me.[10] "The most excellent Observators of our Time," *saith he*, "have made me beleeve, that Animals, or any other organized Beings, do not begin to exist, when we think they do, and that their Seeming *Generation*, is only an *Unfolding*, and a kind of *Augmentation*.[11] But the greatest Quæstion still remains; What becomes of those *Souls* or *Forms*, after the Death of the Animal, or the

---

(1678), bk 1, cap. 3, pp. pp. 97–182; Henry More's *The Immortality of the Soul* (1659), lib. 3, cap. 12, pp. 449–58; and John Ray's ever popular *The Wisdom of God Manifested in the Works of Creation* (1691). See also W.B. Hunter (1950), R.A. Greene (1962), J. Crawford (2005), and A.J. Inglehart (2016).

9 Although an admirer of Descartes' mechanical philosophy, Mather frequently registered serious reservations when pure mechanism seemed to conflict with God's providentialism. See for example *BA* (1:349–52, 355, 393, 399). For Robert Boyle's famous experiments with an air pump, see *Nova Experimenta Physico-Mechanica* (1661), pp. 103–20, and his revision in light of his debate with Henry More and other critics, in *A Continuation of New Experiments Physico-Mechanicall, touching the Spring and Weight of the Air* (1680, 1682). See also J.E. McGuire (1968) and S. Shapin and S. Schaffer (1985).

10 Gottfried Wilhelm Freiherr von Leibniz (1646–1716), renowned German philosopher and mathematician, addresses the philosophical implications of preformationism in the context of René Descartes' dualist concept of the separation of the body and soul, and its entanglement with matter and motion, in Leibniz' essay "Sisteme Nouveau de la Nature et de la Communication des Substances, aussi bien que de L'Union qu'il y a entre L'Âme & le Corps. Par M. D. L[eibniz], *Le Journal des Sçavans* XXV (Juin 27 1695), pp. 294–306, esp. pp. 297–98.

11 Mather explores the popular concept of "preformationism," the time-honored notion that all biological organisms develop from completely pre-formed, miniscule embryos (homunculi) in the male sperm (spermism) or female egg (ovism), or seed (plants), visible only with the help of the most powerful microscopes of the day. Thus the gestation process, so physicians and natural philosophers of the day believed (Swammerdam, Malpighi, Leeuwenhoek, Malebranche), is merely an unfolding and expansion of the preformed tiny body (animalcules) in the seed or sperm and passed down from its progenitors through emboîtement. Preformationism was a popular theory of biological generation, which was superseded by more modern concepts of epigenesis, the progressive development of an embryo from an undifferentiated egg cell – as the English natural philosopher William Harvey had argued in his *Exercitationes de Generatione Animalium* (1651), esp. cap. 31, 33, 45, 50–51, esp. 62-63. See also Mather's CA 1.5 "Monstrous Births"; CA 2.3 "A Monster"; CA 3.2 "A Monstrous Calf"; and *VT* (1718), pp. 8–9. Among the most readable modern studies of the debate are J. Needham (1959), E.B. Gasking (1966), J. Farley (1977), E.G Ruestow (1983) and, more recently, C. Pinto-Correia (1997).

Destruction of the organized Being? The only Way left, is to say, That not only the *Souls*, but also the *Animals* themselves, and their *organized Machines*, are preserved, tho' the Destruction of the Gross Particles ha's reduced them to a Smallness, that is as Imperceptible as it was before they were born."[12]

And I remember, *Poiret* is of the Opinion, that the primitive Subject, unto which our Soul is united, goes out of our Body together with it, when we come to dy.[13]

But oh! the palpable Darkness under which we are languishing! While I was pleasing myself with Speculations, which I thought were growing towards a fair *Hypothesis*, about the *Generation* carried on in the *Animal World*, I found myself compelled by Dr. *Drake*,[14] to Destroy all again, or at least suspend it, until a, *Melius Inquirendum*, should be served upon it.[15] He minds me, That my *Hypothesis* won't account fairly & fully for *Mix'd Generations*. For besides the Vast Variety of *Mungril Dogs*, Animals of Remoter Kinds have had their prolific Mixtures. The Mixture of a *Bull* and a *Mare*, produces a *Gimar*.[16] But

12 This paraphrastic excerpt is from the French philosopher and lexicographer Pierre Bayle (1647–1706), who reviews and cites (by way of translation from the French original) Leibniz's above-mentioned essay in "Rorarius," *An Historical and Critical Dictionary* (1710), 4:2611, col. 2; see also the summary of Leibniz's position in Bayle's essay on Daniel Sennert (Bayle 4:2712, col. 1). A similar argument appears in Leibniz' later disquisition on the soul, in *Monadology* (1714, 1721), §§63–77.

13 Mather's summary paragraph of *Cogitationes Rationalis (de Deo, Anima & Malo)*, "Appendix," numero 1 (1685), p. 611, by Pierre Poiret (1646–1719), a learned French Protestant theologian and radical millenarian, originates in Bayle's biographical essay "Rorarius" (above mentioned), 4:2611 (col. 2). See also *BA* (1:596–97).

14 James Drake, M.D. (1667–1707), fellow of the College of Physicians of the Royal Society, published *Anthropologia Nova*, 2 vols. (1707), which (among other things) explored new theories of generation and hybridization. Mather's paraphrastic citation is from Drake, vol. 1, bk 1, ch. 25, pp. 334–35, 335–36 [misnumbered 352]. Reasonable and convincing as the idea of preformationism appeared to be to the natural philosophers and physicians of the day, Mather quickly realized that this neat theory did not explain such common phenomena as hybrid animals (mules or gimars) which bore the traits of the parentage of two distinct species. However, he was not quite prepared to abandon preformationism for epigenesis simply because preformationism neatly entailed traducianism as well – that like the infinitely small homunculus and its soul were encased (emboîtement) in the first human pair and passed on through propagation to their offspring along with the imputation of Adam's original sin.

15 A better inquiry (hypothesis).

16 Gimar (jumart): "A hybrid animal, erroneously believed to be the offspring of a bull and a mare or she-ass, or of a horse or ass and a cow" (*OED Online*, accessed May 22 2022). According to the *OED*, "Gimar" first appeared in the English language in 1690, in John Locke's *Essay Concerning Humane Understanding in Four Books* (1690).

the most common of all mix'd Breeds, is, the *Mule*, begotten by an *Ass* upon a *Mare*. He also finds Weight, in the Similitude of Children to their Ancestors; whereof some are like the Ancestors in the Fathers Line, and some in the Mothers; For which, to make the Efficacy of a Different Matrix a Solution, he thinks to be as unphilosophical, as to fancy, that an *Orange-tree* transplanted from *Spain* to *England*, would become an *Apple-tree*. Upon all which he concludes; "However old & exploded the Opinion of a *Plastick Power* be, I must however embrace it; even tho' I know not exactly wherein it lies."[17]

An odd Operation then, of this *Plastic Power*,[18] if it must be so, is what I am now to entertain you with.

We have a little Animal called a *Musquash*,[19] pretty frequently lodging about the Brooks that water our Countrey. A small Beast, that lives in shallow Ponds where it builds notable Houses of Earth and Sticks, & feeds upon *Calamus Aromaticus*.[20] It has a strong Scent of Musk; especially about the Month of *May*; and its Testicles laid up in Wool, are of a lasting Fragrancy. It is near the bigness of an ordinary Catt; and the *Mus Muscatulus* in the LXVII Table of *Johnston*, in his, *Historia Naturalis, De Quadrupedibus*,[21] is the nearest of any thing I have seen, unto the Figure of it.

In the Town of *Roxbury*, which adjoins to the City of *Boston*, an honest Man living somewhat solitary in the Woods, had a Catt, which was observed sometimes to visit an adjacent Brook, where a *Musquash* was known to have his Residence. The Catt having received an Imprægnation there, brought forth a Litter, of many Kittens; and each of the Kittens, were of a mixed Form; some had most of the *Musquash*, and some had most of the *Catt*, conspicuous in them. The Mistress of the House had such an horror, of this *unnatural Conjunction*, that she was for destroying the whole New Brood of so monstrous an Aspect. All they that had the Head of a *Musquash*, being destroy'd, the Life of

[17] The quoted passage is from Drake's *Anthropologia Nova*, p. 336 [misnumbered 352].

[18] Mather wrestled with the vitalist idea of "*Plastick Power*" for a long time, yet he was reluctant to let go of the time-worn concept. For it reified God's hand in Nature in the face of pure mechanism. See *BA* (1:103, 104–05, 375, 414, 593, 598–99; 2:463; 3:320; 5:37–38; 9:49–50, 183).

[19] Muskrat.

[20] *Acoraceae*, genus *Acorus*. Sweet flag or muskrat root, used for medicinal purposes.

[21] Joannes Jonstonus, aka. Johnston (1603–75), a Polish natural philosopher, who received his M.D. from the University of Leiden. He is best remembered for his *Historiae Naturalis de Quadrupetibus Libri* (1657). The work went through numerous Latin reprints and translations into English, Dutch, French, and German. Mather here refers to lib. 1, cap. XVII, art. II, fol. 116: "De Muribus Aquaticis"; Tabula LXVII (between fols 104 and 105) includes a depiction of the *Mus Muscatula*, a rodent similar to a beaver.

one that had the Head of a *Catt*, and only the Feet & the Tail, & something of the Fur of a *Musquash*, was upon Intercession for it, reprieved a little While. The Reprieve of this also was but short; for the good Woman could not long bear the Sight of what appear'd such a Monster unto her, but gave Order for the Execution of it.

Lett this Matter of Fact, be admitted into the Number of those, which you will please to consider, when that obscure Subject of *Generation*, shall have your more sagacious Philosophy employ'd upon it.

And when your Contemplations are proceeding upon that Subject, you will allow me to throw in an *Egg* also, (contrary to the ancient Custome) for the finishing of your Entertainment.[22] The rather, because you may remember, that when the Controversy, *De Anthropogeniâ de Ovario humano*,[23] was much agitated, such a thing as I am now to offer, was very much taken into Consideration. The *Naturæ Curiosi*,[24] have thought it worth the while, to præserve the Memory of, *Ova prægnantia*.[25] *Elsholtius*[26] reckoned, that an

22 Mather's witticism alludes to the ancient custom of egging a person as a sign of good luck (fertility), perhaps reminiscent of the cosmic egg from which all life sprang.

23 Of the study of generation in the human ovary. Mather alludes to the preformationist debate between spermists and ovists – each side asserting that the unborn, yet fully pre-formed, progeny (homunculi) exist in miniscule form in the male sperm (animalcule) or female egg. Later, William Harvey, in his *Exercitationes De Generatione Animalium* (1651), esp. cap. 50–51, 62–63, posited that the fetus is formed from unorganized matter in the egg after fertilization. The controversy is reviewed in several contributions to the *PT*, most notably in "A Discourse concerning the Modern Theory of Generation, by Dr. George Garden of Aberdeen," *PT* 17.192 (February 28, 1691): 474–83; "An Objection to the new Hypothesis of the Generation of Animals from Animalcula in Semine Masculino," *PT* 20.244 (September 30 1698): 337, by Martin Lister (F.R.S.); "Some Reflections on Generation," by Daniel de Superville, *PT* 41.456 (January 1, 1740): 294–307; "A Summary of some late Observations upon the Generation, Composition and Decomposition of Animal and Vegetable Substances," *PT* 45.490 (January 1, 1748), n.p., by John Turberville Needham; and in Jan Swammerdam's *Miraculum Naturae sive Uteri Muliebris Fabrica* (1729). A helpful historical survey is provided by J. Needham (1959), B. Glass (1968), P.J. Bowler (1971), and S.A. Roe (1981).

24 Natural philosophers, those prying into nature.

25 Impregnated or fertilized egg.

26 Johann Sigismund Elsholtz, aka. Elßholz, Elsholtius (1623–88), German botanist, chemist, naturalist, and court physician to Elector Friedrich Wilhelm von Brandenburg. Elsholtz relates in his "Observatio LXXX. *De Ovo Praegnante*" that in a clutch of broken hens' eggs, "in albumine unios prope vitellum, ovum aliud parvum jacebat aliud parvum jacebat μαλαχόδερμον, sive molli cute aut membrane tectum" ("in the white of one [egg] near the yolk, another small egg lay

Egg found within an *Egg*, tho' it were only μαλακοδερμον,[27] covered, with a fine Skin, worthy of some Commemoration. But *George Jung*,[28] reckon'd his Egg much more so, wherein *Perexiguum latitabat ovum, cortice candido ac duro aliorum ovorum more munitum*.[29] You shall give me Leave to assure you, That tho' I have never seen such fœtiferous Lemons & Oranges as *Ferrarius* tells of,[30] yett I have newly had in my Hand, a little *Egg*, with an hard Shell, about as big as a *Pigeons Egg*, which was found in the midst of an *Hens Egg*; but judged it not worth the While to transmitt unto you the Egg it self; My Affirmation being with you a sufficient assurance for the Truth of it; As I hope, it will be, that I am, with very great Respect,

[Boston, N. England   Your sincere Servant,

July 2. 1716.][31]

Cotton Mather

μαλακόδερμονi, i.e., either covered with a soft skin or a membrane"), *Miscellanea Curiosa Medico-Physica Academiae Naturae Curiosorum.* Annus sextus et septimus (1677), pp. 115–16.

27 "Thinly covered."

28 Georg Sebastian Jung (1643–81), member of the Leopoldina (National Academy of Science) and court physician to Austrian Emperor Leopold I (1640–1705) of the House of Habsburg.

29 The Latin passage is quoted from Jung's "Observatio CCL. Ovum Ovo Praegnans," *Miscellanea Curiosa Medico-Physica Academiae Naturae Curiosorum.* Annus secundus (1671), pp. 348–49. Jung relates that his maidservant upon breaking eggs to cook them in butter, "opened one in whose white, beside a yolk of very beautiful color, another very small egg was hidden, protected on the outside by a shell or hull, like other eggs."

30 Via Elsholtz's article "De Ovo Praegnante," in *Miscellanea Curiosa* (p. 115), Mather refers to Joannes Baptiste Ferrarius (1584–1655), Italian Jesuit, orientalist, and botanist, whose *Hesperides sive de Malorum Aureorum Cultura et Usu Libri Quatuor* (1646), lib. 3, cap. 19, pp. 263–75; and lib. 4, cap. 8, pp. 403–07, also contains carefully drawn depictions of lemons and pomegranates and their respective seeds to which Mather alludes. Mather's "foetiferous" is a play on words, here suggesting fructifying or generative lemons and oranges that contain within their body embryonic lemons and oranges, like Mather's egg within an egg – a kind of emboîtement.

31 Closing taken from the copy in BL, Sloane 3340, fol. 297.

Figure 10. The Muskrat. In Joannes Jonstonus, *Historiae Naturalis de Quadrupetibus Libri* (Amstelodami, 1657),Tabula LXVII (between fols 104 and 105). Courtesy of the Smithsonian Libraries.

3.2. A MONSTROUS CALF.

July 3, 1716.

To Dr. John Woodward.

{Sir,}

Among the Stores, of a more than Royal Treasury, which are amass'd in the Stock of the Royal Society[1] from the Communications of Inquisitive and Ingenious Men, you have allow'd a particular Class to the *Monsters*, whereof a Relation communicated now & then, ha's found a Good Reception. And indeed *Fortunius Licetus*[2] is not the only Gentleman, who ha's thought Accounts of *Monsters* to be worth collecting; Men of a much finer Philosophy than he, have thrown in many Accounts to the General Collection that ha's been made of such Things: And this from a just Opinion, that an attentive Consideration

1 Lest this letter of Mather to the Royal Society seem too pedestrian if not credulous, the famous Robert Boyle published an "Account of a very odd Monstrous Calf," in *PT* 1.1 (March 6, 1665):10, and its depiction, between pp. 78–79. Similar accounts of malformations – animal and human – can be found in the pages of this journal throughout the first century of its publication. See for instance, Robert Southwell's "Account of a Monstrous Calf with Two Heads," *PT* 20.238 (March 31, 1698): 79–80; John Floyer's "A Relation of Two Monstrous Pigs, with the resemblance of Humane Faces," *PT* 21.259 (December 31, 1699): 431–35; John Craig's "A Description of the head of a Monstrous Calf," *PT* 27.333 (March 31, 1712): 429–30; Andrew Cantwell's "Extract of a Letter … giving an Account of a monstrous Boy," *PT* 41.453 (June 30, 1739): 137–38; and William Gregory's "An Account of a monstrous Foetus, resembling a hooded Monkey," *PT* 41.461 (December 31, 1741): 764–67.

2 The Genoese physician and philosopher Fortunius Licetus, aka. Fortunio Liceti (1577–1657), variously practiced medicine and taught philosophy at Pisa, Padua, and Bologna. He was a friend and colleague of Galileo Galilei (1564–1642) and published numerous works on medicine, philosophy, and astronomy, including a work on monsters, *De Monstrorum Caussis, Natura et Differentiis Libri Duo* (1616, 1634), reprinted in Amsterdam (1665) and Padua (1668). The Mather library included the new Amsterdam edition of 1665, *De monstris. Ex recensione Gerardi Blasii, M.D. & P.P. Qui Monstra quaedam nova & rariora ex recentiorum scriptis addidit. Editio Novissima.*

of these Curiosities, might very much assist our Enquiries into that obscure Work of Nature, Generation; than which, What more mysterious?[3]

The Reasonable Philosopher, will not be too ready to imagine, that *Monsters* carry *Omens* in them, and that the Name of *Monsters* does for that Cause belong unto them.[4] Even *Fortunius Licetus* himself, will furnish Two small Reasons, to confute that Superstition. The one; Because *Monsters* are no where more frequent than in the Solitudes of *Africa*,[5] where there are few People capable of receiving Admonition from them. The other; Because they often are of too criminal an Original,[6] to have it supposed, that the Holy God would employ them to be our Monitors. And yett, *Monsters* may no doubt be sometimes attended with such Circumstances, that they who are more nearly concerned may do well, to be sensible of a Voice from Heaven therein unto them.

It seems, the Denomination of *Monsters* is given to these *odd Things*, *Quià ob sui Novitatem et Enormitatem, quam primum visa cum admiratione, ab unoquoque unicuique monstrantur*.[7] And so, the very Etymology of the

3 Apart from the contemporaneous accounts of monsters, Aristotle's account of how malformations of animal and human bodies can occur at insemination and during gestation is strikingly modern, in *De Generatione Animalium* (4.4.770–73). Mather provides a working definition of the term "monster," from Latin "monstrare," i.e., "to show," "exhibit," or "reveal" some peculiarity in nature that something unusual was about to occur, in CA 1.5 "Monstrous Births." See also CA 2.3 and CA 3.1. Among recent academic studies of prodigious monsters are L. Daston & K. Park (1998, 2001), P.G. Platt (2000); L.L. Knoppers and J.B. Landes, eds (2004); A.W. Bates (2005); J. Crawford (2005); J. Spinks (2009); and Fabian Kraemer (2023).

4 Mather appeals to the authority of *A Discourse Concerning Prodigies* (1663, expanded 1665), by the distinguished classical scholar and antiquarian John Spencer (1630–93), Master of Corpus Christi, Cambridge, who ridiculed the common belief that malformed births, comets, and unusual occurrences in nature were prodigies or signs of God's wrath. See especially K. Park and L.J. Daston (1981). M.P. Winship (1994) examines Mather's changing attitude toward omens in nature.

5 Licetus, *De Monstris* (1665), lib. 1, cap. 2, p. 7; see also lib. 1, cap. 1, p. 4; lib. 2, cap. 53, p. 169; cap. 54, pp. 173–74.

6 Licetus, *De Monstris* (lib.1, cap. 2, p. 7). Mather alludes to the common belief in his time that bestiality occasionally engenders a monstrous offspring with both animal and human features – amply illustrated in Licetus' book of monsters (pp. 13, 57–58, 181–83, 185, 190–91, 194, 234, 236, 256) as well as in Licetus' main source (among others) *Prodigiorum ac Ostentorum Chronicon* (1557), by Conradus Lycosthenes, aka. Konrad Wolffhart (1518–61), a German encyclopedist and professor of dialects and grammar at Basel.

7 Licetus (lib. 1, cap. 2, p. 7). They are called monsters, "because, on account of their strangeness and violation of the norm; as soon as they are seen with amazement, they are pointed out by everybody to everybody else." In his above-mentioned

Name, will Justify my Disposition, to *Show* you what I meet withal, when a *monstrous Birth* occurrs in my Neighbourhood.[8] Yea, it would be *monstrous* in me, if I should not be so disposed; for, *Gestit Natura hominis quod Rarum alteri monstrare.*[9]

It were, I confess, a Shocking Spectacle, to look upon the *Monsters*, reported by Credible Historians, to have had something of *Humane*, mixed with a *Brutal* Shape.[10] Tho' perhaps to a Vertuous Mind, the Depravation too often seen upon Mankind, in the *Bestial* Humours and Manners, whereto Men sometimes degenerate, were a Spectacle no less horrid & hateful.

Many of these *Monsters*, have had *Humane Mothers*. The Men, in *Arrianus* quoted by Ramusius, born with the Heads of Horses, I suppose were so.[11] What *Lycosthenes* mentions of a Lion with the Head of a Man,[12] was more certainly so.

---

article, Sir John Floyer intends to prove that "the Distortion of the parts of a *Foetus*, may occasion it to present the Figure of different Animals, without any real Coition between the two Species," in *PT* 21.259 (December 31, 1699): 431.

8 See also Mather's letter to the Royal Society, "Curiosa" 3.1 "A Monstrous Impregnation" and 2.3 "A Monster," with the longer title that Mather cancels: "A Monster born at *Boston*, in *New England.* 15.*d* 8.*m* 1713]."

9 Licetus, p. 7. "A man's nature eagerly desires to show what is unusual to another person."

10 See the depictions of several hybrid humans with the heads of various types of animals in Licetus' *De Monstris* (1665), lib. 1, cap. 3, p. 8, 10–13; lib. 2, cap. 58, 181–83, 185, 190, 194, etc., several of which are based on the descriptions in Giovanni Battista Ramusio's three-volume *Delle Navigationi et Viaggi* (1559, 1583) and in Conradus Lycostenes' influential *Prodigiorum ac Ostentorum Chronicon* (1557).

11 The 2nd-century CE Greek historian Lucius Flavius Arrianus (Arrian of Nicomedia) is the author of several important histories, including his *Anabasis*, a semi-fictional panegyric on the conquests of Alexander the Great (356–323 BCE), and *Periplus of the Euxine Sea* (130–31 CE), a travel guide to the Black Sea. Mather may well have in mind the mythological centaurs – equines with a human upper torso and head. The Italian geographer and compiler of travel accounts by famous travelers, Giovanni Battista Ramusio (1485–1557) of Venice, excerpts Arrian at some length in the second of his three-volume *Delle Navigationi et Viaggi* (1583) 2:193^{r}–95^{v}, but does not mention any hybrids.

12 See the depiction of the human-headed lion in Licetus, lib. 2, cap. 58, p. 181. It is based on the collection of prodigies and human and animal malformations in *Prodigiorum ac Ostentorum Chronicon* (1557), by Conradus Lycosthenes, aka. Konrad Wolffhart (1518–61).

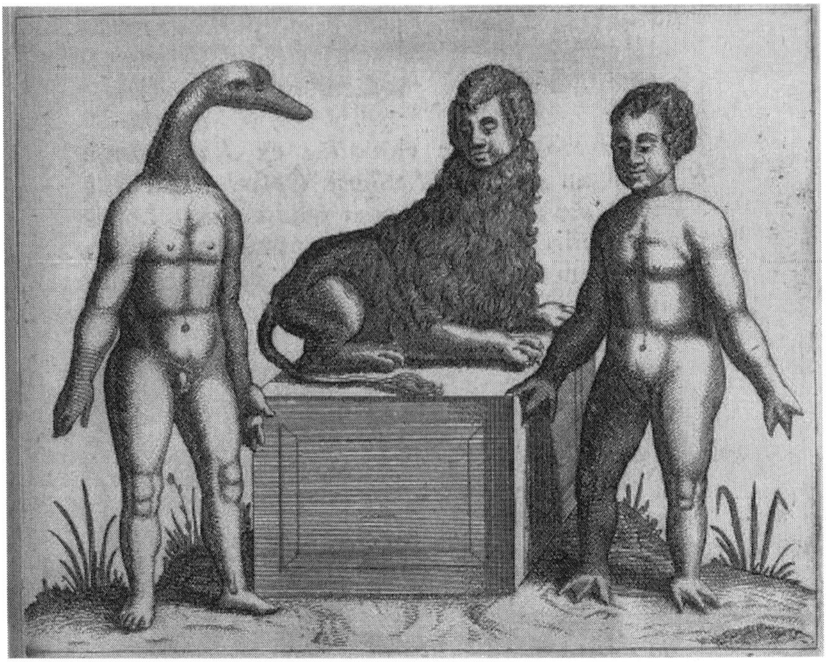

Figure 11. Fortunatus Licetus, *De Monstris* (Amstelodami, 1665), lib. 2, cap. 58, p. 181. Courtesy of the Francis A. Countway Library of Medicine.

So no doubt was the Man mention'd by *Vincentius* in his *Speculum Naturæ*, who had the Head of a Dog.[13] And *Gualtherius* the Chronicler,[14] mentions an Humane *Fœtus*, united by the Back to a Canine. I remember not what the

[13] Via Licetus, p. 181, Mather refers to the monstrous account in *Speculum Naturale* (1494), lib. 31, cap. 126, pp. 408[r–v], a compilation of universal knowledge, by the French Dominican scholar and encylopedist Vincentius Bellovacensis, aka. Vincent de Beauvais (c. 1190–1264). This work went through several editions, including the four volume *Bibliotheca Mundi, Speculum Quadruplex, Naturale, Doctrinalė, Morale, Historiale* (1624).

[14] Again relying on Licetus's *De Monstris* (1665 ed.), lib. 2, cap. 58, p. 182, as his main source, Mather refers at second-hand to Gualtherius Raleigh, better known by his English name, Sir Walter Raleigh (c. 1552–1618), Elizabethan explorer, statesman, and author of his famous *History of the World* (1614). Via Licetus (p. 182), Mather refers to a woman in Saxony, who in "854," during the reign of Lothar, Duke of Saxony, gave birth to what is depicted above. If the given year in the margin of Licetus' book is accurate, then the attribution of the event during the reign "*Lotharij Caesaris*" mistakenly refers to the Holy Roman Emperor Lothair III (1075–1137) rather than his earlier namesake.

Half-Dogs mentioned by *Volaterran* and *Cardan* were.[15] Nor that mention'd by *Magius*; *Humano quidem Capite prœditum, sed Auriculis, Collo, Brachijs, Manibus, partibusque, alijs quibusdam Caninis*.[16] But I well-Remember what *Petrus Damianus* reports concerning *Robert*, a King of *France*; That upon his Coupling with a near Kinswoman, she brought forth a Son, that had a perfect *Head* & *Neck* of a *Goose*.[17] To which may be added, what *Lycosthenes* reports, of an Infant near *Lauffenburg*,[18] whose hands and feet perfectly {matched} the Claws of the *Fowl* that was last mentioned.

But then, *Monsters* have often been brought forth in the Four-footed Part of the World. Yea, and such *Monsters* (and the more hideous for being such) as have had something too much *Humane* upon them. The famous *Horse* of *Julius Cæsar*, whose fore-feet were of the Humane Shape, ha's been celebrated by many Authors.[19] *Paræus* is not the only Writer, who reports a *Sow*, to have brought forth a *Pig*, *Facie, pedibus, manibusque humanis, reliquo autem corpore Suem referentem*.[20] In The same Writer there occurrs, *Equæ*

[15] The Italian humanist Raphael Volaterranus, aka. Raffaello Maffei of Volterra (1451–1522), compiled the encyclopedic *Commentatoriorum Urbanorum libri XXXVIII* (1506), which went through numerous editions and reprints until the 17th century. The reference is to the 1603 edition, lib. 24, title: "Vesica, genitalia, genitura" (col. 903–06), in which chapter Maffei mentions numerous malformations of fetuses, including one born by a "virgin" who had copulated with a canine (col. 906*D*). The Italian physician and mathematician Girolamo Cardano, aka. Hieronymous Cardanus (1501–76), *De Rerum Varietate Libri XVII* (1558), lib. 7, cap. 31, p. 239, dismisses the idea that cohabitating humans and canines can procreate offspring.

[16] Mather's second-hand reference (Licetus, p. 186) is to *Hieronymus Magii, Variarum Lectionum, seu Miscellaneorum Libri IIII* (1564), lib. 1, cap. 20, p. 61^{r}, by the Italian polymath scholar Girolamo Maggi, aka. Hieronymus Magius (c. 1523–72), who describes the malformation as follows: "Endowed, it is true, with a human head, but with the little ears, neck, arms, hands, and other parts of a dog."

[17] Licetus, p. 186, refers to the Dominican Petrus Damianus (c. 1007–c. 1072), Cardinal of Ravenna during the reign of Pope Leo IX (1049–54), whose anecdotal account of King Robert of West Francia (9th–11th century) is here mentioned. It is unclear which of the several French West Francian monarchs by that name the venerable cardinal had in mind.

[18] Laufenburg is a small town in the Swiss canton Aargau. Licetus (p. 181) embellishes the depiction of this claw-footed infant in Lycostenes's *Prodigiorum ac Ostentorum Chronicon* (pp. 502 and 656).

[19] Licetus, p. 181. Suetonius, *Lives of the Twelve Caesars* (1.61), relates that Roman Emperor Julius Caesar (101–45 BCE) bred a horse whose divided hoofs resembled human toes, which prodigy prompted his soothsayers to predict that he would become ruler of the world.

[20] Ambrosius Paraeus, aka. Ambroise Paré (c. 1510–90), French surgeon and physician during the reigns of French Kings Henry II to Henry III (1536–89). The

partus, quadrupes, humano Capite. The Daughter of *Aristonymus*, from an infandous Mixture, called, *Onoscilis*, makes an Article in the Records of Antiquity. *Plutarch* also speaks of a Colt, the upper Parts whereof were of a Form frightfully *Humane*.[21]

But, you certainly begin to reckon it high time, for me to come nearer unto the Subject of my Letter. So, passing by what *Fincelius* relates of a *Cow* that brought forth a *Calf, Vitulum capite humano;*[22] (which it seems, was little

second-hand quote from Licetus (p. 183) is based on Paré's magnum opus, "De Monstris et Prodigiis," in *Opera Chirurgica* (1594), lib. 24, caps 7 and 12, pp. 731 and 735: "with human face, feet, and hands, but resembling a pig in the rest of its body." Likewise, Licetus (p. 183), referring to Paré's work (lib. 24, cap. 1, p. 718), describes the depiction of "the offspring of a mare, four-footed, with a human head." In Paré's depiction, the creature appears to be endowed with the face of a young woman. It is telling that virtually all of Mather's sources on monsters and prodigies share the same tabloid depictions.

21 In his *Parallela Minora* (sec. 29), the Greek philosopher historian Plutarch (c. 46–c. 110 CE) relates the story of Aristonymus of Ephesus (son of Demostratus), who – bored with women – had copulated with a she-ass, whose offspring was a beautiful infant named Onoscelis ("the girl with ass's legs"). Much the same fate befell the offspring of Fulvius Stellus and his mare, whose beautiful offspring was named Epona, protectress of horses – if Plutarch's account (sec. 29) can be taken at his word. Not to forget, according to the apocryphal Testament of Solomon (4:4–7), Onoscelis was a beautiful female demon and shapeshifter.

22 Via Licetus (p. 189), Mather refers to Jobus Fincelius, aka. Hiob Fincel, Finzel (d. 1582), German professor of philosophy at Wittenberg and practicing physician at Zwickau and elsewhere, who is mostly remembered for his *Wunderzeichen. Warhafftige beschreybung und gründlich verzeichnuß schröcklicher Wunderzeichen und Geschichten* (1556). For the year 1523, Fincelius briefly mentions the birth of a calf with a misshapen head, whose tongue hung out of its mouth (n.p.). However, there is no mention of a human face in the story. Depictions of monstrous hybrids ("calf with a human head") were all the rage throughout the ages, and can be found in virtually all books dealing with such prodigies, including Ulisse Aldrovandi's (1522–1605) *Monstrorum Historia* (1642), "Capitis Alieni Procreatio in Foetibus Brutorum," p. 433 (G), sec. II., who acknowledges Fincelius as his muse and provides a rendition of the calf with a woman's face (p. 424). Likewise, a pig with a human face, born in Brussels, in 1564, can be ogled in Ambrosius Paraeus's *De Monstris et Prodigiis* (1582), lib. 24, cap. 12, p. 752; and in Fortunius Licetus, *De Monstris* (1664), lib. 1, cap. 3, p. 13. No less unflattering is the account by Sir John Floyer, communicated by Dr. Edward Tyson, F.C.P., F.R.S.: "A Relation of two Monstrous Pigs, with the great resemblance of humane faces, and two young Turkeys joined by the Breast," *PT* 21.259 (December 31, 1699): 431–35. Significantly, Tyson tries to *disprove* that the monster with its eerie resemblance to "a Humane Face" could have been "produced by the Copulation of two Species." Rather, he speculated that this likeness resulted from "the pervertion of the compression of the Womb, or *Placenta*, or other Pigs in the same part of the Womb" (pp. 431, 432).

Figures 12a and 12b. Depictions of Horses with Human Faces. Courtesy of the Smithsonian Libraries.

other, than what another Author affirms born of a Woman in *Pomerania*;)[23] And, passing by the *Cow* that brought forth, *Vitulum Semi-hominem*, in the Dayes of *Albertus Magnus*,[24] who delivered the *Cow-keeper* from the Suspicion of his Neighbours, by perswading them, from his wonderful Skill in Philosophy, that the *Stars* were concerned in the Business; That whereof I am now to give you a short Narrative, is; That this Year a *Cow* in our Vicinity, brought forth a *Calf*, which had so much of an Humane Visage, as to make the attentive Spectators apprehensive, that the poor Animal had been imprægnated by a Beastly Negro. My shortest & plainest Account of the *Monster*, will be in a lively *Icon* of it; which you have here annexed.[25]

That my Letter may not for its *Bigness* grow into that which ha's been the *Subject* of it; I here finish it; and subscribe,

Sir,

Your {…}

{C. Mather.}

[23] A region in northwest Poland.

[24] Licetus, p. 190; St. Albertus Magnus (c. 1200–80), German Dominican philosopher and scientist of Cologne, is said to have delivered the poor man from the mob. Stories and depictions of animals with human features were all the rage and can be found in numerous works capitalizing on monsters and prodigies (see above). On Albertus Magnus's views on monsters, see *De Animalibus XXVI*, Bd. 16, vol. 2, lib. XVIII, tract. 1, caps 4–6, §§30–54.

[25] Mather's racial slur is indicative of his time. Most likely, he provided a drawing of the calf's head – no longer extant. But see the depictions of such prodigies in Licetus, *De Monstris* (1665), p. 190, and appendix. Mather tells of a similar case in his execution sermon *WFD* (1693) of an animal born in southern Italy (1641) "that had something of an Humane Shape. This Monster had a Blemish in one Eye, just like what a loose Lew'd Fellow in the Town, was known to have. This greater Monster, being upon this Account Suspected, was Examined upon the suspicion, and upon his Examination Confessed his Guilt of most infandous Bestialities, for which he underwent a deserved Execution" (p. 44). Mather reports of yet another example "at the Southward" of Boston: "There have been *Devilish Filthinesses* committed among us; But, Oh! how strangely, have the *Sins of men found them out!* At the *Southward*, there was a *Beast*, which brought forth a Creature, which might pretend unto something of an *Humane Shape*. Now, the people minded, that the Monster had a Blemish in *one Eye*, much like what a profligate Fellow in the Town was known to have. This Fellow was hereupon Examined; and upon his Examination, Confessed, his Infandous *Bestialities*, for which he was deservedly Executed" (*TD*, pp. 50–51) and *MCA* (6:35). See also J.M. Murrin (1998).

3.3. THE NIDIFICATION OF PIGEONS.

July 4, 1716.

To Dr. John Woodward.

Sir,

You were not so cloy'd with a small Treat of my New English *Pigeons*, which I endeavoured for you, in a former Letter,[1] that you will be unwilling to hear any further of them.[2]

Among the Entertainments of your *Ornithology*, and in the vast field of Wonders, for which the *Feathered Tribes* have winged your Curiosity, you have allow'd a particular Consideration to the surprizing Skill with which their *Nidification* is managed.

It surprizes us to see, what *secure* Places they find out, and what *proper* ones, where their Young may ly safe & warm, & have their Growth promoted. But, with what an Artificial Elegancy are some of their Nests præpared; such, that *Humane Skill* could hardly imitate it![3]

We celebrate the *Nests* of the Indian Bird, composed of the Fibres of certain Roots, which are so curiously interwoven, that they cannot be beheld without Astonishment. These *Nests*, the sagacious Bird (which therefore with good Reason they call, a *Subtle Jack*,) hangs on the Ends, of the Twigs of the Trees over the Water, to secure its Eggs & its Young, from the Ravage of *Apes*, & other Beasts, that else would prey upon them.

And what shall we say of the *Flamingo's*? They build their *Nests* in shallow Ponds, wherein there is much Mud; which they scrape together into little Hillocks, like Islands appearing out of the Bottom. They make the Foundation of these Hillocks broad, bringing them up tapering to the top, where they leave a small hollow Pitt, in which they lay their Eggs;

1 See Mather's letter to John Woodward (June 1714), CA 2.5 "The Pigeons."

2 Mather integrated his epistle on nidification in his *CP* (1721), Essay XXX, pp. 188, 189, 190–91, 197. See also *CPS* (1994), p. 201ff.

3 This and the following paragraph also appear in *CP* (1721), pp. 188, 189.

and when they either lay or hatch their Eggs, they stand all the While, not on the Hillock, but close by it, with their Legs on the Ground, and in the Water, resting themselves on the Hillock, and covering the hollow Nest upon it, with their Bodies. Their Legs are very long, and building, as they do, upon the Ground, they could neither draw their Legs conveniently into their Nests, nor sitt down upon them otherwise than by resting their whole Bodies, to the Prejudice of their Eggs, or Young; were it not for this rare Contrivance.[4]

But what my *Pigeons* do, is as worthy to have Remarks made upon it, as any of the rest. They build their *Nests* with little sticks laid athwart one another, at such Distances, that while they are so near together, as to prevent their Eggs falling through, they are yett so far asunder, that their Eggs may feel the cool Air coming at them. Now the REASON for this *Architecture*! Tis this; Their Bodies are much hotter than those of other Birds; and their *Eggs* would be perfectly addled by the Heat of their Bodies in the Incubation, if the *Nests* were not so built, that the cool Air might come at them to temper it.[5]

If Dr. *More* in his Triumphs over Atheism; took the *Eggs* of Birds, for a considerable Ingredient of his *Antidote* against that Madness;[6] and if Dr. *Cheyne*, from the subject of these *Eggs*, did well to take this noble Flight, *It is impossible duely to consider These things, without being rapt into admiration of the Divine Architect*;[7] You will give me Leave, to add, that the Management of our Birds cannot but compell us into the Sense of an Intelligent Being, who has imprinted on these little Animals, a Disposition to Actions of so agreeable a Tendency.

I know not what well to make of an odd Relation, published among you, a few Years ago, but so well attested, that a very pious & worthy Man wrote a large Treatise upon it, entitutled, *Vox Corvi*: – which affirms, That a *Raven* perching on a Steeple, & thence turning towards a Quarrelsome

4 *CP* (1721), pp. 188–89.

5 For the same, see *CP*, p. 188.

6 *CP* (1721), p. 190. Mather refers to the discussion of bird eggs as evidence of divine creation in *An Antidote against Atheism* (1655), lib. 2, ch. 9, pp. 126–28, by Henry More (1614–87), renowned Anglican clergyman, philosopher, and Cambridge Platonist.

7 *CP* (1721), pp. 190–91. The italicized passages are extracted from *The Philosophical Principles of Religion Natural and Revealed* (1705–15), vol. 1, ch. 3, §40, p. 233, by George Cheyne (1672–1743), English physician, dietician, and esteemed natural philosopher.

Neighbourhood, was heard very audibly & articulately to utter those Words, *Look into the Third of the* Colossians, *and the fifteenth*.[8] But this is very certain, *Ask the fowls of the Air & they shall tell thee*;[9] There needs no Genius to take a Possession of our Birds, that we may hear the Admonitions of Piety, and Exhortations to Beleeve and Adore an Infinite GOD, intelligibly enough proceeding from them.[10]

I am glad, that from the Wing of one of them, I am furnished with that Engine; wherewith I may now assure you, that I am, with very great Respect,

Sir,

Your {…}

{C. Mather.}

[8] *CP* (1721), p. 197. The italicized reference is to the title page of Alexander Clogie (1614–98), *Vox Corvi: Or, The Voice of a Raven* (1694). Clogie was an Anglican clergyman and biographer in Wigmore (Herefordshire). Col. 3:15 reads, "And let the peace of God rule in your hearts, to the which also ye are called in one body; and be ye thankful."

[9] See Job 12:7.

[10] *CP* (1721), p. 197.

3.4. A TRITON.

July 5, 1716.

To Dr. John Woodward.

Sir,

Long was I no other than a *Pyrrhonian,*[1] upon the common Tradition of *Tritons*, or *Fishes* in a Shape near *Humane*, sometimes coming from the *Watery World* unto us. *Bellonius* indeed saies, *Vix fabulosa esse quidam putant, quae de Tritonibus tradunt Veteres*.[2] But I was not among the *Quidam*.[3] Nor did I any more beleeve them, than I did *Centaurs* and *Sphinges*;

1 A skeptic who prefers to doubt and suspend judgment rather than believe and embrace certainty – as taught by the ancient Greek philosopher Pyrrho of Elis (c. 360–c. 270 BCE). Those who are not disciples of Pyrrho may enjoy V. Scribner (2017 and 2020).

2 Petrus Bellonius, aka. Pierre Belon (1517–64), was a French traveler, diplomat, and naturalist, who authored several works on marine animals, including *L'Histoire Naturelle des Estranges Poissons Marins* (1551), *De Aquatilibus, Libri duo* (1553), and *La Nature et Diversité des Poissons avec leurs Pourtraicts* (1555) – all of which served as valuable sources in Conrad Gesner's *Historiae Animalium liber IV* (1558, 1604), which went through several editions in the 16th and 17th centuries. The Latin passage, however, is not from Bellonius, but from the subsection "De Pisce Episcopi Habitu, Rondeletius" ["Rondeletus on the Bishop Fish"] in Gesner ([1604] lib. 4, fol. 439, lines 9–10), and signifies, "There are those who believe that the accounts the ancients hand down concerning Tritons are not mythical at all." The bishop fish (see image below) – its head shaped like a bishop's conic mitre – was so named by the French physician Guillaume Rondelet (1507–66), whose work Mather quotes at second-hand. The Mather library included *Historia Animalium* (1588–90) and *De Omni Rerum Fossilium Genere, Gemmis, Lapidus, Metallis, et Huiusmodi* (1565), by Konrad Gesner (1516–65), a Swiss physician, encyclopedist, and natural historian at the University of Zurich.

3 "Some." In other words, Mather claims he did not believe in the reality of tritons – at least not in the past.

nor take a *Triton* to be any other, than what *Suidas* tells us, it meant; the *Neptune* of the Pagans.[4]

I did as little know, as *Ælian* himself saies he did, what to make of the Story, which he quotes from *Demostratus*, of a Dried *Triton* that was once & long to be seen at the Town of *Tanagra*.[5] When he concludes, that he was convinced of the Matter by a *Divine Testimony*, because, forsooth *Apollo Didymæus*[6] had mention'd a *Triton* under the Name of, Θρεμμα Θαλαττιον, *A Beast of the Sea*; this afforded unto me no Conviction at all; Tis true, *Pausanias* commemorates, the taking of this *Triton*, by the *Tanagræans*, who trepann'd him, with leaving an Intoxicating Draught of Wine for the Monster. But all the Use of *Pausanias*'s Commemoration,[7] was only to explain unto me the Meaning of another Fable, which they had added unto this; That *Bacchus* kill'd him.[8]

Theodor Gaza, and *Alexander ab Alexandro*, and *Scaliger*,[9] made me begin to think the Stories of *Tritons*, worth some small Attention, when they

4 Centaurs and sphinxes are mythological creatures: the former in the shape of a horse with a human upper torso; the latter, a lion's body (often feline) with a human head and eagle's wings. Suidas, or Suda, a 10th-century encyclopedic dictionary of the ancient Mediterranean, mentions the mermaid-like monster Triton (Greek god of the sea) in *Suidae Lexicon* (3:507), *sub voce*: Τρίτωνος.

5 Mather here paraphrases Greco-Roman sophist Claudius Aelianus' (c. 170–c. 235 CE), *De Natura Animalium* (13.21), as rendered in Gesner, *Historiae* (1604), lib. 4, fol. 999, lines 2–3. Aelian himself refers to the Greek Demostratus (an obscure Greek author), whose treatise on fishing mentions a "Triton in pickle" that Demostratus claims to have seen at Tanagra (a Boeotian town north of Athens). This pickled triton looked similar to what is depicted in the statue of Neptune, yet its head and scaly body had greatly deteriorated – so Demostratus.

6 Gesner, *Historiae Animalium* (fol. 999, line 3). The vast oracular temple of Apollo at Didyma was near Miletus.

7 The Greek geographer Pausanias of Lydia (fl. mid-2nd century CE) in his *Description of Greece* (9.21.1) claims he had seen a smaller Triton at Rome, with hair on its head (in color like a marsh frog). The bodies of Tritons, according to Pausanias, are scaly like sharks, yet instead of ears they have gills, and the nose of a man with a mouth broader and with the teeth of a beast. "Their eyes seem to me blue, and they have hands, fingers, and nails like the shells of the murex. Under the breast and belly is a tail like a dolphin's instead of feet" (Loeb).

8 The preceding paragraph is a paraphrastic extract from Gesner (fols 998–99).

9 The Greek humanist Theodore Gaza (c. 1398–c. 1475) is renowned for his translation of Aristotle into Latin. The Neapolitan attorney Alexander ab Alexandro (1461–1523) supplied Gesner with his accounts of aquatic creatures (see below). And *Exotericarum Exercitationum Liber Quintus Decimus, De Subtilitate, Ad Hieronymum Cardanum* (1557), Exercitatio CCXXVI, sec. 12 ["Pisces humana figura. Nereides. Tritones, Cippae"], pp. 298–99, by the distinguished Franco-Italian

so affirm Fishes of *Humane Lineaments*; to have been seen in the Dayes of *Augustus*, & of *Tiberius*, and of *Neros*.[10]

Alexander ab Alexandro almost conquered me, when after he had related wonderful Stories of *Tritons*, he adds, *Nostro ævo haud absimile factum comperimus*; One who came ashore on the Coast of *Epirus*, and siezed upon Women for Libidinous Purposes, & ran away into the Water with them. It seems, he was catch'd at last; but starv'd Himself to Death.[11]

Pliny mentioning these *Tritons*, thinks he has good Authority for it; *Autores habeo in Equestri ordine splendentes*.[12] I am not sure, that *Virgil* was one of his Knights who would swear for him, that there was an Animal so to be described,

Frons hominem præfert, in piscem desinit Alvus.[13]

But *Plinyisms* are of no great Reputation in our Dayes.[14] And People may sail thro' the Straits of *Gibralter*[15] safely enough, if their only Danger were from

---

humanist Julius Caesar Scaliger (1484–1558), engaged in famous debates on natural philosophy with the Italian polymath Hieronymus Cardanus (1501–76).

10 Roman Emperors Augustus Caesar, aka. Octavian (BCE 63–14 CE); Tiberius Julius Caesar Augustus (BCE 42–37 CE); Nero Claudius Caesar Augustus Germanicus (37–68 CE).

11 The entire paragraph is based on Gesner's *Historia* (1604), lib. 4, fol. 999, lines 36–39. Gesner himself appears to rely on the edition of Andreae Tiraquelli, *Semestria In Genialium Dierum Alexandri ab Alexandro* (1586), lib. 3, cap. 8, p. 328, lines 1–2, by the Neapolitan attorney Alexander ab Alexandro (1461–1523). Mather claims he was almost sold on the reality of monstrous Tritons when he read Alexandro's Latin affirmation, that "in our time, we have learned of something quite similar that happened"; i.e., sea creatures of this type have also been discovered in our time, like the amorous one on the shores of Epirus on the Ionian Sea.

12 Pliny the Elder (c. 23–79 CE) relates that the governor of Roman Gaul reported to the late Caesar Augustus the sighting of a number of dead Nereids (Tritons) on shore, and that "I offer as authorities distinguished members of the Equestrian order" (*Natural History* 9.4.10).

13 Mather here quotes from Gesner's heading "De Pisce Monachi Habitu, Rondeletius," in *Historia Animalium IV* ([1604], lib. 4, fol. 439, line 41); the standard version (as it appears in Vergil's *Aeneid* (10.211) reads, "*frons hominem præfert, in pristim desinit alvus*" ("the forehead of a man [whose body] ends in a fish") and vividly describes the Triton figurehead at the helm of Aulestes' ship; quoted with a minor misspelling, it appears in Gesner (fol. 728, line 33).

14 Known for its many inaccuracies and exaggerated accounts, Pliny's *Natural History* gave rise to the moniker "Plinyism" or "unreliable assertion."

15 Or rather, "the Pillars of Hercules" – the Gulf of Gades.

such a *Man of the Sea*, as he tells us was very troublesome to Sailors in the Night cruisi{ng} there.

I grew more suspicious of the Matte{r} when I readd, what *Menas* the Præfect of *Mauritius*, avers with all possible Solemnity, That he with many others were Ey-witnesses, of both a Male and Female *Triton*, or Fish of an Humane Form, appearing on the *Nilus*, from Sun-rise to Nine a clock.[16]

My Suspicions had yett more Strength given to them, from what *Bellonius* mentions of a Fish in an Humane Shape, that an Infinite Number of People saw walking on the Shore, for a good While together, on the Coast of *Norway*.[17]

Gillius further strengthened them, when he letts me know, that he himself beleeved what he reported, of such a *Triton*, (by him called, *Marinus homo*) coming out of the Sea, at *Spalato*, to sieze a Woman on the Shore. Yea, that several such had been taken, of whose Hides, there had been made Shoes, that lasted for fifteen Years. *Non igitur, quod de Tritonibus et Nereidibus hominibusque Marinis iam usque ab Heroicis Temporibus in Sermonem hominem venit, vanum esse constat.*[18] For, which he appeals to many Proofs in modern Memory.

I am aware, that among the *Arabians*, the Existence of a *Triton*, (or, a, *Piscis habens formam humanam*, is beleeved; and the Name of *Alchusi* given to him.[19] However, *Arabia* is too far off, to have Relations that come from thence well examined with us.

16 Gesner, in *Historia* (1604), lib. 4, fols 999, lines 58–67, points to the Prefect Menas (probably a governor of Egypt under Maurice), who wrote to the Emperor Maurice about the pair of Triton Menas had seen on the banks of the Nile.' Gesner (fol. 440, lines 39–40) points out the event occurred "sub Gregorio & Mauricio" (under Pope Gregory I [590–604]) and Eastern Roman Emperor Maurice [582–602]).

17 A summary of Gesner (lib. 4, fol. 440, lines 9–11), whose source is Bellonius, *De Aquatilibus*, lib. 1, cap. 7, p. 38 – though the latter seems less forthcoming about the Monkfish's libidinous meanderings on shore.

18 The paragraph, including the Latin citation, appears in Gesner (fol. 440, lines 11-20), who quotes Petrus Gyllus' *Ex Aeliani Historia per Petrum Gyllilum Latini Facti ... accessionibus aucti libri XVI* (1533), lib. 10, cap. 10, pp. 278–79. Mather's Latin passages via Gesner (fol. 440, lines 19–20) is to assert the real existence of these mythological creatures: "Thus, it is established what has entered into the discourse of men concerning Tritons and Nereids and mermen from way back in the time of the heroes is not nonsense." The likely story is told by the French natural scientist Petrus Gillius, Gyllius, aka. Pierre Gilles (1490–1555), who in his *Ex Aeliani Historia* (lib. 10, cap. 10, p. 278), asserts that the attempted abduction by a "sea man" occurred on the shore of the Dalmatian island Spalato, modern Split (Croatia).

19 Gesner (fol. 440, lines 6–7). According to the Italian physician and Arabist Andreas Alpagus Bellunensis, aka. Andrea Alpago of Padua (c. 1450–c. 1522), the Arabians call this piscatorial Triton in human shape "Alchusi."

And so is *Brasile*, where *Purchas*'s Travellers tell us, *Mearmaids* often catch, and kiss, and embrace the Indians, & crush them to Death. *Sofala* is as distant, where *Purchase*'s Travellers also tell of the same *Fish*, that *Horace* alludes to.[20]

A Nameless Author, who in the *Italic* Tongue writes, *De Terra Sanctâ*;[21] celebrates a Fish, of Humane Face, Mouth, Teeth, Nose, Eyes, & Part of the Neck; piscine all the rest. But I did not see myself bound to beleeve him. *Italians* don't alwayes speak True!

One, who is, *Author Incertus*, mentions a Remote River of *Muscovy*, wherein they sometimes take a Fish, which is of an Humane Form, in all the Parts of it. But I find him styled, not only, *Author Incertus*, but also *Fide parum dignus*.[22] However, seeing an Army of *Tritons* like to be raised upon my Incredulity, it began to cry for Quarter.

I feared, I must submitt unto the Truth of the Story, that a *Sea-monster* in the Habit of a *Bishop*, was exhibited unto the King of *Poland*. But then, *Rondeletius* kept my Incredulity in heart, by his representing it as a *Fabulous Business*.[23]

My *Submission* was yett more cogently demanded, by the Famous Tradition of a *Sea-monk*, two ages ago, taken among a Net full of Herrings on the Coast of *Norway*, which was Four Cubits high, lived three dayes, had no Language but that of bitter & heavy Sighs, and in the Midst of them expired.

20 The English encyclopedic compiler of travel accounts, Samuel Purchas (c. 1577–1626), tells his stories in *Purchase His Pilgrimes. Part 4 In Five Bookes* (1625) and describes the city of Sofala (Mozambique) and Brasil (bk 6, ch. 1, p. 1148). In his *Treatise of Brasil* (Pt 4, bk 7, ch. 1, §6, p. 1315), he gives the account of the deadly kisses of the mermaids, whom the natives of Brazil call "*Ypupiapra*." Mather probably has in mind *Ars Poetica* (lines 3–4), where Horace (65 BCE–8 BCE) imagines a painter creating a weirdly hybrid creature: "ut turpiter atrum/ desinat in piscem mulier Formosa superne" ("so that a woman who is beautiful above might terminate in a hideous fish").

21 Gesner (fol. 440, line 30), "Of the Holy Land."

22 According to Gesner (fol. 440, lines 33–36), the remote river is called "Tachnin." Gesner's original "*Verum is parum fide dignum*" can be rendered, "It is true that he [uncertain author] is untrustworthy." Gesner paraphrases a dismissive passage in Sigismundus Liber Baro, i.e., Sigismund, Freiherr von Herberstein (1486–1566), who wrote *Rerum Moscoviticarum Commentarii* (1549), sec. 2, fol. XII^{r-v}, a classic account of Russia.

23 Gesner (fol. 439, lines 50–63) refers to the French professor of medicine at Montpellier, Guillaume Rondelet (1507–66), author of the richly illustrated *Libri de Piscibus Marinis* (1554), which supplies Gesner with many picturesque illustrations, including the Monkfish, which the Flemish physician Gisbertus, practicing in Rome, showed to the Polish king, in Rondelet's *Libri de Piscibus Marinis* (1554), lib. 16, p. 492; and cap. 21, p. 494. A French translation of the Latin original appears as *L'Histoire entière de Poissons* (1558).

Figures 13a & b. "De pisce monachi habitu." In Guilielmi Rondeleti, *Libri de Piscibus Marinis* (Lugduni, 1554), lib. 16, cap. 19, p. 492; cap. 20, p. 494. Courtesy of the Smithsonian Libraries. (See also Petrus Bellonius, *De Aquatilibus* (1553), lib. 1, cap. 4, p. 39. for another embellished version of Monachus piscis, the monkfish.)

But my Historian, still afforded me room for a little Breathing in Suspense, with saying, *Quædam præter veritatem a Pictoribus addita fuisse puto, ut res mirabilior haberetur.*[24]

If the Matter of Fact be true, The *Monachus Maris* mentioned by *Albertus*, as appearing sometimes in the Britannic Sea, & flattering People till he d{evou}red them, shall go for a Kinsman of the *Norwegian.*[25]

24 Mather's historian is Rondelet (*Libri de Piscibus Marinis*, lib. 16, cap. 20, p. 493) – yet quoted at second-hand via Gesner ([1604] fol. 439, lines 20–23). Mather's second-hand Latin quote via Gesner signifies, "I believe that certain features that go beyond strict truth were added by the illustrators in order that the thing might be considered more astonishing." Gesner does not further identify Gisbertus, except to say, "Ejusdem monstri picturam mihi ostendit Gisbertus medicus ex eadem Nortuegia Roman ad se missam, quae pictura nonihil à mea differebat" ("The doctor Gisbertus showed me [Gesner] a picture of the same monster, which had been sent to him at Rome from Norway – a picture that differed to some extent from mine."

25 Gesner, fol. 441, lines 9–11; see fols 546–47. The "Monachus Piscis" or "Monachus Maris" are a species of monk seals often found in the Mediterranean and

My Incredulity gave yett more way when I readd what *Cardan* relates, of a Woman-Fish carried by great Inundations into a Lake of *Pomerania.* And when I found the *Belgic* Records,[26] affirm, That in the Year, 1404. some Women of *Edam*, often saw, at the Ebbing of the Water, a Sea-Woman playin{g} up and down; which they at Length siezed upon, & brought her away with them.[27] And that the Monster anon grew familiar, and lived on ordinary

other hemispheric regions. The "Britannic Sea" is better known by its moniker "North Sea." Mather facetiously dismisses this tall tale just as much as he does the one about the creature caught on the coast of Norway (previous paragraph). The Dominican bishop and medieval philosopher Albertus Magnus (c. 1200–80) describes his "Monachus Maris" (Monkfish) in *De Animalibus* (24.45.85, vol. 2:1540), and elsewhere, as having "white skin on the top of its head, around which is a dark circle, like the head of a monk who has been recently tonsured. It has, however, the mouth and jaws of a fish, flattering people till he devoured them," in Albertus Magnus, *De Animalibus Libri XXVI* (1920).

26 Jacobus Meyerus Baliolanus (1492–1552), a Flemish humanist whose history of Flanders (*Commentarii, Sive Annales rerum Flandricarum* [1561]) was incorporated in *Annales, Sive Historiae Rerum Belgicarum, à Diversis Auctoribus ... ad Haec Nostra Usque Tempora conscripta deductaque* (1580), tom. I, cap. XIIII, p. 251. Meyerus mentions that during the time of Margaret II, Countess of Flanders, in the year 1403 (MCCCCIII), rather than 1404, during a great storm, a marine woman (mermaid), naked and mute, was captured and carried to Edam and from there to Harlem. She even learned to spin, and though mute, she showed honor to the Holy Cross. (Thanks to David Lupher for clearing up this nugget of a likely story). Says Mrs. Jonah to her husband's coming home dripping wet, "For crying out loud, Jonah! Three days late, covered with slime and smelling like fish. What story have I got to swallow this time?" (Gary Larson, *The Far Side*).

27 The reference is to the so-called "Mermaid of Edam," also known locally as the "Purmer Zeewijf," for she was supposedly caught in 1403 (not 1404) in the Purmermeer, a former lake just a few miles west of Edam (Netherlands). In Mather's main source (Gesner, 1604, p. 440), this lake's unfamiliar name had become distorted into the more familiar but geographically absurd Pomerania – an error that seems to have begun with Belon (*De Aquatiibus,* 1553, fol. 38), then was picked up by Rondelet (*De Piscibus Marinis,* 1554, p. 495), who passed it along to Cardano's earlier edition of his *Operum Tomus Tertius; Physica* (1563), lib, 7, cap. 38, p. 143, by the Italian polymath Hieronymus Cardanus, aka. Girolamo Cardano (1501–76). Gesner mainly used the version of Belon, but interpolated bits from Cardano. The persistence of "Pomerania" into Mather's account is a vivid demonstration of how early modern scholarly traditions often resembled modern internet "memes." Mather's other source for the Mermaid of Edam is more elusive. There is an account of her under the year 1403 in Jean de Meyere's chronicle of Flanders (1561), reprinted in the 1580 compendium *Annales, Sive Historiae Rerum Belgicarum* (vol. 1, p. 251). But Mather's account offers details not found in de Meyere's chronicle.

Meats; and being brought unto *Harlem*,[28] lived about fifteen Years; not without often seeking to return unto the Sea. But in this different from our Women; she never spake a Word.

Captain *Whitburn* almost made me yeeld, when he affirms, in his Description of *New-found-land*, That in the Year 1610.[29] He himself saw, a Fish, whose Face, Eyes, Nose, Mouth, Chin, Ears, Neck, & Forhead, were like a Woman; It was very beautiful, & in those Parts well-proportioned, & had Hair hanging down round about the Head. It came within a Pikes Length of him, then dived under Water, from whence rising up again, he had a fuller View of it; and saw the Back of it, white square, smooth, like that of a Man; Below it resembled a broad hooked Arrow. It would afterwards have come in to the Boat, where some of his Men were, till one of them struck it a full Blow on the Head. And others had afterwards a full View of it. I suppose, This was not the *White-hatt*, which the *New-found-Landers* often tell us of; who appears & haloo's unto them, when a Storm is approaching.

That which came in, to make the Conquest of my Credulity, more likely to be carried, was, this; That I knew not well how to rebell again[st] the Authority of our English Chronicles; which affirm, That in the Reign of K. *John*, there was taken a Fish, near *Orford* in *Suffolk*, in all Parts like a Man;[30] & was kept for six Months in the Castle, whence he afterwards escaped, & returned into the Sea. *Fabian*[31] adds, He could by no Methods, no, nor by Torments,

28 I.e., Haarlem, the capital city of North Holland.

29 Mather extracts his material from *A Discourse and Discovery of New-found-land* (1620), Conclusion, sign. N^{r-v}, by Capt. Richard Whitbourne (1561–1635) of Exmouth, Devonshire, who served as governor (1618–20) of Newfoundland, for the island's proprietor Sir William Vaughan (1575–1641) of Golden Grove, Carmarthenshire.

30 Mather refers to the curious story of the wild merman of Orford (Suffolk), whom "*some fishermen hapen'd to catch … in their nets. All the parts of his body resembl'd those of a man; he had hair on his head, a long picked beard, and about the breast was exceeding hairy and rough. But at length he made his escape privately into the Sea, and was never seen* more," in *Camden's Britannia, Newly Translated into English, with large additions and Improvements* (1695), pp. 373–74, by the famous English Chronologer William Camden (1551–1623), whose acknowledged source is Chronicon Anglicanum, by the English chronicler and Cistercian Abbot Ralph de Coggeshall (fl. 1207–26). According to Coggeshall, the wild man was caught in a net by local fisherman (c. 1167) who kept him in Orford Castle for nearly six months. See G.R. Varner (2007), p. 78.

31 *The Chronicles of Fabyan, whiche he hym selfe nameth the concordaunce of histories* (1559), by Robert Fabyan (d. 1513), an English draper who rose to fame as a sheriff and alderman of London. Fabyan's *Chronicle* went through at least five printed editions. Son of Henry II of England, King John (1166–1216) ruled

be brought ever to speak a Word, He would retire to his Lodging at Sun-sett, & gett up again at Sun-rise. One Day they brought him unto the Haven, & lett him go into the Sea; only that they might prevent his getting away, they sett three Rows of Strong Netts before him, to catch him again at their Pleasure; But he diving to the Bottom crept under their Netts, & shewed himself again unto them; & so often diving, & then raising himself unto their View, it seemed as if he mocked them. Nevertheless at Length, of his own Accord, he returned unto them, & for two Months more he remained with them. Afterwards not being very carefully look'd after, he left 'em altogether; went away to Sea, & was never heard of any more.

But now at last, my Credulity is entirely conquered; and I am compelled now to beleeve the Existence of a *Triton*; For such an one, ha's just now been exhibited in my own Countrey; and the Attestations to it, are such that it would be a Fault in me at all to quæstion it.[32]

On 22.^{d} XII.^{m} 1715/16.[33] Three honest & credible Men, coming in a Boat from *Milford* to *Brainford* (which Towns ly near one another on the South-shore, in our Colony of *Connecticot*,) when it was young flood,[34] about the Hour of Two in the afternoon; as they drew near the Shore of *Brainford*-Harbour, they were surprised, with the View of a Creature that seem'd a *Man*, lying on the Top of a Rock, about a dozen foot from the Water. When they were about Three or Four Rods from the Rock, the Animal seem'd affrighted; lifted up his head; look'd forwards & backwards; then clap'd his head close to the Rock, as willing to hide himself; and after this; lifting up his Head again to discover what was approaching. The Three Men rowed unto the Lee-ward of the Rock, & two of the Three went out of the Boat, with a Design to take him; the one in the Front, the other in the Reer. He that went behind the Rock, had a Purpose to strike him with a Pole, and was come within a Rod of him; the other was two Rods off, & the third, in the Boat, might be three. The Person who approached, purposing to strike him, trod on the Ice, the breaking whereof made a Noise, which frighted the Animal, so that he Jump'd off the Rock, & with all possible Expedition flounced into the Water. They had a full View of him; and saw, his *Head*, and *Face*, and *Neck*, and *Shoulders*, & *Arms*, and *Elbows*, and *Breast* and *Back*, all of an *Humane Shape*; only his *Arms*,

England from 1199–1216, popularly remembered for granting the Magna Carta (1215), the royal charter of rights.

32 Mather leaves blank the bottom one-third of his MS. page, resuming at the top of the next.

33 I.e., February 22, 1716.

34 I.e., incoming tide.

were little more than half the Length of a Mans. He wanted not for *Hair*, which was of a *Grayish* Colour. However, – *definit in piscem*;[35] His lower Parts were those of a *Fish*, and coloured like a *Mackarel*. His Tail was forked, and he had Two Fins about half a Foot above the Tail. The whole Animal was about Five or Six Foot in Length. He scrabbled with his Arms, and wrought with his Tail, to reach the Water; and being safely lodged there, and having dived out of the Sight of his Persecutors, until he was near Twenty Rods from them, he there he lift up his Head again to look at them, as high as his Arms reaching down to the Water would allow; and having thus taken a View of them, he took his Leave.

Tho' Honest *Francis Leguat*, has a Remark upon *Rondeletius*; "I wonder" (*saies he*) "at the might{y}Reputation the famous *Rondelet* acquir'd; for {I} never consulted him concerning any thing within my knowledge, but I found him very false & very dry."[36] Nevertheless, I cannot but pay a Just Regard unto a Remark on this *Rondeletius*; *Ea est hominum Vanitas, ut Rei per se satis mirabili prœter verum, plura etiam, affingant.*[37] And therefore the[38] Account I have given you of this American *Dagon*,[39] is exactly True & Plain; & without any flourishes upon it.

You know better than I, what Credit is to be given unto the public Account you have lately had, of a Sea-Monster, in the Shape of a Man, which came ashore this Winter, a few Miles above the City of *Ragusa*,[40] successively for three Dayes together, at so many different Places, tw{o}or three Leagues distant from each other; And every time he came out of the Sea, he walked for three Hours along the Shore, lifting up his hands now & then, & letting them fall again, with such hideous Cries, & from so terrible an Aspect, as to frighten

35 "[H]e ended in a fish."

36 The French Huguenot naturalist and explorer François Leguat, aka. Le Guat (c. 1637–1735), is the author of *A New Voyage to the East-Indies* (1708), a translation of Le Guat's two-volume *Voyage et Avantures de Francois Leguat* (1708). The quote appears on p. 89.

37 Mather agrees with the dismissive remark which Gesner (fol. 439, lines 60–61) attributes to Rondeletius: "For such is the folly of men that when a thing is quite sufficiently astonishing in itself, they foist upon it even more things in utter violation of the truth."

38 MS.: "therefore in y^{e}."

39 An ancient Semitic deity worshiped in the Middle East, Dagon (also Dagan) was the father of the god Baal, and was revered by the Philistines (Josh. 19:27; Judg. 16:23).

40 In the early eighteenth century, Ragusa, a city-state on the Adriatic Sea, was renamed; it is now known as Dubrovnik.

several People to Death; His Head very large; his Body well proportioned; but about fifteen foot high; and having so strong & loud & piercing a Voice, as to be heard at the Distance of a Couple of Leagues.

If the Story in your Account be *Grubstreet*,[41] mine is not so. All that I have to regrett is, that my honest Friends miss'd of taking my *Triton*. For tho' I should not have thought fitt then to have trusted him, with the Conveyance of my Letter to you, yett I should I have been able to have given you a more punctual & entertaining Account of him.

What I have given, I pray may be received among the Testimonies, of the just Concern to communicate unto you all New & Rare Occurrences of Nature, in these Parts of the World, which will alwayes be maintained by,

Sir,

Your— {...}

{C. Mather.}

41 I.e., gossip, taken from the popular tabloid journalism published in Grub Street, London. In fact, the monstrous story was told in the *Flying Post* (February 21, 1716), the *British Weekly Mercury* (Feb), and Read's *Weekly Journal* (February 25). For this monster, see P. Roger (2016).

3.5. A SERPENT SECURELY HANDLED.

July 6, 1716.

To Dr. John Woodward.[1]

Sir,

If you find, *the Incantation of Serpents*, to be the Subject of the Letter that is now address'd unto you, you will allow it for a Subject that is not without its Curiosities; nor will you read the Letter, with the horror of one treading among *Serpents*, especially when you are sensible, that it shall terminate in the discovery of a Surprizing Way to be præserved from their Stings, & one that will please you beyond any *Enchantment*.

One knowes not how well to deny, the very many Passages of *Antiquity*, that mention the force of Charms upon *Serpents*, & imply the Energy of *Dæmons* introduced with them, especially when we read in one of our Inspired Psalms, concerning, *The Deaf Adder, that stoppeth her Ear, & will not hearken to the Voice of the Charmer, charming never so wisely.*[2]

And if one should conceive, That in *Serpents* being more liable to *Incantation* than many other Creatures, there is a sort of Intimation bringing to Remembrance, our old Seduction from God, by the *Divel* possessing of a *Serpent*; he might support the Conceit with a Citation from *Austin* for it.[3]

1 See also Mather's Letter to Richard Waller, November 27, 1712, CA 1.11 "On the Rattlesnake."

2 Psal. 58:4–5. In his subsequent paragraphs, Mather employs much of the same material as in his commentary on Psal. 58:5–6 (*BA* 4:524–28). The Italian physician and naturalist Ulisse Aldrovandi (1522–1605), whose *Serpentum, et Draconum Historiae Libri duo* (1640) charmed his readers with popular depictions of serpents and dragons.

3 Mather alludes to *De Genesi contra Manichaeos* (2.26, secs 39, 40) [*PL* 034.0217], by St. Augustine (354–430), who (like St. Paul before him) cries out against the Manichean devotion to the dualistic serpent-god (Gen. 3). St. Paul (as snake charmer) sings his spell when he warns against false apostles (2 Cor. 11:3) who seduce Christ's followers with their heretical opinions just as the arch marplot of Eden did Eve. Mather's principal source for this epistle to Woodward is

If it should be true (what is commonly said,) That there are Three or Four sorts of *Serpents* (called, *The Deaf Serpents*, because) Magicians cannot make Impressions upon them, as they can upon the rest, one would enquire, whether there may not be intended in it, such a Triumph of the *Divine Power* over the *Diabolical*, as there was when the *Lice* were exempted from the command of the *Magicians* in *Egypt*.[4]

The great *Bochart* allowes not this Distinction, but proposes the Success of Attempts upon the same sorts of *Serpents*, at several times to be various, *Prout variè Deus Diabolo habenas, aut adducit, aut remittit*.[5]

You know, Syr, that some have thought the fame of *Orpheus*, in more early Antiquity,[6] to have risen from his Art in *Enchanting* of *Serpents*. Well-known are his Verses, (if they be his, for *Suidas* assigns them to another,) which mention the Stopping of the hissing of the *Dragon*, – ηδ' Οφεων ιον σβεσαι ερπυστηρων· *et Serpentum repentium venemum extinguere* – To which tis likely our Apostle may allude, when he speaks of, *Quenching the Fiery Darts of the Divel*.[7]

Twould be a perfect Rudeness, to tire you, with citing what occurs in *Plato*, in *Aristotle*, in *Strabo*, in *Gellius*, as well as in *Ovid*; about the *Fascination* of *Serpents* in general.[8]

Hierozoicon Sive Bipartitum Opus De Animalibus Sacrae Scripturae (1692), pars secunda, lib. 3, cap. 6, cols 384–94, by the distinguished French Huguenot Samuel Bochart (1599–1667), a renowned Protestant scholar, whose *Hierozoicon* (1663) and the two-volume *Geographia Sacra* (1646) went through multiple reprints in the 17th and 18th centuries. Better yet, they retained the page-and-line references of the *editio princeps*. Here, all references to Bochart's *Hierozoicon* are to the Dutch 1692 edition.

4 Exod. 8:18.

5 Mather draws on his old mainstay, *Hierozoicon* (1692), lib. 3, cap. 6, col. 394 (lines 68–69); the Latin quote reads, "Just as God in different ways tightens or loosens the devil's reins."

6 The Thracian poet singer Orpheus serves as the mythical patron of music, whose singing could make animals, trees, and even rocks dance to his tune, in *Orphica* (*Lithica* 48–49).

7 Both the Greek passage from the 10th-century Byzantine dictionary-encyclopedia *Suda* and its Latin translation are from Bochart, *Hierozoicon Animalium* (lib. 3, cap. 6, col. 386, lines 73ff). Mather here quotes from *Orphica* (*Lithica*, line 49): "and to quench the poison of the creeping serpent." By implication, Mather (via Bochart) suggests that the Apostle Paul (Eph. 6:16) alludes to this ancient Orphic myth (Bochart, 387, lines 1–2).

8 Mather's cursory references to the Greek philosophers Plato (c. 428–c. 347 BCE). References to Plato's student Aristotle (384–322 BCE); to the Greco-Roman geographer, philosopher, and historian Strabo (c. BCE 64–21 CE); to the Roman

Descending to particular Wayes of *Incantation*, I will take the Leave to cite but one or two Passages on each of the Heads. *Virgil* mentions the Performance of it, by the *Touch* of the *Hand.*[9] And in *Silius Italicus*, we have this thing more than once; – *Tactuque graves Sopire chelydros*. And again, *Tactuque domare venena.*[10] *Arnobius* mentions the Performance of it, by *Talismanical Figures.*[11] *Attalus* in *Pliny* affirms, That a *Scorpion* will not bite a Man, if he cry, *Duo*![12] And *Alcamus* affirms, That at the Sound of, *Os*, or, *Oso*, a *Serpent* letts a Man alone.[13]

But the main Force for performing an *Enchantment* upon *Serpents*, alwayes lay in *Verses*;— but these also were to be *murmured*, or *whispered*; For which Reason *Sorcerers* of old were called, *Whisperers.*[14] *Ælian*, as well as *Pliny*, tells us, That *Serpents enchanted with certain Verses*, επαοιδαις τισι, would be fetched out of their lurking Places.[15] And

---

grammarian Aulus Gellius (c. 125–post 180 CE); and the Roman poet Publius Ovidus Naso (BCE 43–17 CE) can all be found in Bochart's chapter *De Aspide surda* (*Hierozoicon*, pars post. lib. 3, cap. 6, col. 384–94).

9 Bochart (387, line 33); Vergil (70–10 BCE) tames a serpent in his epic *Aeneid* (7.753–55), when King Archippus sent his Marsian priest Umbro, whose charm and touch put vipers and water snakes asleep.

10 The Roman poet, senator, and orator Tiberius Catius Asconius Silius Italius (c. 26–c. 101 CE) of Padua is remembered for his epic poem *Punica* – the longest in Italian literature – on the Second Punic War (218–201 BCE). Mather (via Bochart, lib. 3, cap. 6, col 387, lines 36 and 38) cites from Silius' *Punica* (1.412 and 8.499), which translate (respectively): "to tame poisonous serpents by handling them" and "put venomous water-snakes to sleep by handling them."

11 Bochart (col. 387, lines 58–64). The early Christian apologist Arnobius of Sicca (d. c. 330 CE), in his polemic against the heathens *Adversus Gentes* (2.32.2), reports that pagan warriors tried to protect themselves against the sting of poisonous snakes with amulets made from thin plates (*ANF* 6:446 and [*PL* 005. 360–61]).

12 Bochart (col. 388, lines 24–26). In his *Natural history* (28.5.25–26), Pliny affirms that King Attalus (of Pergamon) could check a scorpion from striking by pronouncing the word "Two."

13 Bochart (col. 388, lines 27–41) quotes from the celebrated Arabic lexicon *Qāmūs al-muḥit* (*Al-Qamus al-Muhit* (aka. *Alcamus, Qamus*), a compilation of several earlier Arabic dictionaries, by Muhammad al-Fayruzabadi, aka. al-Firuzabadi (1329–1414), a distinguished Persian lexicographer who settled in Mecca, whose prized volume is *Qāmūs al-muḥit, sive, Thesaurus Linguae Arabicae et Monumentis Arabum Manuscriptis* (1632). Mather, via Bochart, states that according to the Lexicon *Alcamus*, the whispered sound of "Os" or "Osoh" stops the serpent in its tracks.

14 The Hebrews called *incantations* "*lahashim*," i.e, "whispers."

15 Pliny (*Historia Naturalis* 8.21), "etiam serpentes extrahi cantu cogique in poenam" ("even snakes were drawn out with a song and forced into punishment"), as

Seneca mentions it in his *Medœa.*[16] *Jerom* saies, The Man who did this Trick, was called, *Marsus.*[17] And *Scaliger* affirms himself to have been an *Ey-witness* of it; *Nos aliquandò vidimus cantationibus è Cavernis exciri Serpentes.*[18] Many think, a *Magician* to be for this Cause called, *Chober*, or, *Consociator*; (as R. *Solomon* saies) *Qui consociat Serpentes in Locum unum.*[19]

By Verses too, they would put *Serpents* to flight; else *Lucan* would not have said, *Verbisque fugantibus Angues,*[20] or, at least, they pacified them & stupefied them. *Tibullus* has it, *Cantus et iratœ detinet Anguis iter.*[21] And *Columella,–Magicis Sopitum Cantibus Anguem.*[22] *Seneca* too has the Same.[23] And

well as his fellow Roman author, the rhetorician Claudius Aelianus (c. 170–c. 235 CE), assures his readers in his dual-language *De Natura Animalium* (6.33) that the Egyptians are able to "bring down" birds from the sky with some spells and to bewitch snakes to come out of their hiding places.

16 Bochart (388, lines 55–57); the Roman philosopher playwright Seneca the Younger (c. 4 BCE–65 CE) in his tragedy *Medea* (684–86) has the daughter of King Aeëtes of Colchis, by her magic incantations, pull a brood of poisonous serpents out of its lair for her revenge against Jason's new bride.

17 Mather (following Bochart, col. 388, line 64–66) refers to St. Jerome's annotations on Psal. 57, in *Breviarium in Psalmos* [*PL* 026. 0990C], an exegetical work of uncertain authorship, commonly attributed to St. Jerome, aka. Hieronymus Stridonensis (c. 342–420), theologian, historian, and translator of the Hebrew Bible into Latin. Mather appears to misunderstand Ps.-Jerome, who says, "that man who makes an incantation is called a Marsian." The Marsians, a people in ancient Latium, were noted sorcerers, witches, and snake-charmers, who are said to draw serpents from darkness into light with their incantations.

18 Via Bochart (col. 389, lines 4–7), Mather quotes from an unidentified work by either Joseph Justus Scaliger or Julius Caesar Scaliger, which translates, "we have sometimes seen serpents driven out of caves by singing."

19 Bochart (col. 389, lines 10–13) cites from the dual Hebrew-Latin *Commentarius Hebraicus in Pentateuchum Mosis Latine Versus* (1710), p. 1409, on Deut. 18:11 (Vox חובר [chober]), by the famous medieval French R. Salomon Jarchi, aka. Shlomo Yitzchaki (1040–1105), also known by his acronym Rashi. According to Rashi, "he [a charmer] brings serpents together in one place.

20 Bochart (col. 389, lines 27–30). The Roman poet of Cordoba M. Annaeus Lucanus, in his *Pharsalia* (9.914) has the Roman soldiers clear the terrain "with chanted words to put snakes to flight."

21 Bochart (col. 389, lines 33–34) adds evidence from the Roman poet Albius Tibullus (c. 55–c. 19 BCE), *Elegiae* (1.8.20), that "spells stop the path of the angry snake."

22 Columella of Gades (fl. first century CE) in his book on agriculture, *De Re Rustica* (10.367), told how "the snake guarding the Golden Fleece was lulled to sleep by Medea's incantation."

23 The Nurse in Seneca, *Medea* (704), calls upon the dragon of the Colchians, "come thou to my aid, thou who through my incantations wast first lulled to slumber."

Plutarch, and *Philostratus*, yea, and our *Theodoret* also, have notable Hints to this Purpose.[24]

Yea, by this Method of *Enchanting* all the Poison of *Serpents* ha's been extinguished. *Petrus Chieza*, in *Delrio*, reports, that the *Serpents* on all the vast *Andian* Mountains, are this very Way become Innoxious.[25] The same Way did they that were stung with the *Poison* of *Serpents* use to be healed; *Strabo* finds this usual among the Indians.[26] – And so *Lucan* sings, – *Pestis nigris inserta medullis, excantata perit* [fugit].[27]

Yea, and they *burst* the *Serpents* too. Saies *Virgil, Frigidus in pratis cantando rumpitur Anguis*,[28] And *Ovid, Carmine dissiliunt abruptis faucibus Angues*,[29] Again, *Vipereas rumpo verbis et carmine fauces*.

You will pardon me, Syr, if I ask, That it may not be judged altogether *unphilosophical*, to suppose, That these Tricks, were often performed, by

24 Bochart (col. 389, lines 57–65); Plutarch, *Vitae Parallelae: Cato the Younger* (56.3), relates that the Libyan tribe called the Psylli charm and mortify the serpents with their incantations. Philostratus, in *Life of Apollonius* (3.8), tells of Indian magicians who by placing charmed runes in front of the dragon's burrow make the serpent fall asleep. In his *Interpretatio in Psalmos* (Psal. 58:4–5) [*PG* 082. 1298], Theodoretus of Cyrrhus (c. 393–c. 466), Syrian theologian and bishop of Cyrrhus, relates that David "compared him [Saul] to an asp, which is not naturally deaf but blocks its ears and prevents the onset of the charm" (Psalm 58:4–5), in *Commentary on the Psalms* (1:332).

25 Bochart (col. 390, lines 10–17); the Spanish conquistador Pedro Cieza [Cieça] de León (1518–54) authored an early chronicle of Peru, *Dela chronica del Peru* (1553); however, the citation appears in *Disquistionum Magicarum Libri Sex* (1608), lib. 2, quest. 13, p. 81(*E*), by Martin Antonio Delrio (1551–1608), a Dutch Jesuit theologian of Antwerp.

26 Bochart (col. 390, lines 20–27); Strabo (*Geographica* 15.1.45) explains that according to Nearchus (c. 360–c. 312), Alexander the Great's admiral in the Indian campaign (325 BCE), a multitude of reptiles, including vipers sixteen feet long, would pester the inhabitants near rivers especially during high tide; if stung by serpents or otherwise diseased, charmers would go around and heal the afflicted with their medicine. However, Aristobulus (a Macedonian officer) is incredulous about the size of the reptiles but admits that there are many smaller vipers, asps, and scorpions that are dangerous. Especially the small ones – no more than a span long – are particularly vicious: they hide in crevices, tents, vessels, and bushes, and those who are stung bleed profusely unless an antidote of Indian roots and drugs is administered.

27 Bochart (col. 390, lines 29–31); Lucan, *Pharsalia* (9.930–31): "Often, the pestilence is expelled from the blackened marrow by incantation."

28 Bochart (col. 390, lines 36–37); Vergil's *Eclogae* (8.71): "With song the cold snake in the meadows is burst asunder."

29 Bochart (col. 390, lines 38–42); Ovid, *Amores* (2.1.25): "Spellbinding songs can burst the serpent's jaw." And again, Ovid, *Metamorphoses* (7.203): "With my incantations I break the viper's jaw."

Dæmons who being after some sort *Invoked* by the *Charmer*, stept in to answer his Intention.

And there ha's been known, a strange Effect upon our Modern *Dogs*, as well as upon the ancient *Snakes*, from Applications to *Dæmons* for it; For the *Indians*, in hostility against us, having ceremoniously sacrificed some *Dogs* unto the *Divel*, it ha's been remarkable that when these *Indians* approached our Houses in the Night, upon any Designs of Deprædation, our *Dogs* would not *bark* at them; tho' at another time, the Approach of an *Indian*, even at a considerable Distance, would sett our *Dogs* a barking, so that there could be no quiet for them.

Nevertheless, a little Acquaintance with *Philosophy*, will inform us, of Methods to manage *Serpents*, which have been purely *Natural* and *Innocent*, when *Ignorance* would have been ready to cry out of something *Diabolical*.

An Ingenious Gentleman, who writes, *The Agreement of the Customes of the East Indians, with those of the Jews*,[30] affirms, that he knowes it commonly practised among the *Indians* at this day, to take *Serpents*, and make them *Dance* at the Sound of the *Flute*. They commonly carry about several Sorts of *Serpents*, in Hampers, & make them *Dance*, when any body will give them any thing for the Spectacle. And by their *Flute*, with a *Song*, they fetch the *Serpents* out of their Holes, & can take them up without hurt, & associate them with their Brethren.

What I have now to add, is an *Incantation* upon *Serpents*, that I beleeve will find Some Approbation with you; If there is nothing of the *Divel* in it.

We have a Sort of an *Hawk*, which we call a *Kite*. This *Kite* very much makes his Living upon *Rattle-Snakes*. And our Indians have with the *Fatt* of this *Kite*, used both Outwardly & Inwardly wrought happy Cures on People that have been bitten with the *Rattle-Snakes*. But what ha's appear'd a little strange to the Spectators, ha's been, to See an Indian take a *Rattle-snake*, and handle the fierce Poisoner with as much Freedom, & as little Damage, as if he had only had an *Eel* in his hands.[31] Anon the Indian discovered the Mystery. He had anointed himself with the *Fatt* of the *Bird* aforesaid; and that gave him a greater Security, than what *Alcimus*, writing, *De Peccato Originali*, sings of; *Mox impunè manu coluber tractatur inermi*, *Et morsus*

[30] Mather synopsizes this paragraph from *The Agreement of the Customs of the East-Indians, With those of the Jews, and other Ancient People* (1705), art. 27, p. 93, an epitome of various travel accounts and comparisons of indigenous customs, by M. De La Créquinière (fl. 17th–18th century), variously attributed to John Toland and Daniel Defoe. The French edition appeared in Brussels in 1704.

[31] Mather cancels an unassigned, incomplete sentence written vertically in the left-hand margin: "For now y^{e} *Rattle-snake* would not so much as attem{pt}."

tantum, non virus in Angue timetur.[32] For now the *Rattle-snake* would not so much as attempt the least Bite upon him; or proceed so far as the *Fathers & Poets* generally agree to carry the Teeth of the *Viper*, which fastened on the hand of our Great Apostle;[33] but was there, as they wittily said, *Non ut læderet, sed ut ornaret.*[34]

You have none of our formidable *Rattle-snakes* on your Island. The Animals that most answer to them, are the *Defamers*, which among your Contending Parties, are as Mischievous, as they are very Numerous. But your happy Experience ha's given you to find your *Innocence & Usefulness* and generous *Patience*, a greater Defence than our *Axungia Milvina*,[35] against the Bites of such contemptible Reptiles.

With a just Esteem for your excellent Character in those & all Regards, I am,

Syr,

Your most –

{C. Mather.}

[32] Bochart (390, lines 2–6); Avitus Viennensis, *De Mosaicae historiae gestis libri quinque* (lib. 2, De Originali Peccato) [*PL* 059. 0335D], by the poet Alcimus Ecdicius Avitus, aka. St. Avitus (c. 450–c. 519), Bishop of Vienne (Gaul). See for the same in Alcimus's *De Origine Mundi, De Peccato Originali* (1545), lib 2, p. 28 (lines 21–22): "Soon the snake is harmlessly handled with unprotected hand, / And only the bite, not the venom, is feared in the snake."

[33] Bochart (*Hierozoicon*, pars 2, lib.3, col. 373, lines 30–39). St. Paul, while gathering brushwood to build a fire after his shipwreck on Malta (Acts 28:3); Mark (16:18).

[34] "Not to injure, but to adorn."

[35] "Kite fat." See also *RWSV* (1720), Appendix, xiv.

3.6. A STRANGE MISCHIEF TO THE EYES.

July 7, 1716.

To John Woodward.

Sir,

In reading the Journal of, *An Embassy from* Muscovy *over Land into* China, in the Years, 1693, &c.[1] I find the Mention of a People, called, *Ostiacky*, which inhabit between *Tobolsko* and *Jenokisko*, which have this Curiosity related concerning them;[2] "Men and Women are constantly subject unto a certain *Weakness in their Eyes*; the Cause of which they attribute unto *Want of Bread*; which being a scarce Commodity among them, they seldome are Masters of it, unless they are by Chance furnished with it by Travellers."[3]

This Passage raised in me some Speculations, concerning the *Eye*, and the Things that may have a Tendency either to præserve, or to annoy, that wonderful Organ, and how far it may be affected by the *Food* which we subsist upon.[4]

Where the first Thing that offered itself unto Consideration, was an Experiment, which I have seen made; with a Success which was a little marvellous,

1 Mather refers to Adam Brand (d. 1713), secretary to Everard Isbrand Ides, aka. Evert Ysbrants [Ides] (1657–1708), Danish merchant, traveler, and diplomat, who recorded *A Journal of the Embassy From Their Majesties Iohn and Peter Alexievitz, Emperors of Muscovy* (1698), p. 39. The Dutch translation appeared as *Driejaarige reize naar China, te lande gedaan door den Moskovischen afgezant* (1704).

2 The author of *A Journal of the Embassy* opines that the Ostiaky (aka. Ostiaks, or Ostyaks), a Finno-Ugric people in the basin of the River Ob in western Siberia, are "a lazy people and idle people" (34, 41–42). Tobolsk, the capital of the Tyumen Oblast, is located at the confluence of the Tobol and Irtysh rivers, east of the Ural Mountains of Siberia (31, 39). Ienokisko, aka. Jeneskoi, Ienska, is a regional capital in the Yenisei (Jenisei, Ienska) River valley, "near the River *Tobol*, about 3,000 Miles from the City of Musco," aka. Moscow (30–31, 33, 39).

3 *Journal of the Embassy*, p. 39.

4 See Mather's Essay XXXII. "The Ocultist. Considering, Diseases of the Eye, Especially, *Blindness*," in *ABJ*, pp. 155–60. Mather's admiration for the human eye appears in *CP* (1721), pp. 235–44.

of Recovering the Sight of such as have been struck Blind with Lightning, by feeding much upon Raw Oisters, in the Season of them.[5]

I shall not offer you any of my mean Thoughts on the Subject which I have thus begun upon; but instead thereof I will entertain you with a short Relation, which I have lately received from one of my Neighbours, wherein I find a Curiosity of *Mischief to the Eye*, which I know not that ever it ha's had before any Notice taken of it; A *Mischief to the Eye*, whereof rather *too much Bread* of one sort, than any *Want of Bread*, was the Occasion.

A Neighbour of mine lately on board a Vessel, which was loaden with Pease, & other Commodities, was by a violent Storm driven off our Coast, & forced unto a Flight for Barmudas. The Voyage attended with a thousand Surprizing Difficulties, which it was a Wonder how they could have been out-lived by any Mortals. But one was this; The Vessel took in abundance of Water, which rotted the Pease; and the Steam of the Rotten Pease had a Strange Effect upon the Eyes of the People aboard; producing in them, not only a Blindness, which unhappily unfitted them for much of their Business, but also a Torment that was Insupportable. To releeve themselves in this Distress, they employ'd some hands, to throw the Pease out of the Hold; but e're they had thrown out fifty bushels, these were struck so totally Blind, that they could proceed no further. By keeping at as much Distance as they could from the wett Grain, they gott some Abatements of their Darkness & their Dolour; and yett it remained such, that they could hardly see the Island, when they were just upon it. But when they were gott ashore, they had their Eyes restored unto them.

The Occurrence I thought might be worth mentioning; unto one whose *Eyes* are forever clear-sighted enough to discern what Improvements may be made of it, for the Guard, or the Cure, of an *Organ*, the fullest of Wonders of any thing under the Sun. At least for this Cause I will count it so, in that it gives me an Opportunity of Repeting the Expressions & Assurances, of the very great Esteem wherewith you are alwayes considered by

Sir,

Your most unfeigned

Friend & Servant.

{C. Mather.}

[5] One of Mather's remedies for dim eyesight is the consumption of "*Raw Oysters* Three or Four Every Morning for some time" (*ABJ* [1972], p. 158).

3.7. STRENGTH OF IMAGINATION.

July 9, 1716.

To John Woodward.

Sir,

The Surprizing Force of Imagination ha's doubtless appeared unto you, in so many Instances, as to sett your Philosophy to work, upon some Solution for its very strange Appearances; And I make no doubt, that the Insipid Philosophy of a *Fienus*[1] writing on that Subject, which seems to ly very much in a few School-terms, which are a meer Jargon invented for an ostentatious Cover of Ignorance, give to your sagacious Enquiries very little of Satisfaction.[2]

There are some odd Instances, of what ha's been called the *Potentia Imaginativa*,[3] & People acting therewith upon others at a distance from them. *Avicen* asserts very unaccountable *Powers* to the *Imagination*,[4] until he carries

1 Mather refers to *De Viribus Imaginationis Tractatus* (1608), a Ramist dialogue by Thomas Fienus, aka. Fyens, Feyens, of Antwerp (1567–1631), Flemish professor of medicine at the University of Louvain. Mather also draws on Fienus's work in his fourth letter to John Woodward (November 20, 1712), our CA 1.4 "Antipathies & the Power of the Imagination." For a useful assessment of Fienus's work and the concept of the imagination, see L.J. Rather (1967), E. Fischer-Homberger (1979), P.K. Wilson (1992), M.-H. Huet (1993), and H. Hirai (2011, 2017). On the imagination in the early 18th century, see L.S. King (1978), pp. 152–81.

2 Fienus' syllogistic-dialogic reasoning embedded in his categorical distinctions, propositions, and deductions – reminiscent of medieval scholasticism – had become outmoded by the end of the 17th century.

3 "The Force of the Imagination." See especially the philosophical debate on the power of the imagination of pregnant women and their pervasive impact on the development of the fetus, in *De La Recherche De La Vérité* (1674–75), by the distinguished French philosopher Nicolas Malebranche (1638–1715), whose influential work was translated into English as *Father Malebranche his Treatise Concerning the Search after Truth* (1700). See esp. bk 2, pt 1, ch. 7.

4 Avicenna, aka. Ibn Sina (980–1037 CE), a renowned Persian physician, astronomer, and philosopher, also known as the Father of early modern medicine. Avicenna locates the cognitive-imaginative faculty in the "mid-portion of the brain,"

the Matter so far, as therewith, *Hominem longè remotum ex equo insidentem dejicere, et in puteum prosternere*:[5] – and makes the Souls of some, to be as powerful in their Operations, as the *Supernal Intelligences*. This Gentleman ha's had many Disciples, besides *Marsilius Ficinus*.[6] And, if I should say, that were the *Powers* of *Imagination* well understood, some things that have been ascribed unto that *Witchcraft* which lies in a Confœderacy with *Evil Spirits*,[7] would be

in *Canon of Medicine* (1973), pp. 136–37, §179. Mather's source is Fienus, *De Viribus Imaginationis* (Questio IIII [IV], pp. 40–41), who in his turn refers to *Liber Quartus Naturalium De Actionibus et Passionibus Qualitatum Primarum* (N.D), 4.6 Naturalium, cap. 4; and to *Lectionum Antiquarum* (1516), lib. 20, cap. 15, by the Italian humanist Lodovicus Caelius Rhodiginus, aka. Lodovico Ricchieri (1469–1525). My reference is to the 1620 edition of Rhodiginus' opus, published in Coloniae Allobrogum, lib. 20, cap. 15–17, col. 1115–18. On Avicenna's views of the Imagination, see D. Black (2013).

5 Fienus, *De Viribus Imaginationis* (Quest. II, Concl. XI, p. 40), reads as follows: "to throw a man far distant from the horse he is sitting on and chuck him into a well"; i.e., to mislead an attentive reader who winds up lost in the opaque verbiage of the author. Mather's odd adaptation is his circumlocution for casting aspersions on Avicenna's philosophy of the imagination as being full of misleading jargon.

6 Fienus, *De Viribus Imaginationis* (pp. 25, 40–41), points out that Avicenna had many disciples, including Algazel, Albertus Magnus, Marsilius Ficinus, Pomponatius, Paracelsus. The Neoplatonist philosopher Marsilius Ficinus (1433–99) was the foremost humanist scholar of the early Italian Renaissance. Via Fienus, Mather refers to Ficinus' *Theologia Platonica. De Immortalitate Animorum* (1559), lib. 13, cap. 1, p. 208; cap. 4, p. 224^{r}, arguing that the imagination – via the vibration of the spirits and the nerves – imprints its mark on the unborn fetus.

7 Credulous as Mather's references to the devil's power may strike us today, his allusion is entirely consistent with his contemporaries' physiological understanding of the force of the imagination. The devil, as the argument goes, though an immaterial spirit without physical extension or dimensions, does have the kinetic power to manipulate the imagination of individuals. By operating on the subtle fluids of the mind, the devil can make an impression on the nerves, perturb the humors (animal spirits), thicken the blood (especially in melancholics), and thus distort his victim's perception. Thus the devil's power of prestidigitation can engender in his victims an illusion of reality, arouse the affections (fear, desire, love, hate), thereby agitate the humors, and affect the physical body. Significantly, Mather, though eschewing supernaturalism, here provides a mechanical explanation for how the afflicted manifest signs of bewitchment that *cannot* be blamed on the accused "witches." Analogically, then, just like a mother's power of the imagination can impact her unborn fetus, so an afflicted person's natural imagination – once stirred up – can make impressions on her own body. In this way, it comes as no surprise that when Mather's new-born infant Joseph died from membranous blockage of his intestines (April 11693), the doting father suspected that the birth defect was caused by the terrified imagination of his wife Abigail, who had been accosted by "an

so applied hither, that we should go no further for them; I perceive, I should have *Andreas Cataneus*, writing, *De Causis Mirabilium Effectuum*,[8] besides, *Cornelius Agrippa*,[9] and *Jacobus de Forlivio*,[10] and *Cælius Rhodiginus*,[11] of the same Opinion with me. Not that I would go so far as those Gentlemen do, who equip the Diabolical *Apollonius Tyanæus* with no further Powers.[12]

Whether, the Feats done, by those who are called, *Salutatores*,[13] among the Spaniards, may be ascribed unto those Powers alone, as *Chr. à Vega*[14] thinks they may, I do not know. But if *Pomponatius* press it hard upon me, That the

horrible *Spectre*, in our Porch, which Fright caused her Bowels to turn within her" (*D* 1:164). See *also* Malebranche, *The Search after Truth* (bk 2, pt 3, ch. 6, pp. 99–101).

8 Fienus, p. 41; the Italian scholar Andreas Cataneus, aka. Andrea Cattaneo of Imola (fl. 1500–26), is remembered for his *Opus de Intellectu et de Causis Mirabilium Effectuum* (1502), a brief analysis of Avicenna's metaphysics.

9 Heinrich Cornelius Agrippa (1486–1535), a German physician, scholar-theologian, kabbalist, and Neoplatonist, whose *De Occulta Philosophia Libri Tres* (1531–33) is among his most influential works. It attempts to synthesize the natural world, the occult, and the divine to create a holistic vision of nature.

10 Jacobus de Forlivio, aka. Foroliviensis (d. 1413), Italian physician and philosopher, who authored commentaries on Hippocrates' *Aphorismata* (1477) and on Avicenna, in *Expositio in Primum Avicennae Canonem* (1508).

11 Fienus, *De Viribus Imaginationis* (Quæst. X, p. 97); Caelius Rhodiginus, aka. Lodovico Ricchieri (1469–1525), Venetian professor of Greek and Latin at Bologna and Milan, whose *Lectionum Antiquarum Libri triginta* (1516) is the work in question. Among his most famous students was Julius Caesar Scaliger.

12 Fienus, p. 41, argues that while some ascribe weird actions to demonic intervention, Algazel, Albertus Magnus, Marsilius Ficinus, Pomponatius, and Paracelsus believe these strange actions occur solely through the power of the imagination. Thus they suppose that Apollonius of Tyana (c. 3 BCE–c. 97 CE), a Greek neopythagorean philosopher of Cappadocia, performed astonishing feats through his imaginative faculty. Apollonius' travels, powers of divination, and supposed miracles are celebrated in the semi-fictional *Vita Apollonii*, by his Greek biographer Philostratus (c. 170–c. 245 CE).

13 Fienus, pp. 41, 198. "Salutatores" are clients who call on their patrons for support. In this context, Mather appears to refer to the henchmen who enabled Apollonius Tyanaeus to perform his pretended miracles.

14 Mather, via Fienus, *De Viribus Imaginationis* (Quaest. IV, p. 41; Quaest. VI, Concl. 25, p. 69; and Quæst. XIII, Concl. XL, p. 114), refers to Christopherus à Vega, aka. Cristòbal de Vega (1510–73), a distinguished Spanish physician philosopher at the University of Alcalá, whose *Opera Omnia* (1626) is a huge commentary on the art of medicine as practiced by Galen and Hippocrates. He discusses the effects of the passions on the imagination and humors in *Opera*, lib. 2, sec. 5, cap. 6, fol. 232–33.

Bones of a *Dog*, will work as great *Miracles* as the *Bones* of a *Saint*, if the Imagination of the Worshipper take the former to be the latter, I shall rather than Quarrel with him, allow it to him.[15] Some have proposed this; *Quid miri est, si anima humana, quæ est tam nobilis, et excellens, et Divina, Spiraculum à Deo Spiratum, aliquam habeat potentiam, per quam admirabilia possit præstare? præcipuè Animæ aliquæ nobiliores et præstantiores?*[16]

Who can determine how far the *Powers* of *Imagination* may have their Efficacy, in the Gentlemen-Strokers,[17] who by a Touch of their Hand cure certain Maladies, as King *Pyrrhus* did Splenetic Distempers, with the Touch of his great Toe upon them; and as our Kings do the Kings Evil;[18] and, as they

15 Via Fienus (Quæst. XI, p. 98), Mather alludes to the dismissive statement in *De Naturalium Effectuum Causis, sive de Incantationibus* (1556), cap. 12, pp. 249–50, by the Italian humanist and philosopher Pietro Pomponatius, aka. Pomponazzi (1462–1525), who ran afoul of the Roman Catholic Church for contending that the power of the imagination to restore the health of a worshiper of holy relics would be just as efficacious if the bones of a saint were replaced with those of a canine. Deemed grist for the mills of Protestant polemicists, Pomponazzi's work was subsequently placed on the *Index Librorum Prohibitorum* of forbidden books.

16 Mather truncates his quote from Fienus (Quaest. IV, p. 42), which can be rendered, "Would it be so remarkable if the human soul, which is so noble and excellent and divine, a breath breathed forth by God, were to have a certain power through which it would be able to perform astonishing things – especially certain souls which are nobler and more exceptional than others?"

17 Mather alludes to miracle healers like Valentine Greatrakes, aka. Greatraks, Greatreats (1629–82), an Irish faith-healer, aka. "the Stroker," who cured invalids through a kind of mesmeric (sympathetic) healing or physical massage. See Greatrakes' *A Brief Account of Mr. Valentine Greatraks, And Diverse of the Strange Cures By him lately Performed. Written in a Letter Addressed to the Honourable Robert Boyle Esq.* (1666). Increase Mather's elder brother, the Rev. Samuel Mather (1626–1671) of Dublin (Ireland), discussed Greatrakes's notoriety and dismissed it as superstition. Cotton Mather excerpted their discussion in his uncle's biography "Fructuosus, or the Life of Mr. Samuel Mather," in *MCA* (1702) 4.2.13, pp. 151–52. See *also* Cotton Mather's medical handbook *ABJ* (1972), Cap LXII, p. 298; and "*A Letter Communicated from Mr.* [Ralph] *Thoresby ... concerning the Cures done by Mr. Greatrix the Stroker*, in *PT* 21.256 (September 30 1699): 332–34. A.B. Laver (1978), B.B. Kaplan (1982), and P. Elmer (2013), esp. pp. 66–67, 159, provide useful background and analyses.

18 The Greek King Pyrrhus (319–272 BCE) was the ruler of Epirus to whom divine powers of healing were attributed, as indeed they were to most kings even during Mather's time. The "King's Evil" (scrofula), or tuberculosis of the lymph glands, has been a common disease throughout the ages. Even King Charles II of England (1630–85) reportedly healed the sick by touching the patient. If that didn't accomplish the deed, then the gift of a gold coin, bearing the emblem of St. Michael, certainly worked miracles. The practice fell out of favor after the death of Queen Anne

say, *Vespasian* did some other Infirmities;[19] and as the Lad, whereof *Michael Medina*, in his Book, *De rectâ in Deum Fide*,[20] writes, That he had what was called, *Gratiam Sanitatis*, and, *Multos homines Solo attactu gravissimis morbis liberabat*.[21] We have read of whole Peoples, whose *Eyes* have been strangely fascinating. And of many, who by uttering of *Words* have done *Wonders*, both Beneficial and Pernicious. *Varro* and *Cato*, as well as *Pliny* have given us Multitudes of Instances. And *Benivenius* mentions a Dart pull'd out of a wounded Souldier, by nothing but a Verse recited over it.[22] Perhaps, the

(1702–14). As a matter of fact, Mather was ahead of his English colleague William Beckett, F.R.S. (1684–1738), a London surgeon and antiquarian, who rejected the belief in the efficacy of the king's touch in curing scrofula, in *A Free and Impartial Enquiry into the Antiquity and Efficacy of Touching for the King's Evil* (1722). In his medical handbook *ABJ* (1972), cap. LXII, pp. 293–301, Mather dismissed such practices as superstition, along with the belief in the healing powers of the Seventh Son. See our CA 6.6 "The Seventh Son Examined." Not completely unrelated is the story told in *Boswell's Life of Johnson* (1:42, 43), in which Johnson's biographer points out that the future compiler of the dictionary was one of the last few who "was actually touched by Queen Anne." Alas, her touch was "without any effect." As J.L. Clifford relates this incident, in *Young Sam Johnson* (1955), p. 10, "A specific incitement for the Johnsons in the spring of 1712 was a statement in the *London Gazette* of March 1 that the Queen intended to 'touch' on March 19 and weekly thereafter through Lent." (Thanks to David Lupher for bringing this interesting titbit to our attention).

19 Roman Emperor Vespasian (9–79 CE), in the best tradition of Egyptian Pharaohs, was thought to be endowed with miraculous powers to heal cripples and restore sight to the blind with his spittle.

20 *Christianæ Parænesis, sive De Recta in Deum Fide Libri Septem* (1564), by the Spanish Franciscan theologian at the University of Alcalá, and member of the Concilium Tridentinum (1545–63), Michael Medina, aka. Miguel de Medina (1489–1578).

21 Fienus (Quaest. IV, p. 93; and Quæst. XI, p. 99) paraphrases Medina's story of a boy of Salamanca, Spain (*Christianæ Parænesis*, lib. II, cap. 7, p. 59[V]), who was endowed with the "gift of healing" and "freed many people from the most severe diseases by mere touch." Slyly (perhaps), Mather left out the rest of the story: the disease recurred when the powers of their imagination cooled off.

22 Mather relies on Fienus' discussion (Quæst. IIII [IV], p. 44) on how the imagination animates the soul, by invoking Pliny's *Natural History* (28.4), who in his turn mentions Marcus Terentius Varro's (116–27 BCE) potions for the gout, in *Antiquitates Rerum Humanarum et Divinarum* (2.111; VI.5), and Marcus Porcius Cato's (95–46 BCE) formula for curing sprains (Pliny, *Natural History*, 17.47). To this collection of "memorables," Fienus adds the magical cure of a soldier as described in *De Abditis Nonnullis ac Mirandis Morborum et Sanationum Causis* (1507), cap. 26, sign. ciii, by the Italian physician Antonio di Paolo Benivieni, aka. Benivenius (1443–1502) of Florence.

Powers of Imagination, may in many of these Cases have had their Operation. Especially, if such an Enquirer as an *Adrian*, have the *Hydropic* under cure.[23]

But upon the Allowance of this, it is easy to allow, that the *Potentia Imaginativa* in a Man, may act upon himself, such Things as one would hardly have imagin'd. There is no Room to Doubt of the Efficacy which the *Imagination* of the prægnant Mother, often ha's upon the unborn Infant. We have the *marked Children*, so many of them every day before our Eyes, that any Doubt of it, must proceed from a very singular Obstinacy of Incredulity. We need not go so far as *Hippocrates's* delivering the white Mother of a Black Infant[24] from Suspicion of Adultery (mention'd by *Jerom*)[25] for the Proof of a Matter so very

[23] Via Fienus (pp. 44, 99–100), Mather refers to Hadrian (76–138 CE), whose recitation of certain formulas was rumored to cure the Hydropic (dropsy, edema), a degenerative disease which swells the body by storing excessive fluids – if the lengthy account on chronic hydrops in *De Morbis Acutis et Chronicis* (1709 ed.), lib. III, cap. VIII, pp. 468–91, by the Greco-Roman physician Celius Aurelianus of Sicca (5th century CE), can be relied upon. Mather offers his recommendations on the cure of the dropsy in his *ABJ* (1972), cap. XXIII, pp. 125–29. See also CA 1.7 to Dr. John Woodward (November 24, 1712).

[24] Hippocrates' "De superfoetatione" (see Fienus, pp. 111, 167, 170, 176). M.-H. Huet (1991): 723–24; and in (1993), pp. 22–23, retells the popular story of Hippocrates saving the life of a white princess by proving that the mother's imagination (viewing the portrait of an Ethiopian), not her adultery, resulted in the birth of her black child. Robert Burton's *Anatomy of Melancholy* (1621), Pt I, Memb. 3, subsec 2: "Of the force of Imagination," p. 124, relates that according to Heliodorus' *Aethiopica*, the Ethiopian Queen Persinna "was brought to bed of a faire white child," when during her conception she beheld the picture of Perseus and Andromeda. So, too, we learn from Burton (pt 1, sec. 2, Memb. I, subsec. VI: "Parents a cause by Propagation," p. 84, of the causes in which the imagination of mothers, perhaps frightened or angry during pregnancy, left a permanent mark on their children; thus, "*If a great-bellied woman see a Hare, her childe will often have a Hare lip.*" It appears collectors of "memorables," like Burton, Fienus (Quaest, XIII, Concl. XL, p. 111), and Fortunius Licetus, in *De Monstris* (1616), lib. 2, cap. 66, pp. 112–13; (1665), p. 210, drank from the same fountainhead, for all of them tell the story of a black Ethiopian woman who was suspected of adultery when she gave birth to a white (albino) child. Such stories were perennially fascinating to our ancestors and were widely retold in the literature of the day. Their immediate source is *Opera* (1582), lib. 24, cap. 7, p. 747, by the distinguished French physician Ambrosius Paraeus, aka. Ambroise Paré (c. 1510–90). An English translation by Thomas Johnson appeared as *The Workes of that famous Chirurgion Ambrose Parey* (1649).

[25] Fienus (Quaest. XIII, Concl. XL, pp. 111, 113, 115); St. Jerome, on Gen. 30:33, *Liber Hebraicarum Quaestionum in Genesim* [*PL* 023. 1035], relates the story of a woman who, when experiencing the strongest sexual pleasure during conception, can produce an offspring in the likeness of an image or object she had been

evident, nor to *Fienus*, who goes beyond, – one & thirty in his Enumeration of {some} amazing Exemples for it. The Milk-white Indians, (the Offspring & the Parents of Tawnies,) conceiv'd by Mothers looking on the Full-Moon, at the Time of their Conception, may be added unto the Number; The Things which Men have done upon their own *Bodies*, either for the Bringing or for the Helping of Diseases, or the Performing of unaccountable Motions, Yea, for the Inducing of Death itself, (as in the famous Relation of the poor Man frighted to Death upon the Scaffold,) would afford a Story as ample as curious;[26] and we should hardly acquiesce in any of Aristotles {Commentators?} for the Reason of them; or to tell us, why our Sight of such & such Actions done by others, does powerfully Dispose, and even Enrage us, unto the doing of the Like; or why we wake so præcise at the Hour which we have sett unto ourselves. The Melancholy Gentleman mention'd by *Marcellus Donatus*,[27] who Imagined himself grown so big that he could not pass the Door, but upon the Physicians ordering him to be carried out of the Door, imagined himself to be Inwardly so bruised, at such a rate, that he died of it, was in his Case a little singular.[28] And so was the Gentleman mentioned by *Fulgosius*, who dreaming that a Catt bitt his foot had immediately an Ulcer there.[29] There

---

admiring. "At least this is the argument that Quintilian used to defend a married woman against charges of adultery when she gave birth to an 'Ethiopian.'"

[26] This well-known story also appears in Michel de Montaigne's (1533–92) essay, "*Of the Force of Imagination*," *Essays of Michael Seigneur de Montaigne* (1685), bk 1, ch. 20, p. 135: "Some there are who thorough Fear prevent the Hangman; like him whose Eyes being unbound to have his Pardon read to him, was found stark dead upon the Scaffold, by the Stroak of Imagination." The same story is cited in Fienus (Quaest. X, p. 92) as an example how the imagination when suddenly and strongly aroused can trigger death.

[27] Marcellus Donatus, *De Medica Historia Mirabili libri sex* (1586), lib. 2, cap 1, 34^{r}, relates the tale of a certain Vincent whose imagination (as it turns out) was larger than his physical body and thus came to his tragic end. The same story appears in Fienus (Quaest. X, p. 92).

[28] Mather's illustration is based on the account in Fienus (Quæst. X, pp. 92, 93), who in his turn relies on Marcellus Donatus (1538–1602), *De Medica Historia Mirabili libri sex* (1586), lib. 2, cap. 1, 34^{r}. Donatus identifies this melancholic gentleman as "Vicentius" and lists as his source Antonius Fracantianus (1506–67), *De Morbo Gallico Tractatus Antonii Francanciani* (1564), lib. 2, cap. 1, an Italian physician and professor of medicine at the University of Padua. In this case, as in many others, Mather and his learned Renaissance colleagues on both sides of the Atlantic relied on sources at second, third, even fourth remove – a common enough phenomenon given the scarcity of books.

[29] Via Fienus, *De Viribus Imaginationis* (Quæst. X, p. 92; Concl. XXXVII, p. 97), Mather refers to the learned Genoese Baptista Fulgosius, aka. Battista Fregoso (1453–1504), an Italian compiler of memorable quotes from the classics, who

have been many Exemples of *Stage-Players*, who have Imagined themselves into the very Conditions which they have represented; and the Death of *Le Molier* the famous Comœdian of our Time, ha's been sufficiently talk'd of.[30] In *Seneca* there is a Mention of One *Bibius Gallus, Imitando stultum, ipsemet stultus factus est*.[31] The admitting, & also the Conquering of *Antipathies*, is a fruitful Subject of Thought in this *Field of Wonders*.

Among these, the Instances whereof are Numberless, I will not mention that of *Scholzius*, who did love & eat *Apples* well enough, but if he saw Apples putt into a Rolling Motion, fell into a *Lipothymie*.[32] But I will introduce my

---

relates the story of Arnoldus (Arnaldus), the philosopher, who dreamed that a cat had bitten his toe and promptly developed an ulcer on the spot the next morning. To be sure, Fulgosius, in his *De Dictis Factisque Memorabilibus Collectanaea* (1509), lib. 1, cap. 5: "De Somniis," p. 29^{V} (1588 ed.), insists that the man's dream was not the cause of the ulcer; rather it was already developing in his foot when he dreamt of a cat biting his toe.

30 The renowned French playwright and actor Jean-Baptiste Poquelin, aka. Molière (1622–73), was all the rage in Europe and was talked about even in Puritan New England – where (mind you) any form of theatrical performances was proscribed and the reading of plays and novels frowned upon. Mather here alludes to Molière's final comedy, *Le Malade Imaginaire*; i.e., *The Imaginary Invalid* (1673), in which Molière in the title role of Argan played an inveterate hypochondriac, who is mortally afraid of doctors and of death. In his struggle with his unruly family, Argan lies down on the couch and pretends to be dead to test the loyalty of his family (Act 3). Perhaps Molière's (Argan's) pretended death was so convincing (if not foreboding), Mather seems to imply, that the playwright turned actor collapsed on stage during the fourth performance of the comedy and died later that day.

31 Fienus (Quaest. X, p. 92). "By imitating a fool, he himself has become a fool." Mather's Latin quote is a second-hand paraphrase from the Elder Seneca's *Controversiae* (2.1.25), in which the speaker laughs at Vibius (Bibius) Gallus, who pretending to be insane fell into insanity. Mather's source for Vibius Gallus and Catanaeus is Fienus, *De Viribus Imaginationis* (Quæst. XI, pp. 92 [Bibius]; 98, 114) and Andreas Cataneus, *Opus de Intellectu et de Causis Mirabilium Effectuum* (1502).

32 Probably a reference to the German physician and botanist Laurentius Scholzius, aka. Lorenz von Rosenau (1552–99), whose *Aphorismorum Medicinalium cum Theoricorum, Tum Practicorum* (1589) was among his most significant medical works. A lipothymy is a fainting or swooning episode. A similar apple story is told in Increase Mather's *ERIP* (1684), ch. 4, p. 101. See also ch. 5, p. 155. No doubt, Ben Franklin's fictive Susquehanna Indian felt reassured when the Swedish missionary told him about "the Fall of our First Parents by Eating an Apple": "It is indeed bad to eat Apples," commented the sage Indian orator; "It is better to make them all into Cyder" (*Remarks Concerning the Savages of North-America* [c. 1784]).

Story, with what is related by *Francis Hertodius*; of One who could not have eaten any Bread, tho' he had been as hungry as *Ægon* in *Theocritus*.[33] But falling into a Quartan Ague,[34] he longed for Bread, & with nothing less than a Canine Appetite even raved for it. And yett, when his Ague was gone, his Horror of Bread returned upon him. And what is related by *Johannes Faber*,[35] of a Woman who so abhorred Cheese, that a Knife which had only touch'd it, was an Insupportable Abomination to her. But when she was with Child, then she longed for Cheese. And yett after the Childbirth, all Cheese again became as horrible as before.[36]

My Story now is this; and I have it well attested unto me. A Gentleman of *Carolina*, whose Name is Mr. *Edwards*, in his Repentance of the Errors, which the too free Use of Wine had once led him to, came to entertain an Aversion for that Liquor, æqual to what *Lancelott's* Instance[37] in the German *Ephemerides* has told us of. The least Approach of it wonderfully distressed him. The Minister of the Place (who is my Author for my Story,) perswaded him to present himself unto the Communion, for which he thought so good a Pænitent was well prepared. He objected, His Incapacity to partake of the Cup, which he had so much of a *Calixtine*[38] & a *Protestant* in him, as to

[33] Matthaeus Franciscus Hertodius (aka. Matthias Franz von Hertodt), provincial physician of Bruna, in Moravia, and a frequent contributor to the German *Ephemerides*. In this case, the venerable physician tells the story of a certain citizen of Nicolspurg, named "*Käsmanbueber*," who had a lifelong aversion to eating bread, in "Observatio CXLIV. D. Francisci Matthaei Hertodii. *Panis Antipathia*," in *Miscellanea Curiosa Medico-Physica Academiae*. Annus Secundus (1671): 228–29. In Theocritus' *Idylls* (4.29), Corydon praises the town Croton, "where boxer Milon one fine morn made fourscore loaves his meal" to prepare himself for his match with Aegon.

[34] An intermittent fever recurring every four hours.

[35] Johannes Matthaeus Faber (1626–1702), German physician in Heilbronn, personal physician of the Dukes of Württemberg-Neuenstadt, who contributed several articles to the German *Ephemerides*.

[36] Mather tells the same story in his Letter to John Woodward (November 20, 1712), CA 1.4 "Antipathies & the Power of the Imagination." See also Increase Mather's *ERIP* (1684), p. 101.

[37] Joël Langelott (1617–80), German alchemist, chief physician of Frederick IV, Duke of Holstein-Gottorp. Mather appears to refer to Langelott's "Observatio XIV: De Mirando Antipathiae Exemplo," in the medical journal *Miscellanea Curiosa Medico-Physica Academiae*. Annus Sextus et Septimus (1677): 24–25.

[38] A believer – like the early Protestant Reformer John Huss (1369–1415) – who insists that communicants must partake both of the Chalice (calix: cup; hence wine) and the Host (bread).

think essential unto the Eucharist. The Minister exhorted him, & overcame him, to make the Experiment; which he did, & found not the least Inconvenience in the World: The Wine satt well upon him; and he continued a Religious Communicant. And yett, after this, the Minister seeing the Wife of this Gentleman making of a famous Dish which they call a Pudden, prevailed with her utterly unknown unto her Husband, for to putt a few drops of Wine into it. When it came to the Table, tho' no Mortal else would have been apprehensive of any Wine in it, the Quantity being so very small, yett he was extremely sensible of it, extremely disturbed with it; it nearly {made} him very sick; & produced very ill Effects upon him. – How a Gentleman of such an Antipathy for *Wine*, should so inoffensively drink it at the Table of the Lord! – whether something superiour to meer Imagination may not operate in so rare a Case? –

Could I by the Force of Imagination transport myself, I would immediately Try to converse with you on this & some other Subject. In the mean time, I must be content with subscribing

Yours in the hearty {...}

that the {. ..}[39]

{C. Mather.]

[39] The conclusion of this draft was written by Mather perpendicularly in the left margin; the left edge of the page is badly damaged, resulting in the loss of most of his usually witty closure.

3.8. THE STONE MISTAKEN.

July 10, 1716.

To John Woodward.[1]

Sir,

The Subject I am going to treat you withal, is one upon which I may say,

Animus meminisse horret luctuque refugit.[2]

But yett, I hope, the Communication may be not altogether unserviceable, & therefore not unacceptable.

If That formidable Scourge of Heaven upon Mankind, which we suffer in the *Stone*, come into our Thoughts, without something of horror,[3] we shall

1 The editors have elected to use the MHS holograph as the copytext; the scribal copy at the BL (Sloane) is imperfect.

2 The Roman poet Vergil (*Aeneid* 2.12) did not suffer from a case of bladder stones when he had his hero sing, "My mind shudders to remember, and has recoiled in grief" at the fall of Troy.

3 Mather remembers only too well when he witnessed his grandfather, Richard Mather (1596–1669), die of an excruciating bout of bladder stones accompanied with a blockage of urine, when Cotton was six years old, in 1669. As Increase Mather reported, his father's death was sudden and unexpected: "Onely the two last years of his *Life* he was sorely afflicted with that Disease which some have called *Flagellum Studiosorum*. viz. *The Stone*, which at last brought him to an end of all his Labours and Sorrows," in *LDRM* (1670), p. 26. See also Cotton Mather's lecture, *TT* (1701), pp. 28–29; Increase Mather, too, prayed to be spared from such a painful death, for he knew first-hand the agony of his father's and grandfather's affliction, as Cotton Mather relates in *Parentator* (1724), art. 2, pp. 8–9; art. 15, pp. 65–66; art. 32, pp. 210–11. Samuel Sewall also reports how one day he stopped by Increase Mather's house (July 30, 1723) and heard the aged minister's agonizing cry to God about the pain of "the stone": "Pity me! Pity me!" (*Diary of Samuel Sewall* 3: 325–26). Given the preponderance of the stone in his family, Cotton Mather felt that "the *Stone* is most certainly One of the most Horrible to be thought upon," or so he recorded in "*Flagellum.* The Stone [And other Diseases of the *Kidneys* and *Bladder*]," in *ABJ* (1972), cap. XIV, pp. 82–92. See also Samuel

a little Justify the Account the Poet gives of us, (tho' not his Reason for it) *Genus durum Sumus*,[4] A Malady, which as long ago as *Galen's* time,[5] was pronounced curable by nothing but a manual Operation: And indeed, Since a Cure by all other Medicines, even *Millipedes* themselves cried up for such a powerful Dissolvent in the Case, by others as well as *Hartman*,[6] is despaired of, the manual Operation is now so Successfully practised, that the Horror of the Malady is much abated, tho' even the Remedy itself is horrible.[7]

What shall be done, when the Stone is of such a Magnitude, as that mentioned by *Hildanus*, which weigh'd no less than Two & Twenty Ounces, I know not. I know a *Celsus* would not have the Cure undertaken, when so vast a Discouragement is to be encountred.[8]

Mather's biography, *LCM* (1729), p. 63. Cotton Mather reports of a similar case in his neighborhood: "A poor Boy tortured with the Stone, must be made an Object of singular Compassions. The Expence of the *Lithotomie*, must be Part of my Care" (*D* 2:396).

4 Adapted from Ovid's *Metamorphoses* (1.414): "the toughness of our race." See also Vergil, *Aeneid* (9.603).

5 The Greco-Roman physician and philosopher Claudius Galen of Pergamon (129–c. 216 CE) is among the most distinguished anatomists and physiologists of antiquity. During Mather's lifetime, apart from the voluminous Latin tomes on medicine, the *PT* offered numerous accounts of lithotomy and nephrotomy. See also J. Scarborough (1976); D.E. Greydanus (2016): 101–08.

6 *The Family Physitian* (1696), ch. 12, art. 19, p. 172, by the obscure English medical author George Hartman (fl. 17th century), who is described as a "Phylo Chemist," traveling companion, and physician in the service of Sir Kenelm Digby (1603–65), English philosopher, physico-theologian, and courtier. Much the same appears in Hartman's earlier *The True Preserver and Restorer of Health* (1682), chs 4–5, pp. 46–59. Hartman recommends a quantity of millipedes ("vulgerly [*sic*] call'd Woodlice"), washed, fermented, and powdered, and mixed with white wine. According to Hartman, Sir Kenelm Digby testifies to the nostrum's efficacy (*Family Physitian*, p. 172).

7 See also *ABJ* (1972), p. 86.

8 See *ABJ* (1972), p. 86. Wilhelm Fabry von Hilden, aka Fabricius Hildanus (1560–1634), a German itinerant physician, surgeon, and author of his oft-reprinted *Observationum et Curationum Chirurgicarum* (1606), one of the century's best collections of surgical case studies; and of *Lithotomia Vesicae* (1626). An English translation of *Lithotomia Vesicae* appeared in London, 1640. Mather alludes to the discussion of kidney stones in *De Medicina* (2.7.14–15; esp. 7.26), by the Roman physician and encyclopedist Aulus Cornelius Celsus (c. BCE 25–c. 50 CE). Increase Mather recounts several similar stories of lithotomy and petrified fetuses (mola) remaining in the bodies of women, stones growing under the tongue, or kidney stones voided, in *ERIP* (1684), ch. 9, pp. 304–10. These examples were

However, The Practice of *Lithotomy*, especially Since, that called the *Franconine*, has been so improved by that called the *Collotian*, comes now to be reckoned among the Favours of Heaven to a miserable World.[9]

And yett even here, what Cause for Caution in the Practice?[10]

A Good Man, who lived not far from me, (named Mr. *Call*,) was for diverse Years miserably tortured with the *Stone* in all the most evident Symptoms of it.[11] He first from his Bloody Urine, & other Circumstances, imagined only *Gravel* in his Case. But anon, & for above {a} Year together, his insupportable Pains determined it for the Stone; In all which time there came scarce any Urine from him, except now & then a drop distilling from him. His Dolours were inexpressible; the Houses of the Neighbours at some distance from his, were day & night for many Months, the Witnesses of the grievous Ejulations, whereof the doleful Accents reached them. The most expert *Cheirurgeans* we had, & such as had cutt People for the Stone, with a very uncommon Dexterity, & Reputation; these when they had bestowed the most Examination upon the Seat of his Anguish, pronounced it the *Stone*; and the Operation was urged for. He had not read the Advice of honest *Groeneveldt* for a Desperate Case, "Lett the Patient be advised, not to abbreviate his Life, thro' Impatience,

---

generally gleaned from the pertinent medical publications of distinguished European physicians, from the *PT* and the German *Ephemerides*, including from such mundane tabloid publications as the anonymous *Weekly Memorials for the Ingenious* (1683). See also Ambroise Paré, *On Monsters and Marvels* (1982), ch. 15, pp. 52–53.

9 See also *ABJ* (1972), p. 86. "Lithotomy" is the surgical removal of stones from human organs. Hildanus describes several ways to surgically remove the stone, including "*Lithotomia Franconiana*," in his *Lithotomia Vesicae* (1640), ch. 16, pp. 105–10. A "Franconine lithotomy," named for its originator Pierre Franco (1505–78), a French surgeon, is one in which an incision is made but the stone not removed on the principle that it will come out through the incision (Hildanus, *Lithotomia Vesicae*, ch. 17, pp. 110–13). The term "Collotian" (a Matherian neologism which has not yet made it into the almighty *OED*) appears to be derived from a surgical method of the Parisian lithotomist at the Hôtel-Dieu, Laurent Colot (1520–90), whom King Henri II named official Operator of Stones in 1556. Removing the kidney or bladder stone through a collotian section was yet another way of removal pioneered by Jan Groeneveldt (c. 1647–c. 1710), a native of Holland, who practiced medicine in London.

10 A collection of nostrums for curing the stone are offered in Mather's *ABJ* (1972), pp. 86–92.

11 Mather's neighbor appears to be the same whose petition, "*Lord, Lett us not go Roaring to Heaven!*" was eminently fulfilled, for when "applying himself to the *Urinary Excretion*, he fell at once into a Swoon, and Suddenly and Silently went away to a *Better World*" (*ABJ* [1972], p. 85).

in this or any other Manner; but with Fortitude & Constancy becoming his Christianity, expect Deliverance from the Supreme Physician, who only by his Almighty Power can sett Limits & a Period to all Misery."[12] But yett he took it. The Good Man felt an Aversion on his Mind, for the Proposal; and so waited with an exemplary Patience, till Mortality gave an Opportunity for his Dead Body to come under a Dissection. And then; Behold, unto the Surprise of the Spectators, it was found, that there was nothing at all of a *Stone* in the Case. But what was found, I am now to tell you.

His Bladder was grown entirely Schirrous,[13] uncapable of any Distention or Contraction, & come to be of an enormous Thickness. It had several small fleshy & fungous Excrescences growing in it, and one in the Neck of the Bladder, which was a very large one, & had a considerable Share in the Torment given to him. What a Disappointment must the *Ish Calaphot* have undergone, if the proposed *Lithotomy* had been proceeded in![14]

That Heaven may graciously præserve so valued a Life as Yours, from all such Embitterments, is the Prayer of,

Sir,

Your —

{C. Mather.}

12 The same citation appears in Mather's *ABJ* (p. 85) and is an embellished paraphrase of Groeneveldt's *A Treatise of the Stone and Gravel* (1677, rpt 1710), pp. 88–89. The 1710 London edition has a slightly different title and Anglicizes Jan Groeneveldt's name as John Greenfield, M.D. of the College of Physicians. Our page reference is to this edition.

13 Or "scirrhous": hardened. "A malignant tumor having a preponderance of fibrous tissue" (Merriam-Webster).

14 "*Ish Calaphot*" is – according to Mather – "the Original and Etymology of the Name *Æsculapius,*" aka. the Greco-Roman god of medicine, also known as "*The Master of the Knife*" (*ABJ* [1972], p. 86). Mather uses the same reference in his letter to John Woodward (November 24, 1712), CA 1.7 "Remarkable Recoveries from Wounds," where he acknowledges his original source as Samuel Bochart (1599–1667), who derives "Ish Calabi" (or rather, "is-calibi") from the Phoenician איש־כלבי, in *Hierozoicon* (1712), pars prior, lib. 2, cap. 55, col. 663, which Mather translates as "*A Man of the Dog,*" because "he" [Aesclepius] "carried about a *Dog* with him, to lick the Sores of his ulcerous, or wounded Patients."

3.10. CURIOSITIES OF THE SMALL POX.

July 12, 1716.

To John Woodward.

Sir,

The History of that grievous and wondrous Disease, the *Small-Pox*, would no doubt, be as grateful as the Distemper itself is loathsome.[1]

I shall præsume, at this time, to entertain you with no more than Two or Three American Curiosities upon it.

The Famous *Fernelius* in his Treatise, *De abditis rerum Causis*, makes the *Small-Pox* to proceed, *Ab Occulto quodam Cœlesti influxu*.[2] We have been almost ready to think this, and even suspect a peculiar Agency of the Invisible World, in the Infliction of the Small-Pox upon our City of *Boston*; when we saw, that from the first Foundations of it, in the Year, 1630, down to the Year 1702. it observed the precise Period of Twelve Years, in its mortal Visits unto us.[3] Now & then, a Vessel would in the Intervening

1 In one of his earliest extant letters (November 1678) to his uncle John Cotton (1638–89), Cotton Mather, barely fifteen, describes the mortality from that year's smallpox epidemic in Boston: "To have coffins crossing each other as they have been carried in the street; – To have, I know not how many corpses following each other close at their heels, – To have 38 dye in one week, – 6, 7, 8, or 9 in a day. Yet thus hath it lately been; and thus it is at this day. Above 340 have died of the Small Pox in Boston since it first assaulted the place. To attempt a Bill of Mortality, and number the very spires of grass in a Burying Place seem to have a parity of difficulty and in accomplishment," in "Cotton Mather to John Cotton" (c. November 1678), in *MP* (1868), p. 383; and *SLCM* (1971), p. 7; *CJCJ* (2009), p. 243. See also J. Hull, *Diaries*, pp. 243–45; and M. Peterson, *Boston* (2019), p. 160.

2 Jean François Fernel, aka. Joannes Fernelius (1497–1558), French court physician to Francis I (1494–1547). Mather's citation is from Fernelius' *De abditis rerum causis libri duo* (1548), lib. 2, cap. 11, p. 184, and reads, "by some hidden heavenly influence." The Latin quote appears to be a paraphrase of *ex cœlestium corporum viribus & influxu* ("from the powers and influence of the heavenly bodies").

3 S.B. Woodward (1932) provides a highly useful survey of the periods of smallpox outbreaks and the rate of mortality in 17th and 18th-century New England.

Space, bring in the Distemper among us. However it would not spread. But on the Twelfth Year still no Præcaution would keep it off.[4] In the Twelfth Year, it must be epidemical. So Raging, so Reaching, that it has come at unborn Children; They have been born full of it upon them.[5] But at last, when a Seventh Period for a variolated Twelfth Year was arrived, our Observation has mett with an Interruption. We had not the usual Return of that Contagion; The Compassion of Heaven would not add that Calamity, unto what we suffered the Year before, {1713.} in the *Measles*, which at once arrested almost the whole City & proved so strangely mortal to a Multitude of People, that we buried above an hundred in a Month. Of this Disease, at this Time, there were no less than Five of my own Family, to whose Tomb,[6] I might justly transfer the Words which are on one of yours

---

Isolated cases occurred much more frequently; Mather's calculation of outbreaks every twelve years appears to be based on the epidemics of 1677–78, 1689–90, and 1702–03. *D* 1:451; see J. Duffy, *Epidemics* (1953), pp. 44–50; D. Levin, *Young Life* (1978), pp. 67–69.

4 Mather's experiential insights were crucial to introducing prophylaxis as a means to combat epidemics.

5 See William Derham's "The Case of a Woman big with Child, who recovered of the Small-Pox, and was afterwards delivered of a dead child full of the Pustules of that Distemper," *PT* 28.337 (December 31, 1713): 165–66.

6 *D* 2:248-54, 252, 256–58, 260, 291. In the month of October 1713, Cotton Mather's whole family came down with the measles, and his second wife, three of his youngest children, and maid died from this illness. Just as he would later do during the smallpox epidemic of 1721, he sprang to action during the measles epidemic of 1713: "A good Regiment of the sick under it, might save many Lives, which are like to be lost thro' mismanagement" (November 5, 1713). "I am thinking, whether it would not be a service unto the public, to insert in the News-paper, a brief Direction for the management of the sick. I will advise with a Physician or two, whether I had best offer such a Thing or no" (*D* 2:256). About four weeks later (December 4, 1713), he records, "That which I proposed unto my Physicians, I have now done myself, upon their Declining (tho highly Approving) of the Action. I have given to the Printer a Letter about the *Right Management of the Sick under the Distemper of the Measles*; which is now spreading and raging in the Countrey. I purpose to scatter it into all parts, and propose with the Blessing of Heaven, to save many Lives. Tho' doubtless my Action may expose me to some Invectives, yett my Conformity to my dear Savior, in what He did for the sick, will be my inexpressible Consolation" (*D* 2:272). See Mather's *LMDM* (1713) and *WFS* (1713), "Preface." His *LMDM* is probably the missing document that he had planned to incorporate into his medical handbook *ABJ* (1972), item XIX, "The *Inevitable*. or, The *Measles managed*," p. 2. See also Samuel Sewall's *Diary* (2:407) and the valuable article by P.E. Kopperman et al. (2019): 30–37.

in *Westminster-Abbey, Morbillorum perfidia Sublatæ, videatur ex Igne et Tunicâ Molestâ, evolasse ad Cœlos*.[7]

Of those *Measles*,[8] by the Way, I will take Leave to mention one Remarkable, which I know not how far it may confirm some of Monsieur *Andry's* Thoughts, in his curious Treatise, *Of Worms in the Humane Bodies*.[9] Tis, That a great Part of our People, both young and old, voided *Worms*, when the Infection of the *Measles* was upon them.

I return to the *Small-Pox*, which has usually proved a *Great Plague* to us poor *Americans*; and getting among our *Indians*, hath swept away whole Nations of them, & left not Living enough to bury their Dead.[10]

7 The Latin epitaph reads, "Having been carried off through the treachery of small-pox [measles], he seems to have winged his way from fire and the grievous shirt to the heavens." It refers to the demise of the Greek mythical hero Hercules (Herakles), whose death by a poisoned tunic and self-immolation in his own funeral pyre come to mind. The epitaph appears on the sepulcher of John, Edward, and Stephen Fox of Farley, in whose tomb were interred the ashes of the three brothers along with those of James Fox, age 12, who died on December 13 1677 – all sons of Sir Stephen Fox (1627–1716) of Farley (Wiltshire), Paymaster-General of Charles II's armed forces. The ornate epitaph appears on "a small Monument against the North wall of Alabaster and black Marble," at Westminster Abbey. Mather's source is *Monumenta Westmonasteriensia* (1682), by H.[enry] K.[eepe] (1652–88), p. 352 (Epitaph 162). It is our pleasure to acknowledge Professors Patricia Roberts-Miller (UT at Austin), Edward White (Tulane) and Ava Chamberlain (Wright State U) for identifying the source and location of Mather's reference.

8 See also Mather's *A Letter, About a Good Management under the Distemper of the Measles* (1713). He had every reason to be sensitive about the topic; his second wife and three of his youngest children (including new-born twins) succumbed to the disease in 1713 (*D* 2:248–50, 252–54).

9 The French physician Nicolas Andry de Bois-Regard (1658–1742) made a significant contribution to the study of parasitology in his *De La Génération Des Vers Dans Le Corps De L'Homme* (1700). The English translation was published as *An Account of the Breeding of Worms in Human Bodies* (1701). Mather has much to say on the topic in his medical handbook, "Cap. XLIV. The *Vermine-Killer*. Upon, Worms," in *ABJ* (1972), pp. 203–07. See also CA 4.4 "A Prodigious Worm" (below), and E.H. Christianson (1993). Useful, too, is Edward Tyson's "Lumbricus Latus, or a Discourse … of the Joynted Worm," *PT* 13.146 (April 10 1683), pp. 113–41.

10 John White of Dorchester describes the devastation of the Native populations in the Plymouth and Boston areas, in *The Planters Plea* (London, 1630), p. 25. So, too, Governor William Bradford of Plymouth reports how the settlers found whole Indian villages wiped out and many bodies lay unburied, in his famous log *Of Plimoth Plantation* (2020 ed.), bk 2 ("Anno, 1634"), pp. 418–19. On the decimation of Native Americans by smallpox, see J. Duffy, "Smallpox" (1951).

When our last Visitation was beginning, and I was full of Distress about the Miseries I saw coming on my Neighbourhood, I accidentally litt on an Hint in the *Philosophical Transactions*, That the famous Dr. *Sydenham* had saved more Lives than the Wars of *Ireland* had destroy'd, by teaching the True Method of managing the Small-Pox.[11] I presently consulted his Method; & press'd our Physicians to come into it; and the Success was answerable. Few died in Comparison of the Numbers we formerly destroy'd, when by our unseasonable *Alexipharmacks*,[12] we kill'd one another with kindness, and putt Nature out of the Course, which ought not to be disturbed.

But, why may not they whom the Small Pox ha's once visited, be a Second Time obnoxious to it? Will you repair to the *Material Original* of the Distemper, with which the former Ages were satisfied? or, will Dr. *Drakes* Account of the Matter content you?[13] One Surprising Accident, relating to this Matter, I will now treat you withal. A certain Man, among us, in the Year, 1678. had, all the Symptoms of the *Small Pox*, which at Length appear'd in a Cluster, of a few Inches Square on his Back, & no where else. This he thought was enough to secure him from ever having it any more; But in the Year, 1690. the Man was again taken Sick with all the Symptoms of the *Small-Pox*; which then indeed he had sufficiently, from head to foot; only in that Part of his Back, where he had it before, he had nothing of it.

All that I shall now add, will be my Thanks to you, for communicating to the Public in Dr. *Halley*'s Transactions, the Account which you had from Dr. *Timonius*, at *Constantinople*, the Method of obtaining and preventing

11 The distinguished English physician Thomas Sydenham (1624–89), founder of clinical medicine and epidemiology, devotes his longest discussion to the common distinct and confluent smallpox (variola) along with methods of easing patients through their deadly affliction, in his *Methodus Curandi Febres, Propriis Observationibus Superstructa* (1666), sec. quarta, pp. 123–56; and his *Observationes Medicae circa Morborum Acutorum Historiam et Curationem* (1676). Cotton Mather owned a copy of the latter work. His signed and dated copy (August 26 1679) is at the AAS. To be sure, Sydenham's treatment of the smallpox – before the proliferation of inoculation and later Edward Jenner's cowpox vaccination (from vacca, i.e., cow) – was largely confined to palliative methods, especially bleeding, infected patients.

12 Antidotes.

13 James Drake, F.R.S (1667–1707), English physician, member of the Royal Society, and fellow of the College of Physicians, published *Anthropologia Nova; or, a New System of Anatomy*, 2 vols. (1707), which went through several editions and reprints. Significantly, Drake has little to say about smallpox, in *Anthropologia* 1:25–27.

the *Small-Pox* by *Insition*;[14] which I perceive also by some in my Neighbourhood lately come from thence, ha's been for some time successfully practiced there. I am willing to confirm you, in a favorable Opinion of Dr. *Timonius*'s Communication; And therefore, I do assure you, that many Months before I mett with any Intimation of treating the *Small-Pox*, with the Method of Inoculation, any where in *Europe*; I had from a Servant of my own, an Account of its being practised in *Africa*. Enquiring of my Negro-Man *Onesimus*,[15] who is a pretty Intelligent Fellow, whether he ever had the *Small-Pox*; he answered, both, *Yes*, and, *No*; and then told me that he had undergone an Operation, which had given him something of the *Small-Pox*, & would forever præserve him from it; adding, That it was often used among the *Guramantese*, & whoever had the Courage to use it, was forever free from the fear of the Contagion.[16] He described the Operation to me,

14 Mather refers to the English astronomer and mathematician Edmond Halley (1656–1741), who served as secretary of the Royal Society and editor of the Society's *Philosophical Transactions* (1686–93). Mostly remembered for the Comet of 1682 which bears his name, Halley served as editor of the *PT* at the time and dedicated vol. 29 of the *Transactions* to Sir Isaac Newton, the president of the Royal Society. This volume included an important article by the Greek physician Emanuel Timonius (1669–1720), who was an eyewitness to the inoculation (vaccination) performed by elderly Greek women in Constantinople. Timonius published his "*Account, or History, of the Procuring the Smallpox by Incision, or Inoculation ... Being the Extract of a Letter from Emanuel Timonius ... dated at Constantinople, December, 1713*," in *PT* 29.339 (June 30, 1714): 72–82, more than a decade before the practice was gradually adopted in England and on the Continent. See R.P. Stearns (1950).

15 Given his awareness of Timonius's experimental practice of inoculation, Mather had ample time to draw up his own action plan for the next smallpox epidemic which was bound to arrive in Boston sooner or later. His evolving plan was further strengthened by eyewitness accounts, especially those from Onesimus (fl. 1690–c. 1750), an enslaved African who served Mather for at least a decade (December 1706 to August 1716). Onesimus (Greek: ὀνήσιμος, signifying "helpful" or "useful") was a member of the Guramantes (Garamantes) people of southern Libya. His verbal account of smallpox vaccination practiced in Africa was central to corroborating Mather's thoughts about inoculation. Mather also credits Onesimus in CA 6.2 "*Variolæ trimphatæ*. The SMALL-POX Encountred;" and our CA 6.1 "*An Account of the Method and Success of Inoculating the Small-Pox in Boston in New-England*" (1722). For pertinent background, see M. Minardi (2004), K. Koo (2007), K. Wisecup (2011), R. Tindol (2011), A. Boylston (2012), and A.M. Kass (2012).

16 Years later, in 1721, Mather's colleague across town, the Rev. Benjamin Colman, too, had learned from "a *poor Negro*," at work for one of Colman's parishioners, that he "had been Inoculated" and that it was common practice in his home country

and show'd me in his Arm the Scar, which it had left upon him; and his Description of it, made it the Same, that afterwards I found related unto you by your *Timonius*.

This cannot but expire, in a Wonder, & in a Request, unto my Dr. *Woodward*.[17] How does it come to pass, That no more is done to bring this Operation, unto Experiment & into Fashion, in *England*? when there are so many Thousands of People, that would give many Thousands of Pounds, to have the Danger and Horror of this frightful Disease well over with them. I beseech you, Syr, to move it, and save more Lives, than Dr. *Sydenham*. For my own Part, if I should live to see the *Small-Pox* again enter into our City, I would immediately procure a Consult of our Physicians, to Introduce a Practice, which may be of so very happy a Tendency. But could we hear, that you have done it before us, how much would That embolden us?

We are ambitious to learn of you, and of none more, than of Dr. *Woodward*: And none is more so, than,[18]

Syr,

Your most sincere Servant.

{C. Mather.}

in Africa for people to be presently inoculated when the pox began to spread in their town, in Colman's *Some Observations* (1721), p. 15. See also R.P. Stearns (1950); Beall and Shryock (1954, 1979), p. 99.

17 The English natural philosopher, geologist, paleontologist, and physician Dr. John Woodward (1665–1728), who served as secretary and correspondent of the Royal Society. Mather owned Woodward's *Naturalis Historia Telluris Illustrata & Aucta* (1714), a Latin translation of Woodward's popular *Essay toward a Natural History of the Earth* (1695).

18 Penciled, in another hand: "July 12, 1716," probably the date when Mather's missive was read at the Royal Society.

3.II. THE *FAGIANA*.

July 13, 1716.

To John Woodward.

Sir,

Since I have had the Honour to be taken into the Correspondence, which the Royal Society would have to be maintained with these Parts of America, I have thought it my Duty to make my small Annual Presents, as I have proceeded in my Collection of American Curiosities. And if since my first much larger Exhibition, the Presents have not in their Number much exceeded a *Decad*[1] of Observations, per Year, you will easily pardon one that ha's upon him so uncommon a Variety of Employments, as what this your Servant is oppress'd withal; and what seem rather to call for a Giant of an hundred Hands well to manage than one of my Tenuities. And yett, if every Member of that Illustrious Body, would oblige himself to such a Number of Communications, would you not in a little While have the Repository of your Stores, pretty well replenished? How it is I know not. This I know, That I must entreat your Pardon for my Slothfulness; and yett I will not expect it, except you find me with a Quickened Pace of Industry, endeavouring to Deserve it.

Unto the *Decad* of Letters wherewith I have now address'd you, I shall add this as a Sort of a *Postscript*, which, with a Promise of more, shall give you some Account, of a Plant, which you must impute it unto my Ignorance, if you do not find it, a pure *American*.[2]

1 Or "decade": a group of ten items.

2 The genus *Fagus Sylvatica* identifies the beech tree, under which the beech-drop wildflower (which is really a root parasite) grows. See *Culpeper's Complete Herbal* (N.D.), pp. 48–49, by Nicholas Culpeper (1616–52), the famous English herbalist, botanist, and physician. His *Complete Herbal* was first published in 1653 and went through innumerable editions since then. Mather's trove of plants – along with its Indian medicinal uses – was very likely collected for him by the Rev. Nehemiah Walker (1663–1750) of Roxbury, who was married to Cotton Mather's sister Sarah Mather. In his diary for September 23, 1716, Mather reminded himself to "Employ my Kinsman at *Roxbury* to make a Collection of *Plants*, peculiarly *American*" (*D* 2:371). See also G.L. Kittredge, "Scientific," pp. 13, 26, 27.

It is a short Sprig, of about Seven or Eight Inches in heighth; of a brown Colour; having only little Knotts, rather than of Leaves; and usually two or three Stalks upon a Root; of which you will Judge better, by the Substance of the Plant itself, whereof I now send you a Specimen, than by my Imperfect *Icon* of it, which I have here given you in the Margin.[3]

The Plant appears not until about the Month of *August*; and growes no where upon Earth, but under the Shade of the *Beech-tree*; nor do we find it under every *Beech-tree*, but only in low, dark, swampy, & well-shaded Places.

The first Verse in *Virgils* Eclogs, invited some fanciful Sparks, to putt upon it, the Name of *Tityrus*;[4] But, if you please, we will now call it, *Fagiana*.[5] It was first communicated unto us, by a famous *Indian*,[6] who did very strange Cures upon *Cancers*, by a Decoction of it inwardly taken, & a Cataplasm[7] of the boiled Plant at the same time laid unto the Place affected. And, I am told, that it was not without some Indirect Methods, that the Secret was obtained from him: I mean those upon which, – *Exit Arcanum*.[8]

It is but very lately come in Use among us, And one of the first Instances, I have been informed of, was this.

A Gentleman in the Countrey, fallen into a very Ill-habit of Body, & over-run with Scabs that were thought a Sort of a Leprosy, was cured on the Terms of all your Three Adverbs, with our *Fagiana-Tea*.

I understand, our People begin to use it in *Fevers*, and even in *Hecticks*,[9] with a success that surprizes them.

All that I yett know from any Experience of my own, is, That in a *Tea*, it has a mighty Exhilarating, & Invigorating Vertue; It refreshes the Spirits, with an uncommon Brightness; It prevents and removes Lassitudes, when great Fatigues are to be encountred.

I present you with a little Bundle of it; which I pray you to communicate unto our Industrious & Inquisitive Mr. *Petiver*; and with my hearty Service to

3 In the sent copy, Mather included both a "Specimen" and a drawing or "Icon" of it. Unfortunately, the drawing is not included in James Petiver's catalog (see below).

4 In the character of Tityrus, Virgil has Melioboeus bewail the turbulent changes in Rome as follows: "You, Tityrus, lie under your spreading beech's covert, wooing the woodland Muses on slender reed, but we are leaving our country's bounds and sweet fields" (*Eclogues* 1.1).

5 See also our CA 3.12 "*Curiosa Botanica*," Mather's Letter (September 24, 1716) to James Petiver.

6 Unidentified.

7 A poultice or dressing plaster made from planter material to reduce inflammation.

8 "There goes the secret."

9 "A fever that fluctuates during the day, as in tuberculosis or septicemia" (*AHMD*).

him.[10] Lett him know, that I hope, shortly to present him, with a Collection of *American Plants*,[11] wholly unknown as yett unto you *Europæans* which I have not yett finished.

That, while you know the medicinal Vertues of the vast Vegetable Kingdome, you may be so well as not to want them, is the hearty Desire, of,

Sir,

Him that perpetually wishes

Your Health & universal Prosperity.

{C. Mather.}

[10] James Petiver, F.R.S (c. 1663–1718), an English botanist, pharmacist, herbalist, and voluminous author, illustrator, and collector of plants, herbs, fossils, and minerals. See R. Coulton (2020). Mather's "*Fagiana*" is one of seven plant specimens he sent to Petiver, who listed Mather's contributions in Petiver's catalog *Petiveriana III, seu Naturae Collectanea* (1717), p. 12 (sep. pag.), item # 294, in *Jacobi Petiveri Opera* (1764), vol. 2, Petiveriana III (1717), p. 12 (sep. pag.), Com. V, "Some *American Plants*, with their *Specifick Vertues* and Wonderful *Effects,* lately sent me by the *Reverend* and Learned Dr. *Cotton Mather*, at *Boston* in *New England*, and Fellow of the *Royal Society*, London. Item # 294: *Fagiana* or *Cancer* Rape, *C.[ottonus] M.[atherus] Orobanche* ramosæ *N. Angliæ,* sub *Fagis* nascens. *Nob.* A Pultiss of this Plant laid to the Part, and a *Tea* of it inwardly, does Wonders in *Cancerous* and *Scrofulous* Cases, as also in *Malignant Fevers*; it strengthens and invigorates after *Fatigues* and Weariness. *Grows* only under *Beech Trees* in *shady Places* from *Midsummer* till *Autumn*."

[11] They included (in short order) "*Ophiophuga,* Cottonis Mather"; "*Caltha* Sylvatica, or Wood Marygold"; "Great *Bur* Comfry"; "Small *Bur* Comfry"; "*Formosa,* C. M. by the *Indians* Satanschoo"; "*Taututtipoag*, so called by the *Indians*"; and "*Fagiana* or *Cancer* Rape, *C.M. Orobanche* ramosæ *N. Angliæ*" (*Petiverinana III*, p. 12). Each of these items is accompanied by a succinct description of their locale, season, and medicinal uses.

3.12. *CURIOSA BOTANICA.*

September 24, 1716.[1]

To Dr. James Petiver.[2]

Sir,

Tis high time for me, to make you some Return, that may express my sense of the Obligations, which your Letters with what accompanied them, have laid upon me.

Tis a vast Load, & a somewhat uncommon Vanity of Employment, which my various Capacities at home, and Correspondence abroad, impose upon me; Whereof my Friends, are so sensible, that their Candour forever ha's a Pardon ready for me, when at any time I fail of keeping Time in my Compliance with their Expectations.

But I hope now to make some Reparation for my Delay of my Duty, and endeavour an Entertainment, that may find some Acceptance with you.

I perceive that Botanic Studies, are those to which you have more singularly applied your Inquisitive & Ingenious Mind. And now, that I may treat you with something for which you may have a Relish, after that I have mentioned One or Two curious Observations upon the Vegetable Kingdome, I will fulfil a Promise which I made unto my most valuable Dr. *Woodward*;[3] a few Weeks ago, concerning certain American Plants, which I was collecting for you.[4]

1 24.^d 7.^m [September] 1716. Docketed, "Red. ô Jan. 15. 1716/7." Enclosure: samples of "six or seven plants."

2 On James Petiver's achievements in the field, see R. Coulton (2020).

3 John Woodward (1665–1728), English natural philosopher, geologist, antiquarian, and Mather's frequent correspondent at the Royal Society. Mather owned a copy of Woodward's *Naturalis Historia Illustrata* (1714).

4 On September 12, 1716, Mather records in his diary: "G.D. What shall I do, to render my Kinsman at *Newton*, considerably useful? Employ my Kinsman at *Roxbury* to make a collection of *Plants*, peculiarly *American*" (*D* 2:371). See also *D* (2:208).

In a field not far from the City of *Boston*, there were lately made these Two Experiments.[5]

First; My Friend planted a Row of Hills of Indian Corn, that was coloured Red and Blew, the rest of the field being planted with Corn, of the Yellow, which is the most usual Colour. To the Windward Side, this Red and Blew Row, so infected three or four whole Rows, as to communicate the same Colour to them; and part of the Fifth, & some of the Sixth. But to the Lee-ward Side, no less than Seven or Eight Rowes, had the same Colour communicated unto them; and some small Impressions were made on those that were yett further off.

Secondly; The same Friend, had his Garden robbed ever now & than of the Squashes growing there. To inflict a pretty little Punishment on the Theeves, he planted some Guords among the Squashes, which are very like them, at certain Places which he distinguished with a private Mark; that he might not be himself imposed upon. By this Method the Theeves were deceived, & discovered, & ridiculed. But yett the honest Man saved no Squashes by the Trick, for they were so infected & embittered by the Guords, that there was no eating of them.

Several useful Hints, relating to Vegetation & Agriculture, may be taken from these Experiments; which I wholly leave to your Sagacity, and that of the Philosophers to whom you may see Cause to mention them.

What remains is, to make you a Tender of Six or seven Plants, which are here esteemed peculiar to *America*. I cannot find them in any *Europæn* Herbals;[6] and if you find that I have been mistaken, you will pardon my Ignorance, and still accept my Intention. I have præserved the Plants, as well as I can, in a Book of Brown Paper, and have præsumed so far as to Impose Names upon them; and have added Schedules declaring the Vertues & Uses for which our Indians have employ'd them, and in which our English have been also confirmed by their Experience.

I have committed the Book unto the Care of an Honourable Friend, Mr. *Edward Loyd*,[7] (One of our Judges,) who will do me the Honour to deliver it with his own hand.

This is but a Part of the Collection that I intend for you; because this Year I unhappily miss'd the Season of securing several, which, if I live to another Season, I will do my best, that I may be a Master of them.

5 The following two paragraphs also appear (nearly verbatim) in Mather's *CP* (1721), Essay XXVI, pp. 124–25. On Mather's cutting-edge discovery of plant hybridization, see C. Zirkle's *Beginnings* (1935), pp. 104–07; and his subsequent article, "Hybridization" (1969). Zirkle reprints Mather's letter to Petiver in both works. See also R.P. Stearns, *Science* (pp. 416–17), and B.E. Goulet et al. (2017).

6 See for example Nicholas Culpeper's *Complete Herbal* (1653) and John Ray's *Historia Plantarum Species*, 3 vols. (London, 1686–1704).

7 Most likely Edward Lyde (d. 1724), judge of the Suffolk Court of Common Pleas (1715–23). See Sewall, *Diary* (3:337 *n.*).

In the mean time, I shall not be altogether wanting; in my Essayes to do the best I can in Obedience to your Commands. And I hope, annually to treat the Royal Society also with such a Number of Communications, that if every Member of that Illustrious Body, whose Name stands in the Catalogue (an Honour not yett granted unto mine,)[8] will do but half as much the Stores in your Collection will soon grow considerable.

That the Blessing of Heaven may attend all your Endeavours to do Good in the World, is the hearty Prayer of,

Sir,

Your most affectionate

Friend & Servant.

Mr. James Pettiver.

C. M.

The Vertues of the
Caltha Sylvatica.[9]

It is the greatest Vulnerary in the World. Make a Balsam of the Green Leaves with Oil-Olive, it cures the most grievous Punctures of the Nerves; and Cutting of all sorts within an admirable Efficacy & Celerity. Ordinary Wounds

8 Mather's self-congratulatory fishing for his election to the RSL is noteworthy here, for even though he acknowledges in his *D* (2:245–46) that the "Secretary of the Royal Society" ruled that Mather "shall be made, a Fellow of the Royal Society," Mather was anxious to see his election confirmed in the published membership role of the RSL. See G.L. Kittredge's "Election."

9 The following list of Americana (appended to the MHS MS. N-527 as 355–56) is missing from Mather's original in the Sloane Collection 4065, fol. 255. The two-page list was likely included in the "Book of Brown Paper," along with the plant specimens (see above), but separated from the letter, when James Petiver needed the list for inclusion in his *Petiveriana III* (1717), p. 12 (sep. pag.), "Com. V, Some *American Plants*, with their *Specifick Vertues* and Wonderful *Effects,* lately sent me by the *Reverend* and Learned Dr. *Cotton Mather*, at *Boston* in *New England*, and Fellow of the *Royal Society*, London" (items 288–94), reprinted in *Jacobi Petiveri Opera, Historiam Naturalem Spectantia* (1764), vol. 2:12 (sep. pag.). The English natural philosopher, botanist, and zoologist John Ray (1627–1705) was a distinguished member of the Royal Society of London. Among his noteworthy publications are his three-volume *Historia Plantarum Species* (1686–1704), which Mather is likely to have consulted. All three volumes include various subtypes of the genus *Caltha* (Buttercup flower), Mather's "*Caltha Sylvatica*" is not among them.

are cured with it in four or five Hours unto the astonishment of the Patient; and extraordinary ones in a few Dayes at farthest. The Indians are extremely venturesome in their Fighting, if this Traumatic be near.[10]

Consolida Lappata{?}[11]

For Internal Wounds, it will do Wonders. It releeves Oppressions from clotted Blood in the Stomach. For the Spitting of Blood it affords a Releef hardly to be parallel'd.

Ophiophuga[12]

Bruise it, and lay the Cataplasm to the Part bitten by the Rattlesnake; it Infallibly & Immediately fetches out the horrible Poison, which else would in a few hours bring a miserable Death upon the Patient. It is a marvellous Gift of Heaven to the Countreyes, which have these tremendous Vipers in them. Tis remarkable, that if a Man putt it in his shoes, no Serpent will dare to come anear him.

The Decoction of it, is also a good Ophthalmic.[13] –

Fagiana.[14]

It growes no where but under Beach-Trees, in shady-Places. It appears, in the Summer and expires in the Autumn. An Infusion taken Inwardly, and a

10 Apart from supplementing the missing scientific classification of plants not provided in Mather's description, Petiver does follow Mather's syntax closely. However, the last sentence as printed in Petiver's document bears a peculiar Matheran idiom that does not appear in the MHS draft: "The *Indians* will fight like *Milton's Devils*, if this *Traumatick* be near them, *C.M.*"

11 This herb appears to be what Petiver lists as "Small *Bur* Comfry," in *Petiveriana III*, p. 12 (sep. pag.), item # 291: "It does Wonders in internal Wounds, relieves *Oppressions* in the Stomach, from clotted Blood, not to be parallel'd, *C.M.*" Petiver comments on Mather's description: "This seems another sort of *Avens*, but being without Flower or Seed, I cannot determine it, *J.P.*" See also John Ray's *Historia Plantarum* for several subspecies of the genus *Consolida*.

12 Also known as "Snakeweed." Mather's old nemesis Paul Dudley offered a similar antidote in his "An Account of the Rattlesnake," *PT* 32.376 (April 30, 1723): 292–95. Dudley's designation was "Blood-root," because its juice is "red like Blood" (p. 295). See especially G.L. Kittredge, "Letters," p. 184; and Stearns, *Science* (1970), pp. 463–64.

13 Relating to diseases of the eye.

14 996 See also Mather's more detailed account in his "Curiosa" Letter to John Woodward (July 13, 1716), our CA 3.11 "The *Fagiana*." And *Petiveriana III*, p. 12 (sep. pag.), item # 294.

Cataplasm used Outwardly in Cancerous & Scrophulous Maladies does Things to be wondered at. A Tea of it, will strangely refresh under & against Weariness: And is of grand Use in malignant Feavours.

Formosa.[15] Or, *Virgan aurea americanum.*

It growes near six foot high. Its Red Flowers on the Top give it a Lovely Aspect. The Water in which it is boiled, cures Inflammations of the Eyes that are washed with it. Two Spoonfuls of the said Water, will cure the Malady we call, *The Stomach-ake.*

Geranium Americanum.[16]

Call'd by our Indians, *Taututtipoag* Inwardly the Infusion, and Outwardly the Cataplasm; tis Grand Medicine for the Foul Disease.[17] The whole Plant with the Root boil'd & applied unto the Face, it cures what we call the Ague there; as also any Tumour. A Tea of the Root, is good in Feavers and the Palpitation of the Heart.

15 *Petiveriana III,* p. 12 (sep. pag.), item # 292. John Ray's *Historia Plantarum* lists several subspecies of the genus *Formosa.*

16 Ray's *Historia Plantarum* lists several subspecies of the genus *Geranium.* See Kittredge, "Letters," p. 186. *Petiveriana III,* p. 12 (sep. pag.), item # 293; and Mather's second letter to John Woodward (November 18, 1712), our CA 1.2 "On American Plants & Indian Cures."

17 Venereal disease. See CM's Cap. XXI; Kibroth Hattaavah, or Some Clean Thoughts on the Foul Disease (*ABJ* [1972] 116–20).

3.13. A SINGULAR LIMESTONE.

September 24, 1716.[1]

To Dr. John Woodward.

Sir,

A few Weeks ago, I endeavoured in my poor Way to entertain you with more than a Decad of American Curiosities; – convey'd by the Hand of Mr. *Samuel Woodward* our Secretary,[2] whom I hope ha's before now reached yours.

I now perform what I then promised you, of certain Botanic Entertainments for Mr. *Petiver*;[3] and the Hand of one of our Judges *Edward Loyd* Esqr,[4] who obliges me, with conveying my *Hortus Siccus*[5] to him, does now also convey to you, a further Token that I have not forgotten you.

I blush to think, that I have made no better Progress, in the Collection of Fossils, you proposed unto me.

But I am now going upon it as far as my Variety of Employments will allow me.

In the mean time, I tender to your Acceptance, a very singular sort of a Fossil, whereof the Account I am to give you is this.

1 "24.^{d} 7.^{m} [September] 1716."

2 Samuel Woodward (1660–1751), secretary, commissioner, and treasurer of the Province of the Massachusetts Bay (1715–17). See A. Matthews, ed., "Massachusetts Royal Commissions 1681–1774" (315–16). And Mather's "My most honor'd Doctor," appended to CA 2.1 "A Woollen Snow" (December 1, 1713), *n.* 26.

3 James Petiver (c. 1665–1718), an English botanist, pharmacist, herbalist, member of the RSL, and voluminous author, illustrator, and collector of plants, herbs, fossils, and minerals. See also Mather's letter to Woodward (July 25, 1717) CA 4.1 "Fossils."

4 Most likely Edward Lyde (d. 1724), judge of the Suffolk Court of Common Pleas (1715–23), d. 1724; in Sewall's *Diary* (3:337 *n.*).

5 A collection of dried botanical specimen.

When my Friend Sir W. P.[6] discovered the Spanish Wreck, which had been more than forty Years, diverse Fathom under Water, from which he carried 300000£ home for *London*, – he found that the huge Heap of Peeces of Eight which were stow'd all together, had a thick huge Crust of two or three Inches Thickness, grown over it. It putt him upon the Trouble of using Iron tools with a great Force, to break open this Crust; from which they knock'd out several Bushels of Peeces of Eight[7] as they brought it up in Parcels.

A Fragment of that Crust with a Peece of Eight adhæring to it, I now pray you to allow a Room among your Curiosities.[8] It ha's hung about Eight & twenty Years in my Library & so ha's the Complexion a little changed. But perhaps it may afford you some Speculations, beyond what I may presume to suggest unto a Person of your Sagacity.

May our gracious God prosper your Studies, and bring into the Service of Mankind your noble, & much wished for Præparations.

I am,

Syr,

Your most hearty Servant,

{C. Mather.}

Dr. Woodward.

6 Sir William Phips (1651–95), first royal governor (1691–95) of the province of Massachusetts Bay under the new Charter, farmer, mariner, treasure hunter, and friend of the Mathers. Cotton Mather anonymously published *Pietas in Patriam* (1697) and included it prominently in his *MCA* (2.35–75). In 1687, mariner William Phips of Maine headed an expedition to Hispaniola and recovered thirty-four tons of sunken Spanish treasure that was brought back to England, for which Phips was knighted. See also A. Lounsberry (1941) and E.W. Baker (1998).

7 Spanish silver coins widely circulating in the American colonies and Europe. Pieces of Eight (*real de a ocho*) was worth eight Spanish reales and served as the first international currency of trade.

8 See also *MCA* (2.6.41).

Series Four

4.I. FOSSILS.

July 25, 1717.[1]

To John Woodward.

My most honoured Friend.[2]

The Communications, with which I have endeavoured some Entertainments for you, since I have had the Honour to be related unto your Illustrious Society,[3] have been so many, that my Letters have pretty near as I suppose, æqual'd the Number of Months that ha's passed over me, besides what I might say of other Performances, & Remittances. But for these two or three last Years, I have never been so happy, as to know that my Packetts have not miscarried; and particularly, whether what I addressed unto you, near

1 25.[d] V.[m] 1717. Enclosure: a box containing twenty to thirty fossil samples.

2 John Woodward, M.D., F.R.S. (c. 166–1728), professor of physics and natural history at Gresham College, was particularly keen on marine fossils from the New World to settle his dispute with his confreres at the RSL, especially with John Ray (1627–1705) and Edward Lhwyd (1660–1709), both distinguished naturalists. In their debate on the origin of fossils as sports of nature (lusus naturae) formed by the "plastic virtue" of the earth – a theory widely held by the likes of Robert Hooke, Robert Plot, Martin Lister, and Edward Lhwyd – Ray embraced a hybrid theory of fossils as petrified remnants of living organisms as well as the remains of germinating seeds of plants and animals (seed theory), which (so Lhwyd) grew deep in the earth and sometimes assumed the appearance of sea organisms in a more or less imperfect manner. Woodward, however, firmly held that fossils are the remains of once-living organisms now deeply buried in the earth. When Noah's flood dissolved the crust of the antediluvian earth, Woodward maintained, the remains of plants and animals (because of their differing gravities) sank down in the mud and settled in the earth's different strata, where they became fossilized. No wonder that Woodward encouraged Mather to supply him with American specimens. See Appendix 8.

3 Mather became a member of the Royal Society of London in 1713. For the controversy surrounding his election to the Society, see G.L. Kittredge, "Election" and "Further Notes."

a Year ago, and the *Hortus Siccus*[4] of Plants, peculiar to *America*, which I address'd unto Mr. *Petiver*, have reached you.[5]

Nothing shall discourage me from doubling my Diligence, to add what I can unto your Stores, from what I may think proper for them, in these *American* Regions; whereof a few Weeks hence, I intend you a more ample Testimony.

But I could not lett this Opportunity pass, without Beginning to make some Amends, for my Delay to answer your Desire, to be furnish'd with some Collection of our Fossils.[6] The Truth is, the Load of my various Employments, oblige me to wish and wait for the Assistences of a young Gentleman, whose name is Mr. *John Winthrop*, and whose Grandfather (the Governour of one of our Colonies) was a *Virtuoso*, in the very Beginning of the Royal Society.[7] This Ingenious Gentleman ha's promised me his Assistences,[8] to accomplish what you have desired of me; and as a Specimen of more to follow, he enables me now, by the hand of one Captain *Venteman*,[9]

4 An arrangement of dried plants. Mather refers to his CA 3.12 "*Curiosa Botanica*," to James Petiver, September 24, 1716.

5 James Petiver (c. 1665–1718), an English botanist, pharmacist, herbalist, member of the RSL, and voluminous author, illustrator, and collector of plants, herbs, fossils, and minerals. Mather's shipments of plants are recorded in Petiver's *Petiveriana III* (1717), p. 12 (sep. pag.). See also CA 3.11 "The *Fagiana*" and 3.12. "*Curiosa Botanica*."

6 See also Appendix 12.

7 John Winthrop, F.R.S. (1681–1747), grandson of Governor John Winthrop the Younger, F.R.S. (1606–76) of Connecticut, Harvard class of 1700, Connecticut magistrate, and afterward a fellow of the Royal Society. On the original membership in, and contributions to, the Royal Society of the governor, who was the first colonial American elected to membership on January 1 1661/62, see H.G. Lyons (1940). M. Denny (1947), and R.P. Stearns, *Science* (1970), p. 121.

8 See Mather's letter to Winthrop (15[d] VIII[m] [October] 1716), asking him for specimens of "the various Treasures of Natural Knowledge wherewith you are enriched, & are daily growing Richer & Richer… . A True *Winthrop*, cannot but be generous!" (*MP*, [1868], pp. 419–20). For Winthrop's promise to supply the requested fossils, see his letter to Mather (November 5, 1716), in *CMHS* (sixth series), 5 (1792): 331–34 (ref. at 332).

9 John Venteman, master of the ship *Hannah & Elizabeth*, which sailed mostly between Boston, Jamaica, Barbados, and Bermuda. *The Boston News-Letter* of April 26, 1714, lists the ship as recently arrived in Boston; and the November 12, 1716 issue identifies her as bound for Barbados. Venteman was married to Elizabeth Smith by Mather on July 13, 1699, in Boston. See W.M. Clemens (1926), p. 198.

to transmitt unto you a Box, which contains between Twenty and Thirty of such Things as you have asked for.[10] I am therewithal to give you his most hearty Services, and adds, that if you please to take some Notice of him, when your Pen shall again command the Liesure of writing for N.E.[11] it may still more encourage his Industry, to help me in executing from time to time, all of the Commands you may lay upon us. He is a Man of honour & erudition.

We long exceedingly to see your larger Work, whereof the Harbinger ha's been so grateful & useful to us;[12] & whereof my learned & worthy Friend Dr. *Edwards*,[13] in one of his last Essays before he died, added yett more unto our Expectation.

10 This box likely contained curious American seashells, for in his letter to Mather (April 3, 1721), Woodward thanked him for "your *Curiosa Americana*," yet pointed out that "I hear so litle of Marine Bodyes, Remains of the Deluge, digd up in New England. Pray be more inquisitive on all Occasions of Digging, Mineing, &c." (Appendix 8). Woodward was eager to receive fossilized shells in his ongoing debate with Elias Rudolph Camerarius, Jr. (1673–1734), professor of medicine at the University of Tübingen, who in his *Dissertationes Taurinenses Epistolicae, Physico-Medicae* (1712) had sparred with Woodward in the paleontological debate over the question whether seashells found deep in the earth and even on top of mountains were of marine origin and were thus proof of Noah's flood, or whether they were sports of nature (*lusus natura*e) and had naturally grown in the places where they were found. As early as 1713, Mather had promised Woodward to "make a Collection of our Fossils," but had so far been "overwhelmed with Employment" (Appendix 12). See also G.L. Kittredge, "Election," pp. 83–84, 101. Years earlier (July 10, 1706), Governor Joseph Dudley of Massachusetts sent Mather a description of a giant tooth found at Claverack, New York; Mather incorporated it with additions in his commentary on Gen. 6:4 (*BA* 1:591–94) and in CA 1.1 "Describing Biblia Americana & Giant Fossils."

11 Woodward complied with Winthrop's request for acknowledgement (April 3, 1721) but encouraged him to "search after Fossils of all Sorts," which Woodward desired for "the Perfecting my Nat. Hist. of ye Earth," in *PrMHS* 13 (1873): 110.

12 Referring to John Woodward, *An Essay towards a Natural History of the Earth* (1695), which appeared in a revised third edition in 1723. Woodward's work went through five editions and a translation into Latin. Mather owned Woodward's Latin edition (1714), but here he appears to refer to the anticipated enlarged and illustrated edition of 1723 and 1726. See also Woodward's two-volume *An Attempt Towards a Natural History of the Fossils* (1729).

13 John Edwards (1637–1716), English Calvinist divine, referring to *Some New Discoveries of the Uncertainty, Deficiency, and Corruptions of Human Knowledge and Learning* (1714), in which Edwards describes Woodward's *Essay of the Natural History of the Earth* as "the best Performance of that Kind that I have

May our glorious God continue you long a Blessing to the World, and an Ornament unto our Nation; and prosper your Studies for the Good of Mankind, which must ever acknowledge you one of its most meritorious Benefactors,

Boston, N.E.

25 July 1717.

I am, with your great Esteem,

Syr,

Your most hearty Friend

& humble Servant

{C. Mather.}

Dr. John Woodward.

met with, and hath not the Faults that some other *Theories* are justly charg'd with" (p. 79). Edwards looked forward to when the work would "become a *compleat* Body," since Woodward had allowed him to view the notes towards the larger treatise of 1726.

4.2. AN ECLIPSE OBSERVED.

October 1, 1717.[1]

To Dr. Edmond Halley.[2]

Sir,

The Services you have done the World, by the vast Improvements & Advancements you have made, especially in the mathematical Sciences,[3] which have that peculiar Property of *Certainty*, to recommend them unto the more polished sort of Men, will render your Memory as glorious & as lasting as the Stars, which you touch in your Attainments.

Your Communications in the *Philosophical Transactions*, oblige the curious Part of Mankind.[4] And those of us, who are at a mighty distance below you, for skill in the Mathematicks, are yett more sensibly obliged, when you favour us with Points, as you sometimes also do, which are more within the reach of our lower Capacities.

I am getting ready for you, some select & more Curious Passages, from a Manuscript Journal of a late Voyage round our Globe, which I hope to transmitt by the next Opportunity.[5] In the mean time, something relating to one that every day visits every Part of it, shall entertain you.

1 1.^{d} VIII.^{m} 1717.

2 See G.L. Kittredge, "Scientific," pp. 44, 46.

3 See Edmond Halley F.R.S. (1656–1742), royal astronomer, author of *The Elements of that Mathematical Art, Commonly called Algebra* (1709) and, more famously, *A Synopsis of the Astronomy of Comets* (1705), in which he computed the appearance of the comet that would bear his name.

4 Between 1682 and 1717, Halley authored or co-authored some fifty articles for the *PT* on a variety of topics. For instance, Halley's "Observations of the late Total Eclipse of the Sun," *PT* 29.343 (May 31, 1715): 245–62, may well have inspired Mather to send in his observations.

5 Mather appears to refer to the now-lost document "A Journal of a Voyage round the World," which would become part of his 1720 shipment to the Royal Society (G.L. Kittredge, "Scientific," p. 46).

A Gentleman, who has travelled as you have done, on Purpose to bring home Stores from whence the Commonwealth of Learning might be enriched, has a Right unto all that the Travels of others may give Instruction in.

I could not think of any more agreeable Entertainment for you, than an Observation of a last Eclipse; wherein the principal Manager, was my worthy Friend Mr. *Thomas Robie*, an Ingenious young Gentleman, and the greatest Proficient in Astronomical Studies, which I know in these Parts of the World.[6]

Some Haziness of the Heavens, was a little disadvantage to the Accuracy of our Observation; But yett such as it was the New Astronomical Tables, which we call *Whistonian* received a good Confirmation from it.[7]

I have nothing to add, but, a, *Serus in Cœlum redeas*,[8] and that I am, with very great Respect,

Boston, N. England, Sir,

1^{d}. VIII^{m}. [October] 1717.

Your most affectionate Friend

& Servant.

{C. Mather.}

---

6 Thomas Robie, F.R.S. (1689–1729), Boston native, Harvard class of 1708, college tutor and later physician, who from 1715–22 kept a record of meteorological conditions and published astronomical observations. Robie also published a series of almanacs (1708–20). See F. Kilgour (1939): 473–90, and R.P. Stearns, *Science* (1970), pp. 426–35.

7 Mather refers to his erstwhile friend and correspondent William Whiston (1667–1752) for whom he held an unfeigned admiration. Whiston, an English theologian, outstanding scientist, and Newton's successor to the Lucasian professorship of mathematics at Cambridge, was roundly dismissed from that distinguished position in 1710, because he dared to go public with his antitrinitarian Arian heterodoxy. In the present context, Mather alludes to Whiston's *Praelectiones Astronomicae Cantabrigiae in Scholis Publicis Habitae* (1707), translated into English as "*For the Use of young Students in the University* as *Astronomical Lectures, Read in the Publick Schools at Cambridge ... Whereunto is added a Collection of Astronomical Tables* (1715). The tables, separately paginated, begin after p. 387. See also Whiston's *The Calculation of Solar Eclipses without Parallaxes* (1724).

8 Invoking Horace's benediction of Augustus Caesar (*Carmina* 1.2.45), Mather concludes, "May it be long before you return to heaven."

Observations of the Suns Eclipse, Sept. 23. 1717.
at Cambridge N. England: Lat. 42 = 25.

	h	*l*		
Beginning at	12 —	23 —	00 }	
Eclipse Sensibly decrease, —			}	
an Haze being in the Atmosphere.	1 —	47 —	00 }	P.M.
A Spott in the Eastern Limb of the ʘ Emerg.	3 —	02 —	23 }	
Eclipse ended.	3 —	05 —	10 }	

At the Beg. of the Eclipse, the		
ʘ Altitude —	43 —	21.
And Zenith Distance,	46. —	39.
At the End, the ʘ Altitude —	27. —	44.
And Zenith Distance —	62. —	16.

4.3. AN HORRID SNOW.

December 10, 1717.[1]

To Dr. John Woodward.

Sir,

Tho' we are gott so far onward as the Beginning of another Winter, yett we have not forgott the Last; which at the latter End whereof, we were entertained & overwhelmed with a *Snow*, which was attended with some Things, that were uncommon enough, to afford matter for a Letter from us.[2]

Our *Winter* was not so bad as that wherein *Tacitus* tells us that *Corbulo* made his Expedition against the *Parthians*.[3] Nor like that which proved, so fatal to the Beasts & Birds, in the Days of the Emperour *Justinian*[4] that wherein the very Fishes were killed under the Freezing Sea, when *Phocas* did

1 10.^{d} X.^{m} 1717.

2 For informative background, see T. Wickman (2017) and E.P. Fisher, *Mighty Storms* (2021), esp. ch. 3. In his diary entry for May 24, 1717, Mather declares, "I would make some useful Collections of the rare Occurrence, which were in the Storms and Snows, that we had in the Conclusion of our Winter. I may make a good Use of them. And communicate them [to] the Royal Society" (*D* 2:439–40, 455, 506; quote, p. 455). He spiritualized the climatic challenges of snow and ice of a New England winter in his *WM* (1693), pp. 4, 25–26, 36–37, 40–45, 63, 81, and in his *WP* (1712), esp. pp. 11–15, 33.

3 The Roman senator and historian Publius Cornelius Tacitus (56–c. 117 CE) records Gnaeus Domitius Corbulo's military exploits in Rome's Parthian campaign (58–63 CE), and describes the severity of the winter in northern Armenia: "The entire army was kept under canvas, notwithstanding a winter of such severity that the ice-covered ground had to be dug up before it would receive tents. As a result of the bitter cold, many of the men had frost-bitten limbs, and a few died on sentinel-duty" (*Annals* 13.35).

4 The Byzantine historian Procopius of Caesarea (c. 500–565 CE) reports that, during the Vandalic War, the year 536 CE was so extreme that winter never seemed to end, and the snowstorms were so terrible that vast quantities of birds perished. In the tenth year of Emperor Justinian I (536–38), Procopius writes, "the sun gave forth its light without brightness, like the moon, during this whole year, and it

as much to the Men whom Tyrants treat like the Fishes of the Sea.[5] But the Conclusion of our *Winter* was hard enough, & was too formidable to be easily forgotten; and of a Peece with what you had in *Europe*, a Year before.[6]

The *Snow* was the chief Thing that made it so. For tho' rarely does a *Winter* pass us, wherein we may not say with *Pliny*; *Ingens Hyeme Nivis apud nos copia*;[7] yett our last *Winter* brought with it a *Snow* that excelled them all. A *Snow*, tis true, not æqual to that which once fell and lay Twenty Cubits high, about the Beginning of *October*, in the Parts about the *Euxine Sea*.[8] Nor to that, which the *French Annals* tell us, kept falling for twenty Nine Weeks together. Nor to several mentioned by *Boethius*, wherein vast Numbers of People, and of Cattel, perished;[9] Nor to those that *Strabo* finds upon *Caucasus* and *Rhodiginus* in *Armenia*.[10] But yett such an one, and attended with such Circumstances, as may deserve to be Remembred.

---

seemed exceedingly like the sun in eclipse, for the beams it shed were not clear nor such as it is accustomed to shed," in *The History of the Wars* (4.14.5).

5 Eastern Roman Emperor Phocas (547–610 CE), who usurped the Byzantine throne (602–10 CE), was so brutal in his purges that he in turn was overthrown and executed by his successor Heraclius (610–41). The cruelties exacted upon Phocas' enemies and those of his successor Heraclius are illustrated in *Chronicon Paschale 284–628 AD* (1989), pp. 145–89; and in *The History of Theophylact Simocatta* (8.10.4–15).

6 The winter of 1715–16 was a very cold and severe one in England, with bitter frost from November to February, and the River Thames frozen over. See the anonymous ballad, *A Winter-Wonder, Or, The Thames Frozen Over* (1684); as well as *Saint Patrick's Purgatory ... With a Poetical Description of the frozen River of Thames* (1716); and W. Andrews, *Famous Frosts* (1887), pp. 40–44.

7 "We have a great deal of snow in winter." Although attributed to Pliny, whose *Natural History* was written in Greek, Mather's Latin phrase is most likely a summary reference, not a specific quotation. The Mather family library contained a copy of Pliny's *Historiae Naturalis Libri XXXVIII* (1635).

8 I.e., the Black Sea. *Periplus Maris Euxini* (c. 130 CE), by the Greek historian and Roman senator Arrian of Nicomedia (c. 87–c. 145 CE), is a likely source for this snowy winter.

9 Anicius Manlius Severinus Boethius (c. 477–524 CE), a Roman politician, philosopher, and translator of Plato and Aristotle. Boethius's most famous work is his *De Consolatione Philosophiae* c. 523). Mather's specific reference remains unidentified.

10 Strabo (c. BCE 63–c. 24 CE), a Greek geographer, philosopher, and historian, whose *Geographica* (11.5.6) describes the severe winters and the acclimation of the people who live in the Caucasus. The Venetian Ludovico Ricchieri, aka. Caelius Rhodiginus (1469–1525), Latin and Greek scholar; the Geneva (1626) edition and his principal work, *Antiquarum Lectionum*, in sixteen books, was in the Mather library.

On the Twentieth of the last *February*,[11] there came on a *Snow*, which being added unto what had covered the Ground a few Days before, made a Thicker Mantle for our Mother than what was usual; And the Storm with it, was for the following Day so violent, as to make all Communication between the Neighbours every where to cease. People for some Hours could not pass from one side of a Street unto another, and the poor Women, who happened at this critical time to fall into Travail, were putt unto Hardships which anon produced many odd Stories for us. But on the Twenty-fourth Day of the Month, comes *Pelion upon Ossa*.[12] Another *Snow* came on, which almost buried the Memory of the former: with a Storm so furious; that Heaven laid an Interdict on the Religious Assemblies throughout the Countrey, on this Lords-day, the like whereunto had never been seen before. The Indians near an hundred Years old, affirm, that their Fathers never told them of any thing that æqualed it. Vast Numbers of Cattel were destroy'd in this Calamity; whereof some that were of the Stronger Sort, were found Standing Dead, on their Legs, as if they had been alive, many Weeks after, when the Snow melted away. And others had their Eyes glazed over with Ice at such a rate, that being not far from the Sea, they mistooke of their Way, drowned them there.

One Gentleman,[13] on whose Farms, there were now lost above eleven hundred *Sheep*, which with other Cattel were Interr'd shall I say, or, *Inniv'ed*

[11] Mather *Diary* records (*D* 2:439) that the blizzard began on February, 28 1716–17: "As mighty a Snow, as perhaps has been known in the Memory of Man, is at this Time lying on the Ground; and as mighty a Thaw must be look'd for. Would it not be a seasonable, may it not be a serviceable Action, to entertain the People of God, at the Lecture, with a Discourse on Psal. CXLVII. 18. *He sendeth forth His Word, and melteth them*." He added a week later, "Tis a Time of much Rebuke from Heaven upon us, in the Season. Such Storms and Heaps of Snow, visit us in the approach of the Spring, as were hardly ever known in the Depth of Winter. A great part of the Assemblies in the Countrey have been inter{dic}ted their public Sacrifices {Church services}. A Multitude of Cattel have perished. The Business of the Countrey has an uncommon Stop upon it. Many Difficulties grow upon us. I would procure a Day of Humiliations and Supplications to be kept on these occasions" (*D* 2:440). By contrast, his gadabout friend Judge Samuel Sewall, though noting its intensity, does not spend much time in describing this nor'easter (Sewall, *Diary* 3:120, 121, 122, 123).

[12] I.e., to heap one challenge upon another. In Greek mythology, Pelion and Ossa are two mountains in Thessaly, where Otus and Ephialtes, the two rivaling giants, tried to scale Mount Olympus, the abode of the gods, to overthrow the immortals. Alas, Zeus intervened and put an end to the wannabe usurpers. The story can be found in Homer's *Odyssey* (21.310-16) and Vergil's *Georgics* (1.281–83).

[13] For much of the following description, Mather leisurely draws on the letter (September 12, 1717) of his friend and correspondent John Winthrop (1681–1747), of

in the Snow; writes me, That there were Two *Sheep* very singularly circumstanced. For no less than Eight & Twenty Days after the Storm, the People pulling out the Ruines of above an hundred Sheep, out of a Snow-bank, which lay Sixteen foot high drifted over them, there were Two found alive, which had been there all this time, & kept themselves alive by eating the Wool of their Dead Companions. When they were taken out, they shed their own Fleeces, but soon gott into good Case again.

Sheep were not the only Creatures, that lived unaccountably for whole Weeks without their usual Sustenance, entirely buried in the *Snow-drifts*. The *Swine* had a Share with the *Sheep* in Strange Survivals. A Man had a Couple of young *Hogs*, which he gave over for Dead; But on the twenty-seventh Day after their Burial, they made their Way out of a *Snow-bank*, at the bottom of which they had found a little Tansy to feed upon.

The *Poultry* as unaccountably Survived as these. *Hens* were found alive, after *Seven Days*; *Turkeys* were found alive, after *five & Twenty Days*; buried in the *Snow*, and at a Distance from the Ground; and altogether destitute of any thing to feed them. The Numbers of Creatures, that kept a *Rigid Fast*, shutt up in *Snow*, for diverse Weeks together, & were found Alive after all, have yeelded Surprizing Stories to us.

The Wild Creatures of the Woods, (the *Outgoings of the Evening*)[14] made their Descent as well as they could in this Time of Scarcity, for them, towards the Sea-side. Here a vast Multitude of Deer for the Same Cause taking the Same Course, & the Deep Snow Spoiling them of their only Defence; which is, *To Run*, they became such a Prey to those Devourers, that it is thought, not one in Twenty escaped.

But here again occurr'd a Curiosity.

These Carnivorous Sharpers, and especially the *Foxes*, would make their Nocturnal Visits, to the Pens, where the People had their *Sheep* defended from them. The poor Ewes big with Young were so terrified with the frequent Approaches of the *Foxes*, & the Terror had such Impression on them, that most of the *Lambs* brought forth in the Spring following, were of Monsieur *Reinards* Complexion, when the Dams were all either *White* or *Black*.[15]

New London, Connecticut (*MP* [1868], pp. 681–82), excerpted in K. Silverman's *SLCM* (1971), pp. 247–48. Winthrop – like his famous grandfather, Gov. John Winthrop, Jr. (1606–76) of Connecticut, before him – was elected a member of the RSL on January 10, 1733/34 (*MP* 683). See also M. Denny (1947).

14 Psal. 65:8.

15 *Renard* being the French word for fox. Superstition had it that pregnant women (as well as animals) will give birth to offspring that resemble the creature or person who frightened them. See CA 3.7 "Strength of Imagination."

It was remarkable, that immediately after the Fall of the Snow, an infinite Multitude of *Sparrows*, made their Appearance; but then after a short Continuance all disappeared.

It is incredible, how much Damage was done to the *Orchards*; For the Snow freezing to a Crust, as high as the Boughs of the Trees, anon splitt them to Peeces. The Cattle also, walking on the crusted Snow, a dozen foot from the Ground, so fed upon the Trees as very much to damnify them.

The Ocean was in a prodigious Ferment, and after it was over, vast Heaps of little Shells were driven ashore, where they were never seen before. Mighty Shoals of Porpoises, also kept a Play-day in the Disturbed Waves of our Harbours.

The odd Accidents befalling many poor People, whose Cottages were totally covered with the Snow, & not the very tops of their Chimneys to bee seen, would afford a Story; But there not being any Relation to Philosophy in them, I forbear them. And now, *Jam Satis Terris Nivis*,[16] – And here is enough of my Winter-tale. If it serve to no other Purpose, yett, it will give me an Opportunity, to tell you, That Nine Months ago,[17] I did a thousand times, wish myself with you in *Gresham-Colledge*,[18] which is never so horribly Snow'd upon. But instead of so great a Satisfaction, all I can attain to, is the Pleasure of talking with you in this Epistolary Way, and subscribing my self,

{Boston, New England.} Syr,

10*d*· X*m*· 1717.

Yours with an Affection

that knows no Winter –

{C. Mather.}

Dr. Woodward.

16 Perhaps alluding to the tempestuous storm in the wake of Julius Caesar's murder, the Roman lyric poet Horace (65–8 BCE) addresses Augustus Caesar with the opening line (*Ode* 1.2.1): "Enough of snow [and hail at last / The sire has sent in vengeance] upon the earth" (*Odes and Carmen Saeculare*), p. 2.

17 During the blizzard of February–March 1716/17 (*D* 2:439, 440).

18 Gresham College, London, had been the headquarters of the Royal Society until 1710, when it moved to Crane Street; John Woodward (1665–1728) was a professor of physics at Gresham, an elected F.R.S in 1693, and secretary of the Society.

4.4. A PRODIGIOUS WORM.

December 12, 1717/18.[1]

To John Woodward.

Sir.

Among the many things, that have passed under your curious Observation, you have doubtless allow'd some Thoughts unto the *Worms*, which find out so many of the Bodies in the Animal World, for their Lodging.[2] And among these, you have more particularly considered, the *Jointed Worms*, which are found in the Bowels of Humans Bodies; and examined, whether, what *Hippocrates*[3] asserts be true, That they must be bred in the Infants before they are born; and grow up with them; and whether, what *Spigelius*[4] affirms, be right, That a Man can have no more than one, tho' it may appear

1 12.^{d} X.^{m} 1717/18. Kittredge ("Scientific Communications," p. 29) notes that the letter is dated December 12, "but possibly it should be assigned to Jan. 12, 1717, since the occurrence took place on Jan. 6, 1717."

2 See Mather's *CPS*, "Essay XXVII. *Of Insects*," pp. 168 and 357–58; and his detailed discussion in his medical handbook, *ABJ* (1972), "Cap. XLIV. The *Vermine Killer*. Upon, WORMS," pp. 203–07. In his *Brief Instructions For Making Observations in all Parts of the World* (1696), Woodward gives directions on how to preserve the entrails of animal specimens being sent to the Royal Society, so as to "fence off *Worms* and *Insects* from preying upon them" (19). However, no publications by Woodward on the jointed worm have been found. See also E.H. Christianson (1993).

3 Mather alludes to the discussion of intestinal worms (lumbricus latus) in children, in *De Morbis*, lib. 4, cap. XV, pp. 613–14, in *Operum Hippocratis Coi* (1679), by the Greek physician Hippocrates of Kos (c. 460–c. 375 BCE), the "father of medicine." Mather's source for this reference is *An Account of the Breeding of Worms in Human Bodies* (1701), ch. 2, pp. 24–25, by the Parisian physician Nicholas Andry de Bois-Regard (1658–1742).

4 Adriaan van den Spieg(h)el (1578–1625), or Adrianus Spigelius, Flemish anatomist and physician, and author of *De Humani Corporis Fabrica Libri X* (1627). In the present context, Mather refers to Spiegelius' work on tapeworms, *De Lumbrico Lato Liber* (1618), cap. X, p. 31.

more, by the breaking of it. The Unsuspected & Incredible Mischief done by these Destroyers to the Children of Men, will render it very Reasonable, that such Awakening Treatises about them, as that of Monsr. *Andry*,[5] should be more Thankfully received & studied. But it may be wondred, that they do no more Mischief, and that they dispatch not their poor Brethren who harbour them with a more speedy Execution, if they have so many Mouths as Joints, which I perceive is the Opinion of some very Ingenious & Inquisitive Writers concerning them;[6] especially, when they are of such enormous Dimensions as many that we have mett withal. You need not me for a Remembrancer, That *Alexander Trallianus* long ago finds a Worm, as long as all the Gutts;[7] That *Schultzius* mentions one which was Nine Yards long;[8] *Platerus*,[9] one

5 Nicholas Andry de Bois-Regard, French physician and author of *De la Génération des Vers dans le Corps de L'Homme* (1700), translated as *An Account of the Breeding of Worms in Human Bodies* (1701).

6 See "*Lumbricus Latus, or a Discourse read before the* Royal Society, *of the* Joynted Worm, *wherein a great many* Mistakes *of former* Writers *concerning it*, are remarked" *PT* 13.146 (April 10 1683), pp. 113–41, esp. Tables 1 and 2, by Edward Tyson, M.D., F.R.S. (1650–1708), an English physician and anatomist. Examining the joints of the tapeworm, Tyson believed to see "*orifices*, which I take for mouths" (p. 118).

7 Via Edward Tyson's article "*Lumbricus Latus*," p. 119, Mather's second-hand reference is to Alexander of Tralles, aka. Trallianus (c. 525–c. 605), a Byzantine physician, whose Greek *Twelve Books on Medicine* (Latin translation published in Paris, 1548), was highly esteemed. Trallianus' *De Lumbricis Epistola* (1570) is a treatise on worms of all kinds, including tapeworms. Tyson's own source is the oft-reprinted *De Morbis Puerorum Tractatus Locupletissimi* (1583), by the distinguished Italian physician of Padua and Bologna, Girolamo Mercuriale (1530–1606).

8 Via Tyson, p. 121, Mather refers to "Observatio CCXXIX: *De Lumbrico Lato*," in *Miscellanea Curiosa Sive Ephemeridum*. Decuria 3, pars 2, Annus Tertius (1673), pp. 417–19, by Dr. Simon Schultzius, aka. Simon Schultzens (1622–79), German-Polish physician of Thorn and frequent contributor to the German *Ephemerides*. In his article, "Observatio CXX *De Usu Argenti Vivi in Vermibus*," in *Miscellania Curiosa Annus Quartus et Quintus* (1676), p. 130, Schultzius reports his observation of a little girl's puppy dog which voided a tapeworm "duodecim ferme ulnarum longitudinis" (about 12 span long, i.e., 9 feet) after the canine had received one drachma (⅛ ounce) of mercury.

9 Tyson, p. 119, cites *Praxeos Se de Cognoscendis, Praedicendis, Praecauendis, Curandisque affectibus homini incommodantibus* (1608), lib. II, cap. XIII, pp. 962–91, by Felix Plat(t)er (1536–1614), Swiss physician, anatomist, botanist, and author of several works on human anatomy and diseases. In lib. II, cap. XIII, pp. 962–91, Platter defines the nature of worms, explains their causes, and describes the range of medicine to cure tapeworm infections. The Basil 1625 quarto edition of this work is listed in *Catalogus Librorum Bibliothecae Collegij Harvardini* (1723), p. 63.

that was forty foot; *Clusius*,[10] one that was fifteen Yards; *Oethæus*,[11] one that was eighteen; *Al. Camerarius*[12] one that was above Twenty; *Thaddæus Dunus*,[13] the Like; *Tulpius* one that was Forty.[14] Yea, *Borrichius*[15] relates, that a Patient of his voided eight hundred foot of the Jointed Worm in about a Year; but in several Peeces. And our *Tyson*[16] had a Patient, who voided so many Peeces of diverse Yards long, that being putt all together, would much exceed that of *Borrichius*.

An Instance of this *Worm*, I have not long since had in my Neighbourhood; which, because of my being too much Indisposed in my Health at the Time of its being exposed, either to take the exact *Measure* of it, or to bring it under the Examination of my *Microscope*; I requested a skilful & faithful Physician, to enquire into the Circumstances of it, and make me an exact Report. The Account with which he obliged me, you shall here have in his own Expressive Terms.

10 Tyson, p. 121. Mather refers to Carolus Clusius, or Charles de l'Écluse (1526–1609), physician, pioneering botanist, and commentator on *De Simplicibus Medicamentis ex Occidentali India Delatis* (1574), by the Spanish herbalist Nicolás Bautista Monardes (1512–88).

11 Mather, via Tyson, pp. 121–22, refers to Jakob Oethaeus (c. 1530–86), German physician and pharmacist of Eichstätt (Bavaria). His *Gründtlicher Bericht* (1574) is a handbook for physicians and nurses on the administration of nostrums for the sick.

12 Alexander Camerarius (1696–1736), German physician, botanist, professor of medicine, rector of the University of Tübingen, and member of the *Leopoldina Academia Naturae Curiosorum*, founded in 1652.

13 Tyson, p. 119. Mather refers to the Swiss physician of Zurich Thaddaeus Dunus (1523–1612), whose *Miscellanea Medica* (1592), cap. 15: *De Lumbrico Lato mirae longitudinis*, pp. 155–57, is the specific reference in question.

14 Mather alludes to *Observationes Medicae*. Editio Nova (1652), lib. 2, cap. 42: "*Caput Lati Lumbrici*," pp. 170–73, by Nicolaes Tulpius, aka. Tulp (1593–1674), a Dutch surgeon, anatomist, and mayor of Amsterdam. A depiction of Tulpius' tapeworm appears on p. 171. Perhaps most famous is Rembrandt van Rijn's painting *The Anatomy Lesson of Dr. Tulp* (1632).

15 The document in question is "*De Lumbricis Latis, & Cucurbitinis*," by Ole Borch (1626–90), aka. Olaus Borrichius, or Olaus Borrichus, a Danish physician and scientist, published in *Acta Medica & Philosophica Hafniensia Anni 1673* (1675), vol. 2, Observ. XXXXVII, pp. 148–49, by the renowned Danish physician Thomas Bartholinus (1616–80). *Acta Medica et Philosophica Hafniensia* was a short-lived Danish scientific journal (1673–80), founded by its editor Bartholinus.

16 Mather refers to Tyson's "*Lumbricus Latus*," in *PT* 13.146 (April 10 1683), pp. 113–41.

Sir,

In Obedience to your Commands I have viewed the prodigious *Worm*, that ha's fill'd the Town with Talk and Wonder. I must also confess my own Surprize at the uncouth Object; and I verily beleeve, that scarce any Man alive ha's ever seen its equal.

Mr. *Smith* of this Place,[17] bought a Negro Six Months ago, that came from the *West-Indies* but a few Months before; where he had lived Six Years. The Fellow was Tall, big-boned, but Thin, & Meagre; and from his first Coming, of a dull Aspect; heavy-breathed, & often Stomach-sick; and yett his Appetite very keen, to gratify what he would eat excessively. In this Condition, he could work at times, for diverse Months; yett the aforesaid Symptoms increased apace upon him, especially, his Sick Fitts, his voracious Appetite, & his Difficulty of Breathing: So that at last he was uncapable of Business, & wholly laid aside. Upon much Vomiting of yellow Choler, he was putt into the hands of a Doctor. The first thing he did, was to give an Emetick; but without any relief, it much increase'd the Symptomes, & seem'd much to his Prejudice. The Next Things given were *Anthelminticks*,[18] and chiefly *Mercurials*.[19] A few Days after which, growing suddenly Sick, & reaching to vomit, Behold, a Monstrous Worm broke Prison; thrust up one End of it self, and hung out of the Negro's Mouth. Immediately he siezed upon it, and with much Toyl pull'd it out of its Nest, much as Fishermen hale their Lines in catching their Fish. This happened on the *Lords-Day*, while the Family were at the publick Assembly, so that some Circumstances cannot be exactly related. After it was haled out, and received in a Pott, while it was turgid with Chyle,[20] and coyled up together, it composed an Heap of the Bigness of a Lads Head, in the Judgment of the Spectators. The Doctor was then called, who wash'd it and cleans'd it, & hung it out on a Fence to dry. But the Neighbours, that came flocking to see it at night, took away Peeces of it, unto the Quantity of Twenty foot at least, as was judged by the Marks left on the Fence where it hung; and the Head among the rest, if it had any: I conclude: For which I came the next Day, to take a Survey of it, I could

[17] According to the *Boston Weekly News-Letter*, no. 662 (January 7–14, 1717/18), which published the sensational news, this is Thomas Smith, a sawyer, who consulted "Sebastian Henry Swetzer a German" about the unnamed negro's condition. After ingesting "some Powders," the patient vomited a worm ("a hundred and twenty eight Foot"), almost as thick as "ones little Finger" with a head "like a Snakes" of a "whitish Colour all full of Joynts." See E.H. Christianson (1993), p. 73.

[18] A substance to destroy or eliminate parasitic worms in human intestinal helminths (tapeworm or any parasitic worm). See also Mather's essay in his medical handbook, *ABJ* (1972), p. 205.

[19] Drugs involving the highly toxic mercury as an emetic.

[20] A milky fluid consisting of fat droplets and lymph that drains from the lacteals of the small intestine into the lymphatic system during digestion.

find no Head, but the End where I look'd for it, appear'd unto me, as if a Part had been broken off; while the other End remained entire, & ending in a Small Point. What remain'd being measured before several Witnesses, extended unto the Surprizing Length of *One hundred and twenty Eight* foot.[21] This *Gigantic Worm*, was of the Substance & Colour, that is usual in *Jointed Worms*; of a strong & lothsome Smell, a little resembling the Eggs of Serpents.[22] The Body was Flatt, and Jointed from one End to the other; but the Bigness of the Parts, and the Distance between the Joints, remarkably various. At the End, which I supposed the Head, it was near the Bigness of a Goose-quill; and the Joints above a Third of an Inch asunder, and so it increased very gradually to the Belly, or Middle, which was near thrice that Dimension, as well as Distance of Joints; And so again it lessened to the End of the Tayl, much in the Same Proportion; only the End which I call the Tail, was five times the Smaller of the Two, and the Joints were scarce the Tenth Part of an Inch. – There have been various Accounts of these *Worms* in several Authors. But the most accurate is what the learned & curious Dr *Tyson* gives in the *Philosophical Transactions*;[23] which ours mightily agrees with, only ours, had not, that I could find, those *Orifices* in the *Joints*, which he found in his. Tis not without a singular Pleasure that I have this Opportunity, to entertain your curious Thoughts; being sensible of your Inquisitive Genius, which suffers nothing Remarkable to escape your extensive Survey, & being favoured with your Microscopes, to help me in my Examination. I make no Remarks, but leave them to your Judicious & Penetrating Observation; while, with the greatest Regards, I am,

Syr,

your most obedient Servant;

John Perkins.[24]

21 Mather alludes to the same account in *ABJ* (1972), p. 203 – except that the length of the tapeworm pulled from the man's throat was (as Mather there remembered it) "about one hundred and fifty foot long."

22 See also Increase Mather's *ERIP* (1684), ch. 9, p. 310; Nathaniel Hawthorne's Roderick Elliston may have had such a tapeworm in mind, in his short story "Egotism; Or, The Bosom Serpent" (1843). See S. Bush, Jr.'s intriguing article (1971).

23 See Tyson's "*Lumbricus Latus*," *PT* 13.146 (April 10, 1683): 113–41; his article "Lumbricus Teres, or some Anatomical Observations on the Round Worm bred in human bodies," *PT* 13.147 (May 10 1683): 154–59; as well as his "*Lumbricus Hydropicus;* or *An Essay to prove that* Hydatides *often met with in morbid Animal Bodies, are a Species of Worms, or Imperfect Animals*," *PT* 17.194 (September 30, 1691): 506–10.

24 John Perkins (1698–1781), a native of Lynn, Mass., Boston physician, and friend and neighbor of Cotton Mather (*D* 2:609, 611, 620), who shared his interest in smallpox inoculation but does not appear to have supported Mather's efforts to

This is the Account, with which my Ingenious Friend has complemented me. It is known, that *Worms* of another sort, many Yards long, are often found in the *Legs* of Men.[25] In how many other Parts of our *Bodies* have *Worms* been mett withal, you are well-apprised. But what *Verzascha*[26] relates of his finding them in the *Brains* of Men, deserve a more special Remark, & has occasion'd some Enquiries. I entertain you with none of my poor *Scholia* on these things, lest I incur the hazard of being too much in the Metaphor of Master *Verzaschas* Discovery. Wherefore, wishing you a Deliverance from all other Annoyers as well as from those we have been discoursing of, I subscribe,

[Boston, New England.] Syr,

12d. Xm. 1717/18. Yours with very great Sincerity.

{C. Mather.}

promote variolation during the smallpox crisis of 1721 (*D* 2: 627, 639, 664) – given that Zabdiel Boylston (1679–1766) was the only one to heed Mather's call for vaccination.

25 *Dracunculus medinensis*, or guinea-worms, a parasite transmitted in stagnant water. In his *BA* (2:922–23), Mather speculates that the "fiery serpents" that stung the murmuring Israelites in the Sinai desert were perhaps the quite commonly occurring guinea-worms. See also his *CP* (1721), ch. XXVII, pp. 157–58.

26 Bernard Verzascha (1629–80), Swiss physician and author of several books on the medicinal benefits of plants. In his *Observationum Medicarum Centuria Prima* (1677), "Observatio XXX," p. 81, Verzascha describes the "Convulsiones à lumbricis ortae."

SERIES FIVE

5.2. THE QUADRATURE OF THE CIRCLE.

{February 1720.}

To John Chamberlayne.

In February 1720, Mather's sent a bundle of twelve Curiosa letters (Fifth Series) to his former student and Harvard graduate Henry Newman (1670–1743), secretary of the SPCK, to be communicated to John Chamberlayne (c. 1668–1723), secretary of the Royal Society, having heard that John Woodward had died. (See Mather's Letter to Henry Newman, February 17 1720, MS draft, AAS [*SLCM*, pp. 301–2]*; and Newman's reply, September 17 1720, MS, AAS).*

John Chamberlayne acknowledged their safe arrival, allegedly forwarded them to Sir Isaac Newton and to Edmond Halley, and responded to Mather (August 3, 1720) with profuse praise:

> I thank you for your Noble Entertainment with which so many of my [Chamberlayne's] Friends were Regaled, before I could gett a Snap for myself, who hungred & thirsted for it, that I had not the pleasure thereof till very lately, and indeed too late to communicate the same pleasure to your Illustrious Brethren, the Gentlemen of the R.S. who have always a long Recess at this time of the year. As soon as I had perused your learned Letters, I Immediately communicated them unto Sr *Isaac Newton*, our Excellent Præsident; And he has referred them to the most Learned Mathematician and Great Traveller, Dr Halley, whose Report I expect with no small Impatience. But, as I beleeve, you expect to be informed of their Destiny, with respect to their coming to my hands, I was resolved not to slip the opportunity of telling you, They are safe arrived.[1]

1 See John Chamberlayne as quoted in Mather's Letter to John Winthrop, 26*d.* X*m*. [December] 1720, in *MP* (1868), p. 444, and Mather's follow-up letter to Winthrop (12$^{d.}$ 1$^{m.}$ [March] 1722), in *MP* (1868), p. 455. See also G.L. Kittredge, "Scientific Communications" (1916), p. 45 (43–44 *n.*).

In another epistle to John Winthrop (Boston, 12ᵈ 1.ᵐ [March] 1722), Mather expressed his frustration that even though Chamberlayne had communicated to Mather that his "XII Letters to him [Chamberlayne] have been published[,] And yett I have never to this Hour seen them: nor have they that have published I know not how many other things of mine, ever given the sight of them; (or else they have miscarried.) I have done, I have done, with Expectations from This World." *Curiously, none of the documents were ever published, and the holograph manuscripts – including Mather's draft copies were lost.*

Yet not all was lost. For instance, we get a glimpse of what was contained in at least one of Mather's lost essays, CA 5.2. "The Quadrature of the Circle." *And it comes from an unexpected source: An odd passage extant in William Douglass'* The Abuses and Scandals Of some late Pamphlets In Favour of Inoculation of the Small Pox (1722) *provides further evidence of their arrival in London. Douglass wrote,*

> Tho' unacquainted with the first Principles of the *Mathematicks,* [Mather] sends Home a Solution of **Two** the most abstruse Problems or *Desiderata* in Mathematicks, viz. the SQUARING OF THE CIRCLE, which is only to be expressed by Approximation, or an *infinite Æquation*, whose Nature is known, but whose Value cannot be reduc'd to any determined Lines or Numbers; the other is the LONGITUDE at Sea, the Discovery of which for many Years has been endeavoured by the best *Mathematicians*, encouraged by large *Præmiums*; and is generally thought will continue a Secret, till some Comet or other Cælestial Body appear within our *System,* and perform his Revolutions in a N. and S. Line, or some great *Circle* not much declining from a *Meridian,* as at present we find the *Latitude*, by the Suns Revolution in a Circle not much declining from the *Æquinoctial*, or E. & W. Line, or till an *Automaton* (Clock-work) can be contrived, which shall for some considerable Time move *æqually* or *æquably*, without being affected by the Weather or external Motion; or till we find some Contrivance, notwithstanding the Motion of the Ship, to take the *Altitudes* of the *fixed Stars*, and the *Eclipses of Jupiter's Moons*. A Man guilty of such *Absurdities*, is no good Voucher for an Experiment of Consequences.[2]

No doubt, Douglass' stinging ridicule found its target and served its purpose well enough; Mather must have been chagrined with puzzlement at how his antagonist had found out about the content of the missing documents. Be that as it may, Douglass' invective did serve an unintended purpose – in hindsight.

2 William Douglass, *Abuses and Scandals*, pp. 6–7.

His summary provides a glimpse of what Mather's lost "Curiosa" letter contained: The Quadrature of a Circle and Determining the Longitude at Sea.

Cotton Mather also tried his hand on discovering a way to establish a ship's longitude at sea. And why would he want to bother in the first place? The Longitude Act of 1714 and a chance at a monetary reward of £20,000 for anyone who devised a simple method of determining the longitude of a ship's location at sea within half a degree.

It is well known that before telegraphs, radios, and satellites the navigation of ships, especially when intercontinental seafaring became vital for trade in the sixteenth and seventeenth centuries, depended on a navigator's ability precisely to calculate the whereabouts of his vessel at sea within a degree of the globe's prime meridian east or west of a north-south axis. Doing so required accurate observations, which in the presence of incomplete astronomical tables depended on a fixed time scale at the port of departure and arrival, a ship's unwavering chronometer, a telescope, and knowledge of the position of fixed stars and lunar distance at sea. Easy enough? No doubt, determining a vessel's position relative to its port of embarkation and desired point of entry was fraught with surprising, often deadly, inaccuracies.

Did anyone win the award during Mather's lifetime? No. Not even the English carpenter turned clockmaker John Harrison (1693–1776), who designed the most accurate maritime chronometer over a period of half a century when he published The Principles of Mr. Harrison's Time-Keeper (1767). *He received only a portion of the grand prize.*

5.3. THE WHALE.

{February 1720.}

To John Chamberlayne.

{Sir,}[1]

Pliny mentions the *Balænæ* of the *Indian Sea*, which were nine hundred and sixty Foot long; and he mentions *Whales* that were six hundred Foot long, and three hundred and sixty broad, which came into a River of *Arabia*. In the

[1] No MS. extant. The likely content is reproduced from *CP* (1721), Essay XXIX, pp. 176–78; 2), *TC* (1717), pp. 26–28; and *BA* 6: Jonah 1:1, 17. The editors agree with G.L. Kittredge's argument ("Scientific," p. 46) that Mather's letter "'The Whale'" [CA 5.3] "is doubtless preserved in part in *CP* [1721], pp. 176–178." Winton U. Solberg, however, begs to differ but provides no counterevidence, in his *CPS* (1994), p. 360 *n.*, 188:50–189:58. A case in point: CA 7.4 "*Carcharias Mirabilis*" (June 5 1723), which Mather sent to James Jurin of the Royal Society, shows Mather's methodology of repurposing published and unpublished material: For instance, the first four paragraphs of 7.4 find their reincarnation (nearly verbatim) in his glosses on (*BA* 6 Jonah 1:1, 17). Likewise, portions of CA 1.1 "Describing *Biblia Americana*"; of CA 1.10 "Joseph Beacon's Vision or Dream"; and of CA 2.5 "The Pigeons," are repackaged and repurposed in, for instance, his *BA*, his *CP*, his medical handbook *ABJ* (1972), his treatise on eschatology *TP* (1995), even on witchcraft *WIW* (1693). More often than not, Mather reuses historical and scientific narratives from his published sources to maintain his audience's attention for lively delivery in his sermons. Nothing ever got lost – so it seems. On the topic of whales, Mather's erstwhile nemesis Paul Dudley also submitted a discourse to the Royal Society (roughly at the same time as Mather sent his). Dudley's was published as "An Essay upon the History of Whales, with a particular Account of the Ambergris found in the Sperma Ceti Whale," *PT* 33.387 (April 30, 1725): 256–60. Mather's was lost – although John Chamberlayne, secretary of the Society, did acknowledge receipt of Mather's twelve "Curiosa" letters (Fifth Series), which included "The Whale" and apparently assured Mather that his "XII Letters to him [Chamberlayne] have been published. And yett I [Mather] have never to this Hour seen them" (*MP* [1868] 444, 455). See also G.L. Kittredge, "Election," pp. 94 and esp. 105 *n.* 8 (qt. appears on p. 106); and Kittredge, "Lost Works" (1912), pp. 440–41, 463 *n.* 2. On Dudley's essay, see Stearns, *Science* (1970), pp. 466–67.

Second Chapter of his ninth Book he offers a Reason why the *largest Animals* are bred in the Sea.[2]

But I love to pass from him to a more trusty and modern *Pliny*, our industrious *Ray*; and we will now see something of his Remarks upon these *Belluæ Marinæ*:[3] The *Tail* in these has a different position from what it has in all other *Fishes*; it lies parallel to the Horizon in these, and it is perpendicular in the rest; hereby it supplies the use of the hinder pair of *Fins*, which these Creatures frequently lack; and it serves both to raise and sink their Body at their pleasure. It is necessary that these Creatures frequently ascend to the top of the Water to *breathe*, and therefore they should be furnished with an Organ, by which their ascent and descent might be facilitated. The turning of their Bodies in the Water they perform like the *Birds*, by the motion of one of their *Fins*, while the other is quiescent. It is very remarkable that their whole Body is compass'd round with a copious *Fat*, which we call *the Blubber*, whereby their Bodies are poised, and rendred equiponderant to the Water, and the Water also is kept off at some distance from the *Blood*, the immediate Contact whereof might else have had some chilling force upon it; it serves likewise, as our Clothes do for us, to keep the *Fish warm*, in reflecting the hot Steams of their Body, and so redoubling the Heat thereof: hence they can abide the greatest Cold of the *Northern Seas*, to which they chiefly resort, not only for the Quiet which they enjoy there, but because the *Northern Air*, which is more fully charged with the Particles which we suppose to be *nitrous*, and that are the Aliment of Fire, is fittest of all to maintain their vital Heat in that Activity, which may be sufficient for to move such an unwieldy Bulk as theirs. The *stupendous Magnitude* of these Animals! Thou *Antitype*[4] thereof, among the *Poets* which adorn our Age, describe them to us.

2 Pliny, *Historia Naturalis* (9.2.4). Mather translates Pliny's Roman units of measurement into English measurements: "*ex quibus balaenae quaternum iugerum, pristes ducenum cubitorum, quippe ubi locustae quaterna cubita impleant*" ("among them whales covering three acres each, and sharks 100 ells long; in fact in those regions lobsters grow to 6 ft. long"). "The *iugur* was about two-thirds of an English acre, the *cubitus* or ell about 1 ½ ft" (Pliny, *Natural History* 9.2.4, p. 166 *n.*).

3 "Wild beasts of the sea." Based on the detailed notes of Francis Willughby (1635–72), an English ornithologist and ichthyologist, his friend John Ray, F.R.S. (1627–1705), an influential English naturalist philosopher, published *Historia Piscium libri quatuor* (1686) after Willughby's death. Mather refers to lib. 2, caps 4–5, pp. 35–40 of this work. Both Willughby and Ray are greatly indebted to Guillaume Rondelet's *Libri de Piscibus Marinis* (1554), a popular, richly illustrated work which went through several editions and reprints. See G. Lewis (2012).

4 In biblical typology, Jonah's three-day interment in the belly of the whale serves as the OT type of Jesus Christ's burial in the tomb, the antitype.

While the vast Whale takes in the Deep his place,
Prince of the Waters and the finny Race;
Rolling in Sport, the Billows he removes,
And, like a floating Isle, the Ocean shoves:
Now in his weedy Court he lies at ease,
Now spouts against the Skies exhausted Seas.[5]

And yet one says very well concerning him; he is *minima quaedam Operum Dei, particula ac velut mica.*[6]

Let what I gave you of the *nine hundred and sixty Foot* pass for a *Plinyism*;[7] and so what *Basil* in his *Hexameron* reports of *Whales* equal in bigness to the greatest Mountains,[8] let the Censure of *Brierwood* pass upon it, as *an intolerable Hyperbole:*[9] We will write more sober things. Passing by what *Ælian* affirms of the *Whale* being five times beyond the largest *Elephant*, we find *Rondeletius* assigning him sometimes *thirty-six Cubits* of length, and *eight* of height.[10] *Dion* is a grave as well as an old Writer, and he reports a

[5] Lines extracted from the epic poem *Prince Arthur. An Heroic Poem. In Ten Books* (1697; 4th ed., 1714), bk II, p. 44 (*bottom*), by Sir Richard Blackmore (1654–1729), an English poet-physician whom Mather greatly admired and frequently cited (D 2:105). Blackmore was the personal physician of William III and Mary.

[6] "It is only a small particle of God's work and a mote, as it were." Mather here cites from *Theologia Naturalis Exhibens Augustissimam Naturæ Scholiam* (1615), pars II, cap. XXVII, sec. 11, p. 530, by the German Calvinist clergyman and polymath Johann Heinrich Alsted (1588–1638), who has been called the "father of all encyclopedias." His four-volume *Scientiarum Omnium Encyclopædia* (1630, 1649) became a standard work of encyclopedism.

[7] The OED identifies Mather's *WIW* (1693), pp. 57 [58], 142; and *MCA* (1702) 3.2.19.122 as the first occurrence of the noun "Plinyism," signifying "A statement or account of dubious correctness or accuracy."

[8] Mather's primary source for this and the following three references is *Enquiries Touching the Diversity of Languages, and Religions* (1614), ch. 13, p. 109, by Edward Brerewood (c. 1565–1613), an English antiquarian and professor of astronomy at London's Gresham College. St. Basil the Great (331–79), archbishop of Caesarea, was the author of *Hexaemeron*, a collection of nine homilies on the cosmogony of the creation account in Genesis. St. Basil insists that "'God created great whales' [Gen. 1:21]. Scripture gives them the name of 'great' not because they are greater than a shrimp and a sprat, but because the size of their bodies equals that of the largest mountains. Thus when they swim on the surface of the waters one often sees them appear like islands" (*Hexaemeron* 7.6 [*PG* 029. 0162*bc*] *NPNFii* 8:94).

[9] Brerewood derides the "intolerably hyperbolical" Talmud indulged in by St. Basil in likening whales as "equal to the greatest mountains, & their backs when they shew above the water, like to Islands" *Enquiries* (1614), p. 109.

[10] Brerewood, *Enquiries*, p. 109; Aelian, *On the Characteristics of Animals* (16.12); Regius professor of medicine and chancellor of the University of Montpellier,

Whale coming to Land out of the *German Ocean* sixty Foot in length, twenty in breadth.[11] But *Gesner*,[12] a later, affirms a *Whale* to have landed near our *Tinmouth-Haven*, in the Year 1532, which was *ninety Foot* in length, and the breadth of his Mouth six Yards and an half, and his Belly of such a compass, that one standing on the Fish, and slipping into his Belly, very narrowly escap'd being drown'd there.

But then, if we may take *Hartenius* for a Voucher, among the *twenty* several Kinds of *Whales* by him enumerated, he reckons one sort that is *thirty Ells* long, and hath more than *seventy Teeth*, so large as to make Handles of Knives and other Instruments. He reckons another sort that is *forty Ells* long, and overwhelms *eighty Ells* long, and some of *ninety*.[13]

All these proclaiming the *Grandeur* of their *Great Creator!*

Even in the *cold Sea* too, what a *Warmth* of Parental Affection do the *old ones* express for their *young ones*, and how distinguishing! When the *Seals* are

---

Gulielmus Rondeletius, aka. Guillaume Rondelet (1507–66), published his oft-quoted *Libri Piscibus Marinis* (1554). Mather refers to lib. 16, cap. 11 "Balaena vulgo dicta sive de Musculo," pp. 475–82; ref. on p. 475. See also Petrus Gyllius, aka. Pierre Gilles (1490–1550), *Descriptio Nova Elephanti* (1614), cap. 6, pp. 20–24; and the breviary *Ex Aeliani Historia per Petrum Gyllium Latini Facti* (1533), libri XVI, cap. XCIX, pp. 585–86.

11 Mather's muse is George Hakewill (1578–1649), *An Apologie of the Power and Providence of God* (1627), lib. 2, cap. 8, sec. 2, p. 127, by the Anglican clergyman and rector of Exeter College (Oxford), whose work allows Mather to directly access both Dio Cassius and Gesner. The Greek historian and senator Dio Cassius (*Historiarum Romanarum* 54.21) reports that as the Romans sailed down the Rhine in old Germania, a "sea-monster" some "twenty feet broad and three times as long, resembling a woman except for its head" had come from the North Sea and "become stranded on the shore."

12 Conrad Gesner (1516–65), Swiss naturalist, physician, and bibliographer whose *Historia Animalium* (1554) is a major achievement of Renaissance scientific knowledge. Relying on Hakewill, Mather refers to what appears to be the spermaceti whale which beached (1532) near Tynemouth Haven near Newcastle, mentioned in Gesner's *Historia Animalium* (1558), lib. IV, sec. "De Ceto Britannico," p. 251. Gesner's account is from "A Transcript from a letter to Polydore Virgil, from the Town of Tynemouth in the Northern Parts of Britain."

13 Hartenius (unidentified), apparently a German historian and author of *Chronica Bremensia*, a manuscript chronicle of the Hanseatic city of Bremen, identifies twenty different types of whales. The "Buhrhwal," among them, fits the description: it is "triginta ulnarum, habens multos dentes tam magnos, ut inde fiant tabulae ac manubria cultellorum, habet ultra septuaginta dentes" ("of 30 ells, having many teeth so large that boards or knife-handles are made from them; it has more than 70 teeth"). The Latin quote is from *Ichtyologia et Nomenclaturae Animalium Marinorum* (1624), pp. 27, 28, by Stephanus à Schonevelde, aka. Stephan von Schonefeld (fl. early 17th century), a physician of Hamburg and compiler of his vademecum.

hundreds of thousands of them lying in a Bay coming out of the Sea, they bleat like Sheep for their Young; and tho they pass thro hundreds, yea, thousands of other young ones before they come to *their own*, yet they will suffer none but *their own* to suck them. *Even the Sea-Monsters draw out the Breasts, they give suck to their young ones. Monstrous Parents*, that are *without natural Affection!*[14] These Inhabitants of the *Sea* with open Mouth cry out against you.

Some[15] have disputed, Whether it were a *Proper Whale*, that swallow'd the Fugitive *Jonas*.[16] They pretend, That proper *Whales*, tho' of a Dimension so great, that their Tongue alone has afforded near Twenty Tuns of Oyl, yet have a *Swallow* so narrow, that the Single Arm or Leg of a Man could scarce go through it. They propose, that *Jonah*'s Fish might rather be a *Shark*; [the *Carcharias:*][17] which is found sometimes Twenty foot long, and has a Throat large enough; with a Maw capable of Lodging a Man. The *Statehouse* at *Breme*, has the Picture of this Fish devouring *Jonah*.[18] This is a *Fish*, which is by some annumerated unto the *Whale-kind*; and the *Mediterranean Sea*, is not so much a Stranger to it, as it is to that which more usually bears the Denomination. 'Tis the *Fish* which the *Grecians* make to seize upon their *Hercules*, in their Story, which doubtless was a *Phœnician* Disguise of what befell our *Jonas*.[19] To pass by, what is remark'd by *Rondeletius*, 'tis affirm'd by *Gillius*, that some of these *Fish* have been taken, that Weigh'd four thousand Pound, and had armed Men in their Stomachs. The Great *Bochart* inclines to this Opinion.[20] But others, from the Words of our Saviour insist upon it, that a *Proper Whale* must be the

14 Lam. 4:3; Rom. 1:31.

15 The following paragraph is reconstructed from *TC* (1717), pp. 26–28. Mather's target audience is Boston's whalers who had safely returned from a bountiful whaling season.

16 Jonah 1:17.

17 See Mather's "Curiosa" letter to James Jurin, our 7.4 "*Carcharias Mirabilis*."

18 See below.

19 Bochart, *Hierozoicon* (1663), pars 2, lib. 5, col.743, lines 50ff, cites Lycophron and others that Heracles (like Jonah) was swallowed by a dogfish "canis carcharias" sent by Neptune, and subsequently vomited out. Says his interpreter Theophilus Gale, *The Court of the Gentiles* (1672), pt I, bk 3, ch. 8, p. 86 (§4), Jonah was not swallowed by a whale but "a dog fish, called *Carcharias*. Therefore, in the Grecian fables, *Hercules*, is said to have bin swallowed up of a *dog*, and to have layen three dayes in his *Intrals*. Which fable sprang from the sacred historie, touching *Jonah*, the Hebrew Prophet, as 'tis evident to al." See Lycophron's *Alexandra* (lines 31–34), below.

20 Gulielmus Rondeletius (Guillaume Rondelet, 1507–66), *Libri de Piscibus Marinis* (1554), lib. 13, cap. 12, p. 392. Petrus Gyllius, *Ex Aeliani Historia per Petrum*

Swallower of the Fugitive. And suppose his Throat never so narrow, our Dr. *Edwards*, pleads that there is no necessity of the Poor Mans descent into the Maw of the *Fish*.[21] For, in the Hebrew, the Term of *Belly*, signifies all the *Inward* and *Unseen* parts of any thing; and in the *Greek*, any *Deep, Hollow, Middle* Place. In short, It notes any *Internal Receptacle*. Now, tho' a *Whale* has a small *Throat*, yet he has a vast *Mouth*; and what if *Jonah* should be lodg'd in some Cavity of *that*? And yet, While they make such a Difficulty about the *Swallow* of the *Proper Whale*, there comes the Ingenious *Leguat*,[22] with Demonstrations, that they are mistaken about it; and that the monster sometimes has a Prodigious Throat; and that there were Eye-witnesses to very terrifying Instances of it. And then *Fournier* in his *Hydrography* affirms,[23] Two Men to have been found in the Maw of a *Proper Whale*, that was thrown ashore on the Coast of *Spain*. But this is what I did not propose to dwell upon; I dismiss it, for the Reflection which I did propose it as an Introduction to.

What then was the *Great Fish*, that swallowed up *Jonas*?[24] Some will tell you, Not a proper *Whale* sure;[25] and indeed, when our Lord sais, in Math. 12.40. *Jonas was in the Whales Belly*; the Name there is what other Fishes of the greater Sort may challenge. *Whales* visit not the Sea, where *Jonah*'s Disaster

Gyllium Latini Facti (1533), cap. XCIX: "De Carcharia pisce," p. 586. Samuel Bochart, *Hierozoicon* (1663), pars 2, lib. 5, cap. 12, col. 743, lines 20–44.

21 John Edwards (1637–1716), Anglican bishop of Lincoln, and Mather's much-admired friend and correspondent. Here, the reference is to Edwards' *A Discourse Concerning the Authority, Stile, and Perfection of the Books of the Old and New Testaments*. 3 vols. (1693), vol. 2, ch. 8, pp. 275, 279–80, 281–82, 284.

22 A French naturalist Frances Leguat (1637–1735), whose *Voyage et Avantures de François Leguat* (1708) was translated and published as *A New Voyage to the East-Indies* (1708). He describes his encounter with a pod of whales (pp. 22–24) and affirms with reference to George Fournier's *Hydrography* (1667) at his side that "two Men were found in the Belly of a Whale" on the shores of Spain (p. 23).

23 The French Jesuit priest Pere Georges Fournier (1595–1652) authored *Hydrographie. Contenant la Theorie et la Practique de toutes les Parties de la Navigation* (1667), liv. 4, ch. 28, p. 183: "si grand que sept hommes y pouvoient enterer, & un homme à cheval se tenir dans sa gueule; on trouva deux hommes morts dans son ventre." During the reign of Phillip II of Spain (1556–98), a monstrous whale was found stranded near Valence. Its head was so large that seven men could enter it, and a man on horseback could stand in her mouth.

24 Reconstructed from *BA* 6: Jonah 1:17. Ava Chamberlain, volume editor (*BA* 6: Lamentations to Malachi), kindly shared her transcriptions of and annotations on the book of Jonah.

25 Mather adapted this paragraph from Thomas Bartholin's chapter, "De pisce, in quo sepultus Jonas," in *De Morbis Biblicis Miscellanea Medica* (2nd ed., 1672), 56–57.

Figure 14. "Großes Walbild" (1669). Franz Wulfhagen (1624–70). 3.55 m x 9.55 m. Town Hall of Bremen. Northern Minke whale (Balaenoptera zcutorostrata). Courtesy of Konrad Elmshäuser, Hans-Christoph Hoffman, Hans-Joachim Manske Bremer Rathaus, Bremen.

befel him; and even *Whales* of so vast a Dimension, that their *Tongue* alone ha's afforded eighteen Tun of Oyl, yett have a *Swallow* so narrow, that the single Arm or Leg of a Man, could scarce go thorough it. It is most likely, that *Jonahs* Fish was, *A Shark*; the *Canis Carcharia* which is found sometimes Twenty foot long; & ha's a most patent Throat, and an *Oesophagus* capable of lodging a whole Man therein. I have the Authority of *Breme* for this Exposition; for in their *State-House*, you shall see painted, This very Fish devouring of poor *Jonah*, with open Mouth.[26] *Pliny* also, & before him, *Theophrastus*, tell us, that the Sea's there-away were well stock'd with these Dangerous Animals.[27]

But what saies your Friend *Bochart*, upon *Jonah*'s Whale?[28] Why, He saies, That a *Whale* has but a *little Swallow*; And whereas *Dionysius Periegetes*, and *Festus Avienus*, talk of whole Vessels devoured by a *Whale*,[29] and *Priscianus* has Fabulized,

[26] Bartholin, *De Morbis Biblicis* (1672), p. 56. In 1669 a twenty-nine-foot whale beached on the shore of the Lesum, a tidal river in northern Germany. Franz Wulfhagen (c. 1624–70), German Baroque painter, a pupil of Rembrandt residing in Bremen, was commissioned to paint a life-size portrait of a beached whale, and the painting was hung in the Bremen City Hall ("Obere Halle"), where it still remains on display. See J. Pätzold (2021), p. 74.

[27] Bartholin, *De Morbis Biblicis* (1672), pp. 56–57. Pliny, *Natural History* (13.50); Theophrastus, *Enquiry into Plants* (4.7).

[28] Mather adapts the following entries from Samuel Bochart, *Hierozoicon* (1663), pars 2, lib. 5, cap. 12, "De Ceto Jonæ," cols 742–46.

[29] Dionysius Periegetes, *Orbis Descriptio* (lines 603–04), in *Geographi Græci Minores* (1882) 2:142. The 4th-century CE author Rufus Festus Avienus translated

Quippe solent pariter Navem sorbere Virosque;[30]

he chastises their Fables; adding that *Homer* himself, (tho' he be, *Mendaciorum Pater*) was more modest than these Men, when his Monster *Scylla*, was to be described.[31] *Vincentius* is rather to be credited, who saies, *Cetus habet Oris Meatus strictos, undè non-nisi parvos Pisciculos deglutit*.[32]

Yett he thinks, that *Jonah*'s Fish, was one of the *cetaceous* kind: And particularly, the *Carcharias*, or *Shark*, whereof *Ælian* saies, That some of them are of such a *Bigness*, as to be Numbred, Κητων εν τοις αλκιμωτατοις *cum Cetis robustissimis*.[33] As the *Red-Sea* was well-stock'd with this Animal, so was the *Mediterranean*. They are called, *Lamiæ*, or *Lamnæ*, απο του εχειν μεγαν λαιμον, from the *Wideness* of their *Throats*; which *Oppian* therefore mentions as terrible.[34] *Rondeletius* affirms, That he saw a *small one*, which yett had so wide a Throat, that a *fatt Man* might easily be swallowed up into it.[35] But *Petrus Gillius* tells of those taken (and he tells you where,)

this 2nd-century work into Latin under the title, *Descriptio Orbis Terræ*; the corresponding passage is located in *Geographi Græci Minores* (2:184, lines 792–93).

30 Mather's hexameter line reads, "Truly, they are accustomed to swallow down a ship and its men together." The early 6th-century Latin grammarian Priscian (Priscianus Caesariensis) also produced a translation of Dionysius's geography. The cited passage appears in *Periegesis* (line 602), in *Geographi Græci Minores* (2:195).

31 Bochart, *Hierozoicon* (1663), pars 2, lib. 5, cap. 12, col. 742. Homer claimed that the sea monster, Scylla, was capable of swallowing "dolphins and sea dogs and whatever greater beast she may happen to catch," including sailors. *Odyssey* (12.95). Herodotus, not Homer, is commonly called "the father of lies," following the Spanish Renaissance scholar Juan Luis Vives (1493–1540), who may first have used the phrase "mendaciorum patrem" in reference to the Greek historian, in *De Disciplinis libri XX* (1551), tom. 1, lib. 2, p. 87. But the term was also applied to Homer, who was similarly accused of representing fabulous stories as accurate history. And Vives mentions Cicero, Quintilian, Juvenal, who damned the Greek historians as mendacious. See also John Twyne, *De Rebus Albionicis* (1590), lib. 1, p. 13.

32 Bochart, *Hierozoicon*, pars 2, lib. 5, cap. 12, col. 743 (lines 1–4). "The whale has a narrow passage in its mouth through which it swallows nothing but little fish." Via Bochart, Mather's second-hand quote is from the Dominican friar Vincentius Bellovacensis (Vincent of Beauvais, c. 1184–1264), *Bibliotheca Mundi. Seu Speculum Maioris* (1624), tom. 2, *Speculum Doctrinalis*, lib. 15, cap. 128, col. 1462.

33 "Among the most powerful sea monsters," Ælian, *Characteristics of Animals* (1.55).

34 Suda (Suidas), *Suidae Lexicon, Graece & Latine. Tribus Boluminis* (1705), vol. 2, pp. 412–13 (*voce* λάμια). See Oppian, *Halieutica, or Fishing* (1.370).

35 Gulielmus Rondeletius, *Libri de Piscibus Marinis* (1554), lib. 13, cap. 12, p. 392.

that weigh'd Four Thousand Pounds, & had Armed Men in their Mawes.[36] *Archestratus* calls this Monster θηριον ανθρωποφαγον·[37]

The old Pagans had a Tradition of their *Hercules*, That καρχαρος κυων· a *Shark* sent by *Neptune*, swallow'd him up. You have this Tradition in *Lycophron*.[38] And the old Scholiast upon him adds, That he lay *Three Dayes* in the *Sharks* Belly; coming out at Length with the Loss of his Hair; and that he had the Name of Τριεσπερος for the *Three Dark Dayes* which he underwent in this Condition.[39] *Phavorinus* takes Notice of this very thing, at his Mention of the Name Τριεσπερος: And *Cyril* observes, That the *Greeks* had a Story of *Hercules*, not unlike the story of *Jonas*; and so does *Theophylact*.[40] *Bochart* thinks, That the *Phœnicians* carrying the Story of *Jonas* to the *Græcians*, they

36 Petrus Gyllius (Pierre Gilles, *Liber Summarius de Gallicis et Latinis Nominibus Piscium Massiliensium, libri XVI*, cap. 99; Appendix to *Ex Aeliani Historia per Petrum Gyllium Latini Facti* (1533), p. 586.

37 Bochart, *Hierozoicon*, pars 2, lib. 5, cap. 12, col. 743 (lines 43–44). The ancient Greek poet, Archestratos of Gela (4th century BCE) wrote *Hedupatheia*, or "Life of Luxury," a humorous poem exploring foods and eating habits in the Greek world. Fragments of this work were preserved in the *Deipnosophistae* ("The Learned Banqueters") composed half a millennium later by Athenaeus of Naucratis. Isaac Casaubon produced an edition of Athenaeus's work in the 16th century, which became the standard text until the modern period. For the passage referenced by Mather, see Isaac Casaubon, *Athenæi Deipnosophistarum* (1657), lib. 4, p. 163d; lib. 7, p. 310e. The Greek words mean "man-eating beast."

38 Bochart cites the Greek poet-scholar Lycophron of Chalcis (3rd century BCE), in his *Alexandra* (34–40), the story of the Trojan War, in which Cassandra's prophecies foretell the Trojan War, as reported by a slave to her father Priam. The Greek passage signifies "jagged-toothed dog," implying the dogfish-shark.

39 Bochart, *Hierozoicon*, pars 2, col. 743. According to Lycophron, *Alexandra* (line 33), Heracles was begotten "in a triple night"; Lycophoron (lines 33–40, in Callimachus, pp. 496–97) tells one version of Heracles' tale, according to which "old Triton's hound of jagged teeth devoured [Heracles] with his jaws. But he [Heracles], a living carver of the monster's liver, seething in steam of a cauldron on a flameless hearth, shed to the ground the bristles of his [hound's] head." In short, Heracles killed Poseidon's hound from the inside. According to another version told by Theophilus Gale, Heracles/Hercules was swallowed by a dogfish (carcharias)—thus merging the story of Jonah and the whale with that of Heracles, in Gale's *Court of the Gentiles* (1672), bk 3, ch. 9, p. 86.

40 Bochart, pars 2, lib. 5, cap. 12, cols 743–44. *Dictionarium Varini Phavorini* (1523, 1538), compiled by the Italian lexicographer and librarian Varino Favorino Camerte (1445–1537); Cyril of Alexandria, *Commentarius in Jonam prophetam*, cap. 2, vers. 1 (1:17 KJV) [*PG* 71. 615–16]. Theophylact of Ochrida in modern Bulgaria (c. 1050–c.1109), *Expositio in Prophetam Jonam*, cap. 2, vers.1 (1:17 KJV) [*PG* 126. 931].

chose to apply it unto their *Hercules* rather than another, *Quòd videbatur esse summi roboris argumentum*.[41]

And he then confutes the insinuation of *Josephus*, That *Jonas* was cast ashore upon the Shore of the *Euxine* Sea; proving by many Authors, That no such *Fishes*, ever go thereabouts; and in the *Syriac Sea*, there is *Alexandretta*, as near to *Ninive*, as any Place on the *Euxine Sea*.[42] But it betrayes a gross Ignorance of *Geography*, in *Sulpicius*, to say, That the Fish, cast *Jonah* forth upon the Shore of *Ninive*; as if in Three Dayes time, the *Fish* could swim round all the vast Regions of *Africa*, & no small Part of *Asia*, and arrive to the Banks of the River *Tigris*: *Talos* himself could not have performed such a Task![43]

Jonahs being *Three Dayes* and *Three Nights* in the Belly of the Fish, what is there, that appears Remarkable in this Portion of *Time*? Besides what there was herein, *Typical* of our Lord Jesus Christ, there was in it a Remarkable Correspondence between the Transgression & Chastisement of the Prophet. It is noted of *Ninive*, in Joh. 3.3. *It was a City of Three {Dayes} Journey*.[44] *Jonah* refusing to spend *three {Dayes}* in an evangelical Perambulation of that {Length} must therefore undergo *Three Dayes* of an horrible Imprisonment. *Jonah* that would {not} spend *Three Dayes* in *Preaching* among *Men* {on} *Earth*, must spend *Three Dayes* in Silence & {thus?} among *Fishes* in the *Sea*.

But then, lett a modern Traveller, come in, with some Remarks upon the Matter. The Ingenious *Francis Leguat*, in his Travels, on the Occasion of their Meeting with some *Whales*, bestowes a Smile on the Fright which *Aloysio Cadamusto* discovers in his History on such an Occasion. And he saies, their Seamen laugh'd at *Patricius Vartomanni*'s Relation of *Whales* that piss'd as high as the Clouds.[45] He then proceeds;

41 Bochart, pars 2, col. 744. "which seems to be an argument of the greatest force."

42 Bochart, *Hierozoicon* (1663), pars 2, cols 744–45. Josephus reports that the whale vomited Jonah onto the shore of the Euxine, or Black Sea, in *Antiquities* (9.10.3).

43 Bochart, pars 2, col. 745. Sulpicius Severus claims that Jonah was disgorged near Nineveh. *Historia Sacra* (1.48) *NPNFii* 11:94 [*PL* 020. 124*BD*]. In Greek mythology Talos, a giant man made from bronze, performed a variety of super-human feats to protect the island of Crete, including circling the island's perimeter three times a day.

44 See Mather's annotations in *BA* 6: Nahum 1:1 and Zeph. 2:13.

45 Mather draws this entry from the travel narrative of French naturalist Frances Leguat (1637–1735), *A New Voyage to the East-Indies* (1708), pp. 22–24. Leguat describes his encounter with a pod of whales; the seamen's laughter about the whale's exhaling water from their spout is mentioned on p. 23. The Venetian

But if they had Read *Pliny* and *Solinus*, venerable for their Antiquity, with their *Whales* Nine hundred & sixty foot long, their Inclination to *laugh*, had been changed into a mortal *Fright*, for fear of being all swallowed.[46] For tho' those that are commonly called *Naturalists*, have for a long time used to affirm, that *Jonas* could not be swallow'd by a *Whale*, because the Throat of those Animals is so straight, that hardly a *Pilcher* can go down it; yett every body does not believe what they say, as a Matter beyond all Dispute. There are few Men, who have themselves Dissected *Whales*, and have seen with their own Eyes, what sort of Throats they have; Besides, we must consider, there are several different Kinds of these marine Monsters. And as I cannot refuse believing P. *George Fournier*, a very curious Man, and very knowing in every thing which belongs to the Sea, who assures us in his *Hydrography*, that Two Men were found in the *Belly of a Whale*, that was thrown ashore at *Valentia*, on the Coast of *Spain*, a Jaw of which is præserved in the *Escurial*;[47] so I am satisfied, a *Whale* of the Size that *Solinus* speaks of, would have swallow'd us up as easily as they do *Sack & Loaches* in *England*. The *Animal*, that in all Languages is called, *A Whale*, is the biggest of all Fish, and even of all Animals. And from thence Mr. *Bochart* writes, that the Word, *Balœne*, or, *Whale*, is a Syriac Word, which signifies, *Lord of Fishes*.[48]

{Sir,

a most humble

& obedient Servant.

C. Mather.}

Aloysio Cadamusto (Alvise Cadamosto, c. 1432–88) and Ludovico Patritii Romani di Varthema (1470–1517), aka. Luigi Varthema, aka. Vartomannus, were Italian explorers who wrote important accounts of their voyages. Both *Cadamusto's Navigatio ad Terras Ignotas* and Varthema's *Navigationis Aethiopiae, Aegypti utriusque* are included in *Novus Orbis Regionum ac insularum vereribus incognitarum* (1532), lib. 1, cap. 50, pp. 1–89; 189–297, ed. Simon Grynaeus. Varthema describes a terrifying encounter with some enormous fish that were as large as houses (lib. 7, cap. 4, p. 295).

46 Pliny (*Natural History* 9.2) claims that whales in the Indian Ocean grow so large they span "three acres." Solinus (*Polyhistor*, §52.42), following Pliny, notes that the whales in the Indian Ocean are "ultra spatia quatuor iugerum" ("bigger than four acres in size"), in *Collectanea Rerum Memorabilium* (1895), p. 191. Pliny (*Natural History* 9.5) also describes whales "of an enormous mass of flesh with savage teeth" hunting for prey in the Mediterranean.

47 Leguat, *New Voyage* (1708), p. 23. The French Jesuit priest Pere Georges Fournier (1595–1652) authored *Hydrographie* (1667), liv. 4, ch. 28, p. 183. The jawbone is preserved in the Spanish royal palace El Escorial near Madrid.

48 "Piscium dominus." Bochart, *Geographia Sacra* (1692), pars post. Chanaan, lib. 1, cap. 1, col. 347 (lines 3–5).

5.9. A WATER-DOVE.

{February 1720.}

To John Chamberlayne.[1]

{Sir,}

We are annually visited in our Countrey of *New England*, at the Beginning of Winter, with a *Water-Dove*, that stays for a little while on our Shore near the Sea, and then goes off again to the Sea, without Returning any more, till the Return of the Year.[2] Tis one of the loveliest Birds that fly; and a *Dove* is every thing, except what qualifies it for the Element it is designed for. The *Cock* and *Hen* are paired, as our *Doves* use to be. They are of a Deep Azure-Colour; spotted and streaked with White. The Cock has a fine Topple Crown, & feathers of a Deep Yellow under his Wings. The Hen is blue, & full of white Spotts all over her Body. Their Noise is a little superiour to that of our Doves on the Land. Their Cooings differ a little from these; about as much as the Voice of a Wild Goose from that of a Tame One. Their soft, yett sprightly Murmurs, are more like the gentle Cadence of Waters, than the melancholy Notes of our common Turtle.[3] They differ from the Doves of the Land, in this; that their Bill is a little flatted: and they have the Membranes between their Toes, wherewith Water-fowl use to be accommodated. When they swim about our shores, they often come upon the Land and they are very frequently seen to have in their Bills, a Small-Branch of our long-lived Cedar, which is green with us all the Winter, & growes on our

1 Kittredge, "Scientific," p. 45, gives this date of 1720.

2 In his curious commentary on Jer. 8:7 ("The Stork in Heaven, knowing her Appointed Times, & the Turtle, and the Crane…") in BA 5:879–83, Mather posits that certain migratory birds like the stork and the crane might fly off to the moon or to certain planetoids between the earth and the moon, where they winter until spring. Mather's source is *An Essay Toward the Probable Solution of this Question: Whence come the Stork and the Turtle, the Crane, and the Swallow, when they Know and Observe the appointed Time of their Coming* (1703), pp. 36–39, by Harvard's own Charles Morton (1627–98). No doubt, Bishop Francis Godwin's (1562–1633) *The Man in the Moone: or A Discourse of a Voyage thither by Domingo Gonsales* (1638) had something to do with this extraterrestrial solution. See M. Dopffel (2010).

3 I.e, turtledove (Streptopelia turtur).

Beaches as well as on our Mountains. They are as delicate a Dish, as we can be Regaled withal; Sweet and Fatt; esteem'd a Royalty among the Natives. Our Indians call them *Coecoe-on-Sogelande*; which is as much as to say, *The Messenger of the Great Rain*. This is a little surprizing. And my [Ingenious] Friend. [Mr. *John Winthrop*,][4] from whom I have this Communication, together with the agreeable Sight of the Bird, concludes with this Remark, *That be sure, they are one sort of Dove, that found a Shelter, in the Ark; and it is Rational to suppose, that so wise a Pilot as Admiral* Noah *might send forth a Fowl, that could Swim as well as Flie, on the great Occasion we have been apprised of.*

Upon the whole, I confess myself proselyted unto the Opinion, That *This was the Dove*. *This*, that I have now in my Study before me.[5]

{Sir,

a most humble

& obedient Servant.

C. Mather.}

4 John Winthrop, F.R.S. (1681–1747) of Connecticut, son of Wait Still Winthrop and Mary Browne Winthrop, supplied Mather with a specimen of this unidentified water-dove, which became the material basis for Mather's CA 5.9. In his letters to John Winthrop (30*d* IX*m* [November] 1719), Mather had requested from him "as punctual a Description, as you can of your *Water-Dove*, that so my Account of it may be Authentic as well as punctual" (*MP* [1868], p. 435). In his follow-up letter to Winthrop (11*d* XI*m* [January] 1719/20), Mather acknowledged that he had "assigned our dear *Soogelande*, a very fair Lodging in our *Biblia Americana* [*BA* 1: 643–44], with an honourable mention & character of the generous Hand, which like another *Noah*, sends him forth unto us" (*MP*, p. 436). In yet a later letter to Winthrop (14*d* 1*m* [March], 1720–21), Mather mentioned that he had sent Winthrop's account to "our Masters in the *Royal Society*" and signed his letter "with a pretty little Quill taken from the wing of Your *Water-Dove*" (*MP* [1868], p. 447). John Woodward apparently never received the specimen, for in his letter to John Winthrop (April 3, 1721), Woodward complains that "Dr. Mather has said nothing, as yet, about the Water Doves that you Sent Him" (Appendix 8). Oddly, in his entry on Gen. 8:8 (*BA* 1:644), the only extant source of CA 5.9, Mather reneged on his promised acknowledgement and canceled the [bracketed] reference to Winthrop (above). Nonetheless, as Woodward's letter to Winthrop reveals, Woodward was indeed cognizant of Mather's source (Appendix 8). See K. Silverman, who severely takes Mather to task for his alleged dishonesty, *SLCM* (1971), p. 241.

5 An earlier specimen that Winthrop had promised his friend in Boston apparently fell prey to Winthrop's cat; see Winthrop's unpublished MS. letter to Mather (December 29, 1719) at the MHS, in Silverman, *Life* (246–47). A little more than one hundred years ago, G.L. Kittredge tried to track down Mather's disquisition on "The Water Dove" in the "Biblia Americana," but could not locate it ("Scientific," p. 46). Well, it is in *BA* (1: 643–44). See also CA 2.5 "The Pigeons" and CA 3.3 "The Nidification of Pigeons" (July 4, 1716).

5.10. THE LOCUSTS.[I]

{February 1720.}

To John Chamberlayne.[2]

{Sir,}[3]

There was in the Plague of *Locusts* upon *Egypt*, something extraordinary; even what *neither their Fathers, nor their Fathers Fathers had seen, since the Day they were upon the Earth.*[4]

I suppose, there is no need of our concluding, That the *Locusts* were of a Dimension more than ordinary; like those *Indian Locusts*, whereof some, according to *Pliny*, where Three Foot long, & some Four Cubits; and when they were Dead, & Dried, their *Legs* were used instead of *Sawes*.[5]

1 Evidence of swarming locusts in modern times was of great interest to theologians and historians just as much as it was to entomologists. Mather validated the biblical narratives of horrific locust invasions in his commentary on Nahum 3:16 (*BA*, vol. 6), in which he excerpted a long article by the Irish physician Dr. Thomas Molyneux (1661–1773), M.D., F.R.S., published as "A Letter from Dr. Thomas Molyneux ... concerning Swarms of Insects" infesting some parts of Ireland in 1697, in *PT* 19.234 (November 30 1697): 742–54.

2 See R.P. Stearns, *Science* (pp. 468–71). Edward Floyd's letter "*To Dr.* M. Lister, *giving an Account of* Locusts *lately observed in Wales*," *PT* 18.208 (February 28 1694): 45–47, is the type of scientific description the editors of the *Philosophical Transactions* were likely to have expected from Mather's epistle.

3 The lost manuscript is reconstructed from Mather's commentary on Exod. 10:6 and Num. 11:33 (*BA* 2:184–85, 868–70). For Mather's views on the Saracen locusts of the apocalypse, see his glosses on Rev. 7:4; 9:3–11, 17 (*BA* 10:511, 527–31, 536, 779).

4 Exod. 10:6.

5 Pliny, *Naturalis historia* (11.35.103) introduces his Plinyism with the caveat, "In India, it is said." Perhaps inspired by Joel 1:2, 4, 6, Gilbert Gaulmin (*De Vita Mosis*, p. 30, # 8) believes the locusts of the eighth plague had teeth very similar to those of lions. Gerard Vossius details the voracious appetite of the locusts and their devastating effect on any and all plants (*De Theologia Gentili* [1641]), lib. 4, cap. 94–95, pp. 1625–32; Samuel Bochart (*Hierozoicon* [1663], pars 2, lib. 4, cap. 1–8, cols 441–96) obliges us with their biblical significations in all their variety.

It is enough, that they were more *Numerous*, and more *Mischievous* than ordinary.

Unto what, *Mischievous*? For we read, in the præceding Chapter, The *Hail* [Exod. 9:25] *had smote every Herb of the Field, & broke every Tree of the Field*… .

Consider then a little further, That it is here expressly said, (v. 5, 12, 15.) That the *Locusts* did *Eat the Residue of that which had escaped, and remained unto them from the Hail*; – even, *what the Hail had left*. And what was that? We read, ch. 9.31, 32. *The Wheat and the Rie were not smitten, for they were not grown up*. According to *Pliny*, *Wheat* in *Egypt*, comes a Month after the *Barley*.[6]

The *Wind* which brought in the *Locusts* upon *Egypt*, (which we translate, *An East-Wind*,) was probably, *A South-Wind*. The Greek renders it, Νοτον, and Λιβα, and Καυσωνα.[7] And so the Wind probably brought them from *Ethiopia*; a Land very Fruitful of these horrible Plagues. Whole Nations there, as *Agatharchides* tell{s} us, are, *Acridophages*, or *Locust-Eaters*, for the Plenty of them. The *Winds* thus employed in bringing on, & carrying off, these *Locusts*, is but agreeable to what we find in *Pliny*, and in *Orosius*, and others.[8] And very particularly in *Jerom*; who saies, (on *Joel*. C. 2.) *Etiam nostris temporibus vidimus agmina Locustarum terram texisse* Judæam; *quæ posteà misericordia Domini – vento surgente, in mare Primum et Novissimum præcipitatæ sunt*.[9]

6 Pliny, *Naturalis historia* (18.10.60).

7 LXX Exod. 10:13; 26:20: Νότον respectively signify "south [wind]" and "the south." Exod. 27:9 and 37:8: Λίβα signifies "the south." Ezek. 17:10, Jer. 18:17, and Hos. 12:1: Καύσωνα signifies "east [wind]."

8 The Greek geographer and historian Agatharchides of Cnidus (c. 215–post 145 BCE), in *De mari Erythraeo* [excerpta] (sec. 58, lines 4–6), mentions unimaginable swarms of locusts. Pliny (*Naturalis historia* 11.32.92) relates that grasshoppers ("tettigonia") rather than locusts are much appreciated as food among "the Eastward races, even the Parthians." The Spanish historian Paulus Orosius (early 5th-century CE), in *Historiae contra Gentes* (lib. 5, cap. 11, sec. 312–15) [*PL* 031. 0941–0942], also describes the horrific swarms of locusts that devastate vast regions in N Africa. But it is Strabo (*Geographia* 16.4.12) who relates that the Ἀκριδοφάγοι, the locust-easters of Ethiopia, are of small stature and rarely live longer than four decades.

9 In his *Commentarium in Joelem Prophetam* (lib. 2) [*PL* 025. 0970, sec. 195], St. Jerome (c. 342–420), the great Latin Father and biblical scholar, comments, "Even during our time [4th–5th century CE], we saw that swarms of locusts had covered the land of Judaea, which afterwards, by the mercy of God, with the wind rising, were plunged into the Dead Sea and the Mediterranean Sea." Mather's principal source for the preceding paragraph is Bochart's indispensable *Hierozoicon* (pars 2, lib. 4, cap. 3, col. 463).

It seems clear, that *Quails* were twice afforded unto the murmuring Israelites. They had *Quails* on the fifteenth Day of the Second Month, in the Desert of *Sin*, a little after their coming out of *Egypt*, & before their coming to *Sinai*… . But after all; what if there should be no *Quails* at all in the Case.[10] Dr. *Patrick* thinks, That no body he ha's mett withal, hath laboured so much to give a clear Explication of this History, as *Job Ludolphus*, in his learned Commentary upon his *Ethiopic History*.[11]

There he ha's a long Discourse to show, That the *Hebrewes* take not the Word *Selav* (used here) to signify *Quails*; but we take the Translation only from *Josephus*.[12]

Notwithstanding all that is produced by the most learned *Bochart*, there seem to be several Things in the History, which by no means can agree to *Quails*. *Ludolphus* will therefore have *Selav*, to signify, *Locusts*; by which, it is easy to give a plain Explication, of all that is here concerning them.[13]

Tis very certain, That *Locusts* were not only used for *Food*, in those Parts of the World; but that some of them were very *delicious Meat*, in several Countreyes. They that have eaten them, compare them to Young *Pigeons*; or to a Fresh *Herring*; or to a *Crab*, or *Lobster*, (which they resemble in their Figure:) And they are several Wayes præpared, and accounted very *wholesome Food*, when they have thrown away the Heads, and Wings, and Legs. *Pliny* saies, That some Parts of *Ethiopia* lived upon them; & that they were præserved, *Fumo et Sale*, by being dried in the Smoke, and salted, for their nourishment, throughout the whole Year.[14]

10 Reconstructed from Num. 11:33 (*BA* 2:868–70).

11 *BA* (2:863, 866). Simon Patrick, *A Commentary on the Fourth Book of Moses, called Numbers* (1699), p. 198, enlists *Historia Æthiopica* (1681), lib. 1, cap. 13, §16, by Hiob Ludolphus, the German Orientalist and grammarian of the Ethiopic language Amharic.

12 Hiob Ludolphus, in his *Appendix Secunda ad Historiam Aethiopicam Iobi Ludolfi continens Dissertationem de Locusti* (1694), pars 2: *De Locustis*, cap. 1, p. 21, §§1–5, argues that the Hebrew noun שְׂלָו which he transliterates as "*Selav*" (Exod. 16:13, Numb. 11:35) should *not* be rendered "*coturnicibus*" (quails) but "*locustis*" (locusts). Flavius Josephus (*Antiquities* 3.1.5 and 3.13.1), however, holds fast to "quails" – at least in Whiston's (1733) translation.

13 Bochart, *Hierozoicon* (1663), pars 2, lib. 1, cap. 15, col. 107 (line 70) – col. 108 (line 23) speculates (with reference to the Vulgate's and the LXX's Sapientia 16:2; 19:10–11) that God miraculously created a "new creature" and flesh "with a strange taste" such as a bird that migrates *with* the quails ("ortygometra") to feed the Israelites. However, Ludolph believes they fed on locusts (not quails) – even without a miracle.

14 Patrick (*Numbers* 199); Pliny (*Natural History* 6.195) has groups of Ethiopians solely living on "smoked and salted" locusts; alas, "these people do not live beyond the age of forty."

Now all that occurs in the History before us, (he thinks) will have an easy Meaning, if we follow this Interpretation. But not, if we take them for *Quails*, or *Pheasants*, or *Sea-fowl*. As for Exemple; what we read concerning a mighty *Wind* sent by the Lord, is not hard for to be understood, if we suppose it spoken of *Locusts*, which all Authors tell us, are brought with a *Wind*. But it was never heard, That a *Wind* brought *Quails*, which cannot fly High nor Far; much less, so far, as from the Sea, to the Middle of *Arabia Petræa*.

Probably, they came out of *Africa*, where they abound.

Which Way soever the *Israelites* went, for *Sixteen* or *Twenty* Miles together, (which at least, were the Dimensions of the Camp,) there lay *Heaps* of them upon the Ground. If we understand this, of *Quails*, it cannot be conceived without *Heaps* of *Miracles*. And if we resort thither (as Dr. *Patrick* thinks,) what need was there of any *Wind*, for the bringing of them, when it must be supposed, That God miraculously created them, as He did the *Manna*. But the thing will be found less Wonderful, & very Natural, if it be understood of *Locusts*, who come in thick & vast Clouds, which darken the Sky, as all Authors tell us.[15]

That Clause, of their being *Two Cubits high upon the Face of the Earth*, is by Interpreters look'd upon as impossible. The *Quails* must have been unavoidably choaked & stiffled, if they had been heaped so deep one upon another. And therefore they have devised the Addition of a New Word, & refer it not unto their falling upon the Ground, but their *flying in the Air* two Cubits high above the Ground, that so they might the more easily be taken by their Hands. But there is nothing of this in the Text; and it is contrary to what the Psalmist saies, *They fell in the Midst of their Camp*; and they *came down like Rain*, which alwayes falls upon the Ground.[16] There are some other Difficulties in it, also mention'd by *Ludolphus*.[17] And therefore, it seems better to expound it of *Locusts*; who, tho' they fall one upon another, to a great

[15] Numb. 11:31. Patrick, too, is weary of resorting to miracles to explain these events (*Numbers* 200). Whereas the classical Hebrew commentators – Nachmanides, Kimchi, and Bachya ben Asher – remain silent on this issue, Ibn Ezra, Abarbanel, and Bekhor Shor do not seem to take exception to a literal signification of this quantity. See Nachmanides (*Commentary: Numbers* 112), Bachya ben Asher (*Commentary* 6:2006), and *Miqra'ot Gedolot* (4:83). The famous English non-conformist theologian and biblical commentator Poole (1624–79), in his *Synopsis Criticorum* (1:602–4), mentions the disagreement among interpreters, yet he ultimately follows Bochart's authority, in *Hierozoicon* (1663), pars 2, lib. 1, cap. 15, cols 104–08, which discourages miracles.

[16] Psal. 78:27, 28.

[17] Ludolphus, *Appendix Secunda ad Historiam Aethiopicam* (1694), pars 2: *De Locustis*, cap 48–49: "Sexta Controversia," pp. 48–49. See *BA* (2:869–70).

Depth, are not thereby suffocated, by reason of their *Legs* being so *Long*, & their *Wings* being so *Thin*.

It is evident, that what the People gathered, was Lying on the Ground, & not Flying in the Air.

Every Master of a Family gathered Ten Homers. A vast Quantity, if they were Quails! It would have served them, not for a Month, but for a Year, or two; as is observed by Ludolphus. Besides, we don't use to Measure Fowl, but Number them. And therefore, Bochart being sensible of this Impropriety, takes the Word, Homer here, to signify, only, An Heap. But Ludolphus ha's confuted him.[18]

Whereas it followes, *They spread them all abroad for themselves*. This is another Indication that they were *Locusts*; which they spread, so to Drie them in the *Sun*. The Spreading of *Quails* would have been præposterous, & but the sooner have made them stink. Interpreters therefore ordinarily pass by this Passage; they don't care to meddle with it; and the Vulgar Latin, omitts the Word, *spread*. Whereas all Authors tell us, This is the principal Way of præparing *Locusts*; & præserving them for a Month or more; which they then boil'd, or otherwise Dress'd, as they had Occasion.[19]

From almost æqual Probability, for *Bochart*, & for *Ludolf*, Monsr. *Saurin*, infers, *The Vanity of Sciences*; And, *How great Men can give themselves much Trouble, & can propose Opinions full of Erudition, upon Matters of very small Moment, & of no less Uncertainty*.[20]

{Sir,

a most humble

& obedient Servant.

C. Mather.}

[18] Patrick (*Numbers* 202); Bochart (cols 105–06); Ludolphus, *Ad suam Historiam Æthiopicam antehac editam Commentarius* (1691), pp. 190–92, §§109–12; and Ludolphus's *Appendix Secunda* (1694), pars 2, caps. 54–55, pp. 51–52. The ancient Hebrew "homer," measuring capacity (volume), amounts to c. 100 liters or c. 26 US liquid gallons (*ABD* 6:903–4) – no doubt a huge quantity if each family gathered 10 homers of quails or locusts. No wonder Mather's reality check bespeaks his enlightenment zeitgeist which spurns miracles if rational explanations can be devised.

[19] Ludolphus, *Appendix Secunda* (pars 2, caps 97–98, p. 69).

[20] The notable French Protestant theologian Jacques Saurin (1677–1730), in his *Discours historiques, critiques, théologiques et moraux sur les événements les plus mémorables du Vieux et du Nouveau Testament* (1720), cap. LVIII, p. 391; *Dissertations, Historical, Critical* (1723), Diss. LVIII, p. 556, rehearses much the same argument as here provided by Mather (via Patrick), but is given the last word on the vanity of human wishes.

5.II. *A VOICE FROM HEAVEN.* THE *AURORA BOREALIS.*[1]

{February 1720}

To John Chamberlayne.

Sir,

We are sensible, that of *Later Times*, there has been much *Discourse*, and some *Wonder*, about a *Meteor*, which the Learned agree to call, *The Northern Twilight.* The *Pyramidal Glade of Light*, observed by *Childrey*[2] and

1 In February 1720, Mather's sent a bundle of twelve "Curiosa" letters (Series Five), via Henry Newman, to John Chamberlayne of the RSL for publication in the *PT* (*SLCM*, pp. 301–02). Curiously, although the recipient acknowledged their safe arrival and communication to Sir Isaac Newton and to Edmond Halley (see Mather's Letter to John Winthrop, 26*d.* X*m*. [December] 1720, in *MP*, p. 444), none of the documents were ever published, and the holograph manuscripts were lost (G.L. Kittredge, "Scientific Communications" (1916), p. 45 (notes 43–44). Mather's lost package included an earlier version of his "Aurora Borealis" manuscript to John Chamberlayne. Fortunately, Mather deemed it imperative to publish his "Aurora" in Boston as well – to allay the anxiety of his fellow Bostonians about this seeming prodigy.

2 The principal source for this and the following paragraphs on Childrey and Cassini is *The Posthumous Works of Robert Hooke* (1705), pp. 193–200, by Robert Hooke, M.D. F.R.S (1635–1703), English physicist, natural philosopher, and professor of Geometry at Gresham College, renowned for his discovery of microorganisms (animalculae), in *Micrographia* (1665). Via Hooke's *Posthumous Works* (pp. 193–94), Mather refers to the English astronomer, Joshua Childrey, D.D. (1623–70), prebend of Salisbury, who is mostly remembered for his brief description of the zodiacal light, described as "a small Black Cloud, lying near the Milkie-Way, between the Swans Tail, & the brightest Star in *Cepheus,* which I presume is some new Apparition; for I never saw it before the 16 of *August* 1652… . It seems to be a foretoken of some new Star, Comet, or other Phænomenon," in Childrey's *Szygiasticon Instauratum* (1653), sign. A^5. Childrey repeats much the same in his *Britannia Baconica* (1661, 1662), pp. 183–84. Mather provides greater detail of Childrey's and Cassini's "*Glade of Light*, in his *CP* (1721), Essay II, Of the Stars, pp. 16–17; this account strongly leans on an extract from a letter of William Derham, F.R.S (1657–1735), an English physico-theologian, about the same "*Glade*

Cassim[3] and others in the Northern Hæmisphere of the *Fixed Stars*, has gone by the Name of *Aurora Borealis*; And an Appearance of Light under various Forms in our *Meteorous Regions* has had the Name applied unto it. It should seem, that the Appearance of it, grows a little more frequent in the *Later Times* than it was in the Former; the cause whereof, if the *matter of Fact* be really so, may be worth Enquiring. Be that as it will, this *Aurora Borealis*, has been accounted a Subject worthy to be Registred in the Writings that are to be transmitted unto Posterity, and pondered by the most consummate Philosophers of the Age. Our *Stow* has in the Reign of Q. *Elizabeth*, more than once counted it an Article worthy of a *Chronicle*, That there were Nights wherein the *Heavens did seem to burn*.[4] And I make no doubt, That some of the *Terrible Blazing Stars* Recorded by our *Hollinshead*, in the Reign of K. *Henry* IV. were no other than the *Meteor*, we are now taking into our Contemplation.[5] The

of Light," published in "Part of a Letter from the Reverend Mr. W. Derham, F.R.S. concerning a Glade of Light observed in the Heavens (March 28, 1706)," in *PT* 25.305 (March 1 1706), p. 2220. On Childrey's discovery, see also Mather's Curiosa letter to Richard Waller (November 24 1712) CA 1.8 Astronomical Observations; as well as "An Extract of several Letters from Cotton Mather," in *PT* 29.339 (January 1, 1714), pp. 62–71. The summary of Childrey appears on pp. 65–66.

3 "*Cassim*" is a misprint for *Cassini*; i.e., Gian Domenico Cassini, aka. Jean-Dominique Cassini (1625–1712), the Italian-born French astronomer, distinguished member of the French Académie des Sciences, discoverer of the zodiacal light, published in the French *Journal Des Sçavans*, vol. 11 (May 10 1683): 121–32, and (more readably) in *Découverte de la Lumiere Celeste qui paroist dans le Zodiaque* (1685). The "*Pyramidal Glade of Light*," i.e., the zodiacal light, is "a band of light in the night sky, thought to be sunlight reflected from cometary dust concentrated in the plane of the zodiac, or ecliptic" (G. J. Flynn, "Micrometeoroid"). If both Childrey and Cassini describe the same phenomenon, then the honor of "firsts" belongs to Childrey.

4 The English historian, chronicler, and antiquarian John Stow (c. 1525–1605) is mostly remembered for his *Chronicles of England* (1565) and *A Survey of London* (1598). In his *Abridgement of the English Chronicle* (1618), Stow records at least two blazing stars during the reign of Queen Elisabeth I (1558–1603): "The seventh of October [1564] at night all the North parts of the Element sæmed to bee covered with flames of fire, proceeding from the Northeast and Northwest toward the midst of the firmament, and discended West" (p. 287). Likewise, on "The 14 of November, about midnight, divers strange impressions of fire and smoake were sæne in the Aire, to procæde forth of a black cloud in the North, towards the South, which so continued til the next morning. The next day following, [November 15 1573], the heavens from the parts did sæme to burne, and over our heads the flames from the Horizon round about rising to meete" (pp. 315–16; misn. 406).

5 Not far behind (though better known), the English chronicler Raphael Holinshed (c. 1525–82) compiled a massive chronicle of England, Scotland, and Ireland,

Miscellanea Berolinensia,[6] have given us ample Relations, of the *Radiations* in the Heavens, which were seen extending from the *North-West* to the *North-East*, in several parts of *Europe*, since the Beginning of the present Century. But *Gaßendous*,[7] has thought it worth his while, to give us a larger Description of such an Appearance; which was not long after the Beginning of the former, seen all over *France*; and *Cambden* says, it was also seen in *England.*[8]

It was very much Resembled, by what was lately a matter of some Observation, (and unto some, of *Consternation*) all over *New-England*; whereof an Account shall presently be given you.

---

published in three volumes and two editions, from 1577–87. In *The First volume of the Chronicles of England, Scotlande, and Irelande* (1577), Holinshed reports that during the reign of King Henry IV (1399–1413), "In the moneth of Marche [1402] appeared a blasing starre, first betwene the East part of the firmament, and the North, flashing forth fier and flames round about it, and lastly, shooting foorth fierie beames towards the North, foreshewing as was thought, the great effusion of bloud that followed, about the parties of Wales and Northumberland" (1133*b*).

6 The *Miscellanea Berolinensis*, aka. *Berlin Miscellanea*, were published in seven volumes from 1710–46, until the dissolution of the Berlin Academy. Mather appears to refer to several articles about the aurora borealis mentioned in *Miscellanea Berolinensis ad Incrementum Scientiarum ex scriptis Societati Regiae Scientiarum* (1710). The first in "VIII. O.R. Descriptio-Luminis Borealis" (1:131–33); the second, "IX. Christophori Matthaei Seidelii, "Observatio Luminis Borealis anni 1707" (1:133–35); third, "X. G. Kirchii, "Brevis Descriptio clari cujusdam Luminis, Anno 1707. die 6. Martii" (1:135–37); and fourth, "XI. G.G.L. "Annotatio de Luce quam quidam Auroram Borealem vocant" (1:137–38).

7 The distinguished French philosopher, astronomer, and chronicler Pierre Gassendi (1592–1655) describes the aurora borealis in his *Opera Omnia in Sex Tomos Divisa* (1658), tom. 2, lib. 2, cap. 7: "De Aurora Borea," pp. 107–11 (esp. p. 108, col.1); Gassendi chronologically itemizes their appearances in tom. 4, lib. 3, "Commentarii de Rebus Cælestibus," p. 88ff. Aristotle's *Meteorologia* (bk 1, chs 4–7) has much to say on the great fiery comets in the sky. Of course, Increase Mather's *Kometographia* (1683) was a handy vademecum for his son's readings on the topic. See also "Mr. Prince's Account of the Northern Lights, when first seen in England, 1716" (pp. 14–16), by Rev. Thomas Prince (1687–1758), Congregational minister of the Old South Church in Boston; and Thomas Robie's "A Letter to a certain Gentleman," both in *CMHS* 2 (1793): 14–16; and 17–20, respectively. See also M.E. Holmberg (2019), ch. 4.

8 The English historian, antiquarian, and topographer William Camden (1551–1623) is remembered for his popular *Britannia* (Latin version, 1586), which was first translated into English in 1610 and reprinted and expanded numerous times throughout the 17th and 18th centuries. It is a prime source for the history of regional kings and rulers, but has little to say about blazing stars or comets.

The Large Accounts, that have been given of the *Phænomena* in the irradiated Sky of the Night, which have lately found work for the *Wise Men of Enquiry* in *Britain*, and over good part of *Europe*, you can be no stranger to.[9] The Ingenious Pen of my worthy Neighbour and Brother,[10] who was an Eye-witness to some of them, has oblig'd us with a punctual Relation of what *he* saw, when the Heavens look'd as if He that is to be *Revealed from thence in flaming Fire*, had been making His Descent unto us.

The Story of what we have just now seen, may then hope to come in among the rest, without any indecency.

Of *our* late *Aurora Borealis*, I must observe what *Gaßendous* does of *His*; That such a Vapour must needs be of a *Vast Altitude*, above the Earth; since the Convexity of the Earth was no hindrance to its being Visible, and in the same Scituation, to Places remotely distant from one another.[11]

But the Account of *Ours*, which I now hasten to give you, is; That on the Eleventh of this *December*, (1719.)[12] In the Evening, we were here at

9 See, for instance, Edmund Halley's "Account of the late surprizing appearance of the lights seen in the air," *PT* 29.349 (March 31, 1716): 406–28; and his "Account of the Phænomena of a very extraordinary Aurora Borealis (November 10, 1719)," *PT* 30.363 (December 31, 1719): 1099–1100.

10 Mather refers to the Rev. Thomas Prince, F.R.S. (1687–1758), Mather's colleague and brother-in-(religious) arms, Harvard graduate (1707), pastor of the Old South Church (congregational), historian, and scholar. While visiting England, Prince encountered what appears to have been an aurora borealis in 1716 and published his account as *An Account of a Strange Appearance in the Heavens on Tuesday-Night, March* 6. 1716 (1719). It was Prince who saw Mather's *Voice from Heaven* (1719) through the press (see above).

11 Gassendi, *Opera Omnia in Sex Tomos Divisa* (1658), tom. 2, lib. 2, cap. 7. "De Aurora Borea," pp. 107–11, speaks of the aurora, "cum materies Phaenomeni fuisse pertenuis vapor videretur, oportuerit talem vaporem esse … immensae diffusionis secundum terrae superficiem, aut immensae certè altitudinis, ut nihil superficiei Terrae devexitas obstiterit quo minus e locis adeo dissitis eodem in situ conspiceretur" (p. 108), which might be rendered, "Though the material of the phenomenon would seem to have been a very thin vapor, it would need to be a vapor … of such altitude that the curvature of the earth's surface would not prevent it from being viewed in the same location from such widely separated places."

12 See for instance, the anonymously published *A Letter to a Certain Gentleman* (1719), by Thomas Robie, F.R.S. (1689–1729), scientist, mathematics tutor at Harvard, and physician, who published several almanacs, his meteorological observations, a treatise on earthquakes, and his observation of a comet (December 11, 1719), or rather, the aurora borealis, the subject of both Mather's and Prince's publications. On Robie's career as a scientist, see F.G. Kilgour (1939), and R.P. Stearns, *Science* (pp. 426–35).

Boston, pretty much surprized, with a Luminous Appearance in the *Northern part* of the Heavens, which extended in the Form of an Arch, from the *North-West* unto the *North-East*; a considerable way. It was a sort of *Cloud*, but so thin, that the *Stars* could be seen through it; and first of a *Lighter*, but anon of a *Redder*, and a more *Bloody Aspect*. The Region of it, was much higher than the ordinary Clouds, which were plainly seen moving below it. Of this, we were soon more fully satisfyed, when we received Accounts from our Friends, Fifty Miles to the Northward, and twice Fifty to the Southward of us; That (as One Letter informs us) at this time the Hæmisphære being very clear, (which it was not at *Boston*,) and not a Cloud in the Sky, they saw a *Glade of Light*,[13] grow from a smaller Bulk, first into paler Flames, and then into Redder, and so into the colour of Blood. And, That (as Another) the *Red* was darker at the *Western* End of it, and Brighter at the *Eastern*. It seemed something to *Expire*, and then to *Revive* again. As midnight came on, it Renewed with what was commonly thought a *more Terrible Aspect*, than in the former part of the Evening. Yea, some Hours after *That*, it so *Revived*, that People at Work about their Saw-Mills, perceived their *Trees* to look *Red* with the reflection of it; and they could see to manage their work by it, as if the Light of the Moon, (which was now set) had favoured them.

It is Remarkable to see, how much we are left in the *Dark*, and how much our *Philosophy*, is at a loss, about the *Lights*, that are ever now and then enkindled in the Heavens that are so near unto us. We may talk some fine Things, about the *Sulphur* and the *Nitre*, and the *Je ne scay quoy*,[14] in the composition of them, and make ourselves be admired for our Learned *Jargon*, among them that have not learned the Language.[15] We may also propound unto Consideration, how far the Origin of such a *Northern Twilight*, as *Gassendou's*, and *Our* late One,[16] may be found, in that constant *Milky way of the Sun*, or, *Glade of*

13 See note 3 (above).

14 Or rather, "Je ne sais quoi," lit. "I don't know what" – a phrase used to express something for which one lacks words to describe.

15 Mather appears to poke fun at the pseudonymous author (ΦΙΛΟΣ ΣΟΦΙΑΣ; [Philos Sophias]) of Robie's anonymous *Letter* (1719), who (so Mather) hides his ignorance of the real causes of the aurora borealis behind bombastic platitudes: this phenomenon "may be fired by what the Philosophers of old called the *Antiperistasis* of the Air," Robie conjectures, for "if we take Niter, Brimstone, and Quick Lime, mix them, and put them into an Egg-shell, as soon as they touch the Water they will fly out in an actual Flame, and such is the Nature of an Accid and an Alkali, as that the Contest between them will heat the Plate or Vessel in which you endeavor to satiate them" (p. 7).

16 The aurora borealis of December 11, 1719 (*D* 2:596). Mather wrote the manuscript of his *Voice from Heaven* on 24*d*. Xm. [Dec.] 1719, in *VH* (p. 16), nine days after Robie's letter was printed by James Franklin, Mather's old nemesis.

Light which every Year strikes from that part of the Horizon where the Sun sets, up towards and almost unto the *Pleiads*, in the latter End of *February*, and the Beginning of *March*, whereof there is in the Post-humous Works of Dr. *Hook*,[17] an Account Endeavoured. But still the Old Philosophers ingenuous cry of, *Darkneß, Darkneß!* will return upon us.

I Don't see, That the *Extent* of our *Atmosphære* is hitherto well determined. For tho' 'tis pretended, That our *Baroscope* has fixed it,[18] for less than Fifty Miles; yet *Hartsoeker* will perswade you that the Reports of that *Mercury* in this matter are not so to be relied upon, but that our *Atmosphære* may extend some Hundreds of Leagues;[19] And, I confess, that for some Reasons, which at present I don't care to mention, I incline to that Opinion.[20]

But how poorly Qualify'd are we then to form a Judgment on many Things that are doing it may be up towards the *Selvidge*[21] of this *Atmosphære*.

17 The principal source for the preceding paragraphs is the Cutlerian lecture, "*This Discourse gives an Explication of a Glade of Light*," by Robert Hook, F.R.S (1605–1703), in *The Posthumous Works* (1705), pp. 193–200.

18 Or rather, "barometer," an instrument that measures air pressure. Mather may have in mind Robert Boyle's "An Account of a new kind of baroscope, which may be called statical," in *PT* 1.14 (July 2 1666): 231–39, which gave rise to his famous air-pump experiments. See also his *New Experiments Physico-Mechanical* (1660). For insightful analyses, see S. Shapin et al. (1985).

19 Mather refers to Nicolaas Hartsoeker (1656–1725), Dutch astronomer, natural philosopher, lens grinder, and distinguished member of the French Académe des Sciences and of the Prussian Academy of Sciences. Hartsoeker's *Principes De Physique* (1704), esp. chap. 9 and 12, on magnetism and meteors, may be what Mather has in mind.

20 Ill at ease to bring up the occupants of the invisible world – devils, demons, angels – in what purports to be a scientific publication originally designed for the readers of the Royal Society, Mather nonetheless believed the atmosphere to be the abode of demons swarming in the air catching at the souls on their way to heaven. As he put it in his *AC* (1704), "The *Sovereign* GOD hath, with infinite *Wisdom & Justice*, confined the *Fallen Spirits* unto this *Atmosphere*; but with their *Confinement*, they have so much *Liberty*, that until the *Second coming* of the *Messiah* into this lower world, they may range and rove about, and molest the poor Children of men. Our *Air* is fill'd with them, as with *Flies* in *Mid-summer*. We draw our breath in the *place of Dragons*. *O good God, How much ought we to adore thy Goodness, in thy Restraints upon them*! And O ungodly Sinner, If thy Soul expire in ill Terms with Heaven, tell me, I pray, how tis possible it should get safe to Heaven? It must pass through a Region that swarms with Devils which are waiting to sieze upon it" (p. 6). See also *MPW* (1689), pt III, p. 16; and *WIW* (1693), pp. 50–85.

21 I. e., seam or fringe.

The known Principles of *Mechanism* failing us in some occurences of *Nature*, some that are little enough tinged with *Enthusiasm* or *Fanaticism*, own themselves compelled hereunto, to consider the Operations of *Angels* Good & Bad; *Intellectual* & *Voluntary* Agents.[22]

'Tis true, There can be nothing so Ridiculous, as the *Mahometan Philosophy*,[23] which makes the Angels, to be the doers of all that is continually done in the works of *Nature*; and particularly, makes those *Meteors* which we call, *Falling Stars*, to be the *Firebrands* with which the Good *Angels* chase away the Bad, when they come too near the Heavens, to Eves-drop the Secrets there. But yet we have all possible Assurance, that there are *Angels* both Good and Bad; Our own Country affords Testimonies enough to overwhelm all the *Sadducees* in the World. And so incontestable it is, That all Ages have believed our *Atmosphære* to have some of the *Angelical Tribes* replenishing it. Our Sacred Scriptures do sufficiently assure us, That the *Angels* both Good and Bad, are sometimes particularly concerned about the *Meteors* in this *Atmosphære*; the *Tempests*, and the *Thunders* raised there: yea, That the *Heavens do Rule*, and the *Invisible World*, has an astonishing share in the Government of *Ours*. As *Unphilosophical* as it may seem, to talk at this rate; the further our Improvements in Philosophy are carried on, the less will it be found *Unreasonable*.

The *Uncommon Occurrences* in the *Heavenly Places* of our *Atmosphære*, have doubtless their *Natural Causes*. And yet they may Rationally enough set the admiring and amazed Spectator a thinking, *What unknown Things may be*

[22] Mather has himself and his fellow pneumatologists in mind. He (like many of his peers) embraced Descartes' mechanistic philosophy even as he invoked God as the ultimate power above mechanism. Mather especially admired Robert Boyle's *Christian Virtuoso* (1690), which tries to reconcile the mechanist philosophy with religion, yet he was not prepared to go as far as Boyle did. In his *CP* (1721) for instance, Mather lays down "those *Laws of Nature*, by which the *Material World* is governed, and which, when we come to consider, we have in the Rank of *Second Causes* [laws of nature], no further to go. All *Mechanical Accounts* are at an end; we step into the Glorious GOD *Immediately*" (Essay I, p. 8).

[23] Mather's pejorative views of Muhammad, the Qu'ran, and Islam in general, are entirely consistent with those of his Christian contemporaries. Shaped by his eschatological interpretation of the Turks as the Second Woe, Mather paid close attention to the wars of the Ottoman Empire, whose formidable power in SE Europe was not broken until the Ottoman defeat at the Battle of Vienna (1683). Mather had access to Johann Zechendorff's Latin translation *Turcarum Alcoranus* (1632) as well as to Alexander Ross' *Alcoran of Mahomet* (1649), a second-hand translation from the 1647 French version by André du Ryer. Mather may have had in mind such passages as Sura 67:5, "We have adorned the heaven, and the world with Stars, we expelled thence the Devils, we prepared for them a great fire, and the torments of hell for unbelievers" (*Alcoran*, p. 356). On the views of celestial bodies in Islam, see D. Cook (2008).

doing among the Rational Inhabitants of those Regions? This we do know, That *there shall be fearful Sights, and Great Signs from Heaven*, and *there shall be Signs where we see the Sun and Moon and Stars*, and the *Powers of Heaven shall be in Commotions*,[24] (and it is possible, be at work in producing some of the *Fearful Sights*) among the *Forerunners* of a *Day* that all Sober Men do look for.

Indeed, it is a Weakness, to be too Apprehensive of *Prodigies*, in all *Uncommon Occurrences*. Yea, some things may be thought *Prodigies*, which may really be *Kindnesses* to the World; among which things we may particularly reckon *Exploded Meteors*. Be sure, People are never more fanciful and whimsical, their Imaginations are never more fertil, than when they have *Uncommon Occurrences* in the *Clouds* to work upon. And it becomes not Serious Christians to be *Dismayed at the Signs of Heaven, as the Heathen are dismayed at them.*[25] Not only the *Prodigies* which the celebrated Roman *Livy*, so fills his Pages withal, but also those which *Lycosthenes* and other later Historians have given us, were very many of them doubtless meer *Fancies* or *Fables*.[26] Nevertheless, a total contempt of all *Prodigies* is an Extreme on the other hand, which is to be avoided by them that would *walk wisely*, and *neither turn aside unto the Right Hand, nor unto the Left*.[27] Nor have *some* of the *Prodigies* related by *Josephus*,[28]

24 Luk. 21:11, 26.

25 Jer. 10:2.

26 The Roman historian Titus Livius (59 BCE–17 CE) is famous for his *Ab Urbe Condita*, the myths and legends from before the founding of the City of Rome (753 BCE) to the time of Caesar Augustus (63 BCE–14 CE). It is filled with very many prodigies and phenomena much like those that can be found in *Prodigiorum ac Ostentorum Chronicon* (1557), translated as *The Doome warning all men to the Iudgmente* (1581), by the German encyclopedist Conrad Lycosthenes, aka. Konrad Wolffhart (1518–61). Like Livy's *Roman History* and Cassius Dio's Greek *Historia Romana* (covering Roman history from before its founding to c. 211 CE), Lycothenes's *Chronicon* is filled with prodigious creatures, malformed human beings, and signs in heaven. See L. Daston & K. Park (2001) and S. Satterfield (2022).

27 Mather certainly knew what he was talking about, for many of his early works – including his father's *ERIP* (1684) and *Angelographia* (1696) – were part and parcel of such endeavors. On Mather's ambidextrous views of prodigies, celestial or terrestrial, see M.P. Winship (1994 and 1996).

28 Deut. 2:27; 17:11. The Jewish historian and priest Flavius Josephus (37–100 CE) identified seven signs foreboding the destruction of Jerusalem, in *The Works of Josephus ... Arnauld D'Andilly* (1701), "Of the Wars of the Jews," bk 7, ch. 12, p. 753; "Of the strange Signs and Tokens that appeared before the Destruction of Jerusalem: (1) "a comet like unto a Sword," (2) "a bright light shined about the Altar and Temple by Night, (3) "a Cow that brought forth a Lamb in the midst of the Temple, (4) "the Brazen Gate of the Temple opened it self," (5) "armed

as foregoing & foreboding the Destruction of *Jerusalem*, found a general Disdain among the most Judicious of his Readers.

The Learned and Famous Dr. *John Spencer*, Entertained the World, with a Treatise full of Erudition, on a Design to sink the Opinion of *Prodigies*.[29] But the Venerable Dr. *Increase Mather*,[30] visiting of him, on 27d. IVm. [June] 1689. at *Bennet*-College in *Cambridge*, took the Opportunity to Enquire of him, whether he still continued of the *same Opinion* concerning *Prodigies*? To which he made a modest Answer: "That he was a very Young Man, when he wrote his Book on that Subject; and had not since much considered it; But that he believed the *Dœmons* had *prænotions* of many things, and might give strange *præmonitions* of them in the way of *Prodigies*. And, that he did not

---

chariots, and Men seen in the Air," (6) "a voice in the inward Temple," (7) "*Jesus* a Country-man's Cry and Death" (p. 753).

29 John Spencer (1630–93), English clergyman, antiquarian scholar, and Master of Corpus Christi (Cambridge), dismissed them all in his *Discourse concerning Prodigies* (1663, 1665), ch. 3, p. 60 # 2: "Can any man think God would ever work so ludicrous, so cheap, so insignificant a miracle?" To be sure, Spencer did not reject prodigies outright but tried to discourage the hoi polloi from superstitiously reading natural phenomena. For useful background, see W.E. Burns (1995 and 2002) and, more recently, P.N. Jordan (2002).

30 In his *Angelographia* (1696), "To the Reader," Increase Mather (1639–1723), leading congregational clergyman in Boston, president of Harvard, and father of Cotton Mather, relates that during Increase's visit in England, Spencer confided to him "*That the Evil Angels had* Prænotions *of many Future Things, and did accordingly give strange* Premonitions *of them*" (sign. A^{5r}). And yet, Jonathan Tuckney, Increase Mather' cousin, believed that it was John Spencer's *Discourse Concerning Prodigies* (1663) and the *Discourse Concerning Vulgar Prophecies* (1665), which may well have undermined the impact of Increase Mather's *ERIP* (1684): "Though your peece of Prodigies may possibly have been prevented from efficacy by some we call here Latitudinarians, by a treatise of a Cambridge-man's (Spencer by name) that came out at first some three years after the King's coming in [Charles II]," in "Jonathan Tuckney to Increase Mather" (February 25, 1681/82), *MP* (p. 354). That providential manifestations were particularly called for in New England can be seen in "The Records of the Cambridge Association" (p. 271). They reveal that New England's ministers, at their October 2 1693 meeting, called upon their colleagues to "invigorate the design of recording illustrious and memorable events of Providences throughout this country [New England] ... to be in their names published unto all the ministers of the Country" (p. 271). Increase Mather's *ERIP* (1684) and *A* (1696) were part and parcel of the same endeavor. The proposals were also published in Cotton Mather's *TIC* (1699), "A Design," pp. 71–73, and echo many of the same in the Preface of Increase Mather's earlier *ERIP* (1684), sign. A7r–A8v. They all aimed to combat the mechanist and materialist philosophy of the age that seemingly ruled out the intervening hand of God in the universe.

know, whether he might not err in something of an Extreme, on one side, as others did on the other." Wherefore, tho' I will not say,

Nunquam futilibus resplenduit Ignibus æther,[31]

Yet I will not utterly deny, but that something may be *Read* sometimes by the *Light* of those *Fires*. There is *not* always *Nothing* in them.

In Men of Superiour Wisdom & Goodness, it cannot but move a compassion for a miserable World, when One sees the *Terrors of Death* so generally seizing and frightning People, upon any *Uncommon Occurrences*; and especially upon any *Blazes* in the Heavens over them. *Certainly*, it may be thought, *Certainly the poor Children of Men, are generally conscious to this, that they are in Ill Terms with Heaven! Upon every fiery Eruption seen above, they take it for granted, that the Wrath of God is going to be Revealed from Heaven, against all the Ungodliness and Unrighteousness of Men, which the Earth is filled withal.* People that know themselves Reconciled unto God, would never be so Terrified, as most People are, when they see any *Fires* kindling in the Welkin over them.

In the mean time, it would Vex One to see how ready, & (in their own conceit) skilful, People are to *Prognosticate* upon the Things, which they take for *Prodigies*, where the *Rules of Prognostication* are so Uncertain, Ambiguous & Precarious; But how little Notice they take of those things which are much more *horribly Prodigious*, and on which they may found a *Prognostication* with as much more infallible certainty.

This is very sure; Prodigious *Impieties*, & Prodigious *Divisions*, raging in a Place, are much more *certain Omens* of *Evil to come*, than any *Sights in the Air*, which appear never so formidable. But at those Things, how few can say, LORD, *Horror with a Tempestuous Force has taken hold on me, on the account of Wicked Men who do forsake thy Law.*[32]

The Gentlemen, who had not arrived unto the Skill, (which according to *Seneca's* Prediction) some of our latest Astronomers have now attain'd unto, of Calculating the *Motions* and *Returns* of *Comets*,[33] have reckon'd up (as

31 In his *Astronomica* (1.876), the Roman astronomer Marcus Manilius (fl. 1st century CE) opines, "The upper air has never shone forth with meaningless fires."

32 Ps. 119:53; Prov. 28:4.

33 An allusion to Seneca's *Naturales Quaestiones* (7.25.4) "There will come a time when the passage of time and the hard work of a long age will bring forth into the light those things that now lie hidden... . There will come a time when our descendants will be astonished that we didn't know things so obvious to them." Edmund Halley, who authored "*Astronomiæ Cometicæ Synopsis*," *PT* 24.297 (March 31, 1705): 1882–99, and his "A Synopsis of the Astronomy of Comets," in *Miscellanea Curiosa* (1708), vol. 2, p. 2 (sep. pagin.), is among those "latest Astronomers." See also *CPS* (1994), pp. 53, 329 *n.*; and D.W. Hughes (1987).

I remember *Lubienietzki* does) about Four Hundred & Fifteen *Comets*, of whose Appearing since the Beginning of this World, we find some Footsteps in History.[34] But were so many *Comets* all seen Blazing at once, I must freely say, That tho' such *Worlds in a state of Punishment*, would be a very awful Spectacle, yet I should not be apprehensive of such *horrid Præsages* in them, unto a *Baptized Nation*, as I should, if I saw, in such a Nation, a *Kingdom Divided*; or an Epidemical corruption of Manners in Instances, hardly known among Infidels; or the Institutions of our SAVIOUR prostituted unto very unrighteous purposes; or Impious Attempts to degrade the Infinite & Eternal Son of God, into the Class of *Creatures*, & render Him in all things Different from, & Inferiour to His Glorious FATHER; and the highest Professors of the Christian Religion fall into such a *Laodecean* Temper,[35] as to be shye of appearing to assert the most Vital Point in the Faith of Christianity.[36]

For my part, should I *Really* see, (what many under the prepossessions of a strong Imagination, have *supposed*, when the *Clouds* have an Insolit Aspect upon them; I say, should I *Really* see) the Sky covered with *Phantasms* of *Swords & Spears*, and *Rivers of Blood*, and *Armies* or *Navies* Engaging one another, and the *Canon* with a roaring mouth vomiting out *Fire* upon one

34 The Polish theologian, historian, and astronomer Stanisław Lubieniecki (1623–75) is remembered for his 2-vol. *Theatrum Cometicum, Duabus Partibus constance* (1667–68), which recounts 415 comets from Noah's flood to 1665. The reference is to *Theatri Cometici* (1681), pars posterior, *Historia Universalis Omnium Cometarum* (second title page). At the time of writing his essay on the aurora borealis, Mather's manuscript "The Christian Virtuoso" was being reviewed for publication by the Royal Society (*D* 2:324). It was subsequently published as *CP* (1721); it includes his "Essay X. *Of Comets*," pp. 41–45, republished separately as *An Essay on Comets, Their Nature, the Laws of their Motions* (1744). See also Pierre Bayle's famous *Pensées Diverses sur l'Occasion de la Comète* (1682), translated as *Miscellaneous Reflections, Occasion'd by the Comet which appear'd in December* 1680. 2 vols. (1708).

35 The alleged lukewarm attitude toward early Christianity (Rev. 3:15–16) of Laodicea, one of the seven early churches in Asia Minor.

36 Mather alludes to the apostles of Socinianism, Arianism, and Deism, who (like William Whiston) questioned the Doctrine of the Trinity and argued that Jesus of Nazareth was a mere human being, or at most that he became divine only after his ascent to heaven. Mather was particularly at odds with his erstwhile friend Whiston, Isaac Newton's successor at Cambridge, who insisted that the Johannine Comma (1 John 5:7–8) was a forgery perpetrated by Athanasius of Alexandria and his followers. See esp. Whiston's *Athanasius Convicted of Forgery* (1712) and *Essay Towards the Restoring the True Text of the Old Testament* (1722); and Anthony Collins's *A Discourse of the Grounds and Reasons of the Christian Religion* (1724). See especially G. McDonald (2016).

another; I don't know, that I should look on these things as more *Ominous*, or be more affected with them, than if I should be so unhappy as to see, the *Spirit of Piety* generally lost, & what shall be quite the *Reverse* of it, prevaling among a People that have made an high Profession of it; Or, See a *Spirit of Extortion* generally dispose a People to nothing so much as *Preying* upon one another; Or, See the Blessings of a *Good Education* generally despised & neglected; Or, See a People confessing that such & such *Needleß Expences* ruin them, & yet generally resolving to Retrench none of them; Or, See a People that have but *Few Friends* in the World, madly set upon the Disobliging & the Discouraging of them; Or, See a People Evidently near a dreadful *Convulsion*, & yet no *Men of Sense Uniting* to find out Methods for the Relief of the Publick Distresses, but all generally Alienated from one another, broken into Factions, & Sacrificing all to *Cursed Animosities*: Or, Finally to see a *Faithful Man* torn to pieces, if he tell a People of the Perils he sees them expos'd unto.[37] If our Country have yet the Happiness, to have no such *Symptoms* in any Degree upon it, I should think, we may defy all the *Prodigies* in the World, for any *Prædictions of Evil*, which they may give unto us. I hope therefore we shall Unanimously deprecate 'em*!* A Reign of *Piety*, & *Honesty*, & *Charity*, among us, would be such a *Token for Good*, that the *People in the Wilderness* could see no *Tokens*, that they should have any cause to be *Afraid* of.

But yet, that I may more fully Explain my self, and give you my plain Sentiments on this Point; *What Interpretation is to be made of the* Aurora Borealis, *that Heaven has lately shown unto us*? I will say, That tho' I can do very little by way of *Prognostic*; And I would not say that like the People of *Gibeah*, when we see a *Pillar of Smoke and a Flame ascending in Heaven*, we must conclude, That *Evil is coming upon us*;[38] Nor would I think the *Meteor* to be a *Signal Forerunner* of what happens to *Follow* after it; Like the Honest Old Man upon *Tenderton* Steeple:[39] Nevertheless, No doubt the other *Meteors*

37 Here, Mather's jeremiad seems almost prophetic, for an irate Bostonian tried to assassinate him (November 14, 1721) for advocating smallpox inoculation in Boston during the epidemic of 1721/22 (*D* 2:657–58).

38 Mather may have in mind the Battle of Gibeah (Judg. 20:40, chs 19–21; Acts 2:19) that decimated the tribe of Benjamin for the rape and murder of a concubine of an Ephraimite Levite.

39 A tongue-in-cheek reference to the erroneous linkage of cause and effect, as in the anecdote of an aged man in England who attributed the dangerous shoals of the Goodwin Sands (at the entrance to the Strait of Dover in the North Sea) to the steeple of Tenderton (Tenterton), which had caused the sandbank to form. See also *TP* (1995), pp. 63–64; and M. Braam (1914).

of the Heavens, as well as the *Rainbow*, are designed for *Instructive* ones. The Glorious GOD, who is also our Creator, (even HE, who is also our SAVIOUR,) says of us, whom He calls to be the Spectators of them, *Surely they will Receive Instruction.*[40]

Well Sang the Poet,

> *Qui Meteora*[41] *videt liquido radiantia Cœlo,*
> *Hic videt Æterni facta stupenda Dei.*
>
> Who sees bright Meteors in the Liquid Skies,
> The wondrous Works of the Eternal Spies.[42]

A Religious Mind may even with some *Elegancy* of Devotion, consider some Intimations from Heaven, which our *Aurora Borealis* may, at least, by way of *Occasional Reflection* lead one to think upon. Whether the *Aurora Borealis* were a *Prodigy* or no, the *Man* is *One*, who shall ridicule such an Improvement of it.

The Sieur *Le Peyrere*,[43] has given a most wonderful Account, That in *Greenland*, where the Night in the Winter is excessively long, when they

[40] Prov. 21:11, 24:30–34.

[41] In Mather's day the noun "*meteora*" enjoyed a much wider meaning than it does in our days. The *OED* well captures this difference: meteor: 2a. "any atmospheric or meteorological phenomenon"; 2b. "any of various luminous atmospheric phenomena, such as the aurora, ignis fatuus, rainbow, etc." In short, for Mather and his peers, the term "meteor" also included "Falling Stars" (see below) and was thus a subcategory of "stuff in the sky" ("meteors," *meteora*).

[42] The Latin distich is taken from Marcus Frytschius' "Elegia Authoris," in his *Meteororum, Hoc est, Impressionum Aerearum et Mirabilium Naturæ Operum* (1555), sign. d^{2v}. Translated more literally, it signifies "Stop delaying! Look up at the things in the sky. / They always have something by which you are given a warning." The Latin distich is also quoted in Johann Heinrich Alsted, *Theologia Naturalis* (1615), pars 2, cap. 16, p. 393 # 2. Mather liked it well enough to use it on the title page of his *TDR* (1712), along with his own English translation.

[43] Isaac de la Peyrère (1594–1676), French theologian and controversialist author of *Prae-Adamitae* (1655) and *Relation du Groenland* (1647, 1662), reprinted several times. Most famous for his Pre-Adamite hypothesis and polygenesis, that Adam was the progenitor of the Jews only (Gen. ch. 2), whereas the gentiles stemmed from a different ancestor (Gen. ch 1), La Peyrère was also a geographer of sorts. His *Relation du Groenland* and *Relation d'Islande* (1644) was compiled from travel accounts to Iceland and Greenland he had accessed during his voyages through the libraries of Denmark. Mather refers to *Relation du Groenland*, pp. 99–103, a work which was reprinted at least twice during the 17th century and translated into English and incorporated into other travel accounts.

don't enjoy the *Moon*, there arises a *Light* in the *North*, (called therefore the *Northern Light*,) that shines over all the Country, as if the *Moon* were at the Full; The Darker the Night is, the Clearer the Light; It looks like a Flying Fire; it mounts up like a huge Pole; it passes from one Place to another; it continues the whole Night; and it is of incredible use to them in the Business of their Lives.

We may also render our *Northern Light* of some use unto us, in the Greatest Business of our Lives, if it awaken in us the *Right Thoughts of the Righteous*.[44]

May not the *Fiery Appearance*, make us Inquisitive, Whether we have no *Fires* among us here below, that should be lovingly Extinguished? Or, whether no Rash Doings may threaten to raise *Fires* in our Churches, that will not be easy to be Extinguished! Sollicitous, therewithal to *Put far from our Tabernacles*, those *Iniquities*, that may be punished by *Fires* without a Metaphor laying our Houses in Ashes?

May not the *Bloody Appearance*, admonish us, to beware of that *Wringing*, which we have been told, what it will *bring*, if it be persisted in! And to quicken our Cares and Prayers for our *Eastern Plantations*!

May not *Rare Sights* calling us to look more than without them we should have done unto the *Heavens*, very well put us upon thinking, whether we are not so *Buried* in the Business of the *Earth*, as to need something that may call us off, to converse in a more Divine way, with more *Heavenly Objects?*

It is an Ancient Prophecy, *I will shew wonders in Heaven above, Blood & Fire & Vapour of Smoke*; *before the Great and Notable Day of the Lord come*.[45] Why should not this *Fiery & Bloody* Appearance, in the *Vapour of Smoke* lately before us, put us in mind of the *Great & Notable Day?* A Day, when the *Great GOD our SAVIOUR* shall be *Revealed from Heaven in flaming Fire, with His mighty Angels*;[46] A Day, that shall *Come as a Thief in the Night, and the Elements shall melt with fervent Heat*;[47] A Day, which, O Secure and Sleeping World, *it is near, it is near, and it hasteth greatly!*[48] 'Tis well for us if we *are*, and no little part of our Business, to be, *Ready* for it. The *Pagans* themselves had a Tradition,

[44] Prov. 12:5.

[45] Acts 2:19.

[46] 1 Thess. 1:7–9.

[47] 2 Pet. 3:10.

[48] Zeph. 1:14.

> — *affore Tempus*
> *Quo Mare, quo Tellus, Correptaque Regia cœli*
> *Ardeat, et Mundi moles operosa Laboret.*[49]

We *Christians* have a *more sure word of Prophecy,*[50] which has given us the Warning of a tremendous *Conflagration*, and being *Warned of GOD*, we do in these Oracles find the Methods prescribed, wherein we are to make *Preparation* for it.[51] The whole Work of Christianity*!* *Fires* in the Heavens, how properly do they for this purpose become our *Monitors*!

Finally; What *Extraordinary Spectacles* may be Exhibited in the *Heavens*, before that Grand REVOLUTION, which the *Multitude of the Heavenly Host*[52] making a Descent from thence, is to bring upon the World, I know not. But I am certain of This, That whether any foregoing *Signs* be given of it in the Heavens or no, the Day must be very Near, when the *Stone cut out of the Mountains*, will *break in Pieces and Consume all the Ten Kingdoms*,[53] which it is to fall upon, and the *Kingdom of GOD* shall come on, wherein we shall see *Glory to GOD in the Highest*, and *Good-will among Men:*[54] What cannot be Accomplished, but by Dispensations to be *trembled* at, as well as *longed* for!

But *so much* – [if this may be thought *Enough*!] to satisfy on the present occasion. I am very much of *Avicens* mind; *Quicquid sufficientiæ additur*

49 The second-hand citation (and translation) of Ovid's *Metamorphosis* (1.256–58) is from Thomas Burnet's chapter on the conflagration at judgment day, in Burnet's ever-popular *Sacred Theory of the Earth* (1697), bk 3, ch. 3, p. 10: "A Time decreed by Fate, at length will come, / When Heavens and Earth and Seas shall have their doom; / A fiery doom: And Nature's mighty frame/ Shall break, and be dissolved into a flame." Ovid refers to Jupiter's decision to unleash a great flood upon mankind as a punishment for the sins of Lycaon.

50 1 Pet. 1:19.

51 Mather elaborates at great length on the conflagration at Christ's Second Coming in his *TP* (1995), pp. 155–243.

52 Luk. 2:13.

53 Dan. 2:45. See also Mather's millenarian *SCM* (1716) and his commentary on Rev. 4:6, 17:18 (*BA* 10:443–44, 486, 663, passim), in which Mather associated the apocalyptic "*Ten-Horn'd Beast* with ten kingdoms arising from the collapse of the Roman Empire: the Britains; the Saxons; the Franks, the Burgundians, the Westrogoth, the Suevians and Alans; the Vandals; the Allemans; the Ostrogoths; and the Greeks" (*BA* 10:690–91 and 787).

54 Luk. 2:14.

superfluitati ascribitur.[55] And if the Thoughts to which I have devoted one little piece of a Day prove so *Acceptable* to you, that you may judge, that they may prove also *Serviceable* unto any others, you have my consent unto any Communication, you may think proper for them.

{Sir,

a most humble

& obedient Servant.

C. Mather.}

Written, 24*d*. X*m*.

1719.

[55] The Islamic philosopher, astronomer, and renowned physician Avicenna, aka. Ibn Sīnā ibn Abu Alī al-Ḥusayn ibn Abd Allāh ibn Sīnā (980–1037), a native of Afshana, Persia, was well known among medieval and early modern scholars. His Arabic works were translated into Latin early on, and Mather frequently refers to his *Canon Medicinae* (1490), a medical handbook which enjoyed great popularity among physicians. Avicenna's dictum reads, "That which is tacked onto sufficiency is reckoned as superfluity." Cotton Mather's son Samuel cites the same line in his *Apology for the Liberties of the Churches in New England* (1738), p. 188.

5.13 *THE WORLD ALARM'D. A SURPRIZING RELATION, OF A NEW BURNING-ISLAND.*[1]

{February 1720 and June 10, 1721.}

To John Chamberlayne.[2]
The World Alarm'd.[3]

1 In February 1720, Mather sent a package of twelve Curiosa letters (Series Five) to Henry Newman (Harvard graduate, 1687), an attorney associated with the Middle Temple (London), to be forwarded to John Chamberlayne, F.R.S. (1666–1723), secretary of the Royal Society, because Mather had heard that Woodward (1665–1728) had died (*SLCM*, pp. 301–02). See Mather's letter to John Winthrop, December 26, 1720 (*MP* [1868], p. 444); Newman to Mather, September 17, 1720 (MS. in AAS); and Mather's 1723 list in CA 7.0 Promissory Letter. More than a year later, on June 10, 1721, Mather separately shipped his "A Relation of a New Burning Island" to Chamberlayne (Kittredge, "Scientific," p. 47). The holograph manuscript is lost, but its substance is more than likely a major part of *The World Alarm'd* (1721), which was advertised in the [*Boston*] *News-Letter*, No. 906 (June 22–26, 1721) as "'just published' at the price of fourpence" (G.L. Kittredge, "Scientific," p. 47). Mather was perhaps well advised to have the pamphlet printed in Boston rather than wait for news of its acceptance in London, for the *Philosophical Transaction* (*PT* 32.372, pp. 100–01) published in its August 31, 1722 issue, "Part of a Letter from Thomas Forster [May 12, 1722] ... concerning a new Island lately raised out of the Sea near Tercera." Like Mather's much longer narrative, Forster's brief account closely follows that of Capt. John Robinson, who was an eyewitness of the volcanic eruption near the Island of Terceira (Portuguese Azores), which had occurred on November 20, 1720.

2 In his diary entry for June 15, 1721 (*D* 2:626–27), Mather confesses that "at the very time, when I am writing my Thoughts upon the Subject [subterraneous volcanoes], the Bookseller comes to me, with Desires to have them, that he may give them to the public. So he has them under this Title, *The World Alarm'd.*" Mather also integrates several excerpts of this document in his treatise on eschatology *TP* (1995), pp. 208, 209, 210, 218.

3 Paul Dudley (1675–1751), wealthy son of Governor Joseph Dudley and justice of colonial Massachusetts' Superior Court (1718–45), was Mather's competitor of sorts vying for attention at the Royal Society. He notified John Chamberlayne (April 14, 1721) that Mather had published his *World Alarm'd* in Boston (1721) even though he had submitted it to the Royal Society. Dudley wrote, "Dr Mather

Sir,

'Tis a Scarce Excusable Infirmity, which too often attends me, in Writing on such Subjects as I am now Entertaining and Venturing upon, that I seem for a few Minutes to forget the *Superiour Knowledge* of the Gentlemen to whom I Write, and who Know much more of the *Former Things* which my Letters often begin withal, and have indeed in all Points whatsoever an Erudition much Exceeding all that I dare pretend unto. But yet I cannot always forbear my Freedom with the Candour of the Gentlemen, in Reviving to them the *Remembrance* of things which they are already acquainted withal: Hoping, that (whether it be true or no, that our *Science* is mostly but a *Reminiscence*)[4] if I make not my *Recollection* too tedious for the Length of them, the *Trouble* will not prove much greater than the *Pleasure*. I naturally, and even Unawares fall into the Strains & Frames of a *Camerarius* writing his *Horæ Subcessivæ*,[5] when I can command an *Hour of Leisure* for such Communications. And perhaps the very NAME of that Learned Gentleman being mention'd, I need use no more to obtain my Pardon, where I now address this Imitation of him. Nor am I altogether without Apprehension, that it may be possible for Some of my poor Letters, to be by the Goodness of the Gentlemen communicated unto the View of Others, who may be less Instructed than themselves,

acquaints me that he Sent you the Narrative of the New Island that Rose up among the Azores in the month of November 1720: However to Save my Self the trouble of transcribing it into my Letter I have Inclosed it what Properly you will find in pag. 9. 10 & 11. the Capt. upon his arrival here gave me the Same acct." Dudley included "the observations and draughts [chart] which Capt. Robertson made of this wonderful Island" (RSL MS. EL/D1/75). The accompanying chart on a separate page includes six drawings of the volcanic eruption from different perspectives, "by Captn John Robeson [!] in the Snow Richard and Elizabeth, & by him Communicated to P.[aul] D.[udley] in May 1721" (RSL MS. CLP/9i/67). See also Stearns, *Science*, pp. 462–63, *n.* 160.

4 A tongue-in-cheek rebuff of John Locke's sensationalist philosophy, Mather's comment alludes to Plato's theory of anamnesis (*Phaedo* 73c–76d) that the preexisting soul, before it was trapped in the human body, had innate knowledge, which was lost at birth. Thus the process of education was largely a recollection of this forgotten knowledge.

5 Philipp Camerarius (1537–1624), German humanist, jurist, and polyhistor, was the son of Joachim Camerarius (the elder) and first chancellor of the newly founded Protestant University of Altdorf (1581). Camerarius's *Operae Horarum Succisivarum, sive Meditationes Historicae* (1591, 1599, 1602 etc.), a popular collection of meditations (epistles) on a variety of subjects – geography, natural history, agriculture, medicine, literature, superstition – went through several expansions and was translated into several European languages. An English translation by John Molle appeared as *The Living Librarie* (1621).

and so more capable of being Edified by the Things which may appear to them less usefully or decently Enlarg'd upon.

Hence it comes to pass, that as I am going to Entertain you with a New and a Strange *Tercera* Story, I am siezed with a short Contemplation, on the Tremendous FIRES, which our Globe shakes withal.

Mankind has been generally advis'd and appris'd of a Desolation, which this our World, that has once already perished by *Water*, Expects to Suffer by a *Fire*, which is mention'd under the Term of a *Diluvium Ignis*,[6] in the Discourses of the Ancients. Whether or no the Heathen had some Notion of This, lying under the *Embers* of their Fable, that made *Pyrrha*, (a Name, which you are aware, carries, *Fire*, in it) the Wife of their *Deucalion*;[7] 'tis well known that their Philosophers tell of an Εκπυρῶσις, or *Conflagration*,[8] with which this our World must One day be Visited. The Oracles of our Sacred Scriptures have positively declared it; *The Heavens and Earth which are now, be reserved unto Fire, against the Day of Judgment, and Perdition of Ungodly men*. Then will be fulfilled that Word, *By Fire will the Lord plead with all Flesh, and the slain of the Lord shall be many*.[9] 'Tis true, In the Circumstances of the *Conflagration* to be Expected, there are *Some things hard to be Understood*.[10] The Descent of our Lord REDEEMER, at and for the Destruction of the *Antichrist*, the wicked one, who is to be *Destroyed by the Brightness of His Coming*,[11] and at the Beginning of the *Millennium*, wherein there will be not only *New Heavens*, and there will be in view the *Holy City*, possessed by our SAVIOUR, and the *Raised*, whom He will make the *Teachers* and the *Rulers* of the *Nations*, but also a *New Earth*, occupied by none but *Righteous* Ones, who yet shall *build Houses and Inhabit them*, and *Plant Vineyards and Eat the Fruit of them*:[12] This will not be without a

6 See Mather's *Diluvium Ignis* (1726, 1730), a millenarian pamphlet in Latin describing the Great Conflagration at the Second Coming, which he expected to occur during his own lifetime. On this topic, see his final word in *TP* (1995), pp. 155–294.

7 The popular story of Deucalion and Pyrrha, a Greek version of Noah's flood, appears in Hesiod's *Theogony*, Plato's *Timaeus*, Ovid's *Metamorphoses*, and in others. See also the Sumerian-Akkadian Ur-text of the great flood, *The Epic of Gilgamesh*.

8 The ancient Stoic belief in the universal destruction by fire, in J von Arnim's edition, *Stoicorum Veterum Fragmenta* (2.605, in 2:186).

9 2 Pet. 3:7; Isa. 66:16.

10 2 Pet. 3:16.

11 2 Thess. 2:8.

12 Isa. 65:21–23; see "PrT" (1994), p. 420; *TP* (1995), pp. 50–55, 58, 181, 190–91, 272, 274, 289.

Conflagration. Then, *Our GOD shall come, and a Fire shall devour before Him, and it shall be very Tempestuous round about Him.* Then, *A Fire shall go before the Lord, that shall burn up His Enemies round about, and the Earth shall tremble, and the Hills melt like wax at the presence of the Lord.*[13] But say, Thou Acute Master of Penetration, say, How far this *Conflagration* will be Extended? Say, Whether the first and main Effects of it will not be on the how prepared! – *Italian* Territories:[14] at first reaching (according to the Prophecy,) but a Thousand and Six hundred Furlongs:[15] And whether Dr. *Cresseners* Conjectures, about the Salubrious Effects hereof on the rest of the Earth, may have any thing in them![16] 'Tis plain, a *Total Consumption* of the whole World must not be *Now* accomplished. The Things then to come on are Such as have these Acclamations upon them; *Let the Earth be glad, let the Field be Joyful, and let all the Trees of the Wood Rejoyce.*[17] Surely, a *Total Consumption* of them could not be proposed as that wherein they might *be glad,* and *be Joyful,* and *Rejoyce.* No; a *Total Consumption* of it, cannot be the *Earnest Expectation of the Creature, waiting and groaning for the Manifestation of the Sons of GOD.*[18] Surely the most that can be supposed, is what *Salonius* of old Expressed as the sense of many more, – *Tantum ardebit quantum Malitia Hominum simul ac Dæmonum coinquinare potuit.*[19] And there are more than

[13] Psal. 50:3, 97:5.

[14] *TP* (1995), pp. 219–20, 315.

[15] Rev. 14:20; see also "PrT" (1994), pp. 346–47, 412, 416, 420–21. In his early eschatology, Mather confined the apocalyptic (though partial) conflagration primarily to Italy, from which the combustion would spread outward to awaken Jews and Christians to the speedy return of Christ. When Mather wrote the last essay on the topic (1722–24), he abandoned his belief in a limited and progressively unfolding conflagration for one universal desolation at the beginning of the millennium (*TP* [1995], pp. 41, 155–294). See also his *CPI* (1702), p. 69.

[16] See also Mather's "PrT (1994)," p. 421, and *TP* (1995), p. 315. Drue Cressener, D.D. (c. 1638–1718), English millenarian, vicar of Soham, and prebend in Ely Cathedral. Mather refers to *The Judgments of God upon the Roman-Catholick Church* (1689), ch. 19, esp. 288-301, and to Cressener's *A Demonstration of the First Principles of the Protestant Applications of the Apocalypse* (1690) – both works Mather mines in his commentary on Revelation, in *BA* 10 and elsewhere.

[17] Psal. 96:12.

[18] Rom. 8:19.

[19] Salonius Viennensis (c. 400–75), bishop of Geneva, *In Ecclesiasten Expositio Mystica* [*PL* 53.995A]: "It will certainly burn the heaven on high and the earth below to as great an extent as it is able [to burn] the wickedness of men and the defiled demons." For the same, see *TP* (1995), p. 243.

Twelve Hundred and Sixty Reasons to make One think, that the Century is begun which must see this amazing Revolution.[20]

But then, there seems to be a *Consummation of all things*, that must arrive after a Number of Ages known to GOD alone! when, it may be, this Globe shall become a *Blazing Star*, and come into the Condition of the *Comets*, that seem to be, *Worlds in a State of Punishment*; and be a *Wandring Star*, for which is *Reserved the Blackness of Darkness for ever*.[21] And, if there must come upon This World, a *Substantial Destruction*, it seems reserved until those *Unknown Ages* be rolled about, and the *Great Day of Judgment* has had *Every Work* brought into it. The many passages of this Importance, in the Writers of *Primitive Christianity*; Especially in *Justin Martyr*, in *Irenæus*, and in *Lactantius*, (as well as *Hippolitus*) are what you are no Stranger to.[22] You are too well acquainted with the Original of what is derived under the Name of *Sibylline* unto us,[23] to think any thing of it worth Quoting on this occasion. You also know the Verses that *Justin Martyr* quotes from *Sophocles*, and that *Clemens Alexandrinus* quotes from *Diphilus*, Predicting such an *Utter Destruction* of all things.[24] Nor can you forget *Ovids*,

[20] Dan. 12:11-12; Rev. 11:3-13. After his disappointment of the Annus Mirabilis regarding 1697, Mather expected the millennium to begin in 1716 or 1736. See *TP* (1995), pp. 60–78, 319–47.

[21] Jude 1:13.

[22] Justin Martyr (c. 100–c. 165), Greek philosopher and Christian apologist, in his *First Apology* 20; *Second Apology* 7, *Dialogue with* Trypho 80–81 (*ANF* 1:159, 190, 239–40); Irenaeus (c. 120–c. 203), bishop of Lyon, in *Against Heresies* 5.28.3, 5.26 (*ANF* 1:557, 554–56); Lactantius (250–c. 320), Christian apologist, in his *Divine Institutes* 7.14 (*ANF* 7:211–12); Hippolytus of Rome (c. 170–236), early Christian theologian, in *Appendix*, ch. 21 (*ANF* 5:247), and *Treatise on Christ and Antichrist*, pt 2, chs 43–47 (*ANF* 5:212-14). See also *TP* (1995), p. 326.

[23] *Oracula Sibyllina* (c. 6th century CE), a collection of Jewish and Christian prophecies sometimes attributed to Sibyl, an ancient Greek prophetess. See Mather's "The *Sibylline Oracles*, concerning the CONFLAGRATION," in *TP* (1995), pp. 194–98.

[24] In *The Monarchy* (1.3.291), Justin Martyr (c. 100–c. 165) quotes from the Greek philosopher and playwright Sophocles (c. 496–c. 406 BCE), arguing that "When from the golden ether down shall fall / Fire's teeming treasure, and in burning flames / All things of earth and heaven shall be consumed" (*ANF* 1:291). In his *Stromateis* (5.14.121.1 and 5.14.144.3), the Alexandrian Church Father Clement of Alexandria (c. 150–c. 215 CE), quoting the Athenian New Comedy poet Diphilus (c. 342–291 BCE), says much the same (*ANF* 2:472). The Latin quote from the Roman poet philosopher Publius Ovidius Nasso (BCE 43–CE 17) complements Lucretius with the Words that every schoolboy in Mather's day knew, "It was in the fates [that a time would come when sea and land, the unkindled palace of the sky and the beleaguered structure of the universe should be destroyed by fire]" (*Metamorphoses* 1.256–58).

Esse quoque in Fatis reminiscitur. —

Yea, the wicked *Lucretius* himself, Expects, *Una Dies dabit Exitium* – Indeed it was as you know, with the *Stoicks*, a common Expectation.[25]

The Inhabitants of *Tercera* must have been *Stoicks* indeed, if they had been less affected than they were, upon the Late Eruption in the Sea, at some little Distance from them.

What horrid Commotions, what hideous Convulsions, has this Planet in many parts of it suffered, from Subterraneous Combustions; and such Amassments of those *Igneous Particles* which are *Eternal Fire*, breaking forth at those formideable *Spiracles*, and *Spectacles*, which if they had not been afforded, the Globe would, no doubt, have been torn to pieces![26] If we range over the *Globe*, a little Geography soon informs of many Regions that are Signaliz'd and affrighted with them. Yea, such are the Numbers of them, that there needs another *Varenius* to perfect the Reckoning left Unfinished by the former.[27] The Empires of *Japan*, of *China*, of *Indoustan*, have their Share of Ignivomous Mountains, which give an Inexpressible Consternation to the Beholders of them. The Islands that are *East* of these Countries, and those that are *South*, are in many Places trembling with them. They are in both of the *Java*'s; and *Banda* has the Smoke and the Noise thereof Sensible at a mighty Distance.[28] To the *Northward*, we find the astonishing Eruptions of the *Infernal Fires*, on the Shores of *Tartary*; To the *Southward* we find them at *New Guinea*, and the *Lands afar off*. *Africa* has at least Eight of these Terrifying Eruptions; and the *Islands* which are made such in the midst of the *Atlantick,* by the Sinking of the vast Countries to the *Westward* between the Continent and Them, in the Matchless *Earthquake*, whereof we have such probably Traditions in

[25] Again, in his *Amores* (1.15.23), Ovid quotes *De Rerum Natura* (5.95–98), by the Stoic philosopher Titus Lucretius Carus (c. 94–55 BCE), that the gods "one day shall give the earth to destruction." Mather does much the same in his *TP* (1995), pp. 199–201. See also John Ray's *Three Physico-Theological Discourses* (1713), pp. 326–34.

[26] In his commentary on 2 Pet. 3 (*BA* 10:390–91), Mather paraphrases a long passage from Woodward's *An Essay towards a Natural History of the Earth* [1695], 1723) pt 3, pp. 138–45 on the inner workings of volcanoes and their presence throughout the globe. See also Mather's *CP* (1721), pp. 100–01 and *TP* (1995), pp. 208–11.

[27] Bernardus Varenius (Bernhard Varen) (c. 1621–51) was a German cosmographer, whose *Descriptio Regni Japoniae* (1649) was widely consulted by Mather's peers. The Cambridge 1673 edition includes Varen's description of Siam, Java, and Africa. See also *TP* (1995), pp. 209–10.

[28] The Banda Islands of East India; i.e., Indonesia, in the northeastern Banda Sea.

Monuments of Remote Antequity; these are no Strangers to them. *Fial*,[29] is well-known to Travellers. (What St. *Helens* and *Ascension* have formerly, is Conjectured from what is now found upon them.) *Tercera*, I'l come again to Thee immediately*!* The Flammivomous *Volcano*'s in *America*, are so many, that Some, who might have thought of a better Name for it, have called it, *Vulcans Kingdom*; amazed at the *Estuaries* that have so many and so mighty Chimney's to them. There are no less than Fifteen of them, in the *Andian* Mountains of *Chili*; and if out of *Chili* we pass into *Peru*, they tell us, we shall there find as many more. And the *Terra del Fuogo* is not without them. The *North America* has them as well as the *South*; Diverse besides that at *Nicaragua*. Some of our *Islands* have them. And our *Antigua*, which is *without Water*, is not always *without Fire*, it has a Mountain sometimes Uttering its Admonitions unto the (shall we say, Duely sensible!) Inhabitants. Those in *Europe*, are known to every body. Besides what is Related by *Tacitus*, of One in *Germany* many ages ago, Mount *Carbo* in *Misnia* obliges People sometimes in our Dayes to take notice of it.[30] The Frozen Regions of *Lapland*, and *Iseland*, and *Groenland*, are not so Cold as to admit no *Burning Mountains* there. Visit *Greece*, and there you have particularly the Island *Santorin* in the *Archipelago*. But, *Italy*, *Italy*, is to be *Saluted* with a *doleful Clamour*, as a Country One day to Undergo Stupendous Devastations, from the *Fires* in an awful *Observatory* waiting for it. *Sicily* indeed hears the Roars of *Etna*, and the *Volcanello*'s of the *Liparitan* Islands, particularly *Stromboli*, in its Neighbourhood.[31] But, *Italy*, is the Place, where the *Priests Daughter* is to be *burnt with Fire* for her *Whoredoms*.[32] And here, to omit the Shocking Description that *Kircher* has given of his *Phlegræan* Plains,[33] the Kingdom of *Naples*, has had those Vomits of the *Horrible Pit*, at the Mouth of her *Vesuvius*, which

29 Fial, aka. Fayal (Faial), Pico, Terceira, and St. Michael (Sao Miguel) belong to the Portuguese Azores. See Athanasius Kircher's *Mundus Subterraneus* (1678 ed.), lib. 2, cap. 4, fol. 82; Mather's *D* (2:626–27), and *TP* (1995), p. 209.

30 Tacitus (*Annales* 3). *Carbo* is a dormant volcano in the ancient Roman province of Moesia (modern Bulgaria and Macedonia). Mather may have in mind the dormant volcanoes in the German Eifel region southwest of Cologne, between the Rhine and Mosel rivers. See also Ray's *Three Physico-Theological Discourses*, p, 390.

31 Mt. Stromboli, 22 miles northeast of Liparia in a group of seven volcanic islands northeast of Sicily, is still active.

32 Lev. 21:9.

33 The German Jesuit polymath Athanasius Kircher (1602–80), professor at the Collegium Romanum, and founder of the famous Wunderkammer, Museum Kircherianum (1651). Mather refers to Kircher's *Mundus Subterraneus* (Amstelodami, 1665; 3rd ed. 1678). Kircher describes the Phlegraen Fields (Campi Flegrei), near

have thrown the whole World into the *Terrors of Death*; Such were those of the dreadful Mountain, which are celebrated by several of the ancient Historians; and *Marcellinus* particularly tells us, That all *Europe* was obscured with the Ashes of the Raging Mountain, and the City of *Constantinople* it self (in which the Emperour Leo had not the Courage to continue his Residence,) kept an Anniversary Commemoration of it.[34]

But, behold, as if all these *Vents* for the *Fires* of the Dismal Vault below us, were not enough, behold *New Ones* breaking out, and a very Remarkable One, but a few Months ago.

Looking back to the Year, 1638.[35] We find the Island of St. *Michael*, Shaking with fearful Earthquakes for Eight Dayes together.[36] After which, in the midst of the Sea, about Six Miles from the *Pic* of *Camerino*,[37] in a notable Fishing-Place, at the Depth of an hundred and twenty foot under Water, there broke out a *Fire*, which *many Waters could not quench*, nor the Ocean drown it. It reached thro' the Water up to the very Clouds, carrying up with it Water, Earth, Sand, and Stones, some of which were of Enormous Dimensions, and falling back again into the Water, made what every one trembled at. (And the *Fish*, mighty Shoals of them, were boiled unto Death, in this dreadful Caldron; of the *Deep* thus *boiling as a Pot*.)[38] From a Collection of such things, there began a New Island, which at first was no more than *Five Acres*, but within Four Days, took up the Length of no less than *Five Miles*, and in Heighth about Sixty Fathoms.

An Occurrence much like to this, and not very far from it, has very lately happened: With the Relation whereof, an honest Man who was an Eye-witness of it, newly arrived here, has favoured me.

Naples, in southern Italy, in Kircher's preface (cap. II: *De horrendis* Terrae motibus, *Anno* 1638. *in* Calabria *exortis*) and in lib. 2, cap. 4, fol. 82.

34 The eruption of Mt. Vesuvius in 472 CE was so powerful that the ashes covered much of southern Europe, even as far as Constantinople. Eastern Roman Emperor Leo I, the Thracian (400–74), instituted the commemoration of the catastrophe on the "Twelfth of *November*," in John Ray's *Three Physico-Theological Discourses* (3rd ed. 1713), Discourse II, ch. 5, p. 283. See also *TP* (1995), pp. 209–10; 388 *n.* 25.

35 Mather here exhumes his description from Kircher's *Mundus Subterraneus* (1678 ed.), lib. 2, cap. 4, p. 82. And so does John Ray's *Three Physico-Theological Discourses*, pp. 22–23.

36 St. Michael (Sao Miguel) in the Portuguese Azores. See Kircher, fol. 82.

37 Aka. "Pico delle Camerine" (Kircher, fol. 82).

38 Job 41:31.

Captain *John Robertson*, Commander of a Sneau[39] belonging to *New-England*, arrived at the Island of *Tercera*, Dec. 9. 1720. where he found the poor People in the greatest Consternation imaginable. For about the Twentieth of *November* before, they heard a Prodigious Noise, and felt an Earth-quake, that shook down many Houses, in *Angra*, and the Places adjacent.[40] About the same time, there arose an astonishing Fire out of the Sea, to the South East, near Twenty Leagues distant from the Island, of *Tercera*, where the People now were by *the Sea and Waves* (and Flames) *Roaring*, thrown into *Distress* and great *Perplexity*; and made the Processions which the Miserable *Roman Catholicks* think proper for such occasions. The Government of the Island, Employ'd this honest Man, to make some nearer approaches unto the Formidable Flames, for a more Thorough Discovery. On the Eighteenth of *December*, about Midnight, they made Sail: having aboard a Couple of Priests, and Fourteen other Portuguese, besides their own Ships Company; Steering to the *South-East* from the City of *Angra*. The Government ordered that my Country-man should not be the *first* that went ashore, upon the *New Island*, which they now began to apprehend risen in the Sea; lest our King *GEORGE* might on that pretence lay claim unto it.[41] But he, tells me, He did not think that his KING would be very fond of an *Island all on Fire:* And he had one Reason more, for his not being Ambitious to be the *First* that went ashore: not having (it seems) the Disposition of an *Empedocles* upon him.[42] On Monday by Two a Clock in the Afternoon, they arrived as near as they durst unto their Intended Port; and so near as to be convinced that here was indeed a *New Island* arisen out of the Sea: The Length of it, about Two Leagues; and the Breadth of it the same.[43] They Saw Two Large Hummocks all on Fire, with *Pillars of Smoke* reaching up to Heaven. They Steered on still South East,

39 A type of nautical vessel (aka. snow, snauw, snaw, sneau) with two masts, square-rigged sails, and sharp stern (pink) of various sizes. According to J. Robinson et al. "The 'Snow or Barke Sea Flower, pinke sterned, about 20 Tons,' was built in 1709 at Newbury" (*Sailing Ships* [1922], pp. 21, 30). Capt. John Robertson's vessel is the *Richard and Elizabeth* (RSL MS. CLP/9i/67).

40 Angra de Heroismo, on the Island of Terceira, in the Portuguese Azores.

41 George I of England of the House of Hanover (Brunswick, Germany), king of Great Britain (1714–27).

42 With too much heaven on his mind, the Greek philosopher Empedocles of Akragas (c. 494–c. 434 BCE) is said to have plunged himself into the flames of Mt. Etna to attain his apotheosis – if Diogenes Laertius, *Lives of Eminent Philosophers* (8.69), can be relied upon. The English Victorian poet Matthew Arnold composed a poem on this ancient story, in *Empedocles on Etna, and Other Poems* (1852).

43 A common unit of nautical measurement in Mather's time: two leagues (six nautical miles) of c. 6.9 miles or 11 km.

until Seven at Night; But the Wind being Easterly, they fell something to Leeward; which gave an Opportunity for vast quantities of Dust and Small Stones to fall upon their Decks. The Flames also, (to use my Captains Expression) made a *Noise as if there had been Fifty Canon* at once continually firing. This obliged them to *bear away*, till Day; and then they again *bore towards* it; and being Two Miles to the Southward of it, they made a Good Observation, by which they found it ly in Lat. 38. d. 29. m. and Long. 26. d. 35. m. Then bearing round the Island, that they might find a Place for the Two Priests to Land at, the Wind falling calm, they discerned a Place that looked like an Harbour. But even *there*, and from under the very *Water*, there issued *Fire* & Smoke. Yea, the Island appeared now in a Manner all on Fire; and the Fire had like to have been aboard the Ship, and *Pliny*'s Fate have overtaken the Mariners and Passengers.[44] All Hands betook themselves to their Prayers, for Two Hours. And at Five o Clock, there Sprung up a Breeze, that sufficiently replenished their Decks with Ashes, and Pumice-stones; which also the adjacent Sea is filled withal. But by the Help of this Breeze, they made the best of their way, back for *Tercera*; where they came to Anchor on the Twenty-first of *December*, at Nine o Clock in the Morning; And the Priest ordered their Portuguese Attendents, to wash their Feet, and go barefoot ashore, and make their Prayers in the Kings Church; before they carried their Story to those that sent them. The Island lies to the S.E. Seventeen Leagues distant from the City of *Angra*: Destitute, you may be sure, of all *Vegetables*; and for the Constitution of the Soyl, all that could be discerned was, that the sides of the *Hillocks*, from the Tops whereof there ascend Flames with the Noise of continual Thunder-claps, were formed with pretty Regular Descending Furrows; out of which also the reeking Steams of a Vapour were seen perspiring.

This is the Short Narrative.[45]

[44] The demise during the eruption of Vesuvius (79 CE) of Gaius Plinius Secundus (the Elder) (c. 23–79 CE), Roman historian, biographer, and author of *Natural History*, is recorded by his nephew, Pliny the Younger (c. 61–112 CE), in an epistle to Tacitus (*Letters* 6.16). Pliny the Elder describes the eruption of several volcanoes in the Mediterranean world in *Natural History* (2.106–07). For similar descriptions of volcanic eruptions in Italy, see Thomas Burnet's popular *Sacred Theory of the Earth* (1684; 1697 ed.), bk 3, cap. 7, pp. 35–43. Burnet, like Mather, was certain he had found the scientific key to how the prophesied conflagration at the Second Coming might begin to unfold.

[45] Unfortunate for Mather's submission to the *Philosophical Transactions*, much the same story about the submarine volcano near Terceira was published (at the same time) by Thomas Forster, F.R.S., in "Part of a Letter from Thomas Forster, Esq; F.R.S. to Mr. Machin, Prof. Astron. Gresham, Secr. R.S. Concerning a new Island lately raised out of the Sea near Tercera," *PT* 32.372 (August 31, 1722): 100–01.

And now, shall we *Shew ourselves Men*? It would certainly be a Brutal Stupidity in us, the *Inhabitants of the Earth*, who may justly be in Perpetual *Fear*, lest the Desolations of a *Pit* come with the Surprize of a *Snare* upon us, if such New Eruptions from the Bowels of the *Earth*, should not awaken our Meditations & Expectations of the time, when these *Fires* will prevail to such a *Degree*, as to render this World, a *Fiery Oven*.

I will not Recite the Words of an Eminent *Philosopher*;[46] How little we know when and where we stand upon *Good Ground*; and how much it would amaze the Stoutest Heart, if he could see into the *Subterraneous World*, and how we Walk, and Sleep and Build, sustained only by an *Arched Roof* over a *Dreadful Abyss*, not in all Places of an Equal Thickness. But the Words of a *Divine* may be allowed unto me.

The Description which Dr. *More* gives of the *Conflagration*,[47] which a *Most Sure word of Prophecy* teaches us and Compels us to Look for, is indeed so very Expressive, that I cannot forbear transcribing a little of it.

Our Glorious Lord, – "will cause such an Universal Thunder and Lightning, that it shall rattle over all the Quarters of the Earth; and discharge such claps of Unextinguishable Fire, that it will do sure Execution wherever it falls; So that the ground being Excessively heated, the Subterraneous Mines of Combustible Matter will also take Fire; which inflaming the Inward Exhalations of the Earth, will cause a terrible Murmur under Ground; So that the Earth will seem to thunder against the Tearing and Rattling of the Heavens, and all will be filled with sad remugient Echo's. Earth-quakes and Eruptions of Fire there will be every where; and whole Cities & Countries Swallowed down, by the vast Gapings and Wide Divulsions of the Ground. – And this fiery Vengeance, shall be so thirsty, that it shall drink deep of the very Sea; nor shall the Water quench her devouring Appetite, but Excite it. – The whole Earth shall be Enveloped, in One Entire Cloud of an Unspeakable Thickness, which will cause more than Egyptian Darkness; which added to the choaking Heat and Stench, will compleat this *External Hell*."[48]

46 The eminent philosopher is none other than Thomas Burnet (c. 1635–1715), an English theologian and cosmography, whose popular *Sacred Theory of the Earth* (1697 ed.), bk 1, ch. 9, p. 81, held forth on the hollow recesses or caverns in the earth's crust, which an earthquake might have easily dislodged and plunged the surface into the watery abyss below.

47 A student of the Puritan theologian Joseph Mede (1586–1638) at Christ's College (Cambridge), Henry More, D.D. (1614–87) was a leading member of the Cambridge Platonists, and a favorite of Mather. Mather refers to Henry More, *The Immortality of the Soul* (London, 1659), bk 3, ch. 18, pp. 524–38.

48 Mather's quote is from More's *Explanation of the Grand Mystery of Godliness*, in *Theological Works* (1708), bk 6, ch. 9, sec. 7, p. 164. See also *TP* (1995), p. 218.

How *Livelily Painted*![49]

But as I have my Friends agreeing with me when we have had our View of a caged *Lion*;[50] *That the Lion is really more Fierce than he is or can be painted:* we may certainly say so, of that *Ariel*,[51] which is one day to Devour the Earth; the *Devouring Fire, and the Everlasting Burnings*!

'Tis very sure, PIETY in its due Operation, will dispose us, to Hear the Glorious *GOD Speaking to us from the midst of the Fires*:[52] And there will be nothing of a *Magical* Superstition in it, if we make a Pious Improvement of them, and *Enkindle* our Devotion from them.

And the Almighty GOD having thus fired a *New Beacon* before the *Eyes of the Nations*, it would not be a Folly to be derided, if they should be *Alarmed* from it, into Serious Thoughts on the State of a *World*, which may be nearer to a direful *Catastrophe* than 'tis generally thought for. Had we yet Living and Shining among us, the Admirable *Nieuentyt*, who has by your Excellent Care and Exquisite Skill, become the Instructor of our Nation, I am sure, Such

[49] See also *TP* (1995), pp. 155, 218.

[50] No doubt, the first exhibition of a live lion in Boston was a sensation. Unfortunately, neither Mather nor his friend Samuel Sewall mentions the event in his diary. Odd, too, Boston's news media provide conflicting information about the exhibit: According to *The Boston News-Letter* (Monday, November 26 to Monday, December 3, 1716), the Boston merchant Capt. Arthur Savage (1680–1735) advertised (among the sale of his vendable goods) "at his House in Brattle-Street" the exhibit of "a Lyon of Barbary, with many other Rarities, the like never before [seen] in America"–all "to be shown by William Nichols" (p. 2). However, nearly two years later, the same *Boston News-Letter* announced yet again, "ALL Persons having the Curiosity of seeing the Noble and Royal Beast the Lyon, never one before in America, may see him at the House of Capt. *Arthur Savage* near Mr. *Colman*'s [Brattle Square] Church, Boston, before he is transported for London. But to prevent all disputes with the Negro at the Gate who constantly attends each Person (whether seen him before or not) is desired to pay to the said Negro six pence a piece," *Boston News-Letter* (April 7, 1718), p. 2. (Many thanks to Giovanni Lupia III and Sarah Harwood for tracking down this information). It appears that Savage got tired of hosting Simba in his home, for the king of animals was subsequently relocated to south Boston, the home of Mrs. Martha Adams. According to the *Boston Gazette* (September 26, 1720), "The lyon being the King of Beasts, and the only one of his kind in *America* is removed from Capt. *Arthur Savages*, to Mrs. *Martha Adams's* at the South End of Boston, where attendance will be constantly given, (while he remains in this country,) to any person or persons who has the curiosity to see the same." Admission: "6 *d* per person." See also R.W.G. Vail (1933).

[51] Mather's allusion to the Angel Ariel seems apropos, for the marginal annotation on Isa. 29:1 (KJV, 1611 ed.) the Hebrew name "Ariel" (a metaphor for Jerusalem) signifies "*the lyon of God*."

[52] Deut. 4:33, 5:24. In the burning-bush pericope (Exod. 3:1–16).

a *Religious Contemplation* would not have been judged improper by such a *Religious Philosopher.*[53]

Nor will you, Sir,[54] into whom the Sentiments of *Piety* which inspired that rare Person, have made so Conspicuous a *Transmigration*, judge it at all Incongruous for *me* to make this *Reflection*.

Our Great SAVIOUR having Warned us, that there will be a *Great Furnace of Fire*, which wicked Men shall be cast into; and the *Infernal Fires* that we see growing upon us, being such *Emblems* and *Symptoms* and indeed *Assurances* of it; Can there be any more proper Consequence than this? That the truest *Wisdom* lies in getting into *Good Terms* with Him, and obtaining Him to be our SAVIOUR; Adoring & Obeying of Him as our GOD. The *Burning of the World* is that *Wrath to Come*, from which, our Blessed JESUS will *Deliver* His Followers, as Effectually as the *Ark* did the chosen Patriarchs from the Flood. And it is well for us, that tho' *we* do not yet our selves know of the *way*, wherein it shall be accomplished, *Our GOD knows how to deliver the Godly, and reserve the Unjust unto the Day of Judgment to be punished.*[55]

But, yet more particularly, without espousing an Expression sometimes used, *That the World shall be Burned for a Witch,*[56] I think, the Sight we have of what the *World* is coming to, may very Reasonably dissolve its *Enchantments*, and abate the *Bewitching* Efficacy which it has upon us, and prevent the Madness of an Heart inordinately set upon a World, out of which our GOD is *firing* of us, as out of Another *Sodom*.

I will only add; That, as One would be loath to be Condemned unto a Residence on the *Burning Island* near *Tercera*; so I am Sorry, that the *Fiery Contentions* on Yours, do render it so worthy of such an Appellation: And I wish, that where I am now Writing, there were less of the same *Infælicity* to be complained of.[57]

53 Mather puns on the work *The Religious Philosopher*. 3 vols. (1718–19), by the distinguished Dutch physician and theologian Bernard Nieuwentyt (1654–1718). The Dutch original of Nieuwentyt's *Religious Philosopher* was translated into English by John Chamberlayne, Mather's correspondent to whom Mather sent his CA 5.13 (February 1720 and June 10, 1721).

54 John Chamberlayne.

55 2 Pet. 2:9. No matter his interest in natural philosophy, Mather never fails to turn it into the handmaiden of theology, the queen of science.

56 To be sure, witches were not burned in New England; that was a European custom.

57 Chamberlayne's translation of Nieuwentyt's *Religious Philosopher* (and those of his French and German colleagues) was not well received, for he (like them) took many shortcuts, omitted numerous passages, and misrepresented Nieuwentyt's original. See R. Vermij (2018).

But we are come into the Period, whereto the Unerring Oracles have assigned *That*, as One of its very signal Characters, *The Nations were Angry*.[58] And if we have not more Literal *Volcano*'s very near us, (Tho' they say, you once had something like One in *Scotland*, which burnt for Sixteen Years together) Yet those which a *Metaphor* may call so, will daily multiply. Our best Friends will soon become *Volcanello*'s to us, now *the Love of many waxes cold*,[59] if we do not maintain much *Meekness of Wisdom*, and Study *Patience*.

I have done. I have taken the more of Liberty to do something like what I find Required; *Glorify ye the Lord in the Fires, even the Lord GOD in the Isles of the Sea*;[60] because he is a Devout *Christian* as well as an Acute *Philosopher*, whom I am now Conversing with; and who will always allow to *me* the Liberty of, **The Christian Philosopher.**[61]

On which accounts, were I Master of as many Languages, as were Employ'd by the Learned for the Celebration of the *Peyreskius*,[62] whom you have in so many things made your Pattern; Yea, or of as many Languages as you have lately given us, in a Collection which will Immortalize your Name far more than so many Statues,[63] I should not reckon them too many to Express the Great Regards with which I now Subscribe my self,

Boston, N. E. *SIR,*

June 10, 1721.

Your Most Constant

Friend & Servant.

{C. Mather.}

[58] Rev. 11:18.

[59] Matth. 24:12.

[60] Incorrigible punster that he is, Mather cites Isa. 24:15.

[61] Mather slily refers to his own *CP* (1721), which (as Chamberlayne undoubtedly knew) was then in the press, sponsored by the Royal Society of London, of which Mather was an elected member. Mather's *CP* is also advertised in Samuel Gerrish's imprint of *WASR*, p. 17.

[62] Nicolas-Claude Fabri de Peiresc, aka. Peirescius (1580–1637), French natural scientist, astronomer, antiquarian, and friend and sponsor of the philosopher Pierre Gassendi (1592–1655), who wrote Peiresc's biography, *Viri Illustris* (1641).

[63] Mather flatters John Chamberlayne, who was a veritable polyglot, commanding sixteen languages (*ODNB*).

Series Six

6.1 *AN ACCOUNT OF THE METHOD AND SUCCESS OF INOCULATING THE SMALL-POX, IN BOSTON IN NEW-ENGLAND.*

September 7, 1721.

To Sir *HANS SLOANE*, President. And to the Rest of the College of Physicians.[1]

Gentlemen,

I Receiv'd the following Account of the Method and Success of inoculating the *Small-Pox* in *New-England*, from a Person there, of great Learning and Probity, who desir'd his Name might be conceal'd; having no other View, than a charitable Inclination of doing Good to the World. It seemed to him (as I must freely own it does to me) a Matter of great Importance to the Welfare of Mankind; and if you, Gentlemen, shall be in the same Sentiment, I have no doubt, but it will find your Favour and Encouragement.

Altho' this Practice of ingrafting[2] the *Small-Pox* has been used from Time immemorial among the *Circassians*,[3] and for many Years past in the *Levant*, yet it is a new Thing in these Parts of *Europe*, and still more so in *America:* And as all new Discoveries, however rational in themselves, and beneficial to Mankind, are receiv'd at first with Opposition, none has met with greater than this in *New-England.* A vast Number of honest People quarrel'd with it upon religious Scruples; and they were strongly supported by the Physicians of the Place, for Reasons of a very different Kind, which it is not at all necessary to mention. But as Truth is mighty, and will at last prevail, the great and

1 The English Baron Sir Hans Sloane, F.R.S (1660–1753) was a distinguished Anglo-Irish naturalist and physician, whose huge collection of natural curiosities, botanical specimens, books, manuscripts, and medals, became one of the foundational pillars of London's Natural History Museum. He served as secretary of the Royal Society and editor of its *Philosophical Transactions* from 1693 to 1727. Upon Sir Isaac Newton's death, Sloane succeeded him as president of the Royal Society.

2 Injecting the virulent matter through an incision of the skin.

3 A people originally inhabiting the NW Caucasus between the Black Sea and the Caspian Sea.

surprising Success of this Practice, has convinc'd every reasonable Adversary, and silenc'd the most obstinate. It has been try'd on all Ages, and both Sexes; on every Constitution, the Robust, and the Infirm; on the most contrary Complexions, the Blacks, and the Whites; and in each extreme Season of the Year; the Summer and Winter Solstice. Nor has it failed in any one Instance: Not a single Person has dy'd, or had the *Second Fever*, or, indeed, any considerable Distress. On the contrary, the Patients have been so easy, during the whole Course of the Distemper, that it was hard to perswade them that they were really Patients. I must also observe, that there was no antecedent Preparation of their Bodies; but they have these two Advantages, which are sufficient: One, that they know when they receive the Distemper, (which those who are infected in the common Way can't be sensible of) and from that Time use a regular Diet; the other, that there is such a plentiful Discharge at the Incision-Sores, as prevents any fatal Consequences; which is demonstrable from this Observation, that the Eruptions are more or less, in Proportion to those Discharges. I am inclin'd to think, that the Ease and Security which attended this Operation, may be entirely accounted for from these two Principles, without having Recourse to any other Reasons. But as this is a speculative Point, You, Gentlemen, are the best Judges of it; and to your great Learning and Knowledge it is submitted. I hope some of your Body, will give the World an Account of the late Inoculation of Mr. *Colt*'s Children,[4] who underwent this Operation upon the Relation I gave my worthy Friend, their Father, and his ingenious Lady, of what had been done in *New-England*. They were visited by 15 Gentlemen of the Faculty, who were Witnesses how exactly it answer'd in every Respect. The Pocks came out well, appear'd fair, round, and yellow; and regular with respect to Time, like the true, natural, distinct Kind; and the Children are perfectly well recover'd. It is also particularly to be remark'd, that they were never affected with the usual Disorders, and bad Consequences, that attend this very loathsome and dangerous Discourse.

I AM afraid I have been too prolix, and therefore shall add no more, but that I am, with all imaginable Respect, GENTLEMEN,

Middle-Temple, *Your most obedient Servant,*

Feb. 23, 1721.[5]

JER. DUMMER.[6]

---

4 Unidentified.

5 I.e., 1722 (New Style).

6 Cotton Mather's well-heeled friend Jeremiah Dummer (1681–1739), Harvard (1699), Utrecht and Leiden, Ph.D. (1703), Yale benefactor, and colonial agent

SIR,

A Gentleman[7] well known in the City of *Boston*, had a *Garamantee* Servant,[8] who first gave him an Account, of a Method frequently used in *Africa*, and which had been practis'd on himself, to procure an *easy Small-Pox*, and a perpetual Security of neither *dying* by it, nor being again infected with it.

Afterwards he successively met with a Number of *Africans*; who all, in *their* plain Way, without any Combination, or Correspondence, agreed in *one Story*, *viz*. that in their Country (where they use to die like *Rotten Sheep*, when the *Small-Pox* gets among them) it is now become a *common Thing* to cut a Place or two in their Skin, sometimes one Place, and sometimes another, and put in a little of the Matter of the *Small-Pox*; after which, they, in a few Days, grow a *little Sick*, and a few *Small-Pox* break out, and by and by they dry away; and that no Body ever dy'd of doing this, nor ever had the *Small-Pox* after it: Which last Point is confirm'd by their constant Attendance on the Sick in our Families, without receiving the Infection; and, so considerable is the Number of these in our Neighbourhood, that he had as evident Proof of the *Practice*, *Safety*, and *Success* of this Operation, as we have that there are *Lions* in *Africa*.

After this, he heard it affirm'd, That it is no unusual Thing for our Ships on the Coast of *Guinea*,[9] when they ship their Slaves, to find out by Enquiry which of the Slaves have not yet had the *Small-Pox*; and so carry them ashore, in this Way to give it to them, that the poor Creatures may sell for a better Price; where they are often (inhumanly enough) to be dispos'd of.

Some Years after he had receiv'd his first *African* Informations, he found publish'd in our *Philosophical Transactions*, divers Communications from the *Levant*, which, to our Surprize, agreed with what he had from *Africa*.

First, That very valuable Person, Dr. *Emanuel Timonius*, writes from *Constantinople*,[10] in *December* 1713, That the Practice of procuring the *Small-Pox*, by a Sort of *Inoculation*, had been introduc'd among the *Constantinopolitans*, from the more Eastern and Northern *Asiaticks*, for about

for Connecticut and Massachusetts in London, was instrumental in publishing Mather's essay in London.

7 To protect his anonymity, especially in light of the furor of Bostonians against inoculation, Cotton Mather refers to himself in the third person.

8 Also Guramantee. An African Berber people who lived in what is the southern part of modern Libya. Mather refers to his African servant Onesimus (*D* 1:579; 2:271, 363 *n*.). See M. Minardi (2004) and K. Wisecup (2011, 2013).

9 Coastal region of western Africa.

10 Modern Istanbul, Turkey.

forty Years.[11] At the first (he says) People were cautious and afraid; but the *happy Success* on Thousands of Persons, for (then) eight Years past, had put it *out of all Suspicion*. His Account is, That they who have the *Inoculation* practis'd upon them, are subject to very *slight Symptoms*, and sensible of but very little Sickness; nor do what *Small-Pox* they have, ever leave any *Scars* or *Pits* behind them. They make Choice of as healthy a young Person as they can find, that has the *Small-Pox* of the best Sort upon him; on the *Twelfth* or *Thirteenth* Day of his *Decumbiture*, with a *Needle* they prick some of the larger *Pustules*, and press out the Matter coming from them into some *convenient Vessel*, which is to be stopt *close*, and kept *warm*, in the Bosom of the Person that carries it to the intended Patient. This *Person* ought rather to be some *other*, than he who visited the *sick Chamber* for it; lest he should carry the Infection in the *common Way*, which might prove dangerous. The Patient is to have several *small Wounds* made with a *Needle*, or *Lancet*, in two or more Places of the *Skin*, (the best Places are the Muscles of the *Arm*) and immediately let there be dropt out a *Drop* of the Matter in the Glass on each of the Places, and mix'd with the Blood that is issuing out. The Wound should be cover'd with some little *Concave Vessel*, and bound over, that the Matter may not be rubb'd off by the Garments, for a few Hours. And now, let the Patient (having *Fillets* on the Wounds) keep House, and be careful of his *Diet*. The Custom at *Constantinople*, is to abstain from *Flesh* and *Broth* for twenty Days, or more; and they chuse to perform the Operation, either in the Beginning of the *Winter*, or the *Spring:* The *Small-Pox* begins to appear sooner in some than in others; and with lesser Symptoms in some than in others; but *with happy Success in all*.

Commonly ten or twenty *Pustules* break out; here and there one has no more than two or three, few have an *Hundred:* There are some in whom no *Pustules* rise, but in the Places where the *Incision* was made: And here the Tubercules will be purulent; yet even these have never had the *Small-Pox* afterwards, tho' they have cohabited with Persons that had it on them. No small Quantity of Matter will run, for several Days, from the Places of the *Incision*; but the *Peeks* arising from the Operation, are dry'd up in a little while, and fall of[f], partly in thin Skins, and partly vanishing by an insensible Wasting: The Matter being

[11] Em[m]anuel Timonius (1669–1720), a Greek physician and immunologist, published his observations of smallpox inoculation in the Ottoman Empire as practiced by the common people. The English translation of his report was published as "An Account of history, of the procuring the Small-Pox by Incision, or Inoculation: As it has for some time been practised at Constantinople," in *PT* 29.339 (June 30, 1714): 72–82. See also Zabdiel Boylston (and Cotton Mather), *SAIT* (1721); Mather's *ABJ* (1972), pp. 109–12; A. Boylston (2012).

hardly so thick a *Pus* as in the common *Small-Pox*; tho', at the Places of the *Incision*, 'tis more of the common Sort; and there the *Scars* will always remain, as Remembrances of it. If an *Apostem* should break out in any,[12] (which is more frequent in *Infants*) yet there is no Fear; for 'tis healed safely by Suppuration:[13] In fine, this learned Person assures us, that he never yet observ'd any *bad Consequence* of the Practice, which now so many come into.

After this, we find an honourable Person, whose Name is *Jacobus Pylarinus*, the *Venetian* Consul at *Smyrna*, (who was, I suppose, wholly a Stranger to what had been written by the former) publishing what is entituled, *Nova, ac Tuta Variolas Excitandi per Transplantationem Methodus*.[14]

This Gentleman observes, that this *wonderful Invention*, was first, *à plebeia rudiq; Gente, in Humani Generis adjumentum & in sævissimi Morbi solamen detecta*,[15] found out, not by the learned Sons of *Erudition*, but by a mean, coarse, rude Sort of People, for the *Succour of Mankind*, under and against one of the most cruel Diseases in the World. It was rarely, if ever used among the People of *Quality*, till after the Beginning of the present Century. A *noble Grecian* happening then to try it with happy Success upon four little Sons of his own Family. *Mirum quam multas Nobiliorum Familias ad Imitationem traxit*.[16] It was wonderful to see how many *People of Fashion* presently follow'd the Example; so that at this Day (he says) every one does, *without any Hesitation*, and with all the Security imaginable, practice the *Transplantation*, except here and there a few *Cowards*, that are afraid of their *Shadows*. Indeed, the *Turks* who ascribe all Events to *Fate*, for that Reason come but slowly into it.

The Instructions given by *Pylarinus* about the Operation, differ little from those of *Timonius*, and so there is no Need of repeating them.

He adds, That some do the Business with no more than *one Incision*.

12 Abscess.

13 Discharge of pus.

14 Like his Greek colleague, Jacobus Pylarinus (1659–1718), a Greek physician and Venetian consul in Smyrna, published his observations on smallpox inoculation and attracted wide attention throughout Europe. Pylarinus' *Nova et Tuta Variolas Excitandi per Transplantationem Methodus* (1715) was reprinted as "Nova & tuta Variolas excitandi per Transplantationem Methodus," in *PT* 29.347 (March 31, 1716): 393–99. The Latin title signifies, "A New and Safe Method of Bringing about the Pox through Transplantation."

15 Pylarinus, *Nova et Tuta ... Methodus* (1715), p. 11; *PT* 29 (1716): 393. Mather provides his own translation.

16 Pylarinus, *Nova et Tuta ... Methodus* (1715), p. 17; *PT* 29 (1716): 394. Mather provides his own translation.

The Ferment comes into Action sooner in some than in others; usually 'tis on the *seventh* Day, sometimes on the *first.*

The *Symptoms* prove *remis*, or *intense*, according to the various Constitutions of the Bodies.

The *Small-Pox* proves of the *distinct Sort*, and there are but few of them.

In some few, the *Incision* has produc'd no *Small-Pox* at all; but the Persons have afterwards been taken in the *common Way*, and handled with it like other People.

The *Wounds* made for the *Incision*, prove often very sore; and with some they degenerate into *Apostems*; nay they swell sometimes, and rise and fall, and rise again. There has also happen'd on this Occasion, an *Abscess* with Suppuration, in some Emunctory of the Body; but this is a rare Occurrence.

In fine, *Pylarinus* affirms, it was hardly ever known, that there was any *ill Consequence* of this Transplantation. *Quinimò ritè rectéq; tractata, & in Corporibus per peritum Medicum aptè præparatis, certissimam promittit Salutem.*[17] The Business being well, and wisely manag'd, and the Body being, by a skilful Physician, well prepar'd, you may depend upon it. (He says) in an ordinary Way, there can be *nothing but a good Issue* of it.

After these Communications, and the Thing *establish'd in the Mouths of two such Witnesses*,[18] we met with some ingenious Travellers, who knew so much of the Matter, as very much to confirm the Ideas we had entertain'd of it.

Were one of an ordinary Capacity (for no better is he that is now writing) willing to try a little how far *Philosophy* might countenance the Matter: One might think, the venomous *Miasms* of the *Small-Pox*, entering into the Body, in the Way of *Inspiration*, are immediately taken into the Blood of the *Lungs*; and, I pray, how many *Pulses* pass, before the very *Heart* is pierc'd with them? And within how many more they are convey'd into all the *Bowels*, is easily apprehended, by all who know any Thing how the *Circulation of the Blood* is carry'd on; at the same Time the *Bowels* themselves are infeebled,[19] and their Tone impair'd, by the *Venom* that is thus insinuated. Behold the Enemy at once got into the very *Center* of the Citadel;[20] and the invaded Party must be very strong indeed, if it can struggle with him, and after all entirely expel and

[17] Pylarinus, *Nova et Tuta ... Methodus* (1715), p. 32; *PT* 29 (1716): 398. Mather provides his own translation.

[18] Deut. 19:15; 2 Cor. 13:1.

[19] The English physician William Harvey (1578–1657) described his discovery of the circulation of blood in the body, in *Exercitatio Anatomica De Motu Cordis et Sanguinis in Animalibus* (1628).

[20] See also Mather's CA 6.2 "*Variolae triumphatae.*"

conquer him: Whereas the *Miasms* of the *Small-Pox*, being admitted in the Way of *Inoculation*, their Approaches are made only by the *Out Works* of the *Citadel*, and at a considerable Distance from it. The Enemy, 'tis true, gets in so far, as to make some *Spoil*; even so much as to satisfy him, and leave no *prey* in the Body of the Patient, for him ever afterwards to seize upon: but the *Vital Powers* are kept so clear from his Assaults, that they can manage the *Combat* bravely; and tho' not without a *Surrender* of those Humours in the *Blood*, which the Invader makes a Seizure on, they oblige him to *march out the same Way he came in*, and are sure of never being troubled with him any more. If the *Vermicular Hypothesis* of the *Small-Pox* be receiv'd with us, (and it be, as many now think, an *animalculated Business*) there is less of *Metaphor* in our Account, than may be at first imagin'd.

But to what Purpose is all this *Jargon?* And of what Significancy are most of our *Speculations?* EXPERIENCE! EXPERIENCE! 'tis to THEE that the Matter must be referr'd after all: a few *Empericks* here, are worth all our *Dogmatists*.

About three Months ago, the *Small-Pox* broke in upon the City of *Boston*, where it very much appeared with the *Terrors of Death* to the Inhabitants.[21] On this Occasion, there was address'd a Letter to the *Physicians* of the City, with an Account of the Communications from the illustrious *Timonius*, and *Pylarinus*, entreating them to meet for a *Consultation* upon it, *Whether the new Practice might be introduc'd and countenanc'd among us?*[22] The Writer was perswaded, that herein he did but his Duty, and express'd no other than

21 According to the *Report of the Record Commissioners of the City of Boston* (1883) for May 12, 1721, Boston's selectmen and governor consulted "about the Sea-horse man of war, being Sent down to Spectacle Island and, Pursuant to the Law of this Province to prevent (God willing) the Spreading of the Smal Pox [*sic*] in this Town & Province, Two or three men being Sick of that Distemper on board the Said Ship now in the Harbor" (8:154, entry 399). In his medical handbook *ABJ* (1972), cap. 20, p. 112, Mather is more specific: "About the Month of *May,* 1721, the *Small-Pox* being admitted into the City of *Boston*, I proposed unto the Physicians of the Town, the unfailing Method of preventing Death, and many other grievous miseries, from a tremendous Distemper, by Receiving and Managing the *Small-Pox*, in the Way of *Inoculation*." Mather's worries about the epidemic are detailed in his ledger for the months of May–December (1721), in *D* 2:620–65. A.M. Kass (2012) reviews the main issues of the epidemic. Ola E. Winslow's mainstay (1974) is as useful as ever. J.B. Blake's account (1959) draws the big picture of Boston's health measures during the city's first two centuries, and so do Gareth Williams (2010) and C. Silva (2011).

22 Mather's proposal (June 6, 1721) to Boston's medical establishment first circulated in manuscript among the practitioners (*D* 2:620–21, 624, 628, 662) – was subsequently incorporated in *SAIT* (1721). See G.L. Kittredge, "Lost Works" (pp. 428–40). Folliwing Kittredge, T.J. Holmes argues convincingly that *SAIT* was written by Mather but published under Zabdiel Boylston's name (3:996).

the *Charity of a Christian*, and a proper Concernment and Compassion for his poor Neighbours, whom he saw likely to die by Hundreds about him.[23] His Address found (for what Reasons I know not, or am not willing to know) an *indecent Reception* with our Physitians; all the Return he had, was a Story which they spread about the Town and Country, that he had given an *unfaithful Account* of the Matter to them,[24] tho' they had it in the printed *Philosophical Transactions* before their Eyes to justify it. Then the Story was turn'd, that either Dr. *Halley*[25] had suppress'd Part of the true Account, or *Timonius* and *Pylarinus* were themselves fallacious:[26] Nevertheless, one who had been a more *successful Practitioner* than most of them, and had with a singular Dexterity in his Practice, perform'd Things not attempted by any of them, (namely, Mr. *Zabdiel Boylston*) was prompted,[27] by his enterprizing Genius, to begin the Operation. He thought it most generous to make his first Beginning upon his *own Family*; and here, to make not only two *Slaves*, but a beloved *Son* of his own, (about five or six Years of Age)

[23] On May 26, 1721, Mather revealed his plans to enlist Boston's medical establishment to fight the contagion: "G[ood] D[one]. The grievous Calamity of the *Small-Pox* has now entered the Town. The Practice of conveying and suffering the *Small-Pox* by Inoculation, has never been used in *America*, nor indeed in our Nation. But how many Lives might be saved by it, if it were practised? I will procure a Consult of our Physicians, and lay the matter before them" (*D* 2:620–21). For the composition of *An Account of the Method and Success* and the reconstruction of Mather's lost "Address to the Physicians" (1721), see G.L. Kittredge, "Lost Works," pp. 428–55.

[24] See [William Douglass] *Inoculation of the Small Pox As practised in Boston* (1722), pp 3–4; and Douglass's *Postscript to Abuses, &c. obviated* (1722), pp. 2–4; Zabdiel Boylston's *An Historical Account of the Small-Pox Inoculated in New England* (1726), pp. 3–4.

[25] The English astronomer and physicist Edmond Halley (1656–1742) was elected corresponding secretary of the Royal Society in 1686, a position which entailed the editorship of the distinguished scientific journal *Philosophical Transactions*. Hence the allegation that "Dr. Halley had suppress'd Part of the true Account." Today, his name is most often associated with the famous comet, which bears his name, for in collaboration with Sir Isaac Newton, Halley had mathematically demonstrated that the comet of 1682 (in spite of its orbital variations) was the same one that had previously appeared in 1607 and 1531. Halley published his findings as "*Astronomiae Cometicae Synopsis*," *PT* 24.297 (March 31, 1705): 1882–99.

[26] M.W. Sivils (2011) ably dissects the pamphlet literature of the inoculation controversy. See also C. Silva (2011).

[27] The Boston physician Zabdiel Boylston (1679–1766) was the sole person to heed Mather's call for vaccination. Boylston and Mather collaborated on *Some Account of What is Said of Inoculating or Transplanting the Small Pox* (1721). They describe the standard method of inoculation and respond to each of the main objections. The only biography of Boylston is G. Marvin's (1975).

the Subjects of it: He made the *Transplantation* into them with two or three *Incisions* a-piece, taking the *Leg* as well as the *Arm*, (and in one of them the *Neck*) for the Places of them: He did not use the *Precaution* of sending for the fermenting *Pus* by a third Person: He staid not for what some would have thought more *proper Seasons*; but he did it in the very *Heat* of *Midsummer*; which, with us, is hot enough: He did nothing at all to *prepare* their Bodies; and he chose to leave them to the *Liberties*, which Persons infected with the *Small-Pox* in the *common Way*, do generally take, before their *Decumbiture*, without any Detriment to them.

Under all these Disadvantages, did this Gentleman make his Experiments; but *they succeeded to Admiration*. About the seventh Day the Patients began to grow Feverish, and out of Order; on the third and fourth Day from their falling ill, his Child's Fever grew to an Heighth, beyond his Expectation, which (from the *Novelty* of the Business) did, for a few Hours, considerably terrify him: He had Recourse to the common Remedies of Blisters, and gave the Child a *Vomit*, and presently all the Fright was over: The *Eruption* began; and from the Time of its doing so with the *Child*, and with the two *Slaves*, there was no Occasion for any *other Medicine*; they were easy from this Time; their *Pustules* (which were, tho' not many, yet somewhat more for Number, than what is usual in the *Levant*) grew, and fell off, as they do in the *Levant*; and their *Sores*, which had an agreeable Discharge at them, seasonably dry'd up of themselves; and they all presently became as hail and strong as ever they were in their Lives.

It is incredible, what a *Storm* was rais'd, and very much of it principally owing to some of our enrag'd *Physicians*, on this Occasion.

The Gentleman was threaten'd with an *Indictment* for *Felony*; (tho', in your Countrey, they talk of no such Thing, for them who carry their Children into *infected Chambers*, on Purpose to make them take the *Infection* after the common Way in their Minority) and Words were given out, which had a Tendency to raise the *Mob* upon him.

The *select Men* (an Order of Men, who are the *Overseers* and *Managers* of the *Town-Affairs*) associating with some of the *Justices*, order'd him to appear before them, and severally reprimanded him for spreading the *Small-Pox*; (which was already spreading in the *common Way*) and with high Menaces warned him against proceeding with his Practice any farther. At the same Time, the *Practitioners* of the Town publish'd a Declaration,

> That the *Inoculation* of the *Small-Pox* had prov'd the Death of many Persons, soon after the Operation; and brought Distempers on many others, which have, in the End, prov'd fatal to them; which (they said) appear'd by numerous Instances: *That* the natural Tendency of infusing such malignant Filth in the Mass of Blood, is to corrupt and putrify it; and lay a Foundation for many dangerous Diseases: *That* the Operation tends to spread and

continue the Infection in a Place, longer than it might otherwise be; and *that* the continuing of the Operation among us, is likely to prove of most dangerous Consequence.[28]

There appear'd little to support these *Paradoxes*; but only an *Oath* of an obscure *Frenchman*,[29] (a little known for his *virtuous* Morals) that about five and twenty Years ago, there were thirteen Soldiers in the *French* Army, upon whom this Operation was perform'd; but four dy'd of it; and when they were open'd, horrid Things were found in them; six recover'd with abundance of Trouble, being seiz'd with Tumours, and large Inflamations; the other three found no Effect of it upon them. He swore likewise, that in *Flanders*, twenty Years ago, a Captain, under his Cure for the *Small-Pox*, told him, that ten Years before this, he had been *inoculated five or six Times*, without that cursed Invention's taking Effect upon him; and he was so violently seiz'd, that he had several Ulcers on his Body, especially one on his *Arm*, which was incurable. He swore likewise, that at the *Battle* of *Almanza* in Spain,[30] two *Muscovite* Soldiers had the Operation perform'd on them; and that one recover'd, and the other felt no Impression; but six Weeks after, a Frenzy seized him, and he was swell'd all over his Body, and thought to be poison'd; but his Body being open'd, his Lungs were found ulcerated, which the Physicians concluded was the Effect of the *Inoculation*. Things which Master *Apella* himself would scarce imagine to be credible![31] This notable Testimony was corroborated with one or two more, which amounted to little more than this, That a Man in the *Mediterranean*, twenty Years ago, was told by *somebody*, that *somebody* told him, that *somebody* heard, that the Doctors in the *Archipelago* warned People against *Inoculation* of the *Small-Pox*, as a dangerous Practice.

At the same Time, there was publish'd in our *News-Letter*, a bitter Satyr upon Dr. *Boylston* for his late Action;[32] and his *putting this far fetch'd*, and

[28] "A Resolve upon a Debate held by the Physicians of *Boston*, concerning inoculating the Small-Pox, on the twenty first Day of *July*, 1721," reprinted in Boylston's *Historical Account* (1726), p. 61.

[29] See Boylston's *An Historical Account* (1726), Appendix, pp. 58–60.

[30] The Battle of Almanza, in 1707, pitted Spanish, French, and Irish regiments against an equal number of English, Portuguese, German, Dutch, and French Huguenot forces during the War of the Spanish Succession (1707–14).

[31] An allusion to Horace's *Satires* (1.5.100–01): "credat Judaeus Apella, non ego" ("Let the Jew Apella believe it. I don't"). See W. Schleiner (1995), pp. 82, 87, 94.

[32] A likely reference to Isaac Greenwood's anonymously published *Friendly Debate; or, A Dialogue, Between Academicus* [Dr. Boylston]; *and Sawny* [Dr. William Douglass] *& Mundungus* [John Williams], *Two Eminent Physicians, About some*

(it was pleas'd to say) *not well vouch'd Method, into Practice*. They compar'd it to the Infusion of the Venom of *Serpents*, or the Oil of *Tobacco*, by Injection, immediately into the Mass of *Blood*; and affirm'd, that it would produce only an *eruptive Fever*: but *nothing Analogous to the* Small-Pox; (tho' they inveigh'd against the Gentleman for spreading the *Small-Pox* in what he did) and that the Sufferers might, notwithstanding, receive the *Small-Pox* afterwards in the ordinary Way: They also asserted, that the Thing had been among the Learned universally known in *England*, above *twenty Years*; (and more than twice seven Years before, the Secretary of the ROYAL SOCIETY first recommended it as a *new Thing*, inviting the Thoughts of the Curious in the *English* Nation upon it) but it had never been practis'd, because it was deem'd *wicked* and *felonious*; and then they referr'd the CASE to the *Divines*, which was express'd in these emphatical Terms: *How the trusting more the groundless* Machinations *of Men, than to our Preserver in the ordinary Course of Nature, may be consistent with that Devotion and Subjection we owe to the All-wise Providence of God Almighty*.[33] Six of our Ministers thought themselves bound in Duty to bestow a *publick Rebuke* upon this indecent Satyr; and answer'd the CASE not much to the Satisfaction of some that offer'd it.[34] But the Rage of the People, was, with a very powerful *Pharmacy*, boil'd up to a very great Extremity; and for the more effectual enraging of them, there was a strange Use made of a Passage, that one must have had the *Sense of three Men* in him, to have made such a Construction of it.[35] *Pylarinus* had said, *That sometimes an Abscess happen'd in some Emunctory of the Body*;[36] *which yet* (he said) *was a rare Occurrence*. Now, tho' our Gentlemen knew, that this is a Thing which *daily occurs*, among the People recovering from the *Small-Pox* in the *ordinary Way*; (and we had never seen any Thing of it in our *incisious*

of their Later Performances (1721); and the anonymous *A Friendly Debate; or, A Dialogue Between Rusticus and Academicus About the late Performance of Academicus* (1722).

33 Boston *News-Letter* (July 24, 1721). See also J.B. Blake (1952), p. 493.

34 Mather alludes to the anonymous, *A Vindication of the Ministers of Boston, From the Abuses & Scandals, lately cast upon them, in Diverse Printed Papers* (1722).

35 Mather alludes to *Several Arguments, Proving That Inoculating the Small Pox Is not contained in the Law of Physick, either Natural or Divine, and therefore Unlawful* (1721), and *An Answer To a Late Pamphlet, Intitled, A Letter to a Friend in the Country* (1722) – both by John Williams (1664–1729), a Boston tobacconist and medical practitioner.

36 Any organ of the body, like the kidney or liver, responsible for clearing the bloodstream of toxins. Mather refers to Pylarinus's "Nova & tuta," *PT* 29.347 (March 31, 1716): 398.

Way) yet they rais'd a horrible Cry of *Raw-Head and Bloody-Bones*, that the *Inoculation* of the *Small-Pox* would bring in the *Plague* among us. It must needs be so, because *Plague Sores* were often in some *Emunctory of the Body*; and so *Boston* was going at once to be another *Marseilles*;[37] and they had the *Plague* at *Constantinople*, since *Inoculation* had been admitted there. Our People were scar'd (I cannot say) *out of their Wits*. But the Gentleman having succeeded so well in his first *Experiments*, and being perswaded, that if God would please to make him the happy Instrument of saving the precious *Lives* of his poor Neighbours, it would make amends for all the Obloquies which his *envious Brethren* might raise upon him: He took *little Notice* of the *Inhibition* that had been given him.[38] Divers pious and worthy People consider'd how dreadfully the *Small-Pox* handled many of their Neighbours, and how much the *sixth Commandment* order'd them the Use of Means to preserve and prolong their Lives.[39] They consider'd also, that by managing and governing the Approach of the *Small-Pox*, in the Way of *Inoculation*, they had a Method of securing their *Lives* from the Dangers of it; and that there never arriv'd unto us a *Medicine* of so great a *Recommendation*; it having been used upon so many *Thousands*, and never *one* known to have miscarry'd under it. They therefore apply'd themselves to Dr. *Boylston*, with Desires to come under the Operation; and he charitably gratify'd them in the Thing desir'd; tho' the Objects were most of them, either so *Old*, or so *Weak*, that they would have been the *last* that one would have chosen for it; and one would have apprehended no little Hazard of the *Event:* But they all got *well* and *soon* through it, and so much beyond their Expectation, that they zealously gave Thanks to God, for leading them into it; and seriously profess'd to their Neighbours, that they had rather suffer the Operation *twice* every Year, than *once* to undergo the *Small-Pox*, as it is most *commonly suffer'd*, tho' they should be sure of surviving it.

The *Fever* in these also, was, for a few Hours before the Eruption, more Intense, and the *Pustules* after it, were more numerous, than what is usual in the *Levant*; but in a few Days, and much sooner than what is commonly

[37] Like many other European and northern African cities from the 14th through the 19th century, the French Mediterranean port of Marseille experienced several devastating smallpox epidemics in the 17th century. Mather here alludes to the concomitant outbreak in 1720–22, which killed nearly half of Marseille's population. See "Abstract of a Letter from a Physician at Marseilles, to John Wheake Esq; at London" (September 8 1720), *The New-England Courant*, No. 2 (Monday, August 7–14, 1721), pp. 1–2.

[38] See Mather's *D* (2:445–46, 632, 635).

[39] "Thou shalt not kill" (Exod. 20:13).

done, where they have the *Small-Pox* the common Way, they recover'd their entire Strength, and were, on all Accounts, as well as they were before; and the *Sores* of the *Incision* also heal'd of themselves, rather sooner than they wish'd for. Whereupon, they publish'd a *Declaration* in our *Gazette*, that they might stop *false Reports*, and satisfy the Minds of sober People, in the Midsts of the *Ephesian* Clamours now prevailing;[40] for truly, while these Things were doing, the Town was fill'd with Iniquity, to a Degree which good Men could not observe, without being more than a little *griev'd* at it.

I must say it, I never saw the Devil so *let loose* upon any Occasion. A *lying Spirit* was *gone forth* at such a Rate, that there was no believing any Thing one heard.[41] If the *inoculated Patients* were a little *sickish*, or had a *Vomit* given them, it was immediately reported, That they were at the *point of Death*, or *actually dead*. While the Patients lay blessing and praising Almighty God, for shewing them this *easy Way to escape* a formidable Enemy, it was confidently reported, *That they bitterly repented of what had been done upon them, and would not, upon any Terms, be brought into it, if it were to do again*. When the Patients had their *incisious Places*, either actually and perfectly heal'd in some, or within a Day or two of it in others, it was confidently reported, *That they were perishing under terrible Ulcers, and had their Arms or Legs rotting off*.

These, and the like Things, were asserted with such impudent Confidence, even by such as liv'd in the nearest Neighbourhood, that one was almost ready to fear a Beating, if they durst offer to question them. Then the People would assert, that here were Persons on the very Spot, who underwent the *Inoculation* in *England* a great many Years ago; but afterwards had the *Small-Pox* in the common Way; and, they said, they would *bring these Persons to us*. A few Minutes after, they would assert, that it was never practis'd in England; but there was an *Act of Parliament* which made it *Felony*; and, they said, they would produce the Act to us. But never any Patient had so many *Pustules* of the *Small-Pox*, as there were *Lies* now daily told, and spread among our deluded People.

That which much added to the Misery, was, that the People who made the *loudest Cry*, (who most commonly were what we may not improperly call *of the confluent Sort*, and such also as were past the Dangers of the *Small-Pox* themselves) had a very *Satanic Fury* acting them.

They were like the *possess'd People* in the Gospel, *exceeding fierce*; insomuch, that *one could scarce pass by the Way* where they were to be met

40 Referring to Eph. 4:31, "Let all bitterness, and wrath, and anger, and clamour, and evil speaking, be put away from you, with all malice."

41 See *D* (2:636–37, 641, 651).

withal. Their common Way was to rail and rave, and wish *Death*, or other Mischiefs, to them that practis'd, or favour'd this *devilish Invention*. To inflame them in their Transports, and harden them in their Violences and Exclamations, they pretended *Religion* on their Side; and charg'd all that were not so, with denying and renouncing the *divine Providence*, and I know not how many more *Abominations*; yea, with *going to the Devil*, and the *God of Ekron*.[42] And how strangely they treated the most meritorious Ministers of the Gospel, who did not come into their *Frenzies*, I leave unmention'd. *Father forgive them*.

The View of these Things caus'd some considerate Persons to think, whether the *Angel of Death* promising himself a great *Feast* in this miserable City, *it might not put him that had the Power of Death*,[43] *that is, the Devil*, into a great *Rage*, to see coming in among us, a *Method* of rescuing many *Morsels* from him. They were sure, the *Lying*, and the *Malice*, and the *Outrage*, that manag'd, and carry'd on the Opposition to it, must needs be of a *Satanic Original*; there could be nothing more contrary to the Spirit of a *meek* JESUS.

But the *Opposers* could not hinder the *Proselytes* from increasing. Dr. *Boylston*[44] has his Patients multiplying; and, instead of any one miscarrying, they come off easier, with less *Fever*, and fewer *Pustules*, than those that *led the Van* in his Experiments. He grows more *expert* every Day; and many of the most vehement Opponents, daily come over to an Approbation of the Practice; and People that see their Neighbours, either dying in a tragical Manner about them every Day, or, at best, having their *Loins fill'd with a loathsome Disease*,[45] won't permit themselves to be *talk'd out of their Lives*, by pitiful Impertinencies.

Of what my ingenious Friend hitherto *does*, and *finds*, you will now accept this brief Account.

He makes usually *two Incisions*, (tho' sometimes but one) in the two *Arms* (or an *Arm*, or a *Leg*) of his Patients; and then he puts into them a little Bit of *Lint*, which he has dipp'd into the Quitter, that he had newly fetch'd in a little

[42] Kings 1:1-6. The god of Ekron, aka. Beelzebub, was revered among the Canaanites. Here, Mather appears to allude to the anonymous anti-vaccination pamphlet, *The Imposition of Inoculation As a Duty Religiously Considered* (1721), by Samuel Grainger (c. 1686–1734); and to John Williams's *Several Arguments* (1721) and *An Answer To a Late Pamphlet* (1722).

[43] Heb. 2:14.

[44] The printed text reads, "*Boylstone*."

[45] Ps. 38:7.

Bottle, warm from the Pustules of one who has the *Small-Pox* of a good Sort, now turning upon him. This he covers with a little Plaister of *Diachylon*, to keep it close for two or three Days in its Operation there.

In his first Practice, upon his removing the little *Pledget*, the *Sores* of the *Incision* would sometimes *heal up*, and anon swell and break forth of themselves, when the *Fever* came to be over: But he thought afterwards, that by keeping the Sores *constantly open*, in the Way we do an ordinary *Issue*, the *Fever*, and other Symptoms, presently to come on, are sensibly *moderated*. He allows the Patients, for a while, to go *abroad* about their Business, if the *Weather* be good, and there be no Danger of getting any Cold; but he directs them to regulate their *Diet*, and, particularly to retrench a little their *Carnivorous* Inclinations: And yet he does little this Way, because he finds no great Advantage by thus enfeebling of them.

About the *seventh* Day, (and sometimes a little before, and sometimes a little after) the Patients grow dull, and feel a *feverish* Disposition, and grumbling Pains of the *Head* and *Back*, as in the Invasions of the *Small-Pox* 'tis usual. In some, the *Symptoms* are very *Remiss*, in others they are more *Intense*. But the Doctor governs the *Fever* at his Pleasure, in the Ways that every Body now treats a common putrid *Fever*.

Sometimes on the third or fourth Day, the *Fever* seems, for a few Hours, a little rampant; but then he gives a gentle *Vomit*, and afterward a *Blister*, and the *Eruption* begins without fail immediately.

If you expect now a *long Story* about the Managements in the *Progress* of the Distemper, I shall disappoint you; for *here's the End on't*; there's *no more* to be done; there's no farther *Illness*; no farther *Trouble*; the *Storm* is over; there is not one Atom of that *second Fever*, which, in the *Small-Pox*, is what People generally die of. The Patients have no more to do, but lie *still*, and keep *warm*, and entertain their Friends, and study how to glorify the God of their Lives.

The *Pustules* in some, have been very few, perhaps twenty, or thirty, or so; in others, they have risen to *several Hundreds*; yea, few have come off without *several Hundreds:* In this Point, that is to say, the Number of the *Pustules*, we differ from the *Levant* considerably; and are yet at a Loss for a Reason of it; but in as few Days, as those of the ordinary *distinct Small-Pox*, they are gone.

The *Sores* continue running somewhat longer, (and longer in some than in others) even some Days after their going abroad again. But they dry up of themselves, *Tutò, Citò, ac Jucundè*.[46]

46 "Swiftly, safely, and pleasantly." The Latin dictum was a common formulation in early modern Latin of a maxim attributed by Celsus (c. 25 BCE–c. 50 CE) to the

None of the *inoculated* Patients have yet had any of those *Boils*, which People, recover'd of the *Small-Pox* in the *ordinary way*, are commonly vexed withal. If there had been one Instance of them, our Practitioners would have taught the People to cry, *the Plague! the Plague!* upon it.

Thus Dr. *Boylston* goes on with his Practice *hitherto* successfully. There is one Difficulty, indeed, which very much incumbers him. He has employ'd the *Inoculation* upon one Person, who, having been just before tending two *Patients* that were under it, had very strongly receiv'd the Infection in the *common Way*: The Person was in a few Hours taken down, and prov'd very full of the *Small-Pox*, of the *confluent Sort*, and very narrowly escap'd with Life; to the Preservation whereof, it was yet thought, that the running of the *incisious Sores* might a little contribute.

The Miscarriage of this Person would have been, by the *Mob* as aforesaid, improv'd as a sufficient Subject for their Exclamations. 'Tis therefore no little Distress to the *Doctor*, how they that are *already strongly infected*, may be easily distinguish'd *from those* who are not so, in a Place where the Distemper is become *Epidemical*; for the Condition of every one, will not always allow them to perform the *Quarentine* that would be necessary for it. How he will get over this Difficulty, 'tis yet such *early Days* with us, that I cannot inform you.

Our Practitioners, and the People under their Influences, having had their *other Prejudices* confuted by the happy *Experience* of the Neighbourhood, now chiefly insist upon this, *Well, two or three Years hence you will see the dreadful Effects of this wicked Practice*; *you'll see what happens to the People that are under it*. And the sad Things that were to happen, when the *Lark-time* arrives, are now to terrify us. They are so unreasonable, that one had as good speak *Reason* to a *Post*, or argue with a *Whirlwind*. Were a *Borarius* alive again, what would he say to such *unreasonable Persons*.[47] A Crazy old Man, that is near *Seventy*, having lately enjoy'd the Benefit of *Inoculation*, 'tis thought, that if he should happen to die one Minute before *Ninety*, these People (if not come to their Wits before) will say, *This Inoculation kill'd him*.

Greek physician Asclepiades of Bithynia (early 1st century BCE). In his treatise *De medicina* (3.4), Celsus presented it this way: "Asclepiades officium esse medici dicit, ut tuto, ut celeriter, et iucunde curet" ("Asclepiades says that it is the doctor's job to cure safely, swiftly, and pleasantly").

47 "*Borarius*," i.e., "Rorarius." Girolamo Rorario, or Hieronymus Rorarius (1485–1556), Italian diplomat and Pontifical legate, who in *Quod animalia bruta ratione utantur melius homine* (1644) – a work which went through many editions – asserts that animals make better use of reason than humans. Rorarius' work heavily leans on Plutarch's *Bruta Animalia Ratione Uti* (1st century CE).

It may, perhaps, a little divert you, to see the Humour on't.

A *Negro,* who pretends he can just remember, that he was *inoculated*, when he was a *Pickaninny*, in his own Country, perhaps thirty Years ago, (and was here under Cure for the *French Pox*, a little while ago) lately grew indispos'd, and was laid in a Room where another *Negro* lay, full of the *Small-Pox* upon him. At first it was confidently affirm'd, that this Fellow had also the *Small-Pox* broke out full upon him; and tho', till *now*, the mention of a *Negro* (or of any Thing from *Africa*) was hiss'd at, yet *now*, all on a sudden, a *Negro* was become good Authority, and poor *Inoculation* was like to be knock'd on the Head.[48]

But after the Fellow's lying many Days in the infected Chamber, it is impossible to produce the least Eruption of the *Small-Pox* upon him. What is now to be done? Why, they find something of a *Swelling* under his Arm; and some of our D—-s cry out, that this is the *Small-Pox*, in the Form, you must Note, of a large *Tumour*; and this *pestilential Form* is also owing to his *Inoculation* in his Infancy.

But among all the *Oppressions*, under which this *new Practice* is a Sufferer among us, the chief comes from a *Scruple of Conscience*, which no doubt, in many People, whose *Hearts* are better than their *Heads*, may be sincere and serious.[49]

They plead, That *the Whole have no Need of a Physician*;[50] and that it is not lawful for me to *make myself sick, when I am well*;[51] and bring a *Sickness*

48 In his medical handbook, Mather records, "I was first instructed in it [inoculation], by a *Guramantes*-Servant of my own, long before I knew that any *Europeans* or *Asiaticks* had the least Acquaintance with it; and some years before I was Enriched with the Communications of the Learned Foreigners [Drs. Emmanuel Timonius and Jacobus Pylarinus], whose Accounts I found agreeing with what I received of my Servant, when he showed me the Scar of the Wound made for the Operation; and said, That no Person Ever died of the *Small-Pox*, in their Countrey that had the Courage to use it. I have since mett with a Considerable Number of these *Africans*, who all agree in one Story; That in their Countrey *grandy-many* dy of the *Small-Pox:* But now they Learn This Way: People take Juice of *Small-Pox*; and cutty-skin, and putt in a drop; then by'nd by a little *sicky, sicky:* then very few little things like *Small-Pox*; and no body dy of it; and no body have *Small-Pox* any more. Thus in *Africa*, where the poor Creatures dy of the *Small-Pox* like Rotten Sheep, a Merciful God has taught them an *Infallible Præservative*" (*ABJ* [1972], p. 107). See also CA 7.1 "The Case of the Small-pox Inoculated, further cleared," to John Jurin (May 21, 1723).

49 See Samuel Grainger's pamphlet mentioned above.

50 Luke 5:31.

51 See Cotton Mather, *Sentiments on the Small Pox Inoculated* (1721), p. 2; John Williams' response in *Several Arguments* (1721); and Grainger's *Imposition of Inoculation* (pp. 14, 15, 16, 25–26).

on myself; no, tho' it be to *prevent* a *greater Sickness*. 'Tis to no Purpose to tell them, that they cavil against the Use of all *preventing Physick*; and that they confute themselves as often as they take a *Vomit*, or use a *Blister*; and much more, if they undergo *Salivation*[52] (a Thing a thousand Times worse than the *Dispumation*[53] which is now in Controversy) prevent a Malady which may be fear'd, but is not actually come upon them; and that they are not *the whole*, while they have the Fuel of the *Small-Pox* lodg'd in them; or are in that *Anxiety of Mind* about it, which is, indeed, an *evil Disease*.

They plead, *That what is now done, is a Thing learnt from the* Heathens; *and it is not lawful for* Christians *to learn the Way of the* Heathen.[54] 'Tis to no Purpose to tell them, that *Hippocrates*, and *Galen* were *Heathen*; and that the Gentleman who invented the blundering (but strangely useful) Composition, call'd *Venice-Treacle*,[55] was an *Heathen*, who was Physician to no better a Man than a *Nero:* And from whom is it that we have our *Mithridate?*[56] And how many noble *Specificks* have we *learnt* from our *Indians?* And from whom did they *learn* to smoak *Tobacco*; or drink *Tea* and *Coffee*?

In fine, (tho' there is *no End* of their Follies) while they have been taught the Clamour, *That we take up this new Thing from the* Turks; [A Mistake; for it comes not from *such*, but from the *Greeks*] Their main Cavil against it, were more fit for the Mouths of *Turks*, and is the very thing that keeps the

52 "A secretion or discharge of saliva: esp. the production of an excessive flow by administering mercury" (*OED*).

53 Discharge.

54 See John Williams, *Several Arguments* (1721), p. 14; and his *An Answer To a Late Pamphlet*, p. 6; Grainger, p. 24.

55 Thomas Sydenham, *Entire Works* (1749), pp. 392–93, sec. 102, praises the efficacy of "Venice Treacle" (i.e. Venice Theriaca), a curious concoction of 64 herbal and God-knows-what-other remedies known since ancient times. See Sydenham's *Entire Works* (1749), p. 398, sec. 112; also, "an antidote against the bite of a poisonous animal" (*Medical Dictionary Online*). "Venice Treacle," also known as *Theriacum Andromachi* in honor of its inventor Andromachus of Crete, Nero's physician, was also a standard name in the early modern period for an antidote to poisonous animal bites. See J.P. Griffin (2004).

56 A panacea against all poisons and diseases. Obviously, in referring to the antidote *Mithridate*, i.e. *theriacum mithridatum*, Mather reminds his contemporaries that this famous elixir (developed by a "heathen" doctor) is in common use among British physicians. It is named after the eponymous Mithridates VI Eupator of Pontus (reigned 120–63 BCE), who famously feared being poisoned; he is said to have developed his antidote against any and all poisons. We appreciate David Lupher for drawing our attention to the historical background of this medical potion.

Turks from coming much into it: That God has *decreed when*, and *how* we shall *dye*, and for us to pretend a Remedy that won't fail to *safe our Lives*, and secure us from *Death* by the *Small-Pox*, is to take the *Work of God* out of his Hands. And how do we know, that God will send the *Small-Pox* upon us at all, since there will be some that will escape it, where-ever it comes? 'Tis to no Purpose to attempt their *Instruction*; for if one does, they will quickly, in express Terms, tell him, *You shall never convince me*. But that they may *convince us*, that it is *Religion* which inclines and fixes them, 'tis well if we don't hear them fall to *wishing of Miseries* to them that shall try this *new Practice*, and *railing bitterly* against the most venerable Servants of God in the Land, and giving all the Signs of a *Satanic Energy* upon them. Alas! *Pudet hæc opprobria*.[57]

To quiet the Minds of People that would *think soberly*, the Doctor stated *the Case of Conscience* in these Terms,

> Almighty God, in his great Mercy to Mankind, has taught us a *Remedy* to be used, when the Dangers of the *Small-Pox* distress us; upon the Use of which *Medicine*, they shall, in an ordinary Way, be sure to have it not so severely as in the other Way, and consequently, not to be in such Danger of *dying* by this dreadful Distemper; as also to be deliver'd from the *terrible Circumstances* which many of them, who recover of this Distemper, do suffer from it. Whether a Christian may not employ this *Medicine*, (let the *Matter* of it be what it will) and humbly give Thanks to God for his *good Providence*, in discovering it to a miserable World; and humbly look up to his *good Providence* (as we do in the Use of any other *Medicine*) for the Success of it?[58]

And he thought it Answer enough to say upon it, *It may seem strange, that any Wise Christian cannot answer it*.

Sir, The Reason of my thus giving you the Story of the way to *save the Lives* of People, and *keep them easy too*, from one of the most *formidable Diseases* in the World, thus far practis'd and prosper'd among us, is, because I imagine so considerate and charitable a Mind as yours, may do something to bring it into Consideration, *How far the Thing may be encourag'd in our Nation*, where this Distemper sometimes makes terrible Ravages, and where there are so many Persons of great *Ranks* and *Hopes*, that would be very thankful for some *Assurance of their Lives*, in a Point wherein they will always be *hanging in doubt*, till the *Small-Pox* be over with them.

57 "It is shameful that such reproaches should be cast [upon us]" (Ovid, *Metamorphoses* 1.758).

58 Boylston, *Some Account of what is said* (pp. 14–15).

And then, the Reason of my also giving you the Story of the *Opposition* which the Thing has here met withal, and reporting such *foolish Things* as might be better bury'd in Oblivion, which yet I do without exposing the *Name* of any Person, as if I had been vindictively inclin'd, I might have done; 'tis because it's a thousand to one, that the Thing may meet with the same *Opposition* (for the *Corruptions* and *Weaknesses* of People are the same) on your Side of the Water; and the Knowledge of what has occurr'd, may be of some *Service* to arm you against many of your *Discouragements*. But I shall write a *Treatise*, rather than a *Letter*, if I do not here break off.

I do it with my hearty Wishes, that you may be a great Blessing to the World, and that a World of People may fare the better for you. What better can be wish'd you? By,

Boston, in *New-England*, *Sept.* 7, 1721.

S I R,

Your hearty Friend,

and Servant.

{C. Mather.}

6.2. *VARIOLÆ TRIUMPHATÆ*:[1] OR, THE SMALL-POX INOCULATED[2]

{After December 1, 1721.}

{To Dr. John Woodward.}

It is an *Hard Chapter* that we have now before us. There is a *Great Plague* which we call, The SMALL POX, wherein *the Misery of Man is great upon him*: A Distemper so well *known*, and so much *Felt*, that there needs no Description to be given of it.[3]

So Few among the miserable Children of Men do now escape it, that the *Enquirers after Causes* have suspected the *Original* of this Malady, to be some *Venom Connate* with every Man, (Derived, they'l tell you, from the *Maternal Blood* unto him) which lies Dormient and Buried, until it be fired by *Contagion*, and then furiously breaking out from its unknown Lurking-Place, it mixes with the whole Mass of Blood, and makes the Terrible *Disturbance*, and even *Destruction*, the *Fear* whereof holds Mankind in a very uneasy *Bondage*.

1 "The Smallpox conquered [triumphed over]." The editors agree with G.L. Kittredge, "Lost Works" (pp. 442–44, 455–56) that Mather's "*Variolae triumphatae*" (Sixth Series) sent to John Woodward (c. June 29, 1721), but now lost, is essentially the same as the one Mather incorporated in his manuscript "AB" Cap XX. "Variolae trimphatae. The SMALL-POX Encountred (Part One, MS. 112–41; Part Two, MS. 142–47).

2 See also *ABJ* (1972), pp. 93–116. For our edition, we elected to do a fresh transcription of Mather's "Variolæ triumphatæ" holograph manuscript (MS. pp. 112–41) at the AAS.

3 The Boston broadside *Brief Rule To Guide the Common-People of New-England How to order themselves and theirs in the Small Pocks, or Measles* (1677/78, 1702), by the Rev. Thomas Thacher (1620–78), congregational minister at Boston's Old South Meeting House, is the first medical imprint published in colonial New England. It largely amounts to a short synopsis of Thomas Sydenham's recommendations on treating smallpox, in Sydenham's *Methodus Curandi Febres, Propriis Observationibus Superstructa* (1666), sec. quarta, pp. 123–56; and in *Observationes Medicae* (1676). See the facsimile reproduction of Thacher's *Brief Rule*, by H.R. Viets (1937).

But this *Old Notion* loses much of its Authority, by our considering that it is a *New Distemper*. Tho' the learned *Bartholinus* to that Quæstion, *An deformis hic Morbus Antiquis fuerit notus*? – answers, *Non ausim dubitare*[4] – yett, all that is quoted from *Hippocrates*, & from *Aetius* and *Galen*, will not remove the *Doubt* of other learned Men upon it. Nor will they see, that *Celsus* (the *Tully* of the Physicians) has described so livelily our *Small-Pox*, That one may say, *Ovum Ovo similius non sit*.[5] Tis evident unto us, that the *Ancients* were unacquainted with it: It is one of those *New Scourges*, whereof there are Several, which the Holy and Righteous GOD has inflicted on a Sinful World. It is not many Ages ago, that it was brought into *Europe*[6] on the Wings of those *Arabian Locusts*, which in the *Saracen* Conquests did *spread over the Face of the Earth*.[7]

It seems, as if the Constitution of the *Earth*, and of the *Air*, and of *Humane Bodies*, has altered in Successive Ages; From whence *New Maladies* have

4 The quotation is from a chapter on the origin of smallpox by Thomas Bartholinus (1616–80), a distinguished Danish physician, anatomist, and theologian, whose numerous medical books were widely disseminated. The Latin extract is from Bartholinus' cabinet of medicine, *Cista Medica Hafniensis* (1662), Loculus LVIII, Quæstio 1. p. 592, and asks, "Whether this deforming disease has been known to the ancients?" His answer? "I dare not doubt."

5 Mather agrees with Bartholinus' low esteem of the ancients' medical knowledge of the variola virus (*Cista Medica*, loc. LVIII, Qu. 1, pp. 592–94), for neither Hippocrates (c. 460–c. 370 BCE), the "Father of Medicine"; nor Aetius Antiochenus (fl. 4th century CE), a Syrian theologian and itinerant medical practitioner, in *Aetii Medici Graeci Contractae ex Veteribus Medicinae Tetrabiblos* (1542), Tetrabib, 2, sermo, 1, cap. 129, p. 259; nor the Greek physician Claudius Galen (c. 129–c. 216 CE), in *Method of Medicine* (bk 5, ch. 12); nor the Roman medical writer Aulus Cornelius Celsus (c. BCE 25–50 CE), in *De Medicina* (lib. 5, cap. 28, secs. 1, 4, 6, 15), attribute the different kinds of skin lesions (pustules) to anything like the variola virus (smallpox)–if they had any knowledge of this deadly virus in the first place. To them, the tell-tale pustules all looked alike, though "no egg is like the other." Modern historians of medicine speculate that the smallpox virus originated three to four millennia ago in East Asia and migrated from there to East Africa.

6 In a canceled passage in the holograph manuscript [MS. p. 112], Mather speculates that the variola was brought to Europe "with a *Spice-Trade* from the *East Indies* (whence also the *Pestilence* always comes,) where it still rages more than any where else in the World."

7 Mather here cancels the following sentence: "As there is also Another, that is called (as it is) a Greater [MS. p. 113], which tho' diverse Ingenious Gentlemen besides D[r] Patin, have demonstrated, that it chastised the Wicked in Europe long before the Discovery of America by Colonus, yett most certainly is not of such Antiquity, as they that fancy they find a Mention of it in the Ancient Writers of Medicine, in Hippocrates, yea, and in the Sacred Scriptures themselves, have imagined." Mather refers to the distinguished French physician Gui Patin (1601–72), professor of medicine at the Collège de France and man of letters.

arisen, which the Præceding Ages were Strangers to; (and the *Symptoms* and *Effects* of *Old* ones have mightily altered;) And probably more will yett arise. Hence (as the *Leprosy* grows less common, so) *New Fevers* appear. But the proper *Plague* has never yett visited the vast Regions of *America*: Howbeit *Pestilential* Fevers little better than *That*, have *there* made most fearful Ravages.[8]

It begins now to be vehemently suspected, That the *Small-Pox* may be more of an *Animalculated Business*,[9] than we have been generally aware of. The *Millions* of ——[10] which the *Microscopes* discover in the Pustules, have confirmed the Suspicion. [What would a *Nieuentyt*[11] now say, Reading

8 The genocidal decimation of Native American populations since contact with European arrivals is acknowledged (not always without malicious joy and hatred indicative of his time and age) in Mather's *MCA* (1702). Mather reports how shortly before the arrival of the Plymouth Pilgrims, "The *Indians* in these Parts had newly … been visited with such a prodigious Pestilence; as carried away not a *Tenth*, but *Nine Parts* of *Ten*, (yea, 'tis said, *Nineteen* of *Twenty*) among them: So that the *Woods* were almost cleared of those pernicious Creatures, to make Room for a *better Growth*" (*MCA* 1.2.6.7). See David E. Stannard's discussion of the genocide of America's indigenous populations, especially facilitated by the deliberate spread of diseases, in *American Holocaust* (1992), esp. chs 3–4; S.F. Cook (1973); D.L. Martin et al. (2002); and J. Duffy, "Smallpox" (1951).

9 The term "animalculae" (little animals) designates microscopic organisms (bacteria) and was coined by Antonie van Leeuwenhoek (1632–1723), a renowned Dutch physician, aka. "the Father of Microbiology," and author of *Arcana Naturæ, Ope & Beneficio Exquisitissimorum Microscopiorum* (1696). Mather surveys the discovery of microbes in his chapter "Conjecturalis, or Some Touches Upon a New Theory of Many Diseases," *ABJ* (1972), pp. 43–48. On Leeuwenhoek's perceptions of microscopical discoveries, see E.G. Ruestow (1983).

10 Mather's dash. The missing word is most likely "*Bodies*" (animalcules) since he employs a close paraphrase of Bernard Neuwentyt's *Religious Philosopher* (3:1055) – see next note.

11 The Dutch Cartesian physician, mathematician, and theologian Bernard Nieuwentijt, aka. Nieuwentyt (1654–1718), was a vociferous opponent of Spinoza's materialist philosophy, and several of Nieuwentijt's physico-theological works aimed to provide teleological proof of God's existence. Nieuwentijt's most popular work, *The Religious Philosopher* 3 vols. (1718–21) did just that. It went through several editions and reprints, and Mather profitably used this work in the ten volumes of his *BA* (2010–26) and in his *CP* (1721). Mather's allusion to Job's plaintive "My flesh is cloathed with wormes and clods of dust, my skinne is broken, and become loathsome" (Job 7:5) is validated in Nieuwentijt's *Religious Philosopher* (1:xxxv) praising the microscopical discoveries as follows: "… with this *Microscope* we may be able to see *Millions* of *Bodies*, which we can no wise perceive without such Assistance," and if "we suppose, that our *Eyes* were formed like this *Microscope*, we shou'd cease to wonder at the Minuteness of the Parts of *Bodies*, which now

Job. VII.5. upon it?] And so, we are insensibly drawn into *New Sentiments*, about the *Way* of its *Conveyance*, and the *Cause* why tis convey'd but once; All which, – *Non sunt hujus loci*.[12]

The *Sentiments of PIETY* to be raised in & from this *Grievous Disease*, are what I am first and most of all to be now concerned for.

And now, O *Mankind* in general; Wilt thou not from the View & Sense of this *New Evil Devised* against thee, Humble thyself *under the mighty Hand of GOD*?[13]

Glorious GOD, such a Sharp, and indeed such a New Rebuke of thine upon us, correcting us for our Iniquity, and consuming our Beauty as a Moth, (yea, as a Lion,) why, why must it come upon us? Righteous art thou, O Lord; yett lett us Reason with thee of thy Judgments![14]

The Answer which Heaven thunders down upon us, is; *Ah, Sinful Generation, a People laden with Iniquity, a Seed of Evil-doers, Children that are Corrupters; They have forsaken the Lord! And why are Ye Stricken more, even with Strokes that were unknown to the more early Ages? Tis because Ye Revolt more and more!*[15]

The Distressed *Children of Terror* having such a *Fearful Expectation* of this *Fiery Indignation* to *devour* them; and Seeing that, *Sceptra Ligonibus æquat*,[16] – even *Stars*, yea, *Crowns*, are struck down by it; They are certainly Stupid out of Measure, and more Foolish, than the *wild Asses Colts*,[17] if they do not Immediately Turn & Live unto GOD, and gett into such a *State of Safety for Eternity*, that they may be *Ready* for whatever *Event* this *Distemper* may have upon them, and not be *Afraid with any Amazement*.[18]

The Advice, is; And, Oh! *Behold an Angel with a flaming Sword*[19] over thee giving of it; *Præpare to meet thy GOD*,[20] O thou Traveller thro' an

---

surprize us to perceive, or even think of." So, too, the vermicular infestation of the body is relevant to the question of the purported resurrection of the physical body (*Religious Philosopher* 3:1055–56).

12 "They are not of this place."

13 1 Pet. 5:6.

14 Jer. 12:1.

15 Isa. 1:4–5.

16 The smallpox afflicts "scepters [rulers] and hoes [peasants] equally."

17 Math. 21:2.

18 1 Pet. 3:6.

19 Gen. 3:24.

20 Amos 4:12.

Hatzar-Maveth,[21] a Land where *Fiery Flying Serpents*[22] are hovering every where about thee!

Be Restless, until the *Things that accompany Salvation,*[23] and the *Evident Tokens* of it, be plainly to be found upon thee: until thou art able to say, *My Mind by a New & a Strong Biass given from Heaven unto it, is come to make the Serving & Pleasing of the Glorious GOD, the chief End of my Life: And I now go to GOD, for that which I have heretofore gone to Creatures for. The Blessed GOD is now more to me than this World; The Enjoyment of Him is a Blessedness præferrible with me, to all the Pleasures & Riches & Honours of This World.* Able to say; *Every Thing of a Glorious CHRIST is precious to my Soul; I prize Him in all His Offices; I prize Him with all His Benefits; I would fain have Him to fulfil in me all the Good Pleasure of His Goodness.* Able to say; *There is no known Course of Sin, condemned by the continual Reproaches of my Conscience, wherein I indulge myself; If I am surprised into any gross Act of Sin, I feel my Bones broken, till a Repenting Faith has restored me to the Peace of GOD. The Lusts of Indwelling Sin, I Seek, I Sigh, I Long for a Deliverance from them.* Able to say; *I love my Neighbour; I am glad, when it goes Well with him; I am griev'd when it goes Ill with him. If he Do Evil to me, I durst not so much as wish Evil to him.* And able to say, *It is my Delight and Study to Do Good; And I Labour to be a Faithful Steward, of the Talents wherewith my SAVIOUR has betrusted me.* If thou canst not yett say such Things, *Give no Sleep to thine Eyes, nor Slumber to thine Eylids, until thou canst.*

And then, What wouldst thou do, if the *Summons* were certainly brought unto thee, hearing thyself Summon'd within *One Month* to *appear before GOD the Judge of all*? Surely Thou wouldest without any Delay, sett apart Some Time to go thro' a *Process of Repentance* and *lay hold on Eternal Life.* Instructed from the Gospel, In these Moments of *Agony for Eternity*, thou wouldest cry to GOD, That with Displays of *Sovereign Grace* He would *Quicken* thee to *do what thy Hand finds to do,*[24] that thou mayest be Saved. Thou wouldest *pœnitently confess* the Things, in which {thou} canst upon *Examination* find, that thou hast offended Him. Thou wouldest *carefully with Tears* beg it of the Compassionate GOD, that He would *Pardon* all thy Offences, for the sake of the *Sacrifice*, which His Beloved JESUS has offered up to the Divine Justice for thee. Thou wouldest plead that *Sacrifice*

[21] "A dwelling of death" and "court of death." See also Mather's homily *HM* (1726).

[22] Deut. 8:15, Numb. 21:6–8.

[23] Heb. 6:9.

[24] Eccl. 9:10.

for thy *Atonements*, with Cries that would pierce the Heavens. And then thou wouldest give thyself up unto GOD, with *Dispositions* and *Resolutions* of all possible *Obedience* unto Him; with a *Respect unto all His Commandments*. *All this* wouldest thou do. Then Do it; Even *All this*. Do it *Just Now*. Lett it be done without *any more adoe*.

Finally; Be Inquisitive, what hast thou *left undone*, that if the Pulls of thy *Last Hour* were now upon thee, thou wouldest wish that *it had been done*? My Friend, Suppose thyself within a Few Minutes of thy *Expiration*; Suppose thyself in all the awful Solemnities of a Death-bed; Suppose thy *Life* just agoing, and thy *Soul* presently to appear before the tremendous Tribunal of GOD. Now under the Illuminations of such a Supposal, ask thyself, *What should I wish to have done before I come to this*? Go do it out of hand. Lett it be done with all the *Dispatch*, with all the *Concern* imaginable.

There is an *Epitaph* upon one laid in *West-Minister-Abby*, which has in it such Terms as these; *One who thro' the Spotted Vail of the Small-Pox; rendred a pure & unspotted Soul to* GOD. Shouldest thou *Dy* of the *Small-Pox*, after such a Conversion to GOD, *This* may be thy Happy *Epitaph*.[25] And as it is expressed on the *Grave-Stone* of another there, who dy'd of the like Distemper, the Happiness will be, *Ex Igne ac Tunica Molesta Evolasse ad Cœlos*:[26] To fly away for Heaven, out of as Burning a *Coat*, as what *Nero* putt upon the ancient *Martyrs*.[27]

But I may now suppose the Person thus *Præpar'd*, become a *Patient*.

Being visited with the *Small-Pox*, there are many Exercises of PIETY, which there will be a Call unto; Among which there will be none more pertinent,

[25] The Westminster epitaph reads, "In memory of Mr. *Thomas Smith* of *Elmely Lovet* in the County of *Worcester*, and Batchelour of Arts, late of *Christ Church, Oxford*, who through the spotted vail of the small Pox, rendered a pure and unspotted Soul to God, expecting, but never fearing death, which ended his days *March* the tenth *Anno Dom. 1663¼. ætatis suæ* 27," in Henry Keepe's *Monumenta Westmonasteriensia* (1682), pp. 367–68 (Epitaph 195).

[26] The Latin epitaph reads, "To have flown from the fire and painful tunic to the heavens," and alludes to the death by a poisoned tunic and self-immolation of Hercules (Heracles) in Greek mythology. It is part of an inscription on the sepulcher of three brothers, John, Edward, and Stephen – sons of Sir Stephen Fox (1627–1716) of Farley (Wiltshire), Paymaster-General of Charles II's armed forces. Mather's source is *Monumenta Westmonasteriensia*, p. 352 (Epitaph 162). See also Mather's letter to John Woodward (July 12, 1716), CA 3.10 "Curiosities of the Small Pox."

[27] Roman Emperor Nero (37–68 CE) punished arsonists by burning – a common punishment in ancient Rome, mentioned in Juvenal (Satire 8.235), Martial (Epigram 10.25.5), Seneca the Younger, *Epistle* (14.5). See J. Pollini (2017), part 4, pp. 212–36.

than that of the deepest *Self-Abhorrence*, and *Self-Abasement*, from a Sense of the *Original Sin*, which will obliged us to cry out, *Unclean*! *Unclean*! And confess, *Lord, I am a Filthy Creature*!

My Friend, what a *Loathsome Creature* art thou! *Loathsome* even to Thyself as well as to All that are about thee. Thy *Sin* has rendred thee so unto the Glorious GOD, who is *of purer Eyes than to behold Evil, and cannot look upon Iniquity*.[28] It should render thee so unto Thyself; *Loathing thyself in thy own Sight, for thy Iniquities, & for thy Abominations*.[29]

There is a *Poison within* thee, the *Poison* of an *Evil Heart which departs from the Living GOD*.[30] By *Temptation* as by a *Contagion*, the *Poison* makes horrible *Eruptions*. All the nasty *Pustules* which now fill thy *Skin*, are but little Emblems of the *Errors* which thy *Life* has been filled withal. Make thy Lamentation, *Lord, From the Sole of the Foot, even to the Head, there is no Soundness in me; nothing but putrifying Sores*.[31]

In the *wearisome Nights* that are *Appointed* for thee, thy Complaint is, *My Skin is broken and become Lothesome*.[32] Thy *Bed comforts* thee not; thy *Couch* does not *Ease thy Complaint*; perhaps thou art *Scared with Visions*, and things of a *Frightful Aspect* appear before thy closed Eyes. Now lett *Repentance* have its *perfect Work*.[33] And lett thy Condition lead thee to a *Repenting Sense* of the *Sin* which has provoked & procured this Calamity for thee, and is livelily Resembled in it. Pænitently say, *Lord, My Wounds do Stink and are Corrupt, because of my Foolishness*![34]

Lett this Apprehension of thy *Loathsome Sin*, drive thee to the *Blood* of thy SAVIOUR, for the *Pardon* of it; That thou mayst be *cleansed from all Sin*.[35] At the same time, Importunately Sollicit for the SPIRIT of thy SAVIOUR, to expell the Malignity of *Sin* that is lodged in thee, and send up thy *Groans* unto Him, for His *Purifying Influence*.

While some are very *Sorely* handled by this Noisome and Painful Distemper, others are favoured with a more *Gentle Visitation*. If this be thy Case, Be

[28] Hab. 1:13.

[29] Ezek. 36:31.

[30] Heb. 31:12.

[31] Isa. 1:6.

[32] Job 7:5.

[33] Adapted from James 1:4.

[34] Ps. 38:5.

[35] 1 John 1:9.

very Thankful to a compassionate GOD; Thankfully own, *It is of the Lords Mercies that I am not consumed; because His Compassions fail*[36] *not.*[37]

There is One Circumstance of this Disease, (which an acute and famous Anatomist, methinks, yett gives but a very weak Reason for;[38]) It has hardly ever been seen, that any after having Suffered it *Once*, comes to Suffer it a *Second time.*[39] There are several other *Fevers*, it may be *Six* or *Seven* Sorts of them, that may have the Same Observation made upon them; *No Man undergoes them Twice*. It is to be hoped, *O Man*, That this Observation will be verified in thy *Moral Experience*; and that the *Grosser Sins*, which thou hast once *Repented of*, thou wilt *Never again* fall into them.

But thus having *Sought First*, what is *most* of all to be Sought for, and Serv'd the *Kingdome of GOD & His Righteousness*,[40] from the *Calamity* that is come upon us, we may the more hopefully proceed unto the Work of Encountring and Conquering the *Adversary*.

An *Adversary*, which *wilt thou play with him as with a Bird*?[41] No, He will *fill* thy *Skin* as with *barbed Irons*. And *shall not one be cast down at the Sight of him*?[42] *Who can come to him with his double Bridle*?[43] *Sparks of Fire leap out of his Mouth:*[44]*His Breath kindles Coals.*[45] *When he raises up himself, the Mighty are afraid,*[46] *He spreadeth sharp pointed things, upon* our Clay.[47] He makes our Blood to *boil as a Pott*: He makes our Humours *like a Pott of Ointment*.[48]

Yett lett us be of *Good Courage*; yea, be *very courageous*. There is a Way to manage him![49]

36 The MS. reads "fall."

37 Lam. 3:22.

38 Probably the Danish physician Thomas Bartholin whom Mather quotes above.

39 Mather says as much in item XIV. at the end of this document.

40 Matth. 6:33.

41 Job 41:5.

42 Job 41:9.

43 Job 41:13.

44 Job 41:19.

45 Job 41:21.

46 Job 41:25.

47 Job 41:30.

48 Adapted from Job 41:31.

49 The bottom half of the holograph page is blank.

¶. Unknown is the Number of *Lives*, which the Glorious GOD has made our *Sydenham* the Instrument of Saving, by teaching us a New and a Right Method of treating the *Small-Pox*, and reclaming People from the Madness of *Killing one another with Kindness*, and præposterous Proceedings.[50]

I am willing to treat my Friends, with what I know of the *Sydenhamian* Method. But at the same time I must advise them, That from a Difference of *Seasons* and of *Climates*, there may happen some Circumstances, that may require Proceedings, wherein this *Incomparable Method* must not be strictly in all Points adhær'd unto. The Malady call'd, *Variolæ*, is not without its *Varieties*: Yea, I have myself seen *Anomalies* in the *Small-Pox* wherein a *Sydenham* himself might not be entirely relied upon. There is now {and} then an *Irregular Small-Pox*, in which it may not be easy to fix Rules for an *Unchangeable Regimen*. A *Skilful* and *Thoughtful* PHYSICIAN must be always near the *Patient*, [*My Friend*, Make *Sure*, and then make *Much*, of such an One!] And this *Physician* will take his Measures from the *Indications* as he finds them. Nevertheless, *For the Most Part*, they will *Both* of them find their Account, in Receiving and Remembring such Things, as I will deliver in the ensuing *Aphorisms*.[51]

I. The Distemper we call, *The Small-Pox*, is usually distributed into Two Sorts; The *Distinct* Sort; and the *Confluent* Sort; The *Symptoms* whereof have some Difference; and the *Way of Managing* must be also a little Different.

By not being aware of *This*, what *Mortal Errors* have been often run into!

II. By how much Sooner the *Pocks* come out before the Fourth Day, of the Illness, ordinarily so much the more they will prove of the *Confluent Sort*: However Sometimes there are Accidents that may keep a Restraint upon them.

[50] The famous English physician Thomas Sydenham (1624–89), whose research focused on the origin and treatment of fevers. Sydenham's *Methodus Curandi Febres* (1666) and his *Observationes Medicae* (1676) contain numerous sections on different types of fever, including those of patients infected with the variola virus, he observed during various epidemics in England. The English translation of the Latin original went through numerous editions throughout the 18th century. Cotton Mather's copy of the first edition (1676) at the AAS bears his signature. Sydenham's Dutch colleague, Antonie van Leeuwenhoek (1632–1723), though first in identifying animalcules under the microscope, had not yet fully come to understand how microbes could also bring about immunity, although it was known that patients who recovered from the smallpox became immune to this plague. See his *Arcana Naturae* (1696, 1708).

[51] Conveying knowledge of medicine and medical procedures by aphorism (as a mnemonic device) was first introduced by the Greek philosopher physician Hippocrates and emulated by his latter-day disciples, including Claudius Galen. See for instance, Jacobus De Forlivio's *Expositio in Aphorismos Hippocratis* (1477).

III. If the *Face* be very full, tho' the *Body* should not be so, yett the Sick is in as much danger, as if every Member of the Body were crouded with them. And tho' the *Body* be full, yett if there be Few in the *Face*, the Danger is not so great.

IV. Very *Violent Symptoms* at the *First Arrest* of the Disease, and particularly, *Convulsions* in Children, you must not be always frighted at them: They very commonly introduce very *Moderate Effects*, and the most *Comfortable Issue* that can be look'd for.

V. I am going to mention the Enemy, of which I may say, *Thou shalt Fight neither against Small nor Great, so much as This*![52] All possible Care must be used that the *Ebullition of the Blood* may not *Rise too high*;

Either by *Hastening* the Patients too *Soon* unto their Beds, or by *Confining* them too *much* there:

Or, By heaping on too many *Cloathes*, (which ought not to be more, than what they are in their Health us'd unto:)

Or, By keeping the *Air* of the Chamber too Close & too Hott: (which in Summer may allow of a little Ventilation:)

Or, By giving of pretended *Cordials*, and *Expellers*: (vile *Expellers*, which have Millions of times *driven* the poor *Souls* out of their *Fired Mansions*!

Too hasty an Assimulation[53] of the *Variolous Matter* ha's most certainly *Slain its Ten Thousands*; yea, the most of the *Multitudes, Multitudes*! – that ly *Slain* by the Hand of the *Destroyer* in the *Valley of Death.*

VI. On the other side, unseasonable *Vomits*, and *Purges*, and *Blood-Letting*, may happen too much to diminish the Necessary *Ebullition*. However, if there be a Just Suspicion, that the *Small Pox* coming out, will prove of the *Confluent Sort*, an *Emetic* may be useful yea, a *Phlebotomy*[54] may be Needful

O *Wisdome*, How *Profitable* art thou to *direct*! And how Requisite art thou to the *Physician*! How often to be exercised!

VII. Yett the Patients must not be exposed unto Injuries from the *Cold.* And if the Pustules happen to strike in, or the Swelling of the Hands and Face to fall, from *Such* an Accident, (and a Return to the *Warmth* of the Bed be not Enough to do it) there may be some Recourse to the otherwise exploded

[52] 1 Kings 22:31.

[53] I.e., "Simulation, counterfeiting, pretence" (*OED*).

[54] I.e., bloodletting by puncturing a vein.

Cordials and *Expellers*. However still have a Care of being too lavish in them; *Repeted Ebullitions* of the Blood are very Dangerous Things![55]

VIII. When the Distemper appears, Lett the Patients forbear the Use of *Wine* and *Flesh*: And I again say, of *Cordials* and *Expellers*, and all *Inflamers*. Don't lett 'em Swallow *Fire-brands*! Leave *Nature* undisturbed. It is hardly known, that *Nature* fails of doing its Part for Thrusting out the *Small-Pox*. *Forcing* it, is the most likely Way of *Hindring* it.

Lett their ordinary *Drink*, be *Small Beer*, gently warm'd with a Toast. And lett 'em Drink their Bellyful!

Their *Diet*, Lett it be Oat-Meal Gruel, or Barley-Gruel, or Milk-Porridge; (One Third, Milk, Two Thirds, Water.)

Roasted Apples are no Forbidden Fruit.

Nay, *Roasted Apples* with *Milk*, may now & then be allow'd of.

But nothing too *Hott*; nothing too *Cold*; nothing too hard of Digestion.

IX. Tho' I said, what I said; Yett, at the time of Maturation, when the Purulent Particles flowing back on the Blood threaten to poison it, three or four Spoonfuls of *Generous Wine*, Morning and Evening, may, (if your Physician see no Fevourish Prohibitions,) be allowed of.

X. Lett not the Patients be (as I have said once already) too *Soon*, or too *Much*, confined unto their Beds.

Lett them keep up as long as ever they can.

If the *Small-Pox* be of the *Distinct Sort* (in which Case there usually needs little to be done,) I know not why they should ly *Stiffling* and *Baking* and *Roasting* in their Beds; for whole Days together, if there be so much of *Summer* as will permitt them to Rise with *Safety*.

Be it how it will, If they should be *taken out* of their *Beds*, once or twice every day, all the time of their Illness, they might on many Accounts fare the Better for it. [Only, Be sure, The Inconveniencies of taking any *Cold*, must be carefully *watch'd* against!]

And when they are Abed, lett them *change their Place* ever now & then, to curb any *Sweat* they may fall into. The less they *Sweat*, the better.

Some at the very Point of *Death*, yea, Thought actually *Dead*, have been saved by nothing, but being *taken out of Bed*.

Young Persons are sometimes afflicted with a Total Suppression of *Urine*. Lett them only be *taken out of Bed*, and led once or twice cross the Room; they are help'd Immediately!

[55] Indeed, many patients had been bled to death by this widely used panacea of contemporaneous medicine.

XI. Anon, If the *Small-Pox* don't come out well, the Medicines called *Paregoricks*,[56] have a Success to be wondred at. Your *Liquid Laudanum*[57] (Fourteen Drops, for a Grown Person; *Children* will not want it:) or, *Diascordium*;[58] and the Like, mixed in a Small Quantity, with some agreeable Distilled Water, will check the *Boiling Blood*, and Nature will freely then cast out the Morbific Matter.

XII. Strong *Young Men*, and such as have inflamed their Blood, by Free Drinking, may have Occasion to Bleed in the Arm, when things are hereabouts. Lett the *Doctor* See, how it is!

XIII. In the *Confluent Sort* of the *Small-Pox*, there comes on a *Salivation*,[59] just upon or soon after their First coming out. This having preserved the Lives of the Patients, uses to cease about the Eleventh Day. But a *Swelling* of the *Face* and *Hands* must then supply the Place of it; without which, they must *go to their Long Home*. Tho' the Swelling of the *Face* also should a little abate, yett if the Swelling of the *Hands* continues & increases, there can be no surer Sign of Recovery. The Way to keep this *Discharge of Nature*, in an orderly Condition, is, To *Drink freely* what was formerly allow'd, and be sure, such Drink, as will provoke no *Sweat* upon them: And therewithal to Take what shall presently be mentioned. I will only first observe, what has been in Part already observed, and what cannot be too often inculcated; That as a *Temperate Regimen* is generally the *All in All*, for the Management of the *Small Pox* in all the Kinds of it; [Certainly, The Way to bring a great Company of People handsomely & quietly out of a Room, will not be to throw in *Fire-Works* among them!] So, to prevent the too exorbitant *Ebullition of the Blood*, there can be nothing more adviseable, than the free Use of some Innocent Liquor, which will allay the Heat that Scorches & Wearies the Patients. Besides the *Small Beer* aforesaid; a Decoction of *Bread*, & a Small Quantity

56 A narcotic made from an extract of camphorated opium to relieve diarrhea.

57 Liquid opium extract.

58 An oily tincture made from the dried leaves of *scordium*, which grows in mountainous and swampy areas, and has "astringetia & gusta amara" (an "acidic and bitter taste"), used by the Greek pharmacologist Pedanus Dioscorides (c. 40–90 CE), known for his herbal medicines. See "*De Scordio,*" in *Pedacii Dioscorides Anazarbaeii De Materia Medica, lib. V* (1598), lib. 3, cap. 125, pp. 361–62; translated from Greek into Latin by the French physician Jean Antoine Sarrasin of Lyon (1547–98).

59 "A secretion or discharge of saliva: esp. the production of an excessive flow by administering mercury" (*OED*).

of calcined *Hartshorn*,[60] in a Large Quantity of *Water*, sweetened with Sugar, may be very soberly advised unto. And so may a convenient Mixture of *Milk* & *Water*, if the Stomach be not overcooled with it.

XIV. What was just now promised shall now be mentioned. The Benefit of *Narcoticks* in this Case, is inconceiveable. Tho' my *Sydenham* a long while commended *Liquid Laudanum*, yett anon he came to præfer *Diacodium*:[61] (only, where this might happened to be nauseated; and then to the *Liquid Laudanum*:] I say, *Diacodium*: An Ounce of This, in *Cowslip Water*,[62] or some other such Distill'd Water. It should be given about five or six in the Evening, before the *Restless Fitt* usually comes upon the Patients. Give this, Every Day after there appears Occasion for it; yea, it may be done from the very Sixth Day after the first Invasion. To some young Men, of what we call a very *Sanguine Complexion*, on Ounce and Half will be requisite for a Dose. Yea, when the *Small-Pox* do flux very much, this Anodyne may be given Every Eighth Hour. The Efficacy of this Remedy, surpasses Imagination! *We now come to this Leviathan, with a Double Bridle*!

XV. Besides the Intention of Bridling the *Ebullition of the Blood*, there may be something also necessary, to conquer the *Putrifaction*. And for this Purpose, we scarce know of any thing better, than that on the Fifth or Sixth Day, the Patients come to have the *Spirit of Vitriol*[63] dropt into their *Small Beer*, so as to make it a *little Acid*. Lett this be the *ordinary Drink*; and lett them *Drink freely*, I say, *Drink largely* of it. If they Drink not enough of it, (which it would be strange if they should not!) then mix that *Spirit of Vitriol* with a proper Syrup, or some Distilled Water & Syrup; and give them now and then of That![64] It has done wonderful Things! When the Blood of young People, & such as have been too much acquainted with the Bottel, has raged wth such Violence, that it has broke out of their Arteries into the Bladder, & they have made *Bloody Urine*, which is as desperate a Symptom as the *Small Pox* can be attended with; yett even *Then*, this Course has brought all

[60] Smelling salts (ammonium carbonate).

[61] Sydenham, *Observationes Medicinae* (1676), sec. 3, cap, 2, p. 195. *Diacodium* is an herbal nostrum made mostly from poppies.

[62] A tincture of the cowslip flower against sore throat, bronchitis, even neuralgia.

[63] I.e., Sulphuric acid.

[64] Spirit of Vitriol, Sydenham argued, reduces fever, and keeps the blood from putrefying. See his *The Whole Works of that Excellent Practical Dr. Thomas Sydenham* (1697), sec. 5, ch. 4, pp. 168–69; 419.

to rights. My *Sydenham* seems almost in the Transport of an *Archimedes*[65] at this Discovery.

XVI. When *Children* have the *Small-Pox* of the *Confluent Sort*, a *Looseness* usually follows them, as a *Salivation* does People of Riper Years. But it must not be Stopt. Thousands have been kill'd, by stopping this Provision of Nature, for the Evacuation of the Morbific Matter.

XVII. If in the *Confluent Sort* of the *Small Pox*, the *Spittle* be baked so tough by the preceding Heat, that the Patients are nigh strangled; use a Gargarism, [*Small Beer*, or *Barly-Water*, with *Honey of Roses*,] and Syringe the Throat often with it.

If they are at the last Gasp of *Choaking*, a *Vomit* may be given very seasonably.

XVIII. I have known an Instance, That one taken with the *Small-Pox* was thought siezed with only a <u>Fever</u>. They plied the *Soles of his Feet* with *Pigeons*: and the Consequence of it was, that he had no *Small Pox* above his Waste, but enough below it: His *Head* and his *Breast*, was kept also very easy, all the time of his Illness. Whether such Doings are Adviseable, or may be ordinarily practised without Hazard,

— *melius inquirendum.*[66]

However my *Sydenham*, allows Epithems[67] to the *Soles of the Feet*, in the *Small-Pox* of the *Confluent Sort*. From the Eighth Day, he allows to grown People, *Garlick* sliced and wrapped in a Cloath; to be Repeted every day.

XIX.[68] *Experto Credo:*[69] After all the Methods of Medicines, that our *Sydenham* and others rely upon, I can assure you, There is nothing found so Sure and

65 According to legend, the Greek mathematician Archimedes of Syracuse (c. 287–212 BCE) was so concentrated on his work that he would not even allow the invading Romans to interrupt him – until he met his fate. So John Pechey, M.D. (1654–1718), the translator of Sydenham's *Observationes Medicinae*, reminds us in *The Whole Works*, "The Translator's Preface" (sign. A2^{r-v}). But in the present context, the story goes that when Archimedes discovered the law of buoyancy while taking a bath, he was so excited that he ran naked through the streets of Syracuse to share his discovery.

66 "It should better be investigated before the procedure is applied."

67 External applications to the body.

68 The following paragraph is "a later addition to the text" of "AB." See G.L. Kittredge, "Lost Works," p. 477.

69 "I believe by experience."

Safe as *this*. Procure for the Patient, as early as may be, by *Epispasticks*,[70] a plentiful Discharge at the Hand-Wrists, or Anckles, or Both; and keep them running till all the Danger be over. When the Venom of the *Small-Pox* makes an Evident & Violent Invasion on the Nobler Parts, this Discharge *does wonderfully*. If there be no such Danger, there is less Occasion for this, or any other *Anxious Administration*.

XX. When the Patients are on Recovery, and the Pustules are falling off, & they have begun to feed upon Flesh, (especially, if they have been violently handled) *Bleeding* in the Arm, will be seasonable, and may prevent very Ill Effects of a Depraved Blood. And a little *Purging* will now also come in Season: So all is concluded,

— *Si quid novisti rectius istis*.[71]

¶ But we will not content ourselves with *One Doctors Opinion*. Wherefore, tho' I will not oppress my Friends, with an ostentatious Heap of *Collections*, which it were easy to introduce on a Subject handled by so many Writers, I will call in Two more, who may be enough with our *Sydenham* to constitute, a *Council of Doctors*, on the present Occasion.

The Ingenious *Woodman*,[72] observes, That in the Assaults of the *Small Pox*, the *Constitution* of the Patients, is very much to be considered, whether it be what we call, *Flegmatick*, or *Cholerick*. *Blood Letting* is fatal in the *former*; but in the *latter* it may be very Expedient, yea, Necessary, at the Beginning of the Distemper. Yett, if great *Malignity* attends it, *Blood Letting* must be omitted, whatever Symptoms may seem to demand it.[73]

If the *Stomach* be clogg'd with Gross Humours, a *Vomit* may be given at the Beginning, which may prevent a Future *Looseness*.[74]

A *Costive* Belly, is much præferrible to a *Loose* One. Yett there may be Occasion for a *Clyster*.[75]

If a *Looseness* happens before the Eruption, it must be suppressed. But, if it happen after the Suppuration of the Pustules, and be not very violent,

70 An agent that causes inflammation and blisters to induce a discharge.

71 "If you know how to do it correctly."

72 Philip Woodman, "Practitioner in Physick and Chirurgery" (fl. early 18th century), *Medicus Novssimus* (1712, 1722).

73 Woodman, ch. 29, pp. 219–20.

74 Woodman, p. 217.

75 Woodman, pp. 219–20.

there needs little notice to be taken of it. While it continues, or if it be fear'd approaching, lett the Drink be a Decoction of *Hartshorn*.[76]

No Drinks whatsoever must be given actually *Cold*. The *Small Beer* may have a little *Saffron* infused in it.[77]

Opiates are very good; not only to check a *Looseness*, but also to promote all Intentions.[78]

If the *Pocks* retreat, because of a *Looseness*, Take *Venice-treacle* one Dram;[79] Oyl of *Cinamon* Six Drops, *Laudanum* one Grain; Mix.

Give it every Two or Three Hours.[80]

If *Choaking* be fear'd, give a Vomit out of hand.[81]

If the *Eyes* do suffer very much, often drop into them *White Rose-Water*, wherein *Saffron* ha's been infused.[82]

If a *Cough* be very troublesome, lett the Drink be a Pectoral Decoction of *Hyssop*, *Colts-foot*,[83] *Liquorice*, and the Like, sweetened with Syrup of *Poppy-heads*.[84]

If a *Cold* be taken after the Eruption, so that the Pustules retire, and Languors follow, the Party must be covered very warm, and *Cordials* be given, and *Blisters* be applied unto the Legs and the Feet; From whence if an *Heat of Urine* arises, it may be cured with Draughts of *Whey*, which ha's a few Grains of *Gum Arabic* dissolved in it.[85]

He breaks off so. "More People are lost, thro' a præposterous Use of Needless Remedies, which destroy the Regular Ferment, and check the Expulsive Efforts of Nature, than thro' the Vehemence of the Disease."[86]

76 Woodman, ch. 4, p. 36; ch. 29, p. 218.

77 Woodman, p. 218.

78 Woodman, pp. 254–55, 275.

79 On the strengthening qualities of chalybeates, see Sydenham's *Entire Works* (1749), pp. 392–93, sec. 102; and p. 398, sec. 112. He praises the efficacy of "Venice Treacle" (i.e. Venice Theriaca), a curious concoction of 64 herbal remedies known since ancient times.

80 Woodman, pp. 219–21.

81 Woodman, p. 222.

82 Woodman, p. 222.

83 *Tussilago farfara*, a plant in the Asteraceae family, used to reduce coughing.

84 Woodman, p. 222.

85 Woodman, p. 222.

86 Woodman, p. 224.

The celebrated *Pitcairn*,[87] would have the Patient *Lett Blood*, while the *Fever* does last; yea, tho' the *Small-Pox* do begin to come out. But I doubt, he is *Too Universal* in this Direction.

Do Thou, *O Discreet Physician*, determine when to *Bleed*. However, I must say, I have seen an astonishing Success of *Bleeding*, when the *Second Fever* has greatly threatned the Patient. And I have seen many a Life lost, in a Complement unto the *Physicians* who have earnestly forbidden it.

If on the Fifth, Sixth, Seventh or Eighth Day after the *Small-Pox* be come out, it goes in again; This Gentleman sais, A *Vein* is to be again opened, and *Cantharides*[88] in Powder must be laid unto the *Neck*.

When the *First Fever* is over, and the *Small-Pox* is come out, he would have the Patient often drink any of the Simple, Distilled, Insipid Waters, usually sold at the Apothecaries; in which lett *Sheeps-Dung*[89] be infused for some Hours, and then add Syrup of *White Poppy*.

Besides *Water-Gruel*, he commends a Drink of *Barley-Water*, with Syrup of *White-Poppy*, which will mightily assist the important *Salivation*.

¶ But have we no *Præservatives*, to defend us from the *Invasion* of this dreadful Distemper, or from the *Violence* of it when it has Invaded us?

Our *Sydenham* thinks, That *Purges* duely used before the *Infection* be taken, do hopefully præpare the Body, to feel *Fewer of the Small-Pox*, and of the *Better Sort*.

In a Time & Place of much Infection, perhaps there is, [Give me Leave to say, *After a due Improvement of the NINETY-FIRST PSALM*!][90] no Præservative comparable to that of a Bitt of *Myrrh*, carried in the Mouth of Persons who

87 The Scottish physician-poet Archibald Pitcairn (1651–1713) practiced medicine in Edinburgh before being appointed professor of physics at Leiden. Among his applicable medical works are his *Dissertationes Medicae* (1701, 1713), "De Variolis," pp. 179–81, and the English translation, *The Works* (1715), pp. 244–47; along with *The Method of Curing the Small-Pox: Written in the Years 1704* (1715). Pitcairn's *Opuscula Medica* (1716) went through numerous reprints.

88 Aka. Spanish fly (*Lytta vesicatoria*), a fatty substance (terpenoid) secreted by certain species of beetles.

89 At the time, people believed that sheep dung and related human and animal excrements had a medicinal (repellent) quality to combat certain diseases.

90 For a change of pace, Mather's new translation of the Psalms is opportune (Ps. 91:5–6): "Of what's the terror of the night thou shalt not be afraid; *nor* of the arrow which by day does make a *swifter* flight. *Not* of the pestilence which does in darkness take its walk, *nor* of such a destruction as does at the noon-day waste," in *PS* (1718), pp. 244–45. See also Mather's commentary on Psalm 91 (*BA* 4:635–38).

have already had the *Small-Pox*, and fear not the Return of it, yett if they visit many & nasty Sick Chambers, may have their Spirits horribly poisoned with it.[91] It may do well for *Them* therefore to employ this Præservative.

It ha's been thought, that *Infection* commonly siezes first on the *Salival Juices*. If *so*, the less we *Swallow our Spittle*, the better, where we are in danger of being infected.[92]

Mantissa.

To the foregoing Entertainments, that there may be nothing Wanting, it is fitt that I should add at least a few Delibations from what the valuable Dr. *John Woodward* has written, in his Treatise about, *The State of Physic*.[93]

He observes, Dr. *Sydenham* went a little too fast & too far, into the *Cooling Method*, under the *Small-Pox*, and in his later Times a little quitted it.[94]

The Physicians *Chief Care* in this Distemper, must be to steer & rule the Passions, and keep up the *Hopes* of the Patient. The *Stomach* is very much, the Seat of the *Passions*; in which Bowel also is found the *Source of the Matter*; that being sent from thence into the *Blood-Vessels* with which it has a very notable Communication, is the Cause of the *Small-Pox*: And a Disturbance here, is quickly felt in the Blood, with pernicious Consequences.[95]

If the Physician be consulted in the Beginning of the Distemper, he has an Opportunity of Superseding the Use of almost all other Medicines, by casting out much of the Morbid Matter with a *Vomit*. This Operation being dexterously managed & effectually pursued, the Patient marvellously Releeved: All the Symptoms, not only of the Stomach, which indicate for this Operation, but of the whole Body, Remitt unto Admiration. Yea, There have been Instances, wherein the vitious Matter in the Stomach happening to be little, and not over-boisterous, a *Vomit* seasonably interposed, has discharged almost all the Matter, and the Pustules that began to show themselves, have disappeared. And in other Instances, where such an Evacuation has been made in Season,

91 Mather expressed this legitimate concern numerous times in his *D* (1:365, 451, 547; 2:105, 265, 621, 650, 654; 3:60–61), for as a pastor he visited many sick parishioners.

92 Following a blank space of about seven lines, Mather cancels the following transition: "But I have a further Story to tell you."

93 John Woodward, F.R.S. (1665–1728), English naturalist, geologist, and fellow of the College of Physicians, to whom Mather sent more than a dozen of his "Curiosa Americana" letters. Mather refers to Woodward's *The State of Physick* (1718).

94 Woodward, *The State of Physick*, sec. X. pp. 78–79.

95 Woodward, sec. V, p. 69.

the Pustules have not only proved very Few, but also begun to turn a Day or two sooner, than even the most Favourable Sort use to do.[96]

A *Putrid Phlegm*, in Conjunction with a *Biliose Matter*, seems to be the Cause of the *Small-Pox*; which by the Luxury of the later Times, is now more generated in us, than it was in our Ancestors.[97]

In this Distemper, the *Diet* must be Plain, Thin, Light, Exactly Temperate. *Liquids* must not be too plentifully taken. High *Cordials*, must not be given in great Quantity, & without great Caution. If the Fever lower too much, then some Small Drops of such things may be given.[98]

The *Acids* commonly præscribed, are too much of the same Nature with the *Salt*, that constitutes the chief Peccant Principle in this Disease. And *Absorbents* don't answer Expectation.[99]

In case of a *Diarrhœa*, the main Care of the Physician, must be, to support the Patient and Remove Obstacles, that the Passage of the Peccant Matter downwards, may not be hindred, and so stopped in the Intestines as to be turned back into the Blood; and appease the Tumults of the Stomach, so that it may cease any longer to send the Matter down.[100]

To the Gentlemen who are for *Purging* in the *Small-Pox*, we may address the Words of *Horace*;

Periculosæ Plenum Opus Aleæ
Tractas, et incedis per Ignes
Suppositos Cineri doloso.[101]

My Doctor finds, that of all the Remedies to encounter the *Small-Pox*, there are none comparable to the Sweeter *Vegetable Oyls*, especially the *Oyl of Sweet Almonds*, and such Unctuous Medicines, which may be mixed with Pulps, or Conserves, or Mucilages, for such as can't well take them alone. These wonderfully contribute unto the Frustrating and Subduing of that sharp and hott Matter, which all the Mischiefs of the *Small-Pox* may be ascribed unto.[102]

96 Woodward, sec. XII, pp. 83–84; XLIX, pp. 210–11.

97 Woodward, sec. XIII, p. 89.

98 Woodward, sec. VI, p. 71.

99 Woodward, sec. X, p. 80.

100 Woodward, sec. XIX, p. 104.

101 Horace, *Carmina* (Ode 2.1, lines 6–8): "A work of danger and distrust, / You treat, as one on fire should tread / Scarce hid by treacherous ashen crust" (Horace, *Odes*, Bennett translation).

102 Woodward, sec. XI, pp. 80–81.

But all along, in the several *Stadia* of the Distemper, incredible would be the Benefit of Seasonable *Vomits*. Only, they must be Sufficient Ones, and well-adapted. And among these, the *Ipecacuanha*[103] appears one of the most manageable, & of the most easy Discipline.[104]

In the End of the Disease, these will be of great Consequence, to prevent very grievous Consequences.

But[105] I have now a further Story to tell you. What if we should find out a Way, that the Contagion of the *Small-Pox*, may not (by the *Salival Juices*, as tis commonly thought,) enter the *Stomach*, & make a furious & fatal *Combustion*, in the Phlegmatic and Biliose Matter there, nor enter the *Lungs* more immediately, as with many perhaps it may; but enter by the *Outworks* of the Citadel, and carry off what it has to sieze with very gentle Symptoms, & when it reaches the Stomach in that Way yett be presently conquered with an easy *Emetic* there? This is the Story, which I have now to tell you. And hundreds of Thousands of Lives will be soon Saved, if my Story may be hearken'd to.[106]

Appendix.[107]

There ha's been a *Wonderful Practice* lately used in several Parts of the World, which indeed is not yett become common in our Nation.

I was first instructed in it, by a *Guramantee*-Servant of my own,[108] long before I knew, that any *Europæns* or *Asiaticks* had the least Acquaintance

[103] Known as *Psychotria ipecacuanha* of the *Rubiaceae* family, it is a tropical plant native to South and Central American tropical regions, and is used as a medicine against loose stool and dysentery.

[104] Woodward, sec, LII, p. 220, 222.

[105] Approx. eight lines blank at top of page.

[106] The remainder of the manuscript page is left blank.

[107] The editors agree with G.L. Kittredge's "Lost Works" (pp. 433–35, 444) that MS. pp. 134–41 "substantially" constitutes Mather's "Address to the Physicians," his circulated invitation to Boston's medical establishment to discuss the novel inoculation practice of Pylarinus and Timonius as a means to combat the smallpox epidemic in Boston. See also the anonymous *A Vindication of the Ministers of Boston* (1722), pp. 7–8; and Mather's diary entries for May 26, June 23, July 16, July 27, 1721 (*D* 2:619–20, 624 *n.,* 628, 631–32, 634, 641).

[108] The Guramantes (Garamantes) of southern Libya were an ancient civilization, whose regional capital was at Fazzan. Mather relates the same story in his letter to John Woodward (July 12, 1716), CA 3.10 "Curiosities of the Small Pox"; and in 6.1 *An Account of the Method and Success of Inoculating the Small-Pox*. The enslaved person in question is Onesimus, a young African, gifted to Mather by

with it; and some Years before I was enriched with the Communications of the learned Foreigners, whose Accounts I found agreeing with what I received of my Servant, when he shewed me the Scar of the Wound made for the Operation; and said, That no Person ever died of the *Small-Pox*, in their Countrey that had the Courage to use it.

I have since mett with a considerable Number of these *Africans*, who all agree in One Story; That in their Countrey *grandy-many* dy of the *Small-Pox*: But now they learn This Way: People take Juice of *Small-Pox*; and *Cutty-Skin*, and putt in a Drop; then by'nd by a little *Sicky*, *Sicky*: then very few little things like *Small-Pox*; and no body dy of it; and no body have *Small-Pox* any more. Thus in *Africa*, where the poor Creatures dy of the *Small-Pox* like Rotten Sheep, a Merciful GOD has taught them an *Infallible Præservative*. Tis a *Common Practice*, and is attended with a *Constant Success*.[109]

But our Advice of this Matter, as it comes from Superiour Persons in the *Levant*, is what may have most Attention given to it.[110]

Our first Communication comes from Dr. *Emanuel Timonius* R.S.S. who writes from *Constantinople*, in *December*, 1713. To this Effect.[111]

"Some Gentlemen of our Church" in December 1706 (*D* 1:579). A decade later, Onesimus purchased his own freedom, in August 1716 (*D* 2:363); his place in Mather's household was supplied by "A new Negro servant, (a little Boy)" (*D* 2:384). On the contributions of African slaves during the smallpox inoculation crisis in Boston (1721–1722), see T.H. Brown (1988); M. Minardi (2004); K.S. Koo (2007); R. Tindol (2011); C. Silva (2011); A. Boylston (2012); A.M. Kass (2012); K. Wisecup (2013), pp. 97–126; and K.S Gruesz (2022).

109 See also Mather's *SAIT* (1721), p. 9.

110 Approximately five lines left blank.

111 Emanuel Timonius, "An Account of history, of the procuring the Small-Pox by Incision, or Inoculation: As it has for some time been practised at Constantinople," *PT* 29.339 (June 30, 1714): 72–82. The Ottoman-Greek physician Timonius (1669–c. 1720) studied medicine at Oxford and Padua, and became the court physician of the sultan in Constantinople. From here on out Mather embedded several paragraphs of his CA 6.2 "*Variolae Triumphatae*" in his CA 6.1, which he published in London as *An Account of the Method and Success of Inoculating the Small-Pox in Boston in New-England* (1722), pp. 2ff. Much the same appears in his collaborative work with Zabdiel Boylston, in *SAIT* (1721), pp. 1–4. Although Mather's name does not appear anywhere in Boylston's account, it is widely argued that the majority came from Mather's pen, since he redeploys the same material in several of his other writings on inoculation. In fact, Mather says as much in his diary entry of August 4, 1721, where he acknowledges his own authorship of the document (*D* 2:636).

The Practice of Procuring the *Small-Pox*, by a Sort of *Inoculation*, has been introduced among the *Constantinopolitans*, by the *Circassians* and *Georgians*, and other *Asiasticks*; for about Fourty Years.

At the first, People were Cautious and Afraid. But the *Happy Success* on Thousands of Persons, for Eight Years now past,[112] has putt it out of all Suspicion.[113] The Operation has been performed on Persons of all *Ages*, both *Sexes*, differing *Temperaments*, and even in the worst Constitution of the *Air*; And none that have used it, ever died of the *Small-Pox*; tho' at the same time, it were so malignant, that at least *Half* the People died, that were affected with it in the *Common Way*.

They that have this *Inoculation* practiced on them (*he sais*) are subject unto very *sleight Symptoms*, and hardly sensible of any Sickness; nor do what *Small-Pox* they have, ever leave any *Scars* or *Pitts* behind them.

They make Choice of as Healthy a Young Person as they can find, that has the *Small-Pox* of the best Sort upon him; On the Twelfth or Thirteenth Day of his Decumbiture. With a *Needle* they prick some of the larger Pustules and press out the Matter coming from them into some convenient Vessel of *Glass* (or the Like) to receive it; which ought first of all to be washed very clean with Warm Water. A convenient Quantity of this Matter being thus collected, is to be stop'd *Close*, and kept *Warm*, in the Bosom of the Person that carries it (who ought rather to be some *other Person* than what visited the Sick Chamber for it, lest the Infection of the *Small Pox* be convey'd in the *Garment* as well as in the *Bottel*, and the intended Operation be hurt by the Infection being

[112] Mather refers to the 1713 account given by Timonius in *PT* 29.339 (June 30, 1714): 72–82. For the same, see Mather's CA 6.1 *An Account of the Method and Success of Inoculating the Small-Pox* (1722), p. 3.

[113] See James Jurin's "Letter to the learned Dr. Caleb Cotesworth … containing a Comparison between the Danger of the Natural Smallpox, and that given by Inoculation," *PT* 32.374 (December 31, 1722): 213–27; on p. 215, Jurin includes a summary of some data that Mather's had included in his letter to John Woodward (March 10, 1721–22); Jurin repeats the same in his separately published reprint (1723), p. 6. Also see Jurin's *An Account of the Success of Inoculating the Small-Pox in Great-Britain, for the Year 1726. With A Comparison between the Miscarriages in that Practice, and the Mortality of the Natural Small-Pox* (1727). See also John Gaspar Scheuchzer, F.R.S., *An Account of the Success of Inoculating the Small-Pox in Great Britain, for the Years 1727 and 1728. A Comparison between the Mortality of the Natural Small-Pox, and the Miscarriages in that Practice* (1729); and E. Huth, "Quantitative evidence" (2006).

first convey'd *Another Way,*)[114] and so it should be convey'd as soon as may be, to the Person that is waiting to be the *Patient.*

The *Patient* being in a *Warm Chamber*, is to have several *Small Wounds*, made with a Surgeons Three-Edged *Needle*, or with a *Lancett*, in Two or more Places of the Skin; (the best Places are in the *Muscles* of the *Arm*:) till some Drops of Blood follow: And immediately lett there be dropt out a *Drop* of the Matter in the Glass, on Each of the Places; and mixed well with the *Blood* that is issuing out. The *Wound* should be covered with half a *Walnutt-shell*, or any such *Concave Vessel*, and bound over, that the Matter may not be rubb'd off by the Garments, for a Few Hours. And now, lett the Patient (having *Fillets* on the Wounds) keep *House*, & keep *Warm*, and be careful of his *Diet*; The Custome at *Constantinople* is, to Abstain from *Flesh* and *Broth*, for Twenty Days, or more.[115]

They chuse to perform the Operation, either in the Beginning of the *Winter*, or the *Spring*.

The *Small-Pox* begins to appear Sooner in Some than in Others, and with Lesser Symptoms in Some than in Others: But, *with Happy Success in all.* Commonly *Ten* or *Twenty* Pustules break out: Here and there One has no more than *Two* or *Three*: Few have an *Hundred*. There are some, in whom no *Pustule* rises, but in the Places where the *Insition* was made; And here the *Tubercles* will be purulent. Yett even These, have never had the *Small-Pox* afterwards, tho' they have cohabited with Persons having of it. No Small Quantity of Matter will run for several Days, from the Places of the *Incision*. The *Pocks* arising from this Operation, are dried up in a short time, & fall off; partly in thin Skins; and partly vanishing by an Insensible Wasting.

The Matter is hardly so thick as *Pus*, as in the common *Small-Pox*; but a thinner kind of *Sanies*;[116] whence it rarely pitts; except at the Place of the *Incision*, where the *Cicatrices*[117] are never worn out, and where the Matter is more of the Common Sort.

If an *Apostem* should break out in any, (which is more frequent in *Infants*,) yett there is no fear, for tis heal'd safely by Suppuration.[118]

114 Mather (and those who practiced inoculation) may have learned by then that convivial gatherings of inoculated persons and of those observing the procedure performed risked infection in the common way.

115 The above paragraphs appear – nearly verbatim – in Mather's *AMSI* (1722), pp. 3–4.

116 A thin fluid discharged from an ulcer.

117 Scar tissues.

118 The process of pus formation. The six preceding paragraphs appear nearly verbatim in Mather's *AMSI* (1722), p. 4.

They scarce ever use the Matter of the *Insitious Small-Pox*, to serve the Designs of a *New Insition*.

The *Inoculation* being tried on such as have had the *Small-Pox* before, it had no Effect upon them.

Dr. *Timonius* affirms, That he never yett observed any *Bad Consequence* of the Practice which now so many do come into.[119]

But it is in the *Mouth of Two or Three Witnesses*, that the Thing must *be established*.

We shall again see, this *Leviathan* is not so *Fierce*, but that there are some who *dare to stir him up*.[120]

Since this Communication from Dr. *Timonius*, we have another from an eminent Person, whose Name is *Jacobus Pylarinus*, the Venetian Consul at *Smyrna*.[121] Tis entitutled, *Nova et Tuta Variolas Excitandi per Transplantationem Methodus*.[122]

This Gentleman observes, That this *Wonderful Invention* was, first, *à plebeian rudique Gente in Humani Generis adjumentum, in sævissimi Morbi solamen detecta*,[123] found out, not by the *Learned* Sons of Erudition, but by a Mean, Course, Rude Sort of People, for the *Succour of Mankind* under and against One of the most Cruel Diseases in the World. He seems to look on it, as a marvellous Gift of a Good GOD, unto a miserable World. It was rarely, if ever, used among *People of Quality*, until after the Beginning of the present Century. A Noble *Græcian* then in distress for his Four Little Sons, lest the *Small Pox* might bereave him of them, consulted with *him*, about using the *Inoculation* upon them.[124] At first, his *Ignorance* of the Matter, made him

[119] The preceding paragraphs also appear in Mather's *SAIT* (1721), pp. 1–4.

[120] Remainder of page (c. 8 lines) left blank.

[121] The Greek physician Jacobus Pylarinus, aka. Giacomo Pylarini (1659–1718), served as the Venetian consul in Ottoman Smyrna. Pylarinus studied medicine and law at Padua and practiced medicine in Smyrna and Constantinople.

[122] Jacobus Pylarinus, *Nova et Tuta Variolas excitandi per Transplantationem Methodus* (1715), reprinted as "Nova & tuta Variolas excitandi per Transplantationum Methodus, nuper inventa & in usum tracta: Per Jacobum Pylarinum, Venetum, M.D., in *PT* 29.347 (March 31, 1716): 393–99. Mather's Latin citation can be rendered, "A new and safe method of bringing about the pox through transplantation."

[123] Pylarinus, *Nova et Tuta ... Methodus* (1715), p. 11; *PT* 29.347 (March 31, 1716): 393. Loosely, "By a common person and rude nation discovered and made useful to the human race in the cure of the disease."

[124] The case of Lady Mary Wortley Montague (1689–1762), whose husband served as British ambassador to the Ottoman Empire, is well known. She witnessed the

decline giving him any Advice upon it; But a *Græcian Woman* who was a notable *Inoculatrix* happening to come in while they were, discoursing of the Matter, told them so much about it, that the *Experiment* was Resolved upon. The Woman managed, in her Way, upon all the *Four Sons*. The Three *Younger*, all of which were under Seven Years of Age, felt a very gentle Illness, had a very *few Pustules*, and in about a Week all *Fever* and *Hazard* was over with them. The *Eldest*, about *Eight* Years old, was taken with a *Malignant Fever*: and (tho' he had not many *Pustules*) narrowly escaped with his Life. *Pylarinus* imputes this, to an Atrabiliarious[125] and otherwise Humourous & Unhealthy Constitution of the Lad, and a *Neglect* of using such *Preparatory Expiation* of his Body, as they had been advised unto. But upon this *Happy Success*, *Mirum quam multas Nobiliorum Familias, ad Imitationem traxit*![126] – It was wonderful to see; what a Multitude of *People of Fashion* presently followed the Exemple. So that at this Day, every one, does without any Hæsitation, and with all the Security imaginable, practice the *Transplantation*; except here & there a few *Cowards* that are afraid of their *Shadows*. Indeed, the *Turks*, whose Faith in Fate, is as we know, and who are a more Indocible Sort of Animals, do not yett much come into it.[127]

Pylarinus, instructed by his Greek *Operatrix*, directs; To take a *Proper Season* for the Insition. She would use it only in the *Winter*; but he thinks, the *Spring* may do as well.

The Fermenting *Pus*, must be taken from the Mature *Pustules* of a *Good Sort*, in a Young Person; of a Good Constitution, kept warm in a close Vial, and hastened unto the Application.

The *Air* of the Chamber must be kept very Temperate.

---

practice of inoculation during her travels in Turkey in 1717, and had her young son undergo the procedure with the help of the Scottish surgeon Charles Maitland (c. 1668–1748). Lady Montague related the success of the treatment in several of her letters to her friends in England and became a vociferous advocate of inoculation, which she called "engrafting." Nonetheless, the practice was highly controversial and opposed by the medical establishment in England – even though Lady Mary had the backing of Princess Caroline at the court of George I. See R.P. Stearns (1950), T. Heffernan (2012), and H. Esfandiary (2022).

125 Affected by an excess of black bile, at the time believed to cause indigestion and melancholy.

126 Pylarinus, *Nova et Tuta ... Methodus* (1715), p. 17; *PT* 29.347 (1716): 394. Mather provides his own translation.

127 The preceding paragraphs appears also in Mather's *SAIT* (1721), pp. 5–8; and *AMSI* (1722), p. 5.

The Greek *Operatrix*, prick'd more Places, and less Fleshy Ones, than *Pylarinus* approved of: With an oblique Stroke pricking the Places with an Iron or Golden *Needle*; and with the same Needle dropping and thrusting the *Pus* into the Wound; and so binding all with *Fillets*. Her Way was thus to prick, the Forehead, the Chin, both Cheeks, both Wrists, & both Insteps. This was doubtless over-doing. *Pylarinus* affirms, That Some have done the Business, *Unico duntaxat Vulnusculo ad Brachium inflicto*;[128] with no more than *One little Insition* in the Arm, and it has done very well. [*So it has been with such Africans, as have shown me the Marks of their Inoculation.*]

They must not keep their *Beds* more than is necessary.

Wine, *Flesh*, *Broth* must be laid aside.

The *Ferment* comes into Action, Sooner in Some than in Others. Usually the *Small Pox* appears on the *Seventh* day: sometimes on the very *First*.

The *Symptoms* prove Remiss or Intense, according to the various Constitution of the Bodies.

The *Small Pox* proves, of the Distinct Sort; And there will be but *few* of them; It may be *Ten* or *Twenty*; rarely an *Hundred*.

In some few, the *Insition* has produced no *Small Pox* at all; but the Persons have afterwards been, in the *Common Way*, taken & handled with it, like other People.

The *Wounds* made for the *Insition* prove often very Sore. And with some they degenerate into *Apostems*;[129] yea, These do swell sometimes, & rise, & fall, & rise again. There has also happened on this Occasion, an *Abscess* with Suppuration, in some Emunctory[130]

of the Body: But this is a very Rare Occurrence.

In fine: *Pylarinus* affirms, It was hardly ever known, that there was any *Ill Consequence* of this Transplantation: *Quinimo rite recteque tractata, et in Corporibus per peritum Medicum aptè præparatis, certissimam promittit Salutem.*[131] The Business being well and wisely managed, and the Body being

[128] Pylarinus, *Nova et Tuta ... Methodus* (1715), p. 29; *PT* 29.347 (1716): 398. Again, Mather provides his own translation.

[129] A swelling filled with pus, an abscess.

[130] An organ in the body that eliminates waste.

[131] Pylarinus, *Nova et Tuta* (1715), p. 32; *PT* 29.347 (1716): 398. Mather supplies his own translation.

by a *Skilful Physician* well-prepared; you may depend upon it (*he sais*) in an ordinary Way, there can be *nothing but a Good Issue of it.*[132]

But, I remember, I spoke of *Three Witnesses*. I will therefore add: *Kennedy*[133] says, That in the little Time that he was at *Constantinople*, he was assured of *Two Thousand* that had lately undergone the Method of the *Small-Pox Inoculated*; and there were no more than Two who died under it; & the *Death* of these was entirely owing to their own Ill Conduct in exposing themselves.

Hitherto you have nothing but *History*. But a little *Philosophy* and *Speculation* may be now asked for; and an *Enquiry into Causes* a little endeavoured. No doubt, among the *Wise Men of Enquiry*, there may be found, *So many Men so many Minds*. Every Gentleman may form his own *Hypothesis*; and some of the Later and more Modern Curiosity, will try how the *Vermicular Scheme* will carry them thro' a Solution of these and all *Appearances* in this Distemper.[134]

I have seen the Point, after this *Pothecary* manner talk'd about. The venomous *Miasms* (Lett *That* Word serve at the Present) of the *Small Pox*, entring into the Body, in the Way of *Inspiration*, are immediately taken into the *Blood* of the *Lungs*: And I pray, how many *Pulses* pass before the very *Heart* is pierced with them? And within how many more they are convey'd into all the *Bowels*, is easily apprehended by all that know any thing, how the *Circulation of the Blood* is carried on.[135] At the same time, the *Bowels* themselves are enfeebled, & their Tone impaired, by the *Venom* that is thus insinuated. Behold, the Enemy at once gott into the very *Center* of the Citadel: And the Invaded Party must be very Strong indeed, if it can struggle with him, and after all Entirely Expel & Conquer him. Whereas, the *Miasms* of the *Small-Pox* being admitted in the Way of *Inoculation*, their Approaches are made only by the *Outworks* of the Citadel, and at a considerable *Distance* from the Center of it. The Enemy, tis true, getts in so far as to make some *Spoil*, yea, so

132 This and the preceding paragraphs appear nearly verbatim in Mather's *SAIT* (1721), pp. 5–9; and *AMSI* (1722), pp. 6–7.

133 Peter Kennedy (b. 1685), an English surgeon, published *An Essay on External Remedies* (1715), ch. 37, p. 155. It is remarkable that Kennedy's report affirming the success of smallpox inoculation appeared in press just a year after Emanuel Timonius's "An Account" had been published in *PT* 29.339 (June 30, 1714): 72–82.

134 Mather cancels the transition in the following paragraph. "All that I had of my poor Penetration may venture to say, will be That"

135 The English physician and anatomist William Harvey (1578–1657) discovered the circulation of the blood and published his account in his *Exercitatio Anatomica De Motu Cordis et Sanguinis in Animalibus* (1628).

much as to satisfy him, and leave no *Prey* in the Body of the Patient, for him ever afterwards to sieze upon. But the *Vital Powers* are kept so clear from his Assaults, that they can manage the *Combats* bravely and, tho' not without a *Surrender* of those Humours in the Blood, which the Invader makes a Siezure on, they oblige him to *march out the same Way he came in*, and are sure of never being troubled with him any more.[136] But perhaps, the Few Words, that I wrote, in my Introducing of the Story, may be as much to the Purpose, as all of this *Jargon*. I'l have done with it.[137]

I am now able,[138] as an *Eywitness*, (& more than so) to give a more full Account of the *Practice*, which until *Now* I could only propose as a Matter at a greater Distance.

About the Month of *May*, 1721, the *Small-Pox* being admitted into the City of Boston, I proposed unto the Physicians of the Town,[139] the unfailing Method of preventing Death,[140] and many other grievous Mis-

[136] The preceding paragraph also appears in *AMSI* (1722), pp. 7–8.

[137] Mather cancels the following paragraph: "I durst not engage that the Success of the Trial here, will be the Same, that ha's been in all the other Countreys where it has been tried hitherto: tho' we have seen it succeed well in very different Climates. Nor am I sure that if it should be made upon a Body, where the Blood is already just upon the Point of some unhappy Fever, this may not help to set Fire to such a Thing. But I am very confident, no Person would miscarry in it, but what would most certainly have miscarried upon taking the Contagion in the Common Way. Wherefore, if it be made at all, (and all the Scruples that some have, about the Tempting of Providence, be also gott over,) I advise, that it be never made but under the Management of a Physician, whose Conduct may be much relied upon; and who will wisely Præpare the Body for it, before he Perform the Operation. I have done." The preceding phrase indicates the end of "Part One" of the holograph manuscript (MS. 112–142). See G.L. Kittredge, "Lost Works," pp. 434–35, 442.

[138] Here begins Part Two [MS. 142–147] of the holograph manuscript "AB" (written *after* November 14, 1721) as it mentions the attempt on his life (D 2:657). See G.L. Kittredge, "Lost Works," pp. 442–44, 464–67.

[139] Mather recorded in his diary for May 26, 1721: "The grievous Calamity of the *Small-Pox* has now entered the Town. The Practice of conveying and suffering the *Small-Pox* by *Inoculation*, has never been used in *America*, nor indeed in our Nation. But how many Lives might be saved by it, if it were practised? I will procure a Consult of our Physicians, and lay the matter before them" (*D* 2:620–21). See also Mather's "Curiosa" letter to John Woodward (July 12, 1716), our 3.10 Curiosities of the Small Pox.

[140] Mather's syntactical manuscript revision [~~what I have after~~ the ~~never~~ <un>failing Method of preventing Death] bespeaks his hesitancy rather than his certainty about the practice.

eries, from a tremendous Distemper, by Receiving and Managing the *Small-Pox*, in the Way of *Inoculation*. One of the Physicians had the Courage to begin the Practice upon his own Children & Servants;[141] And another[142] express'd his Good Will unto it. But the rest of the Practitioners treated the Proposal with an Incivility and an Inhumanity not well to be accounted for. Fresh Occasion I saw, for the Complaint of a great Physician; *Heus, Quanto Dolore auger, dum video Naturæ ministrum Medicum, Hostem ejus devenisse*![143] The vilest Arts were used, and with such an Efficacy, that not only the Physician, but also the Patients under the *Small-Pox Inoculated* were in hazard of their very Lives from an Infuriated People. And I myself had thrown into my House in the Dead of the Night, a fired *Granado*, charged with Combustible Matter, and in such a Manner, that upon its going off, it must probably have killed them that were near it, & would have certainly fired the Chamber & speedily have laid the House in Ashes. But the merciful Providence of GOD our SAVIOUR so ordered it, that the *Granado* passing thro' the Window, had by the Iron in the Middle of the Casement

141 I.e., Zabdiel Boylston (1676–1766) of Brookline, Massachusetts, who was the only physician to accept Mather's call to debate the new method. Boylston inoculated his own son and two of his servants (as well as Mather's son Samuel) to test the procedure. In all, he vaccinated c. 248 Bostonians in the epidemic of 1721–22. See also Mather's *D* (2:621–25). In August 1721, Mather recorded that "I will allow the persecuted Physician [Boylston], to publish my Communications for the *Levant*, about the *Small-Pox*, and supply him with some further Armour, to conquer the Dragon (*D* 2:636). Mather refers to *Some Account Of what is said of Inoculating or Transplanting the Small Pox. By the Learned Dr. Emanuel Timonius, and Jacobus Pylarinus*, published by Dr. Zabdiel Boylston" (1721). Mather also collaborated with Boylston on the newspaper article, "A Faithful Account of what has occur'd under the Late Experiments of the Small-Pox managed and governed in the Way of Inoculation," in *Boston Gazette*, No. 101 (October 30, 1721). Almost a decade later, Boylston, F.R.S. (now an elected member of the Royal Society of London and celebrated in England), published *An Historical Account of the Small-Pox Inoculated in New England* (1726; 2nd ed. 1730) in which he narrated the events from his own point of view. Valuable background is provided (among many others) by R.P. Stearns (1950), J. Blake (1959), T.H. Brown (1988), E. Huth (2006), W. Garth (2010), and A.M. Kass (2012).

142 John Perkins (1676–1740), Mather's friend and neighbor. See T.J. Holmes (3:1170, *n.* 3). See also Mather's letter of introduction to James Jurin, CA 7.10 (August 3, 1723). Perkins appears to have been supportive of Mather's drive to enlist Boston's physician in practicing inoculation, yet there is no evidence that he actually did so beyond moral support (*D* 2:609, 627, 664).

143 "Alas, how greatly I am filled with pain, when I see that Nature's minister, the physician, has become her enemy."

such a Turn given to it, that in falling on the Floor, the fired Wild-fire in the Fuse, was violently shaken out some Distance from the Shell, and burnt out upon the Floor, without firing off the *Granado*.[144]

The Opposition was carried on with a *Folly*, and *Falsehood*, and *Malice*, hardly ever known to be parallel'd on any Occasion;[145] And in the Progress of the Distemper many Hundreds of Lives were lost, which might have been Saved, if the People had not been Satanically filled with Prejudices against this *Method of Safety*. However, the Practice went on, and tho' the Physician was under extreme Disadvantages on more Accounts than one, yett he was attended with vast *Success*. The Experiment has now been made on *Several Hundreds* of Persons; and upon both *Male* and *Female*, both *Old* and *Young*, both *Strong* and *Weak*, both *White & Black*, at all Seasons, of *Summer & Autumn & Winter*: And they have generally professed, *They had rather undergo the Small-Pox Inoculated once every Year, than undergo the Small-Pox once in their Lives after the Common Way, tho' sure to live.*

I shall now communicate our *Way of Proceeding*, in the Practice.[146]

[144] At this point in the holograph manuscript, Mather obliterates six and a half lines of text by blotting them out completely. An attempt was made on Mather's life (November 14, 1721) in retaliation for his advocacy of smallpox inoculation. The "Granado," a crude Molotov cocktail of sorts, did not fire: "There was found a Paper so tied with String about the Fuse, that it might out-Live the breaking of the Shell, which had these words in it; COTTON MATHER, *You Dog, Dam you: I'l inoculate you with this, with a Pox to you!* (*D* 2:657–58). The incident was newsworthy enough to make headlines in the "Foreign Affairs" section of *The London Journal*, Numb. CXXXI (Saturday, January 27, 1721), pp. 3–4. See also Mather's letter to James Jurin (May 21, 1723), CA 7.1 "The Case of the Small-pox Inoculated, further cleared."

[145] M.W. Sivils (2011) ably dissects the pamphlet literature of the controversy.

[146] *D* (2:661–62). The following thirteen recommendations originally constitute part of Mather's letter to John Woodward (November 30, 1721) "A Further Account, or the Method and Success of the Small-Pox Inoculated" (CA 6.3), and were the only section of the longer letter published separately as "*The way of proceeding in the* Small Pox Inoculated *in* New England. *Communicated by* Henry Newman, *Esq*; *of the* Middle Temple," in *PT* 32.370 (March 31, 1722): 33–35. Newman (1670–1743), Harvard graduate (1687), Harvard's librarian (1690–93), and secretary of the Society for Promoting Christian Knowledge in London, was at the time Mather's correspondent and representative in London for various transactions, including transmitting several of Mather's "Curiosa Americana" letters to John Chamberlayne of the Royal Society (L.W. Cowie [1956]). Mather's authorship of "The way of proceeding in the Small Pox Inoculated" is well established by G.L. Kittredge, "Lost Works," pp. 461–67.

I. We make usually a Couple of *Incisions* in the *Arms*, where we usually make our *Issues*; But somewhat larger than for *Them*, (Sometimes in an *Arm* & in a *Leg*.)

II. Into these we putt Bitts of *Lint*, (the Patient at the same time turning his Face another Way, and guarding his Nostrils,) which have been dipt in some of the *Variolous Matter*, taken in a Vial, from the Pustules of one (if we can find such an one) that has the *Small Pox* of the more laudable sort now turning upon him; And so we cover them with a Plaister of *Diachylon*.[147]

III. Yett we find the *Variolous Matter* fetched from those that have the *Inoculated Small Pox*, as Agreeable and Effectual as any other. Yea, and so we do, what is taken from them that have the *Confluent Sort*.

IV. In four & twenty Hours, we throw away the *Lint*; and the Sores are dressed once or twice, every four & twenty Hours, with warmed *Cabbage-Leaves*.

V. The Patient continues to do things, as at other Times; only he does not expose himself to the Injuries of the *Weather*, if *That* be at all Tempestuous. But we find, the *Warmer* he keeps himself, he afterwards finds himself no Loser by it.

VI. About the *Seventh Day*, the Patient feels the *Usual Symptoms* of the *Small-Pox* coming upon him; and he is now managed as in an ordinary *Putrid Feaver*. If he can't *hold up*, he goes to *Bed*. If his *Head* ake too much, we putt a common *Poultis* to his *Feet*.[148] If he be very Qualmish at the *Stomach*, we give him a gentle *Vomit*, yea, we commonly do those things *almost of Course*, (especially give the *Vomit*) whether we find the Patient want them or no. If the *Fever* be too high, in some Constitutions, we Bleed a little. And, finally, to hasten the *Eruption*, if it come on too slowly, we putt on an *Epispastic*.

VII. Upon or About the *Third Day* from the Decumbiture, the *Eruption* begins. The Number of *Pustules* is not alike in all. In some, they are very Few. In others, they amount unto an *Hundred*. Yea, In many they amount unto *Several Hundreds*. Frequently, unto more than what the Accounts from the *Levant* say is usual there.[149] But in some, there is not what may be fairly called, *A Decumbiture*: The *Eruption* is made without their suffering one Minute of any sensible *Sickness* for it. *Young Children*, even such as are dandled on the knee, & hanging on the Breast, seem to fare the best of any under this Operation.

147 An adhesive plaster made from litharge, olive oil, and lard.

148 A poultice (cataplasm) generally consists of a warm, medicated ointment on a piece of cloth and placed on the skin.

149 I.e., the accounts given by Timonius and Pylarinus.

VIII. The *Eruption* being made, all *Illness* vanishes; There's an End on't; except there should be something of the *Vapours* in those that are troubled with them. There is nothing more to do, but keep Warm; Drink proper *Tea*'s; Eat *Gruel*, and *Milk-Porridge*, and *Panada*,[150] and Bread & Butter, and almost any thing æqually Simple and Innocent.

IX. Ordinarily the Patient Sitts up every day, and entertains his Friends; yea, ventures upon a *Glass of Wine* with them.[151] If he be too Intense upon Hard *Reading* and *Study*, we take him off.

X. Sometimes, tho' the Patient be on other accounts easy enough, yett he *can't sleep* for diverse Nights together. In this Case, we don't give them *Opiates* or *Anodynes*; because we find that they who have taken these in the *Small-Pox* are generally pestered with miserable *Boyls* after their being Recovered. So, we lett 'em alone; Their *Sleep* will come of itself, as their *Strength* is coming on.

XI. On the *Seventh* Day, the *Pustules* are all usually come to their Maturity; (some, on the *Fifth*:) And soon after this, they go away, as those of the *Small-Pox* in the *Distinct Sort* use to do.

XII. The Patient getts abroad Quickly; and is most sensibly *Stronger*, and in better Health, than he was before. The *Transplantation* has been given to a *Woman* in Childbed, Eight or Nine days after their Delivery and they have gott rather Earlier out of their Childbed, & in better Circumstances than ever in their Lives. Those that have had ugly *Ulcers* long running upon Them, have had them healed on & by the *Transplantation*. Some very Feeble, Crazy, Consumptive People, have upon the *Transplantation* grown Hearty, & gott rid of their former Maladies.

XIII. The *Sores* of the *Incisions* do seem to Dry a little in the Three or Four Days of the Feavourish Præparation for the Eruption. After this, there is a *Plentiful Discharge* at them. The Discharge continues for some Days after the Patient is quite well on other accounts. But the Sores dry up soon enough of themselves; we count the *Later* the *Better*. If they happen to be Inflamed, or otherwise Troublesome, we presently help them in the Way we do any ordinary Sores.

[150] A mush made from bread.

[151] N.B. The contagious patient was not quarantined but freely socialized with friends and family, ignorantly spreading the contagion.

XIV. The *Transplantation* has been tried on such as have gone through the *Small-Pox* formerly in the Common Way; and it has had no Effect upon them; except perhaps an Hour or Two of harmless Indisposition, about the Time when the *Eruption* should otherwise have been made upon them.

It has been unhappily given to some few, that have already Newly received the Infection in the Common Way. The Eruption has then been presently made in Two or Three days after the *Incision*, and they have undergone the *Small-Pox* in the Common Way; hardly escaping with their Lives: Tho' some have thought, the Running of the Sores in these, has been some Advantage to them.

Two, or three have died under, or soon after the *Inoculation*, from a Complication of *other mortal Distempers*.

—[152] But of all the Hundreds that have been under a *Regular Management*, we know not of one, but what Rejoices in their having undergone the Operation.[153]

[152] Mather cancels the following sentences at the opening of the new page: "An Indian Servant getting a violent Cold, fell into a *Pleuretic Fever*, that kill'd her. Another Person that had long been under a Crazy *Melancholy* & *Consumption*, utterly refused all Sustenance, & *starved himself to Death*."

[153] Following a blank space equal to several lines, Mather cancels four lines and renders them illegible by blotting them out.

6.3. A FURTHER ACCOUNT, OF THE METHOD AND SUCCESS OF THE SMALL-POX INOCULATED.[1]

{After December 1, 1721.}

{To Dr. John Woodward.}

[The Way of Proceeding in the Small Pox inoculated in New England. Communicated by Henry Newman, Esq; of the Middle Temple.][2]

1 In his diary entries for November 24 and 30, 1721 and December 1, 1721, Mather relates that he was "draw[ing] up the Method of Proceeding in the Inoculation of the *Small-Pox*, and communicat[ing] Copies of it, that so Physicians about the Countrey may know how to manage it." Roughly a week later he records, "I write a further and more distinct Account of the *Small-Pox Inoculated*, the Method and Success of it among us, and the Opposition to it ..." as well as "The Way of Proceeding, in the *Inoculation* of the *Small-Pox*, I communicate Copies of it unto the Physicians and others, in several Parts of the Countrey" (*D* 2:660, 661–62). See also G.L. Kittredge, Scientific" (21, 48); "Lost Works" (460–67).

2 Mather sent his "Further Account," along with his instructions "The Way of Proceeding," to Henry Newman in London for delivery to the RSL's secretary John Woodward, who published the latter of the two documents in *PT* 32.370 (March 31, 1722): 33–35. Henry Newman (1670–1743), Harvard graduate (1687), served as Harvard's librarian (1690–93). After moving to London, he became the secretary for the Society for Promoting Christian Knowledge, and colonial Agent for New Hampshire (L.W. Cowie [1956]). It was Newman, who "communicated," i.e., delivered Mather's packet of letters (Sixth Series) to Woodward. Newman was *not* the author. Mather's authorship is confirmed by several entries in his *D* (2:660, 661–62). The editors agree with George L. Kittredge's assessment ("Lost Works," pp. 460–67) that "The Way of Proceeding" was a substantial part of Mather's longer letter (Sixth Series), "A Further Account, of the Method and Success of the Small-Pox Inoculated. To John Woodward (c. November 24–December 1, 1721)." Of this original, Woodward only published Mather's step-by-step instructions ("The Way of Proceeding") of the inoculation procedure as practiced in Boston. The remainder of "A Further Account" (Sixth Series), like virtually all of Mather's letters to John Chamberlayne (Fifth Series), were either scattered among the various RSL reviewers who vetted Mather's bundles

1. WE make usually a Couple of *Incisions* in the *Arms* where we make our *Issues*, but somewhat larger than for them, some times in one *Arm*, and one *Leg*.

2. Into these we put bits of *Lint*, (the patient at the same time turning his Face another way, and guarding his Nostrils) which have been dipt in some of the *Variolous Matter* taken in a Vial, from the Pustules of one that has the *Small Pox* of the more laudable Sort, now turning upon him, and so we cover them with a Plaister of *Diachylon*.[3]

3. Yet we find the *Variolous Matter* fetched from those, that have the *inoculated Small Pox*, altogether as agreeable and effectual as any other. And so we do what is taken from them that have the *Confluent Sort*.

4. Within Four and Twenty Hours, we throw away the *Lint*, and the *Sores* are dressed once or twice every Fourt and Twenty Hours, with warmed *Cabbage Leaves*.

5. The Patient continues to do all Things, *as at other times*, only he exposes not himself unto the Injuries of the *Weather*, if that be at all Tempestuous.

6. About the Seventh Day the Patient feels the usual Symptoms of the *Small Pox* coming upon him; and he is now managed as in an ordinary *Putrid Fever*. If he cannot hold up, the goes to *Bed*; If his *Head ach* too much, we put the common *Poultice*[4] to his *Feet*, if he be very Sick at the

of letters or were lost altogether. Fortunately, Mather rarely discarded drafts of important documents – even if they were published in different venues. Mather embedded the substance of "A Further Account" (along with the published "The Way of Proceeding") in the manuscript of "AB" (MS. 112–47), in CA 6.2 "*Variolae triumphatae.* The *Small-Pox* Encountered." See *ABJ* (1972), pp. 93–116; and Kittredge's discussion of Mather's authorship, in "Lost Works" (pp. 440–44, 460–68). That Mather's letters to the RSL could get scattered, misplaced, or lost among the various reviewers during the vetting process can be seen in John Chamberlayne's missive, which Mather quotes as follows in his own letter to John Winthrop: "As soon as I [Chamberlayne] perused your learned Letters [Fifth Series], I Immediately communicated them unto Sr *Isaac Newton*, our Excellent Præsident; And he has referred them to the most Learned Mathematician and Great Traveller, Dr Halley, whose Report I expect with no small Impatience. But, as I beleeve, you expect to be informed of their Destiny, with respect to their coming into my hands, I was resolved not to slip the opportunity of telling you, They are safe arrived," in "Cotton Mather to John Winthrop. Boston. 26^{d}. X^{m}. [December] 1720" (*MP* [1868], pp. 444, 455).

3 An adhesive plaster made from litharge, olive oil, and lard.

4 I.e., a cataplasm of soft, moist, warm, and medicated piece of cloth placed on the aching part of the body.

Stomach, we give him a *gentle Vomit*, yea, we commonly do these Things *almost of Course*, whether we find the Patient want them or no. And we reckon the *sooner* we do these Things, the *better*. If the *Fever* be too high, in some Constitutions, we *Bleed* a little: And finally, to hasten the Eruption, we put on a Couple of *Blisters*.

7. On or about the Third Day from the Decumbiture[5] the *Eruption* begins. The Number of the *Pustules* is not alike in all, in some they are a *very few*, in others they amount to an *Hundred*, yea, in many they amount unto *several Hundreds*; frequently unto more than what the Accounts from the *Levant* say is usual there.[6]

8. The *Eruption* being made, *all Illness* vanishes; except perhaps a little of the *Vapours* in those that are troubled with them; there is nothing more to do, but to *keep Warm*, drink proper Tea's, eat Gruel, Milk Pottage, Panada,[7] Bread, Butter, and almost any thing equally Simple and Innocent.

9. Ordinarily the Patient *sits* up every Day, and entertains his Friends,[8] yea, ventures upon a *Glass of Wine* with them. If he be too Intent upon hard *Reading* and *Study*, we taken him off.

10. Sometimes, tho' the Patient be on other Accounts easy enough, yet he *can't Sleep* for divers Nights together. In this Case we do not give him *Anodynes* or *Opiates*, because we find, That they who have taken these Things in the *Small Pox* are generally pestered with miserable *Biles*[9] after their being recovered. So *we let them alone*; their *Sleep* will come of it self, as their *Strength* is coming on.

11. On the Seventh Day the *Pustules* usually come to their Maturity; *and soon after* this they go away, as those of the *Small Pox* in the *Distinct Sort* use to do.

5 Confinement to a bed.

6 The reports from Turkey and Greece given by Dr. Emanuel Timonius and Dr. Jacobus Pylarinus, in Mather's published letter to Sir Hans Sloane, *An Account of the Method and Success of Inoculating the Small-Pox*, in *Boston* in *New-England* (1722), our CA 6.14 (below).

7 Bread soup.

8 This is one of the great oversights of the entire inoculation procedure: lack of quarantine! Having so-called inoculation parties in which family, friends, and neighbors dropped in on the prone patient inadvertently contributed to the spread of the epidemic among the unvaccinated. Thus the opponents of the procedure were indeed justified in charging that Boylston and the Mathers were spreading the pox in Boston.

9 Boils.

12. The Patient gets abroad quickly, and is most sensibly *Stronger*, and in *better* Health than he was before. The Transplantation has been given to *Women in Child-bed*, Eight or Nine Days after their Delivery; and they have got earlier out of their Child-bed, and in better Circumstances, than ever in their Lives. Those that have had ugly *Ulcers* long running upon them, have had them healed on, and by this *Transplantation*. Some very feeble, crazy, *Consumptive* People, have upon this *Transplantation*, grown hearty and got rid of their former Maladies.

13. The *Sores* of the *Incision* do seem to dry a little in Three or Four Days of the Feverish Preparation for *Eruption*. After this there is a *plentiful Discharge* at them. The discharge may continue a little while after the Patient is quite well on other Accounts; But the *Sores* will soon enough dry upon of themselves; but the *later*, the *better*, as we think. If they happen to be *inflamed*, or otherwise Troublesome, we presently help them in the ways we do any *Ordinary Sores*.[10]

[10] Mather appended this how-to manual to the end of his manuscript of "AB" 143–47); see also "CAP. XX. *Variolae triumphatae.* The Small-Pox Encountred" (*ABJ* [1972], pp. 113–15).

6.4. *CURIOSA VARIOLARUM.*

March 10, 1721/22.

{To Dr. John Woodward.}[1]

SIR,

So considerable a Part of Mankind fearfully perishing by the *Small Pox;* and many more of us grievously suffering by that miserable Distemper, you will allow me to entertain you with a few more Communications, and writ (I think it's) a fourth Letter upon it.

The *Distemper* has lately visited and ransack'd the City of *Boston*; and in little more than half a Year, of more then five thousand Persons that have undergone it, near nine hundred have died. But how many Lives might have been saved, if our unhappy *Physicians*, had not *poison'd and bewitch'd* our People with a *blind Rage*, that has appear'd very like a *Satanick Possession*, against the Method of Relief and Safety in the way of the *Small Pox* Inoculated.

1 The editors agree with G. L. Kittredge ("Lost Works," pp. 475–77) that Mather's CA 6.4 *Curiosa Variolarum*, addressed to John Woodward (March 10, 1721–22), is the same as the imperfect scribal copy (BL, Sloane 3324, f. 260) at the British Library. Mather's letter to Woodward was first published in Charles Maitland's *Account of Inoculating the Small Pox Vindicated* (1722), pp. 58–61. (Maitland is the same physician who supervised the inoculation of Lady Mary Wortley Montague's son in Constantinople in 1717.) A summary extract of Mather's letter to Woodward (CA 6.4), paragraphs 3 and 4, also appears in Dr. James Jurin's own "Letter to the Learned Dr. Caleb Cotesworth, F.R.S. and Physician to St. Thomas's Hospital; containing a Comparison between the Danger of the Natural Small Pox, and of that given by Inoculation, in *PT* 32.374 (December 31, 1722): 215. Jurin republished the same letter to Cotesworth in 1723 as a separate pamphlet, which includes Mather's data (p. 6). For good measure, Jurin's 1723 reprint also included a "Postscript" (pp. 19–22), by Capt. John Osborne stationed in Boston, which confirms Mather's narrative of the Boston inoculation crisis.

I prevail'd with one Physician,[2] (and for it I have had bloody Attempts made upon *my Life*[3] by some of our *Energumens*) to introduce the Practice; and the Experiment has been made upon almost three hundred Objects in our Neighbourhood, Young and Old; (from one Year to seventy) weak and strong; Male and Female, White and Black; in Midsummer, in Autumn, in Winter, and it succeeds to Admiration.

I cannot learn that any one has died of it; tho' the Experiment has been made under various and marvellous Disadvantages. Five or six have died upon it, or after it, but from other Diseases or Accidents; chiefly from having taken the

[2] Zabdiel Boylston (1679–1766), a Boston medical practitioner who was the only physician at the time to accept Mather's call in May 1721 for a united effort to examine (and administer) the "new" prophylactic method of inoculation (*D* 2:620–25). All to no avail. Roughly three months later (August 1721), Mather recorded that "I will allow the persecuted Physician [Boylston], to publish my Communications from the *Levant*, about the *Small-Pox*, and supply him with some further Armour, to conquer the Dragon" (*D* 2:636). Mather refers to *Some Account Of what is said of Inoculating or Transplanting the Small Pox. By the Learned Dr. Emanuel Timonius, and Jacobus Pylarinus. Published by Dr. Zabdiel Boylston* (1721). Mather also collaborated with Boylston on the newspaper article "A Faithful Account of what has occur'd under the Late Experiments of the Small-Pox managed and governed in the Way of Inoculation," in *Boston Gazette*, No. 101 (October 30, 1721). Almost half a decade later, Boylston, F.R.S. (now an elected member of the Royal Society of London and celebrated in England), published *An Historical Account of the Small-Pox Inoculated in New England* (1726, 1730). For American members of the RSL before the Revolution, see D. Margret (1947) and R.P. Stearns (1951). For the history of medical training in England in the seventeenth and early eighteenth centuries, see P. Allen (1946) and J. Barry (2019).

[3] An irate Bostonian "threw a fired Granado" through the window of Mather's house. The attempt to assassinate him failed (*D* 2:657–58). The attack proved to be sufficiently notorious to attract the attention of *The London Journal*. Numb. CXXXI (Saturday, January 27, 1721), pp. 3–4, which reprinted the following account in its Foreign Affairs section: "They write from Boston in New England, Nov. 24. That upon the Tuesday before, some Hours before Day, a Granado-Shell, loaded with combustible Matter, with a lighted Fuse to it, was by some unknown Person, thrown into the House of the Reverend Dr. Cotton Mather, in the said Town of Boston, through the Glass Window of one of his Chambers, to the Terror and Hazard of the People there, and to the endangering that and other neighbouring Houses; and that his Excellency Samuel Shute, Esq; the Governour, in Detestation of such a villanous Action, had issued out a Proclamation, offering 50*l.* Reward for detecting the Author and Actor of that Wickedness, so that he be tender'd to Justice and convicted of the Crime; offering likewise the same Reward, and a Pardon to any of the Accomplices in the Fact, that shall make Discovery of the principal Actor or Actors as abovesaid."

Infection in the common way, by Inspiration,[4] before it could be given them in this way of *Transplantation*. However at present I need say no more of this, having already given you some Report of our Proceedings in it.[5]

To them who are under the *Inoculation* of the *Small Pox*, we commonly give a *Vomit,* in the time of their *Decumbiture*, a Day or two before the expected Eruption. One of our Patients not vomiting so freely as he would have done, thrust a Finger or two into his Throat, which fetch'd up, what was to be discharg'd from his uneasy Stomach; He had but a few of the *Small Pox*, and the Pustules were sufficiently of the *distinct Sort,* as it uses to be where they have the *Small Pox Inoculated*; but the Fingers that had been thus employ'd, prov'd as full as they could hold, of the *confluent Sort*, which he now thought his whole Body would have been, if we had not in this Way prevented it.

Doctor[6] *Leigh* in his *Natural History* of *Lancashire*,[7] counts it an Occurrence worth relating, that there were some Catts known to catch the *Small Pox*; and pass regularly thro' the State of it, and then to die. We have had among us the very same Occurrence.

It was generally observ'd and complain'd, that the *Pigeon-Houses* of the City continu'd unfruitful, and the *Pigeons* did not *hatch* or *lay* as they used to do, all the while that the *Small Pox* was in its Epidemical Progress: And it is very strongly affirm'd that our Dunghil Fowl felt much of the like Effect upon them.

We have so many among us, who have been visited with the *Plague* in other Countries many Years ago, and who have never been arrested with the *Small Pox* after it, tho' they have been exposed as much as any other People

4 Inhaling.

5 "The Way of Proceeding in the Small Pox inoculated in New England," *PT* 32.370 (March 31, 1722): 33–35. Also see CA 6.3 "A Further Account, of the Method and Success of the Small-Pox Inoculated."

6 The Sloane MS. 3324, fol. 260, at the British Library, reads "Your Dr Leigh."

7 Charles Leigh, M.D., F.R.S. (1662–1701) practiced medicine in Preston (Lancashire) and published *The Natural History of Lancaster* (1700). It contains numerous medical case histories as well. Leigh (Lee) relates the case of two cats which "for the most part lay upon the same Beds with the Patients" (who had died of the smallpox). Soon both cats came down with the same symptoms of the smallpox, which erupted and matured, "with *Pustules* exactly like those in Human-kind; at last those subsided, and soon afterwards both the Cats dy'd. By this instance it is evident the small Pox is not a Distemper peculiar only to Human-kind" (*Natural History*, bk 2, ch. 1, p. 8).

to it; that it now begins to obtain a Belief with us, that they who have had the *Plague*, will never have the *Small Pox* after it.

I will add but one Thing more. For Succour under the *Small Pox*, where Life is in Danger, after all the Methods and Medicines, that our *Sydenham* and others rely upon;[8] I can assure you, we have yet found nothing so sure as this; Procure for the Patient, as early as may be, by *Epispastiks*[9] a plentiful Discharge at the *Hand-wrists*, or Ancles, or both, (I say as early as m{a}y be) and keep them running till the Danger is over. When the Venom of the *Small Pox,* makes an evident or violent Invasion on the Nobler Parts this Discharge *does wonderfully*. I am sorry it was so late before we fell into this Way; but it has *constantly prosper'd:* I know not that it has once miscarried, since we came into it.

March 10. 1721–22. Your most hearty Friend

& humble Servant

C. Mather.}

[8] The distinguished English physician Thomas Sydenham (1624–89), founder of clinical medicine and epidemiology, devoted his longest discussion to the natural (regular) and confluent (anomal) smallpox (variola) along with methods of easing patients through their deadly affliction, in his *Observationes Medicae* (1676). Cotton Mather owned a copy of this work; his signed and dated copy (August 26 1679) is at the AAS. To be sure, Sydenham's treatment of the smallpox – before the proliferation of inoculation and before the discovery of Edward Jenner's cowpox vaccination (from *vacca*, i.e, cow) – was largely confined to palliative methods such as inducing vomits and bleeding infected patients. See also Mather's letter to John Woodward (July 12, 1716), CA 3.10 "Curiosities of the Small Pox"; the letter to John Woodward (1721), CA 6.2 "*Variolæ triuimphatæ. The Small-Pox Encountred*"; and the letter to James Jurin (May 21, 1723), CA 7.1 "The Case of the Small-pox Inoculated, further cleared."

[9] Producing a blister to draw infected fluids from the body.

6.5. *NISHMATH-CHAJIM.*[1] THE PROBABLE SEAT AND GENERAL CURE OF ALL DISEASES.[2]

{After December 1, 1721.}

{To Dr. John Woodward.}

(1.) "Spiritus hilaris Ætatem floridam facit: Spiritus tristis Exsiccat Ossa. / (2.) Eodem allusit Regimen Sanitatis Salerni cum ait: Fol. III."
(3.) "*Si tibi deficiant Medici, Medici, tibi fiant*: / *Hæc tria, mens læta, requies, Moderata diæta*."
(4.) "*Quum Ventus pervadit ipsum, mox non Esse*; / Psal. CIII 16."[3]

1 Gen. 2:7: נִשְׁמַת חַיִּים nishmath chajim (breath of life, soul of life). Mather had lent the Nishmath-Chajim MS. to John Winthrop (1681–1747), grandson of Governor John Winthrop of Connecticut, who paid for its publication in New-London, 1722. Nonetheless, Mather was anxious to have it returned to him safely. See Mather's letter to John Winthrop (April 23, 1722) and (12^{d} 1^{m} [March] 1722/23), in *MP* (1868), pp. 452, 454; and Kittredge, "Lost Works," pp. 469–70.

2 No extant MS. Reconstructed from related passages in the AAS holograph of "AB" (MSS, pp. 29–43) and from *TP* (1995), pp. 122–26, with our copy-text provided by *AB* (1722). See *ABJ* (1972), Cap. V. *Nishmath-Chajim*, pp. 28–38.

3 The Latin epigraph on the published title page reads, (1.) "A cheerful spirit makes the age [heart] flourish: A sad spirit dries up the bones"; (2.) "The Rules of Health of Salerno alluded to the same when it said: (3.) 'If no doctors are available to you, let these three thrings serve as your doctors: a happy mind, rest, and a moderate diet'." It refers to *Regimen Sanitatis Salernitanum* (lines 8–9); i.e., *The Salernitan Rule of Health*, a 12th–13th century Latin poem attributed to the Schola Medica Salernitana, the medical school in Salerno (Italy). The poem appears in *Regimen Sanitatis Salernitanum: A Poem on the Preservation of Heath in Rhyming Latin Verse*, edited by Sir Alexander Croke (1830), p. 103. (4.) Mather's quotation from "Psal. CIII 16." is excerpted from the Latin translation of the Hebrew Scriptures in *Testamenti Veteris Biblia Sacra, sive Libri Canonici* (1593), p. 42^{r}, by the distinguished Italian Hebraist Immanuel Tremellius (1510–80) and the French Reformed scholar Franciscus Junius the Elder (1545–1602). The passage alludes to the brevity of life: "when the wind passes over it, it is soon gone." On the Medieval and Renaissance dialectic teaching methods for medicine and science, see B. Lawn (1963) and (1993).

The Probable SEAT of all Diseases, and a General CURE for them, further Discovered.

More particularly, for *Splenetic* & *Hysteric* Maladies, which make so great a part of our Distempers.

<The Glorious GOD, in whom we *live & move & have our Being*, & who is every where continually at Work, as the *Universal Cause* of all things; does by His Almighty *Will* give a Being to *Spirits*, that are in greater or lesser Degrees of Distance from *Grosser Matter*, as it pleases Him. And unto such *Spirits* He has by His *Laws* assigned such *Powers*, that upon their *Tendencies* He does produce what *Effects* He has by His *Laws* assigned unto such *Tendencies*. The more Progress we make in *Experimental Philosophy*, the oftener we shall find ourselves driven to something so much beyond *Mechanical Principles*, as to show us the Insufficiency of a meer *Corpuscularian Hypothesis*.>[4]

I. THERE is *a Spirit in Man*; A Wonderful Spirit, which from very good Authority may be called, *NISHMATH-CHAJIM*, [Or, *The Breath of Life*:] And which may be of a *Middle Nature*, between the *Rational Soul*, and the *Corporeal Mass*; But may be the *Medium of Communication*, by which they work, upon one another. It Wonderfully receives also *Impressions* from *Both* of them; And Perhaps it is the *Vital Ty* between them. The Scriptural Anatomy of *Man*, into *Spirit*, and *Soul*, and *Body*, seems to favour and invite the Apprehensions, which we are now proceeding to.[5]

[4] This paragraph is interpolated from Mather's "Nishmath-Chajim" essay in *TP*, p. 122, and appears neither in Mather's published *AB* (1722) nor in the AAS holograph manuscript "AB," nor in *ABJ* (1972). To reconstruct the content of Mather's Letter to John Woodward as a whole, we have chosen to establish a diplomatic transcription of Mather's *AB* (1722), our copy-text, but to mark the interpolated passages in the text and list the sources in footnotes. Mather introduces his target audience to his larger concern about redressing the philosophical challenges Cartesian mechanism and Hobbesian materialism posed to the belief in the immortality of the soul. The phrase "*Corpuscularian Hypothesis*" (the transformation of matter through the interaction of divisible particles) is associated with the philosophy of Descartes, Hobbes, Gassendi, Cudworth, Boyle, Newton, and others. For instance, Robert Boyle theorized in his *Sceptical Chymist* (1661) that matter consists of minute corpuscles (particles), which possess size, shape, and motion but are divisible, whereas atoms were believed to be indivisible. Their different amalgamations, in turn, determine their different material properties. See F. Bigotti (2020).

[5] Since matter and spirit are completely separate entities (Cartesian "Dualism"), Mather introduces his concept of Nishmath Chajim as a means to bridge this philosophical disjuncture between body and soul by establishing a basis of

>When our SAVIOUR so excused His drowsy Disciples, *The Spirit is willing, but the Flesh is weak*;[6] Doubtless, By *the Spirit*, he means, what His Apostle afterwards called, *The Mind*, and *The Spirit of the Mind*, and, *The Inner-Man*. But there being also in Man, that which is called, *The Soul*, or, *The Heart*, or that Principle & Passion, which is concerned more immediately for the Præservation of the *Life*, & of the Comforts that may sweeten it; This Principle is called, *The Flesh*. The *Flesh* of the Disciples here, which now so Rebelled against the *Spirit*, was not the Lust of *Pleasures* & *Riches* & the Like; But it was that most *Natural Affection* of the *Soul* which lay in a Desire to shun *Death* & *Grief*. And this *Flesh* here is called, *Weak*, not because it wanted *Strength*, for it was in Truth, too *Strong*: But because it wanted *Health*; it was out of order. For the *Health* of the *Soul*, it lies in its *Obeying* and not *Opposing* the Empire of the *Mind*.<[7]

II. THE *Great GOD who formed all things*, and who after a Singular manner *forms the Spirit of Man within him*, has Endued this *Nishmath-Chajim*, with Marvellous *Faculties*; which yet are all of them short of those *Powers*, which Enable the *Rational Soul*, to Penetrate into the *Causes of Things*; To Do Curious and Exquisite Things in the *Mathematical Sciences*; And above all, To act upon a Principle of *Love to GOD*, and with the *Views* of *Another World*.

III. SOME *Rays of Light* Concerning this *Nishmath-Chajim*, have been darted into the Minds of many Learned Men, who have yet after all remained very much in the *Dark* about it.

FAMOUS have been the Sentiments of *Helmont* (and some other *Masters of Obscurities*)[8] about it; who would Exhibit it under the Name of the

---

contact between corporeal substance (possessing shape, size, weight, and motion) and incorporeal spirit (nonphysical and nonspatial). In this manner, he seeks to explain how the protoplast's sin (spiritual lapsus) causally impacted the physical constitution (postlapsarian disease and death) of both Adam and Eve and their offspring (imputation of sin). For useful discussions of Mather's "Nishmath-Chajim," see S.S. Kottek (1982); M.H. Warner (1981); P.A. Watson (1991); and B.M. Grainger (2012).

6 Matth. 26:41.

7 Mather later added this section in the left-hand margin of the holograph manuscript of his medical handbook "AB." It appears neither in *AB* (1722) nor in *TP* (1995), p. 122, but presumably in his Letter to John Woodward [1721], Mather's Sixth Series of Curiosa Letters to the Royal Society.

8 Jean Baptiste van Helmont (1580–1644), the Flemish Roman Catholic Paracelsian physician and iatrochemist, is remembered for his theory of spontaneous generation, the transmutations of metals, and his highly respected *Ortus Medicinæ. Id est*

Archæus;[9] and with much of Reason Press, that in the *Care of Diseases,* there may be more of Regard Paid unto it.

ACCORDING to *Grembs* (Writing, *De Orta Rerum*) it is, *Medium quid inter Vitam et Corpus, et Veluti Aura nitens Splendensque.* A Sort of *Luminous Air*, which is of a *Middle Nature*, {betwixt} Spirituous and Corporeous.[10]

IT has the Denomination of, *The Astral Spirit*, with some Philosophers, who Trouble the *Stars*, more than there is any Need for.[11]

EVEN the *Galenists* themselves, have not been without some *Suspicions*, yea, some *Acknowledgments*, of our *Nishmath-Chajim*, and have given very *Broad Hints* Concerning it. And no doubt, they may thank the Old *Platonists* for Instructing of them. The Great *Fernelius*, one of the most Illustrious Men that ever shone among them, (Writing, *De Abditis rerum Causis*) gives a very Lively Description of it.[12] Yea, He finds it in the, *TO ENORMON*, or, *Inciter*, of *Hippocrates*; and having a great *Power of Incursion*, like the *Wind,* he allows it some Affinity with the Nature of *Body*; But inasmuch as it is *Invisible*, it must also have some Affinity with what is *Incorporeal*. So, he will have it of a *Middle Nature* between *Both*. But he supposes it, the *Vehicle*, and proper Seat of the *Soul*, and all its Faculties; and if we call it, their *Body*, we shall have his Permission for it.[13]

Initia Physicae Inaudita (1648). See esp. his discussion of the archeus in "Causae, et Initia Naturalium" (pp. 26–32) and in "Archeus Faber" (pp. 33–34) et passim, in *Orto Medicinae* (1652). In his Letter to John Woodward (July 9, 1716), Mather complains about "the Insipid Philosophy" of certain individuals "which seems to ly very much in a few School-terms, which are a meer Jargon invented for an ostentatious Cover of Ignorance." See CA 3.7 "Strength of the Imagination."

9 "Archaeus," also "archeus," denotes "a spirit that presided over and governed bodily processes" (*Medical Dictionary*).

10 The Latin quotation is from *Arbor Integra et Ruinosa Hominis* (1657), lib. 1, ch. 7: "De ortu rerum," §2 "De Archeo," p. 34, a medical handbook by the German physician and alchemist of Munich Franz Oswald Grembs (1621–58), an enthusiastic follower of the Flemish alchemist Jan Baptist van Helmont. The citation translates, "Some middle between the soul and the body and just as a shining and bright air." See also G.D. Hedesan (2013).

11 Mather's Platonic "*Astral Spirit*" is an aetherial vehicle from heaven that surrounds the human soul like a garment. See M. Heyd (1995), p. 106.

12 The French physician and professor of medicine at Paris Jean Fernel (1497–1558) authored *De Abditis rerum Causis Libri Duo* (1548). Mather refers to lib. 2, cap. 7, pp. 150–57.

13 Mather here paraphrases Fernel's discussion of the concept of the middle nature between body and soul (p. 150), as conceived by the famous Greek physician and "father of medicine" Hippocrates (c. 460–c. 370 BCE).

AND indeed, the Old *Platonists* had a Notion, of a certain Excellent *Body*, Pelluced, and Ethereal, Subservient unto the Faculties of the *Soul*, and Uniting it unto the more Terrestrial *Body*.[14]

HEURNIUS, whom some Reckon and Value next unto *Fernelius*, (in his *Institutions*) describes it as, *A kind of Ethereal Spirit, Elaborated out of the purest part of the Blood, and changed into the Substance of a very subtil Air; and the prime Instrument of the Soul, for the performance of its Functions*.[15]

IV. OUR *Nishmath-Chajim* seems to be Commensurate unto our *Bodies*; and our *Bodies* are Conformable to the Shape which GOD our Maker gives to that *Plastic Spirit*, (if we may call it so:)[16] But by what *Principle* the Particles of it, which may be finer than those of the *Light* it self, are kept in their *Cohæsion* to one another; tis a thing yet Unknown unto us.

V. AND how it fares in the Case of *Amputations* on our *Bodies*; Whether like a Flame Violently Struck off, what is so, may not nimbly, as by a sort of *Magnetism*, Re-unite with what it belongs unto: But then, how far it becomes for the Present folded up into it: Or, Whether it be not Entirely Lost, but what remains, may have the Power to Produce a *Recruit*, when there shall be a Lodging again Provided for it; This also is yet Unknown unto us.[17]

14 In his *Timaeus* (sec. 35), Plato's friend Timaeus explains that god made the soul of three components: the first, an "indivisible and always changeless" soul; second, a divisible corporeal body, subservient to the soul; and a third component (placed between the indivisible and divisible components) an "intermediate form of being, derived from the other two" and sharing both their qualities (*Complete Works*, p. 1239). This third component approximates Mather's Nishmath-Chajim. See also *TP* (1995), p. 370, *n.* 42.

15 Jan van Heurne (1543–1601), Dutch professor of medicine at Leyden, mostly remembered for his *Institutiones Medicinæ*, of which more than 27 editions appeared between 1593–1666. This reference is to the edition published by his son as *Institutiones Medicinæ. Editio altera* (1638), lib. 3, cap. 1, pp. 57–70.

16 The idea of the "*Plastick Spirit*" of nature is derived from Plato's concept of *anima mundi* (*Timaeus*) – the theory that physical nature is endowed with a spiritual or plastic (formative) power (Aristotle's entelechy) that orders and shapes the size and growth of all biological life forms. This vitalist theory made a brief comeback but was largely discredited by the end of the 17th century. In Mather's time, the English theologian and Cambridge Platonist Ralph Cudworth, in his huge *The True Intellectual System of the Universe* (Cambridge, 1678), and the English botanist and natural philosopher John Ray (1628–1705), in his *The Wisdom of God Manifested in the Works of the Creation* (1691), esp. pp. 31–40, were two of the principal proponents of "plastic nature."

17 See also Mather's commentary on Gen. 6: *BA* (1:595–96); *CP* (1721), pp. 131–37; and *TP* (1995), p. 123. Useful explanations are provided by W.B. Hunter (1950) and L.S. King (1978).

VI. THE *Nishmath-Chajim* is the *Spirit* of the Several *Parts*, where it has a Residence; and it is the *Life* by which the Several Parts have their *Faculties* Maintained in Exercise. *This* tis, that *Sees*, that *Hears*, that *Feels*; and Performs the *Several Digestions* in the *Body*. And the *Animal World*, having *Animam Pro Sale,* if it were not for *This*, would quickly *Putrify.*[18]

VII. WE have sometimes been led by our *Microscopes*,[19] into some Apprehensions, That our *Bodies* are Originally folded up, in Inconceivably Minute *Corpusculicuncules*;[20] and that Generation is nothing but the *Evolution* of the *Stamina* so Involved: Which Operation is carried on, by *filling them up* with a matter agreeable to them, till they have an Augmentation to the utmost Extent of the Dimensions, that they can reach unto: And that the *Resurrection of the Dead*, which is in the Sacred Scriptures called, *A Filling of the Dead Bodies*,[21] will find out the Old *Stamina* of the Forsaken Body, again Shrunk up into its first Parvity, and Replenish it with a more *Etherial Matter*, fit for the Cœlestial Employments and Enjoyments intended for it.[22] But this Hypothesis is Encumbred with Difficulties, which drive us unto a

18 "Spirit to serve as salt" (Cicero, *De Finibus Bonorum et Malorum* 5.13.38).

19 See Mather's theory of germs in his microscopical examination of animalculi (little animals), in *ABJ* (1972), Cap. VII. *Conjecturalis*. or, Some Touches upon, A *New Theory* of many *Diseases*," pp. 43–48.

20 Mather alludes to the concepts of preformationism and epigenesis, early theories of embryology, that fully formed yet miniscule embryos (homunculi) are present in the vesicle of the male sperm or female egg. Upon fertilising the undifferentiated matter of the egg, the animalcule (Mather's "*Corpusculicunculus*") merely unfolds during the gestation process. His hesitation about preformationism is informed by the unresolved debate between preformationist and ovists, given that hybrid animals such as mules evince the characteristics of both male and female progenitors and could therefore not have embodied the sole characteristics of the preformed animalcule of one parent only. Mather was familiar with George Garden's "Discourse concerning the Modern Theory of Generation, in *PT* 17.192 (February 28 1691): 474–83, which neatly synopsizes the debate. See Needham (1959); M-H. Huet (2004), and C. Pinto-Correia (1997). Mather's tongue-twisting neologism "*Corpusculicuncules*," perhaps derived from the merger of "corpuscle" (a minute cell in an organism) and "homunculi" (little man inhabiting a sperm cell), has not yet found its lodging in the *OED*.

21 Jer. 33:5.

22 The venerable Robert Boyle similarly believed that microscopically small corpuscles of the dead body remain, even if completely consumed by wild animals and mixed with its cells, "and yet retain their own Nature." From this remnant, God could revive the lost body and rejoin it with the incorporeal soul at the resurrection, in *Some Physico-Theological Considerations about The Possibility of the Resurrection* (1675), pp. 18, 33–39.

Nishmath-Chajim, either to Support and Perfect the Hypothesis, or to yield us a Better upon the failing of it.[23]

VIII. THE *Nishmath-Chajim* is indeed, *Generationis Faber ac Rector*;[24] and as it Leads to the Acts Requisite in *Generation*, without any further *Instructor*, So it is the *Spirit*, whose *Way we know not, for Shaping the Bones*, and other Parts, *in the Womb of her that is with Child.*[25]

IX. THERE are indeed many Things in the Humane Body, that cannot be solved by the Rules of *Mechanism*. Our *Nishmath-Chajim* will go very far to help us in the Solution of them. Indeed we can scarce well Subsist without it.

X. THERE is an astonishing Operation, & indeed some Illustration and Explanation, of the *Nishmath-Chajim,* in *Prægnant Women*; Whose Imaginations frequently makes Impressions on the *Unborn Infants*,[26] that would Exceed all Belief, if we had them not continually in View before our Eyes. The Instances are so Numerous and so Various that One might Compile a Large Volumn of them; and almost ask a *Palæphatus* to afford a Title for it. But in what other way to be accounted for?[27]

XI. FOR the *Nishmath-Chajim* we may safely be *Traducians*. It is a Flame Enkindled in, and so Derived from the *Parent*. And this *Traductian* (which is *Luminis e Lumine*) may help us considerably, in our Enquiries, *How the Dispositions of our Original Sin are Convey'd & Infus'd into us?*[28]

23 See also Mather's Letter to John Woodward (July 2, 1716), CA 3.1 "Monstrous Impregnations."

24 Derived from Jan Baptiste Van Helmont's *Ortus Medicinae* (1652), "Archeus Faber," p. 33, and signals "the artificer and director of begetting," an early precursor of our modern concept of DNA. See also Silverman (*LT*, p. 252).

25 Eccles. 11:5.

26 The archaic belief in the transformative power of the imagination impacting the four humors and the foetus of a pregnant mother. For this archaic belief, see Robert Burton's *Anatomy of Melancholy* (1621), Pt 1, Sec. 2, Mem. 1, Subs. 6 ("Parents a Cause by Propagation"). See also Mather's Letter to John Woodward (July 9, 1716), our CA 3.7 "Strength of Imagination."

27 I.e., Palaiphatos (Palaephatus), mentioned in *Suda*, a 10th-century CE, a compendium of abstracts and sayings from the classics. *Suidae Lexicon, Graece & Latine* (1705), 3:8 lists several writers by that name. Though used as a pseudonym, Palaiphatos is believed to have been the fictitious name of a 4th-century BCE Athenian mythographer to whom all sorts of fictitious works are attributed (*Pauly* 2449–55). Hence Mather's facetious reference.

28 "Traducianism" argues that the human soul, along with Original Sin, is passed from Adam and Eve to their offspring during procreation. Instead of creating each soul individually at conception, God created only one human soul – that of Adam – and Eve and their offspring received their respective souls from their original

XII. IT was of Old, yea, it is at this day, a prevailing Opinion, among the Strangers to the *Glorious Gospel of the Blessed God;* That the *Manes,* which remain after *Death,* have still an *Humane Shape,* and all the Parts both External & Internal, which there were in the *Body* that is now deserted:[29] yea, That there is a *Food* which this *Departed Spirit* craves for & Lives on. *Homer* inflicts Punishments on the Wicked after *Death,* which there must be a Sort of *Bodies* to be the Subjects of. And *Plato* Speaks of those that are Punished in *Hell,* as having such *Members* and *Faces* as they had once upon the Earth. Indeed *Justin Martyr* argues from it, that these old Gentlemen must needs have some Knowle{d}ge and Belief of our Doctrine of, *The Resurrection of the Dead.* But what shall we say, when our Glorious LORD-REDEEMER, in His *Parable of the Rich Man,* supposes his *Body* in the *Grave,* and yet being in *Hell,* he cries out of a *Body,* and particularly, of a *Tongue,* <[the Member which the *Wealthy* much abuse in their *Table-Talk*]> that is tormented there?[30] Many of the Ancients thought, there was much of a Real *History* in the *Parable*; <[yea, the *Talmuds* have such an *History*, and make the Name of the *Rich Man* to be, *Nimensis*;]>[31] and their Opinion was, That there is, *DIAPHORA KATA TAS MORPHAS, A Distinction* (and so a Resemblance) *of men as to*

---

progenitor. Condemned by Pope Anastasius II (498), traducianism was upheld by Gregory of Nyssa, Tertullian (*On the Soul* 26–27), and St. Augustine, for whom it signified *spiritual* generation. The Latin passage translates “Of light from light.” On the 17th-century debate on Traducianism vs Creationism of the soul, see Joseph Glanvill's *Lux Orientalis* (1682); Henry More's *The Immortality of the Soul* (1659), Edward Warren's *No Præexistence* (1667); and Henry Hills' *A Short Treatise Concerning the Propagation of Souls* (1667). See also P.C. Almond (1994), ch. 1.

29 Derived from the gods of the Underworld (*Dii Manes*), the term *Manes* signifies the souls of the dead separated from their bodies. It is in this context that Mather refers to Homer's *Odyssey* (10.563–80, 11.30–104), Plato's *Phaedrus* (249, 256), *Republic* (3.386, 10.613–14); to Justin Martyr's *First Apology* (18–21), *ANF* (1:168–70); and to the evangelist's reaction to Plato and Aristotle, in *Exhortation to the Greeks* (6), *ANF* (1:275–76).

30 Luke 16:23–25. Mather interpolated the bracketed phrase in his version of Nishmath-Chajim, in his *TP*, p. 124.

31 The bracketed interpolation is Mather's and appears in *TP*, p. 124. The Greek goddess Nemesis metes out retributive justice. Her name does not appear in the Talmud, yet in the Talmudic tractate Tamid (32a), the interlocutor asks his students, “Who is called a rich man? They replied: Who is rich? He who rejoiced in his lot. He said to them: What shall a man do to live? They replied: Let him mortify himself.” A person's obsessive preoccupation becomes his undoing (Soncino Classic Collection, Davka Corporation).

their Shapes after Death.[32] We find This was the Opinion of *Iræneus*; who proves, from what our SAVIOUR Speaks of the *Dead Man*, that the *Souls* which have put off their *Bodies*, do yet *Characterem Corporum Custodire*, Preserve the *Shapes* of the Bodies, to which they were United. And from the same Speech of our SAVIOUR, *Tertullian* does infer, *Effigiem Animæ et Corporales Lineas*, the *Shape* and *Corporeal Lineaments* of the Soul, I will say nothing of what *Thespesius* returning to Life reported about the, *TA TONE PSEUCONE CROMATA*, *The Colours of Souls*, & the Ulcers by their *Passions* left upon them.[33]

On this Occasion, the Words of M. *Dacier*, in her Notes upon *Homers* Odysses, are not unworthy to be considered:[34] which give us the Observation, That a Fine Subtil Sort of *Body*, accompanied the *Intellect*, after the Separation of the *Soul* by Death from the Grosser *Body*.[35]

XIII. THE *Nishmath-Chajim* is much like the *Soul* which animates the *Brutal World*; Even that *spirit of the Beast, which goeth downward unto*

[32] Mather transliterates the Greek passage ΔΙΑΦΟΡΑ ΚΑΤΑ ΤΑΣ ΜΟΡΦΑΣ (*TP*, p. 124) to accommodate his target audience. The passage appears in several places, including Pseudo-Justinus Martyr's *Quaestiones et Responsiones ad Orthodoxos* (Moral page, 426, sec. A, line 8), and in Theodoretus of Cyrus, *Quaestiones et Responsiones ad Orthodoxos* (page 7, line 23; p. 71, line 24).

[33] Mather refers to Irenæus, bishop of Lyon (c. 130–200), *Against Heresies* (5.6.1; *ANF* 1:531). The Latin quote from the same work reads, "To preserve the body's form" (*Against Heresies* 2.34.1; *ANF* 1:411). Tertullian's "Image of the soul and its corporeal lines" is from *Against Marcion* (2.9; *ANF* 3:304). The vision of Thespesius, a character in Plutarch's *De sera numinis vindicta: On the Delays of the Divine Vengeance* (Stephanus 565C) is indeed an unpleasant one, since the souls of his heroes return with the ulcers of their passions.

[34] The reference is to *L'Odysée d'Homere, Traduite en François, avec Des Remarques Par Madame Dacier*. 3 vols. (1708), by the distinguished French classical scholar, translator, commentator, and editor Anne Lefèvre Dacier (1647–1720). In his holograph manuscript "AB," Mather cancels his own translation of the quoted passage from Dacier: "The Egyptians beleeved, 'That after Death, that is to say, after the Separation of the Soul from this mortal Body, there was yett another Separation of Two Parts of the Soul; that is, Of the Intellect or Understanding, which Homer calls, ΦΡΗΝ, and that of the First and subtil Body, wherewith the Intellect was cloathed, and which the same Poet calls the ΕΙΔΩΛΟΝ, or Image; and that y[e] Intellect goes to Heaven, but the Image to Hell." Yet Mather liked the passage sufficiently to include a close paraphrase of the same in *TP* (1995), p. 124. See also *BA* (4:346) and below.

[35] Mather added this paragraph (along with its cancelled sentences) in the left-hand margin of his medical handbook "AB."

the Earth;[36] but is by the Hand of the Glorious Creator impregnated with a *Capacity*, and *Inclination*, for those Actions, which are necessary for the *Preservation* of themselves, & the *Propagation* of their *Species*. The *Nidification* of *Birds*, the *Mellification* of *Bees*, and a Thousand such Things, how surprizing Works done in the Brutal World, without any *Rational Projection* for them![37] And hence, there are also many Actions done by us, that have a Tendency to our Safety & Welfare, which are not the Effects of any *Rational Projection*; but such as we do by what we call, *A meer Instinct of Nature*, fall into. The *Sucking Infant*, yea, and the *Nursing Mother* too, do very Needful and Proper Things, without Consulting of *Reason* for the doing of them.[38] <It is a thing, which who can observe without Astonishment? In every other *Machin*, if any thing be out of Order, it will remain so, till some Hand from abroad, shall rectify it. But the *Humane Body* is a *Machin*, wherein if any thing be out of Order, presently the whole Engine, as under an Alarm, is awakened for the helping of what is amiss, and other Parts of the Engine, strangely putt themselves out of their Way, that they may send in Help unto it, whence can this proceed, but from a *Nishmath-Chajim* in us, with such Regards for the *Law of Self-Preservation*, by GOD imprinted upon it.>[39]

XIV. HAVING at some time or other felt a Considerable *Smart*, or been Considerably *Sick*, from something that we have met withal, we have an *Abiding Horror* for that thing perhaps all our Days. Tho' we *Certainly Know* that the Thing will now do us *No Hurt*, but rather do us *Much Good*, yet no Conviction of *Reason* will Overcome our *Abiding Horror*, We cannot Swallow the *Pill*, or take the *Meat* or the *Drink*, & do an hundred Things, which we have heretofore been horribly frighted at. Our *Nishmath-Chajim* has an *Incurable Aversion* for them.

36 Eccles. 3:21.

37 See Mather's Letters to John Woodward (July 4, 1716), CA 3.3 "The Nidification of our Pigeons." Unfortunately, his Curiosa essay on "Melissologia; with a new Method of Bee-hunting" (see CA 7.0) has perished. On the act of building a nest (nidification) and the queen bee's governance of her hive and of making honey, see Mather's *CP* (1721), Essay XXX, pp. 188–89; and respectively, Essay XXVII, pp. 152–54. See also Paul Dudley's "An Account of a Method lately found out in New-England, for Discovering where the Bees Hive in the Woods, in order to get their Honey," *PT* 31.367 (April 20, 1721): 148–50.

38 In his *CP* (1721), Essay XXXII, "Of Man," p. 256; and *TP* (1995), pp. 124–25, Mather expresses his admiration for nature's providence in postponing the growth of teeth in nursing infants.

39 The two preceding sentences are restored from Mather's "Nishmath-Chajim," in *TP* (1995), p. 125.

XV. TIS the *Nishmath-Chajim*, that is the *Strength* of every part in our *Body*, and that gives *Motion* to it. Here perhaps the Origin of *Muscular Motion* may be a little accounted for. <This is that *Aura Motiva*, which in the Proportion of its *Force* to the *Pondera Elevanda*, so admirably observed and adjusted by the Rare *Borelli*, (and others after him, carries in it what will *force* the Greatest Astonishment.>[40] And this is the *Spirit,* and the *Balsome*, & one might almost say, the *Keeper*, of each part, which is Occupied and Befriended with it. Yea, what Construction shall we make of it, when People have Lived without any *Brains* in their *Heads*, & after the Destruction of almost all the *Bowels* in their *Bodies?*[41] We are supplied and surprized with many most Credible Relations of such Things. And I Question, whither any thing will do so well, or go so far, as our *Nishmath-Chajim*, to account for them?

XVI. THE Principal Wheel in the *Animal Oeconomy*, is the *Stomach*. And we shall now find that which above all things the *Digestion* there, is to be ascribed unto. Dispute, *O Philosophers, and Physicians*, How *Digestion* is Performed in the *Stomach*. Tis the *Nishmath-Chajim* after all, that is above all, the *Main Digester.* Else, how could a *Stomach* that is actually *Cold*, and has in it no very *Tastable* or *Notable* Humour for this purpose, *Digest* the very *Stones* that are taken down into it?[42]

THE taking of some *Repast*, is in our Sacred Scripture sometimes called, *The Establishing of the Heart*. The *Heart*, is not seldom, a term for our *Nishmath-Chajim*.

XVII. IT is the *Nishmath-Chajim*, that is more Eminently the *Seat* of our *Diseases*, or the *Source* of them. To pass by what they quote of *Herophilus*, we find *Plato* Elegantly Demonstrating, That all *Diseases* have their Origin in the *Soul*. Yea, as long ago, as the Days of *Hippocrates*, the *Essentials* of *Diseases*

[40] The preceding sentence is restored from Mather's "Nishmath-Chajim," in *TP* (1995) p. 125. "Motive Breath" and "Weights to be lifted up" – both Latin phrases appearing in *De Motu Animalium* (1680), pars prima, cap. 1, pp. 3–4, by Alfonso Giovanni Borelli (1608–79), an Italian physician and biologist who developed a mechanical interpretation of the process of life (iatromechanism).

[41] For accounts of this nature, see CA 1.4 "On Antipathies" and CA 1.7 "Remarkable Recoveries."

[42] Mather's February 1720 Curiosa essay "A Strong Digester" (Sixth Series) to John Chamberlayne has perished; however, since Mather was in the habit of repurposing useful information, some portions of it may have been digested in his discussion of the stomach, in "The Main Wheel scoured and oiled. or, Help for, the Stomach depraved. [and, Vomiting], in *ABJ* (1972), pp. 195–201, as well as in *CP* (1721), pp. 206–07, 266–68.

began to be Discovered; and the Pacifying & Rectifying, of the *Enforcing Spirit* was proposed as the most ready way to Cure them.[43]

QUÆRE; How far the Decays of *Old Age* are to be found in the Circumstances of the *Nishmath-Chajim* falling under Impairments? And Whence it came to pass that when *Moses* was very Old, yet *his Eye was not dim, nor his Natural Force abated?*[44]

IS not this the true, *Humidum Radicale,* they use to talk of?[45]

AND is not this the *Microcosmic Air*, whereto *Tachenius* ascribes the Cure of the *Gout* by a strong perturbation of the Mind; upon which he concludes it animated?[46] <Most[47] certainly, The *Physician* that can find out Remedies, that shall have a more immediate Efficacy to Brighten, and Strengthen, and Comfort, the *Nishmath-Chajim*; Especially, if he can irradiate the Spirit in the *Stomach*, and also fortify the *Blood*, and restore a *Volatil Ferment* in *That*, when it grows Vapid & Languid; and in a Word, if he can *Confirm the System*, which our *Sydenham* supposes, of our *Spirits*, wherein proper *Anodynes*, and *Chalybeates*, are often very potent: Such an one will be the most successful *Physician* in the World. He will be yett more so if as *Baglivi* advises, he can

43 Plato enlightens his interlocutor Charmides, explaining that "Because … the soul is the source both of bodily health and bodily disease for the whole man, and these flow from the soul in the same way that the eyes are affected by the head. So it is necessary first and foremost to cure the soul if the parts of the head and the rest of the body are to be healthy." How is the soul cured? Says Charmides, "by certain charms … consisting of beautiful words" (*Charmides* 156e–157a), in *Complete Works* (642–43). Herophilus (c. 4th–3rd century BCE) was one of the most notable physicians, anatomists, and physiologists of antiquity – if we consult Tertullian's *Treatise On the Soul* (10.4, 25.5), in *ANF* (3:189, 206).

44 Deut. 34:7.

45 This "root moisture" is alleged to give the body its agility and youthful appearance. That this concept has not lost its sap is shown in a recent article by Stanford University researchers Hamilton Se-Hwee Oh et al., "Organ aging signatures in the plasma proteome track health and disease." *Nature* 624 (December 6 2023), pp. 164–72.

46 The German iatrochemist Otto Tachenius (1610–80) is mostly remembered for his *Hippocrates Chimicus* (1666) and his *Antiquissimæ Hippocraticæ Medicinæ Clavis* (1697). For Mather's reference to Tachenius's explanation and treatment of the gout, see *Otto Tachenius His Hippocrates Chymicus* (1696), ch. 15, pp. 46–51, and ch. 19, pp. 65–72, esp. p. 68. Mather has two hefty chapters on the cure of the gout and of rheumatism (the gout's "Younger Brother") in his medical handbook *ABJ* (1972), chs XII and XIII, pp. 66–79, 79–82.

47 The long segment beginning with "Most" to the end of the paragraph is restored from Mather's "Nishmath-Chajim," in *TP* (1995), pp. 125–26.

Skilfully Quiet the *Passions of the Mind*, the *Griefs* and *Cares*, by which his Patient is disordered. Wonders have been done, in the *Art of Curing by Expectation*; & there would be yett greater Wonders done in the, *Art of Curing by Consolation.*>[48]

XVIII. IT is Probable, that when we Dy, the *Nishmath-Chajim*, goes away, as a Vehicle to the *Rational Soul*; and continues unto it an Instrument of many Operations. Here we have some Solution for the Difficulties, about *Place*, and the *Change* of it, for such an *Immaterial Spirit* as the *Rational Soul*; And some Account for *Apparitions* of the *Dead*: The *Spectres*, which are called both, *Spirits*, and *Phantasms*, in our *Gospel*. Yea, We are certain of it, That Persons before they have *Died*, upon strong Desires to Visit and Behold some Objects at a Distance from the *Place* to which they were now Confined, have been thrown into a *Trance*, wherein they have lain some Considerable While without *Sense* or *Breath*; and then Returning, have reported what they have seen. But incontestible Witnesses have deposed, that *in This Time*, they were actually Seen at the *Place*, which they affirmed they had gone unto.[49]

AND here also we do a little understand, how our Apostle in the Raptures (which the Scoffing *Lucian* derides him for) wherein he supposes he might be *Out of the Body*, yet he *heard Words*; he was yet *Sensible* of *Occurrences*.[50]

<In reading of *Homer*, (as has been already in Part observed) we find his Notion to be the Same with what was in the *Egyptian* Philosophy; which supposed, that *Man* is composed of Three Parts; An Intelligent *Mind*, called, φρην, or ψυχη· A *Vehicle*, called, ειδωλον· The *Image*, or the *Soul*: And a Gross *Body*, called, Σωμα· The *Soul*, in which they look'd on the *Mind* as lodged, they supposed exactly to resemble the *Body*, in Shape, & Bulk, &

[48] Anodynes administered in a chalybeate liquid are pain-relieving drugs that have an iron taste. Georgio Baglivi (1669–1707) was for Italy what the renowned Thomas Sydenham (1624–89) was for England, whose *Observationes Medicinae* (1676) went through numerous editions. As a physician of the soul and body, Mather was thoroughly at home in the art of curing by consolation, for his third wife, Lydia Lee George, was given to bouts of insanity. See Mather's *D* 2:583–84, 723, 735, 742, 749–50, 755; and his *ABJ* (1972), chs 24–25. That New England's clergy often suffered from "*crazy melancholy*" is documented in Mather's *D* 1:327.

[49] See especially the story of Sir Charles Lee's wife, in Mather's CA 1.10 "Joseph Beacon's Vision or Dream."

[50] The Syrian satirist Lucian of Samosata (c. 125–80 CE) was known for poking fun at all manner of religious beliefs, esp. in his *Dialogues of the Gods*. Whether or not Lucian referred to St. Paul's vision on the road to Damascus (2 Cor. 12:2) is hard to tell.

Features; being in the Body as the Statue in the Mould, & so after its Departure keeping the *Image* of the *Body*.[51]

Plutarch very distinctly delivers this Doctrine; and sais, when the *Soul* is compounded with the *Understanding*, it makes *Reason*; & when compounded with the *Body*, it makes *Passion*. The one Composition, is the Principle of *Pleasure* & *Pain*, the other of *Vertue* & *Vice*. He adds, Man dies *Two Deaths*: The *First Death* makes him Two of Three: The *Second* makes him One of Two.>[52]

XIX. <IN the Indisputable and Indubitable Occurrences of *Witchcrafts* (and *Possessions*) there are many things, which, because they are *Hard to be Understood*, the *Epicurian-Sadducees* content themselves, in their Swinish Manner,[53] only to Laugh at. But the *Nishmath-Chajim* well understood, would give us a marvellous Key to let us into the Philosophy of them.>[54]

XX. AND now, for some Important Consequences.

MOST Certainly, the Physician that can find out Remedies (particularly in the *Mineral* or *Vegitable* Kingdom) that shall have a more Immediate Efficacy to Brighten, and Strengthen, & Comfort, the *Nishmath-Chajim*, will be the most Successful Physician in the World. Especially, if he can Irradiate the Spirit in the *Stomach*, he will *do Wonderfully*.[55]

51 Homer's *Odyssey* (10.563–80, 11.30–104), Plato's *Phaedrus* (249, 256), *Republic* (3.386, 10.613–14). Mather paraphrastically translates this passage from *L'Odyssée D'Homere, Traduite en Francois, avec Des Remarques. Par Madame Dacier* (1708), tome seconde. My reference is to the Paris 1756 ed., vol. 2, pp. 415–16. Also see Dacier (footnote, above).

52 This and the preceding paragraph are restored from Mather's holograph manuscript of "Nishmath-Chajim," in "AB." Plutarch, *De placitis philosophorum: Sentiments Concerning Nature* (4.2–7. 23) – said to be an epitome of a work by Aetius (fl. late 1st century CE)

53 In *TP* (1995), p. 126, Mather has "in their *Bekkerian* Manner."

54 Both the Epicureans and Sadducees were philosophical materialists, who denied the existence of the soul and of spirits. For Mather's views on demonic possessions, see his *MPW* (1689), *WIW* (1693), and "Thaumatographia Pneumatica," in *MCA* (1702), 6.7.66–83. Mather sneers at the Jewish sect of the Sadducees (formed c. 200 BCE) who, like the Epicureans, denied the immortality of the soul and the existence of spirits. For similar reasons, the Dutch theologian Balthasar Bekker (1634–98), a Cartesian minister in Amsterdam, was removed from office (1692) when his *De Betoverde Weereld* (1691) gave a rationalist explanation of the "enchanted world," insisting that neither devil nor demon – though existing in their own spheres – can possess man or incite him to evil.

55 Mather cancels the following sentence from the holograph manuscript of "Nishmath-Chajim," in "AB": "Here, Saffrons, do thou assert thy Vertues."

THOSE things also, which Fortify the *Blood*, and restore a *Volatil Ferment*, in the Vapid and Languid *Blood*, will do Wonders for us. It is impossible to Kill a Man (the *Nishmath-Chajim* will never leave him) till the Circulation of his *Blood* be ruined.[56]

>He who will best keep in Heart, the *Nishmath-Chajim*, will be,

ΙΗΤΡΟΣ ΑΝΗΡ ΠΟΛΛΩΝ ΑΝΤΑΞΙΟΣ ΑΛΛΩΝ·<[57]

XXI. WE read, *Heaviness in the Heart of man makes it stoop; but a Good Word makes it Glad.* We read, *A Chearful Heart does good like a Medicine; but a broken spirit dries the Bones.* The Invigoration, or the Debilitation of the *Nishmath-Chajim,* is that wherein these ancient Observations are accomplished. >Dr. *Auerbachius*, after Forty Years Practice, made this Declaration, *Reipsa comperi plures homines Mœstitia ac Dolore Animi mori, quam violenta Morte.*<[58]

IT is a Remark of *Baglivi*, but it may have been made by Ten Thousand more; "That a great part of our *Diseases*, Either do Rise from, or are Fed by, a Weight of *Cares*, lying on the *Minds* of men. *Diseases* that seem Incurable, are easily Cured by agreeable *Conversation*. Disorders of the *Mind*, first bring Diseases on the *Stomach*; and so the whole Mass of *Blood* gradually becomes infected. And as long as the *Passions of the Mind* continue, the Diseases may indeed change their *Forms*; but they rarely quit the Patients."[59]

56 The English physician and anatomist, William Harvey (1578–1657) offered his break-through theory on the circulation of blood in animal bodies in his *Exercitatio Anatomica De Motu Cordis et Sanguinis in Animalibus* (1628).

57 The Greek quotation and the sentence preceding it are restored from the holograph manuscript of "Nishmath-Chajim, in "AB." The line from Homer's *Iliad* (11.514) appears in other places as well: Galen's *Thrasybulus sive utrum medicinæ sit an gymnasticæ hygieine* [*On whether Hygiene belongs to Medicine or Gymnastics*] (Kühn 5:869, line 10), and Plato's *Symposium* (Stephanus, 214*b* line 7). Says Homer's Idomeneus to the noble Nestor, "A healer is worth many other men."

58 The preceding reference to Dr. Auerbach is restored from the margin of Mather's holograph manuscript of "Nishmath-Chajim," in "AB." The German physician and alchemist Heinrich Stromer von Auerbach (1479–1542), professor of medicine at Leipzig, is mostly remembered for his *Saluberrime adversus Pestilentiam Observationes* (1516), *De Morte Hominis Decreta aliquot Medica* (1531), and *Decreta Medica et Senectute* (1531) (*ADB* 1:638). Whichever primary or secondary work Mather relies on for his citation, the Latin passage seems particularly apropos for his present context: "In fact, more men die from sadness and pain of mind than by violent death." Words of wisdom, indeed!

59 Mather's quote is a pastiche of passages from the English translation of Giorgio Baglivi's *De Praxi Medica ad Priscam Observandi Rationem Revocanda* (1696).

A *BONIFACIUS* heretofore Address'd the Physicians on this Occasion, in such Terms as these.[60] "*Tranquility of Mind*, will do Strange Things towards the Relief of *Bodily Maladies*. Tis not without Reason, that *Hofman* in this Dissertation, *Des Moyens de vivre longtem{p}s*, does insist on *Tranquility of Mind* as the chief among the *Ways to Live Long*; and that this is the cause why we read, *The Fear of the Lord tendeth to Life*.[61] They that have Practised, *The Art of Curing by Expectation*, have made an Experiment of what the *Mind* will do towards the Cure of the Body. By Practising, *The Art of Curing by Consolation*, you may carry on the Experiment.[62] I Propound then, Let the *Physician* with all possible Ingenuity of *Conversation*, find out, what matter of *Anxiety*, there may have been upon the Mind of his *Patient*; what there is, that has made his Life *Uneasy* to him. Having Discovered the *Burden*, let him use all the ways he can devise to take it off. Offer him such *Thoughts*, as may be the best *Anodynes* for his Distressed Mind; especially, the *Right Thoughts of the Righteous*,[63] and the Ways to a Composure upon *Religious Principles*. Give him a *Prospect*, if you can, of some *Deliverance* from his Distresses, or some *Abatement* of them. Raise in him as *Bright Thoughts* as may be, and Scatter the *Clouds*, remove the *Loads*, which his Mind is Perplexed withal: Especially, by Representing and Magnifying the *Mercy* of GOD in CHRIST unto him."[64]

XXII. IT is well known that if *one Third* of our Diseases, be those which we call, *Chronical*, more than *one Half* of this Third, will be those, which in *Men* go under the Name of *Splenetic,* and in *Women* go under the Name of *Hysteric*; tho' the *Spleen* and the *Womb* are often enough unjustly accused in

---

Mather's immediate source is Baglivi's *The Practice of Physick, Reduc'd to the ancient Way of Observations* (1704), bk 1, ch. 14, secs 1–4, pp. 177, 179, 180, 181. The same citation is also put to good use in Mather's *B* (1710), p. 132, and in *MAM* (1726), §19, p. 136.

60 Bonifacius, or doer of good. See also Mather's *B* (1710).

61 Mather refers to the same French work on redressing psychosomatic diseases in what appears to be a French translation of a work by the distinguished German physician and voluminous author Dr. Friedrich Hoffmann (1660–1742) of Halle, whose *Medicina rationalis systematica* in 6 vols. (1718–34) is among the best-known of his more than four hundred publications. Mather employs the same citation reference in his *B* (1710), Essay 18, p. 132; and in his *MAM* (1726), §19, p. 136. Baglivi's *The Practice of Physick* (1704), bk 1, ch. 14, pp. 184–89, offers similar remedies for mental distress.

62 See also Mather's "Cap. XXIV. De Tristibus, or the Cure of Melancholy," in *ABJ* (1972): 132–37.

63 Prov. 12:5.

64 As in the preceding extract from Baglivi, Mather's citation is from his own *B* (1710), pp. 132–33.

these Denominations. It is marvellous to see, in how many *Forms* we undergo *Splenetic* and *Hysteric* Maladies; The very *Toothache* it self often belongs unto them: And Marvellous will be the *Success*, Marvellous the *Esteem*, of the Physician that can Discover 'em & Encounter 'em.[65]

THE Sagacious Dr. *Sydenham*, seems to have the Scent of our *Nishmath-Chajim*, when he tells us, That as the *Outward Man* is fram'd with Parts, obvious to *Sense*, thus the *Inward Man* does consist of a due *Series*, and as it were a *Fabric*, of *Spirits*, to be view'd only by the Eye of Reason: And as this is united with the Constitution of the *Body*, so the *Frame* of it is more or less easily Disordered, by how much the *Constitution of the Spirits* is more or less Firm within us. And that the Origin of the *Splenetic* and *Hysteric* Ataxy in the *Body* is a *Feeble Constitution* of the *Spirits*, and the breaking of their *System*, so that they are easily Dissipated, or have an Unequal Distribution.[66]

THESE Maladies have many Symptoms, which may serve as *Diagnosticks* for them; Especially these Two: That the *Urine* is Clear, Limpid, & Copious. And, That the Patient is *Chiefly affected* with his Indispositions, when he has just had his *Mind* under some Disturbance and Affliction.[67]

IT is Plain, that these Diseases are not mainly in the *Humours*;[68] inasmuch as *Evacuations* do not Relieve, but fearfully Produce & Increase the Diseases. Only indeed, when the Ataxy of the *Spirits*, has by its Continuance at last Considerably Vitiated the *Humours*, – *Then* a little *Purging & Bleeding* may be allow'd of.[69]

THE Cure is; First, *Quiet* the *Spirits* with Proper *Anodynes*. Then bring them to Rights, and Revive them & Refresh them, & bring a *New Strength* into them. In short, *Confirm the System. Chalybeates* do Wonders this way;

65 See also Mather's "Cap. XI. Dentifragibus, or the Anguish, and Releef, of the Tooth-ache," in *ABJ* (1972): 61–66.

66 See Thomas Sydenham's "An Epistolary Discourse to the Learned Doctor William Cole, concerning some Observations on the confluent Small-pox, and of Hysterick Diseases," in *The Whole Works of that Excellent Practical Physician Dr. Thomas Sydenham* (1696), pp. 454, 463: "*Hysteric* Ataxy" is an emotional disorder affecting physical movement, including vertigo.

67 See Sydenham's *Compleat Method*, 4th ed. (1710), p. 6; and Baglivi, *Practice of Physick* (1704), pp. 89–90.

68 The ancient concept of an imbalance of the four metabolic humors – phlegm, blood, black bile, and yellow bile – causing disease was still regnant in Mather's time.

69 Sydenham, *Whole Works* (1696), p. 454. An "A*taxy* of the S*pirits*" amounts to a neurological loss of coordination (*Miller-Keane*). On vertigo, see Mather's "Vertiginosus, or How to Steer Under Dizziness," *ABJ* (1972), pp. 148–50.

And usually the Steel in Substance, more than many of the Common Preparations. *Corroborating Plants*, infused in Generous Wine, or a *Tea* of them, have also done Wondrously. Yea, *Venice-Treacle* alone (our *Sydenham* says) as Contemptible as it may seem, yet if often used, and a long while, it is a Great Remedy in This, as well as very many other Diseases, "and perhaps *the most Effectual that has hitherto been known in the World.*"[70]

THE Force of the *Peruvian Bark* Regularly administred, for giving a Vigour to the *Blood,* and so to the *Spirits,* has also been very Surprising And for the Recovering of People to an Healthy Constitution out of *Splenetic* and *Hysteric* Diseases.[71]

BUT there is nothing like the Exercise of *Riding on Horse-back,* Every Day, when the Weather will allow it; and increasing the Journey by Degrees till one comes to a Score of Miles in a Day.[72]

XXIII. UPON the Whole;

OF all the Remedies under Heaven, for the Conquering of *Distempers*, & for the Preservation of *Health* & Prolongation of *Life*, There will now be found none, like Serious PIETY. *Many* Remedies *have done Virtuously* (& had their Virtues) *but thou Excellest them all.*[73] The *Rational Soul* in its Reflections has Powerful and Wonderful Influences on the *Nishmath-Chajim*. Now in the Methods of PIETY, get a *Soul* into the *Peace of GOD*, with Assurance of a *Reconciliation* to Him; and *walk in the Fear of GOD*, & the *Comfort of the Holy Spirit*; Keeping always in, and Filled always with, His *Love*; and Indulge none of those *Lusts*, which render the *Wicked like the Troubled Sea.*[74] Keep a *Conscience*, which in a *Continual Aim* at what is Right shall make a *Continual Feast*. Be not Anxious about *Futurities*, nor Disturbed upon *Provocations*: But let the *Strong Faith* of a Faithful SAVIOUR *Performing the Thing that*

70 On the strengthening qualities of chalybeates, see Sydenham's *Entire Works* (1749), pp. 392–93, sec. 102. He praises the efficacy of "Venice Treacle" (i.e. Venice Theriaca), a curious concoction of 64 herbal and God-knows-what-other remedies known since ancient times. See Sydenham's *Entire Works* (1749), p. 398, sec. 112.

71 "Peruvian Bark," aka. Jesuit's bark, contains quinine and was widely considered the most effective nostrum against malaria, fevers, and agues. See Sydenham's *The Whole Works* (5th ed.), corrected (1712), pp. 224, 360.

72 Sure enough, Sydenham also recommends horseback riding (in combination with light chalybeates, i.e., liquids containing salts-of-iron minerals) as an efficacious nostrum, one that encourages bowel movement (*Entire Works* [1712], p. 32, *n.*(*s*), and pp. 401, 402, 439 etc. Mather follows suit in his advice for young clergymen, in his *MAM* (1726), p. 131.

73 Prov. 31:29.

74 Isa. 57:20.

is Appointed for us in all that happens,[75] Produce a perpetual Tranquillity & Serenity in the Soul. Go on *Singing in the Ways of the Lord*, & Casting all *Burdens* on Him, & *Rejoice in the Hope of the Glory of GOD*.[76] Thus, *I show you a most Excellent Way*.[77]

XXIV. LET this be Remembred; Moderate *Abstinence*, & Convenient *Exercise*; and some Guard against Injurious *Changes of the Weather*, with an HOLY & EASY MIND, will go as far, in Carrying us with *Undecay'd Garments* thro' the Wilderness, to the Promis'd & Pleasant Land, which we are Bound unto, as all the *Prescriptions* with which all the *Physicians* under Heaven, have ever yet Obliged us.[78]

{Your most hearty Friend

& humble Servant

C. Mather.}

[75] Job 23:14.

[76] Ps. 138:5, Rom. 5:2.

[77] 1 Cor. 12:31.

[78] In the published version of *AB* (1722) – on a separate page – is printed the Latin poem (19 lines) "De Flatibus Humanum Corpus Molestantibus." [On the vexations of breathing in the human body" (i.e., flatulence)]. Probably added by Winthrop as a humorous conclusion to Mather's weighty subject of Nishmath Chajim, the breath of life, the poem is taken from *De Flatibus Humanum Corpus Molestantibus, Commentarius Novus ac Singularis* (Sanctandreana, 1592), p. 196, by Joannis Fieni Andoverpiani, aka. Joannes Fienus, Jean Feyens (1537–85), a physician in Antwerp. Fienus attributes the poem to Johannes Postius, Aulæ Palat. El. Archiater, candido Lectori S. Johann Postius von Germersheim (1537–97), chief physician of Rhenish Palatine, mostly remembered for his richly illustrated *Tetrasticha in Ovidii Metam. lib. xv* (1563). Fienus' *De Flatibus* appeared in an English translation, titled *A New and Needful Treatise of Spirits and Wind Offending Mans Body* (London, 1668). We have elected to exclude the poem, because the subheading in black letter reads "By another hand."

6.6. THE SEVENTH SON EXAMINED; WITH A TOUCH UPON THE KINGS EVIL.[I]

{After December 1, 1721.}

{To John Woodward.}

Sir,[2]

We have a Fancy among our Common People, That a SEVENTH SON, among Brethren that have not had a Sister born between them, is endued with I know not what, *Power of Healing*, various Distempers, with a Touch of his Hand upon the Part affected.[3] But may we not be afraid lest some *Unlawful*

1 Mather knew the healing arts of his time well enough to dismiss the alleged curative powers of the king's touch for scrofula as pure superstition. However, he recommended a concoction of the Native American "*Boar-thistle*," "*Cancer-root*," and "*Devils-bitt*" as a likely remedy for the disease. See Mather's Letter to John Woodward (November 18, 1712), our CA 1.2 "On American Plants & Indian Cures"; and CA 3.7 "Strength of the Imagination." See also A. Finley-Crosswhite (2003) and Kittredge, "Lost Works," pp. 473–75.

2 It is likely that Mather's epistolary essay was triggered by John Williams' *Several Arguments, Proving, That Inoculating the Small Pox is not contained in the Law of Physick, either Natural or Divine, and therefore Unlawful* (1721). Williams (1664–1729), a Boston tobacconist, apothecary, and medical practitioner, had charged that inoculation was not only dangerous but criminal. To expose Mather's felonious ignorance, Williams insisted that he knew of infallible practices that Mather had condemned as superstitious nonsense. For instance, to cure agues, Williams held forth, the patient should "wear a Spell about the Neck," written on a piece of paper "by the hand of a Seventh Son" and sticking it in the bark of an "Elder-Tree" (like the one on which Judas had hanged himself). Likewise, he knew of "Children that have had the King's Evil," who were cured by the King's touch (lawful or no), and the likes of Cotton Mather "ought not to deter them from it, lest the Disease may not go off" (pp. 11, 16).

3 The Elizabethan writer Thomas Lupton (fl. 1572–84) reported in his ever-popular *A Thousand Notable Things, of Sundry Sortes* (1579) that "It is manifest by experiēce that the seventh Male Chyld by just order (never a Gyrle or Wench being borne betweene) doth heale onely with touching through a naturall gyft, the Kings

Fellowship with the Invisible World may ly at the Bottom of it? If we suppose the *Matter of Fact*, I would enquire;

First; can any Reasonable Man imagine, that a *Seventh Son* should be distinguished *Naturally*, with such a Sanative and Balsamic Vertue in his Constitution, that he should be like the *Miraculous Pool*, when an Angel made his Descent into it?[4] Why should the Intervening of a *Sister* deprive him of this *Distinguishing Prærogative*? Why should the *Seventh* have it any more than the *First-born*, to whom there belongs *the Excellency of Dignity & the Excellency of Power*?[5] Yea, Is not the *First-born* as really the *Seventh* as the *Last*? Or, Is it requisite, that all the *Seven* have the same Father and Mother? Or, may the Father have the *Seven Sons* by different Wives? Or, The Mother of these *Maccabees*,[6] by different Husbands? Or, *How*?[7]

Or, Secondly; Is it possible to resolve the Matter into any other than a *Sacramental Original*?

Either we must suppose, that the Glorious GOD, before whom we are *as the Clay before the Potter*,[8] has endued the *Seventh Son* with a *Gift of Healing*,

---

Evyll: which is a special gyft of God geven to Kings or Queenes. As daylye experience doth withnesse" (bk 2, p. 25, §2). See also S. Brogan, *The Royal Touch* (2015).

4 The Pool of Bethesda (John 5:1–4).

5 Gen. 49:3.

6 2 Macc. 7 tells the story of a mother who bore seven sons, all of whom were martyred.

7 This reductio-ad-absurdum approach throughout Mather's Letter to John Woodward (1721) seems refreshingly rational and modern; however, it must be seen in the context of Mather's persisting occult views of the invisible world. For in his medical handbook *ABJ* (1972), cap. LXII, "*Fuga Daemonum*, or, Cures by CHARMS" (pp. 293–97) he not only adds a long section on the demonological origin of charms (hence their danger and sinfulness), but also willingly embraces cures from the invisible world, revealed by angels and especially in dreams (Cap. VIII: "Raphael, Or, Notable Cures, from the INVISIBLE WORLD"). This seeming contradiction bespeaks less his much-touted "transitional" place in the history of medicine and its allied arts than his endeavor to reconcile the old with the new sciences of his day, as well as to keep an open mind about the harmony of all pansophic knowledge. In fact, Mather's writings on natural philosophy (*CP* [1720]), on medicine (*AB, ABJ* [1722, 1972]), and on science and its allied arts (CA [1712–24]) establish him as the first American to publish a unified theory of all the sciences and the best flowering of the venerable tradition of pastor-physicians this side of the Atlantic. See O.T. Beall, Jr. and R.H. Shryock (1954), and P.A. Watson (1991).

8 In Gen. 2:7, God is likened to a man who forms mankind from the dust of the ground. See also Job 10:8–9; Lam. 4:2; Jer. 29:16: 45:9.

like that which in the primitive times of Christianity,[9] recommended some *Favourites of Heaven* among the First Preachers of the Gospel unto the World; or at least, like that which the famous *Greatreaks*, the *Stroker*[10] in *Ireland*, was for a while, an unaccountable Instance of. But where have we been *Taught of GOD* any thing of this importance? Or, what if the *Seventh Son* should be a *Vile Person*; a *Child of Belial*;[11] a *Devil Incarnate*? and one belonging to the Tribes that are the *Abhorred of the Lord*?[12]

Or, we must suppose, (what is most likely) That in the *Kingdome of Darkness*,[13] there is a surprizing Imitation of many Things that occur in the *Kingdom of GOD*. And as the Number SEVEN, has been strangely considered in the affairs of the Heavenly World, (as well as the Number, *Twelve*,) so, in the *Kingdome of Darkness*, it is affected, for the Number *Seven* to be brought into a peculiar Consideration, (as well as the Number, *Twenty five*?)[14] And more particularly, an Application to a *Seventh Son* for a Cure, shall be a *Sacrament*, wherein the *Dæmons* that attend him, shall reckon themselves acknowledged, as the Dispensers of the *Cure*, that is here looked for?[15]

That the *Dæmons* are very officious to convey *Cures*, unto the Patients, that with *Cæremonies* approved by *Them*, do seek unto them, is notorious from the Experience of all Ages, & of all Places, as well as from the *Dreams* of old obtained in the Temple of *Esculapius*.[16]

9 1 Cor. 12:8–9.

10 Mather alludes to miracle healers like Valentine Greatrakes, aka. Greatraks, Gratrakes, Gratreaks (1629–82), an Irish faith-healer, aka. "the Stroker," who healed invalids through a kind of physical massage or strokes. See Greatrakes' *A Brief Account of Mr. Valentine Greatraks* (1666). Increase Mather's elder brother, the Rev. Samuel Mather (1626–71) of Dublin (Ireland), discussed Greatrakes' notoriety and dismissed his healing methods as superstition. Cotton Mather excerpted their discussion in his uncle's biography, "Fructuosus, or the Life of Mr. Samuel Mather," in *MCA* (4.2.13, pp. 151–52); and alludes to "the Stroker," in CA 3.7 "Strength of the Imagination." See also Ralph Thoresby's eyewitness account of Greatreaks' healing method, in "A Letter … concerning the Cures done by Mr. Greatrix the Stroker," *PT* 21.256 (September 30 1699): 332–34; B.B. Kaplan (1982); P. Elmer (2013).

11 1 Sam. 2:12–36; 10:27; 30:22; Prov. 6:12; 2 Cor. 6:15.

12 Prov. 22:14.

13 Col. 1:13; Eph. 6:12.

14 See J.J. Davis (1968).

15 The idea that the devil apes God.

16 Aka. Aesculapius (Asclepius), ancient Greek god of medicine and healing, worshiped in the Temple of Asclepius at Epidaurus, in Rome, and wherever his cult

If the *Healing Vertue* be not in the *Natural Constitution* of the *Seventh Son*, must it not needs be in the Operation of some assisting *Dæmons*?[17] And the Homage, or the Regard, that is paid unto the Number *Seven* in this Way, is the *Sacrament* that brings the Patient under the Efficacy of it.

But what an *Enchanted Field* are we now brought into? And what a Door is opened unto Ten Thousand *Sorceries*? If *Numbers* are to be so esteemed for the Healing of Diseases, why should not *Figures* be so too? And from *Figures* lett us pass to *Letters*; yea, to all the *Ephesian* Ones. And why should not the *Seventh Son* pronounce *Words*, and use his *Lip* as well as his *Hand*, for a Cure? And why not also manage the *Motions* of his *Hands*,[18] in the Way of *Magical Cæremonies*? And would not a Cure be wrought as effectually, if he should only pronounce *Words*, and make no Use of his *Hand* at all? Would not a *Spell* from him, do as well as a *Touch*? I doubt it not!

Indeed, I have read concerning the *Seven Sons* of one *Sceva* a Jew,[19] among certain Exorcists and Vagabonds, who did some *Notable Things*. But among these, *all the Seven*, for ought we see, were as potent Knaves as the *Seventh*. And I make no Doubt, That if People would once take up the Fashion of applying to a *Sixth Son*, they would find *him* do as *Notable Things* as now they think the *Seventh* can do for his Ill-advised Patients.

I pray, lett us not foolishly leave the *Ordinary Ways of Cure*, and run unto Impertinent, yea, very *Unwarrantable* Superstitions.

In my Opinion; One setting up for, *An Healer*, no otherwise Qualified and Empowered for it, than as being, *A Seventh Son*, deserves little better Usage than a *Quack* was want formerly at *Montpelier* to meet withal. *Is apprehensus*

---

flourished in the ancient world. Asclepius (like the Greek Hermes) was generally portrayed with his caduceus and writhing serpents twined around it (see Moses' copper serpent [Num. 21:4–9]). Like the God of the Hebrews, he revealed his messages through visions, voices, and in dreams. As Mather put it, "It is well-known, That it was the ancient Custome, for Patients under various Infirmities, upon the Observation of certain Cæremonies, to fall asleep in the Temple of *Æsculapius:* and in their Sleep, to enjoy *Dreams*, which gave them Directions how to obtain the Releef of their Infirmities" (*ABJ* [1972], cap. 8, p. 48). See also 1 Sam. 3:1–15.

17 In his letter to John Woodward (November 22, 1712), CA 1.6 "Remedies Revealed in Dreams," Mather grants that good angels may be employed for good purposes, including healing individuals. Yet he warns that even "*Evil Spirits* may *Do Good*" as well, but only "upon an *Evil Design*."

18 Prestidigitation.

19 Acts 19:11–20: While Paul cured the diseased with miracles given by God, the chief priest Sceva and his seven sons in Ephesus invoked an evil spirit for their magic.

Asino Imponitur, et urbe pellitur, ovis putridis et alijs Sordibus, sit venia dicto, quae in ipsum conjiciunt, donec per urbem magno dedecore deductus sit, alium locum petiturus.[20]

I will at this time trouble you with but one Story more.[21] We have a Sort of a *Thistle*, which we call, the *Boar-thistle*. It is the shortest, and the most prickly Sort, in all the Tribe of Thistles;[22] And it ha's a large & long *Root*, which is the thing that is to bear its Part, in the Medicine that is now to be compounded. With this, we join a Root which we call, the *Cancer-root*; and add unto it a little of what you call *Divels-bitt*.[23] Of these three Roots, we make a Decoction, or Infusion. Lett a Person languishing (yea, tho' very far gone) in the *Kings Evil*,[24] take every day a little Draught of this Liquor. At the

20 The poor quack of Montpellier (southern France) was probably one who ran afoul of his colleagues at the famous medical school (founded 12th century), the oldest in Europe: "He is caught and put on a donkey, and driven out of the city, with rotten eggs and other filth [cast upon him]; let it be said that they are forgiven, for what they cast upon him, until he passed through the city with great dishonor, that he might seek another place."

21 Mather's recommendation for treating King's Evil – missing from his Cap. LXII in his medical handbook *ABJ* (1972) – is here restored from his CA 1.2 "On American Plants & Indian Cures" (November 18, 1712). See also *D* 2:204.

22 Boar thistle, also known as purple-milk thistle, is a species of annual herb in the family *Asteraceae* found in the North Atlantic region.

23 American cancer-root is a parasitic plant with yellowish flowers that grows on the roots of some species of oak; Devil's-bit scabious is a perennial flowering plant in the honeysuckle family, with purple lobed flowers, which grows in damp ground.

24 Scrofula, a form of tuberculosis of the lymph nodes of the neck (lymphadenitis). According to an unpublished draught letter (AAS), Mather responded to "Mr. Jos. Web," who had sought advice on the treatment of scrofula for (probably) his own child. Mather's reply was straight forward: "ye Opinion, & Expectation of Curing ye King's Evil, by ye Touch of a Seventh Son, is not only a Vulgar Error, but also a very Foolish One." Mather dismisses it as "a very unjustifiable Superstition" but recommends as a treatment "a decoction of the 'Boar Thistle," – "ye Lowest of ye Tribe; very prickly; and of a Large Root" (quoted in Kittredge, "Scientific," p. 49). As a matter of fact, Mather was ahead of his English colleague William Beckett, F.R.S. (1684–1738), a London surgeon and antiquarian, who rejected the belief in the efficacy of the king's touch in curing scrofula, in *A Free and Impartial Enquiry into the Antiquity and Efficacy of Touching for the King's Evil* (1722). The venerable custom, apparently first introduced by the Anglo-Saxon King St. Edward the Confessor (c. 1003–66), fell out of favor during Queen Anne's reign (1702–1714).

same time, lett the Patient lay the bruised Roots, on the Scrophulous Ulcers or Tumours, and every day Renew the Cataplasm. Tis astonishing to see, how soon that *Reproach of the Physicians* will fly before this Remedy. It works a swift and a strange Cure; and I don't think, that the Gentleman who ha's lately published at *Frankfort*, what is by him called, *Praxis Medicinæ infallibilis*, ha's a more *Infallible*, among all his boasted Medicines, fetched, *è concilijs ac secretis Archiatrorum nostro tempore fœlicissimorum*.[25] It will do more good, than some Cæremonies, which from *Edward* the Confessors time to ours, have imposed on the Credulity of {the} World, for the Cure of this grievous Malady.

{Your true Friend,

& Servant.

C. Mather.}

[25] "From the councils and secretaries of the most successful archiators [chief physicians] of our time," part of the title of *Praxis Medicinae Infallibilis* (1711), and "Prologus" (sign 1^{V}), a voluminous collection of medical studies by different physicians on different diseases, edited by Michael Bernhard Valentini (1657–1729), professor of medicine and rector of the University of Giessen (Germany). The infallible Praxis Medicinae went through numerous reprints. See I. Veith (1978).

6.II. AMBERGRISE DISCOVERED AND DETERMINED.

{1721.}

Ambergris *found in* Whales. *Communicated by Dr.* Boylston *of* Boston *in* New England.[1]

The most learned Part of Mankind are still at a Loss about many Things, even in medical Use; and, particularly, were so in what is called *Ambergris*, until our *Whale* Fishermen of *Nantucket*, in *New-England*, some three of four Years past, made the Discovery. Their Account to me is this.

Cutting up a *Spermaceti* Bull *Whale*, they found accidentally in him, about twenty Pound Weight, more or less, of that Drug. After which, they, and other such Fishermen, became very curious in searching all such Whales they kill'd; and it has been since found in lesser Quantities, in several Male Whales of that Kind, and in no other, and that scarcely in one of an Hundred of them. They add further, that it is contained in a Cyst, or Bag, without any inlet or outlet to it, and that they have sometimes found the Bag empty, and yet entire.

[1] *PT* 33.385 (December 31, 1724): 193. Boylston is identified as the one who "communicated," *not* who has *written*, the piece. Boylston is *not* the author. We believe it was written by Cotton Mather – though based on first-hand accounts of whalers in Mather's congregation; its rhetorical style is within the scope of Mather's other pieces, and the circumstances of Boylston's close collaboration in the smallpox epidemic in Boston may well support our contention. G.L. Kittredge explains that the term "communicated" as used in the *Philosophical Transactions* is "a formula especially used when a letter or essay was handed in or read [to the assembly] by some person other than the author" ("Lost Works," p. 461). The same caveat applies to CA 6.3. Mather's old nemesis Paul Dudley, F.R.S. (1675–1761), chief justice of Massachusetts and son of Governor Joseph Dudley, published "An Essay upon the Natural History of Whales, with a particular Account of the Ambergris found in the Sperma Ceti Whale," *PT* 33.387 (April 30, 1725): 256–69. Also see our Introduction.

The Bag is no where to be found, but near the Genital Parts of the Fish. The *Ambergris* is, when first taken out, moist, and of an exceeding strong and offensive Smell.

Whether or not (from the Account above) the Ambergris *be naturally, or accidentally produced in that Fish, I leave to the Learned to determine.*

{Your most hearty Friend

& humble Servant

C. Mather.}

Series Seven

7.0. LETTER TO DR. JAMES JURIN.

May 21, 1723.

Syr,

You are better known to me, than I can be to you; And I congratulate unto myself as well as unto the World, the fælicity of your Succession in the Office of *Secretary*, to the *Royal Society*. But in order to a mutual Better Knowledge, I owe you, & must now give you, some very short Account of myself; more particularly, with relation to that Illustrious Body, whereof, I hoped, I had the Honour of being esteemed a Member.

Having the Pleasure of some Correspondence with your excellent Predecessor, Mr. *Waller*, I did communicate unto him (and unto my valuable Dr. *Woodward*) a great Number of American, and philosophical, Curiosities; with an Intimation of my Purpose to continue the Communications. These Gentlemen putt the, as unexpected as undeserved Respect upon me, of proposing me for a Member of the *Royal Society*; and they both wrote unto me, That I was chosen accordingly both by the Council and Body of the Society, on the Anniversary Day for such Elections, in the Year 1713.[1] – Adding, that the only Reason of my not having my Name in the *Printed List* of the *Society*, was, because of my being beyond-Sea, and yett a Natural Born Subject, & so not capable of being inserted among the Gentlemen of other Nations. Your Secretary also, Dr. *Halley*, in the *Philosophical Transactions* of 1714. printed my Name, with an F.R.S. annexed unto it.[2] Mr. *Petiver* did the Like, in his

[1] See G.L. Kittredge, "Election" (1912), pp. 90–91. *D* 2:245–46. R.P. Stearns has examined the issue further in his article, "Colonial Fellows" (see esp. pp. 184–85, 192, 199–203, 208), and in his *Science* (1970), pp. 403–26.

[2] Alas, the title of Mather's published Extract (or summary, rather) of the twelve "Curiosa Americana" (First Series) lists his name as "Cotton Mather, *D. D.*" – no F.R.S. there – in Appendix 6, "An Extract of several Letters from Cotton Mather, D.D. to John Woodward, M.D. and Richard Waller, Esq; S.R. Secr.," *PT* 29.339 (January 1, 1714): 62–71. However, Mather's distinguished F.R.S. does appear in the Table of Contents of *PT* 29.339 (item IV, p. 51).

Naturæ Collectanea;[3] And in his Letters to me, he had these Words, "Your Election succeeded without Opposition, and you was elected after the usual Method of Balloting. The Reason of your being out of the *Printed List*, is your not being personally here, to subscribe to the Orders that should be tendred you." To which he added Congratulations & Complements not proper to be by my Pen transcribed.

A distinguished, & a diminutive Crue of odd people here, when they could find no other Darts to throw at me, imagined their not finding my Name in the printed List of the *Royal Society*, would enable them to detect me of an Imposture, for affixing an F.R.S. unto my Name, on some Just Occasion for it. And an Infamous Fellow, whose Name is *John Checkley*, a sorry *Toyman*,[4] (that yett had the Impudence to write as a *Divine*) wrote a Letter full of scandalous Invectives against me, which was publickly read in the *Royal Society*. This wretched Man, ambitious to do the Part of a *Divine*, printed here some Rapsodies, to prove, *That the God whom* K. William, *and the Christians* of New England, *have worshipped, is the D—l—*. A Young and a Bright Kinsman of mine,[5]

3 James Petiver (c. 1665–1718), an English botanist, pharmacist, herbalist, member of the RSL, and voluminous author, illustrator, and collector of plants, herbs, fossils, and minerals. Mather's list includes 7 plant specimens (items 288–94) he sent to Petiver, who listed his contributions in Petiver's catalog, *Petiveriana III, seu Naturae Collectanea* (1717), p. 12 (sep. pag.), in *Jacobi Petiveri Opera* (1764), vol. 2, *Petiveriana III* (1717), p. 12 (sep. pag.), Com. V, "Some *American Plants*, with their *Specifick Vertues* and Wonderful *Effects*, lately sent me by the *Reverend* and Learned Dr. *Cotton Mather*, at *Boston* in *New England*, and Fellow of the *Royal Society*, London."

4 Or "toy man," or "one who sells trinkets, knick knacks, or fashion accessories" (*OED*); unless a deliberate misspelling of "Toryman," a supporter of Toryism, a politically conservative, and member of the Church of England (see Kittredge, "Election," pp. 96–102); John Checkley (1680–1754), sometime apothecary and bookseller in Boston, inveighed against Mather's dishonesty in styling himself F.R.S., for his name could not be found on the Society's roster of elected members. In his letters (1720, 1721) to Lt. General Francis Nicholson (1655–1728) and to the famous Edmond Halley (1656–1742), secretary of the Royal Society, Checkley tried to expose Mather's imposture. A thorn in Mather's flesh, Checkley authored several controversial works, including his anonymously published *Choice Dialogues Between a Godly Minister And an Honest Country-Man, Concerning Election & Predestination* (1719) and *A Discourse concerning Episcopacy* (1723). On Mather's views on Episcopacy, see his election-day address of the General Court, *The Serviceable Man* (Boston, 1690), esp. 31–52; see also Mather's *D* (2:605, 703 *n.*, 716 *n.*). On Checkley, see E.F. Slafter (1897), 1:1–139; T.C. Reeves (1965); and more recently, J.B. Bell (2008), pp. 33–41.

5 Mather's nephew and protégé Thomas Walter (1696–1725), who responded to Checkley's *Choice Dialogues* (1719), pp. 2, 21, 31, with his own (most likely coached by Mather) *Choice Dialogue Between John Faustus A Conjurer, and*

bestowed such Castigations on the Blasphemer, that I became thereupon the Object of his Implacable Revenges. But of this Matter, I gave Dr. *Woodward* a more full Account, a Year and half ago: Which, because I know not whether he received it, I now repeat a Little of; & Fly to your Justice, if any further Indignity of this Nature should be offered me.

But, that I may not lay aside any of the meek Patient, Humble Disposition, with which I should address you on this Occasion, I shall keep such Terms, as I used unto my Doctor, when he had what he required of me upon it.

I should never presumed upon affixing an F.R.S. unto my poor unworthy Name, had I not thought, that my Honourable Masters, would have taxed me as guilty of some Ingratitude unto them for their unmerited Favours, if I had always declined it.

The many Treatises (many more than Three Hundred) which Heaven has allowed & assisted me to publish, (in the midst of many other constant & heavy Labours) on various Arguments, and in various Dead as well as Living Languages, added unto some other Circumstances needless to be mentioned, had procured me from some *Europæan* Universities, without my Seeking in the least Measure for them; the Testimonies of the highest Respect, that they could shew to the meannest of Men; and among the rest, a *Diploma*, for the *Doctorate in Theology*.[6] Upon this I was under some Temptation unto the Vanity of Thinking, that it was possible the *Royal Society* also might esteem it no Disgrace, to permitt my Name among their Members. Especially, when my Remittances to their Treasury, might for Number (however not for Value) be æqual to what they might receive of some other Members, whose Correspondence they cast a kind Aspect upon.[7] For the Embellishments, wherewith I studied usually, (after the Manner of the *German Ephemerides*)[8] to make my dry & dull Stories a little

Jack Tory His Friend (1720), pp. viii, 2, 6, 8, under the pseudonym "By a *Young Strippling*."

6 The University of Glasgow bestowed on Mather the honor of a D.D. degree in 1713. See Mather's *D* (2: 63–64, 77); his open letter (25. d. 10. m. 1710), *To the truly Honourable & Venerable ... Professors, of the Renowned University of Glasgow* (1710); and K. Silverman, *LT* (1984), pp. 222–25.

7 In the left margin, Mather wrote, "[The Enclosed Catalogue, —]," apparently referring to the attached list of letters he had previously sent to the Royal Society.

8 *Miscellanea Curiosa, sive Ephemeridium Medico-Physicarum Germanicarum Academiae Caesareo-Leopoldinae Naturae Curiosorum* (Lipsiae, Norimbergae, etc.), published under the aegis of the Academia Leopoldina Naturae Curiosorum (founded in 1652), published numerous volumes (also under different auspices and titles) similar to those in the British *Philosophical Transactions*. See U. Müller (2008); see also *BA* (1:1003, *n.* 30). Mather had access to the German *Ephemerides* at Harvard.

more palatable to Men of Erudition, some of your own Members, as well as Monsieur *Tournefort*,[9] help'd me to some Apology.

But if after all, it be the Pleasure of those Honourable Persons, who compose or govern the *Royal Society*, that I should lay aside my Pretensions to be at all Related unto that Illustrious Body, upon the least Signification of it by your Pen, it shall be dutifully complied withal. I will only continue to take the Leave of still communicating Annually to you (as long as I live) what *Curiosa Americana* I can become the Possessor of. For (my Jewish Rabbi's having taught me, to *love the Work*, and have *little Regard unto the Rabbinate*,) it is not the *Title*, but the *Service*, that is the Heighth, & indeed the Whole, of my Ambition.

As a Token of my Purposes this Way, and as an Earnest, of a much greater Variety, which I propose to send you by another Hand, about a Fortnight hence, I now present you with a tedious Account of Sentiments & Occurrences, relating to a Subject, about which I perceive you are sollicitous to have the exactest Informations.

At this time, I add no more, but, with hearty Prayers, to Heaven, that you may be continued long as a great Blessing to the World, I subscribe,

Boston, N. England. Syr,

May. 21. 1723. Your most affectionate

Friend & Servant,

Cotton Mather.

Dr. Jurin.

Besides the Scores of *Curiosa Americana*, communicated in the Letters, to Mr. *Waller* and Dr. *Woodward*, the Extracts whereof were published by Dr. *Halley*, in the *Philosophical Transactions* for the Year 1714.[10] I have since, at several times transmitted these Communications.

9 Joseph Pitton de Tournefort (1656–1708), French botanist and traveler, whose *Relation D'Un Voyage du Levant* (1717) was translated into English and published as *A Voyage into the Levant* (1718). Mather puts this work to good use in the volumes of his commentary on the Bible *BA* (2010–) and in his *TP* (1995).

10 Appendix 6: "An Extract of several Letters from Cotton Mather, D.D. to John Woodward, M.D. and Richard Waller, Esq; S.R. Secr.," *PT* 29.339 (January 1, 1714): 62–71.

A Woollen Snow.
Surprising Influences of the Moon.
A Monster (of Two Children united.)
The Moose of *America.*
The New English Pigeons.
A strange Instance of long Fasting.
An uncommon Sisterhood.
A Rare Exemple of some Idiots.
A surprising discharge of an Ear of Rye at the Shoulder-blade.
A surprising Molestation of the Ear, with Fly-blows.
An American Plant, called, *Fagiana.*
A strange Mischief to the Eyes.
Curiosities of the Small Pox.
A monstrous Calf.
Monstrous Impraegnations, of the *Musquash* on the Catt.
The Strength of Imagination.
The Stone mistaken.
The Nidification of our Pigeons.
A Serpent securely handled.
The Limestone produced in Sir. William *Phips* Wreck.[11]
A Collection of Fossils.
A Triton.
A prodigious Worm.
An horrid Snow.
All these, to Dr. *Woodward.*

An *Hortus Siccus*, of American Plants.
To Mr. *Petiver*.

An Eclipse observed.
To Dr. *Halley*.

A strong Digester.
The Quadrature of the Circle.
The Whale.
Prodigies.
Odd Effects of the Thunder.

[11] Sir William Phips (1651–95), governor of Massachusetts, and friend of the Mathers. See Cotton Mather's biography, *Pietas in Patriam* (1697).

A large Description of *Mississipi*; by one who sojourned several Years in that Countrey.
A Rocquing Stone.
A Sea-Lion.
A Water-Dove.
The Locusts.
The *Aurora Borealis*.
A Journal of a Voyager round the World.
A Relation of a New Burning Island.
All these to Mr. *Chamberlayne.*

An Account of the Method & Success of Inoculating the Small-Pox.
To Mr. *Dummer*.

A Further Account, of the Small-Pox Inoculated.
Curiosa Variolarum.
Nishmath-Chajim; The probable Seat & general Cure of all Diseases.
The Seventh Son examined, with a Touch upon the Kings Evil.
Lacus Mirabilis.
An Horrible Tempest.
Ostreophagi; or, a Matchless Oister-hill.
Absinthium Sempervivum.
Ambergrise discovered & determined.
Nœvi Materni.
Melissologia; with a New Method of Bee-hunting.
All these to Dr. *Woodward*.

Unto these may be added Several Communications in the Book entitutled, *The Christian Philosopher*.

7.1. THE CASE OF THE SMALL-POX INOCULATED, FURTHER CLEARED.

May 21, 1723.

To Dr. James Jurin.

Sir,[1]

It is a Thing well known, to all who know any thing, That the *Small pox* has from the Days that the *Saracens* brought it into *Europe* with them, still proved a *Great Plague* unto the Inhabitants of the Earth, and bin enough, if there were nothing else, to procure the Denomination of a *Wo*, for those woful *Harpyes*.[2] The Numbers of the Slain by this tremendous Malady, have been far more than of those, who have perished by the *Pestilence*; and the Distemper which has been, by way of Eminency called, *The Destroyer*, has not been Such a *Besome of Destruction*,[3] as this Competitor to it, among them, who have had the *Graves waiting for them*. The *perpetual* (and sometimes very strangely *periodical*) Visits,[4] which this Destructive Malady is ever making to all the *Commercial* Parts of the Earth, & even unto them, that are *Afar off*

1 See James Jurin's response to this letter, Appendix 9.

2 For Mather's eschatological Saracen and Turkish "Wo" (Rev. 8:6–13; 11:15–18) as apocalyptic opponents of God's Saved Nations, see *TP* (1995), 241, 394; and *BA* 1:405, 450, 745, 795, 773–74; *BA* 10:537–30, 533, 540, 779–81). See also G.L. Kittredge, "Lost Works" (478–79). For useful background on the history of smallpox in colonial America, see S.B. Woodward (1932), A Boylston (2012), and A.M. Kass (2012). For the history of medical training in England in the seventeenth and early eighteenth centuries, see P. Allen (1946) and J. Barry (2019).

3 Isa. 14:23.

4 See Mather's letter to Dr. John Woodward (July 12, 1716), CA 3.10 "Curiosities of the Small Pox," in which Mather calculated that an outbreak of smallpox occurs in Boston every twelve years. He speculated that lacking protection, a new generation born after the last outbreak had not attained immunity as those who had been previously exposed to the variola virus.

upon the Sea,[5] do hold Mankind in a continual *Bondage*, thro' the *Fear* of being once in their *Life* siezed with it, yea, of having their *Life* extinguished by it. And as the *Leviathan* approaches, from whose *mouth Sparks of Fire leap out*, and who *makes* our Blood *boil as a pot*; how are even the *Mighty*, as well as the Many, *affraid*![6] The apprehensions of Dying a very *terrible Death*, after a Burning for many Daies, in a *painful* as *loathsome* a Malady, or at Best, of having many *weary Nights* roll away under the uneasy circumstances of *Loins filled with a Loathsome Disease*,[7] & recovering with *Boils*, & *Scars* and *Wounds*, not quickly to be forgotten, hold the Children of Men in the *Terrors of Death*,[8] until the *Fiery Trial* be over with them.

The Inventions of the most Skilful Pretenders to the *Art of Healing*, have hitherto signifyed very little to abate the Slaughter. Yea, tho' our *Sydenham* hath been complemented for *Saving more Lives, than some Wars have carried off*, by teaching us the true way of managing the *Small-pox*;[9] yet *his* way too has not given Universal Satisfaction: And, *hâc non Successit, aliâ aggrediamur viâ*,[10] is the Note, which our Physicians are still Harping on.[11]

After all, was the Dire Malady even more mortal, than it is, *at this day*! In how many of your Towns, have One in Three or Four died, that have been taken with it? And in some ultramarine places, I am told, that scarce One in Three of Four have lived. In your *Capital City*, how has it grown upon you, till many more than an Hundred a week dy of it, and the Dead by it make a *Seventh part* of those that fill your *Bills of Mortality*?[12]

5 Ps. 65:5

6 Job 41:19, 25, 31.

7 Ps. 38:7.

8 Ps. 55:4.

9 The founder of clinical medicine and epidemiology, Thomas Sydenham (1624–89), a renowned English physician, started his medical career by devoting himself to the cure of fevers, especially those closely linked to smallpox, in "An Epistolary Discourse to the Learned Doctor William Cole, concerning some Observations of the confluent Small-pox, and of Hysterick Diseases," in *The Whole Works of ... Dr. Thomas Sydenham* (1696), pp. 404–78; and in his *Observationes Medicæ* (1676). See K. Dewhurst (1959).

10 Since "this did not succeed, let us proceed in another way."

11 Nonetheless, palliative practices such as bloodletting and diet – among the standard treatments of the time even in Sydenham's medicine chest – were at best supportive of patients suffering from the smallpox, but never an antidote for the disease, let alone a panacea.

12 See E. Huth (2006). Daniel Defoe (1660–1731), English pamphleteer, novelist, and journalist, provides a telling account of the devastation in 1665 of the bubonic

One would have thought, when the Compassion of Heaven had made unto a miserable World a Discovery of an *Unfailing Method*, not only to *Redeem our lives from the Destruction*[13] threatned by this *Common Enemy of Mankind*, but also to prevent the *Uneasy circumstances*, which make the most, that Out-live it, profess, that Thousands of pounds would not hire them again to undergo them; I say, One would not but think, That the Children of men should have agreed in the most *Solemn Thanksgivings* to a gracious God, and most thankfully have accepted the offered Favour. But now to find, that people should be generally under the *Energy* of an unaccountable Aversion from coming into this *Method of Safety*; yea, that they Should with a Rage, that reache's up to Heaven, malign and revile those that only propose it unto them, and not only *Wish* the *Death* of those that come into it, but also actually *seek* the *Death* of the Friends, that only show them how to save their Lives from a formidable Adversary![14] — What should one make of this? or, can any Account be given, but that, which good old *Baxter*,[15] gives of the constant Opposition, that all *Good* meet's withal, [which, *he sais*, convinced him] *That most certainly there is a Devil*; and there are *Powers of Darkness* alwaies militating against the Welfare of Mankind.[16] If the *Angel of Death* had not a very large Commission, and if *Evil Spirits* could not when they have a permission from Heaven for it, exert amazing Influences, and enchant those

plague decimating the population of London, in his *Journal of the Plague Year* (1722). John Graunt (1620–1674), an English haberdasher turned demographer, preceded Defoe in his *Natural and Political Observations ... made upon the Bills of Mortality* (1662) by providing some of the first statistics to track the effects of the contagion. See also James Postlethwayt et al, *A Collection of the Yearly Bills of Mortality, From 1657 to 1758 inclusive* (1759).

13 Ps. 103:4.

14 See Mather's bitter complaints (*D* 2:634, 639, 641, 657–58, 659).

15 Richard Baxter (1616–91), Scottish presbyterian divine, pastor, apologist, and friend of the Mathers. The Mather library contained several of Baxter's works, but none have a passage that matches the one Mather cites. In his sermon "What Light must shine in our Works," in *Practical Works* (1707), vol. 4, p. 830, Baxter writes, "And what Wonder if all the Powers of Darkness do bend their Endeavours to obscure this Sacred Light… . Christ is the Light of the World … Satan's Work is to stir up all that he can against them, High and Low, Learned and Unlearned, and to put Christ's Lights (both ministers and People) under a Bushel." Obviously, Baxter's last work, *The Certainty of the Worlds of Spirits* (1691) is as much as Mather's own devoted to dispelling the materialist argument against the existence of the soul and invisible world as it is to proving the existence of God and the devil.

16 See Baxter's *Certainty of the Worlds of Spirits* (1691), pp. 236, 237.

that are *fallen under their Dominion like the Beasts that perish,*[17] to become distracted *Suicides*, one could not imagine it possible for people to act so madly, as they have done, when the Method of *Receiving and Conquering the Small-pox in the way of Inoculation, has bin propounded unto them*![18] O *Foolish* Anti-Inoculators, *What has bewitched you*!

The manner, wherein the *Opposition* has been usually carried on, has been so *Satanic*, that he must be very blind who sees not the *Evident Original* of it. The *Railing*, the *Lying*, the *Fury*, the bloody *Malice*, & even the Subordination[19] of *Perjury*, with Attempts of *Assassination*, which has distinguished it, have been Such, as hardly have ever been æqualled on any Occasion;[20] and One must, *after the manner of men*, be engaged, as in a *Fight with Beasts at Ephesus*,[21] who will appear as an Intercessor with people, that they would be willing to have their Lives reprieved and sweetened.

But, if there be any thing of *Argument*, that has ever been produced against this *Method of Safety*, I pray, Let us be apprised of it.

In several parts of the *East-Indies*, the Sun alwaies rise's in *clouds*, which gather about the Horison; And the *Indians* have a Fancy, that these *Clouds* are *Giants*, which combine to stop the Sun from rising to the World: But the Sun every morning, for these thousands of years, it seems has forced his way thorough, and been too hard for them. All the *Objections* that have appear'd unto the Fancy of Weak people, as *Giants*, to obstruct the Passage of this *Healer* into the World, will be presently found but Silly *Clouds*, and I hope the *Light* will force its' way thorough them all.

The *Objection*, drawn from this *Party*, (political or Ecclesiastical,) approving the practice, and that *Party* not approving it, is too mean to have so much as one Thought bestowed upon it. It is with the utmost Indignation,

17 Adapted from Ps. 49:20.

18 Mather alludes to the circular letter "An Address to the Physicians" he sent to Boston's physicians (June 6 and June 23, 1721) to invite them to discuss small-pox inoculation (*D* 2:620–21, 624 *n.*, 628, 631–32, 641). No complete copy of Mather's invitation is extant. The substance of the circular appears in the "Appendix" of our CA 6.2 *Variolae triumphatae* (MS. 134–MS. 147). Another extract survives in the anonymous *A Vindication of the Ministers of Boston* (1722), pp. 7–8, and in Mather's letter of invitation (June 24, 1721) to Zabdiel Boylston, in *MM* 1 (1789): 778. G.L. Kittredge, "Lost Works" (pp. 428, 433–35, 444) provides invaluable insight into Mather's circular.

19 MHS holograph has "Subornation"; i.e., underhandedly.

20 See *D* (2:634, 636–37, 639, 641, 657–58). The attempted assassination occurred in the wee hours of November 14/15, 1721.

21 1 Cor. 15:32. Ephesus, a Greek metropolis in Asia Minor, was famous for bullfighting.

that some have sometimes beheld the Practice, made a meer *Party-business*; and a *Jacobite*, or *High-flying* Party,[22] counting themselves bound in duty to their *party*, to decry it; or perhaps, the *party* disaffected unto such and such persons of publick *Station* & *Merit*, under the Obligations of a *party* to decline it. And yet I foresee, that I shall quickly find it in my way, to touch upon *One party* (mentioned in the Third Chapter of *Genesis*) from whence One small Essay towards an *Argument* for the practice, has been some-times allowed of.[23]

One Sort of *Objections* against the Method of Managing and Governing the *Small-pox* in the way of *Inoculation*, has been fetched from Conscience; and it has been the *Cry of a Multitude*,[24] that they *cant see thorough it*, how One can with a good *Conscience*, bring a *Sickness* on himself, until it shall please the *God of our Life* to send it upon him.[25]

Now and then persons of Serious *Piety*, become the Opposers of it, from a Real *Scruple* of *Conscience*, concerning the Practice. And as these, for the most part, express their Dissent with Modesty, Sobriety, Humility, & without censuring of those that are *otherwise minded*, So, wee ought ever to treat them with the respects of *Brethren*. But it has been somewhat *remarkable* and equally *ridiculous*, that the most *fierce Opposers* of the practice have commonly been Such as have most cast off the Restraints of *Conscience*, in *other* matters, and in their way of talking upon *This* also, show, That *Conscience* has very little awe upon them; *With a mouth full of Cursing*,[26] and the Language of Fiends, at the moment, they vow, *They would not be guilty*

22 Mather associates Boston's opponents to smallpox inoculation (William Douglass and his ilk) with the supporters of the Church of England.

23 Perhaps a sly reference to Gen. 3:15 (*BA* 1:487–89), the seed of the woman (Christ and his church) crushing the head of the serpent. Similar symbolism governs the depiction of the caduceus of Asclepius and the "brazen serpent" of Moses in the wilderness (Numb. 21:4–9). See also Mather's commentary on Numb. 21:9 (*BA* 2:925–37).

24 Possibly an allusion to Act. 25:24, in which Roman governor Festus describes to King Agrippa how "the multitude of the Jews" were "crying that he [Paul] ought not to live any longer."

25 See the anonymous tract *A Letter From one in the Country, to his Friend in the City* (1721); Samuel Grainger, *The Imposition of Inoculation As a Duty Religiously Considered* (1721); John Williams, *Several Arguments, Proving, That Inoculating the Small Pox Is not contained in the Law of Physick, either Natural or Divine, and therefore Unlawful* (1721), and his *An Answer To a Late Pamphlet, A Letter to a Friend in the Country* (1722)–they all argue that employing inoculation as a prophylactic method of inoculation spreads the disease, worse, it stops God's hand.

26 Ps. 10:7; Rom. 3:14.

of Such a Wickedness, for all the World![27] I have known Some good people, that have opposed the Practice, brought over to it by *This*, as the *first* of their motives; They were *ashamed of their Company*. For, as an Eminent Person (One of Weight enough, to weigh against many *Masseys*)[28] writes upon it, "There are Some worthy persons that are not clear about it; nevertheless, it cannot be denyd, but that the known children of the wicked One are generally fierce Enemies to *Inoculation*, It is a grave Saying of Old *Seneca*; *Pessimi Argumentum Turba*.[29] For my part I should be ashamed to join with Such persons; *O my Soul, come not thou into their Secret, unto their Assembly be not thou united*."[30] However let us hear what may be said in Opposition: Verily, Nothing but what may be said against all the *preventing Physick* in the World. It is urged, That it is *unlawful* to prevent our falling into a *greater*, & even a *deadly* Sickness, by useing a Remedy, which will throw us in to a *Lesser*, and more *Gentle* turn of *Illness*, and such an One as will not ordinarily endanger the Life of him that uses it. This is a thing that must be affirmed by all that hold it *unlawful* to seek a Deliverance from the Danger of a Deadly *Small-pox*, in the Gentle way of *Inoculation*. But they who affirm this, *understand not what they say, nor whereof they affirm*.[31] With Such profound Casuites,[32] it is not lawful for a man to undergo so much *Danger* for the saving of his *Life*, as Every Sailor dos, (and often *Sickness* too) to gett a little *money*. Certainly, never till now was that Rule contested; *Of Two Evils, chuse the Least*.[33] How commonly do people in Health, use *Emeticks* and *Cathartticks*, to *prevent* a Sickness, which they are afraid of? And those *Emeticks* and *Catharticks* too, under the Operation whereof, ten thousand more have lost their Lives, than ever there have, under the *Small-pox inoculated*?

27 John Williams, *Several Arguments* (1721), pp. 3–4, 11–12; and Williams' *An Answer To a Late Pamphlet* (1722), pp. 1–2.

28 An allusion to *A Sermon against the Dangerous and Sinful Practice of Inoculation* (2nd ed. 1722), by Edmund Massey, M.A. (1690–1765), Lecturer of *St.* Alban's in Wood Street. This work was reprinted from the third edition at Boston, for Benjamin Indicott (1730).

29 Seneca, *De Vita Beata* (2.1): "The crowd is proof of the worst choice."

30 Gen. 49:6.

31 1 Tim. 1:7.

32 MHS holograph has "Casuists," that is, clever but unsound reasoners.

33 Aristotle, *Nicomachean Ethics* (5.3.1131*b ad finem*): "… the lesser evil is reckoned a good in comparison with the greater evil, since the lesser evil is rather to be chosen than the greater" (*Complete Works of Aristotle* 2:1786).

Yea, how many Millions have only in Learning to *smoke Tobacco*[34] (and this perhaps only to qualify themselves for the pleasure of an Exercise, which their *Health* is little enough consulted in) procured unto themselves much more *sickness* & of a more dangerous *Tendency* too, than is undergone by the most of them, that undergo the *Small-pox inoculated*?

It is an hundred to One but the *Objector* may have been chargeable at Some time or other with One or both of these things; But he would be angry, if you should charge them as *crimes* upon him.

Lett us yet more exactly enquire into the Conduct of the *Inoculated Patient*.[35]

He is in *Hazard* of being smitten with the *Small-pox*, & for ought he knows, it may be of the *Confluent sort*, which is that which people do mostly dy of; Of the *Distinct sort*, there are few {who} miscarry. He is in the *way* of it, *Reason* and therefore *Duty*, bid's him to look for it. If it be not so, The *Inoculation* is unseasonable & impertinent. If it be so, He is not One of *the whole that have no need of a Physician*.[36] And if it be so, why should he forget a very Ancient Oracle, *The Prudent forsees the Evil, and shelters himself; The Simple pass on and are punished*?[37]

Well; He is acquainted with a Medicine, that never was known to fail of preventing, the *Small-pox* of the *Confluent sort*,[38] which he[39] is in *hazard* of. I beseech you, What is there in the Word of the Blessed God (which præscribes and limits the *Whole Duty of Man*)[40] that forbids the Use of this *Medicine*, any more than an *Antidote against the Plague*? It is rather plain,

34 In his medical handbook *ABJ* (1972), pp. 301–11, Mather explicitly warns against the dangers of smoking tobacco and using it as snuff. He repeats his warnings in his handbook for young ministerial candidates, *MAM* (1726), pp. 133–35. No doubt, Mather's nemesis John Williams, a Boston tobacconist, was not happy with Mather's proscription (see below).

35 Mather may have Sammy, his youngest child, in mind, who begged his father (August 1, 1721) to be allowed to undergo the procedure since the smallpox had begun to spread at Harvard (*D* 2:635, 637–40). Not a moment too soon, so it seems, for Samuel's Harvard roommate died of the virus within a fortnight, August 15, 1721 (*D* 2:637–38).

36 Luk. 5:31.

37 Prov. 22:3.

38 Mather distinguished between two types of smallpox: the distinct and the confluent. The latter was the most deadly of the variola virus.

39 Mather added this word in proofreading the copy.

40 Eccl. 12:13. The anonymously published *The Whole Duty of Man* (London, 1658) was an extremely popular sermon in its time; it went through many editions.

that the *Sixth Commandment* require's him to use it;[41] And I alwaies thought, the *Word* of the Blessed God had instructed us; That for our *physick* as well as for our *food, Every Creature of God is good, and nothing to be refused, if it be received with Thanksgiving.*[42]

Certainly our Sacred Bible will not be a *Dispensatory* that shall determine any thing in the *matter* of the Medicine to be *unlawful*. What if it were the powder of *Toads*,[43] or the powder that *Johannes Anglicus*[44] cured agues withal? What if it be a *Variolous Quitter*,[45] or the Galls of *Rattlesnakes*?[46] The Sacred Scripture, no where sais, *Touch it not, Tast it not, Handle it not.*[47]

If there be any thing *unlawful* in the Medicine, it must arise from the *way of it's Operation*. But lett us examine *This*.

It has been found, That a *Vomit* just before the *Invasion* of the *Small-pox* has marvellously abated the Symptoms of it. But when shall I know the precise time when the *Invasion*, is to be looked for? Yea, our Dr. *Woodward* has observed, that a *Vomit* wisely order'd, as the *Decumbiture* has been coming on, has made the Symptoms not only of the *Stomach* (where the Fire first find's its' fuel) but also of the whole *Body*, remit unto Admiration.[48] The *Pustules* have not only proved very few, but also turned a Day or Two sooner, than the most favourable use to do. Now I ask, were such a *Vomit* as this *unlawful*?

41 I.e., "Thou shalt not kill" (Exod. 30:13).

42 1 Tim. 4:4.

43 Used for centuries, "powder of *Toads*," derived from the toxins and skin of poisonous frogs, has been used for a variety of medical conditions – including cancer, tumors, and melancholia (depression).

44 Johannes Anglicus, aka. Joannes Pyxcomiston (1502–77), a German Roman Catholic priest of Strasburg, who turned Protestant, and held the first Protestant church service in Strasburg's Cathedral in 1561.

45 MHS has "Variolus Quittor," an infection of a horse's hoof.

46 Mather details the salvific potency of the gallbladder of snakes in his letter to Richard Waller (November 27, 1712), CA 1.11 "Rattlesnake." Likewise, he describes its metaphoric healing power in his essay *RWSV* (1711, 1720), where he confides that the best way to cure the bite of calumniators is found in his own medicine cabinet: "I have my self made a fine Anodyne and Febrifuge, with the *Gall of Rattle-Snakes*," Mather delights to tell, yet his literal serum here is metaphorically to "get his own Heart & Life healed of all that he can find amiss in it; by the Exercise of PIETY, which the Sibilations of Slanderers, may excite him to" (*RWSV* [1720], Appendix, p. xiii).

47 Col. 2:21.

48 John Woodward's *The State of Physick: and of Diseases; with an Inquiry into the Causes of the late Increase of them: But more particularly of the Small-Pox* (1718), pp. 3, 87–89, 161, 211–38.

Or would our *Medicine* be *unlawful*, if it were only[49] a *Vomit*? Would the Use of it be, To take the *Work of God out of His Hands*, as our Ideots talk? No, surely; and yet how many Thousands have been known to *dy* in & from the Operation of a *Vomit*!

Suppose the *Medicine* were a Strong *Diaphoretic*,[50] &, for the time raised a mighty *Heat* and *Sweat*, upon the patient; would *that* be *unlawful*? None can[51] imagine it![52]

Nay, suppose the *Medicine* would throw the Patient into a *Salivation*:[53] would this be *unlawful*? How many Thousands have brought such a thing upon themselves, and none blame's them for it? And yet how many Thousands have been kill'd in & from the Operation!

But then Suppose the *Medicine* work in the way of what wee may[54] call a *Despumation*,[55] and raise a *Ferment* with *Pustules*, that wee may call a sort of *Small-pox*; and is indeed so much of it, as for ever to secure from any further Arrest from that grievous Distemper? Where is the *Word* of God that forbid's this *Despumation*, when it may be made with Expectation of *Safety* and *Success*? Though the *Medicine* be a *Variolous Quittor*, applied unto an *Incision*, I[56] say again; Where is the Use of it forbidden unto us, any more than the rest, that the *Microcosmus Medicus* advises of?[57] This, at last, is the very case! *Here* is the *Medicine*! He takes a *Medicine*, that operates with the Symptoms of that *Small-pox*, that ordinarily kills no body; and prevents his ever falling into that *Small-pox*, that kills us by Millions; and prevents also the *Distresses*, which many, that are not kill'd, yet commonly meet withal. What shall wee say? The *Word* of God has allowed *Medicines*, that operate in all

49 Mather added this word in proofreading the scribal copy.

50 A drug that induces sweating.

51 In reviewing the scribal copy, Mather changed "Nor then" to "None can."

52 Boylston, *Some Account of what is said of Inoculating or Transplanting the Small Pox. By the Learned Dr. Emanuel Timonius and Jacobus Pylarinus*. Published By Dr. Zabdiel Boylston (1721), p. 18.

53 "A secretion or discharge of saliva: esp. the production of an excessive flow by administering mercury" (*OED*).

54 Mather added this word in proofreading the copy.

55 The act of discharging or removing impurities from the body, by lancing a boil or by bloodletting.

56 In proofreading the copy, Mather deleted "and" and replaced it with "I."

57 Daniel Beckher (1594–1655), German physician of Danzig and Königsberg, authored *Medicus Microcosmus seu Spagyria Microcosmi triplo auctior & correctior* (1622; 1633).

other waies except only that of *Despumation*? Profane Caviller! *Add not unto His Word, lest he Reprove thee, and thou be found a Lyar.*[58]

All that can be pretended is, That the Patient run's an *Evident Hazard of his Life*, in this Operation: And in what peace can any one dy, who shall by this *Medicine* have his *Death* hastned upon him? Now, This is a most *Evident Falshood.*

Suppose it were true, That One or Two in an Age did miscarry under the *Small-pox Inoculated*; Have not some dyed of drawing a *Tooth*? and yet the Operation is practised every day. There is a Blundering sort of a Medicine, which is more used, and has done more good, than any one that ever was in the World; I mean *Venice-Treacle.*[59] A Thousand times more have died in & from the Use of that Medicine, than ever have died from the *Small pox Inoculated.* And yet the use of it, Who question's it?

On the mention of This, If those conceited Christians, who have roared against this Practice; *Wee Learn't it from the Heathen* (which by the way is not true) *and wee must not learn the way of the Heathen*; had been worthy to have been argued with, I would have asked them, whom wee had our *Venice-Treacle* from, and *whose Physician* he was that first invented it?[60]

Briefly, The Thing may be so illustrated, with a very familiar case, That One may be astonished, at the Bias of Intellectuals, that can see any room for Disputation after it.

Suppose, There is a *Bloody Flux* prevailing in the Town,[61] and it proves deadly to a great part of them that have it; they *dy by hundreds in a month*, A Physician is a Master of a *Purge*, which, whosoever takes it, is, in an ordinary way sure of being delivered from the Danger of the Fatal Distemper. But One Ingredient of the *Purge* is a *Drop* taken from the Stool of One that lay Sick of

58 Deut. 4:2; Prov. 30:6; Rev. 22:18–19.

59 Sydenham's *Entire Works* (1749), p. 398, sec. 112; and pp. 392–93, sec. 102, praises the efficacy of "Venice Treacle" (i.e. Venice Theriaca, aka. Mithridatium), a concoction of 64 herbal and other remedies known since ancient times. It was used as an antidote for snake poison and other conditions. *See* J.P. Griffin (2004).

60 Mather alludes to Andromachus the Elder, of Crete (1st century CE), physician to Roman Emperor Nero, who devised his potion, known as *Theriacum Andromachi*. Among those who objected to medicines devised by "pagans" are Grainger, *Imposition of Inoculation* (1721), p. 24; Williams, *Several Arguments* (1721), p. 14; and his *An Answer To a Late Pamphlet Intitled* (1722), p. 6. See Boylston's response, in *Some Account* (1721), pp. 9, 21.

61 I.e., dysentery. Mather appears to allude to the outbreak in September 1711 (*D* 2:105–06). See his handbill "Sentiments on the Small Pox Inoculated" (1721), p. 2, col. 1.

this *Dysentery*. An *Artificial Purge* seasonably taken (which also, for the time, considerably indisposes him) saves him from Death by the *Natural Purge*, which he is exposed unto. Will any scruple the taking of this *Artificial Purge*? Surely, none but such as want a purge of *Hellebore*.[62] The Man most properly *takes Gods Time to fall sick*; He does it seasonably, and in the *time*, when God has commanded him to do it.

In short, I have never seen any *Scripture* brought against this Practice, but what has been brought with as foolish and faulty a Violation of the *Third Commandment*, as it would be to bring that *Scripture* against *Blood-letting*; *He that sheds man's Blood, by Man shall his Blood be shed*:[63]

But I am now drawn to make some Remark upon another Sort of *Objections*, against the method of *Saving our Lives from the Horrible Pitt*;[64] and this they pretend is fetched from NATURE; They alledge, Tis a *dangerous Practice*.

Now to confute this Allegation, there can be no Answer comparable to that of CONSTANT EXPERIENCE; And by this it is abundantly answered, and so victoriously, that it is amazing, there should be heard the least *Sibilations* of it any more.

Some[65] Physicians may talk like *Pothecaries* (and prove themselves true φαρμακεις)[66] as long as they please, and amuse people with their Learned and Senseless *Jargon* at their pleasure, and abound in their *Hypothesimanies*,[67] till their poor Patients are sacrificed by Thousands to their Obstinacys. If CONSTANT EXPERIENCE do shew that *Lives are generally saved* by a Method, which the gracious and wondrous Providence of our Good God has leadd us to, people of sense will, by degrees, more and more come into it, Let their *Doctors* never so much assume the Character of *Butchers* upon them.

62 Helleborus plants are toxic. Woodward, *State of Physick*, pp. 177–78

63 The Third Commandment, Exod. 20:7, forbids taking the name of the Lord in vain. The quote beginning, "He that sheds man's blood," is from Gen. 9:6. Here, Mather singles out John Williams' contention in *Several Arguments* (1721). Rev. William Cooper (1694–1743), minister of Boston's Brattle Street Church, addressed the common scruples against inoculation in his *Letter to a Friend in the Country* (1721), which John Williams tried to refute in *An Answer to a Late Pamphlet* (1722).

64 Adapted from Job 33:24, 28.

65 Mather added this word in proofreading the copy.

66 A sly dig at physicians and apothecaries whom Mather dubs Φαρμακεις (fem., pl. of φαρμακίς, sorceress, witch). Thus, the "Learned and Senseless *Jargon*" of these drug-making φαρμακεις (pharmakeis) sounds like the mumbo-jumbo of sorceresses.

67 Mather's preponderance for neologisms, here suggesting "mania of hypothesizing" has not yet been discovered by the editors of the venerable *OED*.

Now This is what I maintain; That CONSTANT EXPERIENCE has declard for this practice, That it is *a safe and sure Method* of delivering ourselves and our children from *going down into the Pit*, by the stroke of a Distemper, which dispatches Mankind thither by Millions. Yea, *there never* did *arrive to us* (no, not in the boasted *Cortex* itself,[68] or any of it's Rivals) any Medicine, recommended with such a CONSTANT EXPERIENCE, of a Divine Blessing upon it.

The CONSTANT EXPERIENCE, to which I make my *first Appeal*, shall be fetcht from *forreign Countreys*. And there wee find men of Honour and Learning & incontestable Veracity, not One or Two, but a Considerable *Constellation* of them, all agree, in the Relation they give us, of the practice being used with *perpetual Success* in the *Levant*. It has been used upon vast Multitudes, even upon many Thousands, and for some scores of years; and when regularly used, it appears not, that ever *One person* miscarried of it, or had the *Small-pox* after it. Indeed a Scotch Practitioner asserted, in our Paltry *Boston-News-Letter*,[69] That in the Relation given by Dr. *Timonius*,[70] there were Two Children {who} died *from Inoculation*. But the Man of *Brass* knew his Assertion to have no Truth in it; He knew that the Doctor sais; There had been such a Thing reported, but he took the pains to go to the Houses, and inform himself, and found the *Report absolutely false*. They also prevailed with a *French* Fellow of a very vicious Character,[71] to swear, that a Trial was made of the *Inoculation* upon many in the *French* Army, six or seven and twenty years ago, with tragical Effects upon such as it was tried upon; but who in his Wits can believe any other, but that the *wretch forswore himself*?

68 Referring to the famous "Jesuit powder" made from Peruvian bark, as described in Francis Metcalfe, *Cortex Peruvianus* (1663). See Antoni van Leeuwenhoek, "Microscopical observations on the Cortex Peruvianus," *PT* 25.312 (December 1, 1708): 2446–55.

69 Colonial America's first newspaper published continuously (1704–26). Douglass's article, over the pseudonym "W. Philanthropos," is "To the Author of the Boston News-Letter," *Boston News-Letter* (July 24, 1721), p. 2, in which he states that the Greek physician Emanuel Timonius (1669–1718), in his account of the smallpox in Constantinople, recorded that "*Two Children* three Years old *Dyed* some short time after Eruption *from Inoculation*." See also *Boston News-Letter* (May 22–29 and June 9–12, 1721).

70 Timonius (1669–1718) published his "An Account, or History, of the Procuring the Small Pox by Incision," in *PT* 33.339 (June 30, 1714): 72–82. On the historical background, see T. Kyrkoudis (2021).

71 Lawrence Dalhonde (or Jonathan Delhond), died in Boston (1746), a Huguenot refugee who served in the French army in Italy and Flanders. https://primaryresearch.org/medical-training/

But wee find likewise an Army of *Africans* among our Domestick Slaves to be our[72] *Auxiliaries* in this cause, who had the Operation performed upon them, while they were yet in *Barbary*; and as they shew the *Marks* of it, so wee as well as they reap the *fruits* of it, in their secure Attendance upon our sick. And wee understand, That in *Barbary*, the common Usage is this; when about half a Dozen in a Village fall sick of the *Small pox*, presently all the *sound* at once repair unto them to be furnished with the *pus*, for the *Inoculation* of the *Small-pox* upon them; All of whom, after a small Indisposition, have a few *pustules*, which for ever secure them from the Distemper; whereas, the first half a Dozen generally dy; yea, the poor *Negroes* generally dy like rotten Sheep, when the *Small pox* getts among them and this *Method of Safety* is not practised.[73]

But then, *My own Countrey*, if I may call it so; or that where I was born and bred, & in the *Tents* whereof I am a *Sojourner*, shall be what I will with yet more Assurance *appeal* unto.

When I first addressed the Physicians of *Boston* (as the *Small pox* was entring the city, two years ago) with the Account which *Timonius* and *Pylarinus*[74] gave of this practice, One Word (of the *fœminine Gender*.)[75] lett fall in that account, cause some to foretel, That the *Physicians* would never come into it, but would sett themselves to decry it with all their might. *I will not mention the Reason* given for this *prædiction*, because I am far from casting a

72 In reviewing the copy, Mather deleted "the" and inserted "our."

73 See Mather's *Diary* (2: 620–21, 624, *n.*1) and his anonymous *An Account of the Method and Success of Inoculating the Small-Pox* (1722), pp. 1–3; and his medical handbook, *ABJ* (1972), p. 107. See also M. Minardi (2004) and K. Wisecup (2011, 2013).

74 Mather refers to his "Address to the Physicians" (*D* 2:620–21, Kittredge, "Lost Works," pp. 428–44, 467). Jacob Pylarini (1659–1718), a Greek physician and consul of the Republic of Venice in Smyrna (Ottoman Empire), published *Nova et Tuta Variolas excitandi per Transplantationem Methodus* [*A New and Safe Method of Beginning Smallpox by Transplantation*] (1715), reprinted the following year as "Nova & tuta Variolas excitandi per Transplantationem Methodus, nuper inventa & in usum tracta," *PT* 29.347 (March 31, 1716): 393–99.

75 The likely word is "Africa" – because Mather was ridiculed for trusting the accounts of enslaved Africans. In his *Sentiments on the Small Pox inoculated* (November 23, 1721), Mather justified the introduction of inoculation as follows: "*We have an army of Africans among our selves, who have themselves been under it, and given us all the Assurance, which a Rational Mind can desire, that it has long been used with like Success in* Africa" (p. 2, col 1). This statement was parodied and earned Mather the ridicule of his opponents. We gratefully acknowledge Dr. David Lupher of the University of Pudget Sound for drawing our attention to this somewhat puzzling reference. See also Kittredge, "Lost Works" (pp. 437–40).

Blemish, on that Honourable Profession, from the Humours of Some, that are gott into it. But this I will say, The *Prædiction* was accomplished unto Admiration. And how far they can comfort themselves, in seeing above a Thousand of their Neighbours within a few months killed before their eyes, when they knew a Method, that, in Ordinary way, would have saved them — *They know better than I!*

Two Pretenders to Instruct people in the *Art of Physick*,[76] published their Pamphlets against the Practice: Both of them stuffed with gross *Lies* in matters of Fact, (which indeed are ever used and needful to be so, for the Support of the Cause) and both them full of a *Non-sense & Folly*, which a Text, in the *Twenty Sixth of the Proverbs*, forbids an *Answer* to.[77]

The first of these was a sorry *Tobacconist*, who could hardly spell a word of *English*, (even the word *English*, from his acute pen was *Engleche*) and could not read his own Manuscript, but pray'd the Printer to find out the meaning, & make *English* of it. This hideous fellow, who is more known by the Name of *Mundungus*, than that of *John Williams*,[78] directed his Read-

[76] John Williams and William Douglass.

[77] Prov. 26:4, "Answer not a fool according to his folly, lest thou also be like unto him."

[78] "*Mundungus*" signifies "tobacco of poor quality; bad-smelling tobacco" (*OED*. Mather refers to John Williams (1664–1729), "tobacconist" (apothecary), who participated in the fray with his *Several Arguments* (1721) and *An Answer To a Late Pamphlet* (1722). See Kittredge, "Lost Works," pp. 471–74. To be sure, the medical education at Oxford, Cambridge, and Edinburgh, did not differ substantially from that at Harvard, for until the early 18th century, medical education was largely based on "reading medicine"; i.e., perusing the medical tomes of the ancients – Hippocrates and Galen – and only later in the century works on anatomy and on specific diseases published in England and on the Continent. Hard to believe nowadays, hands-on practical training including diagnosing a patient's disease or performing surgery was *not* part of the medical curriculum, neither was preparing and administering medicines (see P. Allen [1946], O.T. Beall and R.H. Shryock [1979], and J. Barry [2019]). The only physician in Boston with a medical degree, William Douglass, M.D. (c. 1691–1752), was born in Scotland and settled in Boston (1718), where he felt slighted by Mather for not having given Douglass sufficient credit for lending Mather a copy of Timonius's *Philosophical Transactions* article (see above) on inoculation as practiced in Constantinople. In his *Abuses and Scandals* (1722) and his *Postscript To Abuses, &* (1722), Douglass – eager to elevate himself above self-taught practitioners and preacher-physicians in the colonies – attacked Mather for meddling in business for which he lacked qualifications. Probably coached by Mather, Isaac Greenwood (1702–45) came to Mather's aid and responded with his satire *A Friendly Debate; or, A Dialogue, Between Academicus* [Zabdiel Boylston] *And Sawny* [William Douglass] *& Mundungus* [John Williams], *Two Eminent Physicians,*

ers to *Studde Sempeti and Anthepeti*,[79] and he forbad his *prates*, because, to *Specke for hoomain Invenecions in Fisecke* is not *allowabel*. (This is a Specimen of the *Orthography* in his Manuscript.)[80] But because the Sott abusively brought some Clauses from the *Sacred Scriptures* (which he spelt *Scakered Scripters*) 'tis incredible, how much the Mob about the Countrey were enchanted by this poor *Smoaky Conjurer*, The *Smoke*, that came out of a *Tobacco-cellar* made a Strange Impression on them, and this *Hans en kelder*[81] was a meer *Apostle* (when doing the part of an *Apollyon*)[82] to his wretched Votaries.

But however he was thought an *Author*, worthy to be quoted by another, who value's himself upon his being a *Scholar*, and whose name I would not have mentioned, if you had not among you published some of his *Leasings* under his Name unto the World; And I am under a necessity of telling you, That the First Letters of his Name are, *William Douglas*.[83] I shall say nothing about the too well known *morals* of the man. But if his Rapsodies have

About some of their Late Performances (1722). The title of Greenwood's anonymous satire is based on that of the anonymous tract published by Ben Franklin's brother James Franklin (see below).

79 I.e., "Study Sympathy and Antipathy." [Greenwood], *Friendly Debate* (1722), p. 21, Mather could give as good as he got, when he alludes to the anonymous *A Friendly Debate; or, A Dialogue Between Rusticus an Academicus About the late Performance of Academicus* (Printed and sold by James Franklin, 1722). In his *Postscript To Abuses, &c. obviated* (1722), sign. A, Douglass bitterly complained about the aspersions cast upon him for his Scottish accent.

80 Mather added this parenthetical notation in the left-hand margin. It quotes Greenwood's *Friendly Debate* (1722), p. 21.

81 "HANS IN KELDER, jack in the cellar, i.e. the child in the womb: a health [toast] frequently drank [sic] by breeding women or their husbands," in *A Classical Dictionary of the Vulgar Tongue* (1785), sign. M^2.

82 I.e., Abaddon, or destroyer (Rev. 9:11).

83 William Douglass (c. 1691–1752), M.D., Scottish-born Boston physician, *Inoculation of the Small Pox As practised in Boston, Consider'd in a Letter to A–S–M.D. & F.R.S. In London* (1722), and *The Abuses and Scandals Of some late Pamphlets In Favour of Inoculation of the Small Pox, Modestly obviated, and Inoculation further consider'd in a Letter to A–S–M.D. & F.R.S. In London* (1722); the abbreviations "A–S–" in the titles of Douglass's pamphlet apparently refer to Alexander Stuart (1673–1742), a Scottish natural philosopher, physician of Aberdeen, and correspondent of William Douglass in Boston. Less than a decade after the controversy, Douglass eventually came to recognize the benefit of inoculation and adopted the procedure as well. See his *A Dissertation Concerning Inoculation of the Small-Pox* (1730).

reached you, I shall be guilty of a *Fault*, & be wanting to the *Truth*, if I do not inform you & assure you, that there are many *gross Falsities* in them; and scarce a word that he sais, is to be relied upon. This he has here been very publickly convicted of! For the rest you will not need any Reviews of mine; For (besides an inveterate spite expressed against the Ministers of the Countrey, whom he Scoffingly calls *Inoculating Parsons*)[84] there you will read with your own eyes, how, in a Complement, no doubt, unto the *Royal Family*, or, at least, unto their Physicians, he proclames, The *Inoculation* of the *Small-pox*, to be a *Felony*, and within the *Act against poysoning*,[85] and – presse's that the *Physician*, who use's it may be *tried for* his *Life*. There you will also be taught, That whatever *Chronical Distempers there may be in the person, from whom the pocky matter is received for Inoculation, are communicated unto the Inoculated person.*[86] And then presently he yet moves, That *Inoculation may be ordered by Act of Parliament,*[87] *as an adæquate Succedaneum to the Genuine small-pox, to purify the blood from the remaining Fæculency*. Much such a Speculation and Contradiction, as he had in his first Libel (for he shot more than One or two of his fiery *Granado's*) where he rail'd at an *Inoculation* for *propagating the infection of the Small-pox*;[88] and in a few Lines after,

84 Douglass, *Inoculation*, p. 2. *Abuses and Scandals* (1722), "Introduction," pp. 1–2.

85 Douglass, *Inoculation*, p. 13.

86 Douglass, *Inoculation*, pp. 14, 15.

87 Douglass, *Inoculation*, p. 14.

88 A reference to the character-assassination in print. A quite literal attempt to assassinate Mather occurred in the night of November 14/15, 1721, in retaliation for his advocacy of smallpox inoculation. He recorded in his diary, "Towards 3 a Clock in the Night, as it grew towards Morning of this Day, some unknown Hands, threw a fired Granado into the Chamber where my Kinsman lay, and which uses to be my Lodging-Room. The Weight of the Iron Ball alone, had it fallen upon his Head, would have been enough to have done Part of the Business designed. But the *Granado* was charged, the upper part with dried Powder, the lower Part with a Mixture of Oil of Turpentine, and Powder and what else I know not, in such a Manner, that upon its going off, it must have splitt, and have probably killed the Persons in the Room and certainly fired the Chamber, and speedily laid the House in Ashes. But *this Night there stood by me the Angel of God, whose I am and whom I serve;* and the merciful Providence of GOD my SAVIOUR, so ordered it, that the Granado passing thro' the Window, had by the Iron in the Middle of the Casement such a Turn given to it, that in falling on the Floor, the fired Wild-fire in the Fuse was violently shaken out upon the Floor, without firing the Granado… . There was found a Paper so tied with String about the Fuse, that it might out-Live the breaking of the Shell, which had these words in it; COTTON MATHER, *You Dog, Dam you: I'l inoculate you with this, with a Pox to you!*" (*D* 2:657–58, 663, 658). See also the report in the *Boston News-Letter* of November 20, 1721. He confided to his diary

saie's, *The Inoculation produces nothing Analogous to the Small-pox*. But he compares it unto the Injection of the *Oil of Tobacco* into the Mass of Blood,[89] which, he sais, may be received by the *Stomach* or *Lungs* without prejudice. It was in that Libel, that he affirmed, The Practice of *Inoculating* the *Small-pox*, had been *among the Learned universally known* in *England*, above twenty, (now twenty two) years; and more than twice Seven years before the communication of it from *Timonius* (albeit your Dr. *Harris* more truly tells us,[90] *That* was the first communication of it, unto the Western parts of *Europe*) but that it had never been practised, because it was deemed *wicked*, & *Fellonious*. I hope I have not at all trespassed upon the *Innocency of the Dove*,[91] in these few Touches (and they are very few to what I might have justly given) upon the performances of a person, upon whose esteem with you, a great Injury cannot but be done to some whose *Innocency* ought thus to be defended.

Wee had but *One Physician* in the city, who ventured on the practice;[92] whose Courage with the Blessing of God upon his Endeavours, triumphed over the Attempts of his Adversaries to ruine him. His *Method* you have already received and imparted. The *Success* of it has been such, that wee need nothing from the other side of the *Atlantic*, to embolden our proceeding in the *Method*, when there shall be a new Occasion for it; which with us is very *rare*, but with you is never wholly intermitted.

The Experiment has now been successfully made on several (more than One or Two) Hundreds of persons among us: upon both *Male* and *Female*, both *Old* & *Young*, both *strong* & *weak*; *Whites* & *Blacks*, on *Tawnies*, On Women in *child-bed*, and (unknown to the Physician) on Women *with* child, at

---

that he would willingly embrace martyrdom for his cause (*D* 2:659–60). See also CA 6.2 "*Variolæ Triumphatæ*. The SMALL-POX Inoculated."

89 Douglass, *Postscript To Abuses, &c. obviated* (1722), p. 8. See J.R. Sanchez-Ramos (2020).

90 In reviewing the scribal copy, Mather added "more truly." Walter Harris, F.R.S. (1647–1732), an English physician and author of *De Peste Dissertatio Habita Apr. 17. 1721… . Cui accessit Descriptio Inoculationis Variolarum* (1721), pp. 21, 41–43, 48. See also Increase Mather's *Some further Account from London, of the Small-Pox* Inoculated. 2nd ed. (1721), p. 1; and Mather's extract from Harris published in the *Boston Gazette* (February 5, 1722).

91 Math. 10:16.

92 Zabdiel Boylston (1679–1766), a largely self-taught physician (common for the period) who learned his trade by being apprenticed to a local practitioner. At Mather's hand he published *Some Account of what is said of Inoculating or Transplanting the Small Pox. By the Learned Dr. Emanuel Timonius and Jacobus Pylarinus*. Published By Dr. Zabdiel Boylston (1721), and *An Historical Account of the Small-Pox Inoculated in New England* (1726).

all Seasons of *Summer* & *Autumn* & *Winter*, and under all the Disadvantages imaginable;[93] yea, when the Patients under the *Inoculation* have been by an Infuriated *Mob* compelled unto Such *Removes*, in cold & bleak winds & rains, for several miles upon the Water, as would have probably murdered any that were in the least invalid circumstances. But they have *still done well*, and had generally Such an easy time of it, that they profess, *They had rather undergo the Small-pox inoculated once every year, than undergo the small-pox once in their Lives, as they see their Neighbours handled with it, in the common way, tho' they should be sure of out-living the trouble of it.*[94] Tho' indeed *Infants*, even such as are dandled on the knee, and hanging on the breast, and can profess nothing about it, are such as wee have usually found, come off the easiest. Wee have also found, in a variety of Instances, That, if the *Inoculation* be used, on such as have had the *Small-pox* before, it produces *nothing at all* upon them.

There are two Towns contiguous to *Boston*. The *Small-pox* entred the Town to the Northward; where the people were poisoned with Outcries against the *Inoculation*. There they died, by *scores*; they died in *Shoals*; the place was an *Alcedema.*[95] The *Small pox* entred the Town to the Southward; and of the first fourteen or fifteen men that were taken with it, about eleven died. But the Survivers after the Example of their wiser Pastors, coming at once into the *Inoculation*, there died *not one man* after it. One would think, here was an Experiment enough to instruct a *Countrey*; yea, to instruct a *Nation*.

Of the *Five* or *Six* in all this Numerous Company of Patients, which died before the Operation of the *Inoculation* was over with them, 'tis undoubted, that there were two or three, that came not under it, until they were strongly first infected in the *common way*.[96] Of the *other Three* I must give you this faithful Account.

The First was an *Indian*, who was kept in a cold Garret, without any Fire in the Depth of our[97] Cold Winter; where she could not be well tended, because the whole family, was at the same time under the *Inoculation*, They *all did well*, but *she* took a violent *cold*, which threw her into a *pleuretick Fever*, a Distemper; that had formerly much endangered her Life; And she died of *that*.

93 Boylston, *Some Account*, p. 11.

94 Boylston, *Some Account*, p. 13.

95 MHS draft has "Alcedama," i.e., "Field of blood," a burying ground for the poor in Jerusalem, which according to legend was purchased with the money that Judas Iscariot took for betraying Christ.

96 See Edward Huth, "Quantitative evidence," pp. 262–66.

97 Mather added this word to the copy.

The Second was a Woman, who was languishing of a *Consumption*, and had not been allowed by sickness, to be above once or twice at the Publick Worship for several years together. She came under the *Inoculation*, when she had an immoderate flux upon her; which thereupon grew to an Extremity; And she died of *That*.

The Third was a man, who had formerly been recovered out of an *Hectic*;[98] and had[99] now been, for diverse months under violent symptoms of a Relapse into his former Weakness, and was under such a *Complication of Infirmities*, that few who knew him, expected him to live many more weeks in the world. He came under the *Inoculation* with some Hopes, that it might prove unto him, as it had unto some others, *A Relief of his other Infirmities*. But a *distemper of Mind* came upon him, whereof some touch had been before observed in him; which brought upon him an obstinate Refusal of Medicine or of[100] *Sustenance*. The Abstinence was to a Degree, that might have killed any man of so weakly a Constitution, should he have had *nothing else* to have grappled with. A Physician, coming from him, told me, One porringer of *Water-gruel*, if there could be any perswadeing him to take it, would certainly save his Life. This crazy *Abstinence* (with what had been many months upon him) was what he died of.

The Hundreds, that have been under the happy Experiment with us, have now seen all of them a year, most of them a year & half, Others yet more, since the *Inoculation* was used upon them. And though our Quacks and their Disciples kept ever threatning of them with the *Dismal_Things*, they should find in their Health, before many months were out; The *Raw Head and Bloody Bones* is all come to nothing; They are all *alive* to this day (except only two, that have been *killd*, the One by a *Fall*, the other in the *Field*;) yea, all in *health*, and some that had been very feeble people, more *healthy* & *hearty*, than ever they were in their Lives before.[101]

Upon the whole, One, that shall have the Effrontery to utter an Intimation, That it is[102] confessed, as many dy under the *Small-pox Inoculated*, as there do of the *small pox* taken in the *common way*, ought certainly to have the penance of a very long[103] *Silence* imposed on him. Instead of insinuating so vile a Falshood, it may, for ought any thing, that I have ever yet mett withal, be

98 A regularly recurring fever, often accompanying tuberculosis.

99 Mather added this word to the copy.

100 Mather deletes "Abstinence" and interlineates "Medicine or of."

101 Mather added this word to the copy.

102 Mather added this word to the copy.

103 Mather added this word to the copy.

very truly declard; *It yet remains to be proved, that of the many Hundreds and Thousands, that have been under the Small pox Inoculated, there was ever any One person, who miscarried, that the Operation was regularly used upon.*[104] There never was a more *successful Operation* brought into the World! And the Reception, which it has among those, that should be wiser, administers to Heaven too much cause of that Complaint, *Wherefore is a price put into the hands of the Fools, that have no Heart unto it?*[105]

If your physicians will discourage the Practice, they will do well to examine their Hearts; *What principles they act upon*; and consider, How they can answer to God, for the Loss of so many Hundreds of Thousands of Lives, as would have been saved, *if THEY had not hindred it*! But, *Meliora Speramus*.[106]

As for the other vehement Opposers of this Practice, if they are not so *possessed*, that all talking to them, would be only to argue with a Whirlwind! I would ask them, whether it be not a bold *Præsumption* in them to make *That* a *Sin*, which God, *the Judge of all* has never made a *Sin*, but really commanded, & enjoined and required as a *Duty*?[107] And I would ask them, whether They can count it a *Trusting* in God, & not a *Tempting* of Him,[108] to depend on Him for the prolongation of their Lives, while they neglect the *Duty* of *Doing what they can* in the Use of *means*, that He has kindly shewn them, for it.

I would also ask them, Whether it be not a most criminal *Ingratitude* unto the *God of Health*, when He has acquainted us with a most *Invaluable-Method*, for the Saving of our Lives from *So great a Death*, to treat it with neglect, and contempt, and multiply Abuses on them, who Thankfully and in a Spirit of Obedience to Him, embrace His Blessings!

I would finally ask them, Whether they have no dread at all of being accessory to the *Innumerable Deaths*, which may be, in part, oweing to their boisterous Opposition under the *Method of safety*?

For my part, I cannot lay aside my Sentiments, That the Sense of the *Fascination* with which, the Great Adversary of Mankind, makes these unhappy men his Instruments of helping the *Curse to devour the Earth*, and Those, who *dwell therein to be desolate*;[109] so that anon, *few men*, to what might have

[104] See Boylston, *Some Account*, pp. 18, 19, 22.

[105] Prov. 17:16.

[106] "We hope for better things."

[107] Williams, *Several Arguments* (1721), p. 17.

[108] Anonymous [Francis Archibald], *A Letter From one in the Country, to his Friend in the City: In Relation to their Distresses occasioned by the doubtful and prevailing Practice of Inoculation of the Small-Pox* (1721), p. 5. Williams, *Several Arguments*, p. 5; Grainger, *Imposition of Inoculation*, pp. 11–15, 22.

[109] Isa. 24:6.

been, *shall be left*, may call them to walk humbly all their daies: That ever they should be *so left of God*!

In the mean time, *Wee*, that cry with a loud voice to them, *Do yourselves no Harm*,[110] and shew them, How to keep themselves from the *pathes of the Destroyer*, are conscious of nothing, but of a *pity for Mankind* under the *Rebukes of God*; A concern to see the *madness of the people*; A desire to have our Neighbours *do well*; and a sollicitude for a *Better State of the World*. And all the Obloquies, and Outrages wee Suffer, for our Charity, Wee shall entertain as *Persecutions* for A *Good Cause*, which will not want its Recompences.

I write unto a Person of so much Goodness, that I am sure he will pardon the Fatigue, which the perusal of this Long Epistle may give him; upon an Affair, the Importance whereof, will make it's Apology; and he will with his usual Candour, accept the Intention of,

Boston, New England.[111] Syr,

May. 21. 1723.[112]

Your most hearty

Friend & Servant,

Cotton Mather.

Dr. Mather

May 21st 1723.

read Dec: 19: 23.[113]

110 Mather here adapts the Hippocratic oath "Primum non nocere" or, "First do no harm" to his own purpose. Generally attributed to the famous Greek physician Hippocrates (c. 460–c. 375), the oath is now believed to have been coined by Sydenham – though no textual evidence has been found in any of his published works. See C.M. Smith (2005) and, especially, H. King's demystification of the oath in *Hippocrates Now* (2020), pp. 68–72. We thank David Lupher for drawing our attention to King's work.

111 The location and date are in Mather's hand, as is his signature.

112 Mather MHS draft copy bears the date of May 4, 1723. The date of "May. 21. 1723." appears to be the date when Mather received and corrected the scribal copy he sent to the Royal Society.

113 The docketing information is written in another hand; the latter date indicates when Mather's document was read to the assembled members at the RSL.

7.2. A MONSTROUS DRAGON.[1]

June 3, 1723.

To Dr. James Jurin.[2]

Sir,

The Relation wherewith I am going to Entertain you, will require a Little Introduction from *Antiquity*, that you may not Imagine me a *Continuator* of the Ancient Work that has given you a Collection, *Of Incredibles*.[3] Thô I cannot mention any thing from thence that will add to Your Erudition; from which hardly any thing has escaped, or quote any thing which you *did not know*: Yett I may Stumble on a Quotation of Something which Perhaps you *did not just now think upon*. To make way for a Story of a SERPENT for which I bespeak a Room amongst Your Curiosity; thô I will allow that Some report we have had about the Enormous Dimensions of the *Serpent* which we call a DRAGON,[4] may be Fabulous,[5] Yett it is Possible we shall anon Subscribe to Father *Austin* who not only Sais, That they are *Magna Animantia*; but also adds, *Majora non*

1 Read to the members of the RS (December 5, 1723). See Jurin's Letter to Cotton Mather (December 16, 1723), Appendix 9.

2 In the sent MS. copy, the heading (but not the date) is in Mather's hand, who would never let a good story go to waste if it could be repackaged and utilized for different purposes. So he did with this letter to James Jurin, which he integrated in his *BA* 10:582–85 as a commentary on Rev. 13:3. An extract of Mather's discussion of dragons also appears in his letter to Richard Waller (November 27, 1712), CA 1.11 "On the Rattlesnake."

3 A reference to the Greek mythographer Palaephatus of Paros, Περὶ ἀπίστων. *De Incredibilibus* (c. 4th century BCE), a collection of well-known myths, among which is the story of Daedulus and Icarus.

4 Mather's references to the ancients – including to the Church Fathers – are borrowed from Samuel Bochart's *Hierozoicon Sive bipertitum opus de Animalibus Sacrae Scripturæ* (1663), pars post. lib. 3, cap. 14, cols 430–33.

5 This clause is interlineated by Mather.

Sunt Super Terram.[6] *Ætius*, and *Avicenna* will so Assure us, that we need not go to *Pliny* for the Assuring of it, That these *Dragons* are from *Five Cubits*,[7] to *thirty Cubits* in their Dimensions;[8] *Ælian*, and *Pausanias*, and *Philostratus*, tell us of *Dragons* that were *thirty Cubits* Long;[9] And Such an one was brought alive unto, (a Brother of his) *Ptolomæus Philadelphus*, in *Alexandria.*[10] *Agath-*

6 Bochart, pars post. lib. 3, cap. 14, col. 430 (line 37); St. Augustine's *Enarrationes in Psalmos* (Ps. 148:7) [*PL* 36. 1943] (*NPNF* 8:675) believes that Dragons are "gigantic animate beings," adding, "there aren't any larger ones on earth." See for the same quote in the popular bestiary, *The History of Four-footed Beasts and Serpents* (1607, 1658), by the English theologian Edward Topsel (c. 1572–1625). Illustrations of winged dragons appear on pp. 701 and 705.

7 Mather inserts this word.

8 Bochart, col. 430 (lines 40–49); the Byzantine physician Aetius Amidenus (fl. 5th–6th centuries CE) describes his dragons in *Contractae ex Veteribus Medicinae ... id est sermons XVI* (1542), Quartae sermo I, cap. 34, p. 700, insisting that those in Ethiopia and Lycia – although of varying colors – are from five, ten, and thirty cubits long (2.3 m, 4.6 m, 13.8 m). The Islamic philosopher, astronomer, and renowned physician Avicenna (980–1037), a native of Afshana, Persia, was well known among medieval and early modern scholars in Western Europe. His Arabic works were translated into Latin early on, and Mather frequently refers to Avicenna's *Canon Medicinae*, a medical handbook which enjoyed great popularity among physicians. The Latin translation was published in Venice in 1490. This translation is generally attributed to Gerardus Cremonensis (c. 1114–87) rather than Gerard de Sabloneta (fl. 13th century), an Italian translator who specialized in translating Arabic medical treatises into Latin. Pliny has much to say about dragons in his *Natural History*. In 8.12, for instance, he relates that dragons are so large that they can wrestle with elephants and swallow them whole.

9 Bochart, col. 430 (lines 43–55); The Greco-Roman author and rhetorician Aelian (c. 175–c. 235 CE) relates that the Phrygian dragon was "decem orgylas" long (50–60 feet or 15–18 m), and that of his relatives around the Red Sea forty cubits (*On the Nature of Animals* 17.1); in India, conquering Alexander the Great discovered a hissing serpent in its lair measuring at least seventy cubits long, although only its head poked out (Aelian, *On the Nature of Animals* 15.21). Likewise the Greek geographer Pausanias of Lydia (146–76 CE) knowingly relates that dragons in India and Libya were up to thirty cubits long, in *Description of Greece* (2.28.1); so, too, the Athenian biographer and sophist Lucius Flavius Philostratus (c. 170–c. 245 CE) knows of an Indian dragon thirty cubits long (*Life of Apollonius*, lib. 3, caps. 6–9). See also Mather's letter to Richard Waller (November 27, 1712), our CA 1.11, "Rattlesnake."

10 Bochart, col. 430 (lines 55–63); Diodorus Siculus, in his *Library of History* (3.36.1–5), tells of a dragon thirty cubits long (but dismissing those of 100 cubits as mere falsehoods). As the story goes, since Ptolemy II Philadelphus' reign (284–246 BCE) promised rich rewards to anyone who could bring such a monstrous dragon to Alexandria, many set out to do just that, but succeeded only after the loss of life of some who were gobbled up by the dragon.

archides affirms their having Seen *Dragons* of that extent, and *Strabô* brings the Authority of *Artemidorus* for the like.[11] My Admirable *Bochart* is willing therefore to grant a thing, that has been Confirmed, with so many Testimonies of Antiquity; But then, he can Scarce tell what he shall think (*I Beleeve, I must help him*!) of those in *Ælian*, that[12] grow to *Forty Cubits* and those that in *Philostorgius* grow to *forty five*; thô that Author Asserts That he *himself Saw* the Skins of those *Dragons* brought unto *Rome*.[13] And what shall we then Say, to the *Dragon* of no less than *fifty Cubits* long; which *Suetonius* reports exposed by *Augustus*? Yea, *Dio* Sais, That in *Hetruria*, there arrived a *Dragon* that was *four score and five foot* long, which after he had made fearful Devastations, was Killed with a thunderbolt,[14] and it is a Wonderful thing related by *Strabo*, That in *Cœlo-Syria* there had been a *Dragon*, that was an *hundred foot* Long, and of Such a *Thickness*, that a Couple of Men on horseback, on Each side of him, could not See one another, and of a *Mouth* so Wide that a Man might Sit on horse-Back in it, and every *Scale* of His Hide was as big as a *Shield*. Yea, One of an *Hundred and Twenty foot* long was Killed near *Utica* by the Army of *Regulus*; and the Skin & Jaws of it were Kept at *Rome* for a Sight; And yet after all, tis Affirmed; That the Dragons of *Ethiopia*, and of *India* do exceed all of these.[15] (To recite nothing of any *Norway* Stories:) They are so *Big* that in

11 Bochart, col. 430 (lines 63–74) and col. 431 (lines 1–3); the Greek historian and geographer Agatharchides of Cnidus (fl. 2nd century BCE) affirms this tale in his *On the Erythraean Sea* (bk 5, fragm. 80a). The indispensable Greek geographer Strabo of Pontus (c. 64 BCE–c. 24 CE), in his *Geographica* (16.4.16), has his reptile – of a mere thirty cubits – overpower bulls and elephants, though Strabo's dragons are puny when compared to the fabulous ones in India. He relies on his Greek predecessor, the geographer Artemidorus of Ephesus (fl. 100 BCE), whose geography (11 books) is extant mostly in fragments by Strabo, Pliny, and Stephanus of Byzantium.

12 Mather interlineated this word.

13 Bochart, col. 431 (lines 3–13); Aelian (*On the Nature of Animals* 17.1) forty cubits. The Cappadocian Philostorgius (368–c. 433 CE) affirms the existence of monstrous serpents forty-five cubits long, or "fifteen fathoms" (75–90 feet), and bodies "thick as beams," so he insists, having seen their skins brought to Rome with his own eyes (*Ecclesiastical History*, lib. 3, cap. 11, p. 450).

14 Bochart, col. 431 (lines 13–21); the Roman historian Suetonius Tranquillus (c. 69–c. 130 CE) reports in his *De Vita Caesarum: Divus Augustus* (43.3) of a serpent no less than fifty cubits long, yet the Greco-Roman historian Dion Cassius (164–c. 235 CE), in his *Historiae Romanae* (50.8.4), reportedly knows of a two-headed serpent in Hetruria (central Italy), eighty-five feet long.

15 Bochart, col. 431 (lines 21–38); to top it all off, Mather invokes the authority of Strabo (*Geographica* 16.2.17), whose fallen leviathan in the Plain of Macra

Figure 15. Olaus Magnus, *Historia de Gentibus Septentrionalibus* (Romae, 1555), lib. 21, cap. 43, p. 771. Courtesy of the University of Ghent.

Truth I durst not bring them into my Letter.[16] — But what is a *Dragon* akin to the *Serpent*, of which I am now going to make a_Spectacle? Or is the famous *Dragon* any other than a *Mighty Serpent*? It is thought that a vast Age (*Longus Annorum Decursus*) is that which turns a *Serpent* into a *Dragon*.[17] Thô *Paulus Jovius*, and *Scaliger* assign Feet, like those of a Goose unto *Dragons*; Yett *Austin* more truely Sais, *Dracones Sine Pedibus*;[18] And *Lucan*, – *Serpitis Aurato nitidi*

(Coele-Syria) was no less than a "plethrum," or one-hundred feet long, allegedly slain in the First Punic War, at Utica (near Carthage), by the Roman general Marcus Atilius Regulus (fl. 3rd century BCE). See also Pliny, *Natural History* (8.11.32). For a shorter version of it all, see Mather's *CP* (1721), Essay 28, p. 168.

16 Mather's tongue-in-cheek remark appears to allude to the "Great Norway Serpent," or sea monster, by the Swedish cartographer and archbishop of Upsala Olaus Magnus, i.e. Olof Månsson (1490–1557), in his folkloric *Historia de Gentibus Septentrionalibus* (1555, 1558), bk 21, caps 43–44, pp. 771–73. Olaus Magnus' depiction of the "Great Norway Serpent" attracted much attention (and imitators) at the time. See especially A.C. Oudemans (1892), ch. 4.

17 So it may be (Mather waxes facetious) that just any old snake may – in time – transmogrify into a monstrous dragon.

18 Bochart, col. 432 (lines 40–47), col. 433 (lines 15–17); the Italian physician, historian, and prelate Paulus Jovius, aka. Paolo Giovio (1483–1552), in his *Historiarum*

fulgore Dracones. *Ovid* assigns *Combs* and *Crowns* unto *Dragons*;[19] But *Pliny* more truely Corrects it, Saying, *Draconum Cristas qui Viderit non reperitur*.[20] *Lucan*, and *Jerom*, and *Austin*, and *Jovius*, and *Scaliger*,[21] and all the Painters of St. *George*, and of our *Signs*, allow *Wings* unto the *Dragons*.[22] But these are Certainly Deceived; The *Mystical Dragon* in the *Apocalypse*, has no *Wings*; but it is by the Advantage of *Wings*, that the Woman Escapes him.[23] The *Wings* of *Dragons* probably grew the Fancies of the Ancient *Mythologists*, who Intended

sui Temporis Tomus Primus (1550), tom. 1, lib. 18, p. 1084, endows his dragon with feet like waddling geese. And the Italian French classical scholar Julius Caesar Scaliger (1484–1558), in his *Exotericarum Exercitationum liber XV* (1582), Exercitatio 183, sec. 7, p. 602, knows of four-footed serpents in Calicut (India) with a pig's head and a long, hairless tail. Moreover, says Scaliger in Exercitatio 189, sec. 4, p. 620, "Draconibus autem crura esse, notum est" ("but it is known that dragons have legs"). However, St. Augustine, in his *De Genesi ad Litteram* (3.9) [*PL* 034. 0284] contradicts all those who claim to know of two or four-footed serpents: "Dragons," the great Father of the Church insists, "lack feet."

19 Bochart, col. 433 (lines 19–21); the Roman poet Lucanus (39–65 CE) sings in his *Pharsalia* (9.728) "you dragons crawl shining with a golden gleam." The Roman poet Ovid (BCE 43–17 CE), as in latter-day regal coats of arms, endows his dragon (sacred to Mars) "cristis praesignis et auro" (*Metamorphosis* 3.32) with crests and gold. See also Aelian's *History of Animals* (11.26), who bestows a crest and beard on male dragons.

20 Bochart, col. 433 (lines 23–25); in spite of all who are knowledgeable about dragons, Pliny (*Natural History* 11.44.122) knows best: "None one can be found who has ever seen serpents' crests."

21 Bochart, cols 432 (lines 40–46 and 71–75); col. 433 (lines 1–18); Lucan (*Pharsalia* 9.729–30), St. Jerome (c. 347–419), in his Commentary on Isa. 5:13, 20–22 [*PL* 024. 159C] speaks of "Cristati sunt, et volantes" ("crested and flying") dragons; St. Augustine, *De Genesi ad Litteram* (3.9.) [*PL* 34. 284]: "Dracones autem, &c. in aërem sustolli perhibentur" ("But the dragons, etc., are said to be lifted into the air"). So in his *Enarrationes in Psalmos* (148.7) [*PL* 36. 1943]: "De speluncis procedunt, feruntur in aëra" ("They come from caves and are borne off into the air"); Paulus Jovius, *Historiarum sui Temporis Tomus Primus* (1550), tom. 1, lib. 18, p. 307, says, "Asserunt Gogiani [Georgiani] regni incolæ his in vallibus dracones gigni alatos" ("The inhabitants of the kingdom of Georgia claim that in these valleys dragons are born with wings"); Scaliger (Exerc. 189, sec. 4, p. 620): "Draconibus autem crura esse, notum est" ("but it is known that dragons have legs").

22 "Our signs" may refer to the signs hung outside of taverns and other businesses. In his letter to Richard Waller (June 21, 1714), Mather dismisses the effigy of St. George and the dragon as pure fable, of a kind with the Unicorn: "We are now satisfied, That the *Unicorn* which pretends to support our [Coat of] Arms, is as Fabulous for a *Beast*, as our Patron St. *George* is for a *Saint*," in CA 2.4 "The Moose."

23 Rev. 12:7, 14.

Frygates[24] in the Stories which they told of *Dragons*; and this the rather because they had the *Carved Images* of *Dragons* upon them. The *Serpent* which I am agoing to Show you, will not Serve to Decide Some of these things; because I can shew you Nothing but the Skeleton of him; Never the Less that Stupendous Article of the *Dimensions* found in Some of the Given Instances, may be a Little Confirmed from what I am now to tell you.

A few Months Ago, near the Falls of *James-River* in *Virginia* Some digging a Water Course for a Mill, by a Small run of Water, under the side of an Hill ten foot under ground, they found the *Back Bone* of a *Snake* lying in Such a form as this . They dug along from the *Head* to the *Tail* of it, both of which were very much Decayed; but the *Earth* about the Decay'd Parts was of a *Different Colour* from all the Earth about it. *These*, with all the *Bones*, were found exactly of the Same Depth under ground. It was about a Mile, and an half Distant from the Great river, and about One hundred and fifty miles from the Sea; But now for the *Prodigy*! The Monster extended above *One hundred, and fifty foot* in Length! A Credible Person who measured the Trench out of which it was taken, brings the Account unto us, and brings with him one *Joint* of the *Back* Bone, which the Workmen gave him, and which is[25] Five Inches and a Quarter in *Depth*; Five Inches & an half in *Width*; Four Inches & an half *Diameter* backwards & forwards; and tho' it be grown very Dry, and some of it be worn off, yett it weighs Thirty three Ounces & an half. *Monstrum horrendum*![26] —

I shall not go to trouble You with any of my Poor Speculations on the Subject, chusing and wishing rather to be Instructed from *Yours*. And so Praying for Your Deliverance from a *Greater Dragon*, than that whose Bones were lately found in *Virginia*, I Subscribe,

Sir,

Boston N. England. Your most hearty Servant.

Jun. 3. 1723.

Cotton Mather.[27]

[24] Mather overwrites this word.

[25] The remainder of the paragraph is in Mather's hand.

[26] Vergil's *Aeneid* (3.8), referring to Aeneas recounting to Dido how he and his shipmates caught sight of the blinded Polyphemus on his island.

[27] The location, date, and signature are added in Mather's hand.

7.3. A NEW RATTLESNAKE.[I]

June 4, 1723.

To Dr. James Jurin.

Sir,

Lett me not incurr any Danger of being listed among the *Ophitæ*, much less of Procuring to my self the Name of an *Ophioneus*;[2] if after the Various Treats, which I have given to you or my former Correspondents, upon Our *Serpents*, I præsume once more to treat you with them.

Tis Horrible, and hideous to See, what an *Ophiolatry*[3] has been Practised by Several Nations of Mankind, whom the *Old Serpent* has Miserably deceived, and enslaved, and in this way triumphs over them. There will be no Part of *it*, in our Inquiries after the *Knowledge* of the Animal; It shall be Consigned rather to the *Rattle Snakes* I am now going to tell you of.[4]

We have lately discovered a *New Snake* among us, which in all things has the Resemblance of a *Rattle Snake*: Only it has not *Rattles*, but an hard *Bone* instead of them. Our People have not yett had the Time to make much, and Long Observation of his Conduct among the *Rattle Snakes*; But it has been Observed, That he Exercises a Sort of a *Government* among them; And if any

1 Read to the members of the RSL (December 5, 1723). See Jurin's Letter to Cotton Mather (December 16, 1723), Appendix 9.

2 According to St. Epiphanius of Constantia (c. 315–403), the Ophites were snake worshipers who held that Eve in the Garden of Eden believed the serpent "as if he were God's son" (*Epiphanius* 2:56.9–10). See also *BA* 1:480. Mather's characteristic wordplay – on Ophioneus as a worshiper of snakes, as well as on Ophioneus (the elder Titan) and his wife Eurynome, who were cast out from Olympus by Cronos and Rhea – may allude to the song of Orpheus in Apollonius of Rhodes's *Argonautica* (1.496–511), who recites their fate.

3 The adoration of serpents.

4 See also 1.11 "On the Rattlesnake."

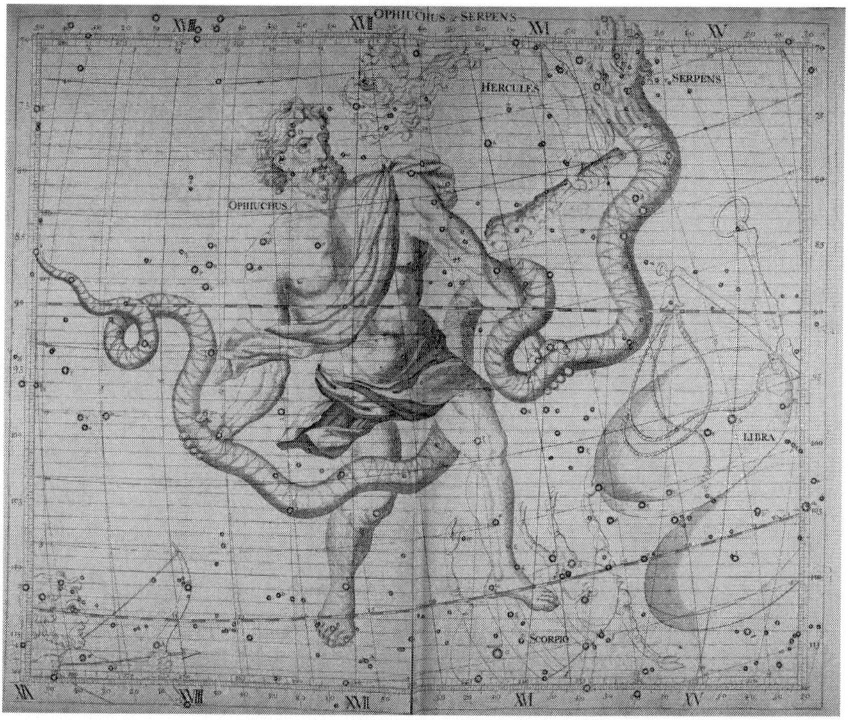

Figure 16. Ophiuchus, the Serpent Bearer. In John Flamsteed, *Atlas Coelestis* (London, 1729). Courtesy of the Wellesley College Library.

of them do not Conform to his Expectation, with his Tail he Strikes them dead upon the Spott.[5]

I know not whether he would have presumed to do so, (or, whether a Venerable *Old Age* would have been any Protection) unto a *Old Rattlesnake*, which I am informed was not long ago Seen at a Place called *Cowass*,[6] and which had hairy Bristles on his Back, and no less than *Seventy two Rattles*; whereas we rarely find any with above *Sixteen*; and their *Age* in Our Common Tradition Computed from them.

It ha's appeared a thing Somewhat Marvellous unto many of our People, That thô our Woods have Abounded with *Rattle Snakes*, and their Bites are

5 See also Mather's letter to John Winthrop (12^{d} 1^{m} [March] 1722/23), in *MP* (1868), p. 454.

6 Cowass, modern Newbury (New Hampshire), in the territory of the Native American Abenaki tribe of the Cowass (Cowasuck), whose vast territory included most of present-day New Hampshire, Vermont, and southern Quebec.

Terribly Venemous; yet it is a very rare thing for any of us to be bitten with them; And when any are Bitten, Our Indians have so Instructed us, that they are Presently Cured.

Among the Remedies Provided for this Occasion, we have a *Plant*, which among us goes by the Denomination of, *Poor Robins Plantain*.[7] The Story is That an Indian Called *Robin*, Saw a *Snake* bite a *Partridge*; but the *Partridge* escaping, and the Indian following him; he found the *Partridge* Nibbling at a Plant, and with his Bill Applying the Bruised Leaf to his Wounds; On which the Wounded *Partridge* Soon recovered, and flew Away where the *Reptiles* could not Come at him; My Indian took the *Plant*, and found that by laying of it Bruis'd unto the Bite, and by taking the Infusion of it Inwardly, the Bite of the *Rattle Snake* has an Effectual, and a Wonderful Antidote; and upon the Security which his Plant gave him; he became as hardy and as famous an *Ophiuchus*,[8] as any in All the Country.

Sir,

I wish Your Safety from all that may hurt you, and Particularly from the Bites of *Malice*, and *Envy*, wherein Persons of Your Attainments, and Services, often find, the *Teeth of Beasts are Sent upon* them, *with the Poison of the Serpents of the Dust*.[9]

With Such Wishes, I am,

Sir,

Your Perpetual Servant

Jun. 4. 1723.

Cotton Mather.

[7] The Poor Robin's Plantain, also known as *Erigeron Philadelphicus* of the *Asteraceae* family, is a herbaceous plant native to North America.

[8] Punster that he is, Mather summons Ophiuchus (ὀφιοῦχος), i.e., "Serpent-bearer" (constellation Serpens), represented as a man clutching a serpent in the celestial equator.

[9] Deut. 32:24.

7.4. *CARCHARIAS MIRABILIS.*[I]

June 5, 1723.

To James Jurin.

Sir,

You will allow me to Confess, That I have been under an Agreeable Surprize, in finding a Man *Stark Blind*; yea, *Born Blind*, make Discoveries, and those in the way of *Reading* too, which no Man ever made before him.

This ha's been done by my Invaluable Friend Mr. *William Jameson*, a late Professor in the University of *Glasgow*; He has done it in Several Books, full of Criticism, and Quotation, and most Uncommon Erudition; And he has very Singularly done it, in the Treatise he published just before his Death, entitutled; *Spicilegia Antiquitatum*.[2]

Among the Curiosities in that Performance, there is one that very Particularly gratified me. We find in *Mela* a Tradition Concerning the City of *Joppe*, That it was *Ante Diluvium Condita*;[3] and a Mention of the Songs

1 Read to the members of the RSL (December 5, 1723). See Jurin's Letter to Cotton Mather (December 16, 1723), Appendix 9.

2 William Jameson (f. 1689–1720), a Scottish Presbyterian, professor of history at the University of Glasgow, and author of *Spicilegia Antiquitatum Ægypti atque ei Vicinarum Gentium* (1720). For many of the citations below, Mather gleaned his historical material from *Spicilegia* (cap. 7, §2, pp. 139–42). The substance and close wording of this and the following two paragraphs are also integrated in Mather's commentary on Jonah 1:1 (*BA* 6). Mather also cited Jameson in *TP* (p. 177) and throughout the volumes of *BA* (vols. 1, 4, 5, 6, 9). He had hoped to visit Jameson in Glasgow, but Mather's voyage to Europe never materialized. See his letter to Jameson (March/April 1715), in K. Silverman's *SLCM*, p. 179. See also A. Chamberlain (2018), pp. 144–62.

3 Jameson, *Spicilegia*, p. 139; "Founded before the Flood." In his *De Situ Orbis, Libri Tres* (1635), lib. 1, cap. 11, p. 30, by Pomponius Mela (fl. 43 CE), an early Roman geographer, whose geography was employed in tandem with Strabo's *Geographica* (BCE 64–24 CE) until the 16th century. Joppe (or Joppa) is the modern city of Jaffa in Israel. See also F.E. Romer's edition of *Pomponius Mela's*

about *Andromeda* here Saved by *Perseus*, from a Wild Beast of the Sea, the *Vast Bones* whereof were Still to be Seen.[4] *Pliny* also Speaks of *Joppe*, as being, *Antiquior Terrarum Inundatione*,[5] And he Sais, That the *Bones* of the Sea-Monster that would have devoured *Andromeda*, were Carried from *Joppe* to *Rome*, and exposed by *Scaurus* there; That the Animal was *forty foot* in Length, and in heighth Exceeding the Tallest *Elephants* of *India*, And the Joints of the Back Bone a foot, and half in Extent.[6] *Solinus* tells of *Joppe* as being the Oldest City in the World, and built *before the Flood*, and famous for the Rock, that had Footsteps of the Chains that Bound *Andromeda*, Still to be Seen upon it; Adding; That *Scaurus*, in the Time of his *Ædile-Ship*, Exhibited the *Bones* of the Wild Beast, which was to have devoured her; And he agrees with *Pliny* in the Circumstances.[7] The fancy of *Andromeda* exposed unto the

---

Description of the World (1998), p. 53. Although Mather cites Mela's expertise in this letter, he roundly dismissed the myth of Andromeda in his commentary on Gen. 6:4 (*BA* 1:587), even as Mather rehearses numerous other gigantomachiae to bolster his discourse on the giant Nephilim and how they might have attained their tremendous size (*BA* 1:582–99).

4 Via his source in Jameson (p. 139), Mather refers to the story of Perseus (famed slayer of serpentine Medusa), son of Zeus and Danaë, and Andromeda (beauteous daughter of Cepheus and Cassiope of Joppa), who was chained to a rock as a sacrifice to a sea monster, but whisked away by Perseus on the back of his trusty Pegasus. Customary for his time, Mather reads ancient myths through the lens of biblical history and discovers – in the best manner of *Prisca theologia* – the story of Jonah and the whale hidden among pagan historians who drank from the Mosaic fountainhead but distorted the original. Theophilus Gale's huge four-part *The Court of the Gentiles: or A Discourse touching the Original of Human Literature* (1669–82) is among the most noteworthy examples, and like Ralph Cudworth's *True Intellectual System* (1678) posited that the ancient theology, or *Prisca theologia*, was distorted by the postdiluvian descendants of Noah by mixing the original with their own pagan myths.

5 Pliny, *Historia Naturalis* (5.14.69): "more ancient than the flooding of the lands" (Deucalion's flood).

6 Pliny (9.4.11) relates that the monster which Marcus Aemilius Scaurus (c. 92–c. 52 BCE) brought to Rome was indeed the gargantuan marvel to which Andromeda in the story was to be sacrificed. Scaurus, Roman consul, praetor, and politician, served under Pompey the Great (106–48 BCE) in Syria and Palestine during the Third Mithridatic War, and in 58 BCE was elected "curule aedile" (overseer of public buildings and public festivals).

7 Roman geographer and compiler Solinus (fl. 3rd-century CE) retells the hoary tale in his *De Mirabilibus Mundi* (34.1–3) and provides the same measurements.

Whale at *Joppe*, is Reported by *Strabo.*[8] *Apollodores* and *Ovid* have the Fable too; thô they Seem to lay the Scene in *Ethiopia*, which *Pliny* Endeavours a Conciliation for.[9]

Behold here, a Confession of the *Flood* in the Writings of the Ancient *Pagans*; and, behold also the History of *Jonah* Marvellously Disguised, but Confirmed.[10] *Jonah* was cast ashore near *Joppa*, from whence he took his departure. That it was done on the Shore of the *Euxine* Sea, as *Josephus* would have it;[11] or that it was done on the Shore near *Ninive*, as tis fancied by *Sulpicius*, it is most Irrationall to Imagine it.[12] It is Probable that the *Whale* was by the rising Waters Carried so farr up, that he was there Left ashore, which might also Contribute unto *Jonah*'s more Easy Rescue. The *Whale* dies, and rotts there; and the *Skeleton* in Length of time Covered with Earth, was dug up after Many Centuries of Years; and the *Bones* were Exhibited by *Scaurus* at *Rome*, as the Relicks of Some *Antediluvian Monsters*. This was the Occasion of the Vain fancy that *Joppa* existed *before* the *Flood*. *Jonah* in the Mean time, was transformed not only into *Andromeda*, but also in to *Hercules*; of whom they tell us (as tis related by *Cyril* of *Alexandria*) that he was devoured by a

8 Jameson (p. 140). Strabo (*Geographica* 16.2.28.759) attributes the story of Andromeda and Perseus to "certain" mythologists.

9 In *Bibliotheca of Pseudo-Apollodorus* (2.4.3–4), attributed to Apollodorus of Athens (3rd century CE), the circumstances of Andromeda's binding to the rock at Joppa (as compelled by the Ethiopians) and happy escape at the hands of Perseus are rendered vivid. Ovid (BCE 43–c. 18 CE) seconds the tale in his *Metamorphoses* (4.668–750). See also Pliny (5.34.128). These fossils and their fossilized stories are ably described in Adrienne Mayor's *The First Fossil Hunters* (2000), esp. pp. 138–39.

10 Since the likes of the "heretic" Isaac La Peyrère, in his *Men before Adam* (1656) (bk 4, ch. 7, p. 239; and ch. 9, pp. 248–59) had limited the Noahic Flood to a local event in Palestine – not the whole world – Mather was only too happy to employ the discovery of the bones of dinosaurs (a term not established until the early 19th century) and other fossils as evidence of a universal flood. See also the delightful collection, *The Dinosaur Papers 1676–1906*, edited by D. B. Weishampel and Nadine M. White (2003).

11 Jameson (pp. 140, 141). The Graeco-Roman historian, Jewish priest, and scholar Flavius Josephus (c. 37–c. 100 CE) sends Jonah's whale to the shores of the Black Sea, from which likely locale the prophet footed it to the Assyrian Nineveh, on the banks of the Tigris River (*Antiquities of the Jews* 9.10.2).

12 The Christian historian of Aquitaine (S.W. France) Sulpicius Severus (c. 363–c. 425) posits that the whale swam up to the shallows of the River Tigris and disgorged the prophet near the walls of Nineveh (*Chronici*, lib. 1, cap. 48) [*PL* 20. 0124*CD*].

Whale, but then Cast up again, without any other Damage than that the Heat of the Fish, had hurt his *head*, and Stript him of his *Hair*.[13] This Incommoding the *Head* of the Hero, My *Jameson* thinks is Nothing but a Relique of what the Sacred Scriptures Report, about what the Head of Jonah Suffered from the Hott Sun beating upon it.[14] And no doubt, there was in the days of *Josephus* a Noted *Rock*, on the *Joppan* Shore, which has excavations upon it, that Contributed unto the *Andromedan* Fiction.[15]

But, as these Remarks of my learned Professor, brought Something that was *New* about an *Old Whale* unto one of my Small Observation, and Erudition, and we are very Prone to invite our Friends unto what we ourselves are much Pleased withal, which is my Apology for Setting before you these Communications, which are Indeed but a *Postscript* unto what[16] I have written of the *whale* in my former Letters:[17] Thus I shall make them only Serve to Introduce the Mention of a Thing, which my Ignorance had made also *new*

13 Jameson (p. 141). The Patriarch Cyril of Alexandria (c. 378–444 CE), *Commentarius in Jonam Prophetam*, on Jonah 2:1 [*PG* 71.4. 0376], has turned Jonah's metamorphosis into Hercules and Alcmena, not Andromeda (Alcmena's mother).

14 Jameson (pp. 139–40).

15 Flavius Josephus points to the "great stones" in the harbor of Joppa, "where the chains wherewith Andromeda was bound have left their footsteps, which attest to the antiquity of that fable" (*Jewish Wars* 3.9.3, in Whiston's 1733 translation). See also Pliny, *Natural History* (5.14.69 and 9.4.11).

16 The copyist erroneously dropped the word "what," which does appear in the MHS draft.

17 This may be the lost letter to John Chamberlayne (February 1720), CA 5.3 "The Whale," or the lost letter to John Woodward (1721), CA 6.11 "Ambergris discovered and determined." Mather also put the story to good use in his sermon *TC* (1717), pp. 24–29, addressed to New England's fishermen, especially whalers; the story also appears in his book published by the Royal Society, *CP* (1721), Essay XXIX, pp. 176–79. See also R.P. Stearns' *Science*, pp. 466–67. Mather's old nemesis Paul Dudley, F.R.S. (1675–1761), chief justice of Massachusetts and son of Governor Joseph Dudley, published "*An Essay upon the Natural History of Whales, with a particular Account of the Ambergris found in the Sperma Ceti Whale*," *PT* 33.387 (April 30, 1725): 256–69. Dudley's was preceded by "Ambergris found in Whales. Communicated by Dr. Boylston of Boston in New England," *PT* 33.385 (December 1724), p. 193, our CA 6.11. It is highly likely that Boylston's brief account may actually have been Mather's own, since Boylston is not listed as the author but only as the one who had "communicated"; i.e., he presented or read it to the assembled members of the RSL (see Kittredge, "Lost Works," p. 461). It is well known that Mather had collaborated with Boylston on *Some Account Of what is said of Inoculating or Transplanting the Small-Pox… . Published by Dr. Zabdiel Boylston* (1724), yet allowed it to be made public under Boylston's aegis.

to me when I mett with [it] very Lately.[18] Tis this; Off of our *Cape Cod* we ever now and then Kill a SHARK, (And by the Way, you Know what Authority will have *Jonah's* Fish to be a *Shark*;)[19] A *Shark*; which is not only destitute of *Teeth*, a *Shark* wholly *Toothless*; But also has no *Bone* at all in His Body, all the *Bone* he has, is only a *Pristle*.[20] He is Usually thirty foot Long; But our People Catch him only for the Sake of his *Liver*; which alone Usually Yeelds at least a Tun of *Oyle* unto them.

I am ashamed that my Letter *Smells* no more of it; But I have not wholly *Lost my Oil*, if it in any Measure add unto the Testimonies, how much I am;

Jun. 5. 1723. Sir,

Sincerely Studious to Entertain

and Gratify You.

Cotton Mather.

[18] MS. draft, as the copy, reads: "... New to me untill very lately," so this was a change Mather made in reviewing the copy, but forgot to include the indirect object.

[19] Possibly a "toothless" whale shark; their tiny teeth are used to crush crustaceans, not to rip up chunks of meat. Samuel Bochart's disquisition on Job's leviathan (Job 40:15, 41:1–34), in *Hierozoicon* (1663), pars 2, lib. 5, caps. 16–18, cols 770–96, provides the most authoritative interpretation of the Hebrew terms "leviathan" and "tanim," which (if not left untranslated) are associated with the crocodile, even the hippopotamus. Hiob Ludolphus (1624–1704), the German Lutheran Orientalist and grammarian of the Ethiopian language (Amharic), comes to the same conclusion in his *Iobi Ludolfi ... Ad suam Historia Æthiopicam antehac editam Commentarius* (1691), lib. 1, cap. 11, p. 86. See also Mather's gloss on Gen. 1:21 (*BA* 1:328) and Theophilius Gale's *Court of the Gentiles* (1672; second rev. ed.), part I, bk 3, ch. 8, p. 86 (§4), which reiterates the ancient tradition of Hercules (Herakles), who is swallowed up by a dogfish. To Mather and his peers, this version is nothing but the story of Jonah and the whale in Phoenician guise. Bochart, *Hierozoicon* (pars 2, lib. 5, col. 743, lines 50ff) cites Lycophron and others that Hercules (like Jonah) was swallowed by a dogfish "canis carcarias" sent by Neptune, and subsequently vomited out.

[20] Scribal error on RSL copy. Mather's MHS copy reads "Gristle"; i.e., cartilage. See also Mather's commentary on Jonah 1:17 (*BA*, vol. 6, forthcoming).

7.5. MYSTERIOUS RAINS.[I]

June 6, 1723.

To Dr. James Jurin.

Sir,

While we are Admiring the Works of the Glorious One who is the *Father of the Rain*,[2] in exhaling from the *Salt Sea*, by a *Natural Distillation*, a *Water fresh* to such[3] a degree as hardly any *Artificial Distillation* of ours has yet Effected it; and in the *Equal Dispersion* of the Clouds to Accommodate all Parts of the Earth, Except when Heaven will Punish our Crimes with a Drought; and the *Gradual falling* of the Drops from the Clouds to refresh us, when *Streams* would ruin us; The *Strange Things* that Sometimes fall in, and with the *Rain*, Afford a very Particular Matter of Admiration.

Concerning *Frogs* Generated in the *Clouds* there has been a great Noise made, by the Assertions of what they call *Equivocal Generation*.[4] But among

1 Read to the members of the RS (December 5, 1723). See Jurin's Letter to Cotton Mather (December 16, 1723), Appendix 4.

2 Job 38:28.

3 Mather deletes "that" and inserts "such."

4 Mather employs the same material, in greater detail, in *CP* (1721), Essay 27, pp. 145–47. The equivocal generation of frogs – like that of insects – engaged numerous natural philosophers since ancient times; confusion arose in part because the terms "equivocal" and "spontaneous" generation were frequently employed interchangeably. Those following Aristotle (*History of Animals* 5.1) argued that insects and other smaller creatures like crustaceans, eels, and small fish, were the result of spontaneous generation from *organic* matter (putrefying plants, mud, or slime) *without* procreation of the same species. Thus Aristotelians generally embraced René Descartes's philosophy of matter in motion, for it seemed to confirm the ancient philosophy underlying abiogenesis, and by implication, creation *de novo*. By the end of the seventeenth century, however, this sacrosanct belief had become more than suspect, especially when the Italian natural philosopher, biologist, and physician Francesco Redi (1626–97), in his *Esperienze Intorno alla Generazione Degl' Insetti* (1668), seemed to prove

the many things *Poured in* for the *Confutation* of the very Unphilosophical, and Unreasonable Speculation I cannot omitt a Passage of[5] Our Industrious, and Ingenious *Ray*, who Sais; *He* that *can Swallow the raining of Frogs, has made a fair Step, towards the Beleeving, that it may rain Calves Also; For we read, that one fell out of the Clouds in* Avicens *time.*[6] Now that *Frogs* do Sometimes fall in, and with the *Rain*, We have so many Incontestible Witnesses, that I am Under a Necessity of Beleeving it; But then, who Can tell how long the Seed of[7] Animals Carried up with Vapours by the Heat of the Sun into the Clouds, may Continue there, and have their *Generation* Carried on? And who can tell, what force the *Winds* may have, to hurry up very Heterogenous things into the Clouds, and waft them away to Considerable Distances?[8]

---

through experimentation that *no* worms, flies, or parasites could spring from decomposing matter if the substance were *covered* in hermetically sealed jars. However, those which came from the meat, fish, or putrefying plants in *uncovered* jars sprang from the larvae or eggs generated by flies. The debate is nicely caught in the review and summary of Redi's experiment in "An Account of Three Books. I. Ezperienze Intorno alla Generatione Degl'Insetti, fatte da Francisco Redi ..." *PT* 5.57 (March 25 1670): 1175–76. See also Anthony Leeuwenhoeck, "An Abstract of a Letter ... about the Generation by an Animalcule of the Male seed. Animals in the seed of a Frog," *PT* 13.152 (October 20, 1683): 347–55; Richard Waller's "Some Observations made on the Spawn of Frogs, and of the Production of Todpoles [*sic*] therein," *PT* 17.194 (September 30, 1691): 523–24; George Garden, "A Discourse concerning the Modern Theory of Generation," *PT* 17.192 (February 28, 1691): 474–83. Valuable historical background information is provided in E.B. Gasking (1967); J. Farley (1977); and D. Lehoux (2017). The terminological confusion between "spontaneous" and "equivocal" generation is addressed in Peter McLaughlin (2005).

5 Mather inserts the preposition "of."

6 John Ray, F.R.S. (1627–1705), English divine and naturalist, one of the earliest "physico-theologians" in England, who published works on botany, zoology, and natural theology. Among his most popular works are *The Wisdom of God manifested in the Works of the Creation* (1691), which went through numerous editions and reprints in the 18th century; his *Three Physico-Theological Discourses* (1713); and his *Philosophical Letters* (1718). In the present context, Mather refers to Ray's *Wisdom of God*, third edition (1701), Part II, pp. 319–20. Significantly, Ray dismisses the "many and great Authors" who testify to the "raining of Frogs and their Generation in the Clouds." "Avicens time"; i.e., Avicenna, aka. Ibn Sīnā in Abu Alī al-Ḥusayn ibn Abd Allāh ibn Sīnā (10th–11th centuries CE).

7 Mather inserts the preposition "of."

8 Clearly, Mather seems caught between the ancient concepts of spontaneous generation and equivocal generation of insects from the seed of parents. See

Tis an indisputable Matter of Fact, that in the Town of *Rutland* among us,[9] the last Autumn, while Some honest Carpenters were at Work upon the roof of the New Church then Building in that new Plantation there fell down as from the Clouds on the Roof of the Church, a Considerable *Shower of Barley*, Perhaps near a Peck; (a Quantity of which, I have at this time in my hands)[10] And this at a Time, when if the *Barley* Came from the Earth, it must be Brought by the *Wind* from a Town of Some distance from them.

It is likewise Affirmed by Some of the first Planters in *Rutland*, who are People of Unquæstionable Veracity, that before any *English Grain* was raised among them, they found a Sheaf of *Wheat* in the Woods at about four Miles distance from that Village, and more than Six Miles distance from any other, So that they Conclude, it must fall *from Above*.

Showres of Blood, have Produced much Consternation, among the more fancyfull Regarders of *Prodigies*; And the Sons of *Livy* have made them a Notable Article in their Historys.[11] But they have lost their Terror, Since we have upon Enquiry litt[12] upon Such Easy and Likely Accounts of them. However for *Showres of Grain* we must enquire after Another Original; which

especially his commentary on Exod. 16:33 (*BA* 2:45–46, 248–51; and *CP* (1720), Essay 27, pp. 145–47 and his views on epigenesis vs. preformationism (*BA* 1:103; 595–99).

9 Rutland, in central Massachusetts, was settled in 1713 and incorporated in 1722.

10 Mather inserted the parenthetical phrase in the left margin, indicating its location with carets.

11 The Roman historian Titus Livius, aka. Livy (BCE 59–17 CE) authored his famous *Ab Urbe Condita*, the history of Rome from the city's legendary past, its founding in 753 BCE, to the reign of Emperor Caesar Augustus (BCE 63–14 CE). Prodigies – portents of seemingly preternatural events sent by the gods – were of particular interest to ancient and modern historians who tried to discern the will of the gods. For instance, at the defeat of the Sabines (672–640 BCE), Livy reports in his *Ab Urbe Condita* (1.31.1–5), Roman King Tullus Hostilius received news that a shower of stones had been witnessed on Mount Alban. Envoys sent to confirm this prodigy returned with accounts of pebbles (like hailstones) raining down from heaven and piling up in drifts even as a loud voice from the mountain grove commanded Albans and Romans alike to offer sacrifices to the gods lest they be punished for neglecting their ancient rites. In short, such prodigious occurrences in nature as bloody rains, monstrous births, fiery comets, and Parhelia were useful devices for prophets, soothsayers, and historians alike to discern the wrath or favor of the gods – at least until the time when Baconian science focused on the laws of nature governing natural and seemingly supernatural events, thereby demystifying marvelous phenomena in nature. See especially John Spencer's *A Discourse concerning Prodigies* (1663, 1665).

12 Mather deleted the copyist's "little" and inserted "litt."

we may do, without looking for the *Windows of Heaven* which the *Samarian Lord* Sometimes talked of.[13]

Showres of *Locusts*, which *Cluverius*,[14] and others tell us of are Easily Accounted for. But not so Easily, the Showres of *Ducks* and *Geese* falling in *Croatia*, whereof *Leon Clavius*[15] was an Eye Witness, and wherewith many People were Nourished. However very Lately, and in a very Tempestuous *Rain*, there fell in half a Dozen Places of Our *Newton* – a Place about Seven Miles Distance from the Sea Side, Several *Sea fowl*; (as their Claws determined them) So *Young* that they were not able to fly, and of a Shape so Unknown to the People, that I do not understand they Yett Know, what Name to Putt upon them. Their *Early Death* Denied our Sea-Side Fowlers an Opportunity to Examine them.

13 A reference to the account in 2 Kings 7, in which the prophet Elisha predicts to the King of Samaria, who was being besieged by the Syrians, that the following day "a measure of fine flour [would] be sold for a shekel, and two measures of barley for a shekel, in the gate of Samaria" – in other words, that the siege would be lifted. The king's assistant doubted, saying, "Behold, if the Lord would make windows in heaven, might this thing be?" In the night, God routed the Syrians and Elisha's prophecy came true.

14 Mather's cryptic reference may either be to the German geographer and historian Philippus Cluverius (1580–1622), or to his contemporary, the German theologian and historian Johannes Cluverus (1593–1633). Philippus Cluverius, aka. Philipp Clüver (1580–1622), studied classical history under Joseph Scaliger at the University of Leiden, where he subsequently worked as a librarian. Cluverius is remembered as a leading figure in the revival of interest in the geography of antiquity. His *Introductionis in Universam Geographiam Libri VI* (1624) is among his most noteworthy publications. Johannes Cluverus, aka. Johannes Clüver, served as a Lutheran pastor in Marne, was appointed professor of theology at the Danish Academia Sorana, and published his oft-reprinted *Historiarum Totius Mundi Epitome* (1631). Cluverus' *Historiarum* and his posthumously published *Diluculum Apocalypticum Seu Commentarius Posthumus in Apocalypsin* (1647) engaged a number of intellectuals in Europe, most notably the English theologian Matthew Poole and the philosopher Thomas Hobbes. See P. Springborg (1996).

15 Johannes Leunclavius, aka. Hans Lewenklaw, Löwenklau (c. 1541–94), German historian and orientalist, whose *Historiae Musulmanae Turcorum* (1591) was the standard handbook on all things pertaining to the Ottoman Empire. It was deemed dangerous enough by the Roman Catholic Church to be placed on the *Index Librorum Prohibitorum et Expurgandorum Novissimus* (1657), lib. 1, p. 632. In the present context, Mather refers to the swarms of locusts and geese descending on Croatia in December of 1587, as recorded in Leunclavius's *Annales Sultanorum Othomandiarum, A Turcis sua Lingua scripti* (1588), p. 180 (lines 15–29). On Leunclavius, see also Mather's commentary in *BA* (2:932; 10:534).

I am also well assured, That in our Town of *Lyn*,[16] a little while ago, they beheld a *Black Snake*, between five and Six foot-long, fall from the Clouds, in a Violent Showre of *Rain*. The *head* of the Wretch Coming foremost, So Struck upon a *Rock*, as to Stun him; on which the Spectators went, and Killed him.

I Conclude, with wishing you Plentifull *Showres of Blessings* and the favours of Him, who *Comes down like Rain upon the Mowen Grass, and as the Showres that Water the Earth*.[17] And Am

Sir,

Jun. 6. 1723.

Yours in the most

Inviolable Friendship.

Cotton Mather.

[16] Lynn, in northeastern Massachusetts. See A. Lewis and J.R. Newhalt (1865) relate numerous encounters with poisonous snakes in the area.

[17] Ps. 72:6; Ezek. 34:26.

7.6. THE LAND SAIL'D UPON.

June 7, 1723.

To Dr. James Jurin.[1]

Sir,

Among the Various Objects upon which the Witts of Men have been Exercised, You are Sensible, that *Engines* for the Use of Mankind, and *Mechanic*, and *Handicraft* Instruments, have been Considerable, and that as early as before the *flood*, they began upon them.

What *Inventions* have there been, for facilitating, and Meliorating Our *Manufactures*? With what *Inventions* have we releeved our *Ploughs*, our *Mills*, and our *Maltkilns*, or at least offered Releefs unto them? Our *floating Machines*, and our *Diving Machines*, What useful *Inventions* have they been, or might they have been?

If the Common *Adversary* of Mankind, had not Inspired People, with a Disposition rather to *Ridicule*, than to *Cultivate*, well projected *Inventions* for the good of the World, we should before now have Seen the World in Much better Circumstances than it is.[2]

We try for *Machines* to render the *Wind* as well as the *Water* Serviceable to Us; and extend our *Empire* into all the Elements.

To *Sail by Land*, was a Proposal Published by that Eminent Philosopher, in whose Lodging the Illustrious *Royal Society* had it's *first Conception*.[3]

1 Docketed: "Recd Dec. 5. 1723. / Enter'd" (presumably by Jurin). See Jurin's Letter to Cotton Mather (December 16 1723), Appendix 9. The title and name of addressee, minor corrections, and date and signature are Mather's.

2 An allusion to the half-satirical *Essay upon Projects* (1697), by English writer, novelist, and satirist Daniel Defoe (1660–1731).

3 Gresham College (London). John Wilkins (1614–72), bishop of Chester (1668–72), and one of the Gresham College "Group of '45," in which the Royal Society had its origins. Mather refers to Wilkins' popular *Mathematical Magick. Or, The Wonders That may be performed by Mechanicall Geometry* (1648), lib. 2, cap. 2, pp. 154–62, which describes the dynamics of endowing a chariot with sails and

Figure 17. Sketch of a Land-Boat. In John Wilkins, *Mathematical Magic* (London, 1648), bk 2, ch. 2, p. 158. Courtesy of Early-English Books, 1641–1700.

And as he was a *Bishop*, So you will allow me to make you Some report, Concerning a *Minister* in these Parts of the World, who apprehends, that he has Improved that Invention to Such Advantage, as to be worth a Little hearkning to.

There Lives at this time in *Freehold* of *new Jersey* a Person whose Name is Mr. *Joseph Morgan*.[4] He is a *Minister* of the Gospel, an Industrious, and Ingenuous, and a very Religious Man; & one of Such a Character every way, that he will deserve an hearing from us. I am willing that he may tell his own Story, and therefore I will only Transcribe a few Passages of his Last letter to me.

> I have made a Small Tryal of my *Engine* to *Sail upon the Land*. One difficulty I wanted to be Conquered, which was *Sailing against the Wind*; For I can easily make it sail with any Wind a Ship can Sail with; and it will Sail *twice as Swift* as a Ship upon the Water; and Can bear *four times the Sail* (if need be) which a Ship rolling on the Water of the Same Burden can bear; I cannot but think (it being a thing so Easy and Plain) that it would have been long in Use in Clear Countries, if it were not for Some Difficulties; *All which I have now fully removed*. First, The Danger of *Over setting* with high Sailes; Secondly; The Impossibility of *Axle trees* that will be Strong enough to bear a Load that will quit the Cost; Thirdly, The Difficulty of *Steering*; Fourthly, the Danger in going *down hills*; Fifthly, The Impossibility of *Sailing against the Wind*; because we Cannot have roads wide enough to *turn it against the Wind*, as at Sea. Any one of these *five things*, would render the Proposal of no Worth; But I have demonstratively *removed them every one*; And

---

wheels and sailing on long, straight roads. Wilkins does acknowledge several of his predecessors and contemporaries – Giovanni Botero, Hugo Grotius, Johannes Walchius, Pierre Gassendi, Jodocus Hondius – who either witnessed or described the mechanics of the contraption and its experimentation in Holland, Spain, and Italy.

4 Joseph Morgan (1671–c. 1742), native of Norwich (Preston), Conn., Yale class of 1702, ministered in several churches in Connecticut and New York before being settled as the pastor of the Presbyterian church of Freehold, N.J., from 1709 to 1728, when he resigned after the Synod charged him with being an astrologer, a dancer, and a tippler. He was absolved and went on, though not without further controversy, to serve other churches in New Jersey until his death. Beside several sermons, he published *The History of the Kingdom of Basaruah* (1715), based on John Bunyan's *Pilgrim's Progress* (1678). On Morgan, see W.J. Bell, Jr. (1951). According to R.P. Stearns's *Science* (1970), pp. 410–11, and 497–99, Morgan was also a man of science who loved to tinker and experiment, some of which he described in his letters to the RSL, which never published them.

Particularly that of *Sailing directly in the Eye of the Wind*; I caused one to be made twelve foot Long, and eight foot wide; And twenty People with {me} Saw it Sail about a foot Pace (which was the Highest I expected) So right against the Wind, as the Skilfullest of them could Steer it. But they said, it would be of no Use in Our Country, which is so full of Wood. But a Gentleman of our Governours Council, who has lived in *England*, Earnestly advised me to Compleat the Work, which was but Partly done when tried; And to go to *England* with it; For he said, the Carting from Several Inland shires, to *London*, is Exceeding Costly, which makes the Produce very dear, the Way being So long, But yet the Countrey clear & suitable.

It will *Steer with Ease* to an Inch, and turn Clear round in little more than it's length.

It will go in an *Uneven Ground*, where a Cart can go, and go *down an Hill* more Safely, and (if need be) more Slowly.

It may by drawing a few Ropes be made to go with *either End before*.

If it be as large as any Ship, no man need go aloft to Order the Sails.

Five or *Six* men, Can Manage One that will Carry *two hundred Cart loads*, and Can sail as Comfortably in the *Rain* as in *fair Weather*.

Thô twould go but Slowly directly *against the Wind*, Yet *with other Winds* it would in *One Day* (if the Wind be fresh) over take a *Cart* that Sett out a Week before it.

The Charge of Making it will be farr less, than to make so many *Carts* as will Carry its Load, and will Save the *Time* & *Charge* of a Man, and Team to Each Cart or Waggon.

I am afraid of Leaving my *Pastoral Charge*, for any *worldly* gain; and Chuse rather to Acquaint the *Royal Society* with it, in this way of desiring *you* to Transmitt the Matter to them.

If the *Royal Society* will by writing Encourage me to do it, I will (God willing) Send them the whole *Draught* of it. But I fear, that except I be Present at the Making the *first*, Something may fail and a Small Mistake may Spoil it; and a Balk at first may put it quite out of Credit.[5]

[5] In his letter to Cotton Mather (December 16, 1723), James Jurin asked Mather to convey to Joseph Morgan, that "the Royal Society never concern themselves in any undertaking of this kind, how usefull & practicable soever they may seem any otherwise than in making them known to the World, when brought to perfection but that they wish success to that Gentleman, & shall be glad to hear of his having fully accomplished his design, in which case there is no doubt, but he will meet with encouragement" (Appendix 9).

Having tendred you this Transcript from the Letter of the Gentleman, in Compliance with his Orders to me, it is not for me to Pass any Censures of Mine, or to Declare how far I am in the Order of *St. Thomas*, upon it.[6]

You See the Tender that is made you, and I have Nothing to add, but my Wishes that the World may grow more, and more happy in Discoveries of what may be the Welfare of Mankind, and that both *You* and *it*, may be all that *you* desire, which I am Sure will be to be *filled with* the *Blessings of Goodness*.

I am Sir,

Jun. 7. 1723.

{Your} Sincere Friend

And Humble Servant,

Cotton Mather.

6 The Knights of St. Thomas was an ancient military order dating to the Crusades, composed of Englishmen, whose members swore to a life of poverty, chastity, and obedience.

7.7. *SUBTERRANEA.*[I]

June 8, 1723.

To Dr. James Jurin.

Sir.

Had We any *Kirkers* in our Countrey, there Would Soon be Another Volumn of, A SUBTERRANEOUS WORLD, in Our Libraries.[2] Our Countrey alone would afford a Number of Materials; and Verily, Some of them Very Mysterious Ones. But the Family of the *Cicero's* is not found Yett among them, that are *Transplanted* into *America.*[3]

I have not Yett made any Progress, in My Collection of those Curiosities. But Yett in the Civil Law, I find the Terms of, *An Acceptilation,*[4] which is Used when Some *Lesser Thing* is *Accepted,* instead of a *Greater Debt.* I will Confess, that I owe You Some Larger Communications; but at present I Entreat an *Acceptation* of the Little, I have to offer You; And I will pray Your Thoughts on one or Two

1 The title, address, concluding date and signature are in Mather's hand. Read to the members of the RS (December 12, 1723). See Jurin's Letter to Cotton Mather (December 16, 1723), Appendix 9.

2 Mather refers to Athanasius Kircher (1602–80), a German Jesuit, polymath, and author of numerous works on geology, medicine, Egyptology, Sinology, and comparative religion. Kircher's lavishly illustrated *Mundus Subterraneus, In XII Libros digestus* (1665) covered a wide variety of topics on things subterranean – petrified bones, grottos, lakes, and minerals – and along with Kircher's *Linguæ Ægyptiaca Restituta* (1643), *Oedipus Ægyptiacus* (1652–55), and *Arca Noë, in tres libros digesta* (1675), was of great interest to Mather and quoted and extracted numerous times in his commentary on the Bible, *BA* (2010–26). A copy of Kircher's work on pestilence, *Scrutinium Physico-Medicum Contagiosæ Luis, quae dicitur Pestis* (1659), bearing Increase Mather's signature, was in the Mather family library (AAS).

3 Cicero, 1st-century BCE Roman politician and lawyer, considered one of the greatest orators of the age. His writings, particularly his letters, are known for the great detail he provided about his contemporaries and the events of his time.

4 A legal term in Roman civil law, signifying that "a claim of creditor or obligee has been satisfied with or without payment" (*Merriam-Webster Dictionary*). Or, to put it simply, "to accept the will for the deed." See 2 Cor. 8:12.

odd *Phænomena*, Lately Seen in our *Subterraneous World*, whereof I have had within these few Hours a Good Assurance.[5]

In a Town of Ours, Called *Springfield*, both of them Were mett withal.

The first is this. Digging for a Well about Eighty Rod from the Great River,[6] and about Seventeen foot under Ground, They found a *Limb of a Tree*, with Evident Marks Left upon it, of the *Axe*, which it had been hew'd withal.[7]

The Next is Like to it. About Thirty Rod from the River, digging for a Well, they Twenty foot under Ground found a Quantity of *Chips*; with a Parcel of those Fruits they Call by the Name of *Butter-Nutts*.[8]

I Confess, These are not Such Considerable Objects, as the *Ship*, which You tell us, was found a Mighty Depth under Ground, Somewhere, as I remember, in *Switzerland*.[9] But Yett, *When*, and *How*, were these Little Things Convey'd into the *Subterraneous Regions*?

That Nothing of *You*, but Your Sagacious and Penetrating *Searches*, may for Yett a Good While be *there*; And that *then*, for the Better Part – *Serus in Cœlum Redeas*,[10] is the Wish of,

Sir,

Jun. 8. 1723.

A Friend, who wishes

You all that is Good.

Cotton Mather.[11]

5 In his letter (19 d. 7 m. [September] 1712) to Samuel Penhallow (1665–1726), speaker of the General Assembly, treasurer, justice in the superior court, and later chief justice of New Hampshire under Governor Samuel Shute, Mather requested information about a tree found "*at an Unaccountable Depth under ground*" while the people were digging a well "*some years ago*." Specifically, Mather wanted to know "what the Depth; what the Trees; which way they lay; how far from the Sea," and promised to acknowledge Penhallow in his communications with the Royal Society (*D* 2:175–76).

6 I.e., the Connecticut River.

7 Mather could have learned a great deal about how trees become submerged, in the anonymous *The Late dreadful and most admired Calamity of a Parcel of Land, and many great Oaks, and other Trees sunk, many yards under ground* (1657). A similar discovery (along with Roman artifacts) was reported in "Part of a Letter from the Reverend Mr. Abraham de la Pryme to the Publisher, concerning Trees found under Ground in Hatfield Chace," *PT* 22.275 (October 31, 1701), pp. 980–92.

8 Lest such discoveries appear to be too pedestrian for the *Philosophical Transactions*, John Woodward expressed a particular interest in such evidence, especially fossils (Appendix 8).

9 A search in *PT* and other venues for a mention of this ship has not yielded any results.

10 "That you may return late to heaven." Horace (*Odes* 1.2.45).

11 The date and signature are in Mather's hand.

7.8. A SINGULAR CASE.[1]

June 10, 1723.

To James Jurin.

Sir.

You may Expect one day from Your *American* Servants, a Small Cargo of Communications, relating to that noble Profession, in which there are So Many Continually acknowledging You by the Title of, *A Benefactor*.

But how *Lame a Messenger*, do I now Employ, to Acquaint You with my Purposes?

I have Lately Putt into my Hands a Relation of as Miserable a Patient, as most that in Your Extended Practice, You may have mett withal.

Orosius Entituled his History, *De Miseria Hominum*.[2] And I Remember, *Bongars* Observes, how well the Title Suits all *History* in General; *Etenim quid aliud Historia*?[3]

Your Charitable and Honorable Employment Leads You Every day to the Monuments and Materials for Such an *History*. But not Every day to Such

1 Title and addressee, and the concluding date and signature are Mather's. The scribal copy is endorsed after concluding date: "read Dec: 12. 23. / Enterd," the date when Jurin read and recorded it in the RSL Letter Book. See also Jurin's Letter to Cotton Mather (December 16, 1723), Appendix 9.

2 Paulus Orosius (385–420 CE), Roman priest, theologian, and historian, author of *Historiae Adversus Paganos Libri VII*, lib. 1, cap. 1 [*PL* 031. 0669C]. Mather refers to the first chapter, which in some editions is titled *De miseria hominum per pecca tum ab initio*, "The Misery of Man through Sin from the Beginning."

3 "Truly, what else is history?" Jacques Bongars (1554–1612), French scholar and diplomat in the service of King Henry IV of France. Bongars authored *Gesta Dei Per Francos, Sine Orientalium Expeditionum, et Regni Francorum Hierosolimitani Historia. Tomus Primus* (1611), a collection of histories of the crusades. Mather's Latin quote appears in the address to the Reader, "Lectori Historiarum Studiosos" (sign. *d* ^{2v}), which refers to Paulus Orosius' *Historiae* as its source. However, the Latin citation is most likely at second hand, for it also appears in *The Dictionary Historical and Critical of Mr. Peter Bayle* (1647–1706) (orig. French ed., 1697; English translation, London, 1709; 2nd ed., 4 vols, London, 1734–38), 4:424, by Pierre des Maizeaux (c. 1666–1745), Huguenot exile in London, and translator of Pierre Bayle's famous *Dictionnaire Historique et Critique*. 2 vol. (1697–1709). This title is not in the Mather family library.

an One as This; which may have in it a Passage or Two worthy of Your Contemplation.

I shall Give it You in the Words of the Patient himself in his Written Address unto Me.

> We Sailed from *Boston*, the nineteenth day of *December*, bound for *Jamaica*, where I design'd to work at my Trade, which is that of a *Shipwright*. But How Wonderfully has it pleased GOD to disable me! The Same Night that we Sailed, I went unto my Chest, in *Charity* to fetch a Dram for[4] a Poor Man that lay Sick and had none of his Own. I went with a Candle in a Lanthorn; but Some how, My Powder took fire; and I was very much burnt; and for Want of Proper Means I Suffered very much. I was blind for fifteen days, and the Fire So Eat into my Neck, that my Wind-Pipe & my Collar Bone Lay Entirely bare. In Thirty days We arrived at *Jamaica*; where at Length I Recieved a Cure. I went then to work; but in Four Days found Myself too Weak to Proceed. I was Persuaded hereupon to Go Carpenter's-Mate in a Privateer. We were Six Months on the Cruise, and I Gott Nothing, but Languishing Sickness.
>
> I lay Sick of the Flux at *Jamaica*, Sixteen Weeks; and Spent what little I had, and ran Twenty Pounds in Debt. My Landlady turned me out of Doors; and having no Place to reside or to repose in, I betook myself to Another Privateer, in Some Hopes, that tho' I were very Sick, Yett I might recover; and in the Like Hopes they received me. But I Grew worse, and my Sickness fell into my Leg, and I also became Light headed. So, they Going to the Windward of the Land, for Watering, and being Loth to be farther troubled with me, they Laid me, & Left me ashore, on the Bare Ground; and Left by my Side a Little Bread, & a Peice of Meat; hoping that Somebody Coming Quickly to water there, might relieve me. I Can Give No Account of the Small Provision they Left me; But It was thought the Negroes might find it and Steal it. I had No Bed but the Land, and No Covering but the Heavens; and I was unable to Give an Account how Long I had been there, unto them whom the Good Providence of GOD Sent at length to releeve me. But a Sloop Came to water there; and from *them*, who knew the Time of the Privateers Sailing, it was found I was *Eight Days and Nights* in that Condition; and I took *in all this Time No Manner of Sustenance*. They found me Speechless and Senseless; and brought me to *Port-Royal*; where they Putt me ashore without Divulging of my Circumstances, for fear of bringing Charges on themselves. Here I lay all Night in the Streets for Some Time; till the Church Wardens were Informed of my Condition; and they Putt me into The Hospital, where I had Some Care taken of me. Here I Grew worse;

4 Mather inserts "fetch a Dram for."

and had a Very Bad Leg, which broke and many Quarts Run from it; and I was in Extremity of Misery. I had also a Sore bred in my Mouth, which was a Great Addition to My Misery. It is unaccountable, what I underwent! My Jawbone perished and broke away; I have now *Two* Considerable Peices of it in My Hand; with Six of my Teeth.

But thro' the Goodness of GOD, I Gott a Cure of my Leg and Face in Seven Months time; after which a Kind Countreyman of Mine took me aboard, and brought me to *New-England*, where I Now Ly Languishing under a Rheumatism; which obliges me to Entreat, that *Dr. Mather* would Please to take Some Cognisance of my Condition.

Joseph Skeath.[5]

If there be Such objects elsewhere to be mett withal, I wish them the Fælicity of Coming into the Skilful hands of my *Dr. Jurin*, and wish Him the Consolation of doing them all the Good, which his Generous Goodness may dispose him to do unto them;

And am,

Sir,

Jun. 10. 1723.

Your most Affectionate Friend

and Servant,

Cotton Mather.

[5] A Joseph Skeath married Hannah Davise on April 18 1692, in Boston. W.M. Clemens (1926), p. 74.

7.9. LETTER OF INTRODUCTION FOR ISAAC GREENWOOD.

June 10, 1723.

To Dr. James Jurin.[1]

Sir,

A few Days ago, I wrote you a Large Account, of the Success which the *Small Pox Inoculated* has had in these Parts of the World.[2]

In hopes of its coming safe, there Now comes a small Packet of some Communications,[3] which I then promised you, tho' not of all that I then Intended you.

The Bearer, (Mr. *Isaac Greenwood*)[4] is an Ingenious Young Gentleman, & a Graduate in our Colledge; who proposes to make some Addition to

1 This letter is in Mather's hand.

2 Curiosa Letter to James Jurin (May 21, 1723), our CA 7.1 "The Case of the Small-pox Inoculated, further cleared."

3 This packet is most likely the bundle of Mather's letters to James Jurin for May 21–June 10, 1723 (Seventh Series). Given the uncertainty of mail delivery across the Atlantic (including the loss of several of his previous letters to the RSL, Series 5 and 6), Mather made a good choice in sending Isaac Greenwood (1702–45) as his protégé. Mather's letter of recommendation was productive in yet another way: Greenwood, who studied in London under the natural philosopher and clergyman John Theophilus Desaguliers, F.R.S. (1683–1744), was introduced to Thomas Hollis, the great benefactor of Harvard.

4 Greenwood, the recipient of Thomas Hollis' (1659–1731) largesse to become the first Hollisian Professor of Mathematics and Natural Philosophy at Harvard, and published *Arithmetick Vulgar and Decimal with the Application thereof to A Variety of Cases in Trade, and Commerce* (1729). For Greenwood as the first science professor at Harvard, see D.C. Leonard (1981). During the Boston smallpox inoculation crisis, Greenwood joined the fray in support of Mather and his fellow advocates of inoculation in his satire *A Friendly Debate; or, A Dialogue Between Rusticus and Academicus; and Sawny & Mundungus, Two Eminent Physicians, About some of their Late Performances* (1722).

his Accomplishments, by visiting *Europe*, and particularly by visiting of Dr. *Jurin*.[5]

I recommend him to your Favour and Advice; and he having been One of our *Inoculates*,[6] I offer him to your Examination; because I have heard, the *Inoculation* is an Article on which you are very particularly Inquisitive. And considering the Importance of it, for the Saving of hundreds of thousands of precious Lives, I do not wonder at it, if a Person of your Wisdome & Goodness be so.

With humble Supplications to Heaven, for your Prosperity, and especially, that you may have the Fælicity of being so useful as you desire to be, I am,

Sir, Your affectionate

Boston. N.E. Friend & Servant.

Jun. 10. 1723. Co. Mather.

Dr. James Jurin.

5 See James Jurin's Letter to Cotton Mather (December 16, 1723), Appendix 9.

6 One who underwent the inoculation procedure. Mather thus delivers to Jurin's candid observation a "specimen" survivor of Boston's inoculation practices.

7.10. LETTER OF INTRODUCTION FOR JOHN PERKINS.

August 3, 1723.

To Dr. James Jurin.[1]

Syr,

Twice in the Summer now Spending, I have given you the Trouble of some *American* Packetts. In the former of them, I tendred you what the *Inoculation of the Small-Pox*[2] ha's produced. In the latter, there were Seven Letters of, *Curiosa Americana*, Succeeding the more than Seven times Seven transmitted unto some other Correspondents; from Some of whom I have not so much as understood, whether they have received them.[3] I hope these are come unto your hands.

I could not omitt this Opportunity of addressing you once more, before I know any thing of the Reception of the præceding Addresses; Because there now waits upon you, a Gentleman whose Name is Mr. *John Perkins*.[4] He is

1 This letter is in Mather's hand.

2 Letter to Dr. James Jurin (May 21, 1723), our CA 7.1 "The Case of the Small-pox Inoculated, further cleared."

3 Possibly, Mather alludes to the Fifth Series of his Curiosa Letters to John Chamberlayne (February 1720), or those of Mather's Sixth Series to John Woodward (1721). In both cases, most of the shipments appear to have perished in one way or the other. See G.L. Kittredge, "Lost Works," pp. 418–79.

4 Mather's relationship with John Perkins, a medical practitioner in his neighborhood, went back for more than a decade. For instance, in his diary for March 22, 1721, Mather jotted, "G.[ood] D.[one]. I will take my religious Neighbour Dr. *Perkins*, into the Number of my Relatives; I will study his Welfare; I will promote his Practice; I will do for him all that I can do for the dearest Brother!" At the height of the smallpox crises (June 1721), Mather pledges, "Being entirely satisfied in the Skill of my pious Neighbour Dr. *Perkins*, I will take all Occasions to recommend him unto the Use of the Neighbourhood, in the Calamity of the *Small-pox* coming upon us: As well as make my daily Prayers for him." Six months later (December 13), Mather adds, "I am using further Methods to promote the Credit and Practice of the beloved Physician in my Neighbourhood; whom I would embrace as a beloved Relative" (*D* 2:609, 627, 664). Thus Perkins may indeed have joined

a Physician of great Learning & Wisdome & Success; and Inferiour to none that we have in this Countrey. But he is a Gentleman, whose humble Modesty is æqual to his other very bright Accomplishments. His business on your Side of the Atlantic (whither we can ill spare him) is, to possess himself of a considerable Interest there falling to him. While he is pursuing of this, I have directed him to pay you a Visit, that so you may enquire of him, every thing about these Parts of the World, that your generous Curiosity may lead you to. I purpose, (if GOD please) a Continuation of Communication to you, when I shall understand that those I have already offered, have not missed of what I have hoped for.

But at this time, I shall content myself with tendring to your candid Acceptance, a little Book entituled, India Christiana.[5] Upon which I would gladly be informed, whether any further Accounts of our *Aboriginal Natives*,[6] and their Customes, may be any Acceptable Ingredients of my Communications, if you order me to go on in making them.

I commend your Person & Practice, to the Blessing of the Great GOD our SAVIOUR;

And am,

Syr,

Your most affectionate

Boston. N. England. Friend & Servant,

Aug. 3. 1723. Co. Mather.

Dr. Jurin.

Zabdiel Boylston's endeavor to inoculate his patients, while most others of the medical establishment in Boston opposed their methods.

5 Mather's *India Christiana. A Discourse, Delivered unto the Commissioners, for the Propagating of the Gospel among the American Indians which is Accompanied with several Instruments relating to the Glorious Design of Propagating our Holy Religion, in the Eastern as well as the Western, Indies* (1721), a long letter with extracts from his correspondence with the Halle Pietist missionaries in Tranquebar, in southeastern India.

6 For Mather's previous letters on Native Americans' medicinal, cultural, or religious aspects, see his Curiosa letters to John Woodward (November 18 and 24, 1712), our CA 1.2 "On American Plants & Indian Cures"; CA 1.8 "Astronomical Observations" (letters to Waller); our CA 2.5 "The Pigeons"; and (July 6, 1716), our CA 3.5 "A Serpent Securely Handled"; and to John Petiver (September 24, 1716), our CA 3.12 "*Curiosa Botanica*."

Series Eight

8.I. CLIMATICAL INFLUENCES.[1]

October 1, 1724.

To Dr. John Woodward and Dr. James Jurin.[2]

Sirs,

It shall not be for your *Instruction* (I am not so impertinent as to imagine myself capable of doing any thing for *That*,) but for the Use of it which I shall make in the *Conclusion*, that I shall presume to remind you, of the strange *Influence* that *Climates* appear to have on the *Humours* & *Manners* & *Actions* of the People that inhabit them. What should be the Meaning of it, I beseech you, that tho' the *Nations* of some *Climates* have undergone such *Revolutions*, yea, quite *another People* have come in the Room of the *old Inhabitants*, yet the People remarkably retain the *same Qualities*, by which their Prædecessors were *distinguished*, some hundreds above a Thousand Years ago?

Every Body knows, what it was, *Græcari*,[3] a great while ago; and that the Ancient *Greeks* began with *small Cups*, at their *merry Meetings* (in which they would be *as merry as Greeks*) and then call for *large ones*, when they were half-Seas over.[4]

1 The title, addressees, the concluding date, and signature are in Mather's hand.

2 As with his subsequent letters (CA 8.4 and 8.7), Mather's missive responds to Jurin's call for accurate reports on weather phenomena in America. Jurin hoped to develop a theory on how weather impacts people and their environment. In fact, he had sent copies of a handbill to Mather, asking him to distribute them among New England's scientists with access to reliable instruments. To reach a wider audience beyond England, Jurin also issued a call for meteorological information in the *Philosophical Transactions*: "Invitatio ad Observationes Meteorologicas communi consilio instituendas," in *PT* 32.379 (October 31, 1723): 422–27. See also Appendice 10 and 11. On Jurin's interest in meteorology, see A. Rusnock (1996), pp. 27–31. On the development of meteorology as a discipline and the use of meteorological instruments, see Wolf, *History of Science ... 18th Century* (1938), vol. 1, chs. XI and XII. A useful study of environment factors and their influence on human nature is C.J. Glacken (1976), esp. part 1, ch. 2: "Airs, Waters, Places."

3 "To imitate the Greeks."

4 Homer's epics *The Odyssey* and *Iliad* are filled with scenes of libations, drinking, and merrymaking. However, the carousing banqueters in Athenaeus' *Deipnosophists* (esp. books 10–11) are probably what Mather had in mind.

But the modern Travellers tell us, That their *Posterity* are noted for keeping up their old Fashion, *To this very Day*.[5]

Tacitus, you remember, observes concerning the ancient *Germans*, that it was no *Disgrace* among them *Diem Noctemque continuare potando*; and that it was their Custome, *De jungendis Affinitatibus, de Bello denique et Pace, in Conviviis consultare*;[6] From whence a learned Antiquary, will have the present Assemblies of the Circles in the Empire to hold the Name of DIETS. Now, since the Days of *Tacitus*, how often has *Germany* shifted its People with successive Colonies of *Sueves* & *Sclaves* & *Goths* and *Huns* & *Vandals* & *Saxons*, & how many more? And yet we See, in *Germany* they are notoriously & scandalously just as they used to be, *To this very Day*.[7]

You have not forgot the Character that *Julius Cæsar* gave of the *Gauls*, above seventeen hundred Years ago; *Gallorum subita Ingenia*, that they were a quick & rash & harebrain'd People.[8] *Juvenal* also reports, they were so given to *Law-suits*, that they could supply their Neighbours with *Lawyers*.[9] And *Florus* tells how they fought; *Gallorum primus impetus*

5 Several popular travelogues can be mentioned here: Joseph Pitton de Tournefort (1656–1708), a French botanist and traveler, whose popular *Relation d'un Voyage du Levant* (1717) appeared in an English translation as *Voyage into the Levant*, 2 vols. (London, 1718). M. de la Guillatiere, aka. Guillet de Saint-George (fl. 17th century), *Les Voyageurs Français* (1675), translated as *An Account of a Late Voyage to Athens* (1676); Pierre Belon du Mans (1517–64), a French traveler, naturalist and diplomat, *Le Voyage au Levant: Les Observations de Pierre Belon du Mans de Plusieurs Singularités & Choses Mémorables, trouvées en Gréce, Turquie, Judée, Egypte, Arabie & autres pays éstranges* (1555); and Jacob Spon (1647–85), a French physician and Greek scholar, whose *Voyage D'Italie, de Dalmatie, de Grèce et du Levant, fait aux années 1675 & 1676* (1679) was published in two volumes.

6 This and the following two paragraphs and the Latin citations and references are extracted from *A Full Relation Of two Journeys* (1656), bk 2, ch. 1, p. 36, by the English theologian and historian Peter Heylyn, aka. Heylin (1599–1662). The Roman orator and historian Cornelius Tacitus (c. 56–c. 120 CE), in *De Origine et Situ Germanorum Liber* (cap. 22.2–3), reported that it was no disgrace among them "to pass days and nights without intermission [in drinking]... . In their banquets "they generally deliberate ... on family alliances, and on war and peace," in *The Germany and the Agricola of Tacitus* (1897), p. 45.

7 Heylyn, *Full Relation*, p. 36.

8 Adapted from "Gallorum subita et repentina consilia," in Julius Caesar's *De Bello Gallico Commentarius* (3.8.3): the measures of "the Gauls are sudden and hasty." However, Mather's adaptation and translation is extracted from Heylyn (p. 36).

9 Via Heylyn's *Full Relation* (p. 36), Mather refers to the litigiousness of the Gauls and (Celtic) Britons as rendered in Juvenal, *Satyra* 15 (line 111): "Gallia causidicos docuit facunda Britannos" ("Eloquent Gaul has trained the pleaders of Britain").

major quam virorum, Secundus minor quam Fœminarum.[10] Now, tho' the present *French* are of another Original, yet they have stil so much of the *Gauls* remaining in them, that it is not easy to give a truer Portraiture of them, than in the Terms of those old Historians: Those Properties remain in them, *To this very Day*.

Nearer Home! You need not be told who made that Remark upon the *Britons* more than sixteen Centuries ago, *Dum singuli pugnant, universi vincuntur.*[11] Now what is become of the *Britons*? And how many Nations have their Blood running in the Veins of, *A True born Englishman*? But then how remarkably have the *English* lost every where & undone themselves by *Intestine Divisions*? There cannot be a juster or more lasting Brand upon us than this, *Dum singuli pugnant* – They are a *divided People*; Their intestine *Quarrels & Factions* exceed what is ordinarily to be found in other Nations.

I know not whether the Observation be too *Superficial*, for you to think it worthy of your bestowing any Speculation upon it: or whether the incomparable Sir *Richard Blackmore*, will count his noble Poem on, *The Nature of Man*,[12] confirmed in it. But it will Serve to introduce, what I have observed in the Countrey from whence you are now written to.

One very observable Quality of our *Indians*, has always been this; that they have no *Family Government* among them; all *Family Discipline* is a Stranger to them; Their Children are the most Humoured, Cockered, *Indulged Things* in the World. One that has given us a Narrative of what he found among

10 Heylyn, *Full Relation* (p. 36); the Roman historian Lucius Annaeus Florus (fl. 2nd century CE), *Epitomae De Tito Livio Bellorum Omnium Annorum DCC Libri Duo* (1.20): "The first attack of the Gauls is more severe than that of men, the second less than that of women."

11 Heylyn, *Full Relation* (p. 36); Tacitus (*Agricola* 12) opined that "so long as they fight separately, the whole are conquered." Tacitus' dictum is a popular quote in many historical works and can also be found in Don Carlos Garcia's *France and Spain Natural Enemies* (1704), ch. 9, p. 108.

12 Mather had a special fondness for the poetry of Sir Richard Blackmore, F.R.S. (1654–1729), an English poet-physician whose poetic oeuvre is massive. Blackmore's *The Nature of Man, A Poem in Three Books* (1711, 1720) explores several popular 18th-century themes: Book 1 examines the passions and intellectual faculties of people and the advantage of temperate climate on civilization; book 2 holds forth on how the air and soil of European nations shaped the distinct character of their people; book 3 extrapolates on the popular theme of "translatio studii," the westward movement of civilization and learning. Although Blackmore's poetry was later ridiculed by John Dennis and in Alexander Pope's *Dunciad* (1728), bk 2, lines 351–52, Mather believed that as a poet Blackmore has "forever out-done *Homer* and *Virgil*, in two Immortal Composures" (*BA* 4:174). So much for poetasters. See also Mather's *D* (2:105, 141), *BA* (1:143, 643); and *TP* (p. 192).

them, relates, That in one of their *Wigwams* beholding a Boy behave himself refractorily towards his Father, he show'd the Man the Way to employ a *Switch* upon him; and the Man was almost scared out of his Wits, to see what an Operation the *Switch* had upon the Boy, & unto what good Order the little Tool in such an Hand soon reduced him.[13]

Now tis as observable, that tho' the first English Planters of this Countrey, had usually a *Government* & a *Discipline* in their *Families*, that had a sufficient *Severity* in it: Yet, as if the *Climate* had brought us to *Indianize*,[14] the *Relaxation* of it is now such, that it seems almost wholly laid aside; and a foolish *Indulgence* to Children is become an *Epidemical Miscarriage* of the Countrey, & like to be attended with many evil Consequences.

An other very observable Quality of our *Indians* is this; That they are intolerably *Lazy*; They hate *Work*; To keep them at *work* would be a more grievous Punishment unto them than to *scourge* them; *Death* would be more eligible to them than *Slavery*.

13 The story originates in *A Key into the Language of America: or, An help to the Language of the Natives in that part of America, called New-England* (1646), by Roger Williams (1603–83), a controversial Puritan minister, advocate of religious freedom, and founder of Providence Plantations in the Colony of Rhode Island. Mather appears to retell the tale from memory, though it is very similar to how Williams relates it: "I once came into a *house*, and requested some *water* to drinke, the *father* bid his sonne (of some 8. yeeres of age) to fetch some *water*: the *boy* refused, and would not stir; I told the *father*, that I would correct my *child*, if he should so disobey me, &c. Upon this the *father* took up a sticke, the *boy* another, and flew at his *father:* upon my perswasion, the poore *father* made him smart a little, threw down his stick, and run for *water*, and the *father* confessed the benefit of *correction*, and the evill of their too indulgent *affections*" (*A Key*, ch. 5, p. 30). I am grateful to Professor Patrick Cesarini, University of South Alabama, for identifying Mather's source.

14 Mather also intones this Jeremiad theme of declension in his *OT* (1699), appended to *DL* (1699), pp. 211–12. Espousing a contemporaneous racist prejudice that characterize Native Americans as thieves, liars, and downright evil, Mather looks askance at the rising generation of Puritans: "… They [Native Americans] are *Sluggards* to a proverb; they are for any way of Living rather than work. Thirdly, They are abominably Indulgent unto their *Children*; there is no *Family Government* among them. Will you now *Enquire*, Sirs, how far we have *Indianized* in every one, but especially, the last of these *Evil Manners*? If we find these *Indian Vices* to grow Epidemical among us, Oh! don't wonder, that our God hath been with *Indian Hatches* cutting down the *Tree*, that brings forth *Fruits* thus disagreeable to Him that Planted it" (pp. 211–12). See also Mather's *TDP* (1696), p. 11. Mather was not alone in his worries about life on the frontier and its effect on European manners. The French American diplomat and writer Hector St. John de Crèvecoeur (1735–1813) also worried that European settlers become "Indianized," *Letters from an American Farmer* (1782), Letter 12: Distresses of a Frontier Man, p. 305.

Now, I will not complain of my Countrey as if they were not generally, a sober, honest, *Industrious* People; yet I must say I wish we did in this Matter less answer our *Climate & Indianize*.

I am sure, more *Industry* would not be any Harm unto us, yea, more *Industry* would make this quickly become one of the finest Regions in the World. It is noted, That a laborious English Husbandman coming over hither & labouring as they do in *England*, presently grows rich, & outstrips the *Natives*.[15]

But for the Proof of the Article I am now upon, I will rather excuse the rest of my Countreymen, & exhibit *myself* unto You, as an Object for your Censures. For if I had not been inexcusably *Lazy* my Letters this Year to You would not have come short of the Number which in some former Years they have arisen to. My Friends indeed flatter me, as a Man of numerous & ponderous Employments, & Some too wonder at my being able to apply myself unto so many Correspondencies & other Intentions: But you will do me more Justice, if you censure me as a Tame *Indian*,[16] tainted with the vice of the *Climate*, and rebuke me for my *Idleness*. However, the Quæstion will be, Whether such *Climatical Distempers* be not *Incurable*?

And so you will be favourable to one, who tho' he *do little* for your entertainment, yet is glad & proud that he can *do something*; and hopes to *do more*, if you please to animate him with such a kind Reception of what he does, as he has formerly met withal. He is, at least,

Syrs,

Boston, N. England. Your Sincere & well meaning

Oct. 1. 1724. Servant.

Cotton Mather.

[15] Mather here sounds what would become a popular theme in American literature – the Puritan work ethic and the theme of rags to riches. See especially his biography of Sir William Phips, in *PP* (1697), and *CC* (1701) – themes when secularized were proliferated in Benjamin Franklin's "Hints for Those That Would Be Rich (1736), "Advice to a Young Tradesman" (1748), "The Way to Wealth" (1758), and by his many emulators, culminating in the 19th century in Horatio Alger's rags-to-riches dime novels.

[16] Mather's witticism is not entirely groundless. He was perhaps the first European-American, born in America, who called himself "American" (without hyphen), in his *TRR* (1691), "The Dedication" (A^{4r}) – at a time when the designation American signified a "savage Indian" and "cannibal." None less than Jonathan Swift (1667–1745), the Anglo-Irish clergyman, satirist, and political pamphleteer, needled English parliamentarians in his satire *A Modest Proposal* (1729): "I have been assured by a very knowing *American* of my acquaintance in *London*, that a young healthy Child well Nursed, is, at a year old, a most delicious nourishing and wholesome Food, whether *Stewed, Roasted, Baked, or Boyled*; and I make no doubt that it will equally serve a *Fricassee*, or a *Ragoust*" (p. 6).

8.2. AN *AMPHISBÆNA*.[I]

September 21, 1724.

To Dr. John Woodward.

Syr,[2]

An *Amphisbæna*[3] is what I am now to present you withal.

As long ago as *Lucan's* Time, it was the Song,

Est Gravis in geminum surgens caput Amphisbæna,[4]

unto which to make up a Distich, *Georgius Pictorius* in his *Lerna* κακων, long after added;

Serpens qui visu pernecat et Sibilo.[5]

1 Title, addressee, and date in Mather's hand.

2 Mather's youngest son, Samuel (1706–85), appears to be the copyist of this holograph at the MHS.

3 The two-headed amphisbaena – spawned from the dripping blood of Medusa's serpentine head – has been perennially fascinating to mythologists then and now, and depictions of the monster can be found throughout history.

4 Adapted from Lucan, *Pharsalia* (9.719): "Dread Amphisbaena with his double head (rising)" (1896).

5 Quoted in Georg Pictorius (c. 1500–69), German physician and mythographer, author of *ΠΑΝΤΟΠΩΛΙΟΝ, continens omnium fermè quadrupedum, avium, piscium, serpentum* (1563), sec. De Serpente Amphisebena, p. 41 (p. 210, in the 1568 edition). The Latin distich adapted from Lucan (with Pictorius' addition of the second Latin line) also appears in *The History of Four-footed Beasts and Serpents* (1658), 1:700, by Edward Topsel (1572–c. 1625), an Anglican clergyman, who renders Pictorius' second-hand distich as "*This Serpent Double-head, is grievous to be seen / Whose cloven-head doth kill with sight and hissing keen*." It is safe to say that Mather cited Conrad Gesner's *Historiae animalium Liber V. qui est de Serpentium natura* (1587), lib. 5, fol. 25[r-v] (by Gesner [fol. 25r., line 28] attributed to Pictorius' "Lerna κακῶν"). The Mather library contained the Frankfort 1588–90 editions of Gesner's work. Pictorius' "Lerna κακῶν" refers to Hecataeus of Miletus (c. 550–c. 476 BCE), who relates Heracles' task to vanquish the many-headed

Now, tho' the *Latter* be a *Vulgar Error*, I think, I can satisfy you, that the *Former* is not so.

The *Ancients*, your *Ælian*,[6] & *Pliny*,[7] and *Galen*,[8] and Company, beleeve him {to} be a *Serpent*, really with *two Heads*, at the *two Ends*, of him. And many *Moderns*, besides *Mantuan*, (if he may be called one,) go into the Beleef of an *Amphisbœna Biceps*.[9] Indeed *Aristotle* mentions the Animal as a *Monster*, not a *Species*; and produced only by an accidental & unnatural Cohæsion of *two Serpents*, as there have been of *humane Bodies*, in their Conception.[10] Anon,

---

hydra in its lair in the lake of Lerna, rumored to be the entrance to Hades. Much the same is mentioned in Petrus Gyllius, *Ex Aeliani Historiae Animalium per Petrum Gyllium* (1553), lib. 9, cap. 44, pp. 260–62.

6 Aelian, *On the Characteristics of Animals* (9.23), emphasizes that while Hecataeus may brag of the mythic hydra of the Lernaean lake (the second of Heracles' Twelve Labors); and Homer of his tripartite Chimaera, "in the fore part a lion, in the hinder a serpent, and in the midst a goat" (*Iliad* 6.180–82; 16.306), the two-headed amphisbaena, however, is real and does exist in nature.

7 Pliny the Elder, *Natural History* (8.35), is for once not telling a Plinyism: "the amphisbaena has a twin head, that is one at the tail-end as well, as though it were not enough for poison to be poured out of one mouth."

8 Claudius Galenus (129–c. 216 CE), the Greco-Roman physician and philosopher of Pergamon, whose medical works were highly esteemed through the ages. Galen's *Theriaca ad Pisonem* (146–47) mentions both Heraclides of Tarentum and Apollonius Mys, who developed an antidote to snake venom. And yet, he also mentions Sostratos, whose potent mixture of venom facilitated Cleopatra's escape into the underworld (*Theriaca ad Pisonem* 235), in Joannes Juvenus's *De Theriaca ad Pisonem. Interprete & Commentatore, Ioanne Iuvene, Medico* (1575), cap. 9, pp, 79, 83. See also Caius Julius Solinus (27.29), who believes the amphisbaena moves in circular fashion because of its heads on either extremity.

9 Baptista Spagnuoli Mantuanus (1447–1516), Italian Carmelite reformer, poet, and amateur physician whose most popular work is *Adulescentia.* As is the case with Mather's second-hand citation of Pictorius (see preceding note), Mather may well have culled this reference, along with others below, from Gesner, *Historiae animalium* (lib. 5, fols 66ʳ–67ᵛ). See also G.W. Jones (1962).

10 Aristotle, *De Generatione Animalium* (4.4. 769*b*–773*a*) employs Democritus as his starting point, explaining that monstrosities occur when two semina in the uterus start to grow unevenly and in different stages of development, causing the different parts to grow attached to one another and throwing the whole into disorder. Snakes with two heads are similarly formed in the uterus, owing to their shape and the number of eggs gestating at different stages in the uterus (769*b*, 770*a*). The ancients and moderns were perennially fascinated with monstrous births and misshapen animals, and Mather includes in his "Curiosa Americana" several letters on the topic: CA 1.5 "Monstrous Births"; CA 2.3 "A Monster"; CA 3.2 "A Monstrous Calf"; CA 7.2 "A Monstrous Dragon." The vademecum for all those who wrote

Matthiolus and *Grevinus* mightily decry the Tradition;[11] and think, that the Notion is taken up from an heedless Observation of the *Serpents* being able to do (what his Name signifies,) move *backwards* as well as *forwards*.[12]

I find it a Passage occurring in Mr. RAYS *Philosophical Letters*; "Meeting with Mr. *Dale*, & examining him about *Serpents*, he confidently affirm'd to me, that he had twice seen the *Amphisbæna* in this Countrey, & named the places: Not that there is any *Serpent* that hath *two Heads*, or an *Head* at each *End*; but only that it hath a Faculty of going *backward* as well as *forward*; and the *Tayl* is turrit, & somewhat resembles an *Head*."[13]

on the topic was, among others, Fortunius Licetus' richly illustrated *De Monstris* (1665), esp. lib. 1, cap. 4, pp. 17, 21–22, which went through several editions. (The Mather library at the MHS contains the 1665 edition with his nephew Mather Byles' signature). On Aristotle's views on the formation of monsters, see S. Connell (2017) and T.V. Upton (2003).

11 Pietro Andrea Gregorio Mattioli (1501–c. 1577), Sienese naturalist and personal physician of Archduke Ferdinand of Austria. Mather refers (at second hand) to Mattioli's *Opera quae extant omnia: Hoc est Commentarij in VI. libros Pedacij Dioscoridis Anazarbei de Medica materia* (1598), cap. XLVIII, pp. 1022–23. His French colleague, Jacobus Grevinus, aka. Jacques Grévin (1538–70), French poet and Parisian physician, discusses the amphisbaena in his *De Venenis Libri Duo* (1571), lib. 1, cap. 16, pp. 81–83, who states that the amphisbaena, or *Double-marcheur* (biceps), is so called because the Greek and Latin designations are composed of two names. For Mantuanus, Aristotle, Matthiolus, and Grevinus, Mather relied on Gesner (lib. 5, p. 25r).

12 Grevinus, *De Venenis*, lib. 1, cap. 16, p. 81. The Scottish physician and astrologer William Ramesey, aka. Ramsay (1627–c. 1676), who consults both Mattioli and Grevinus on numerous occasions, did the latter author the honor of adopting the same title for his own book, aptly named *De Venenis: or, A Discourse of Poysons, their Names, Natures, & Vertues* (1663). In *De Venenis* (Div. III, sec. 11, pp. 198–99), Ramesey argues that the "*Amphisbæna* moves both forward and backward (whence by some) it is thought it hath its name." Yet, pregnant women must beware, warns the venerable doctor, for if they cross the path of the amphisbaena or its "something thicker" cousin, the "Scytale," they "shall miscarry or be in great danger in Child-bed." The *Potentia imaginativa* at work again!

13 John Ray (1627–1705), distinguished English natural philosopher, botanist, and theologian, father of natural history in Britain. Mather refers to Ray's posthumously published *Philosophical Letters* (1718), p. 243. In the Preface, "*Mr.* Dale of Braintree" is identified as being "*Mr.* Ray's *Neighbour and industrious Friend.*" The *OED* defines the term "turrit" (noun) as a "small or subordinate tower." Dale appears to use the term as an adjective, suggesting that the tail end of the amphisbaena is like a small head raised at an angle.

What I have to do, is to lay before you a well-attested Relation, in the Terms wherein the public among us has received it.

> *Newbury*. Jun. 14. A Serpent was killed here this week, about Two foot long, with TWO PERFECT HEADS, one at each End. In *Each Head* there were *two Eyes*, & a *Mouth*; and in each *Mouth* a forked *Sting*; *Both* which he thrust out at the *Same time* with equal Fierceness. The Manner of his Defence was; Raising up his *Heads* about two Inches from the Ground, he kept one always directed towards his Adversary; thrusting out both his Stings at once. The Lad that killed him, affirm'd, That when running, if his Motion were obstructed one Way, he would run directly the contrary Way, and never turn his Body. *One Head* was something Bigger than the other; And from the bigger to the other, his Body was somewhat Taper-wise: but in a far less proportion than in common Snakes. The *Colour* of the *Snake* was not much differing from that of many common *streaked Snakes*; but neither of his *Heads* any ways like them.[14]

This Relation may be relied upon. Tis here quæstioned by no body. Afterwards I made a more exact Enquiry into the Matter, and I find it very fully Attested; That the Business of *Two Real Heads* was very strictly brought under a Critical & Repeated Examination; And it was found most certainly so; There were *Two Real Heads*, one at each End of the Snake; with *Two Mouths*, which they opened with a little Stick; in each of which there was a *Tongue* or *Sting*.

[14] Mather copied this article from *The New-England Courant* (July 1, 1723), issue 100, p. 2. The following is the account as it appeared in the newspaper of Newbury, Massachusetts, as quoted in Joseph T. Buckingham's *Specimen of Newspaper Literature* (1850) 2:85: "*Newbury, June* 14. – A serpent was killed here this week, about two foot long, with two perfect heads, one at each end; in each head were two eyes and a mouth, and in each mouth a forked sting, both which he thrust out at the same time with equal fierceness. The manner of his defence was, raising up his heads about two inches from the ground; he always kept one directed towards his adversary, thrusting out both his stings at once. The lad that killed him affirmed that when he was running, if his motion was obstructed one way, he would run directly the contrary way and never turn his body. One head was something bigger than the other, and from the biggest to the other his body was somewhat taker-wise, but in a far less proportion than in common snakes. I the subscriber with several others saw the said serpent just after he was killed, and can testify to all above-written, except his motions described by the lad, who only saw him alive. *Nath. Coffin*."

And there {were} *Two Eyes* in each of the Heads. With Master *Dales* Leave,[15] Tis no longer to be doubted.[16]

I am sorry, that my Distance from the *Field of Battel*, where the Lad killed this little *Python*, which is forty Miles off, made me uncapable of such an *anatomical* Proceeding upon the Animal, as would have enabled me to answer many Enquiries, which would immediately offer themselves on this Occasion, and particularly what may relate unto the, *Ore Parere*, which *Claudius Minos*[17] pretends to *recite* out of *Pliny* and *Solinus*;[18] tho' indeed he does only *infer* it from his Authors.[19] But what Intelligence I am owner of, I have now honestly communicated. And it is not fitt I should now *add any more*.

Ælian tells us from a Book of old *Nicanders*,[20] which is not now extant; That the *Skin* of an *Amphisbœna* being lodged in a Staff, he that carries it

15 See the reference to John Ray's *Philosophical Letters* (p. 243) above.

16 Mather adds the preceding paragraph sidewise in the left-hand margin.

17 Claudius Minos, or Charles Mignault (1536–1606), French jurist and philologist, whose reprinted work was his edition of *Omnia Andreae Alciati v.c. Emblemata cum Commentariis* (1577), a rich collection of emblems and Latin verse, believed to have started the genre of emblem books, by the French-Italian humanist and jurist Andrea Alciato (1492–1550). Minos insists that "amphisbaenas ore parere tradunt: his enim caput est geminum, hoc est, & ad caput & ad caudam" (Embl. 1, p. 52). In other words, "Amphisbaenas are said to bring forth [eggs] by the mouth: for with them the head is twin, that is, both at the head and at the tail." Both the Latin citation and references to Pliny and Solinus Polyhistor appear in Minos' edition (1577), lib. 1, Emblema Primum, p. 47, illustrated with a serpent swallowing a man. Mather's source for this citation is Gesner, *Historiae animalium* (1587), lib. 5, fol. 25[v], which also includes the referces to Pliny, *Naturalis Historia* (8.13); i.e, (8.35), and to the 3rd-century CE historian Gaius Julius Solinus, *Polyhistor*, cap. 30; i.e., (27.29).

18 I.e., Pliny, *Historia Naturalis* (8.35); Solinus *Polyhistor* (27.29).

19 Faulting Minos for dressing himself in borrowed robes is rather disingenuous; it was a common practice even in Mather's day – and Mather was not above reproach in such matters. But perhaps he is poking fun at himself as well, for Mather accesses Minos' work at second hand from Gesner!

20 Aelian, *On the Characteristics of Animals* (8.8), invokes the authority of Nicander of Colophon (fl. 2nd century BCE) that the skin of the Amphisbaena, when wrapped around a cane, drives away snakes and other poisonous creatures. Nicander was a Greek poet, physician, and priest of Apollo at Pergamum. His *Theriaca* is the longest of the many fragments of his works extant in Mather's time. Jacobus Grevinus discusses the contributions to medicine derived from venom by Galen, Hippocrates, and Nicander, and includes a long section of Nicander's *Theriaca* in his above-mentioned *De Venenis* (1571), lib. 1, cap. 2, pp. 276–311.

about with him, need fear neither *Serpents* nor any other Creatures, the Bites whereof are venemous. I cannot promise any such Powers to my *Letter*, from the *Amphisbœna* thus inclosed in it. It is dignified with vertues enough if it may find Acceptance with you, according to the hearty Regards with which it comes, from one, who forever wishes your Protection from the *Amphisbœna's* in *humane Shape* among you, and most of all, from a greater & older *Serpent*: And who now subscribes himself,

[Boston, New England.] Sir,

Sept. 21. 1724.[21] Your perpetual Servant.

Cotton Mather.

Mather's mediate sources are Gesner's *Historia Animalium* (lib 5, fol. 66^{v}–67) and Gyllius, *Ex Aeliani Historia* (1533), caps. 44–45, pp. 261–62, and cap. 57, p. 534. See also Topsel's *History* (1:700–01).

21 The date and signature are Mather's.

8.3. AN UNUSUAL VOMIT.[1]

September 22, 1724.

To Dr. James Jurin.

Syr,

Since my Letter is not like to be read, until after you have risen from the *Table*, and perhaps have your *Stomach* well-setled with the *Liquids of the Afternoon*, I will venture to make *Strange Vomits* the Matter of your Entertainment.

The *Strange Vomits* of *Nails* & *Pins* & *Coals* and *Balls of Hair*, and other such *Unaccountable Stuff*, which have been seen in poor *Energumens*, we will say nothing of; because they seem not any otherwise than from the Operations of *Præstigious Dæmons* to be accounted for.[2] What Our *Philosophy* may presume to enquire after, is all that my Pen ventures now to meddle with.

And here I was going to tack the Names of my Authors, to the Relations of *strange Vomits*, which I was thinking with all possible Brevity, to bring as an *Introduction* to the Story, which I am now to add unto them. I was going to remind you, that *Forestus* mentions one who vomited a *Caterpillar*;[3]

1 Title, addressee, date, and signature are in Mather's hand.

2 The paraphernalia of witchcraft and the tell-tale signs of possession here mentioned are all too common in the publications of the day. Vomiting crooked pins, nails, and needles, hairballs, coals, worms, and stones seemed to be the order of the day, and can be found on both sides of the Atlantic. A few examples: Meric Casaubon's *A Treatise Proving Spirits, Witches, and Supernatural Operations* (1672), p. 87; John Webster's *The Displaying of supposed Witchcraft* (1677), pp. 35, 250, 264, 265, 274; John Bell's *The Tryal of Witchcraft* (n.d.), sec. 1, p. 7; Joseph Glanvill's *Saducismus Triumphatus* (1681), pp, 28, 101, 170, 173, 177, 195, 197, 202, 206, 207; Richard Baxter's *The Certainty of the Worlds of Spirits* (1691), pp. 74–79, 93, 95–97, 100, 109; Cotton Mather's *WIW* (1693), pp. 50, 66, 86–90; and Robert Calef's *More Wonders of the Invisible World* (1700), pp. 4, 5, 44, 125. See also William Perkins' important handbook, *A Discourse of the Damned Art of Witchcraft* (1608).

3 Pieter van Foreest, or Petrus Forestus (1521–97), a prominent Dutch physician, author of *Observationum et Curationum Medicinalium Liber Decimus-Octavus* (1590), a huge collection of medical cases and observations of his patients'

another who vomited a *Beetle. Schonfelder*, one who vomited, those which we call *Canker worms;*[4] *Borellus*, one who vomited *Spiders*.[5] To bring *vomited Animals* from another Element, I was going to remind you, how *Cornelius Gemma*, mentions one who vomited an *Eel;*[6] *Segerus*, one who vomited a

---

diseases. Mather refers to lib. 18, Obser. 19: "*De Vomitu atrae bilis, una cum vermibus & scarabeis*" ("On the vomiting of black bile along with worms and beetles"), pp. 105–09. Forestus (pp. 106–07) describes the case of a young woman, who "tamen paulo ante scarabeos vomitu reiecerat… . Referebat & mater vivum scarabeum reiecisse, qui per duos dies supervixerat: referebatur autem ad incantationem hoc malum": "but just a little before I visited her again she had vomited up beetles," adding that he would not have believed it had he not seen it himself. The physician related that the "mother reported that she had vomited a live beetle that had survived for two days; moreover, this malady was attributed to incantation." See also "Scholia," pp. 107–08. The English natural philosopher, arachnologist, conchologist, and geographer Dr. Martin Lister, F.R.S. (1639–1712) of Yorkshire, related the account of the son of Mr. B. (near Ripon, Yorkshire) who "vomited up several strange Worms" (one alive the other dead), which Dr. Lister preserved in Spirit of Wine. They were "*Caterpillars* with fourteen legs, *viz.* six small pointed, the eight middle stumps, and the two hind claspers; something more than an inch long, and of the thickness of a Ducks-quill, thin haired or rather naked, with brown *annuli*, and a black head." These fine specimens were all recorded in "An Extract of a Letter of July 28 1675, by Mr. Lister from York, containing … odd worms vomited by children," *PT* 10.117 (September 1675): 391–95; quotation at p. 395. See also Lister's *Historiæ Animalium Angliæ Tres Tractatus* (1678).

4 Philippus Jacobus Schönfelder (active 1670–90), German physician in Ingolstadt (Bavaria), author of *Historiarum, Enarrationum et Curationum: Medicarum Liber Primus* (1681), Historia II. Dolor ventriculi à verme vomitu excreto causato, pp. 5–6, 9. Mather also reports the devastating effect intestinal worms have on the human body, in "CAP. XLIV. The *Vermine-Killer*. Upon, Worms, *ABJ* (1972), pp. 203–07.

5 The French physician, chemist, and botanist Pierre Borel (c. 1620–71), in his *Historiarum, & Observationum Medicophysicarum Centuriae IV* (1670), Cent. I, Obser, LXXXI: Aranea deglutita, p. 87, relates the observation of his colleague Dr. Josias de Fregeville, that "Corpore aperto, aranaea in stomacho ejus inventa est. Sic assertum mihi ab alio Medicum 7 araneas è corpore humano remediis expulisse," or loosely, "I was likewise assured by another person that a doctor had expelled with his remedies seven spiders from a human body," in the same work, Cent. III, Obser. XIX: Aranea mira, vomitu viva rejecta, & cirara de araneis (pp. 212–14).

6 Cornelius Gemma (1535–78), Flemish physician and astronomer of Leuven, who in 1552 published the first illustration of a human tapeworm, and later published *De Naturae Divinis Characterismis; seu Raris & Admirandis Spectaculis* (1575), which compiled accounts of medical marvels. He mentions a girl in Leuven, in whose excrements were found "anguillae cum rebus aliis maximè peregrinis":

Gudgeon;[7] *Rhodius*, one who vomited a *Crab*.[8] I was going with a, *paulò majora canamus*,[9] and besides those, who have affirmed *Catts & Weasels* to have been vomited, to have reminded you of the *Two Puppies* vomited by the Taylor of *Hanover*, mentioned by *Paulinius* in his *Cynographia*.[10] But fearing lest my Quotations be already, *ad nauseam*, surfeiting, I laid aside all Thoughts of quoting my *Authors*, for the *strange Vomits* of *Serpents*, &

"eels with highly unusual things" (p. 22). A fourteen-year-old girl voided (through her lower parts) live eels: "anguillae vivae per inferiora secessus" (lib. 1, cap. 6, p. 103; see also lib. 2, cap. 4, p. 202). See for the same in Francesco Maria Guazzo's *Compendium Maleficarum* (1626), lib. 2, cap, 9, pp. 188–89, who identifies her as Catharina Gualteri (age 15), who voided her live fishy "anguillam vivam" in 1571.

7 Georg Seger (1629–78), Polish-German physician in Gdansk (Danzig), *Miscellanea Curiosa Medico-Physica Academiae, Annus Tertius, Anni scilicet 1672* (1673), Observatio CCXXVII: De Rene Dextro Monstroso Ejusque Portionibus cum Urina Excretis, et Calculo Renis Sinistri insolitae Magnitudinis," pp. 412–15: Georgius Mochingerus "vomitu rejecerit Gobium fluviatilem, vulgò Fundulum, verum & vivum, minimi digit longitudinem aequantem, atque eo rejecto, pristinam sanitaem recuperaverit": Georg Mochinger "vomited out a river gudgeon – popularly called a *Gründling* – the length of one's little finger, and after it had been voided, he recovered his former state of health" (p. 415). A gudgeon is a small edible European freshwater fish, often used as bait by anglers.

8 Joannes Rhodius, aka. Johan Rode (c. 1587–1659), a Danish physician in Kemberg (Saxony), authored *Observationum Medicinalium Centuriae Tres* (1657), Cent. III, Observ. 66, p. 183: "Gulielmo de Haert Batavo Serenissimae Reipublicae Venetae tribune, vermis ab eo secessit multiceps, cauda longa, cornu gemino, cancro similis, flavi & nigri coloris, palmi longitudine": "Upon the death of the Dutchman Willem de Haert, representative at the Republic of Venice, a multi-headed worm came out of him, with a long tail and a double horn, similar to a crab, yellow and black in color, a span in length."

9 Indeed, "Let us sing of things a little more elevated" (Vergil, *Eclogues* 4.1).

10 Christianus Franciscus Paullinus, aka. Christian Franz Paullini (1643–1711), German physician, polymath, poet, and author of *Cynographia Curiosa seu Canis Descriptio* (1685), sec. 1, cap. 3, §58, p. 60. Here we learn that "Anno 1580 sartor Hanoveranus, Albert. Hencke, duos *catulos albos, vives, repentes & nondum videntes* evomuit, qui in patina singular in templo S. Georg expositi fuerunt, ut ab omnibus viderentur": "In the year 1580, the Hanoverian tailor Albert Hencke vomited out two white, live puppies, crawling and not yet sighted, which were exposed in St. Geroge's Church on a broad earthenware dish so they could be seen by everybody." Mather might have added the hoax of Mary Toft of Godalming, Surrey, who allegedly gave birth to several rabbits, in 1726! William Hogarth famously poked fun at the story in his engraving *Cunicularii. The Wise Men of Godliman in Consultation*. London, 1726.

Leeches & *Toads* & *Frogs*, &, finally, several Sorts of *Lizards*:[11] And only observe, that I find them generally agree in Suspicions, That the *Vomiters* had either *drunk* unawares the *Sperm* of these *Animals*, or swallow'd them with their *Drink* when they were very little ones: Or, that the *Animals* grown larger had stolen down the *open Throats* of the Patients, while they were *asleep*.[12]

So that I shall no longer defer my Story of a Gentleman at our *Newbury*; who has lately vomited a *Lizard*;[13] Upon which addressing myself to his worthy *Physician* (who is also the *Minister* of the Place)[14] for Satisfaction, I shall transcribe the words of his Letter to me.

> As for the *Serpent* vomited by a *Gentleman* in my Neighbourhood, it was what is commonly called an *Evet*;[15] which differs but little, if at all, from a

[11] See Mather's *CP* (1721), Essay XXVII, p. 158.

[12] Mather's speculation (and that of his peers) about the origin of insects, semi-aquatic amphibians, and tapeworms in the stomachs of unwitting carriers seemingly bespeaks the credulity of the age – if viewed from our modern perspective. To be sure, the underlying issues are rooted as much in the debate about preformationism, spontaneous generation, and epigenesis as they are about the discovery of microscopically small animalcules (parasitic worms, microbes) in water, menstrual blood, and sperm – topics that engaged the leading minds of the period such as Leeuwenhoek, Boyle, Harvey, Redi, Schwammerdam, Malpighi, Spallanzani, and many others (E.G. Ruestow (1983). As a natural philosopher in his own right, equipped with his own microscope, Cotton Mather in faraway New England was very much ensconced in the progress of the debate as is most evident in his chapter on animalcules, "VII. Conjecturales, or Some Touches Upon a New Theory of Many Diseases," *ABJ* (1972), pp. 43–48. See also Mather's letter to John Woodward (July 2, 1716), our CA 3.1 "Monstrous Impregnations." For helpful background, see John Farley (1977), esp. pp. 8–30; and C. Pinto Correia's delightful book on the ovaries of Eve (1997).

[13] Increase Mather tells a similar story of a woman who drank water from a pond in which frogs had laid their eggs. And yes, you guessed it, sometime after she complained of hearing croaking frogs in her stomach, her physician "gave her a strong *Cathartick*, whereupon she cast up two living Frogs pretty large ... ." To be sure, the poor woman did not croak and had nothing on a Frenchman from Lyon, who in 1680 "voided a Worm seven Ells long, scaly like a Serpent, and hairy," as related in Increase Mather's *ERIP* (1684), ch. 9, p. 310. Popular stories like these could be found in the medical literature of the day as well as in such tabloids with scientific aspirations as the anonymous *Weekly Memorials for the Ingenious ... with Other Accounts relating to Arts and Sciences* (1682), No. 9, p. 67; No. 17, p. 100. See also our CA 4.4 "A Prodigious Worm."

[14] Christopher Toppan (1671–1747), native of Newbury, Mass., graduate of Harvard, 1691, and pastor of the First Church from 1696 until his death.

[15] The common newt or eft, or one of several species of aquatic salamanders.

Lizard. There are several Sorts of them; Some that keep in Wells & watery Places; Some under old Logs & rotten Wood; – and they are of different Colours. That which keeps in stony moist Places, is Reddish about the Head; and so was this; It was for Bulk about the Bigness of an ordinary Mans Thumb, at the Head; but lesser backwards; & about Four Inches long. It had four Legs under its Belly a little backward of its Head. It was vomited alive, but soon killed & thrown away. The Person from whom it came, has been under an ill Habit of Body some years, more especially for two or three Years past, being exercised with great Pain, occasioned by the *Stone & Strangury*:[16] often sick at his *Stomach*, with frequent *Vomiting*; able to eat little or [2r] No meat; living chiefly, if not only, on Liquid Food. Since the Coming away of this Creature from him, there has been no great Alteration as to his State of Health, except that he is Somewhat more free from Pain, and not so sick at his Stomach as he was.

Concerning this Odd Production, if I may give my Opinion, I imagine, it was bred in the Gentleman from an *Evets* Egg drank in water;[17] He, the last Summer drinking abundance of Water, taken out of a stoned Well, where tis likely, some *Evet* might lay its Eggs; and his *Stomach* being much vitiated, might afford suitable Nourishment, for its Growth, till it came to that Bigness, in which it was vomited up. Or, it might be taken down in Water, when it was very Small, soon after it was hatched (if it be proper to term it so,) & not easily percieved; For it was his Custome to drink Water by Night as well as by Day.

Thus my Candid Correspondent.[18]

16 A condition caused by blockage or irritation at the base of the bladder, resulting in severe pain and a strong desire to urinate.

17 This likely story parallels the case of a melancholic man in Switzerland who accidentally swallowed the spawn of frogs when he fell into a pit filled with water. His Swiss countryman, the famous physician Felix Plater (1536–1614), in his *Observationum, In Hominis Affectibus plerisque, corpori & animo, libri tres* (1614), Observ. lib. 1, pp. 40–43, relates that he had much ado to undeceive the hapless man from his melancholic conceit that he heard frogs croaking in his stomach. At least the English scholar Robert Burton (1577–1640) deemed the story sufficiently entertaining to excerpt it in his encyclopedic *Anatomy of Melancholy* (1621), pt 1, subsec. 2, pp. 258–59.

18 In his *MCA* (1702), "A General Introduction" (fol. C^{4v}), Cotton Mather mentions the story of a bosom "serpent with diverse tails found in the left ventricle of [a young man's] heart" – a story that may well have given rise to Nathaniel Hawthorne's "Egotism; Or, The Bosom Serpent" (1843), at least if we agree with the scholars of numerous articles on the subject, including S. Bush, Jr. (1971); J.W. Shroeder (1959); or D.R. Barnes (1972).

Garmannus, who in the *German Ephemerides*, brings a Woman vomiting of a living *Mouse*, passes no other Judgment upon it but this. *Nihil habeo quod dicam, nisi illud; Forsan Ore dormienti {mulierxulae} in ventriculum descendit.*[19] And you see how far my Friend agree with him, in judging of his *Lizard*.

Having thus contributed a little *Evet*, unto the Stores in your more curious Collections, I take Leave only to add, that the Blessings of a good & sound *Stomach*, that shall have nothing in it but what Shall agree with it, and answer its offices, tho' they are truly *great Blessings*, yet they are among the *Least*, that are wished you, by

Syr,

Sept. 22. 1724.[20]

Your most hearty Servant,

Cotton Mather.

[19] "I have nothing to say, save this; perhaps it went down into the woman's stomach through her mouth when she was asleep," so commented Christian Friedrich Garmann (1640–1708), city physician of Chemnitz. *Ephemerides Germanicae* was a periodical first published in Frankfurt in 1670 by the *Academia Naturae Curiosorum*, an early scientific society. The venerable doctor contributed his "Observatio CLVII: De Musculo vomitu rejecto," pp. 341–47, to the distinguished *Miscellanea Curiosa sive Epheremidum Medico-Physicarum…* . Decuriae II. Annus Octavus 1689 (1690). Mather's Latin citation appears on p. 347. Significantly, virtually all of Mather's medical authorities cited above are also referenced in Garmann's article.

[20] The date and signature are in Cotton Mather's hand.

8.4. A TIDE & STORM OF UNCOMMON CIRCUMSTANCES.[I]

September 23, 1724.

To Dr. John Woodward.

Syr,

The *Reading* of a Storm is not so bad as the *Feeling* of it. I shall therefore think it no Trespass on *Civility*, to entertain you with a short Relation of a *Storm* & a *Tide*, wherein these parts of the world, saw what no man alive remembers to have been seen before, & suffered Incomputable Damages.[2]

It was on Febr. 24, 1723,[3] when our little *American Philosophy* observed an uncommon Concurrence of all those Causes, which an *High Tide* were to be expected from.[4]

The *Moon* was then at the Change, & both *Sun* & *Moon* together in the *Meridian*.

The *Moon* was in her *Perigee*;[5] and the *Sun* very near his; having passed it but a little before.

1 Title, addressee, date, and signature are in Mather's hand.

2 The same letter appears in *MP* (1868), pp. 456–57; and, along with the description by John Winthrop (1681–1747), of a similar storm occurring seven years earlier (1717), in *CMHS* 2 (1793): 11–12 and 12–13, respectively. As with his subsequent letter (CA. 8.7), Mather responds to Jurin's call for accurate reports on weather phenomena in America. Jurin hoped to establish a theory of meteorological formations and sent copies of a handbill to Mather, inviting him to distribute them among New England's scientists with access to reliable instruments. Jurin also published a call for meteorological information in the *Philosophical Transactions*: "Invitatio ad Observationes Meteorologicas communi consilio institutendas," in *PT* 32.379 (October 31, 1723): 422–27. See also Appendice 10 and 11. On Jurin's interest in meteorology, see A. Rusnock (1996), pp. 27–31.

3 I.e., Old Style.

4 For an account of the high tide and flood, see *The Boston News-Letter*, no. 996 (February 21–28, 1723), p. 2, col. 2, "*Boston, Febr. 25.*"

5 At a point when the moon in its orbit is closest to the earth.

Both the *Sun* & *Moon* were near the *Equinoctial*;[6] & so fell in with the *annual* & *diurnal* Motion of the Terraqueous Globe.

There was a great fall of *Snow* & *Rain.*

The Temper of the *Air* was cool & Moist, & such as contributed unto a mighty Descent of Vapours.

A *Cloudy Atmosphœre* might also help something to swell & raise the Waters.

Finally; The *Wind* was high & blew hard & long; & first from the Southward, it threw the Southern seas in a vast Quantity to the Northern Shores; Then vearing Easterly, it brought the eastern Ocean also upon them; and then stil veering to the Northward, it brought them all with yet more Accumulations upon us. They raised the *Tide* unto an Height, which had never been seen in the Memory of Man among us.

The *Tide* was very high in the Night; But on the Day following (aforesaid,) it being the *Lords day* at Noon, it rose two foot higher than ever had been known, into the Countrey; and the City of *Boston* particularly suffered from it incredible Mischiefs & Losses. It rose two or three foot above the Famous *Long Wharf*; & flowed over the other Wharfs & Streets to so Surprizing an Height, that we could sail in Boats from the *South Battery* to the Rise of the Ground in *Kingstreet*; & from thence to the Rise of the Ground ascending towards the North-*Meeting-house.* It filled all the Cellars, & covered the Floors of the lower Rooms in the Houses & warehouses in the Town. The Damage inexpressible!

In the Countrey on the inside of *Cape-Cod*, the Tide rose four Foot, & without rose ten or a dozen Feet higher than ever was known.

At *Rhode-Island* and *Piscataqua*, they Fared as we did at *Boston.* At *Hampton*, the Sea broke over its *Natural Banks*, for many Miles, and continued running over, for several Hours. Almost all over the Countrey the *Artificial Banks* of the Sea were broken down; the Marishes were overflown & overwhelm'd; mighty Stacks of *Hay*, some Remov'd, Some Destroy'd; Many Acres of *Marish* ruined, being either torn up thro' the Rage of the *Water*, or covered with the Sand from the *Beach.*

This is the Sum of the Story.[7]

If there be nothing in it worthy to be *Remembred* any otherwise than *as Waters that pass away*,[8] (or any thing like the Memorable *November-Storm*, which filled the English World with Horror Twenty Years ago, and whereof a

6 When they are at a state equal to the time of day and night.

7 Significantly, neither the diary of Samuel Sewall nor that of Cotton Mather has much of anything recorded on this auspicious storm.

8 Job 11:16.

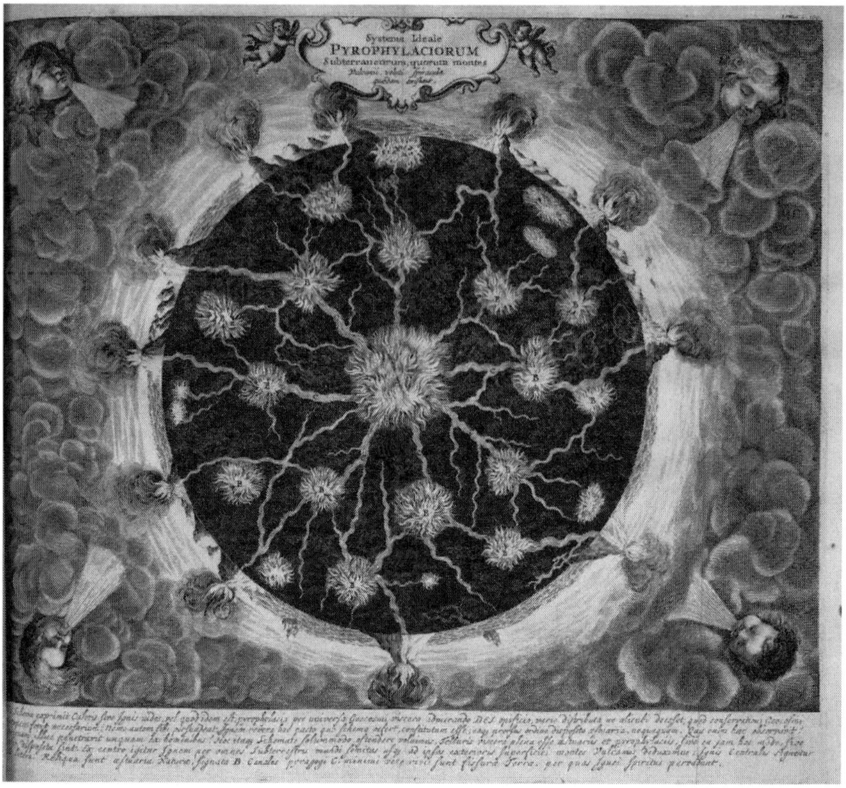

Figure 18. "Pyrophy Laciorum," Athanasius Kircher, *Mundus Subterraneus, in XLL Libros digestus* (Amsterdam, 1665), lib. 4 (insert following 180). Courtesy of the Library of Congress.

large Book was written, asserting, that *it had never been parallel'd before,*)[9] yet it may lead a Person of your Sagacity to some considerable Speculations.

[9] Mather refers to the great storm (November 26–27, 1703), in which (by modern estimates) some 8,000–15,000 people were killed in England and the Netherlands. The "large book" is by his sometime-correspondent Daniel Defoe (1660–1731), *The Storm: or a Collection Of the most Remarkable Casualties and Disasters Which happen'd in the Late Dreadful Tempest, both by Sea and Land* (1704). See also the anonymous *An Exact Relation of The Late Dreadful Tempest* (1704); and "A Letter for the Reverend Mr. William Derham, F.R.S. Containing his Observations concerning the late Storm," *PT* 24.289 (February 28, 1704): 1530–34; and Anthony van Leeuwenhoek, F.R.S. "Part of a Letter … giving his Observations on the late Storm," in *PT* 24.289 (February 28, 1704): 1535–37. For a modern assessment, see D. Ludlum (1963), pp. 19–20. Mather also alludes to the same storm (1703) in his impromptu *VGT* (1723), p. 7.

And more particularly, Tho' I have mentioned what *our Small Philosophers* here may dream for the Cause of such Occurrences, Yet you will also consider, how far the subterraneous *Heats* and *Steams*, below the Bottom of the *Ocean*, rising thence, & passing thro' it, & causing the *Deep to boil as a Pot*,[10] may further contribute unto them.[11]

However, as for a *Tempest*, so for a *Letter* about one, You may think, The *Shorter* the *Better*. It shall therefore now be *over*. I will add no more; and you shall be sensible of Nothing but a *Swelling Tide* of Esteem & Affection for You, in the Breast of,

Syr,

Sept. 23. 1724. Your hearty Friend

& Servant,

Cotton Mather.

[10] Job 41:31.

[11] Mather here leans on *Mundus Subterraneus in XLL Libros digestus* (1665), by the German Jesuit polymath Athanasius Kircher (1602–80), author of many significant works on comparative religion, Egyptian hieroglyphs, geology, and medicine. Mather was well familiar with Kircher's work and quoted it throughout *BA* (2010–26) and in his "Curiosa Americana." See CA 1.1 "Describing Biblia Americana"; CA 1.12 "Thunder & Lightning"; CA 7.7 "*Subterranea*"; and in *WASR* (1721). A copy of Kircher's medical work (bearing Increase Mather's signature), *Scrutinium Physico-Medicum Contagiosæ Luis quæ dicitur Pestis* (1659), still remains in the Mather Family Library (AAS).

8.5. A MOVEABLE ROCK.

September 24, 1724.[1]

To Dr. James Jurin.

Syr,

You may judge me akin to *Johannes de Rupe Scissâ*,[2] by my obtruding upon *You*, another Story of a *Rock*, after my having transmitted unto our [Alas, *deceased*] Friend, an Account of the *Rocquing Stone*,[3] to be seen at the Town of *Glocester* in this *American* Colony.[4]

But tho' the thing be little above the Quality of a *Trifle*, yet because there is a little *Curiosity* in it, and some other Things as *Trivial* have been commemorated among *observable Rarities*, I will now give a Relation of another Large *Rock* in the very same Town of *Glocester*; the Circumstances whereof are a little singular.

There stands on the Shore there, between Low and High Water Mark, a *Rock* upon a *Rock*; where it would require many Team of Oxen to draw it from its Basis.[5] The *Rock* is about Nineteen or twenty foot long; about nine or ten foot Wide; and about Seven or eight foot deep.

1 Title, addressee, date, and signature are in Mather's hand.

2 Joannes de Rupescissa, aka. Jean de Roquetaillade (c. 1310–c. 1366), French Franciscan monk, alchemist, and millenarian visionary, who authored numerous works on alchemy. Mather puns on Rupescissa's name, meaning "split rock," hence a moveable one in Mather's studied joke.

3 Mather's sent his Curiosa letter "A Rocquing Stone" to his correspondent at the Royal Society, John Chamberlayne (c. 1668–1723), English courtier, translator, contributor to the *Philosophical Transactions*, and member of the Society for the Propagation of Christian Knowledge. Unfortunately, all but one of Mather's twelve Curiosa letters (February 1723) appear to have perished.

4 Referring to CA 5.7 "A Rocquing Stone." To John Chamberlayne, [February 1720], which is no longer extant. For a recent account of the "Moving Rock," see *Gloucester Times*, August 27 2010 issue.

5 According to the town's nineteenth-century historian J.J. Babson (1860), p. 6, the boulder is one of several such ice-age monoliths dotting the landscape around Cape

Tis not *Length-wise*, but *Side-wise*, that it faces The Sea. But when a Storm arises, and the Sea beats Tempestuously upon the *Rock*, the Spectators, have a Sight that a little surprizes them; They see the *Rock*, notwithstanding the vast *Bulk* & *weight* of it, move backward & forward three or four Foot, and in less than a Minute or two perform its motions with continual Repetitions.

The *Rock*, whereon it stands is worn into a most regular Smoothness, as far as this Motion extends; and a *Rock* on which the Back of it rubs is also thereby worn considerably.

If it were possible for the Friendship established, and cultivated in our present Correspondence to suffer any Little *Vacillations*, yet there will be no moving of it from the *Basis* on which it stands firm, in a very high Esteem for your Person & bright Accomplishments, always maintain'd inviolable, with

Syr,

Your unshaken Friend & Servant,

Sept. 24. 1724. Cotton Mather.

Ann, and weighs "about seventy tons, resting on a rock on a level with the surface of the ground, and may be set in motion by one person; so that, after the withdrawal of the force, about fifteen vibrations may be counted."

8.6. A NEW-FOUND-LAND.[I]

September 25, 1724.

To Dr. John Woodward.

Syr,

Some that have written, *De Arte Combinatoria*,[2] have amused us with a pretended Computation of no fewer than *one hundred & seventy nine millions, one thousand & sixty*, different Sorts of Earth. By what proceedings, they formed their Enormous Computation I cannot imagine. They talk, as if almost Every *Clod of Earth* were a Species different from its Neighbours: or do they mean the *Children of Men* at one Time upon the *Earth*?[3]

I am aware, that the old *Geoponic* Writers,[4] are content with a Reckoning sufficiently short of theirs. Nor do they themselves acquaint us with a *Sixth*

1 The title, addressee, date, and signature are in Mather's hand.

2 See, for instance, *Dissertatio De Arte Combinatoria, In qua Ex Arithmeticæ fundamentis Complicationum ac Transpositionum Doctrina novis* (1666), by the German philosopher and polymath Gottfried Wilhelm von Leibniz (1646–1716). In this work, Leibniz tried to develop a mathematical formula for all possible combinations of human ideas expressed in simple terms and in grammatical syntax, subject and predicate – a kind of computational linguistics that informs the most recent developments of artificial intelligence (AI) chatbots.

3 Biblical commentators frequently tried to entertain their readers by providing mathematical calculations of the ante- and post-diluvian population numbers that descended from a single human pair were wiped out in Noah's flood, or multiplied exponentially from the Patriarch Jacob, whose family numbered seventy at the beginning and six-hundred thousand males along with a mixed multitude at the end of their three-hundred-year Egyptian sojourn. See, for instance, Mather's commentary on Exod. 12: 37 (*BA* 2:208–12); see also his *TP* (pp. 242–43, 275–76). Lest Mather appear singular, compare his statistical extrapolations with those of Thomas Burnet (c. 1635–1715), *The Sacred Theory of the Earth* (1681; third ed. 1697), bk 1, ch. 3, p. 16; and of William Whiston (1669–1752), Lucasian chair of mathematics at Cambridge, *A New Theory of the Earth* (1696), bk 3, ch. 3., pp. 175–81.

4 Writers like Columella and Pliny (fl. 1st century CE), Vindonius Anatolius (fl. 4th century CE), and Cassius Bassus (fl. 6th–7th centuries CE), collected all

part of the *Sixty*, which it pleases them to add to the *Thousand*, that they tack to their incredible *Millions*.

But I know not whether I may not raise it unto *Sixty one*. Indeed, my Studies carrying me mostly *another Way*, I have so little Skil in the *Earth*, as to be ignorant, whether I shall now exhibit any thing, that is *wholly new* to you or no. But my *Intention* will make *Atonement* for my *Ignorance*: And I will give you this Relation of something, that has appeared unto us *Americans*, what is with us esteemed, a *Curiosity*.

A Man in the Vicinity of *Woodstock*,[5] about Seventy miles from our City of *Boston*, digging a Ditch for the Defence of his Field, about three foot below the Surface of the Earth, came to a *Lay of Earth* of the Consistence and Cohæsion of *Clay*, about the Thickness of two feet;[6] next below which there lay a *Stratum* of the Usual Clay. The lay of Earth, which he found, as I said, three foot below the Surface, is as *black* as *Jet*, or any thing, that never so much absorbs all the Rays of Light: and of a *Weight* vastly heavier than any other *Earth* yet known unto us. By being long exposed to the Air it grows not only *paler* but also *Lighter* and indeed arises to a very surprizing *Levity*. But that which was most new to us was; That this honest *George*,[7] planting an Hedge upon the *Earth*, which he thus threw up out of the Ditch, and this *Hedge* happening to take fire, this *black Earth* was by the Force of the *Fire* turned as *white* as the Plaister of *Lutetia*,[8] or as any thing under the Sun. Upon this accidental *Experiment*, the people get into the Way of burning this *black Earth*, by which (losing its *Weight* likewise with its *black*) tis changed into a *Chalk*, and they use it as *Chalk*; and also make a *Wash* with it, with which they *white wash* their Houses.

---

available knowledge on agricultural pursuits (*Geoponica*) and compiled handbooks such as Lucius Junius Moderatus Columella's *De Re Rustica Libri XII*. Modern soil-science taxonomies, like those of the German chemist Justus von Liebig (1803–73), or of the Russian scientist Vasily V. Dokuchaev (1846–1903), classified soil types, their chemical composition and formation, and their most efficient utility for plants.

5 In the northeast corner of Connecticut, formerly part of the Massachusetts Bay Colony.

6 Probably "Gloucester loam" (a composite of sand, silt-humus, and clay) prevalent around Woodstock.

7 Mather puns on the Greek etymology of the name *George* from γεωργός; i.e, γῆ (soil) and ἔργον (work), hence *georgos* (planter, farmer, gardener).

8 Lutetia (the Roman name for Paris), Plaster of Paris (gypsum), Paris limestone, is unique to the area around the French capital.

I cannot hear, that this *black Earth*, has yet been found any where else, but in one place more, not far from this; where, a Company of *Indians* digging in a Swamp, about Five or six Foot under Ground have met with it.

May our GOD & Saviour, præpare us for & provide for our Safety in, the approaching Day, when the Element which turns our *black Earth* into *white*, shall bring upon the *whole Earth* incomprehensible Changes, and introduce[9] the *New Earth*, wherein shall *dwell the Righteousness*, which good Men are wishing for.

In a Friendship as *candid* as any *Fire* can accomplish, I am,

Syr,

Sept. 25. 1724. Your cordial & constant Friend

& Servant,

Cotton Mather.

[9] Mather deleted "bring on" and interlineated "introduce."

8.7. MORE AND STRANGE EFFECTS OF THUNDER.

September 26, 1725.

To Dr. James Jurin.[1]

Sir.

The *Historians* in all Ages have bestow'd such Remarks on the *Effects of Thunder*, that another *Zuinger* might now exhibit as many more as those, which the old one has upon his *Theatre*.[2] The *Philosophers* as well as the *Historians* offer such Things unto Consideration; and you also have them in your *Transactions*.[3] This encourages me to hope, that I shall not be thought impertinent, if I now make some Additions to the Relations, which I have tendred in some former Letters,[4] of what the *Thunders* now & then do in these parts of the World. I should be so, if I should go to *Philosophizing* upon the Occurrences, unto a Person of a Skil so much superior to my own; or if

[1] Mather's missive to Jurin is in response to Jurin's call for detailed reports on weather phenomena in America. Jurin hoped to establish a theory of meteorological formations and sent copies of a handbill to Mather, inviting him to distribute them among New England's scientists with access to reliable instruments. Jurin also published a call for meteorological information in the *Philosophical Transactions*: "Invitatio ad Observationes Meteorologicas communi consilio institutendas," in *PT* 32.379 (October 31, 1723): 422–27. See also Appendice 10 and 11.

[2] The Swiss physician and encyclopedist Theodore Zwinger the Elder (1533–88), professor of theoretical medicine, Greek, and Ethics at the University of Basel, is mostly remembered for his *Theatrum Humanæ Vita* (1565–1604), an early encyclopedia of all knowledge arranged on the basis of Ramistic logic.

[3] From 1666–1725, the *PT* had at least twenty articles on the effects of lightning – all demonstrating that Mather's easily held muster to those of his peers.

[4] Mather appears to refer to his letter to Richard Waller (November 29, 1712), CA 1.12 "Thunder & Lightning," which reports on the changing polarities of a ship's magnetic compass (First Series 1712); as well as CA 5.5 "Odd Effects of Thunder," a letter to John Chamberlayne (February 1720). The latter work appears to have perished along with the whole batch of those in Mather's Fifth Series. See also his *BS* (1695).

I should remember you of the *Superstitious Division* which *Pliny* makes,[5] or the more *allowable Division*, that *Seneca* makes of *Thunderbolts*;[6] or entertain you with the Contemplations of a *Kronlande* or a *Tylkowski*,[7] and an Army more, besides the more than Twenty I can just now think on, dogmatizing with their Paradoxes upon the *Lightning*. It was a Tradition among the Ancients, That the *Thunderstruck* tho' they survived the Stroke, yet never after enjoy'd their *Senses*. Tho' you might count me *Thunderstruck*, if I should be so *Senseless* as to trouble you with *Hypotheses* on this Occasion; Yet barely to propose one or two *matters of Fact*, may be excusable.

I shall only transcribe, what I have newly received from a Town, that is not half a Score miles distant from the City, where I am now paying my Duty to you.

5 Pliny's account in his *Natural History* (2.43) does not appear to be all that superstitious given that true to the ancient cosmology of his time, he believed that a ring of fire in the realm of the stars surrounded the earth, proceeded downward, and agitated the air and clouds, causing thunder and lightning. Nonetheless, Pliny did distinguish between several varieties of thunderbolts – a dry flash (without fire) causing an explosion; the smoky ones that merely blacken, but not burn; and the bright ones known for draining casks filled with fluids, melting gold, silver, and copper, while not damaging the containers themselves. To be sure, Pliny does relate that the ancient Tuscans in Italy believed that nine gods were hurling thunderbolts of eleven different kinds. Jupiter sends three varieties during daytime, and Summanus at night. The most mysterious, so the Tuscans, are the ones that issue directly from the ground – not the stars (*Natural History* 2.42–43).

6 Lucius Annaeus Seneca, *Natural Questions* (2.27–29, 57–58), distinguishes thunder according to their different sounds – a deep rumble preceding earthquakes; hoarse when air is trapped within clouds; a sharp crack when the cloud expels the air trapped in itself and makes people's minds deranged and thunderstruck. Lightning flashing across the sky occurs when dense clouds disperse and rarify, and hot air catches fire through friction. Thus, when smoky fumes arising from the earth's cracks and mix with clouds, the accumulated fuel causes bright flashes to occur, or so Seneca explains.

7 Johann Marcus Marci von Kronland (1595–1667), Bohemian physicist and rector of the University of Prague. His Πάν ἕν πάντων *seu, Philosophia Vetus Restituta* (1662), pt II, sec. V, subsect. 1, pp. 145–51, examines how lightning caused by fiery meteors affects the air from which it dissolves mercury, salt, and sulfur. Adalbertus Wojciech Tylkowski (1624–95), Polish Jesuit theologian and polymath, professor of meteorology at the University of Warsaw, author of *Meteorologia Curiosa ad D. Boguslavm Leszcynski* (1669). In pars. I, cap. 10: De Fulmine, pp. 23–36, Tylkowsky holds forth on the origin and effect of lightning.

Newtown. Jun. 24. 1724. About the middle of last Night, We had a very severe Clap of Thunder, and a terrible Flash of Lightning, that fell upon a Red oak Tree, in a peece of Low Meadow Ground. The Top of the Tree seems no otherwise affected with the Lightning, than that the Small Twigs, Moss & Leaves are scorched. Within about Ten foot of the Ground, the Body of the Tree begins to be barked: And at six foot high the Bark is all peeled of, and the Tree quite naked: some of the Peeces are Scattered unto a great Distance. The Tree itself, from the Ground upwards to Six or Seven foot, is shattered & rived into many places with the Grain. From the Root of the Tree, which is also much broken, the Lightning passed into the Earth, and divided its Flame into five or six distinct Lines, some of them visibly reaching more than threescore foot distance, tearing up the Ground like Furrows: The most considerable of these Furrows runs to the Southward of the Tree about five Yards to a dead Stump, which stands just on the Edge of a Boggy Meadow; and the Lightning for ten or twelve foot together made a Trench of near six foot in Width, & near three foot deep. From this great opening the Furrow continues for several Yards two foot deep, and a yard broad, and then divides into two Branches for several Rods, til tis lost. To the South east & North west, there are also two Remarkable Lines where the Lightning pass'd, and the Ground broke up at fifty foot distance from the Tree: But these Lines or Furrows are discontinued in many Places, the Fire passing wholly under ground, sometimes for ten or twelve foot together, and in one Place for near twenty Foot, without opening the Ground at all; but by thrusting a Pole in at the Holes, it is plain the Lightning had a Passage there. To the westward of the Tree, there are two more small Lines, that reach near forty foot from the Tree, where the Lightning here & there has broke up the Ground. Upon the Meadow where the Lightning pass'd, a parcel of Hay, that was there in Cocks, was either destroy'd by it, or scattered & carried quite away. The Place where the Tree stood is clear & open; there was no Hill or Tree near it. And the Ground that was tore up by the Lightning is a loose Soyl, & full of Roots, where it had its greatest Force, the Ground is moist & boggy.[8] —

Having thus related the odd *Effects of Thunder*, I will to this Relation tack another, which I had sent you several Months ago, if I had not supposed that

[8] *Boston News-Letter* (July 9, 1724), p. 2. The article identifies the town as Newton and the property owner as Joseph Fuller, Jr. (b. 1685), a militia lieutenant and later a town selectman, and the son of Capt. Joseph Fuller (1653–1740). The extended Fuller family owned nearly a thousand acres in Newton and the surrounding area during this time. See H.E. Woods (1905), p. 72.

you might have it from another & a better hand. But being uncertain of its having reached You, I will venture to insert it.

It is by meer *Accident*, that in the German *Ephemerides Curiosæ* for the year 1671. The Observation [XXXVII] entituled *Miri fulminis Effectus*,[9] is immediately followed with an observation [XXXVIII] of this Title, *Sal Tartari volatile*.[10] But I shall pass to a *Salt of Tartar*[11] upon the very Intention of reporting the *Effects of Thunder* in it.

A Gentleman of our *Cambridge*, having felled a *white Oak*, his Man that he might serve the Occasions of his *Pipe*, kindled a little Fire about the Stump of the Tree. The odd Bubling & Murmur of the Fire, & the *hard Lumps* of, they know not what, left in the Ashes, awakened their Enquiries into the Matter. One or two ingenious young Gentlemen, soon took the Method of dissolving the Lumps in a convenient Quantity of Water, & So pouring it from the Sediment at the Bottom, & boiling it *ad Siccitatem*;[12] when, behold, it became True & Fine *Salt of Tartar* to the Quantity of perhaps half, more or less, of the Mass that had been dissolv'd. It was not at all to be distinguished from the right *Salt of Tartar*; The *Apothecaries* bought it for such; the *Physicians* used it for such; all the Ends of it were answered. Upon Trial they found a considerable part of the Tree would upon Burning yeild the Same, that they had found in the first Essay. But it was a part of the Tree, that had been wounded, unto all Appearance by the *Lightning*; and had a Fresh Bark so grown over the Wounds of it, as to give, what they please to call the *Nitre*, an opportunity for some Digestion there. They did or might fetch by Computation an hundred weight of *Tartar*, from this one peece of a Tree.[13]

[9] "Observatio XXXVII: Miri Fulminis Effectus, Fulimine Percussorum Cura," by the German physician of Leipzig practicing in Bergen (Norway), Dr. Johann Friedrich Treubler (1626–79), in *Miscellanea Curiosa sive Ephemeridum Medico-Physicarum* 1. Decuria 2. 1671 (Jenae, 1671): 64–69. ("Observation on the wonderful effect of lightning, and the cure of thunder affecting a thunderstruck person").

[10] The follow-up article is "Observatio XXXVIII: Sal Tartari Volatile," by the Swiss physician Dr. Johannes Jacob Wepfers of Schaffhausen (d. 1695), in *Miscellanearum Curiosa* 1. Decuria 2. 1671, pp. 69–70 ("Observation on the volatile tartrate").

[11] Potassium carbonate.

[12] Until [all the water has] dried up.

[13] Alas, Harvard's own Thomas Robie (1689–1729), sometime science tutor, librarian, and physician, beat Mather to the story, in Robie's "An Account of a large Quantity of Alkaline Salt produced by burning rotten Wood. By Mr. Robie of

The *Meteors* we are upon, are what my *Schultzius*[14] has this Note upon, & indeed every body makes it; *Ut omnibus Notissima, ita maximè Terribilia existunt*. May our great Redeemer always preserve us from *Damage* by them; & in the *Fear* of him who creates & governs them. Tis among the Wishes of that *Love*, which is forever *constraining* all the Regards of the most Sincere Friendship for You, from

Sir,

Sept. 26, 1724. Your most hearty

Servant,

Cotton Mather.

Harvard College in Cambridge, New-England. Communicated by the Rev. Mr. Derham, Prebendary of Windsor, F.R.S." *PT* 31.366 (December 31, 1720): 121–24. See also F.G. Kilgour (1939).

[14] Dr. Simon Schultz (1622–79), distinguished German physician in Thorn (Poland) and frequent contributor (of more than 200 medical articles) to the *Miscellanea Curiosa* (aka. German *Ephemerides*). As "they [meteors] are known to all" and engender thunder and lightning, Schultzius seems to argue, "so they prove to be most frightful." See Tylkowsky, *Meteorologia Curiosa*, pars I, pp. 11–15.

8.8. A RARE DISCHARGE OF BULLETS.

September 28, 1724.

To Dr. John Woodward

Syr,

You will not refer yourself to what you may find in *Deusingius* upon that Maxim, *Quia Tu aut Ego, non vidimus, idcircò Rem negari posse.*[1] But I will humbly offer you a *Matter of Fact*, which tho' not *seen* by either of us, has been so *proved* as to be incontestable.

It is astonishing to behold the Provision made in the *Body* of Man, to stave off the *Evils*, which may threaten it. Among the Instances of this Provision, we may reckon the successful Methods & Efforts of Nature in uncommon & very surprizing Ways, to bring all to Rights, & expel & eject an Enemy; when Something is got where it should not be, within us. *Hippocrates* in his Book, *Of Aliments*, is not the only Gentleman, who has admired the Sagacity of Nature, in the finding out *unknown Passages* (more improbable than the *North-west* one, your Navigators talk of,) to discharge things, that are offensive to the Body. Tis an Article well worthy to have the Pen of *The Christian Philosopher*, insist upon it[2]: and they who see nothing admirable in it, may be worthy to have the *Hippocratical Fetters*[3] assigned unto them.

1 Anton Deusing (1612–66), German physician, astronomer, mathematician, and Orientalist, who was appointed professor of medicine at the University of Groningen (Holland). The Latin quotation is from Deusing's *Resurrectio Hepatis Asserta; Contra Socium Larvatum ... Disquisitio Ulterior, de Chyli Motu Atque Officio Hepatis* (1662), p. 59, and cites the adage of Ibn Sinæi, aka. Avicenna, or Abū 'Alī al-Ḥusayn ibn 'Abd Allāh ibn Sīnā (980–1037), a distinguished medical philosopher of Persia. The Latin aphorism translates, "the matter cannot be denied [simply] because you or I have not seen it." The motto is likely buried deeply in Avicenna's *Principis. Ex Gerardi Cremonensis Versione, & Andreae Alagi Belunensis Castigatione* (Venetiis, 1608). The Mather Library (AAS) included Deusing's *De Pulvere Sympathetico* (Groningæ, 1662).

2 Indeed, so it does, and Mather does so as well in his own *Christian Philosopher* (London, 1720), Essay 32, p. 227, where he repackaged the same paragraph.

3 Stephano Roderico Castrensi, *Commentarius in Hippocratis Coi, Libellum de Alimento* (Florentiae, 1635), sect. 2, quaest. 2, Sycoph. 3, §12, p. 264, "Hippocraticis

A learned *Illmerus*, counts it a Matter worth celebrating, That certain *Stones*, made their Way thro' a painful Tumour & Abscess in the *Abdomen* pretty near the *Navel*, of a Woman, in *Hungary*.[4] On this Occasion the *Naturæ Curiosi* seriously enquire, whether there must be no *Magic* in this Occurrence, to secure the *Vessels* that must be broken, if the *Stones* thus discharged, were bred in any of the *Bowels*. But that they may avoid all Danger & Fancy[5] of *That*, they imagine a *Lapidific Spirit* at work among the Muscles of the *Abdomen*.[6]

What I am now to present you withal is a Relation of Things not *bred* in the *Body*, but from abroad *Thrown* into it, and into the very Bowels; and yet making Way thro' the *Abdomen* without any Harm ensuing to the Patient.

The Wife of *Joseph Meader*[7] belonging to our *Dover*, had long been afflicted with that *miserable Distemper* which they call, *the Twisting of the Guts*. Her Physician advised her to swallow a Couple of *Leaden Bullets*; upon which after some time, her Pain was abated and the Use of her Limbs returned unto her. A *Year after this*, one Day walking in a Room she observed a *Bullet* fall from her on the Floor. It gave her an extreme Surprize; for she had no *Symptom* or *Suspicion* of any such Matter; nor did any *Pain* ensue, any more than there did præcede it. But she percieved an *Orifice* a little below her *Navel*, from whence it issued; which yet presently closed & healed without any further Trouble. A

Vinculis alligandi," or "bound by the Hippocratic bonds," such as, "First do no harm," appearing in Hippocrates' *De Alimento* (and in Mather's *CP* [1721], p. 227).

4 Mather leans on an article by the Viennese physician Dr. Ferdinand Friedrich Illmer von und auf Wartemberg (c. 1640–99), professor of medicine at the University of Vienna and personal physician of the last Spanish Hapsburg Emperor Juan Carreño de Miranda (1661–1700). Illmer's article "Observatio XVI: Excretorum lapidum quorundam per Abdomen," in *Miscellanea Curiosa sive Ephemeridum Medico-Physicarum*, 2 Decuria 3 (1684): 59–60, describes the case of a Hungarian woman from whose abdomen were expelled several stones.

5 The copyist wrote, "all Fancy & Danger," which Mather changed to "Danger & Fancy."

6 Illmer hypothesizes, "Manet igitur quod hi lapides ex pituita vitrea vel gypsea, concurrente aliquo spiritu lapidifico, vel acido, fuerint producti; qui cùm panniculum carnosum & musculos abdominis … hunc dolorem, tumorem et abscessum effecerunt" (p. 60): "It remains, therefore, [to be seen] that these stones are produced by vitreous slime or gypsum, combined with some "spiritu lapidifico" (petrifying spirit), or acid, which developed with a fleshy patch and abdominal muscles … and caused this pain, swelling, and abscess." The venerable doctor remembered having read several similar cases of this type of petrification in previous pages of the *Miscellanea Curiosa*.

7 Possibly Abigail Tuttle Meader, wife of John Meader, Jr. (1625–1715); they were married in 1653, in Dover, New Hampshire, and were the parents of at least four sons and three daughters.

little while after, the other *Bullet* also[8] drop'd out much as the former did. But that was at an *Orifice* a little above the *Navel*; and it was attended with more uneasy Circumstances both before & after its coming away; Yet these also were quickly over, & all is well. I am informed, the Woman has had something of an *Hernious Malady*,[9] by which, tis thought, the Discharge might be somewhat facilitated. But the Story needs no further to be enlarged upon.

Attempts to *swallow Bullets* have not always terminated so well; tho' many times they have done well enough.

I remember, many Years ago, I read in Mr. *Hooks* Collections,[10] a tragical Story, related by an honest Chirurgeon whose Name was *Young*, of a poor Man, in whom the *Epiglottis* permitted the Descent of the *Bullet* into his *Lungs*; From which & from other unhappy Experiments, I think, I should endure abundance, before I tried such a Remedy.

One would imagine that *Bullets* entring into the *Abdomen* from the vile Instruments of Death which are daily doing such Execution, should be no more mortal than *Bullets* coming out of it, in such a Manner as has been related.

I wish my Friend may be always out of the Way of all *Bullets*, that may hurt him. The Brace of them with which this Letter is charged, will do him no Hurt. They only come with the *Report* of my being,

Syr,

Sept. 28. 1724. Your Servant, always full of

the best Wishes for You.

[Cotton Mather.]

8 Mather interlineates this word.

9 I.e., a hernia, an abnormal protrusion of an organ, usually the bowel, through the wall of the body cavity.

10 Robert Hooke, F.R.S. (1635–1703), distinguished English physicist, discoverer of the law of elasticity (Hooke's law), professor of geometry at Gresham, and author (among many other works) of *Micrographia* (1665), describing his microscopical discoveries. In his *Lectures and Collections made by Robert Hooke* (1678), pp. 105–12, Hooke reprints "*A Relation communicated to me in a Letter by that ingenious and experienced Chirurgion Mr. James Young* [Yonge] *of* Plimouth, ... *of the fatal Symptoms caused by a Bullet Swallowed into the Lungs*." Yonge (1646–1721), surgeon of the Royal Navy, describes the case of Anthony Williamson of Liscard [Liskeard] (Cornwall), age 65, who swallowed "two Musket Bullets," along with "some Carminative Clyster" in an effort to cure himself of a colic. When his pain returned, he swallowed "three Pistol Shot" by casting them down his throat while lying on his back. Alas, one of the bullets was inexplicably lodged at the opening ventricle of his lungs and restricted his breathing, of which condition he died some years later. An autopsy revealed that one of the bullets had remained lodged in his trachea. Two copies of the 1678 edition of this work are in the Mather Family Library (AAS).

8.9. NATURE AT WORK WITH A NEEDLE.[I]

September 29, 1724.

To Dr. James Jurin.[2]

Syr,

Perhaps it may appear a little *præposterous*; But I will now first entertain you, with what shall be no other than a long *Threed*, only to draw on a *Needle*.

Many Years ago, it was one of my simple Speculations, that suppose all the *morbific Matter* of our Maladies, (whereof we find so many relieved by *Steel-Medicines*,)[3] were a Mass of *Steel Particles*, then a *Magnet* applied unto proper *Emunctories* of the Body,[4] might give a wonderful Relief, & wondrously fetch them off. While I was in the Midst of my silly Speculations, I stumbled on *Angelus Sala*, and that Chapter in him, which describes the *Emplastrum Magneticum Arsenicale*,[5] & reports the astonishing Effects of it in the *Pestilence*. I took Occasion to inform a certain Practioner, how to prepare this *Magnetic Plaister*, & advise him *how* & *where* & *when* to apply it. The venturesome Spark paid little Regard unto my *Cautions*, but used it almost at hap-hazard, & made a *Polychreston*[6] of it, & got more than a little

1 From ἔμπλαστον "to daub." Thus "emblastrum magneticum" signifies "magnetic plaster."

2 Title, addressee, date, and signature are in Mather's hand.

3 Metallic elements that interact with the human body.

4 An organ or body part that expels human waste.

5 The Italian physician and chemist Angelo Sala (1576–1630) practiced medicine in Switzerland and later in Germany. Among his principal works are *De Variis tum Chymicorum, tum Gelenistarum Erroribus* (1602) and, most of all, his *Anatomia Vitrioli, in duos Tractatus divisa* (1609). In tractatus II, cap. 10, pp. 69–73, Sala describes the composition, application, and virtues of "*Emblastrum magneticum in Peste*," which magnetic plaster includes (among other nostrums) gum, ammoniac, wax, turpentine, magnetic arsenic, sulfur, and oil.

6 "Polychreston" is an antidote in the form of a liquid substance, composed of a powder and some juices, syrups, preserves, jams and honey to give it body and consistency.

Money by it. I do not remember all the Cures he wrought by it: But this I do well remember, that he cured obstinate *Heart-Burns* & Pains in the *Stomach* by laying it upon the *Stomach*. And one of his Ventures, I may not forget. He protested vehemently unto me, that he not only found it *forever harmless*, but also in very *malignant Fevers*, when the Patients have been in their *last agonies*, he has clap'd the *Magnetic Plaisters* on their Breasts, and within a few Hours the Plaister would attract all the Malignity of the Distemper & fall off with an intolerable Stench; and the Patients have to Admiration recovered.

A more Cautious Physician than he, having at my Desire, furnished himself with this Remedy to be used upon Occasion, was called unto a Gentlewoman, who had long been tortured & languished under the Miseries of an Arm sorely inflamed & greatly tumified.[7] Many Means had by many Hands been used in vain; But my Physician being entirely at a Loss, what else to do, clap'd on the *Magnetic Plaister*. And when he came a few Hours after, to take off the *Plaister*, Behold, it fetch'd out a small *Needle* with it; and the Gentlewoman who could give no Account how the *Needle* came into her Arm, yet now being thus delivered from it, her Pain was all over, and all come to rights immediately.[8]

All I shall add upon this, is; That from what our Celebrated *Boyl*,[9] as well as what the old Chymist I first mentioned,[10] has written about the *Magnetic Plaisters*, it seems a Pity that no more Experiments have been made upon it. But it is now Time for my *Needle* to come on.

I will not entertain you with the Stories of *Aciculæ deglutitæ*,[11] coming away with the *Urine*, related by *Languis* & *Claudinus*;[12] or coming away

7 Swollen or puffed up.

8 A similar report is given in "The Case of Mary Howell, who had a Needle run into her Arm, and came out at her Breast," *PT* 41.461 (December 31, 1741): 767–68. It demonstrates – at the very least – that Mather's story was hardly out of place for the time. See also Ambroise Paré, *On Monsters and Marvels*, ch. 14, pp. 48–49.

9 Robert Boyle (1627–91), Anglo-Irish natural philosopher, considered the founder of modern chemistry. In his *The Sceptical Chymist* (1661), pp. 78, 113, Boyle takes a skeptical view of his Flemish colleague's *De Magnetica Vulnerum Curatione* (1621), "On the Magnetic Cure of Wounds," by Jan Baptist van Helmont (1579–1644).

10 Angelo Sala (see above).

11 Swallowed needles.

12 Johannes Lange (1485–1565), German physician, surgeon, and author of an oft-reprinted *Epistolarum Medicinalium* (1589) and Julius Caesar Claudinus (unidentified), respondent to Langius's debate about the excretion of a needle in the urine of

thro' the perforated *Stomach*, related by *Benivenius*;[13] or coming out by the side of the *Neck* related by *a Rejes*.[14] If I could be under any Temptation to stop you here, with any such Relations, it should be with what *Kerckringius*[15] has given us, of a *Loadstone* fetching away such a *Needle* thro' the *Throat* where it stuck: And with what *A Castro*,[16] and *Moinechen*[17] have given us, of *Swallowed Needles*, which were at last *E crure erumpentes*.[18] But I shall supersede all, with a Narrative of a *Needle*, that was not *swallowed* indeed; but may be as worth telling as that of *Dorstenius* in the *German Ephemerides*.[19]

---

a patient. Both are listed in Mather's principal source: Johannes Daniel Dorsten's article (cited below).

13 The esteemed Renaissance physician, anatomist, and pathologist Antonio di Paolo Benivini, aka. Benivenius (1443–1502) is now mostly remembered for his *De Abditis Nonnullis ac Mirandis Morborum et Sanationum Causis* (1507), a work on anatomy and pathology based on his pioneering use of autopsy. In chapter 20 (n.p.) of this work, Benivenius relates the case of a woman who had swallowed a bronze needle, which passed through her stomach lining and came out of a small hole in her side.

14 The Spanish physician Gaspar de los Reyes Franco of Carmona (fl. 17th century.) authored his distinguished *Elysius Iucundarum Quæstionum Campus Omnium Literarum Medicis Inprimis* (1661). In Quæstio 92 (p. 693), he relates the case of a 15-year-old girl who accidentally swallowed a needle which a month later came out on the side of the neck near the ear.

15 The Dutch anatomist and physician Theodor Kerckring (1638–93), who practiced mostly in Amsterdam and Hamburg, is remembered for his *Opera Omnia Anatomica: Contenti Specilegium Anatomicum,* editio secundæ (1717). In his Observatio LXIV, pp. 99–101, he relates the case of a young maidservant who (you guessed it) imprudently swallowed a needle, which was extracted with a magnet from her throat.

16 The Portuguese physician Rodericus à Castro (1546–1627) practicing in Hamburg, published his two-volume *De Universa Mulierum Medicina* (1603–04), which went through several reprints.

17 The Danish physician Henricus à Moinichen, aka. Henrik Moinichen (1631–1709), published his *Observationes Medico-Chirurgicae* (1688). In Observatio XXI, pp. 64–66, he relates the case of a Venetian woman, who like her fellow unlucky compatriots accidentally swallowed a needle, but for forty years suffered no pain until at last she felt a stabbing pain in the inside of her leg which was caused by a swelling; yet somehow, the meandering culprit was expelled from her leg.

18 "breaking out of the leg."

19 Johann Daniel Dorstenius, aka. Dorsten (1643–1706), professor of medicine in Marburg (Germany), published more than twenty articles in the *Miscellanea Curiosa*. Here Mather refers to Dorsten's "Observatio LIX. De Aciculâ deglutita," in

An Honourable Member of the Kings Council in our Province of *New-Hampshire*, has lately thus written to me.[20]

> You will permit me to relate a Thing worthy of your Speculation; Of a young Gentlewoman of my own Family (the Truth whereof you may assert,) which is as follows. Being overset in a Chaise, she was Stun'd with the Fall; & coming to herself was sensible of no other Hurt, but a Pain in the outside of her Thigh, where there was a small Rising as big as a Pea; which disappeared in a few Hours; but the Pain continued. She had the Advice of several of the best Physicians & Cheirurgeons in *Boston*, (where she then was) for a Month or five Weeks. But their *Opinions* were various, what should be the Cause; and their *Præscriptions* as various. She found Relief by Nothing. The Pain encreasing, her Friends, being sixty Miles from her, fetched her Home. Riding was exceeding painful to her; the {pained} part being next to the Person that rode before her. When she came Home, she stil complained; But the Pain was removed unto the Midst of her Thigh; & by Degrees more to the Inside; until she could perceive, when she touched it, something that shee thought like the Head of a *Pin* pressing against her Hand. The Cheirurgeons being stil at a Loss, at last advised to lay to the Place something that should be *drawing*. In a Day or two it was felt between the Flesh & the Skin; & in a few days more it plainly appeared the Point of a middling *Needle*; which he drew out with an Instrument; after it had been eight Weeks in the Flesh; & of itself wrought thro' the Thigh. She then remembred, That her *Housewife*[21] (as the Ladies call it) was open in her Pocket, when she recieved her Fall; and that a *Needle* had fallen out of it. Her Cheirurgeon Says, That had the

---

Miscellanea Curiosa Medico-Physicarum Germanicarum Academiæ. Decuriæ 2 Annus Tertius, Anni 1684 (1685): 147–49. It describes the case study of a Marburg tailor's daughter, who swallowed a bronze needle half a finger long, with almost no pain, except that she could hardly swallow. The needle was again extracted from the outside of her neck, below the scutiform cartilage (L-shaped cartilage attached to the auricular muscle of the lower ear canal); she is now alive and well, we are assured. Dorsten's article is Mather's principal source for the above paragraph.

20 Samuel Penhallow (1665–1726), speaker of the General Assembly, treasurer, justice in the superior court, and later chief justice of New Hampshire under Governor Samuel Shute. Penhallow also supplied Mather with information on various other oddities of nature he had discovered. See for instance Penhallow's letter (19 d. 7 m. [September] 1712) to Mather (*D* 2:175–76) and our CA 7.7 "*Subterranea*"; as well as Mather's "Letter to Samuel Penhallow" (24 *d* 10 *m* [December] 1713), requesting information about the twin sisters who died on the same day (*D* 2:291). See our CA 2.7 "A Strange Sisterhood."

21 A small case for needles, thread, and other sewing items.

Needle stop'd against a Bone or Artery, or met with some other Parts, it would have killed her.

I add no more. But if you will please to allow, that in my poor Essays to entertain & gratify you, *Rem Acu tetigi*,[22] it will be an abundant Satisfaction, to,

Syr,

one always glad & Proud of Serving

Sept. 29. 1724. You,

Cotton Mather.

[22] I.e., "I touched the matter with a needle."

8.10. UNCOMMON DENTITION.[1]

September 30, 1724.

To Dr John Woodward.[2]

Sir,

Your *Galen*[3] observes, that we commend the Skil & Sense of him, that shall well marshall a Company of *Thirty two*. But how much ought we to admire the Contrivance of our Glorious Creator, in marshalling the *Thirty two*,[4] which in our *Jaw-bones* have by Him their Station, so regularly, so exquisitely assign'd unto them![5]

The *old Observations* of this Matter, are well-known. But there is to be received among the *new* ones, what the Inquisitive and Ingenious *Raw* affirms unto us;[6] That the *Seeds* of all the *Teeth* ly hid in the Sockets of the Gums or

1 Title, addressee, date, and signature are in Mather's hand.

2 Mather's friend and correspondent John Woodward (1665–1728), an English geologist, natural philosopher, physician, and distinguished member of the Royal Society of London. Woodward had made Mather a present of his popular *Essay toward a Natural History of the Earth and Terrestrial Bodies* (1695).

3 Galen of Pergamum (129–c. 210 CE), Greek physician and surgeon, who provided the basis for the humoral view of the human body. In the present context, Galen deserves credit for arguing that the teeth of a fetus form *in utero* and develop when the skull hardens (odontogenesis), in *De foetuum formatione libellius* (IV.673, 16–674, 1 Kühn) and *De semine* (IV.637, 10–14 Kühn). On Galen's views on dental anatomy, see D.C. Koutroumpas and H. Koetsi-Kounari (2012).

4 Galen, *De usu partium corporis humani* (III.868, 18–869, 2 Kühn) and *De ossibus ad tirones* (II.753, 9–10 Kühn). See also Mather's teleological discussion of human teeth in *CP* (1721), "Essay 32, *Of Man*," pp. 255–57; and in his *ABJ* (1972), ch. 11, pp. 61–62.

5 This and the following paragraph are integrated in Mather's discussion of treating toothaches, in "Cap XI. *Dentifrangibus.* Or The Anguish, and Releef, of the Tooth-ache," in *ABJ* (1972 ed.), pp. 61–66.

6 The Dutch surgeon and anatomist Johann Jacob Rau, aka. Raw, better known by his Latinized name Johannes Jacobus Ravius (1668–1719), professor of surgery

Jaws of a *Fœtus*; and that as many *Seeds* as have lain latent there, so many *Teeth* a Man should have in his Life. He is able to demonstrate unto the Eye, the *Seeds* that were hid in the Gums of a new-born Infant. Every *Fœtus* contains the *Seeds* of the compleat Number of *Teeth*, which he is to have, which are *Fifty two*. Besides the *Thirty two* Teeth which are found in a perfect Man, there are the Tracts of *twenty* more, which, it seems, are to be found in every Infant. How the *New Dentition* of old Men past an Hundred must be accounted for, – *Melius inquerendum*.[7]

But what *Anomalies* do we sometimes meet withal?

The Illustrious *Thomas Bartholinus*, in his *Historiæ Anatomicæ rariores*, has thought the Article of *Dentes Continui*, worthy of a Place among his *Rarities*.[8] He not only mentions it as a *Curiosity*, that *Pyrrhus*,[9] had not any *Distinction of his Teeth*, but he counts it may deserve a mention, that his renowned Grandfather *Finckius*, often fed in his House at *Copenhagen*, a *Groenlander*, who, *unum continuumque; Dentem totam Extremitatem maxillæ implentum curiosis exhibuit spectandum*.[10]

You will then pardon me, if I hope, it may pass muster among the Communications, with which my Letters are ambitious to entertain you; That a Young Gentleman of our *Harvard-Colledge* t'other Day, doing something, which made the friendly standers-by tell him, *He would spoil his Teeth*, he presently convinced them, there would be no mighty Danger of it. For he shewed them, his *Jaw-bones*, which instead of *Teeth* had only a *white, sharp,*

---

and anatomy at the University of Leiden, and famous for his operations of prostatic lithotomy. See E. Riches (1968), p. 192.

7 "Make a formal inquiry."

8 The chapter title in Thomas Bartholin's *Historiarum Anatomicarum Rariorum Centuria I et II* (1656), Centuria I, cap. 35, pp. 48–50, signifying "continuous or interconnected teeth," a rarity which Bartholin found in a "barbarian" from icy Groenlandia (Greenland).

9 The Greco-Roman biographer Plutarch relates in his biography of Pyrrhus of Epirus, famous Greek king and conqueror (319–272 BCE), that "he had not many teeth, but his upper jaw was one continuous bone, on which the usual intervals between the teeth were indicated by slight depressions" (Plutarch, *Vitae Parallelae: Pyrrhus* 3.4).

10 Thomas Bartholin (1616–80), Danish physician and theologian, *Historarium Anatomicarum Rariorum* (Centuria I, cap. 35, p. 49), refers to his grandfather Thomas Fincke (1541–1656), Danish physician and author of a geometry textbook, who entertained the person in question. Most likely an Inuit from Danish Greenland, he had "one continuous [toothlike ridge]; he exposed the entire extremity of the tooth of the jaw for the curious to look at."

continued Edge, and the *Gums* reaching very near the Top. And herewith, he not only chews his ordinary Food, but also gnaws the hardest Bisket, that can be given him.[11]

This *peculiarity* is *hereditary* to the Family; his Father, a Good Man of our *Dedham*, has it; and for how many Generations his Ancestors had it, I have not been informed.

But nothing, that any *Teeth* ever had under their Mastication can be so grateful to me, as the instructive Things, which you can & will give me to feed upon. I beseech you, Lett there be therein a *Continual Feast* allow'd unto.

[Boston, New England.] Syr,

Sept. 30. [1724.] Your [...][12]

[C. Mather.]

[11] This may possibly be referring to Daniel Dwight, born in Dedham in 1707, graduated Harvard class of 1726, who went on to serve as a preacher and merchant in Hanover, Mass., and died in Dedham in 1747. His father was Michael Dwight (1679–1760). See F.L. Weis (1936), p. 75.

[12] Both the MS. draft and copy have this incomplete closing.

8.11. *ANGEL OF BETHESDA* ADVERTISEMENT.

October 5, 1724.

To Dr. James Jurin.

Syr,

You have so encouraged me, by the kind Reception, which my former Communications have had with you, and by your Means with my Illustrious Masters, that I cannot but, in my poor Way, continue them. I wish, that they had been more valuable for Curiosity or Erudition; But they are what I have. And you will have the Goodness to consider me, as a Man exceeding full of Employments; Able but now & then after a mean Manner to express my Zeal for your Noble Design. Tis indeed nothing but that well-meaning Zeal, that can bespeak for me, the Room you are pleas'd to allow me in a SOCIETY, which I esteem as one of the most Illustrious in the World.[1]

Tho' my Studies & Labours most run in a *Theological* Channel, and among the Effects thereof I have lying by me some very considerable Præparations for the Public (and much more valuable Ones, than any of those many more than Three Hundred, which have been on various Arguments and in various Languages, drawn into the Public from the mean Hand that has præpared these,) yett now & then I know not how to deny myself that Satisfaction of some Visits to *Philosophy*. And these have enabled me to gett ready for the Press, a Work of above Thirty Sheets, whereof This is the Title-Page.

THE ANGEL OF BETHESDA.[2]

An ESSAY, upon the COMMON MALADIES OF MANKIND. Offering, First, the Sentiments of PIETY, whereto the INVALIDS are to be awakened in and from

[1] Mather expresses his gratitude to Jurin for seeing to it that the mystery of Mather's (1713) election to membership in the Royal Society was solved. See G.L. Kittredge, "Election" (p. 114). *D* 2:245–46.

[2] His plans for composing a medical handbook with remedies for the most common diseases date back at least to March 1693, when he decided to "collect, at Leisure,

their Bodily *Maladies*; And then, A Rich Collection of Plain, but Potent and Approv'd *Remedies* for the *Maladies*. Accompanied with many very *Practicable Directions* for the *Præservation* of *Health*, to such as enjoy a Good Measure of so Great a Blessing. And many other, curious & grateful & useful *Entertainments* occasionally intermixed.[3]

I may shortly attempt the Publication of this Treasure in this Countrey;[4] And, I hope, I shall have your good Wishes to it. I mention this that so you may see, I am at Work for you, tho' I don't Remitt my, *Curiosa Americana*, so fast as I could wish to do. And yett, I pray, tell me; Have you many Correspondents, whose Remittances are more than this your poor *Americans*?

In the mean time, I should not have been sorry, if my Letter to you, Justifying the *Inoculation of the Small-Pox*, had been published;[5] — might it at all have

---

a fit Number of most parable and effectual *Remedies* for all Diseases, and publish them unto the world; so, by my Hand, will bee done things that the *Angels* love to do" (*D* 1:163). He finished the work nearly thirty years later (February 1723–4): "If my glorious Lord will please to accept of it," Mather confided in his diary, "it may prove one of the most usefull Books, that have been written in the World. I must now apply myself both to Heaven and Earth, to bring on the Publication" (*D* 2:698–99). Mather's huge manuscript of his medical handbook remained unpublished until 1972 – with the exception of three selections: "Nishmath-Chajim. The Probable Seat of all Diseases" (Cap. V), published separately as *The Angel of Bethesda* (1722); "The Inevitable or, The Measles managed" (Cap. XIX), missing from the holograph manuscript of "The Angel" but published separately as "A Letter, About a Good Management under the Distemper of the Measles" (1713); and "Retired Elizabeth. A Long, tho' very Hard Chapter of a Woman Whose Travail Approaches With Remedies to Abate the Sorrows of Child-bearing" (Cap. LIII), previously published as *Elizabeth in her Holy Retirement* (1710). O.T. Beall, Jr. and R.H. Shryock (1954), published a selection from Mather's medical handbook (AAS) prior to G.W. Jones's *ABJ* (1972) edition.

3 The same opening paragraph appears in Mather's "Proposals for Printing a BOOK entituled; The Angel of Bethesda" (MS. at AAS). See the introduction to *ABJ* (1972), p. 1.

4 Since publishers were not likely to risk the expense of printing a large manuscript without subventions from well-heeled benefactors or by subscription, Mather's "Angel" includes a note inviting subscribers to deposit a down payment or pay the full amount (12 shillings) up front. Anyone sending in "Subscriptions for *Six Books*, may expect a *Seventh* gratis," *ABJ* (1972), p. 4.

5 Mather refers to his March 10, 1721/22 letter to John Woodward/James Jurin, our CA 6.4 "*Curiosa Variolarum*" and to CA 6.3 "A Further Account, of the Method and Success of the Small-Pox Inoculated." His complaint that Jurin did not publish the letter in *the Philosophical Transactions* is perhaps legitimate. CA 6.4 appeared in Charles Maitland's *Account of Inoculating the Small Pox Vindicated* (1722), 58–61; but a large extract of CA 6.3 was published in *PT* 32.370 (March

contributed unto the more General Entertainment, of so marvellous, — but, alas, how Satanically-despised! — a Blessing.

As soon as I can find any capable Undertakers, I will prosecute your Invitatio,[6] with all possible Vivacity. But it is not easy for you to imagine, how destitute we are, partly of well-framed Instruments, but very much more, of agreeable Minds & Hands to use them.

We are Objects for your Compassion. — But it may Be, none more so, than,

Sir,

Boston, N. England. Your Feeble, & yett Hearty,

Oct. 5. 1724. Friend & Servant.

Co. Mather.

The Bearer, (Mr. *Howard Wyborn*)[7] is an Hopeful, Modest, Ingenuous Young Gentleman, of a liberal Education. To whom, whatever Civilities you show, they are shown to,

Your Servant,

Mather.

Dr. Jurin.

31, 1722): 33–35, yet without Mather's name. See G.L Kittredge, "Lost Works," pp. 460–67; and Mather's *D* (2:660, 661–62).

6 Mather alludes to James Jurin's "Invitatio ad Observationes Meteorologicas communi consilio instituendas," *PT* 32.379 (October 31, 1723): 422–27, in which Jurin invited his colleagues with meteorological instruments to send in measurements of precipitation, barometric pressure, and seasonal temperatures in America and in Europe. Jurin supplied Mather with a number of printed copies to be forwarded to those with accurate instruments in the colonies.

7 Possibly Howard Weyburn, b. 1720 in Boston, Harvard class of 1721. See S.F. Weyburn (1911), p. 32.

8.12. INTRODUCTION FOR ZABDIEL BOYLSTON.

December 15, 1724.

To Dr. James Jurin.

Sir,

Having lately addressed you, with even some Number of Letters,[1] I have just now nothing to add; but that I have an agreeable Occasion of Introducing to your Knowledge and Kindness, a Friend that has been to me as the Golden Wedge of *Ophir*.[2]

Tis Mr. *Zabdiel Boylston*,[3] the Sight of whom will doubtless be the more welcome to you, because his Name has already reached you.[4]

He is a Gentleman, whose Performances as a *Cheirurgeon*, (and very particularly in *Lithotomy*)[5] have hitherto been æqualled by no Person in these Parts of the World. And as a *Physician*, he has been to an uncommon

[1] Mather's letters to both Woodward and Jurin (September 21–October 5, 1724). See our Curiosa (Eighth Series).

[2] Isa. 13:12; 1 Kings 9:27–28. Ophir is the (unknown) locale from which Solomon's fleet is to have fetched gold for the temple.

[3] James Thacher's brief medical biography of Zabdiel Boylston, in *American Medical Biography* (1828) 1:185–92, is still as relevant as ever. Boylston's family papers are at the MHS. To date, the only biography of Boylston is Gerald Marvin's "Zabdiel Boylston: Medical Pioneer of Colonial Boston" (1975).

[4] Zabdiel Boylston (1679–1766), a Boston medical practitioner who was the only physician at the time to accept Mather's call (May 1721) for a united effort to examine (and administer) the "new" prophylactic method of inoculation (Mather's *D* 2:620–25). Almost half a decade later, Boylston, F.R.S. (now an elected member of the Royal Society of London and celebrated in England), published *An Historical Account of the Small-Pox Inoculated in New England* (London, 1726, 1730). It is exemplary for its scientific approach and thorough detail.

[5] The surgical operation for bladder stone and related ailments from the urinary tract. See Mather's letter to John Woodward (July 10, 1716), our CA 3.8 "The Stone Mistaken," and *D* 2:396.

Degree successful, and so Beloved and Esteemed, that his Absence for a few Months from us, on his present Voyage,[6] is a matter of uneasy Apprehension to a Multitude.

But that which will more particularly recommend him to your Notice, is, That *This* is the Gentleman who first brought the Way of saving Lives, by the *Inoculation of the Small-Pox*, into the *American* World. When the rest of our Doctors, did rather the Part of Butchers, or Tools for the Destroyer, to our perishing People, and with envious and horrid Insinuations infuriated the World against him, this worthy Man had the Courage and Conscience to enter upon the Practice; and (generously beginning with his own Family)[7] he Alone, with the Blessing of Heaven, saved the Lives of (I think) several Hundreds; yea, at One time, he saved a whole Town[8] from a fearful Desolation, after the Small-Pox had begun to do the Execution of a Great Plague upon it. With an admirable Patience, he sleighted the Allatrations[9] of a Self-destroying People; and the Satisfaction of having done Good unto Mankind, made him a noble Compensation for all the Trouble he mett withal.

You having done so much to oblige the Public in your candid Essays,[10] to procure a just Reputation for a Practice, which if Mankind were not obstinately bent upon Self-Destruction, would soon save the Lives of Millions, it cannot be unacceptable unto you, to have an Opportunity for Enquiring of this Gentleman, what has occurred in his own Experiments; and particularly, how far he can Justify the Account I have given you of those Few, who died after the Inoculation.[11]

6 The Anglo-Irish physician and distinguished member of the Royal Society and of the Royal College of Physicians, Sir Hans Sloane (1660–1753), had invited Zabdiel Boylston to meet the members of the Royal Society of London.

7 On June 26, 1721, Boylston began the experiment by inoculating two black servants and his son (13 years old) and, subsequently, his entire family. The story is told in Boylston's own *Historical Account* (1726, 1730), "The Preface," pp. i, 2–3, 7, 10, 13. In this treatise, Boylston gives a complete list of the inoculated patients, their symptoms, condition, and recovery.

8 Roxbury, Mass. See Capt. John Osborne's independent observation while stationed in Boston during the entire episode, appended to Jurin's *Letter to the Learned Caleb Cotesworth* (1723), pp. 19–22 (see below).

9 Barking, from "To bark out, utter with a bark" (*OED*).

10 See for instance Jurin's *A Letter to the Learned Caleb Cotesworth* (1723), which includes an extract of Mather's letter to Jurin, pp. 6, 17–18, and Capt. John Osborne's independent observation while stationed in Boston during the entire episode, pp. 19–22.

11 See Mather's letter to John Woodward (March 10, 1721–22), our CA 6.4 "*Curiosa Variolarum.*" Inoculation was performed "upon almost three hundred Objects in

Yea, perhaps the Prince & Princess themselves,[12] if informed of such an Ones coming to *London*, may not be unwilling to take some Cognisance of a Person, so distinguished by an Operation of so much Consequence.

I add no more; But commending my Friend unto your Favour, I Request the Blessings of GOD our SAVIOUR on both of you; and subscribe,

Boston, N.E. Sir,

Dec. 15. 1724.

Your most hearty Servant

Co. Mather.

Dr. Jurin.

our Neighbourhood… . Five or six have died upon it, or after it, but from other Diseases or Accidents; chiefly from having taken the Infection in the common way, by Inspiration, before it could be given them in this way of *Transplantation*." See also Jurin's *Letter to Caleb Cotesworth* (p. 6) and Boylston's *Historical Account* (1726), pp. iii–iv.

12 George and Caroline, Prince and Princess of Wales, who became king and queen of England in 1727. Boylston dedicated his *Historical Account* (1726; 2nd ed. 1730) to "her Royal Highness the Princess of WALES" (cover).

Appendix 1

Manuscript Descriptions & Locations

Abbreviations

AAS	American Antiquarian Society
BA	*Biblia Americana* (2010-)
BL	(Sloane) British Library, Sloane Manuscript
CJJ	*Correspondence of Jame Jurin*, ed. Rusnock
FLG	Frederick Lewis Gay Manuscript, MHS
LWIHM	Library of the Wellcome Institute of the History of Medicine
MHS	Massachusetts Historical Society
MP	*Mather Papers* (1868)
PMHS	*Proceedings of the Massachusetts Historical Society*
RSL	Royal Society Library
SLCM	*Selected Letters of Cotton Mather*, ed. Silverman
TP	"Triparadisus." *The Threefold Paradise of Cotton Mather*
WIW	*Wonders of the Invisible World*
[ ]	Mather's brackets in MS
{ }	Editor's braces
†	Headings, addressee, date, and signature in Mather's hand.

First Series

1.1. {Describing 'Biblia Americana' & Giant Fossils.} To John Woodward. November 17, 1712.

Sent copy, RSL, EL/M2/21; scribal copy, BL (Sloane) Ms. 3339, fols 99ff.; transcript, FLG, 1–27. Previously published in *SCLM*, 110–12 (partial); transcript, D. Levin (1988); *BA* 1:582–99.

1.2. {On American Plants & Indian Cures.} To John Woodward. November 18, 1712.

Sent copy, RSL, EL/M2/22; scribal copy, BL (Sloane) Ms. 3339, fols 99ff.; transcript, FLG, 27–28.

1.3. {On American Birds. To John Woodward.} November 19, 1712.

Sent copy, RSL, EL/M2/23; scribal copy, BL (Sloane) Ms. 3339, fols 99ff.; transcript, FLG, 39–47. Previously published in *SCLM*, 113–15 (partial).

1.4. {On Antipathies & the Force of the Imagination.} To John Woodward. November 20, 1712.

Sent copy, RSL, EL/M2/24; scribal copy, BL (Sloane) Ms. 3339, fols 99ff.; transcript FLG, 48–56.

1.5. {Monstrous Births.} To John Woodward. November 21, 1712.

Sent copy, RSL, EL/MZ/25.

1.6. {Remedies Revealed in Dreams.} To John Woodward. November 22, 1712.

Sent copy, RSL, EL/M2/26; scribal copy, BL (Sloane) Ms. 3339, fols 99ff.; transcript, FLG, 65–72. Previously published in *SCLM*, 115–18 (partial).

1.7. {Remarkable Recoveries from Wounds.} To John Woodward. November 24, 1712.

Sent copy, RSL, EL/M2/27; scribal copy, BL (Sloane) Ms. 3339, fols 99ff.; transcript, FLG, 72–87. Previously published in *SCLM*, 118–19 (partial).

1.8. {Astronomical Observations.} To Richard Waller. November 24, 1712.

Sent copy, RSL, EL/M2/28; scribal copy, BL (Sloane) Ms. 3339, fols 99ff.; transcript, FLG, 88–98. Previously published in *SCLM*, 119–20 (partial).

1.9. {Rainbows & Parhelia.} To Richard Waller. November 25, 1712.

Sent copy, RSL, EL/M2/29; scribal copy, BL (Sloane) Ms. 3339, fols 99ff.; transcript, FLG, 99–107.

1.10. {Joseph Beacon's Vision or Dream.} To Richard Waller. November 26, 1712.

Sent copy, RSL, EL/M2/30; copy in BL (Sloane) Ms. 3339, fols 99ff.; transcript, FLG, 109–19. Previously published in part in *WIW* (1693), 80–82; *SCLM*, 120–22 (partial); *TP* (1995), 115–16 (partial).

1.11. {On the Rattlesnake.} To Richard Waller. November 27, 1712.

Sent copy, RSL, EL/M2/31; scribal copy, BL (Sloane) Ms. 3339, fols 99ff.; transcript, FLG, 119–30. Previously published in G.L Kittredge, "Letters of Samuel Lee and Samuel Sewall Relating to New England and the Indians" (1912), 174 (partial).

1.12. {Thunder & Lightning.} To Richard Waller. November 28, 1712.

Sent copy, RSL, EL/M2/32; scribal copy, BL (Sloane) Ms. 3339, fols 99ff.; transcript in FLG, 130 –40.

1.13. {Cases of Longevity & Fruitfulness.} To Richard Waller. November 29, 1712.

Sent copy, RSL, EL/M2/33; copy, BL (Sloane) Ms. 3339, fols 99ff.; transcript, FLG, 141–50.

Second Series

2.1. [A Woollen Snow.] {To Richard Waller.} December 1, 1713.

Draft, MHS, Ms. N-527, 315–22. Previously published in *SCLM*, 134 (partial).

2.2. [Surprising Influences of the Moon.] {To Richard Waller or Dr. John Woodward.} 1714.

Sent July 11, 1716.

Draft, MHS, Ms. N-527, 323–24; BL (Sloane) Ms. 3340, fol. 293a. Previously published in *SCLM*, 136 (partial).

2.3. [A Monster.] {To Richard Waller or Dr. John Woodward.} June 21, 1714.

Draft, MHS, Ms. N-527, 327–28. Previously published in *SCLM*, 132 (partial).

2.4. [The Moose.] Containing the Description of the *Moose Deere*. To Richard Waller. June 21, 1714.[1]

Sent copy, RSL, EL/M2/35; draft copy, MHS, Ms. N-527, 327–28; transcript, FLG 169–72.

2.5. [The Pigeons.] {To Dr. John Woodward.} June 1714.

Draft, MHS, Ms. N-527, 331–32, 373–74. Previously published in A.W. Schorger (1938); *SLCM,* 137–38.

2.6. [Long Fasting.] {To Dr. John Woodward.} June 1714.

Draft, MHS, Ms. N-527, 335–36.

1 See Paul Dudley's "A Description of the *Moose-Deer* in America," *PT* 31.368 (August 31, 1721).

2.7. [A Strange Sisterhood.] {To Richard Waller or Dr. John Woodward.} June 1714.

Draft, MHS, Ms. N-527, 337–40.

2.8. [Uncommon Ideots.] {To Richard Waller or Dr. John Woodward.} June 1714.

Draft, MHS, Ms. N-527, 341–42. Previously published in *SCLM,* 138–40 (partial).

2.9. [An Unusual Discharge.] {To Dr. John Woodward.} June 1714.

Draft, MHS, Ms. N-527, 343–44. Previously published in *SCLM,* 140 (partial).

2.10. [An Ear Strangely Molested.] {To Dr. John Woodward.} June 1714.}

Draft, MHS, Ms. N-527, 345–46.

Third Series

3.1. [Monstrous Impregnations.] {To Dr. John Woodward.} {July 2, 1716.}

Draft, MHS, Ms. N-527, 367–70; imperfect scribal copy, BL (Sloane) Ms. 3340, fols 277–79.

3.2. [A Monstrous Calf.] {To Dr. John Woodward.} {July 3, 1716.}

Undated draft, MHS, Ms. N-527, pp. 363–66; dated and addressed scribal copy, BL (Sloane) Ms. 3340, fols 280–82. Previously published in *SLCM*, 209–10 (partial).

3.3. [The Nidification of Pigeons.] {To Dr. John Woodward.} {July 4, 1716.}

Undated draft, MHS, Ms. N-527, pp. 385–87; dated and addressed sent copy, BL (Sloane) Ms. 3340, fols 282b–284a.

3.4. [A Triton.] {To Dr. John Woodward.} {July 5, 1716.}

Undated draft, MHS, Ms. N-527, 397–405; dated and addressed scribal copy, BL (Sloane) Ms. 3340, fols 284b–290a. Previously published in *SLCM*, 210–12 (partial).

3.5. [A Serpent Securely Handled.] {To Dr. John Woodward.} {July 6, 1716.}

Undated draft, MHS, Ms. N-527, 379–83; dated and addressed sent copy, BL (Sloane) Ms. 3340, fols 290b–92b.

3.6. [A Strange Mischief to the Eyes.] [To Dr. John Woodward.] {July 7, 1716.}

Undated draft, MHS, Ms. N-527, 357–58.

3.7. [Strength of Imagination.] {To Dr. John Woodward.} {July 9, 1716.}[2]

Undated draft, MHS, Ms. N-527, 371–74, 332–33. Previously published in *SLCM*, 212–13 (partial).

3.8. [The Stone Mistaken.] {To Dr. John Woodward}. {July 10, 1716.} [3]

Undated draft, MHS, Ms. N-527, 375–77. Lost in MHSM, Ms. 3340.

3.9. [Surprising Influences of the Moon.] {July 11, 1716.}[4]

Draft, MHS, MS. N-527, 323–24; scribal copy, BL (Sloane) MS. 3340, fol. 293a. Previously published in *SLCM*, 136 (partial).

3.10. [Curiosities of ye Small Pox.] {To Dr. John Woodward}. {July 12, 1716.}

Undated draft, MHS, MS. N-527, 359–62; dated and addressed scribal copy, BL (Sloane) MS. 3340, fols 293b–95b. Previously published in Kittredge, "Lost," p. 422 (partial); *SCLM*, 213–14 (partial).

3.11. [The *Fagiana*.] {To Dr. John Woodward.} {July 13, 1716.}[5]

Undated draft, MHS, MS. N-527, 347–49; dated and addressed sent copy; enclosures: plant sample and drawing.

3.12. [*Curiosa Botanica*.] {To Dr. James Petiver.} September 24, 1716.[6]

Draft,[7] MHS, MS. N-527, 351–56; sent copy, BL (Sloane) MS. 4065, fol. 255; enclosure: samples of "six or seven plants." Previously published in *SLCM*, 216–17 (partial).

3.13. [A Singular Limestone.] To Dr. John Woodward. September 24, 1716.

Draft, two leaves, MHS, MS. N-527, 389–91. Enclosure: sample of salt-water mineral accretion.

2 The British Museum sent copy [BL (Sloane) MS. 3340] is lost. George L. Kittredge provides the bibliographical background on this letter in "Scientific," pp. 9, 25.

3 Kittredge "Scientific," p. 40, infers this date.

4 Kittredge, "Scientific," pp. 28–30, 41, claims that this is the same letter, just redated, as CA 2.2 "Surprising Influences of the Moon." The BL (Sloane) copy, now lost, is supposedly incomplete, but carries this date. See Kittredge, "Lost Works," p. 422.

5 Presumably there is a copy of the letter in the Sloane MS. but Kittredge fails to provide the exact citation, even though he gives the recipient and date.

6 See Kittredge, "Scientific," pp. 20, 21, 23, 36, 41–43, for bibliographical comments on CM's letters to Petiver.

7 The editors elected to use the MHS Draft because the sent copy, BL (Sloane) MS. 4065, fol. 255, lacks Mather's description of the samples, which Petiver most likely removed for inclusion in his subsequently published catalogue *Petiveriana III* (1712).

Fourth Series

4.1. [Fossils.] To Dr. John Woodward. July 25, 1717.[8]

Undated draft, MHS, MS. N-527, 393–95; enclosure: a box containing twenty to thirty fossil samples. Previously published in *SLCM*, 233–34 (partial).

4.2. [An Eclipse Observed.] To Dr. Edmond Halley. October 1, 1717.[9]

Draft, MHS, MS. N-527, 407–09. Previously published in *SLCM*, 234–35 (partial).

4.3. [An horrid Snow.] To Dr. John Woodward. December 10, 1717.[10]

Draft, MHS, MS. N-527, 415–18. Previously published in K.B. Murdock (1926), 373–77; *SLCM*, 245–47 (partial).

4.4. [A Prodigious Worm.] To Dr. John Woodward. December 12, 1716 or January 12, 1717.[11]

Draft, MHS, MS. N-527, 411–14. Previously published in E.H. Christianson (1993).

Fifth Series[12]

5.1. A Strong Digester. [To John Chamberlayne.] {February 1720.}

No extant MS.[13]

5.2. The Quadrature of the Circle. [To John Chamberlayne.] {February 1720.}

No extant MS.[14]

8 See also Paul Dudley's "Some further Account of the before-mention'd Dudley Fossils," *PT* 46.496 (December 31, 1750): 600–02. See also CA 7.2 "A Monstrous Dragon." To James Jurin. June 3, 1723 – about the fossilized remains of a likely dinosaur. See also Mather, *WM* (1693), 19.

9 1.^{d} VIII.^{m} 1717.

10 10.^{d} X.^{m} 1717.

11 12.^{d} X.^{m} 1717./18. Kittredge ("Scientific Communications," p. 29) notes that the letter is dated December 12, "but possibly it should be assigned to Jan. 12, 1717, since the occurrence took place on Jan. 6, 1717."

12 Kittredge constructs this list from Mather's letter to Jurin on May 21, 1723, which contains a postscript of all letters Mather had sent thus far (excluding the 1712 Series).

13 For a likely textual parallel, see Mather, *Angel of Bethesda* (1972), ch. XLII.

14 See John Wallis, "An Account of a small Tract, entitutled, Thomae Hobbes Quadratura Circuli, Cubatio Sphaerae, Duplicatio Cubi ..." *PT* 4.55 (1669), 1121–22; and A. Bird (1996).

5.3. The Whale. [To John Chamberlayne.] {February 1720.}
No extant MS.[15]

5.4. Prodigies. [To John Chamberlayne.] {February 1720.}
No extant MS.

5.5. Odd Effects of Thunder. [To John Chamberlayne.] {February 1720.}
No extant MS.[16]

5.6. A large description of *Mississippi*; by one who sojourned several years in the Countrey. [To John Chamberlayne.] {February 1720.}
No extant MS.[17]

5.7. A Rocquing Stone. [To John Chamberlayne.] {February 1720.}
No extant MS.[18]

5.8. A Sea-Lion. [To John Chamberlayne.] {February 1720.}
No extant MS.

5.9. A Water-Dove. [To John Chamberlayne.] {February 1720.}
No extant MS.[19]

5.10. The Locusts. [To John Chamberlayne.] {February 1720.}
No extant MS.[20]

5.11. *A Voice from Heaven*. The *Aurora Borealis*. [To John Chamberlayne.] {February 1720.}

15 Significant portions of CA 5.3 "The Whale" are likely incorporated in *CP* 1721), Essay XXIX, pp. 176–78; *TC* (1717), 26–28; and *BA* 6: Jonah 1:1, 17. S*ee* also Paul Dudley's "An Essay upon the History of Whales, with a particular Account of the Ambergris found in the Sperma Ceti Whale," *PT* 33.387 (March–April 1725): 256–69; and Stearns, *Science in the British Colonies of America*, pp. 466–67. Both Mather and Paul Dudley received their information from the same sources (ibid., 467). See also CA 7.4 "*Carcharias Mirabilis*."

16 See CA 8.7 "More and Strange Effects of Thunder."

17 Based on the account by Joseph Kellogg. See Stearns, *Science* (1970), pp. 460–61; and Stearns, "Joseph Kellogg" (1936).

18 See CA 8.5 "A Movable Rock."

19 The substance of this letter has been reconstructed from Mather's gloss on Gen. 8:8, in *BA* (1:643–44). See also *MP* (1868), 436 *n*.

20 The substance of this letter was incorporated into *BA* (2:184–85, 868–69), entries on (1) Exod. 10:6 and on (2) Numb. 11:33.

No extant MS. Likely published substantially in Cotton Mather, *VH* (1719).[21]

5.12. A Journal of a Voyage round the World. [To John Chamberlayne.] {February 1720.}

No extant MS.[22]

5.13. *The World Alarm'd. A Surprizing Relation, Of a New Burning-Island.* [To John Chamberlayne.] [June 10, 1721.][23]

No extant MS. Likely published substantially in Cotton Mather, *WA* (1721).[24]

Sixth Series[25]

6.1. *An Account of the Method & Success of Inoculating the Small-Pox in Boston in New-England.* Sept. 7, 1721.

No extant MS.[26]

6.2. *Variolæ triumphatæ*: or, The Small-Pox Inoculated. {To Dr. John Woodward.} {1721.}

No extant MS. Incorporated substantially into Cotton Mather, "Angel of Bethesda," Cap XX, "*Variolae trimphatae.* The SMALL-POX Encountred" (MS., AAS, Part One, 112–41; Part Two, 142–47). [27]

6.3. A Further Account, of the Method and Success of the Small-Pox Inoculated. {To Dr. John Woodward.} {1721.}

No extant MS. "The Way of Proceeding in the Small Pox Inoculated in New England" was published in *PT* 32.370 (March 31, 1722): 33–35, appears to have been part of a much longer letter.[28]

[21] CA 5.11 "*A Voice from Heaven.*" (1719), dated "24d. Xm. 1719." on the final page of the pamphlet. See Mather's letter to Thomas Prince, January 25, 1719, in *D* (2:596).

[22] See CM's reference in CA 4.2 "An Eclipse Observed."

[23] Date appears at the end.

[24] See also Thomas Forster's "Concerning a new Island lately raised out of the Sea near Tercera," *PT* 32.372 (August 31, 1722): 100–01.

[25] Based on the 1723 loose-leaf list held at the AAS.

[26] Kittredge claims that this is likely a repeat of CA 5.13 "A Relation of a New Burning Island." See *D* (2:626–27).

[27] Kittredge, "Lost Works," pp. 442–44, 455–56.

[28] Kittredge's assessment ("Lost Works," pp. 460–67) is that "The Way of Proceeding" was a substantial part of Mather's longer letter (Sixth Series), "A Further Account, of the Method and Success of the Small-Pox Inoculated. To John Woodward (c. November 24–December 1, 1721)." See also Kittredge ("Scientific," pp. 21, 48).

6.4. *Curiosa Variolarum*. {To Dr. John Woodward.} {1721.}

Copy text "Mr. MATHERS Letter from *Boston* in *New England*. March 10, 1721–22," in Charles Maitland's *Account of Inoculating the Small Pox Vindicated* (1722), 58–61. Sent copy (scribal), imperfect, BL (Sloane) MS. 3324, fol. 260.[29]

6.5. *Nishmath-Chajim*; The Probable Seat and General Cure of all Diseases. {To Dr. John Woodward.} {1721.}

No extant MS. Incorporated into the holograph of "Angel of Bethesda" (MS., AAS, Part One, 29–43) and in *ABJ* (1972), Cap. V. Nishmath-Chajim. The Probable Seat of all Diseases, and a General Cure for Them, Further Discovered," 28–38; and in *TP* (1995), 122–26; and in the first edition of *AB* (1722).

6.6. The Seventh Son Examined; with a Touch Upon the Kings Evil. {To Dr. John Woodward.}{1721.}

No extant MS. Incorporated into "Angel of Bethesda" (MS., AAS, 378–80) and in *ABJ* (1972), Cap. LXII, "*Fuga Daemonum,* or, Cures by CHARMS considered, And, a SEVENTH-SON Examined," 293–301. The applicable pages pertaining to the "Seventh Son," under "Mantissa," are on pp. 297–99.[30]

6.7. *Lacus Mirabilis*. [To Dr. John Woodward.] {1721.}

No extant MS.

6.8. An Horrible Tempest. [To Dr. John Woodward.] {1721.}

No extant MS.

6.9. *Ostreophagi* or a Matchless Oister-Hill. [To Dr. John Woodward.] {1721.}

No extant MS.

6.10. *Absinthium sempervivum*. [To Dr. John Woodward.] {1721.}

No extant MS.

6.11. Ambergrise discovered and determined. [To Dr. John Woodward.] {1721.}

No extant MS.[31]

[29] Kittredge claims that this is likely the same as 6.15.

[30] See Kittredge, "Lost Works," p. 473, on the debate between CM and John Williams (*Several Arguments*), 16–17.

[31] See Paul Dudley's "An Essay upon the History of Whales, with a particular Account of the Ambergris found in the Sperma Ceti Whale," *PT* 33.387 (March–April 1725): 256–69; and. Stearns, *Science in the British Colonies of America*, 466–67.

6.12. *Noevi materni*. [To Dr. John Woodward.] {1721.}

No extant MS.

6.13. *Messologia*,–with a New Method of Bee-hunting. [To Dr. John Woodward.] {1721.}

No extant MS.[32]

6.14. *An Account of the Method and Success of Inoculating the Small-Pox, in Boston in New-England*. To Jeremiah Dummer. September 7 1721.

No extant MS. Printed as a pamphlet with the same title (1722). The subtitle reads, "*In* a LETTER *from a Gentleman there, to his Friend in* London." Dedication by Jeremiah Dummer.

6.15. Dr. Mather of New-England. To Dr. John Woodward. March 10 1722.[33]

No extant MS. Same as 6.4 [BL (Sloane) MS. 3324, f. 260]?

Seventh Series

7.0. Letter to Dr. James Jurin, May 21, 1723.

Sent copy, RSL, EL/M2/36, four leaves; enclosure: list of previously sent "Curiosa."

7.1. The Case of the Small-pox Inoculated, further cleared. To Dr. James Jurin. May 21, 1723.[34]

Sent copy (scribal), RSL, CLP/23ii/31; docketed: "read Dec: 19: 23." Draft, AAS, MS. 271821, b. 01, f. 16, 3–23; transcript, FLG, 181–86. Extract published in *SLCM*, 360–67; *CJJ*, 15–63.

7.2. A Monstrous Dragon. To Dr. James Jurin. June 3, 1723.†

Sent copy (scribal), RSL, EL/M2/37; draft, AAS, MS. 271821, b. 01, f. 16, 25–28; transcript, FLG, 181–86. Previously published in *CJJ*, 165–67.

7.3. A New Rattlesnake. To Dr. James Jurin. June 4, 1723.†

Sent copy (scribal), RSL, EL/M2/38; draft, AAS, MS. 271821, b. 01, f. 16, 29–31; transcript FLG, 186–89. Previously published in *CJJ*, 167–68.

7.4. *Carcharias Mirabilis*. To Dr. James Jurin. June 5, 1723.†

[32] See also Paul Dudley's "An Account of a Method … for Discovering where the Bees Hive in the Woods, in order to get their Honey," *PT* 31.367 (April 20, 1721): 148–50.

[33] Kittredge mistakenly cites Hans Sloane as the recipient.

[34] Date and signature are in Mather's hand.

Sent copy (scribal), RSL, EL/M2/39; draft, AAS, MS. 271821, b. 01, f. 16, 33–35; transcript, FLG, 189–94. Previously published in *CJJ*, 168–70.

7.5. Mysterious Rains. To Dr. James Jurin. June 6, 1723.†

Sent copy (scribal), RSL, EL/M2/40; draft, AAS, MS. 271821, b. 01, f. 16, 37–39; transcript, FLG, 194–98. Previously published in *CJJ*, 170–72.

7.6. The Land Sail'd Upon. To Dr. James Jurin. June 7, 1723.†

Sent copy (scribal), RSL, EL/M2/41; draft, AAS, MS. 271821, b. 01, f. 16, 41–44; transcript, FLG, 198–203. Previously published in *CJJ*, 172–74.

7.7. *Subterranea.* To Dr. James Jurin. June 8, 1723.†

Sent copy (scribal), RSL, EL/M2/42; draft, AAS, MS. 271821, b. 01, f. 16, 45–46; transcript, FLG, 203–05. Previously published in *CJJ*, 174–75.

7.8. A Singular Case. To Dr. James Jurin. June 10, 1723.†

Sent copy (scribal), RSL, EL/M2/43; docketed: "read Dec: 12:23." Draft, AAS, MS. 271821, b. 01, f. 16, 47–50; transcript, FLG, 205–09. Previously published in *CJJ*, 175–77.

7.9. {Letter of Introduction.} To Dr. James Jurin. June 10, 1723.

Sent copy, RSL, EL/M2/.44; transcript, FLG, 210–11.

7.10. {Letter of Introduction.} To Dr. James Jurin. August 3, 1723.

Sent copy, RSL, EL/M2/45; transcript, FLG, 211–13.

Eighth Series

8.1. [Climatical Influences.] To Dr. John Woodward and Dr. James Jurin. October 1, 1724.†

Sent copy (scribal), RSL, EL/M2/56; draft, MHS, MS. N-527, 421–24; transcript, FLG, 213–19. Previously published in *SLCM*, 397–99 (partial); *CJJ*, 268–71.

8.2. [An *Amphisbœna.*] To Dr. John Woodward. September 21, 1724.†

Sent copy (scribal), RSL, EL/M2/47; draft, MHS, MS. N-527, 425–27; transcript, FLG, 219–22.

8.3. [An Unusual Vomit.] To Dr. James Jurin. September 22, 1724.†

Sent copy (scribal), RSL, EL/M2/48; draft, MHS, MS. N-527, 429–32; transcript, FLG, 223–27. Previously published in *SLCM*, 394–95 (partial); *CJJ*, 271–73.

8.4. [A Tide & Storm of Uncommon Circumstances.] To Dr. John Woodward. September 23, 1724.†

Sent copy (scribal), RSL, EL/M2/49; transcript, FLG, 227–30. Previously published in *MP* (1868), 456–57.

8.5. [A Moveable Rock.] To Dr. James Jurin. September 24, 1724.†[35]

Sent copy (scribal), RSL, EL/M2/50; draft, MHS, Ms. N-527, 431-434; transcript, FLG, 231-232. Previously published in J.J. Babson, *History* (1860), 5-6 (partial); *CJJ*, 273–74.

8.6. [A New-found-Land.] To Dr. John Woodward. September 25, 1724.†

Sent copy (scribal), RSL, EL/M2/51; draft, MHS, MS. N-527, 435–36; transcript, FLG, 233–35.

8.7. [More and Strange Effects of Thunder.] To Dr. James Jurin. September 26, 1724.†[36]

Sent copy (scribal), RSL, EL/M2/52; draft, MHS, MS. N-527, 437–40; transcript, FLG, 236–41. Previously published in *CJJ*, 274–76.

8.8. [A Rare Discharge of Bullets.] To Dr. John Woodward. September 28, 1724.†

Sent copy (scribal), RSL, EL/M2/53; draft, MHS, MS. N-527, 440–43; transcript, FLG, 241–45. Previously published in *SLCM*, 395–96 (partial).

8.9. [Nature at Work with a Needle.] To Dr. James Jurin. September 29, 1724.†

Sent copy (scribal), RSL, EL/M2/54; draft, MHS, MS. N-527, 445–46; transcript, FLG, 245–50. Previously published in *SLCM*, 396–97 (partial); *CJJ*, 277–79.

8.10. [Uncommon Dentition.] To Dr. John Woodward. September 30, 1724.†

Sent copy (scribal), RSL, EL/M2/55; draft, MHS, MsS N-527, 449–50; transcript, FLG, 251–53.

8.11. {*Angel of Bethesda* Advertisement.} To Dr. James Jurin. October 5, 1724.

Sent copy, RSL, EL/M2/56; transcript, FLG, 253–56. Previously published in *SLCM*, 400 (partial); *CJJ*, 279–80.

8.12. {Introduction for Zabdiel Boylston.} To Dr. James Jurin. December 15, 1724.

Sent copy, RSL, EL/M2/57; transcript, FLG, 256–59. Previously published in *SLCM*, 402 (partial); *CJJ*, 284–85.

[35] Same as CA 5.7.

[36] See CA 5.5.

Appendix 2

"Cotton Mather to Richard Waller (December 1, 1712).
RSL, *Letter-Book*, XV, 44–45; R.P. Stearns, *Science* (1970), 405–06.

Appendix 3

"Richard Waller to Cotton Mather" (July 22, 1713).
Draft copy, RSL, MS. EL/W3/75–79.

Appendix 4

"Richard Mather to Cotton Mather" (December 23, 1713).
Draft copy, RSL, MS. EL/W3/75–79.

Appendix 5

"Richard Waller to Cotton Mather" (December 23, 1713).
Draft copy, RSL, MS. EL/W3/75–79.

Appendix 6

An Extract of several Letters from Cotton Mather, D.D. to John Woodward, M.D. and Richard Waller, Esq; S. R. Secr.
In *PT* 29.339 (June 30, 1714): 62–71.

Appendix 7

"Cotton Mather to Richard Waller" (March 10, 1715).
Sent copy, RSL MM.1.18; MDA/A/25/16/57; A5.16.

Appendix 8

"John Woodward to Cotton Mather" (April 3, 1721).
PMHS 13 (1873): 110–11.

Appendix 9

"James Jurin to Cotton Mather" (December 16, 1723).
LWIHM, Wellcome MS. 6146.

Appendix 10

"James Jurin to Cotton Mather" (February 22, 1723/24).

MS copy Wellcome MS. 6146.

Appendix 11

"James Jurin to Cotton Mather" (May 20, 1725).

MS copy Wellcome MS. 6146.

Appendix 12

"Cotton Mather to John Woodward" (c. 1713; sent c. July 17).

MS draft; MHS MS. 319–21. Previously published in *SCM* 135–36 (partial).

Appendix 2

December 1, 1712

To Richard Waller, Esq.[1]

Sir.

For some very good Reasons, which yet there is no need of mentioning, I have addressed certain Letters to you; and joyn'd you with my Excellent friend Dr. John Woodward in a small Collection of *Curiosa Americana*, to the making whereof I devoted near a fortnight a Little before the sailing of the fleet from hence, as far as my many other Employments would give me Leave.

I could have been glad I could have provided a Richer Entertainment for you; But your Candour will accept of these mean things because they were the best, that I could in so little time as I could allow myself to be master of. I despair not that if I live, I may annually entertain you with something or other from these parts of the World that a Secretary of the Royal Society may judge it proper to take some notice of. And tho' I cannot presume so much upon my own merits as to dream of being thought worthy to be admitted a member of that more than illustrious Body but rather to belong to the order of the *fratres Ignorantiae*, yet I will here venture to subscribe the Promise which your Candidates make at their admission, That I will Endeavor to promote the good of the Royal Society of London for the Improvement of Natural Knowledge.

I have Entreated my Dear & Honored Dr. [Woodward] that he would obtain an Interview with you, and there at a perusal of some Testimonies with how much Regard, I am Sir

Yours to Command,

Cotton Mather.[2]

1 See Waller's letter in response, July 22, 1713 (Appendix 3).

2 Royal Society, *Letter-Book*, XV, 44–45 (R.P. Stearns, *Science* [1970], pp. 405–06).

Appendix 3

A Letter from Mr. Richard Waller Secretary
to the Reverend Dr Cotton Mather
upon the Receit of his Letters.
London, July 22^{d}. 1713.

Reverend Sir

It is with all the appearance of Justice, that you may think, I ought to be asshamed to own the Reciept of your obliging Letter dated December the 1st 1712: after so long an omission of returning you a suitable answer, and must beg the ffavour of your accepting for my Excuse, an account of the Truth of the matter that caused the Delay. Sir, your Letter to me mentioning the Curiosa Americana; I apprehended, at first, that they were some Collections of natural Raritys of New-England designed for the Royal societys Repository; till our common friend Dr. Woodward informed me that the Curiosa Americana were several Observations you had made, and communicated in diverse letters you had bin pleased to write partly to him selfe and the Rest to me. The Manuscript of which I had not the Happyness of receiving from him till the 15th instant: when to my surprize I found my selfe obliged in so great a Debt; that at present I must own my selfe unable to make a tollerable satisfaction, and like other unhappy insolvents, desire time for Repayment: I have not since I recieved the MSs had time and leisure to read over all your Letters nor consider duly of all that curious collection of Discoverys and Observations you have therin communicated. Some few of them I read at the next meeting of the Royal Society who were very well pleased and entertained with them, and I was ordered by the first Opportunity, to return the thankes of the Society to you for your Informations, and for the Offer you are please to make of future Obligations of the like Nature: which, as they will bee extreamly acceptable to the Society; so I shall in particular think my selfe happy in cultivating a Philosophical Correspondence with so candid and learned a Person. Sir I tooke the same Opportunity of Proposing you to be a Member of the Society which was recieved to the satisfaction of the Members present. But their being a Formality requisite, of being allowed off by the Council of the Society and of Being chosen in a full Assembly of the Fellows, which at this time of the year can hardly be expected, I doubt the Election will be deferred till after Michaelmass, when I need not question but I shall be able to send you an account, of your being chosen: at which time I shall not fail (God willing) to send you a Letter, which I hope to render more acceptable, by some Informations of several matters of a Philosophical Nature, which have lately bin brought before the Royal Society, and shall from time to time communicate such Discoverys or Observations: Esteeming my selfe happy in being any ways instrumental in promoting an intercourse of such Intelligence, between the old and new World, and for the present desire you would please to believe me

Sir

Your most obliged humble servant

Richard Waller. R.S. Secretary

PS[1]

Sir I was desired to request Some Information from you concerning the Mouse Deer as to its Size Horns Colour Descript and the like. what we have hitherto had being very imperfect & not to be depended on.

Another Quere I have to recommend which is whether in the summer of the year 1711 a Meteor like a globe of fire past one Evening in the Dusk from West to East as is said, with so swift a motion as to pass from One Horizon to the other in the space of two minutes the Truth of which we desire to be informed of it is said Mr. Banister observed it.

[1] Added from Waller's draft copy

Appendix 4

A Letter from Richard Waller Secretary R.S.
To Dr. Cotton Mather at Boston in New England.
Accompanying several Remarks on a Manuscript of Letters.
sent by Dr Mather.[1]
London Dec. 23^{d}. 1713.

Reverend Dr.

Some Months are now past since I sent you by the Conveyance of my good friend Mr. Jeremy Dummer a Letter in answer to your former, which I hope came to your hands, in that I gave you an account of my having recieved your Manuscript of ingenious Letters to Dr. Woodward and my selfe, tho' as I then intimated, they had bin so Short a time in my possession that I could not make a sutable return for the Kindness and the honorable sentiments you are pleased to entertain off me.

As for y^{r} being chosen a ffellow of the Royal Society, that has bin don {.} you have bin approved by the Council and Elected by the Body of the Society, so that only the Ceremony of an actual admission is wanting, which you being Beyond sea cannot be performed, and you being a natural born subject, cannot be inserted among the Gentlemen of other nations in the printed Liste, which is the reason that your name is not printed in the Liste herewith sent you. The chief reason of printing these Listes is for the convenience of the fellows present in choosing the Council President and Officers for the ensuing year.

But to come to the contents of your Letters in your Manuscript at the Desire of the Society I made a large Extract of them which was read at our publique meetings. What Remarks were made by the Gentlemen present with some few of my own I shall take the freedom to send you herewith, desiring your farther Communication, of what you shall meet with relating to Natural Phylosophy, the improvment of which is the present business as well as the first design of the Royal Society nor shall I be wanting in communications of the like nature from hence.

1 In the upper left-hand corner, Waller's notion reads: "Place the Extract of {Dr.} Mathers Letters just before this Letter."

To conclude wishing you may long enjoy the best of terrestrial blessings Mens Sana in corpore sano with leisure for your Philosophical entertainment for as Ovid says of Poetry secessum scribentis et otia quærunt I subscribe my selfe

Reverend Sir &c.

{Richard Waller. R.S. Secretary}

Enter these after the Extract of Cotton Mathers letters which at Present [next to the Letter to him] Mr. Innis has to print in the Philosophical Transactions. They must be procured from the Printers as soon as they are wrought off.[2]

Transcribed.[3]

[2] Written in another hand on a snippet of paper.

[3] In yet another hand.

Appendix 5

Some Notes and Querys drawn up by Mr. Waller
Upon the Reverend
Dr. Cotton Mathers Letters wrote
To Dr. Woodward & Mr. Waller
In November 1712.[1]

In the first place As to the account and Discourse concerning Giants in the first Letter upon the finding the large Teeth and bones found buryed in the Earth.

There may yet remain some doubt whether these were ever human. Wherefore it were very desirable that an exact figure or draught were made of these Teeth and bones if possible to be procured. Especially to the bone 17 foot long supposed to have bin a thigh bone of a Man wherewith the measure or girth of it in its several parts this would determine whether human or not.

A Catalogue and Description of those American Plants, which are strangers to our Climate would be very acceptable to us. But as to this Article I suppose my good friend Mr. James Petiver fellow of the Royal Society will give you a farther Information by Letter as he promised me. Next as to Birds, the largness of your wild Turkeys and Eagles is remarkable an exact figure of these with a scale as to their proportions with what observable in them were very desirable.

{A}s for your Itinerants the vast flights of Pigeons were it not so well attested would be almost incredible. It might be worth notice what winds favour their coming, as likewise with what winds they depart, which might possibly point out their Native country. But as for any other attendant on our Earth except the Moon it could not always so shelter it selfe under the Rays of the sun as to be invisible.[2] To what Species of the Numerous Tribe of the Pigeons these belong, with a figure and description is desired. {I am inclined

1 Holograph manuscript of Richard Waller, RSL Ms. EL/W3/75-79.

2 At this point, Waller's cancels the following passage from his draft. It reveals how Waller struggled with Mather's conjecture that seasonal birds migrate to an unseen satellite orbiting the earth: ~~now will the Laws of gravitation and the circular or elliptical Motion partial motion of allow of such as unseen as if unseen useless Satellite near us~~.

to believe that these Pigeons Quit y^{r} Country as likewise they come to you at some stated seasons in Quest of their food and that there are <some> several yet undiscovered Islands to which they resort and breed in wch by y^{r} acct. I do not find they do with you.}[3]

As to what you mention of their rising up high in the air before their flight, the same is observed of storks & possibly all other Itinerants. The advantage of great heights ~~is~~ <being> very considerable, that higher part of the atmosphere being freer from the fleeting Nebulæ. Mr. Trumballs in his acct of going up to the Pyk of Tenariffe says that looking down from so great a height the Grand Canarys the several miles distant seemed as if just below him. {To this as a hint may be added, that the at least Philosophical use of the Turris Babel & the like el{ev}ated structures of the ancients was for the greater convenience of surveying the subjacent Regions and therby drawing the exactest plans, all beneath lying open and exposed to the exactest mensuration.}[4]

In the Next place as to the Power of the Imagination I allow it to be very great as in the Person imagining so as sometimes to produce wonderfull Effects. as may be seen in the unaccountable actions of distracted persons disorders of a fever performing what might seeme to exceed the very powers of the human Machine & strength of human muscles. But how this vis Imaginationis should act upon the fœtus seemes at least a hard probleme, {and I must own my selfe a kind of an unbeliever as to maculæ Maternæ[5] so frequently mentioned},[6] for I doubt the Maculæ Maternæ upon a due examination will be found to proceed rather from the Imagination of others than the persons affected thus. If a Child happens (as it frequently dos) to be born with any small wartlike excrescence in any part of ye body, the Good women presently cry out the Mother longd for a something for some shall say tis like a Cherry others a Mulberry, Raspberry strawberry or the like when in reality tis only a sort of wart which begging pardon of the Reverend Gentleman who attested it I take to be a true solution of the pea-like excressence on the Childs forehead. Entia neq^{e} miracula sunt multiplicanda alique necessitate.[7] In the Next Place as for Dreams foretelling or discovering Matters after an unaccountable manner; as the matter of fact in many such cases cannot be denyed, so the Notices at least such as may be defended on are too few and imperfect to found a rational or Physicall

3 Waller cancels the passage in braces. Nonetheless, it reveals how he eschewed less-than-likely conjectures.

4 Waller cancels the passage in braces in his draft.

5 Birthmark (lit. mother's mark).

6 Canceled in Waller's draft.

7 I.e., miracles are not to be multiplied as a necessity.

solution of them wherfore as less proper for our disquisition I shall leave them and proceed to the consideration of Monstrous births of which kind we have lately had two very remarkable instances, One was a Pigg without any Gula or passage that way into the stomach yet that pig was the fattest of the whole litter and the contents in the stomach and gutts the same as were in the perfect Pigs, which seems to determin the Question whether the fœtus recieves any nourishment by the mouth whilst in utero or only by the communication with the parent by the umbilical vessells, for at least in this subject that was the only way. Another Instance of Monsters was likewise in a litter of Pigs of which the rest being all perfect only One, in several particulars resembled the Dog kind. The outward hairy skin white with some liver-coloured spotts, the forhead round and prominent like a hound or spaniel the Eye has closed as puppys are with two ears hanging down thin and flat above an Inch long, at the End of the Nose-bone an imperfect beginning of a snout about an Inch long hanging loose, the Tongue and under Chop or jaw, like a dog, the Back round and breast between the forelegs with the tail exactly like a puppy it was also blind the Eyelids being fastened together as puppys are: But the four Legs and feet exactly of the hog kind, as were also all the viscera.

As for wounds which at first seeme mortal yet are happily cured there have bin several extraordinary Instances. I remember a ffrench Surgeon who was employed in the Hospital of the Sick and wounded in the french army in the war between france & Savoy published a little tract about 8 or 10 years since in which he relates several very unusuall cures of wounds particularly I remember one of a Souldier that had bin wounded with a Bayonet which went in a little above the Collar bone and came out at back under the scopula as far as I can now recollect having not the book at present by me to have recourse to. And I my selfe was some years since present with Dr Tyson at the Dissection of a man dying of a fever in Bedlam, but who 7 years before being distracted had attempted to kill himselfe by discharging a pistol against his right Temporal muscle, the Ball entered at that place went under the right Eye tore the optic nerve and the bones of socket and Nose and came out a little above ye corner of ye left Eye which was therby saved, but upon opening the Scull and viewing the brain of this person there was a sort of Callus on the Cortical part of the brain where the bullet had grazed in its passage. as to the Rattlesnake & Magnet needle vide ad finem hujus.[8]

Your accounts of Meteors were very acceptable a communication of all such Appearances being desired. As for the Evening Glade of light about the Equinox, the first account of it was published by Dr. Childrey in his Britannia

[8] "See the end of this [Note]."

Baconica since which some discourses have bin printed of it in the Journal des Scavans, and Dr. Hook wrote a particular Lecture concerning it which is published in his Posthumous works. As for Rainbows & Parhelias Dr. Halley publish an Ingenious account and explication of them in the Transactions several years since. That Meteor you mention appearing like a narrow dark cloud with light in the middle is a subject well worth farther enquiry since it moves in so narrow a space with so great a violence and Noise.

Your Account of the pitts or Receptacle cut time out of Mind in the hard Rocks But especially

The two lines of the unknown Inscription on the Rock are really great Curiositys. If you can as you are pleased to promise procure the Rest of that Inscription it would be very acceptable. There was printed in the Transactions about ye Year 1694 several Inscriptions in unknown Characters taken at the old Persepolis but they were wholy different from your Indian Characters if they are really such, as in all probability they are. which reminds me of what Simon Stevinus has asserted and by many Instances endeavoured to prove that there was formerly even before the fabulous Age, a learned Age in which Arts and learning flourished. and Indeed the Pyramids the Tower of Belus as described by Herodotus and these unknow{n} Inscriptions seem to favour such an assertion.

Thus Sir I have adventured to make some remarks in answer to your ingenious Letters. I had thoughts of troubling you with a severall Inquirys but to a Person so knowing in all the parts of Natural Philosophy they would but unnecessary, since your own genius will readily furnish you with what ever of that kind can be hinted at by any other Especially since you have bin pleased to promise a farther Communication of such matters which we expect to recieve with Satisfaction.

I shall therefore fill up the Remainders of my sheet with an Account of some few of the several subjects that have lately bin communicated to the Royal Society. Dr. James Douglass a very learned and ingenious Physitian and an excellent Anatomist, has made several accurate dissections of the amphibious Genus as of frogs Lizards snakes Vipers &c in which he has made many curious discoverys both as to their generation and manners of life with very accurate designs drawn by the life, which he designs to make publick. The Same Gentleman is now writing a History of Anatome in which he will assert the several Inventions as far as he is able to their true Authors and Inventors, a work really of excellent use and Justice that Every One may have that due praise and honor he deserves, which is the lest we ow to the Memorys of famous men who have laid out their time and pains for benefit of Posterity.

Mr Cheselden an Ingenious surgeon & Anatomist has cmmunicated several curious observations on morbid bodys. viz. of the adhesion of the Pericardice and its being partly ossifyed. Of the Vena Azygos inserted into the right Ventricle Part of ye Dura Mater ossifyed. of a double ureter on one side from the Kidney the other as usual. The Tubæ fallopianæ impervious. A woman having a Child after a very great Prolapsus Uteri, a spleen reaching from the Diaphragm to the Os Ilium in circumference round the Edges 35 Inches weighing above 4 pounds. Pease coming out of an Imposthumation in the groin, the Patient afterwards cured & I omitt giving an account of several Experiments that have bin made and Discourses or other matters that have bin communicated having I fear I already tired the patience as likewise any acc^t^ of new books lately published, only there is One I ought not to omitt it being by far the best that has yet bin published of its kind. the Title is. Physico-Theology: or a Demonstration of the Being and Attributes of God from his workes of Creation being the substance of 16 sermons at the honourable Mr. Boyls Lecture preached in 1711 & 1712 by William Derham Rector of Upminster in Essex. 8vo. pag. 483.

Your account of the Rattle snake & the juice contained in the Gall bladd^er^ is I believe perfectly new as that likewise of the Trochees made therof which you record a good cordial sudorifick. I am informed that at least to me a new sort of medicine has bin given with good success in the intermitting fever viz. 2 or 3 granes of the Common house spiders web which has only caused a kind of Nauseating in the stomach of ye patient but without any vomiting purging or great sensible evacuation cured the Patient.

The Account of the Magnetical Needle altered by a Storme of Thunder we have had before but your confirmation in saying that at your writing you had one of those so disordered Needles before you on the Table is a farther attestation, and possibly Mr. Boyls Experiment of the ignited Magnet may go a great way to the Solving that od Phenomenon.

Enter these papers as is directed in one of them, & get the Extracts from Dr. Mathers Letters of William Innys.[9]

9 Written upside down in a different hand – probably instructions to the printer of the *PT*.

Appendix 6

An Extract of several Letters from Cotton Mather, D.D. to John Woodward, M.D. and Richard Waller, Esq; S. R. Secr. In *PT* 29.339 (June 30, 1714): 62–71.

The first Letter directed to Dr *Woodward*, is dated at *Boston* in *New England*, *Nov*. 17. 1712. In this the Writer gives an Account of a large Work in Manuscript, in two Volumes in Folio, but does not name the Author.[1] This, according to the account of it, is a large Commentary upon some Passages in the Bible, interspers'd with large Philosophical Remarks, taken out of Natural Historians, and the Observations of himself and others, more particularly as to Matters observ'd in *America*, when he entitles the Work, *Biblia Americana*. This Work Dr. *Mather* recommends to the Patronage of some generous *Mecœnas*, to promote the Publication of. As a Specimen of it, he transcribes a Passage out of it, being a Note on that Passage in *Gen. Chap*. 6. *v*. 4. relating to *Giants*; and confirms the Opinion of there having been, in the *Antediluvian* World, Men of very large and prodigious Statures, by the Bones and Teeth of some large Animals, found lately in *Albany* in *New England*, which, for some Reasons, he judges to be Human; particularly a *Tooth* brought from the Place where it was found to *New York*, 1705. being a very large Grinder, weighing 4 pounds and three quarter, with a Bone, suppos'd to be a Thigh-bone, 17 Foot long. He also mentions another Tooth, broad and flat like a fore-Tooth, four Fingers broad, the Bones crumble to pieces in the Air after they are dug up; they were found near a place call'd *Cluverack*, about 30 Miles on this side *Albany*. He then gives the Description of one, which he resembles to the Eye Tooth of a Man; he says it has four Prongs, or Roots, flat, and something worn on the top; it was six Inches high, lacking one eighth, as it stood upright on its Root, and almost thirteen Inches in circumference; it weigh'd two pound four ounces *Troy* weight: There was another near a pound heavier, found under the Bank of *Hudson*'s River, about fifty Leagues from the Sea, a great way below the Surface of the Earth, where the Ground is of a different Colour

[1] In electing to write in the third person, Mather inadvertently confused Waller about the authorship of "Biblia Americana," our CA 1.1 "Describing 'Biblia Americana' & Giant Fossils."

and Substance from the other Ground, for seventy five Foot long, which they suppose to be from the rotting of the Body, to which these Bones and Teeth did, as he supposes, once belong. *It were to be wish'd the Writer had given an exact Figure of these Teeth and Bones.*

The second Letter[2] to the same Person, is dated *Nov.* 18. 1712. from *Boston*, as all the following are. In this he treats of the Plants of *America*; and, in the first place, offers a Conjecture of his concerning the *Shittim Wood*, mentioned in the Sacred Writings to be made use of for the Ark, *&c.* It is said to be not as most other Woods, subject to rot: He judges that it was the black *Acacia*; that the *Gopher* Wood was the *Juniperus arbor tetragonophyllos*, frequent in the *East-Indies, &c.* He observes that the *Indians* often perform very great Cures with their Plants; of which there is a great variety, differing from the *European*, which he promises a Catalogue and description of; and, for the present, instances in some. As, a Plant efficacious in curing *Inflammations*, whence they call it *Antierisypelas*; it grows plentifully in the Woods: A Chymical Oil extracted from it, taken inwardly, does Wonders in absorbing scorbutick Salts. Another Plant, which goes by the Name of *Partridge-berries*, excellent in curing the *Dropsy*; a Decoction of the Leaves being drank several days as a Tea, discharging a vast quantity of Urine, as long as the Disease lasts; after which it may be drank without provoking Urine observably: Gouty Persons drink it with benefit.

The Root call'd the *Bleeding Root*, curing the *Jaundies* in five or six days.

Another Specifick for the Bite of *Rattle-Snake*, and another for *Quinsies*, or sore Throats. A Plant, call'd by the *Indians*, *Taututtipang*; infallible for the *Lues Venerea*, the Root being used in a Decoction, and drank half a Pint; a Cataplasm of the same Root, bruised, apply'd to the Ulcers cures them also.

A Thistle call'd the *Boar-Thistle*; very short and prickly, with a large and long Root. To this they add a Toot, call'd the *Cancer Root*, and a sort of *Devils-Bit*: A Decoction of which three Roots is a Cure for the *King's Evil*, tho' very far gone; a small quantity being drank every day, and the bruised Roots apply'd to the Scrophulous Tumors.

But of these *American* Plants he promises a larger Crop.

The third Letter[3] relates chiefly to the Birds of that Country; where, he says, they have many of the same *Species* with ours in *England*. He mentions very large wild *Turkies*, some weighing 50 or 60 pound, but the Flesh is very tough and hard. He takes notice of a very large *Eagle* with a great Head,

[2] CA 1.2 "On American Plants & Indian Cures."

[3] CA 1.3 "On American Birds."

soaring very high, as all of that *Genus* do. As to the Itinerants; he takes notice of vast Flight of *Pigeons*, coming and departing at certain Seasons: And as to this, he has a particular Fancy of their repairing to some undiscovered Satellite, accompanying the Earth at a near distance.

The next Letter[4] relates to *Antipathies*, and the Force of the *Imagination*. As to the first, he says, a Gentlewoman of his Neighbourhood swoons upon the seeing any one cut their Nails with a Knife; which if done with a Pair of Scissors has no Effect upon her. The Wife of a Person, vomiting upon seeing her Husband take a Vomit; the Patient that took it being not remov'd, but forc'd to take a fresh Emetic.

Some Histories are here related of the *Maculæ Maternæ*. One particularly, of a Woman longing for Peas, but refusing to gratifie her desire, for fear of a sort of Bug, with which at that time most of their Peas were infested; this Woman's Child, when born, had an Excrescence on the Forehead, resembling one of those Peas, with a black Speck, as the buggy Peas had, which after some time, dry'd away, and shell'd out they fancy'd, as the Bugs are observed to leave the Husk of the Pea.

This Letter concludes with an account of a Stone, generated under the Tongue, near the Root.

The fifth Letter[5] gives an account of some *monstrous Births*, but nothing very observable.

The sixth Letter[6] relates the Stories of some Persons that had Informations of Medicines for the Distempers they lay under, in their Dreams; these Accounts relate little to Natural Philosophy.

The next,[7] and last to Dr. *Woodward*, relates the Cures of several Wounds in Persons, which were judg'd mortal. In this little of Philosophical Information.

The next Letter,[8] being the first to Mr. *Waller*, is dated at *Boston*, *Nov.* 24. 1712. In this the Writer observes, in the first place, That the *Indians* have no Division of Time, except by *Sleeps, Moons* and *Winters*. Altho' the *Indians*

[4] CA 1.4 "On Antipathies & the Force of the Imagination."

[5] CA 1.5 "Monstrous Births."

[6] CA 1.6 "Remedies Revealed in Dreams."

[7] CA 1.7 "Remarkable Recoveries from mortal Wounds."

[8] CA 1.8 "Astronomical Observations."

have not divided or distinguished the Stars into Constellations, yet it is observable that they call the Stars of *Ursa Major, Paukunawa*, that is, *the Bear*; and this long before they had any Communication with *Europeans*. He says, there is a Tradition among them, that in *November*, 1668. a Star appear'd below the Body of the Moon, within the Horns of it. In the next place he mentions the *Evening Glade*; first taken notice of by Dr. *Childrey*, to be constantly observ'd there in *February*, and a little before and after that Month; adding, that the Cause of that Appearance must be sought for above the Atmosphere. Then he gives a new Method of his own for finding the *Julian Period*, adding a Table for that purpose; which concludes this Letter.

The next[9] relates the Appearance of several uncommon *Rainbows* and *Mock Suns*. On the 2d of *January*, in a clear Sky, but very cold; the Sun was from Ten o'Clock, for near Three Hours after, attended with four *Parhelia*, in the midst whereof were two Rainbows.

About six Weeks after this, in a Day much colder than used to be at that time of the Year, the Air a little hazy, a little after One o'Clock, for about half an hour, four Mock-Suns were seen.

He observes, that these Appearances with them are of great varieties, each usually differing in some respect from the other.

The next Letter[10] date, *Nov*. 26th. to the same Person, has the Relation of a strange Discovery of the Murder of a Person in *England*, to his Brother *Joseph Beacon*, at that time at *Boston*, in a Dream; the Person wounded appearing with the Wound on his Head; with the Attestation of several Persons, as to the Truth of it. The Information by the Dream was on the 2d of *May*, 1687, about Five o'Clock in the Morning; on the very same Day his Brother dy'd at *London*, of the Wounds he had receiv'd in *April* before: Of which Misfortune his Brother *Joseph Beacon* neither had, nor could have any notice, till the next Communication by Shipping, towards the latter end of *June* following; when he had a Letter of his Brother's Death, and the Cause of it, agreeable to his Dream, but this not directly relating to Natural Philosophy, I omit the Particulars, tho' the Relation seems to be well attested.

The following Letter[11] sent likewise to Mr. *Waller*, treats chiefly of the *Rattle-Snake*, hinting at the occasion of its Name, from the Rattles in its

[9] CA 1.9 "Rainbows & Parhelia."

[10] CA 10 "Joseph Beacon's Vison or Dreams."

[11] CA 1.11 "On the Rattlesnake."

Tail, in which he says are sometimes twenty of those loose Rings, tho' he does not come in with the Opinion, that one is added very Year. Next follow these Observations. That the more Northerly they travel, these Snakes are less numerous, as well as less venomous; nor as it is said, are any seen to the North of *Merimack* River, which is about 40 miles North of *Boston*. Here he relates a Story, as he says, constantly affirmed by the *Indians*, that these Snakes frequently lie coiled at the Bottom of a great Tree, with their Eyes fixed on some Squirril above in the Tree; which tho' seeming by his cries and leaping about, to be in a Fright, yet at last runs down the Tree, and into the Jaws of the Devourer. Then he relates, that the Winter abode of these Snakes is in the Clefts of inaccessible Rocks, from whence in the Spring, they come forth a Sunning themselves, at first very feeble, which is their chief time of destroying them. At this time the *Cystis* or Gall bladder in these Snakes is full of an acid azure coloured Juice, which they squeeze out into a Glass, but it is so Spirituous, that if the Glass be not immediately stopt, it will soon evaporate; this Liquor therefore they fix with a convenient quantity of powder'd Chalk or *Indian* Meal, and use it as a proper Medicine against the venomous Bites of this Snake; some have name it *Trochisci Connecticotiani*, from the Connecticot *Colony*. 'Tis observable when the Summer Heats come on, the Snakes have no longer this azure Liquor in their Gall-Bladders, in which there is only found a black thick Sediment, of no known use, at which time they think the forementioned spirituous Juice is carried to, and lodged in their Gums, and so conveyed or thrown by the hollow of the Teeth into the Wound, when they Bite, having received another Digestion, and higher Exaltation, by passing thro' several Strainers and Glands before it arrives to the Gums; as an instance of the virulence of this Liquor, he tells us, that a Traveller killing one of these Snakes, with the lashes of which he had disabled him; and a Fly by chance disturbing one of his Temples as he rod on afterwards, he rub'd his Temple with the other end of the Switch, which as he relates it, immediately caused hi whole Head to swell to a great excess, the Poison as he supposes permeating the whole length of the Switch. He adds another relation as to the penetrating quality of this Poison, a Person provoking a *Rattle-Snake* to bite the Edge of a broad Axe he had in his Hand; the colour, of the Steeled part bitten, was immediately changed, and at the first stroke he mad with it in using his Ax, the so discoloured part broke out, leaving a gap in his Ax. But to return to the *Trochies* made of the Gall, he says it is a Cordial sudorifick, and so good an Anodine, that he knows some who take 3 or 4 Grains of it to compose them to rest after Travel. 'Tis good in all Fevers, especially the Malignant. 'Tis an infallible Remedy for Obstructions incident to Women upon catching cold in Childbed. Being taken in a convenient Quantity, 12 Hours before the Fit, it certainly cures a Quartan Ague. The Dose is 14 Grains more or less according to the Circumstances of the Patient in any Vehicle.

The next Letter[12] treats of the Effects of *Thunder* and *Lightning* very frequent with them, which from its frequent destroying Animals, without any visible Hurt on the external Parts, he compares to the *Jewish* Punishment of pouring melted Lead down the Throats of the Condemn'd which they call'd *Combustio animæ*. Tho' he likewise observes some have had their Hair singed with Marks on their Skin like those made by small Shot; some have had their Bones made limber like a Gristle. The Captain of their Castle was found Dead in his Bed after a storm of *Lightning* without any apparent Hurt. Here he relates a Passage of which an Account has sometime since been given in the *Philosoph. Transact.* but is here confirm'd That *July* 24th, 1681. A Ship whereof one Mr. *Lad* was Master about 100 Leagues from *New England* in Lat. 38. met with a violent storm of *Thunder*, which did much Damage to the Ship; at which time, a bituminous Matter fell on the Ship burning with that Violence, as not to be extinguish'd with Water till it was all burnt out, smelling strongly like fore Gunpowder; and when they came to observe the Stars at Night, they found the Polarity or Direction of their Sea-Compasses to be changed; the North-Point being turned to the South, and so continued to do for the rest of the Voyage for a 1000 Leagues. He adds farther, that one of these Compasses continues to do so still, and was upon his Table before him at the time of his writing this present Letter. He makes a *Quere* whether this may be accounted for by Mr. *Boyl*'s Experiment of heating a Loadstone red-hot, and by altering the Position in which it was cool'd, he could change its Polarity. Which some may say, might happen tot his Needle, supposing it was made red-hot, and turned upon its Center in the Storm.

From *Thunder* he proceeds to *Earth-Quakes*, which tho' he says they have not done with them the Mischiefs frequent in *Sicily, Italy,* &c. yet they have had several very sensible and affrightning. In the Year 1663, they had 6 or 7 violent shakes in the space of 3 Days: a Town lying on the River *Connecticut*, has had scores of them in a Year, for many Years together. The *Indians* affirm, that several Rivers have not only been stopt in their course and diverted, but some wholly swallowed up by *Earth-Quakes*. He father adds, a passage out of *Josselin* who dwelt in the Neighbourhood, that in the Years 1670, at a place called *Kenebunch*, near the side of the River, a piece of clay Ground was thrown up over the tops of high Oakes, growing between it and the River,

[12] CA 1.12 "Thunder & Lightning."

which it thereby stopt, and left a Hole in the Place from which it was thrown forty Yards square, *&c*. Next as to Storms of Hail, he relates that they have had very extraordinary ones, insomuch that they have lain 3 or 4 Foot thick on the Ground, some as big as Hens Eggs, others five times as big. He mentions, as a Accident sometimes happening to them in the Winter, that it has rain'd plentifully, and at Night frozen so extremely, that the weight of the Icicles has broken the Limbs of Trees, and not unfrequently split their Trunks. Tho' they have not those *Hurricanes* to which the *Caribbe* Islands are subject; yet they have had *Whirlwinds*, or Gusts, drive along a particular narrow Tract, for divers Miles together, with a violence not to be opposed by any thing on Earth; that if their Towns had stood in the way, they must undoubtedly have been destoy'd. Of these, he says, a thick dark, small Cloud has arose, with a Pillar of Light in it, of about 8 or 10 Foot Diameter, and past along the Ground in a Track not wider than a Street, horribly tearing up Trees by the Roots, blowing them up in the Air like Feathers, and throwing up Stones of a great weight to a considerable height in the Air, throwing down all in its passage; the Noise this Cloud made was so great all the while, that the Noise of the Mischiefs done by it, was thereby quite drown'd.

The remaining of this Letter relates to some very ancient Remains, at a Place call'd *Ammuskeag*, a little above the hideous Falls of *Merimack River*. There is a huge Rock in the midst of the Stream, on the Top of which are a great number of Pits, made exactly round, like Barrels or Hogsheads of different Capacities, some so large as to hold several Tuns; the Natives know nothing of the Time they were made; but the Neighbouring *Indians* have been wont to hide their Providions in them, in their Wars with the *Maqua*'s; affirming, God had cut them out for that use for them. They seem plainly to be artificial.

In the next place, he gives an account of a strange Inscription found on a Rock, in these Words. *At* Taunton, *by the side of a Tiding River, part in part out of the River, there is a large Rock, on the perpendicular side of which, next to the Stream, are 7 or 8 Lines, about 7 or 8 Foot long, and about a Foot wide, each of them ingraven with unaccountable Characters, not like any known Character*. He has not yet been able to procure the whole, which he hopes to be Master of before long, and has herewith sent a Copy of two of them, promising the rest; they are as is represented, *Fig*. 8.

The last Letter[13] of this Collection, dates *Nov*. 29. 1712. gives a Calculation of the possible Increase of the Descendants of *Adam*; and from this

[13] CA 1.13 "Cases of Longevity & Fruitfulness."

Introduction proceeds to the Account of some long-liv'd Persons there, as likewise of their Fruitfulness. He says, 'tis no rare thing with them to have an aged Gentlewoman see many more than 100 of her Offspring. He mentions one Woman that had 23 Children, of which 19 liv'd to Man's Estate. Another that had 27; another 26, of which 21 were Sons, one whereof was Sir *William Phips*; another 39 Children. Here he gives several Instances of Persons living, with them, to above 100 Years of Age. One *Clement Weaver* lived to 110, his Wife being upwards of 100. This Man, to the last Year, could carry a Bushel of Wheat to the Mill, above 2 Miles. He relates the Case of an old Man, above 100, that lost the memory of several of the latter Years of his Life, but very well retain'd the Remembrance of what past in his younger days. I do not find, by any of these Relations, that the Persons observ'd any Regularity, or Method, in their manner of Diet, Exercise, or the like.

Appendix 7

To my Honour'd Friend,
Richard Waller Esqe,
Secretary of the ROYAL SOCIETY.
March 10, 1715.[1]

Sir.

Among the valuable Circumstances, which have brought you into Knowledge and Esteem with Men of Letters, as far off as the Regions beyond *Thyle*, and even on the Western shore of the wide *Atlantic*, One is, the Relation you sustain unto a SOCIETY, which is one of the most Illustrious in the World.

It was very much by your favourable Recommendation, that an Admission into that SOCIETY has been granted unto an *American*, of so obscure a Character, that a Place in the Academy of the *Nascoti* at *Milain*, or the *Innominati* at *Parma*, or the *Incogniti* at *Venice*,[2] would by their Titles doubtless have been more suitable for him, than a Room among your Honourable *Virtuosi*.

And for *These*, as well as other Causes, an Inscription to YOU, belongs unto a Performance, which was hereby awakened & animated, and which now prayes for a favourable Reception with you.

No Man of any Reasonable Thought can consider, your *Noble Undertaking*; How worthy it is to be pursued by Gentlemen who would *shew themselves Men*; How useful it ha's already been to Mankind, & capable of being yett more applied unto the *best of Purposes*; a Tendency it has to *Refine* and *Sweeten* the Minds of Men, and Reconcile them unto *Just Regards* for *True Merits* in one another; with an Extirpation of that noxious *Clamour-Wort*, the *Party-Spirit*: And, finally, how generously the more polite Literators of the World go on in it, with a decent Contempt on the Banters of the *Bruitish*

1 RSL MM.1.18; MDA/A/25/16/57; A5.16.

2 The Accademia dei Nascosti (i.e., the Hidden Ones") was founded in Rome, Italy; the Accademia degli Innominati (i.e. "the Nameless Ones") was founded in 1580 in the Italian city of Parma, and the Venetian Accademia degli Incogniti (Academy of the Unknowns) was a learned society of freethinkers, founded in Naples (c. 1540) and Venice (1626–27). The same self-deprecatory reference appears in his [Separate Letter] appended to CA 2.1 "Woollen Snow," and in *DL* (1699), p. 7; see also Kittredge ("Election," pp. 84–87).

among the People; but the Result of his Consideration will be, that it will be a greater Honour to be taken into the List of your Servants, than to be *mixed with the great Men of Achaia*.

One who is entirely of that Opinion, having been so listed with you,[3] ha's been desirous, to discharge his *Obligations* by agreeable *Assiduities*: And therefore, besides what every Year brings you from him, as an Addition of memorable Curiosities to the rich & vast Amassment you are præparing, he has bestow'd a few Hours upon that PHILOSOPHICAL RELIGION which he now humbly tenders to your Acceptance.

He is aware of what ha's ben said about, *The Religion of Learned Men*. But, as the *Religion* here exhibited, under the Advantage and Influence of the most Improved *Philosophy* will be what Men of *Learning*, and of *Breeding* will have a Relish for, so, if they will unite in *This*, the Consequences will be glorious.

If the Author had been a *Man of but one Business*; or, if many Scores of Observations proper for this Collection, had not been lodg'd in several Apartments of the, BIBLIA AMERICANA, (yett unpublished,) from whence he did not think it proper to transfer them hither; or, if he had not been afraid, that a *great Volumn* would have been reckoned a *great Evil*, and that more than a *Sufficiency* would have been reckon'd a *Superfluity*; the Treatise had been much bigger than it is.

But of the Two *Poets*, he who *knew what to say*, wears not so many Laurels, as he who *knew what not to say*. Perhaps, t'will be not much otherwise among *Philosophers*.

However, The *Libri Elephantini*, or a Book that should require *Twelve hundred & Sixty Oxhides for a Covering*, would not be of a Bulk proportionable, either unto the Dignity of the copious and excellent Subject, or indeed unto the Respect I owe, unto the Person, to whom I now make the Dedication.

Of the many Things *left unsaid*, this must not be one; That I am,

Sir,

Your most affectionate Friend

& obedient Servant,

March. 10. 1715.

Cotton Mather.

3 Mather appears to hint that his name has not yet appeared in the RSL Members List. See Kittredge ("Election").

Appendix 8

John Woodward to Cotton Mather.[1]

Gresham College, 3 April 1721.

To

The Reverend Dr. Mather,

At Boston, in New England.

Sir—

By Mr. Gale, returning to New England, I take the Opportunity to Salute you: and pay my Acknowledgements for your very civil Letter of the later End of last Summer. I must likewise assure you that, sooner, or later, I ever answer your Letters: & generally delivered my Answers to Mr. Dummer; so that I was not in Fault for so long Silence & Intermission. Then, as to what you Say of your *Curiosa Americana*, after I had carefully perused them my Self, I delivered them to the Roy. Society; that they might print any thing out of them, that they judgd proper, in the Philosophical Transactions. But the Editors, since Mr. Wallers Death, are very neglectfull & partial; by which the Society suffers not a little: and indeed things are very low with them at present. For my own Part I have not been wanting in Doing you Justice: and makeing the Curious here sensible of your great Diligence there.

As to my own Studyes, as they find great Approbation from the Ingenuous, I am zealous in the Prosecution of them, so far as my Business in Physick will give way: & my *Medical Undertakings*, as well as the *Natural History of the Earth*, are ever going on. By what you write, you seem to have forgot how near I am to supplying all the Defects of my first *Essay*, in my *Observations & Reflections in Answer to Camerarius*, 8°. I much admire I hear so litle of Marine Bodyes, Remains of the Deluge, digd up in New England. Pray be more inquisitive on all Occasions of Digging, Mineing, &c. I am,

Sir

Your very faithfull humble Servant,

J. Woodward.

[1] *PMHS* 13 (1873): 110–11.

***

John Woodward to John Winthrop

For John Winthrop, Esquire,

New England

Gresham College, 3 April 1721.[2]

Sir –

Tho' my Business happens to be very urgent, I cannot prevail with my-self to pass over this Opportunity of Saluteing you, by Mr. Gale, bound for New England: & returning my Thanks for the Shells you Sent,[3] that afforded some entertainment to the Curious here. For my Self, my studies determin me more to the search after Fossils of all Sorts, & such marine Bodyes, Remains of the Deluge, as are found in Digging, Mineing, &c. These will be of Use to the Perfecting my Natural History of the Earth: and I find much Difficulty in Procureing any from your Countryes there. Dr. Mather has said nothing, as yet, about the Water Doves that you Sent Him.[4] I should be glad to see any of the Utensils or Instruments, or any Things used in Religion, by the Pagan Americans there: or to have Samples of their Skulls, Bones, &c. as they may be found, after Battles, in the Fields. If there be any Thing in this Place that may gratify your Curiosity, pray be very free in Laying your Commands upon,

Sir

Your very faithful humble Servant,

J. Woodward.

2 *PMHS* 13 (1873): 110.

3 See Mather's letter to John Woodward (c. late 1713, c. July 1716), CA 2.1; as well as his letter to John Woodward (July 25, 1717), CA 4.1 "Fossils."

4 On the Water-Dove, see Cotton Mather's *BA* (1:643–44).

Appendix 9

James Jurin to Cotton Mather.[1]

London
December 16th 1723.

Reverend Sir,

I am favour'd with two Packets from you, which came to Town, while I was at Tunbridge Wells in Kent during the Recess of the Society, so that I had not the pleasure of seeing Mr. Greenwood,[2] & of receiving from him by word of Mouth a Relation of your success in inoculating the Small Pox in New England, which otherwise I should have been glad to have done. I have read over your *Case of the Small Pox inoculated, farther cleared*,[3] with a great deal of satisfaction. I must needs say, that not only the Publick, but in particular all our Faculty are obliged to you for your Condescension & pains you have taken in refuting the many weak Arguments brought against Inoculation. For certainly if those be of any force, & it be a Crime to incurr, or inflict a lesser natural Evil, or Danger, to avoid a greater, there must be an end of our Practice.

Your Book[4] was Presented to the Royal Society at one of their late meetings, when I likewise read to them your several Letters on the *Curiosa Americana*, & I am commanded to return you the thanks of the Society for the one and the other.

As to the Proposal of your Ingenious Friend, the Reverend Mr. Morgan,[5] I am to acquaint you, that the Royal Society never concern themselves in any undertaking of this kind, how usefull & practicable soever they may seem any otherwise than in making them known to the World, when brought to perfection but that they wish success to that Gentleman, & shall be glad to hear of

1 LWIHM, Wellcome MS. 6146.

2 See CA 7.9 "Letter of Introduction of Isaac Greenwood" (June 10, 1723).

3 CA 7.1 "The Case of the Small-pox Inoculated, further cleared," Cotton Mather to James Jurin (May 21, 1723).

4 CA 7.1; docketed: read Dec: 19: 23.

5 CA 7.6 "The Land Sail'd Upon" (June 7, 1723).

his having fully accomplished his design, in which case there is no doubt, but he will meet with encouragement.

I am sorry, Sir, that you have met with so much trouble on account of the Relation you bear to the Royal Society. Your case is briefly this. You had been proposed to the Royal Society by your Friends to be made a member of it, & had been referr'd, as usual, to the Council for their approbation. The Council had approved of you as a proper Person, after which there remain'd nothing more to be done but to have proposed you again to the Body of the Society, as a Person recommended to them by the Council, upon which the Election always follows of Course. But this it seems, Mr. Waller had either forgot to do, or had neglected to insert your being chosen, into the Journal of the Society; for upon a Search, that I made my self, there was no Minute to be found of your Election.[6]

However this is now rectified; for you have since been proposed to the Society over again as one recommended by the Council, & have been regularly chosen by Balot, & your Election is inserted into the Minutes of the Society. So that you are now in the same case with Mr. Dudley of your Country, & several other Gentlemen, that is, you are an elected Member of the Royal Society, tho' you can not be put into the List of Members for want of your being present to be formally admitted, as is required of all Members, who are the King's Subjects. Notwithstanding which all such Gentlemen, who by reason of unavoidable absence could not be admitted, are looked upon as Fellows of the Royal Society, & have that Title given them by the Secretaries upon publishing any of their Pieces in the Philosophical Transactions, & are at liberty to prefix it before any of their Works published by themselves, if they think fit.

I am, Good Sir,

Your affectionate Friend & much obliged humble Servant

J. Jurin.

[6] On Mather's election, see G.L. Kittredge, "Election" (1912); *D* 2:245–46; on Mather request for membership, see "Cotton Mather to Richard Waller" (December 1, 1712); (RSL, *Letter-Book*, XV, 44–45). Stearns, *Science* [1970], pp. 406–07.

Appendix 10

James Jurin to Cotton Mather.[1]
February 22, 1723/4

London, Febr. 22, 1723/4

Worthy Sir,

Since my last of Dec. 16. I have had the pleasure of seeing Mr. Greenwood,[2] & talking with him about your Success in Inoculation in New England, tho' after the exact & copious account sent over by your self & others,[3] there was indeed little occasion for any farther relation of it. Your account of the Transit of Mercury has been presented to the Royal Society,[4] & I am order'd to make their acknowledgements to you for your constant care in informing them of anything curious that is observ'd in America. But that account being in some degree imperfect, I should be glad of receiving from you the Minutes of the Observations at large, with an account of the Instruments used therein, & the are that was taken of having the true time, &c. It was taken notice of by some Gentlemen at the meeting of the Society, that the duration of the Transit as determin'd by calculation & by actual Observation did not differ a single second: which was through too near an agreement, it being impossible to observe either the beginning or the end of the Transit to so great a nicety.

1 MS copy Wellcome MS. 6146.

2 See Mather's CA 7.9 "Letter of Introduction" (June 10 1723).

3 Jurin refers to Mather's long letter CA 7.1 "The Case of the Small-pox Inoculated, further cleared," including "The Way of Proceeding in the Small Pox Inoculated in New England. Communicated by Henry Newman, Esq; of the Middle Temple," *PT* 32.370 (March 31, 1722): 33–35; and 6.14. "*An Account of the Method and Success of Inoculating the Small-Pox, in Boston* in *New-England*" (September 7, 1721); Thomas Robie's "Part of a Letter from Mr. Thomas Robie, Physician in New-England, to the Reverend Mr. Derham, F.R.S. Concerning the Effects of Inoculation; The Eclipse of the Sun in November 1722; And the Venom of Spiders," *PT* 33 (1724): 67–70.

4 RS Journal Book XII, December 12, 1723; Jurin appears to refer to Mather's CA 5.11 "*A Voice from Heaven*" (Feb. 1720), sent to John Chamberlayne.

This created a suspicion that the Observations had been a little wrested to bring them to agree with the computation.

You will find enclosed some Copies of an Invitation I have publish'd for Observations of the Weather,[5] &c. which I flatter my self you will easily see the consequence of. They have been dispers'd all over Europe, & I shall be glad if you can help me to any Observers in your part of the World. Pray send one of them to Mr. Dudley,[6] who, if he has leisure & Instruments, will be a very proper Person. I am sending other Copies to their Excellencies Governor Nicholson & Burnet,[7] who I hope will favour the design. I heartily wish you all happiness & prosperity both here & in a better World, being with great respect, Reverend Sir,

Your faithfull humble Servant

J. Jurin R.S. Secr.

5 Jurin hoped to establish a network of correspondence on the weather thereby to theorize on meteorological phenomena. He sent a number of handbills to Mather for distribution, and Jurin also published his "*Invitatio ad Observationes Meteorologicas communi consilio institutendas*," in *PT* 32.379 (October 31, 1723): 422–27. See also Appendix 11.

6 Paul Dudley, F.R.S. (1675–1751), Attorney General of Massachusetts, son of Governor Joseph Dudley, and contributor to the *Philosophical Transactions*.

7 Gov. Francis Nicholson (1655–1728), F.R.S., of South Carolina (1720); Gov. William Burnet (1688–1729), of New York and New Jersey (1720).

Appendix 11

James Jurin to Cotton Mather.
May 20, 1725.[1]

Reverend Sir,

Since my last,[2] I am favour'd with two Letters from you, one of Oct. 5th & the other of Dec. 15th by Dr. Boylston,[3] whom I shall be so glad to serve upon all occasions on account of his own Merit, that he stands in no need of your Recommendation, tho' that will always have the greatest weight with me. Besides these, I have received your large Packet of Letters to Dr. Woodward myself, which I have at times communicated to the Royal Society[4], & am commanded to return you their thanks for the several curious relations contain'd therein. I have likewise received several small Pieces publish'd by you upon different subjects, for which I return you my hearty thanks & beg you will accept of my account of Inoculation herewith enclos'd,[5] as a poor return for so many favours. Give me leave to hope for some assistance from you with regard to my Proposal for Observations upon the Weather. If you cannot conveniently take in the whole, any part of it, as the Winds, Weather, and the heights of the Barometer, will be very acceptable.[6] In the mean time, I

1 MS copy, Wellcome MS. 6146.

2 February 22, 1723/24 (Appendix 10).

3 CA 8.11 Mather's "Advertisement for Angel of Bethesda" (October 5, 1724); CA 8.12 Mather's letter of Introduction for Zabdiel Boylston (December 15, 1724).

4 RSL, Journal Book XII. Mather's CA (Seventh Series).

5 James Jurin, "An Account of a remarkable Instance of the Infection of the Small Pox. By Dr., Jurin, Soc. Reg. Secret," *PT* 32.373 (October 31, 1722): 191–93. And "A Letter to the Learned Dr. Caleb Cotesworth, F.R.S. of the College of Physicians, London … containing a Comparison between the Danger of the Natural Small Pox, and of that given by Inoculation. By James Jurin, M.D. R.S. Secret." *PT* 32.374 (December 31, 1722): 213–27 – both publications mention the success of inoculation in Boston.

6 Two years earlier (February 22, 1723/24), Jurin had sent Mather several copies of his invitation requesting detailed weather reports to help Jurin's develop a meteorological theory about weather formation. Jurin asked Mather to distribute his

have orders to make a request to you in the name of the Royal Society, on the following occasion.

In Phil. Trans. Nº 339 is published an Extract of one of your Letters to Mr. Waller, containing the following remarkable passage.

At Taunton, by the side of a tiding River, part in part out of the River, there is a large Rock,[7] on the perpendicular side of which, next to the Stream, are 7 or 8 Lines, about 7 or 8 foot long, & about a foot wide, each of them engraven with unaccountable Characters, not like any known Characters. In a figure to that Transaction is represented a copy of two of these Lines, which you had sent over, not having been able to procure the rest.

This Copy has been lately seen by Mr. De la Croze of Berlin, a Gentleman of uncommon Learning, & particularly skill'd in the Characters & Alphabets of the generally unknown Eastern Languages, used in Japan, but thinks it is somewhat incorrectly taken. Mr. De la Croze offers his assistance towards decyhering the whole, in case a correct copy of it can be procured. This, Sir, the Royal Society hope for from you, & if you can procure it to be done,[8] I assure you it will highly oblige the whole Body, & particularly, Honoured Sir,

Your most obedient and humble servant

J. Jurin.

---

pamphlet among worthy scientists in New England who had the trustworthy instruments to do so. Jurin also published his "*Invitatio ad Observationes Meteorologicas communi consilio instituendas*," in *PT* 32.379 (October 31, 1723): 422–27, for wider distribution. Mather was happy to oblige with CA 8.7 "More & Strange Effects of Thunder" (September 26, 1725). See also Appendix 10.

7 See CA 1.12 "Thunder & Lightning" (November 28, 1712) and Mather's "An Extract of several Letters from Cotton Mather," *Pt* 29.339 (June 30, 1714): 62–71; Appendix 6.

8 Jurin's letter (May 29, 1725) to Mather, however, remained unanswered (if it ever reached him), but the desired information was finally supplied by Harvard's Isaac Greenwood (1702–45), an elected member and active correspondent of the Royal Society, who sent to London a fresh transcript of the mysterious hieroglyphics (April 28, 1732). See R.P. Stearns, *Science* (1970), p. 452. On the Dighton Rock, see especially E.B. Delabarre (1916, 1917, 1928), and L. Barnett (2019).

Appendix 12

{To Dr. John Woodward.}
{Written late 1713; sent c. July 1716.}[1]

My most honour'd Doctor.[2]

For your most acceptable Present, which I received, when you honoured me with your *Defence* of your *Natural History of the Earth*,[3] I return you all the Thanks of a Mind fill'd with Gratitude.

Your Triumphs over the Doctor of *Tubingen*,[4] do but whett our Appetites for that most valuable Work; whereof I find my excellent Friend Dr. *Edwards* in a learned Essay with which he favoured me, raising also our Expectations.[5]

1 G.L. Kittredge, "Election" (1912), p. 83 *n.* 2; MHS MS. 319–321; K. Silverman, *SLCM* (1971), p. 135.

2 Mather's letter is addressed to John Woodward, M.D., F.R.S. (c. 1665–1728), professor of physics and natural history at Gresham College, and secretary of the RSL.

3 Woodward's *Naturalis Historia Telluris Illustrata & Aucta Una cum Ejusdem Defensione* (1714), translated as *The Natural History of the Earth, Illustrated, and Inlarged ... against Dr. Camerarius* (1726). This document is a new edition of Woodward's popular and oft-reprinted and expanded *Essay toward a Natural History of the Earth* (1695), and includes Woodward's 1714 response to *Dissertationes Taurinenses Epistolicae. Physico Medicae* (1712), by Elias Rudolph Camerarius (1641–95), professor of medicine at the University of Tübingen. Woodward and Camerarius had sparred over the question whether seashells found deep in the earth and even on top of mountains were of marine origin or had naturally grown in the places where they were found and were the sports nature (lusus naturae). See also Woodward's letters to Mather and to John Winthrop (April 3, 1721), Appendix 8.

4 Mather alludes to the debate between the Italian physiologist and biologist Marcello Malpighi (1628–94), the English physician and botanist Nehemiah Grew (1641–1711), and Elias Rudolph Camerarius, Jr. (1673–1734), professor of medicine at the University of Tübingen. Woodward, in his *Naturalis Historia* (1714), had taken issue with his German colleague on the debate about Noah's flood, the origin of fossils, petrification, and the sedimentary layers of the earth's crust. Woodward presented Mather with a signed copy of this Latin work (AAS).

5 John Edwards, D.D. (1637–1716), a notable English Calvinist divine, vicar of St. Peters (Colchester), and correspondent of Cotton Mather. Mather quotes Edwards

I am overwhelmed with some Confusion, that I have not all this while yeelded a due Obedience to the Commands you laid upon me, to make a Collection of our Fossils. I can plead nothing for my Delay of my Duty in this Matter, but my being overwhelmed with Employment. For truly, Syr, I am far from what some have putt into their Letany, *Homo unius Negotii.*[6] But I am forming the best Projection I can, in an Infant Countrey, entirely destitute of Philosophers, to-have this, and other Intentions answered.

In the mean time, I pursue my Design, as well as I can, to make my Annual Presents of American Curiosities.

I am very much afraid, whether one of my Collections, did not miscarry; In which I also enclosed a Couple of *Guinea*'s or, *Louis d'ors*,[7] (I forgett which,) to be spent by you, and our dear Mr. *Waller*, when you should (with your Treasurer,) undergo, the Trouble of Reading over my tedious Letters. I therefore look'd out {for} some Imperfect Copies which I kept of those Letters, & have thrown them into the Packett now before me. If you have had 'em, — tis well; There is no hurt in this Repetition.

The last Year, instead of a *Decad* of Letters,[8] I presented you, with a large Composure in Manuscript, entituled, *The Christian Virtuoso*;[9] which likewise, I never heard, whether it ever came to hand.

I now address you with a *Decad* of Letters; to the Reading of which, I pray you to invite what Gentlemen you please, (and your Honourable Treasurer:)[10] and that the Fatigue of the Perusal may be abated, lett the Two Bitts of yellow Dust enclosed, be spent on the Occasion.[11]

---

extensively throughout the volumes of *BA* (2010–). In the present context, Mather appears to refer to Edwards' *Brief Remarks upon Mr.* [*William*] *Whiston's New Theory of the Earth* (1697, 1712). See also *SLCM*, pp. 128–29; *D* 1:550; 2:81, 243.

6 "A man of one business"; i.e., employment.

7 A "Guinea" was a British gilded coin in use between 1663–1814; the "Louis d'or" was a French gold coin first issued in 1640.

8 Collection of ten individual letters.

9 Mather's manuscript copy of *CP*, published by the Royal Society of London, 1721, was originally titled "The Christian Virtuoso." To avoid confusion with Robert Boyle's own *The Christian Virtuoso* (1690), the editors of Mather's manuscript at the Royal Society appear to have opted for its present title. For safekeeping, Mather had sent the manuscript to his younger brother Samuel Mather (1674–1733), pastor of the congregational church of Witney (Oxfordshire), asking him (August 1715) that "if you know no better way to make it public Lett it pass thro' *Dr. Woodward's* hand, into the Repository of the Royal Society" (*D* 2:324).

10 Sir William Ashurst (1647–1720), British MP, treasurer of the New England Company.

11 Mather's two Louis d'or.

But I am willing that one of the Gentlemen then present should be he that is the Honourable Bearer of my Letter; who is the Secretary of this Province,[12] and a Person whose Conduct among us ha's been Such, as to procure him the universal Respect & Esteem of all our People; who will be well-pleased with his Return unto us.[13]

I have the rather made Choice of his Hand, for the Conveyance of this Packet unto yours, because his Name is Mr. *Samuel Woodward*;[14] And now you will have Opportunity to find out, whether you are akin or no.

If Dr. *Halley*, would please to enrich his Transactions, more plentifully with Communications of what we call *Natural Philosophy*, and also of *Medical Curiosities*, as well as of mathematical Rarities,[15] whereof he is so great a Master; but how few able to be so much as Learners from him! – it would much help to recommend the Treasures wherewith Mankind is obliged from him.

At this time, I add no more; but humbly tender you Two or Three of our latest Publications;[16] which are singled out from the best Part of Three Hundred, Treasures on various Arguments, in various Languages, & of various Dimensions; which this mean hand has had extorted from it.

With my hearty Prayers to the glorious God, & our Saviour, that you may be continued as a Rich Blessing to the World. I am,

Sir,

Your most affectionate

Friend & Servant.

[12] Samuel Woodward (see below).

[13] See Mather's *D* (2:324, 332).

[14] Samuel Woodward (1660–1751), secretary, commissioner, and treasurer of the Province of the Massachusetts Bay (1715). See "Massachusetts Royal Commissions 1681–1774," in *PCSM* 2 (1913): 2:315–16.

[15] Mather's ingratiating praise of Edmond Halley (1656–1742), F.R.S., English astronomer, geophysicist, mathematician, and meteorologist, is amply justified by an even cursory review of Halley's many publications in the *PT* and in other learned journals throughout Europe.

[16] Mather published no less than 11 sermons and essays in 1715.

Bibliography

Primary Works

A Castro, Rodericus. *De Universa Mulierum Medicina*. Coloniae, Hamburgiae, 1603–04.

À Vega, Christopher. *Commentaria in Librum Aphorismorum Hippocratis*. Madrid, 1563.

—*Opera Omnia nunc Denuo Publici Iuris Facta*. Lugduni, 1626.

Ab Heers, Henricus. *Most True and Wonderful Narration Of two Women bewitched in Yorkshire ... As it is attested under the hand of that most famous Phisitian Doctor Henry Heers*. N.p., 1658.

—*Spadacrene. Hoc Est, Fons Spadanus; Ejus Singularia, Bibendi Modus, Medicamina Bibentibus Necessaria*. Leodii, 1614.

Abarbanel (Abravanel), Judah. *Perush 'al Nevi'im ahronim,* aka. *Commentarius celeberrimi Rabbi Ishak Abarbanel super Iesaiam, Ieremiam, Iehazkelem et Prophetas XII. Minores*. Amstelodami, 1642.

Acosta, José de. *The Naturall & Morall History of the East and West Indies*. London, 1604.

Adam, Melchior. *Vitae Germanorum Theologorum, qui Superiori Seculo Ecclesiam Christi Voce Scriptisque Propagarunt et Propugnarunt Congestae.* Haidelbergae, 1620.

Aelianus, Claudius. *De Natura Animalium. On the Characteristics of Animals*. Translated by A.F. Scofield. 3 vols. Cambridge, MA: Harvard University Press, 1958.

Aetius Amidenus. *Aetii Medici Graeci Contractae ex Veteribus Medicinae Tetrabiblos ... id est sermons XVI*. Basileae, 1542.

Aetius Antiochenus. *Aetii Medici Graeci Contractae ex Veteribus Medicinae Tetrabiblos*. Basileae, 1542.

Agatharchides of Cnidus. *De mari Erythraeo (excerpta)*. Edited by K. Müller. *Geographi Graci Minores*. Paris: Didot, 1855. 1:111–94.

—*On the Erythraean Sea*. Translated by Stanley M. Burstein. London: Hakluyt Society, 1989.

Agrippa, Cornelius. *De Occulta Philosophia Libri Tres*. Paris and Cologne, 1531–33.

Albertus Magnus. *De Animalibus Libri XXVI*. Edited by Hermann Stadler. (Beiträge zur Geschichte der Philosophie des Mittelalters). Band 15–16. Vols 1–2. Münster: Aschendorffsche Verlagsbuchhandlung, 1920.

Albosius, Joannes. *Portentosum Lithopaedion, sive Embryon Petrefactum Urbis Senonenis,* in *Collectanea De Diuturna Graviditate, Seriem Tractatum, Versa Pagina Exhibet.* Amstelodami, 1683. 11–32.

Al-Fayruzabadi, Muhammad. *Qāmūs al-muḥit, sive, Thesaurus Linguae Arabicae et Monumentis Arabum Manuscriptis, & Impressis Bibliothecae Ambrosianae eruit, concinnavit, Latini iuris fecit.* Quartus volume. Editio Antonius Giggeius. Mediolani, 1632.

Alcimus Vitus. *De Origine Mundi, De Peccato Originali.* Basileae, 1545.

Aldrovandi, Ulisse. *Monstrorum Historia.* Bononiae, 1642.

—*Serpentum, et Draconum Historiae Libri duo.* Editio Bartholomaeus Ambrosius. Bononiae, 1640.

Alexander ab Alexandro. *Semestria In Genialium Dierum Alexandri ab Alexandro.* Editio Andreae Tiraquelli. Lugduni, 1586.

Alexander Trallianus. *De Lumbricis Epistola.* Venice, 1570.

Alsted, Johann Heinrich. *Encyclopaedia Septem Tomis Distincta.* Herborn Nassoviorum, 1630. Lugduni, 1649.

—*Theologia Naturalis Exhibens Augustissiman Naturæ Scholiam.* Francofurti, 1615.

Andry de Bois-Regard, Nicolas. *An Account of the Breeding of Worms in Human Bodies; their Nature, and several Sorts; Their Effects, Symptoms, and Prognostics.* London, 1701.

—*De la Génération Des Vers Dans Le Corps De L'Homme.* Paris, 1700.

Anonymous. "IV. The Case of Mary Howell, who had a Needle run into her Arm, and came out at her Breast." *PT* 41.461 (December 31, 1741): 767–68.

—*An Exact Relation of The Late Dreadful Tempest: Or, A Faithful Account of the Most Remarkable Disaster Which Happened On That Occasion.* London, 1704.

—*A Friendly Debate; or, a Dialogue between Rusticus and Academicus.* Boston: Printed and Sold by J[ames] Franklin in Queen-Street, 1722.

—*The Late dreadful and most admired Calamity of a Parcel of Land, and many great Oaks, and other Trees sunk, many yards under ground.* London, 1657.

—"A Letter for the Reverend Mr. William Derham, F.R.S. Containing his Observations concerning the late Storm." *PT* 24.289 (February 28, 1704): 1530–34.

—*A Letter From one in the Country, to his Friend in the City: In Relation to their Distresses occasioned by the doubtful and prevailing Practice of the Inoculation of the Small-Pox.* Boston, 1721.

—*St. Patrick's Purgatory … With a Poetical Description of the frozen River of Thames.* London, 1716.

—"A Relation of the Effect of a Thunder Clap on the Compass of a Ship on the Coast of New England." *PT* 14.157 (March 20, 1684): 520–21.

—*A Vindication of the Ministers of Boston, From the Abuses & Scandals, lately cast upon them, in Diverse Printed Papers.* Boston, 1722.

—*A Winter-Wonder, Or, The Thames Frozen Over*. London, 1684.

Ante-Nicene Fathers: The Writings of the Fathers Down to A.D. 325. Edited by Alexander Roberts and James Donaldson. 10 vols, 1885. Peabody, MA: Hendrickson Publishers, Inc., 1999.

Apollodorus of Athens. *Bibliotheca of Pseudo-Apollodorus. The Library*. 2 vols. Translated by Sir James George Frazer. London: William Heinemann, 1921.

Apollonius of Rhodes. *Argonautica*. Edited and translated by William H. Race. Cambridge, MA: Harvard University Press, 2009.

Aristotle. *The Complete Works of Aristotle*. Edited by Jonathan Barnes. 2 vols. Princeton: Princeton University Press, 1984.

—*De Generatione Animalium*. In *The Complete Works*. 1: 1111–1218.

—*Meteorologia*. In *The Complete Works*. 1:554–625.

—*Nicomachean Ethics*. In *The Complete Works*. 2:1721–1867.

Arnobius of Sicca. *Adversus Gentes*. *ANF*. 6:405–543.

Arrian of Nicomedia. *Periplus Ponti Euxini. Arrian's Voyage Round the Euxine Sea*. Translated by William Falconer. London, 1805.

Athenaeus. *Deipnosophistae*. Edited by G. Kaibel, *Athenaei Naucratitae Deipnosophistarum Libri XV*. 3 vols. Leipzig: Teubner, 1877–1890.

—*The Deipnosophists*. With an English Translation by Charles B. Gulick (Revised edition). 7 vols. Cambridge, MA: Harvard University Press, 1951.

St. Augustine of Hippo. *Against Julian. Contra Julianum Libri Sex*. [*PL* 044. 0641–0874].

—*De Civitate Dei* [*PL* 041. 009-0804]. *NPNFi* 2:1-511.

—*De Enarrationes In Psalmos*. [*PL* 036. 1591-1967]. *NPNF* 8:1–683.

—*De Genesi contra Manichaeos*. [*PL* 034. 0172-0220].

—*De Genesi Crista ad Litteram*. [*PL* 034. 0246-0486].

Aventinus, Johannes (Thurmayr). *Annales Boiorum Libri Septem*. Ingolstadii, 1554.

Avicenna, Ibn Sina. *Canon Medicinae*. Translated into Latin by Gerardus Cremonensis. Venetiis, 1490.

—*Canon of Medicine of Avicenna*. Translated by Oskar Cameron Gruner. 1930; New York: AMS Press, 1973.

—*Liber Quartus Naturalium De Actionibus et Passionibus Qualitatum Primarum*. Venetiis, n.d.

—*Principis. Ex Gerardi Cremonensis Versione, & Andreae Alagi Belunensis Castigatione*. Venetiis, 1608.

Avitus Viennensis, *De Mosaicae historiae gestis libri quinque*. [*PL* 059. 0323-0382A].

Bablot, Benjamin. *Dissertation sur le Pouvoir de L'Imagination des Femmes Enceintes*. Paris, 1788.

The Babylonian Talmud (Soncino). Brooklyn, NY: Judaica Press, Inc. 1975. In *Judaic Classic Library* (Version 2.2. March 2001) by David Kantrowitz (Davka Corportation).

R. Bachya ben Asher (Bechai). *Midrash Rabbeinu Bachya Torah Commentary*. Translated and annotated by Eliyahu Munk. 2nd Revised Edition. 7 vols, 1998; Brooklyn, NY: Lambda Publishers, 2004.

Baglivi, Giorgio. *The Practice of Physick, Reduc'd to the ancient Way of Observations*. London, 1704.

—*De Praxi Medica ad Priscam Observandi Rationem Revocanda*. Roma, 1696.

Baronius, Caesar. *Epitome Annalium Ecclesiasticorum Caesaris Baronii*. Colonia Agrippinae, 1602.

Barrow, Isaac. *Lectiones Mathematicae*. Londini, 1664.

Bartholinus, Thomas. Editor. *Acta Medica et Philosophica Hafniensia*. 1673–80.

—*Anatome ex Omnium Veterum Recentiorumque Observationibus*. Lugduni Batavorum, 1673.

—*Cista Medica Hafniensis, Variis Consillis, Curationibus, Casibus Rarioribus Vitis Medicorum*. Hafniæ, 1662.

—*De Morbis Biblicis. Miscellanea Medica. Editio secunda correctior*. Francofurti, 1672. Translated by James Willis. *On Diseases in the Bible: A Medical Miscellany. Acta Historica Scientiarum Naturalium et Medicinalium*. Copenhagen: The Danish National Library of Science and Medicine, 1994.

—*De Nivis Usu Medico Observationes Variae*. Hafniae, 1661.

—*Historiarum Anatomicarum Rariorum Centuria I et II*. Hagae Comitum, 1656.

—*Institutiones Anatomicae Novis Recentiorum*. Lugduni Batavorum, 1641.

Basil of Caesarea. *The Hexaemeron. NPNFii*. 8:51–107.

—*Homiliae in Hexaemeron* [*PG* 029. 2–208].

—"Letter XXXVIII. To his Brother Gregory, concerning the difference between οὐσία and ὑπόστασις." *NPNFii*. 8:137–41.

—*De Spiritu Sancto. On the Holy Spirit. NPNFii*. 8:1–50.

Bauhini, Caspar. *Pinax Theatri Botanici Caspari Bauhini, sive Index in Theophrasti Dioscoridis Plinii et Botanicorum*. Basilae, 1623.

Baxter, Richard. *The Certainty of the Worlds of Spirits. Fully evinced by unquestionable Histories of Apparitions, Operations, Witchcrafts, Voices &c. proving The Immortality of Souls, the Malice and Misery of the Devils, and the Damned, And the Blessedness of the Justified. Written as an Addition to many other Treatises, for the Conviction of Sadduces & Infidels*. London, 1691.

—*Of the Nature of Spirits; Especially Mans Soul. In a placid Collation with the Learned Dr. Henry More, In a Reply to his Answer to a private Letter, Printed in his second Edition of Mr. Glanviles Sadduceismus Triumphatus*. London, 1682.

—"The Preface." In Cotton Mather's *MPL*. N.P.

—"What Light must shine in our Works." *The Practical Works of the late Reverend and Pious Mr. Richard Baxter*. 4 vols. London, 1707. 4:818–32.

Bayle, Pierre. *Dictionary Historical and Critical of Mr. Peter Bayle*. Edited and translated by Pierre des Maizeaux. London, 1709, 1734–38.

—*Dictionnaire Historique et Critique*. 2 vols. Rotterdam, 1697–1709.

—*An Historical and Critical Dictionary. By Monsieur Bayle. Translated into English, with many Additions and Corrections, made by the Author himself, that are not in the French Editions*. 4 vols. London, 1710.

—*Pensées Diverses sur l'Occasion de la Comète*. Paris, 1682. Translated as *Miscellaneous Reflections, Occasion'd by the Comet which appear'd in December* 1680. 2 vols. London, 1708.

Beacon, Joseph. "Fraternum Donum / sent to his Loving Brother Nathaniel Beacon / In London / From J[oseph] B[eacon] in Boston. N England/ in the year 1688" (pp. 15–67). In "Miscellanies on various subjects: translations & collections of diverse authors / by Joseph Beacon in Boston, New England": Manuscript, 1688. MS AM 527. Houghton Library, Harvard University, Cambridge, MA.

Beaumont, John. *An Historical, Physiological and Theological Treatise of Spirits, Apparitions, Witchcrafts, and other Magical Practices*. London, 1705.

Becanus, Johannes Goropius. *Origines Antwerpianae. Sive Cimmeriorum Becceselana Novem Libros Complexa. Atvatica, Gigantomachia*. Antverpiae, 1569.

Beckett, William. *A Free and Impartial Enquiry into the Antiquity and Efficacy of Touching for the King's Evil*. London, 1722.

Beckher, Daniel. *Medicus Microcosmus seu Spagyria Microcosmi triplo auctior & correctior*. Wittenberg, 1622; Lugduni Batavorum, 1633.

Bekker, Balthazar. *De Betoverde Weereld*. Amsterdam, 1691.

Bell, John. *The Tryal of Witchcraft*. N.P. N.D.

Bellonius, Petrus. *De Aquatilibus, Libri duo*. Parisiis, 1553.

—*L'Histoire Entière des Poissons*. Lion, 1558.

—*L'Histoire Naturelle des Estranges Poissons Marins*. Lion, 1551.

—*La Nature et Diversité des Poissons avec leurs Pourtraicts*. Paris, 1555.

—*Libri de Piscibus Marinis, in quibus verae Piscium effigies expressae sunt*. Lugduni, 1554.

Bellovacensis, Vincentius. *Bibliotheca Mundi, Speculum Quadruplex, Naturale, Doctrinale, Morale, Historiale*. Duaci, 1624.

—*Speculum Naturale*. Colonia Agrippina, 1494.

Belon due Mans, Pierre. *De Aquatilibus, Libri duo*. Paris, 1553.

—*L'Histoire Nautrelle des Estrages Poissons Marins*. Paris, 1551.

—*La Nature et Diversité des Poissons avec leurs Pourtraicts*. Paris, 1555.

—*Le Voyage au Levant: Les Observations de Pierre Belon du Mans de Plusieurs Singularités & Choses Mémorables, trouvées en Gréce, Turquie, Judée, Egypte, Arabie & autres pays estranges*. Paris, 1555.

Benedictus (Benedetti), Alexander. *De Observatione in Pestilentia*. Venetiis, 1493.
—*Omnium À Vertice ad Calcem Morborum Signa, Causae, Indicationes & Remediorum Compositions utendique Rationes, Generatim Librim XXX conscripta*. Basileae, 1539.
Benivieni, Antonio di Paolo. *De Abditis Nonnullis ac Mirandis Morborum et Sanationum Causis*. Florentiae, 1507.
Beverovicius, Johannes. *Schat der Gesontheyt*. Dordrecht, 1636.
Blackmore, Richard. *The Nature of Man, A Poem in Three Books*. London, 1711, 1720.
—*Prince Arthur. An Heroic Poem. In Ten Books*. London, 1697. Fourth edition, 1714.
Blondel, James Augustus. *The Power of the Mother's Imagination over the Foetus Examin'd*. London, 1729.
Bochart, Samuel. *Geographia Sacra, cujus Pars Prior Phaleg. Pars Posterior Chanaan*. Editio Tertia. Lugduni Batavorum, 1692.
—*Hierozoicon Sive Bipartitum Opus De Animalibus Sacrae Scripturae*. Londini, 1663. Ex recensione Johannis Leusden, Lugduni Batavorum, 1692; Editio Quarta, Lugduni Batavorum, 1712.
Boethius, Anicius Manlius Severinus. *De Consolatione Philosophiae. The Consolation of Philosophy of Boethius*. Translated into English Prose and Verse. By H.J. James. London: Elliot Stock, 1897.
Bolducus, Jacobus (Jacques Bolduc). *De Ecclesia Ante Legem. Secunda Editio*. Lugduni, 1626, 2nd ed., 1630.
Bongars, Jacques. *Gesta Dei Per Francos, Sine Orientalium Expeditionum, et Regni Francorum Hierosolimitani Historia. Tomus Primus*. Hanoviae, 1611.
Borel, Pierre. *Historiarum et Observationum Medicophysicarum Centuriae IV*. Paris, 1657; Francofurti, 1670.
Borelli, Alfonso Giovanni. *De Motu Animalium*. Romae, 1680.
Borrichius, Olaus. "*De Lumbricis Latis, & Cucurbitinis*." *Acta Medica & Philosophica Hafniensia Anni 1673*. Hafniae, 1675. Vol. 2, Observ. XXXXVII, 148–49.
Bouchard, François. "Observatio XII, 'Infante Monstroso Lugduni in viam publicam die V. Martii A. MDCLXXI. exposito. *Miscellanea Curiosa Medico-Physica Academiae Naturae Curiosorum, sive Ephemeridum Medico-Physicarum Germanicarum Curiosarum Annus Tertius*. Lipsiae & Frankofurti, 1673.
Boyle, Robert. "An Account of a new kind of baroscope, which may be called statical; and of some advantages and conveniences it hath above the mercurial." *PT* 1.14 (July 2 1666): 231–39.
—"An Account of a very odd Monstrous Calf." *PT* 1.1 (March 6 1665): 10.
—*The Christian Virtuoso*. London, 1690.

—*A Continuation of New Experiments Physico-Mechanicall, touching the Spring and Weight of the Air, and their Effects*. London, 1680, 1682.
—*The Excellency of Theology, Compared with Natural Philosophy*. London, 1671.
—*Natural Philosophy. Essays of the Strange Subtlety, Determinate Nature, Great Efficacy of Effluviums. To which are annext New Experiments to make Fire and Flame Ponderable*. London, 1673.
—*New Experiments Physico-Mechanical, Touching the Spring of the Air, and Its Effects*. Oxford, 1660.
—*Nova Experimenta Physico-Mechanica de Vi Aeris Elastica*. London, 1661.
—*The Sceptical Chemist: or Chymico-Physical Doubts & Paradoxes, Touching the Spagyrists Principles commonly call'd Hypostatical; As they are wont to be Propos'd and Defended by the Generality of Alchymists*. London, 1661.
—*Some Considerations Touching the Usefulness of Experimental Naturall Philosophy*. Oxford, 1664.
—*Some Physico-Theological Considerations about the Possibility of the Resurrection*. London, 1675.
Boylston, Zabdiel (with Cotton Mather). "A Faithful Account of what has occur'd under the Late Experiments of the Small-Pox managed and governed in the Way of Inoculation." *Boston Gazette*, No. 101 (October 30, 1721).
—"Ambergris found in Whales. Communicated by Dr. Boylston of Boston in New England." *PT* 33.385 (December 31, 1724): 193.
—*An Historical Account of the Small-Pox Inoculated in New England*. London, 1726; 2nd edition; London, 1730.
— (with Cotton Mather). *Some Account Of what is said of Inoculating or Transplanting the Small-Pox. By the Learned Dr. Emanuel Timonius, and Jacobus Pylarinus. With some Remarks thereon. To which are added, A Few Queries in Answer to the Scruples of many about the Lawfulness of this Method. Published By Dr. Zabdiel Boylston*. Boston, 1721.
Bradford, William. *Of Plimoth Plantation. The 400th Anniversary Edition*. Edited and introduced by Kenneth P. Minkema, Francis J. Bremer, Jeremy D. Bangs, and Paula Peters. Boston: Colonial Society of Massachusetts, New England Historic Genealogical Society, 2020.
Brand, Adam. *A Journal of the Embassy From Their Majesties Iohn and Peter Alexievitz, Emperors of Muscovy, &c. Over Land into China, Through the Provinces of Vstiugha, Siberia, Dauri, and the Great Tartary, to Peking, the Capital City of the Chinese Empire. By Everard Isbrand, Their Ambassador in the Years 1693, 1694, and 1695… . Translated from the Original in High-Dutch, Printed at Hamburgh, 1698*. London, 1698.
—*Driejaarige reize naar China, te lande gedaan door den Moskovischen afgezant*. Amsterdam, 1704.

—*Three years travels from Moscow over-land to China*. London, 1706.

Brerewood, Edward. *Enquiries Touching the Diversity of Languages, and Religions through the cheife parts of the world*. London, 1614.

Bridgeman, Orlando. "Part of a Letter from Orlando Bridgman Esq; F.R.S. to Capt. Wine, giving an Account of a Storm of Thunder and Lightning that happen'd at Ipswich, July 16, 1708." *PT* 26.316 (August 1, 1708): 137–39.

Browne, Sir Thomas. *The Garden of Cyrus*. London, 1659.

—*Pseudodoxia Epidemica: Or, Enquiries into Very many Received Tenents* [sic], *And commonly Presumed Truths*. 2nd edition, London, 1650; 3rd edition, London, 1658.

B.[ulwer], J.[ohn]. *Chironomia: Or, The Art of Manuall Rhetorique*. Appended to *Chirologia: Or the Naturall Language of the Hand. Composed of the Speaking Motions, and Discoursing Gestures thereof. Rhetoricke*. London, 1644. 1–146 Sep. Pag.

Bunyan, John. *Pilgrim's Progress from This World, to That which is to come*. London, 1678.

Burnet, Thomas. *The Sacred Theory of the Earth*. London, 1684, 1697.

Burthogge, Richard. *An Essay upon Reason, and the Nature of Spirits*. London, 1694.

—*Organum Vetus et Novum*. London, 1678.

—*Of the Soul of the World, and of Particular Souls*. London, 1699.

Burton, Robert. [R.B.] *The General History of Earthquakes*. London, 1694.

Burton, Robert. *Anatomy of Melancholy*. Oxford, 1621.

Buxtorf, Johannes. *Epitome Radicum Hebraicarum et Chaldaicarum Complectens omnes voces, tam primas quam derivatas, quae in Sacris Bibliis, Hebraeâ & ex parte Chaldaeâ linguâ scriptis, extant*. Basileae, 1607.

—*Lexicon Hebraicum et Chaldaicum Complectens omnes voces, tam primas quam derivatas, quae in Sacris Bibliis, Hebraeâ & ex parte Chaldaeâ linguâ scriptis, extant*. Basileae, 1645.

Cabeus, Nicolaus. *Philosophia Magnetica, in qua Magnetis Natura penitus explicatur*. Ferrara, 1629.

Cadamusto, Aloysio. *Navigatio ad Terras Ignotas, Archangelo Madrigiano Interprete*. In Simon Grynaeus. Editor. *Novus Orbis Regionum ac Insularum Veteribus Incognitarum una cum Tabula Cosmographica*. Basileae, 1532. Fol. 1–89.

Caelius Aurelianus. *De Morbis Acutis & Chronicis Libri VIII. Joannes Conradus Amman recensuit*. Amstelaedami, 1709.

Calef, Robert. *More Wonders of the Invisible World*. London, 1700.

Calvin, John. *Commentaries of The First Book of Moses called Genesis*. Translated by the Rev. John King. (The Calvin Translation Society) 22 vols. Rpt; Grand Rapids: Baker Book House Company, 2005. Vol. 1.

Camden, William. *Britannia; sive Florentissimorum Regnorum Anglicae*. Londini, 1586.

—*Camden's Britannia, Newly Translated into English: with large Additions and Improvements. Published by Edmund Gibson*. London, 1695.

Camerarius, Elias Rudolph. *Dissertationes Taurinenses Epistolicae, Physico-Medicae. Ad Illustr. Ital. Ac German. Quosdam Medicos scriptae, continents Annotationes In Varia Modernorum, Dn. De Noiies cumprimis, ac Dn. Woodwardi Scripta atque Experimenta*. Tubingae, 1712.

Camerarius, Philip. *The Living Librarie, or, Meditations and Observations Historical, Natural, Moral, Political, and Poetical*. Translated by John Molle. London, 1621.

—*Operae Horarum Succisivarum, Sive Meditationes Historicae, Auctiores Quam Antea Editae. Centuria Prima.* Noribergae, 1591, 1599; Francofurti, 1602–09, 1644.

Camerarius, Rudolph Jakob. *De Sexu Plantarum Epistola.* Tubingae, 1694.

Cantwell, Andrew. "Extract of a letter dated at Montpelier, December 27, 1731. N.S. from Andrew Cantwell, M.D. "Monspel. To T.S. M.D. and by him translated from the French, giving an Account of a monstrous Boy." *PT* 41.453 (June 30, 1739): 137–38.

Capitolinus, Julius (Scriptor Historiae Augustae). *The Scriptores Historiae Augustae: Maximini Duo*. Translated by David Magie. 3 vols. Cambridge: MA: Harvard University Press, 1921–32. 2:314–79.

Cappellus, Jacques. *Historia Sacra et Exotica ab Adamo usque ad Augustum*. Sedani, 1613.

Cappellus, Ludovicus. *Commentarii et Notae Criticae in Vetus Testamenti.* Amstelodami, 1689.

Cardano, Girolamo. *Operum Tomus Tertius; Physica*. Lugduni, 1563.

—*De Rerum Varietate Libri XVII.* Avinione, 1558.

—*De Subtilitate Libri XXI.* Parisiis, 1550.

—*The De Subtitlitate of Girolamo Cardano*. Edited by John M. Forrester. With an Introduction by John Henry and John M. Forrester. Tempe: ACMRS, 2013.

Casaubon, Isaac. *Athenaei Deipnosophistarum*. Lugduni, 1657.

Casaubon, Meric. *A Treatise Proving Spirits, Witches, and Supernatural Operations*. London, 1672.

Cassanione (Chassanione), Joannes. *De Gigantibus, eorum que Reliquiis ... Joannis Goropii Error*. Basileae, 1580.

Cassini, Gian Domenico. *Découverte de la Lumiere Celeste qui paroist dans le Zodiaque*. Paris, 1685.

Cassiodorus, Magnus Aurelius. *Expositio in Cantica Canticorum*. In *Opera Omnia Cassiodori*. [*PL* 070. 1055–1106C].

Castrensi, Stephano Roderico. *Commentarius in Hippocratis Coi, Libellum de Alimento*. Florentiae, 1635.

Catalogus Librorum Bibliothecae Collegij Harvardini Quod est Cantabrigiae in Nova Anglia. Bostoni Nov-Anglorum, 1723.

Cataneus, Andreas. *Opus de Intellectu et de Causis Mirabilium Effectuum.* Florence, 1502.

Celsus, Aulus Cornelius. *De Medicina.* With an English Translation by W.G. Spencer. 3 vols. Cambridge, MA: Harvard University Press, 1961.

Chamberlayne, John. "A Relation of the Effects of a Storm of Thunder and Lightning at Sampford Courtney in Devonshire on 7 Oct. 1711." *PT* 27.336 (December 31, 1712): 528.

Cheyne, George. *The Philosophical Principles of Religion Natural and Revealed.* 2 vols. London, 1705, 1715.

Checkley, John. *Choice Dialogues Between a Godly Minister And an Honest Country-Man, Concerning Election & Predestination.* Boston, 1719.

—*A Discourse concerning Episcopacy*. Boston, 1690.

Childrey, Joshua. *Britannica Baconica: Or, The Natural Rarities of England, Scotland, & Wales*. London, 1661.

—"The *Evening Glade*; first taken notice of by Dr. *Childrey*." *PT* 25.305 (March 1, 1706): 2220.

—"An Extract of a Letter, Written by Mr. Joshua Childrey to the Publisher, containing an Accompt of a passage by sea to the East-Indies." *PT* 40.50 (August 16, 1669): 1003–09.

—*Szygiasticon Instauratum. Or, An Ephemeris of the Places and Aspects of the Planets, as they respect the [Sun] as Center of their Orbes*. London, 1653.

Chronicon Paschale 284–628 AD. Translated by Michael Whitby and Mary Whitby. Liverpool: Liverpool University Press, 1989.

Chrysostom, Johannes. *Four Discourses of Chrysostom. Chiefly on the Parable of The Rich Man and Lazarus*. Translated by F. Allen. London: Longmans, Green, Reader, and Dyer, 1869.

—Homily on Genesis. In *The Fathers of the Church* (74, 82, 87). Translated by Robert C. Hill. Washington, D.C.: Catholic University of America Press, 1986–92.

Cicero, Marcus Tullius. *De Divinatione. On Divination*. Translated by W. A. Falconer. Cambridge, MA: Harvard University Press, 2023. 214–568.

—*De Finibus. On Ends*. Translated by H. Rackham. Cambridge, MA: Harvard University Press, 1914.

—*De Senectute*. In *Cicero on Old Age on Friendship on Divination.* Translated by W.A. Falconer. Cambridge, MA: Harvard University Press, 2023. 8–99.

Cieza, Pedro. *Dela chronica del Peru*. Sevilla, 1553.

A Classical Dictionary of the Vulgar Tongue. London, 1785.

Clement of Alexandria. *The Stromata* [*ANF* 2: 299–567]. *Stromateon.* Tom 2. *V–VIII* [*PG* 008.0011–0602].

Cluverus, Johannes. *Diluculum Apocalypticum Seu Commentarius Posthumus in Apocalypsin*. Lubecae, 1647.

Cluverus, Philipp (Clüver). *Introductionis in Universam Geographiam Libri VI*. Lugduni Batavorum, 1624.
Collins, Anthony. *A Discourse of the Grounds and Reasons of the Christian Religion*. London, 1724.
Colman, Benjamin. *Observations on the New Method of Receiving the Small-Pox by Ingrafting or Inoculating*. Boston, 1721.
Columella, Lucius Junius Moderatus. *De Re Rustica Lbiri XII. On Agriculture*. Translated by Harrison B. Ash. 3 vols. Cambridge, MA: Harvard University Press, 1941.
Cooper, William. *A Letter to a Friend in the Country*. Boston, 1721.
Copernicus, Nicolaus. *De Revolutionibus Orbium Coelestium*. Nuremberg, 1543.
Craig, John. "IV. A Description of the head of a Monstrous Calf." *PT* 27.333 (March 31, 1712): 429–30.
Cressener, Drue. *A Demonstration of the First Principles of the Protestant Applications of the Apocalypse*. London, 1690.
—*The Judgment of God upon the Roman Catholick Church*. London, 1689.
Croke, Alexander. Editor. *Regimen Sanitatis Salernitanum. A Poem On the Preservation of Health in Rhyming Latin Verse. Addressed by the School of Salerno to Robert of Normandy, son of William the Conqueror, with an Ancient Translation: And an Introduction and Notes*. Oxford: D.A. Talboys, 1830.
Crooke, Helkia. *Micrographia: A Description of the Body of Man. Together with the Controversies Thereto Belonging*. London, 1615.
À Cruce, Johannes Andreae. *Chirurgiae Iannis Andreae A Cruce, Veneti Medici Libri Septem*. Venetiis, 1573.
—*Quamplurimis Instrumentorum Imaginibus Arti Chirurgicae Opportunis suis locis exornati*. Venetiis, 1573.
Cudworth, Ralph. *The True Intellectual System of the Universe*. Cambridge, 1678.
Culpepper, Nicholas. *Culpeper's Complete Herbal: Consisting of A Comprehensive Description of Nearly All Herbs with their Medicinal Properties extracted from them*. London: W. Foulsham & Co, n.d.
Cyril of Alexandria. *Commentarius in Jonam Prophetam*. [*PG* 071. 0598–0638]
—*Contra Julianum* (lib. 1–2). Edited by P. Burguière and P. Évieux. *Cyrille d'Alexandrie. Contre Julien*. Tome 1. [*Sources chrétiennes* 322]. Paris: Éditions du Cerf, 1985:100–318.
—*Glaphyra in Pentateuchum* [*PG* 69. 009–677].
Danaeus, Lambertus. *Physices Christianae Pars Altera, sive De Rerum Creatarum Natura*. Genevae, 1676.
De Crèvecoeur, Hector St. John. *Letters from an American Farmer*. 1782; rpt New York: Fox, Duffield & Co. 1904.

De Dominis, Marcus Antonio. *De Radiis Visus et Lucis in Vitris Perspectivis et Iride Tractatus Marci Antonii de Dominis. Per Ioannem Bartolum in lucem editus*. Venetiis, 1611.

De Forlivio, Jacobus. *Expositio in Aphorismos Hippocratis*. Padua, 1477.

—*Expositio in Primum Avicennae Canonem*. Venice, 1508.

Defoe, Daniel. *An Essay on the History and Reality of Apparitions*. London, 1727.

—*Essay on Projects*. London, 1697.

—*A Journal of the Plague Year: Being Observations or Memorials, Of the most Remarkable*

Occurrences, As well Publick as Private, Which happened in London During the last Great Visitation In 1665. London, 1722.

—*The Storm: or a Collection Of the most Remarkable Casualties and Disasters Which happen'd in the Late Dreadful Tempest, both by Sea and Land*. London, 1704.

—*A True Relation of the Apparition of one Mrs. Veal*. London, 1706.

De la Créquinière, M. *The Agreement of the Customs of the East-Indians, With those of the Jews, and other Ancient People*. London, 1705.

De La Guillatiere, Georges. *Les Voyageurs Français*. Paris, 1675. Translated as *An Account of a Late Voyage to Athens*. London, 1676.

De La Pryme, Abraham. "III. Part of a Letter from the Reverend Mr. Abraham de la Pryme to the Publisher, concerning Trees found under Ground in Hatfield Chace." *PT* 22.275 (October 31, 1701): 980–92.

De la Roche, Michel. Editor. *Bibliotheque Angloise, Ou Histoire Littéraire de La Grande Bretagne*, tome premier, premiere partie. Amsterdam, 1717.

De Laet, Joannes. *Notae ad Dissertationem Hugonis Grotii, De Origine Gentium Americanarum*: *Et Observationes*. Amstelodami, 1643.

—*Novus Orbis seu Descriptionis Indiae Occidentalis Libri XVIII*. Lugduni Batavorum, 1633.

De Lost Reyes, Gaspar Franco. *Elysius Iucundarum Quæstionum Campus Omnium Literarum Medicis Inprimis*. Bruxellae, 1661.

De Montaigne, Michel. "Of the Force of Imagination." *Essays of Michael Seigneur de Montaigne. In Three Books* ... Newly rendered into English by Charles Cotton, Esq. London, 1685.

De Peiresc, Nicolas-Claude Fabri. *Viri Illustris. Per Petrum Gassendum*. Paris, 1641.

De Superville, Daniel. "Some Reflections on Generation." *PT* 41.456 (January 1, 1740): 294–307.

De Thou, Jacques-Auguste (Thuanus). *Historia Sui Temporis ab Anno Domini 1543. Usque ad Annum 1607. Libri CXXXVIII*. Parisiis, 1604–08.

—*Monsieur de Thou's History of his own Time*. 2 vols. London, 1729–30.

De Tournefort, Joseph Pitton. *Relation D'Un Voyage du Levant*. Lyon, 1717.

—*A Voyage into the Levant*. 2 vols. London, 1718.

De Vale, Godefridus (du Val). *De Arte nihil Credendi*. 1570.

De Zárate, Agostin. *Historia del Descubrimento y Conquista del Peru* (1555). Translated as *The strange and delectable History of the Discoverie and Conquest of the Province of Peru*. London, 1581.

Della Valle, Pietro. *Travels of Sig. Pietro della Valle, A Noble Roman, Into East-India and Arabia Deserta*. London, 1665.

Delrio, Martin Antonio. *Disquistionum Magicarum Libri Sex*. Lugduni, 1608.

Derham, William. "The Case of a Woman big with Child, who recovered of the Small-Pox, and was afterwards delivered of a dead child full of the Pustules of that Distemper. In a Letter from the Rev. Mr. W. Derham, F.R.S. to Dr. Hans Sloane, S.R. Secr." *PT* 28.337 (December 31, 1713): 165–66.

—"A Letter from the Reverend Mr. W. Derham, F.R.S. to Dr. Hans Sloane, R.S. Scr. giving an Account of some Inundations; Monstrous Births, Appearances in the Heavens, and other Obvservables he received from Ireland. With his Observations on the Eclipse of the Sun, Sept. 3. and of the Moon, Sept. 18, 1708." *PT* 26.320. (April 1, 1709): 308–13.

—"Part of a Letter from the Reverend Mr. W. Derham, F.R.S. concerning a Glade of Light observed in the Heavens (March 28, 1706)." *PT* 25.305 (March 1, 1706): 2220.

—*Philosophical Experiments and Observations of the Late Eminent Dr. Robert Hooke ... and Other Eminent Virtuoso's in his Time*. London, 1726.

—*Physico-Theology: or, a Demonstration of the Being and Attributes of God, from his Works of Creation*. 2nd edition. London, 1714.

Descartes, René. *Dioptrique, les Météores et la Géométrie*. In *Discours de la méthode*. Leyden, 1637. 157–294.

—*The Philosophical Writings of Descartes*. Translated by John Cottingham et al. 3 vols. Cambridge, UK: Cambridge University Press, 1985.

Deusing, Anton. *De Pulvere Sympathetico*. Groningae, 1662.

—*Resurrectio Hepatis Asserta; Contra Socium Larvatum ... Disquisitio Ulterior, de Chyli Motu Atque Officio Hepatis*. Groningae, 1662. John Wilson and Son, 1857.

Dickinson, Edmund. *Physica Vetus & Vera: Sive Tractatus De Naturali Veritate hexaëmeri Mosaici*. Londini, 1702.

Digby, Sir Kenelm. *Of Bodies, and of Mans Soul. To Discover the Immortality of Reasonable Souls*. London, 1669.

—*Of the Sympathetick Powder. A Discourse in A Solemn Assembly At Montpellier*. London, 1669.

—*Two Treatises: In the one of which, the Nature of Bodies; In the other, the Nature of Mans Soule, is looked into*. London, 1645.

Diodorus Siculus. *Library of History*. Translated by Charles H. Oldfather. Cambridge, MA: Harvard University Press, 1933–35.

Dion Cassius. *Historiae Romanae: Roman History. In Nine Volumes*. Translated by Earnest Cary. 9 vols. London: William Heineman, 1914–15.

Dionysius Periegetis. *Orbis Descriptio*. In *Geographi Graeci Minores*. 2 vols. Edited by Karl Müller. Parisiis, 1882. 2:103–76.

Dioscorides, Pedacius. *Pedacii Dioscoridis Anazarbaei De Materia Medica, lib. V*. Francofurti, 1598.

Dodonaeius, Rembert. *Medicinalium Observationum Exempla Rara*. Lugduni Batavorum, 1585.

Donatus, Marcellus. *De Medica Historia Mirabili libri sex*. Mantuae, 1586; Francofurti, 1613.

Dorstenius, Johann Daniel. "Observatio LIX. De Aciculâ deglutita," in *Miscellanea Curiosa Medico-Physicarum Germanicarum Academiæ* Decuriæ 2 Annus Tertius, Anni 1684. Norimbergæ, 1685. 147–49.

Douglas, James. "The Natural History and Description of the Phœnicopterus or Flamingo; with two Views of the Head, and three of the Tongue, of that beautiful and uncommon Bird." *PT* 29.350 (December 31, 1716): 523–41.

Douglass, William. *The Abuses and Scandals Of some late Pamphlets In Favour of Inoculation of the Small Pox, Modestly obviated, and Inoculation further consider'd in a Letter to A–S–M.D. & F.R.S. In London*. Boston, 1722.

—"To the Author of the Boston News-Letter." *Boston News-Letter* (July 24, 1721): 2.

—*A Dissertation Concerning Inoculation of the Small-Pox. Giving some account of the Rise, Progress, Success, Advantages and Disadvantages of receiving the Small Pox by Incisions*. Boston, 1730.

—*Inoculation of the Small Pox As practised in Boston, Consider'd in a Letter to A–S–M.D. & F.R.S. In London*. Boston, 1722.

—*Postscript To Abuses, &c. obviated. Being a Short and Modest Answer to the Matters of Fact maliciously misrepresented in a late Doggrel Dialogue*. Boston, 1722.

—"William Douglass' Letter to Alexander Stuart" (September 25, 1721), Royal Society (Records, EL/D2/2 Early Letters).

Drake, James. *Anthropologia Nova; or, A New System of Anatomy*. 2 vols. London, 1707.

Drusius, Johannes van den Driesche. *Ad Loca Difficiliora Pentateuchi id est Quinque Librorum Mosis Commentarius*. Franekerae Frisiorum, 1617.

Du Hamel, Jean-Baptiste. *De Meteoris et Fossilibus Libri Duo*. Paris, 1660.

Du Pin, Louis Ellies. *Dissertationes Historiques, Chronologiques, Geographiques et Critiques sur la Bible*. 2 vols. Paris, 1711.

Dudley, Joseph. "Joseph Dudley to Cotton Mather." (July 10, 1706). *CMHS*. Second Series, vol. 2 (1814) 263–64.

Dudley, Paul. VII. “An Account of a Method lately found out in New-England, for Discovering where the Bees Hive in the Woods, in order to get their Honey.” *PT* 31.367 (April 20, 1721): 148–50.

—“IV. An Account of the Rattlesnake. By the Honourable Paul Dudley, Esq; F.R.S.” *PT* 32.376 (December 31, 1723): 292–95.

—“I. A Description of the Moose-Deer in America. By the Honourable Paul Dudley, Esq. F.R.S. Communicated by John Chamberlayne, Esq.” *PT* 31.368 (January 1, 1721): 165–68.

—“II. An Essay upon the History of Whales, with a particular Account of the Ambergris found in the Sperma Ceti Whale.” *PT* 33.387 (April 30, 1725): 256–60.

Dunus, Thaddaeus. *Miscellanea Medica*. Tiguri, 1592.

Duretus, Ludovicus. *Adversaria, Enarrationes et Scholia in Jac. Holleri Opera Practica*. Parisiis, 1571. Cologny, 1623.

—*Hippocratis Aphorismi, Prognostica, Prorrhetica, Coacae Praenationes, Aliaque Decem Opuscula*. Leiden, 1627.

—*Hippocratis Coacae Phraenotiones*. Parisiis, 1588.

Edwards, John. *Brief Remarks upon Mr.* [*William*] *Whiston's New Theory of the Earth*. London, 1697, 1712.

—*A Discourse Concerning the Authority, Stile, and Perfection of the Books of the Old and New Testaments*. 3 vols. London, 1693. Vol. 2.

—*Some New Discoveries of the Uncertainty, Deficiency, and Corruptions of Human Knowledge and Learning*. London, 1714.

Elsholtz, Johann Sigismund. “Observatio LXXX. *De Ovo Praegnante*.” In *Miscellanea Curiosa Medico-Physica Academiae Naturae Curiosorum, sive Ephemeridum Medico-Physicarum Germanicarum*. Annus sextus et septimus. Francofurti & Lipsiae, 1677. 115–16.

Epiphanius of Constantia. *Epiphanius, Ancoratus and Panarion* [*Die griechischen christlichen Schriftsteller* 25, 31, 37] 3 vols. Edited by K. Holl. Leipzig: Hinrichs, 1915–33.

Epiphanius of Salamis. *On Weights and Measures*. In *Epiphanius' Treatise on Weight and Measures: The Syriac Version*. Edited by James Elmer Dean. Chicago: University of Chicago Press, 1935.

Erastus, Thomas. *Disputationum De Nova Medicine Philippi Paracelsi*. Basileae, 1572–73.

Etheridge, J.W. *The Targums of Onkelos and Jonathan Ben Uzziel on the Pentateuch*. 2 vols. 1862–65; Piscataway: Gorgias Press, 2005.

Eusebius Nierembergius, Joannis. *Historia naturae, maxime peregrinae libris XVI distincta*. Antverpiae, 1635.

Eustachio, Bartolomeo. *Opuscula Anatomica*. Venetiis, 1564.

Eutropius. *Breviarium Historiae Romanae. Abridgement of Roman History*. Translated by John Selby Watson. London: Henry G. Bohn, 1853.

Fabricius ab Aquapendente, Hieronymus. *De Formato Foetu.* In *Opera Omnia Anatomica et Physiologica.* Lugduni Batavorum, 1687.

Fabyan, Robert. *The Chronicles of Fabyan, whiche he hym selfe nameth the concordaunce of historyes*. London, 1559.

Favorino, Varino Camerte. *Dictionarium Varini Phavorini*. Roma, 1523; Basileae, 1538.

Fazelli, Thomas. *De Rebus Siculis Prioris Decades Duae, nunc Primum in Lucem Editae.* Panormi, 1558.

Fernelius, Johannes. *Ambiani de abditis rerum Causis Libri Duo*. Parisiis, 1548.

Ferrarius, Joannes Baptista. *Hesperides sive De Malorum Aureorum Cultura et Usu Libri Quatuor*. Romae, 1646.

Ficinus, Marsilius. *Theologia Platonica. De Immortalitate Animorum*. Parisiis, 1559.

Fienus, Thomas. *De Viribus Imaginationis Tractatus*. Lovanii, 1608.

Fincelius, Jacobus. *Wunderzeichen. Warhafftige beschreybung und gründlich verzeichnuß schröcklicher Wunderzeichen und Geschichten*. Jena: Nuremberg and Frankfurt, 1556.

Firmianus, Lucius Caecilius. *Institutes Divinae. The Divine Institutes*. In *ANF*. 7:9–223.

Flavius, Josephus. *Antiquities of the Jews*. In *Complete Works*. 23–426.

—*The Complete Works of Josephus* (Whiston translation). Grand Rapids: Kregel Publications, 1981.

—*Wars of the Jews*. In *Complete Works*. 427–605.

—*The Works of Josephus ... according to the Excellent Translation of Monsieur Arnauld D'Andilly*. London, 1701.

Florentinus, Thaddeus. *Expositiones in Arduum Aphorismorum Ipocratis Volumen*. Florentini, 1527.

Florus, Lucius Annaeus. *Epitomae De Tito Livio Bellorum Omnium Annorum DCC Libri Duo*. Lipsiae: Teubner, 1872.

Floyer, Sir John. "IV. A Relation of two Monstrous Pigs, with the resemblance of Humane Faces, and two young Turkeys joined by the Breast, by Sir John Floyer, Communicated by Dr. Edward Tylon, Fellow of the College of Physicians, and R.S." *PT* 21.259 (December 31, 1699): 431–35.

Foreest, Pieter van. *Curationum Medicinalium Liber Decimus-Octavus*. Lugduni Batavorum, 1590.

Forster, Thomas. "III. Part of a Letter from Thomas Forster, Esq; F.R.S. to Mr. Machin, Prof. Astron. Gresham, Secr. R.S. concerning a new Island lately raised out of the Sea near

Tercera." *PT* 32.372 (August 31, 1722): 100–01.

Fournier, George. *Hydrographie. Contenant la Theorie et la Practique de toutes les Parties de la Navigation*. Paris, 1643, 1667.

Fracantianus, Antonius. *De Morbo Gallico Tractatus Antonii Francanciani*. Bononiae, 1564.

Franklin, Benjamin. "Advice to a Young Tradesman." Philadelphia, 1748.

—*The Autobiography of Benjamin Franklin*. The Federalist Papers Project, www.thefederalistpapers.org.

—"A Letter from Mr. Franklin to Mr. Peter Collinson, F.R.S. concerning the effects of lightning." *PT* 47 (1752): 567–69.

—"Hints for Those That Would be Rich." Philadelphia, 1736.

—"The Way to Wealth." Philadelphia, 1758.

Frézier, Amédée-François. *Relation du Voyage de la Mer du Sud aux Cotes du Chily et du Perou, Fait pendant les années 1712, 1713 & 1714*. Paris, 1716.

Frytschius, Marcus. *Meteoroum, Hoc est, Impressionum Aerearum, et Mirabilium Naturae Operum*. Noribergae, 1555, 1563.

Fulgosus, Baptista. *Chronique de Baptista Fulgosus*. Parisiis, 1494.

—*De Dictis Factisque Memorabilibus*. Milano, 1509.

—*De Dictis Factisque Memorabilibus Collectananea*. Parisiis, 1588.

Fuller, Thomas. *The History of the Worthies of England*. London, 1662.

Gaffarel, Jacques. *Curiositez Inouyes sur la Sculpture Talismanique des Persans. Horoscope des Patriarches et Lecture des Estoiles*. Paris, 1629.

—*Unheard-of Curiosities: Concerning Talismanical Sculpture of the Persians; The Horoscope of the Patriarkes; And the Reading of the Stars*. London, 1650.

Gale, Theophilius. *The Court of the Gentiles: A Discourse touching the Original of Human Literature, both Philologie and Philosophie, From the Scriptures & Jewish Church*. Part I. 2nd edition revised and enlarged. Oxon, 1672. London, 1669–82.

Galen, Claudius. *Claudii Galeni Opera Omnia*. 4 vols. Edited by Karl Gottlob Kühn. Cambridge, UK: Cambridge University Press, 2012.

—*De dignotione ex insomniis. Operum Hippocratis Coi, et Galeni Pergameni, Midicorum Omnium Principum*. Tomus VI. Lutetiae Parisiorum, 1639. 6:517–19.

—*De foetuum formatione libellius* (IV.673, 16–674, 1 Kühn). In *Claudii Galeni Opera Omnia*. Vol. 4.

—*Method of Medicine*. 2 vols. Translated by Ian Johnston and G.H.R. Horsley. Cambridge, MA: Harvard University Press, 2011.

—*Opera ex Sexta Iuntarum Editione*. Venetiis, 1586.

—*Operum Hippocratis Coi, et Galen Pergameni. Medicorum Omnium Principum*. Editio René Chartier. Lutetiae Parisiorum, 1638–39.

—*De ossibus ad tirones* (II.753, 9–10 Kühn). In *Claudii Galeni Opera Omnia*. Vol. 2.

—*De semine* (IV.637, 10–14 Kühn). In *Claudii Galeni Opera Omnia*. Vol. 4.

—*De Theriaca ad Pisonem. Interprete & Commentatore, Ioanne Iuvene, Medico.* Antverpiae, 1575.

—*Thrasybulus sive utrum medicinae sit an gymnasticae hygieine.* Edited by C. G. Kühn. *Claudii Galeni Opera Omnia.* 19 vols. Leipzig: Knobloch, 1821-1830. 14:210–94.

—*De usu partium corporis humani* (III.868, 18–869, 2 Kühn). In *Claudii Galeni Opera Omnia.* Vol. 3.

Garcia, Don Carlos. *France and Spain Natural Enemies. or, Several Reasons to Prove it Impossible they should long be Friends. Englished with large Additions and Improvements.* By Thomas Brown. London, 1704.

Garden, George. "A Discourse concerning the Modern Theory of Generation, by Dr. George Garden of Aberdeen, being part of a Letter to Dr. William Musgrave, LLD. Reg. So. S. and by him communicated to the Royal Society." *PT* 17.192 (February 28, 1691): 474–83.

Garmann, Christian Friedrich. "Observatio CLVII: De Musculo vomitu rejecto." *Miscellanea Curiosa sive Epheremidum Medico-Physicarum Germanicarum Academiae.* Decuriae II. Annus Octavus 1689. Norimbergae, 1690. 341–47.

Gassendi, Pierre. *Exercitationes Paradoxicae adversus Aristoteleos.* Amstelodami, 1649.

—*The Mirrour of True Nobility & Gentility: being the Life of The Renowned Nicolaus Claudius Fabricius Lord of Peiresk.* London, 1657.

—*Opera Omnia in Sex Tomos Divisa.* Lugduni, 1658.

—*Romanum Calendarium Compendiosè Expositum.* Parisiis, 1654.

Gaulmin, Gilbert. *De Vita Mosis, Libri Tres.* Parisiis, 1629.

Gemma, Cornelius. *De Naturae Divinis Characterismis; seu Raris & Admirandis Spectaculis.* Antverpiae, 1575.

Gersonides. In *The Commentary on the Tanakh* (II Kings 18). *The Book of Kings 2. A New English Translation of the Text, Rashi and a Commentary Digest*, translated by Rabbi A.J. Rosenberg: The Judaica Press, 1998.

Gesner, Conrad. *Historiae Animalium Liber I. De Quadru Pedibus Viviparis.* Tiguri, 1551.

—*Historiae Animalium Liber II. De Quadru Predibus Oviparis.* Tiguri, 1551.

—*Historiae Animalium liber IV. Qui est De Piscium & Aquatilium Animantium Natura.* Tiguri, 1554; Francofurti, 1604.

—*Historiae animalium Liber V. Qui est De Serpentium Natura.* Tiguris, 1587.

—*De Omni Rerum Fossilium Genere, Gemmis, Lapidus, Metallis, et Huiusmodi.* Tiguri, 1565.

G.G.L. "Annotatio de Luce quam quidam Auroram Borealem vocant." *Miscellanea Berolinensia ad Incrementum Scientiarum.* Berolini, 1710. 1:137–38.

Giambattista della Porta. *Magia Naturalis*. Neapolitae, 1558, 1589. Translated into English as *Natural Magick by John Baptista Porta in Twenty Books*. London, 1658.

Gilbert, William. of *De Magnete, Magneticisque Corporibus, et de Magno Mangete Tellure.* London, 1600.

Glanvill, Joseph. *Lux Orientalis; or an Enquiry into the Opinion of the Eastern Sages concerning the Præexistence of Souls.* London, 1682.

—*Saducismus Triumphatus: Or, full and plain Evidence Concerning Witches and Apparitions.* London, 1681, 1689, 1700.

Godwin, Francis (Domingo Gonsales). *The Man in the Moon: or A Discourse of a Voyage thither*. London, 1638.

Goulartius, Simon. *XL Tableaux de la Mort*. Genevae, 1606.

[Grainger, Samuel]. *The Imposition of Inoculation As a Duty Religiously Considered. In a Letter to a Gentleman in the Country, Inclin'd to admit it.* Boston, 1721.

Graunt, John. *Natural and Political Observations Mentioned in a following Index, and made upon the Bills of Mortality*. London, 1662.

Greatrakes, Valentine. *A Brief Account of Mr. Valentine Greatraks, And Diverse of the Strange Cures By him lately Performed. Written in a Letter Addressed to the Honourable Robert Boyle Esq.* London, 1666.

[Greenwood, Isaac]. *Arithmetick Vulgar and Decimal with the Application thereof to A Variety of Cases in Trade, and Commerce*. Boston, 1729.

—*A Friendly Debate; or, A Dialogue, Between Academicus* [Zabdiel Boylston]; *and Sawny* [William Douglass] *& Mundungus* [John Williams], *Two Eminent Physicians, About some of their Later Performances*. Boston, 1722.

Gregory, David. *Astronomiae Physicae & Geometricae Elementa.* Oxford, 1702.

Gregory, William. "An Account of a monstrous Foetus, resembling a hooded Monkey: Communicated by Mr. William Gregory of Rochester." *PT* 41.461 (December 31, 1741): 764–67.

Grembs, Franz Oswald. *Arbor Integra et Ruinosa Hominis; id est: Tractatus Medicus Theorico Practicus in Tres Libros Divisus*. Francofurti, 1657.

Grevinus, Jacobus. *De Venenis Libri Duo*. Antverpiae, 1571.

Grew, Nehemiah. *Anatomy of Plants. With an Idea of a Philosophical History of Plants*. N.p, 1682.

—*Cosmologia Sacra: or a Discourse of the Universe As it is the Creature and Kingdom of God. Chiefly written, To Demonstrate the Truth of the Bible.* London, 1701.

—*Musaeum Regalis Societatis, or, A Catalogue and Description Of the Natural and Artificial Rarities*. London, 1685.

—"Some Observations touching the Nature of Snow, presented to the Society." *PT* 8.92 (March 25 1673): 5193–96.

Groeneveldt, Jan. *A Treatise of the Stone and Gravel, their Causes, Signs, and Symptoms, with Methods of their Prevention and Cure. And Some Account also of the Manner of the Collotion section*. London, 1677, 1710.

Grotius, Hugo. *See* De Laet.

Grynaeus. Ed. *Novus Orbis Regionum ac insularum veteribus incognitarum*. Basileae, 1532.

Guard-Book. Manuscripts from correspondents, Royal Society Archives.

Guazzo, Francesco Maria. *Compendium Maleficarum*. Mediolani, 1626.

Gyllius, Petrus. *Descriptio Nova Elephanti*. Hamburgi, 1604.

—*Ex Aeliani Historia per Petrum Gyllilum Latini Facti ... accessionibus aucti libri XVI*. Lugduni, 1533.

—*Liber Summarius de Gallicis et Latinis Nominibus Piscium Massiliensium*. Appendix to *Ex Aeliani Historia*. 1533.

Hacspan, Theodor. *Notarum Philologico-Theologicarum in Varia et Difficilia Scriptorae loca Pars Prima*. Altdorphii, 1664.

Hakewill, George. *An Apologie of the Power and Providence of God*. Oxford, 1627.

Halley, Edmond. "An Account of the Cause of the Change in the Variation of the Magnetical Needle; With an Hypothesis of the Structure of the Internal Parts of the Earth." *PT* 17.195 (October 19, 1692): 563–78.

—"An Account of the late surprizing appearance of the lights seen in the air," *PT* 29.347 (March 31, 1716): 406–28.

—"An Account of the Phaenomena of a very extraordinary Aurora Borealis (November 10, 1719)." *PT* 30.363 (December 31, 1719): 1099–1100.

—"*Astronomiae Cometicæ Synopsis*, Autore Edmundo Halleio apud Oxonienses Geometriae Saviliano." *PT* 24.297 (March 31, 1705): 1882–99.

—*The Elements of that Mathematical Art, Commonly called Algebra. Compiled by John Kersey*. London, 1709.

—*Miscellanea Curiosa: Containing a Collection of some of the Principal Phænomena in Nature, Accounted for by the Greatest Philosophers of this Age*. Vol. 2. London, 1708.

—"Observations of the late Total Eclipse of the Sun on the 22d of April last past [1714], made before the Royal Society at their House in Crane-Court in Fleet-street." *PT* 29.343 (May 31, 1715): 245–62.

—"II. *Some Queries concerning the Nature of Light, and Diaphanous Bodies. Proposed to the Royal Society by* E. Halley." *PT* 17.206 (December 1, 1693): 998–99.

—*A Synopsis of the Astronomy of Comets*. London, 1705. Rpt in *Miscellanea Curiosa* (1708). Pp. 1–/6 (sep. pagination).

—"A Theory of the Variation of the Magnetical Compass." *PT* 13.148 (June 10, 1683): 208–21.

Harris, John. *Lexicon Technicum: Or, An Universal English Dictionary of Arts and Science*. 2 vols. London, 1704.

Harris, Walter. *De Peste Dissertatio Habita Apr. 17. 1721. In Aphitheatro Collegii Regalis Medicorum Londinensium. Cui accessit Descriptio Inoculationis Variolarum*. Londini, 1721.

Harrison, John. *The Principles of Mr. Harrison's Time-Keeper*. London, 1767.

Hartman, George. *The Family Physitian, or, A Collection of Choice, approv'd and experienc'd Remedies*. London, 1696.

—*The True Preserver and Restorer of Health*. London, 1682.

Hartsoeker, Nicolaas. *Essay de Dioptrique*. Paris, 1694.

—*Principes De Physique*. Paris, 1696; Amsterdam, 1704.

—*Recueil de Plusieurs Pièces de Physique, où l'on fait principalement voir l'invalidité du Système de Mr. Newton et où se trouve*. Paris, 1722.

Harvey, William. *Exercitatio Anatomica De Motu Cordis et Sanguinis in Animalibus*. Francofurti, 1628.

—*Exercitationes De Generatione Animalium*. Londini, 1651.

Hawthorne, Nathaniel. "Egotism; Or, The Bosom Serpent, From the Unpublished 'Allegory of the Heart." *The United States Magazine and Democratic Review* 12.55 (January 1843): 255–61.

Hecataeus of Miletus. *Fragmenta*. Edited by F. Jacoby. *Die Fragmente der griechischen Historiker (FGrH)* # 1. Leiden: Brill, 1923–58.

Heidegger, Johann Heinrich. רָאשֵׁי אָבוֹת [Rashi Aboth] *Sive de Historia Sacra Patriarcharum Exercitationes Selectae*, 2 vols. (1667–71). Tiguri, 1729.

Helwig, Johannes. *Observationes physico-medicae*. Augustae Fidelicorum, 1680.

Herodianus. *The History of the Empire of Marcus*. With an English Translation by C.R. Whittaker. 2 vols. Cambridge, MA: Harvard University Press, 1969.

Hesiod. *Theogonia: Works and Days. Testimonia*. Edited and translated by Glenn W. Most. Cambridge, MA: Harvard University Press, 2007.

Hernandez, Francisco, of Toledo. *Plantarum Animalium et Mineralium Mexicanorum Historia*. Romae, 1650.

Hertodius, Matthaeus Franciscus. "Observatio CXLIV. D. Francisci Matthaei Hertodii. *Panis Antipathia*." In *Miscellanea Curiosa Medico-Physica Academiae Naturae Curiosorum*. Annus Secundus. Jenae, 1671. 228–29.

Heurnius, Johannes (Jan van Herne). *Hippocratis Aphorismi, Prognostica, Prorrhetica, Coacae Praenationes, Aliaque Decem Opuscula, Graecè et Latinè, pleraque ex interpretatione J. Heurnii*. Lugduni Batavorum, 1627.

—*Institutiones Medicinae. Edito altera*. Lugduni Batavorum, 1638.

Hevelius, Johannes. *Selenographia: sive Lunae Descriptio*. Gedani, 1647.

Heylyn, Peter. *A Full Relation Of two Journeys: The One Into the Main-Land of France. The Other, Into some of the adjacent Islands*. London, 1656.

Hildanus, Gulielmus Fabricius. *Lithotomia Vesicae*. Basilae, 1626.

—*Lithotomia Vesicae, That is An Accurate Description of the Stone in the Bladder*. London, 1640.

—*Observationum et Curationum Chirurgicarum Centuriae*. Basileae, 1606, 1641.

Hill, Henry. *A Short Treatise Concerning the Propagation of Souls*. London, 1667.

Hippocrates. *Hippocratis de Alimento Liber*. Tomus VI. In *Operum Hippocratis Coi et Galeni Pergameni*. Edito Renatus Charterius *Vindocinensis*. Tomi XIII. Lutetiae Parisiorum, 1637–39. 6:238–99.

—*Operum Hippocratis Coi. Et Galeni Pergameni, Medicorum Omnium Principum. Tomus VII*. Lutetiae Parisiorum, 1679.

Hippolytus. *Appendix to the Works of Hippolytus*. *ANF*. 5:242–66.

Hobbes, Thomas. *Leviathan or the Matter, Forme, & Power of Common-wealth Ecclesiastical and Civil*. London, 1651.

Hoffmann, Friedrich. *Medicina Rationalis Systematica*. 6 vols. Halle, 1718–34.

Hogarth, William. *Cunicularii. The Wise Men of Godliman in Consultation*. London, 1726.

Holinshed, Raphael. *The First volume of the Chronicles of England, Scotlande, and Irelande*. London, 1577.

Homer. *Iliad*. 2 vols. Translated by A.T. Murray. Rev. by William T. Wyatt. Cambridge, MA: Harvard University Press, 1924.

—*L'Odysée d'Homere, Traduite en François, avec Des Remarques Par Madame Dacier*. 3 vols. Paris, 1708, 1716, 1756.

—*Odyssey*. 2 vols. Translated by A.T. Murray: Rev. by George E. Dimock: Cambridge, MA: Harvard University Press, 1919.

Hooke, Robert. *Lectures and Collections made by Robert Hooke*. London, 1678.

—*Micrographia: or some Physiological Descriptions of Minute Bodies made by Magnifying Glasses*. London, 1665.

—*The Posthumous Works of Robert Hooke … Containing his Cutlerian Lectures, and other Discourses, Read at the Meeting of the Illustrious Royal Society*. London, 1705.

Horace, Quintus Horatius Flaccus. *Ars Poetica, Art of Poetry*. Translated by H.R. Fairclough. Cambridge, MA: Harvard University Press, 1926. 442–90.

—*Carmina. Odes*. In *Odes and Epodes*. Edited and translated by Niall Rudd. Cambridge, MA: Harvard University Press, 2004. 22–261.

—*Odes*. Translated by Charles E. Bennet. New York: W. Heinemann Macmillan, 1914.

—*The Odes and Carmen Saeculare of Horace*. Translated by John Conington. London: George Bell and Sons, 1882.

—*Satires*. Translated by H.R. Fairclough. Cambridge, MA: Harvard University Press, 1926.

Houillier, Jacques. *Opera Practica*. Parisiis, 1571. Cologny, 1623.

Hull, John. *Diaries of John Hull, Mint-master and Treasurer of the Colony of Massachusetts Bay*. Boston: John Wilson and Son, 1857.

Ibn Ezra, Abraham (Aben Ezra). *Ibn Ezra's Commentary on the Pentateuch: Genesis (Bereshit)*. Translated by H. Norman Strickman and Arthur M. Silver. New York: Menorah Publishing Company, 1988.

Illmer, Ferdinand Friedrich. "Observatio XVI: Excretorum lapidum quorundam per Abdomen." *Miscellanea Curiosa sive Ephemeridum Medico-Physicarum*, 2 Decuria 3 (1684): 59–60.

Index Librorum Prohibitorum et Expurgandorum Novissimus. Romae, 1657.

Irenaeus. *Against Heresies*. *ANF*. 1:309–567.

Jackson, Thomas. *A Collection of the Works of Thomas Jackson, D.D.* London, 1653.

—*The Humiliation of the Sonne of God by His Becoming the Son of Man*. London, 1636.

Jameson, William. *Spicilegia Antiquitatum Ægypti atque ei Vicinarum Gentium*. Glasgow, 1720.

Jarchi, Solomon. *Metsudah Chumash/Rashi: A New Linear Translation by Rabbi Avraham Davis*. 5 vols. Monsey, NY: Eastern Book Press, Inc., 2006.

—*R. Salomonis Jarchi, Commentarius Hebraicus in Pentateuchum Latine Versus*, by Johann Friedrich Breithaupt. Gothae, 1710.

Javelli, Giovanni Crisostomo. *Medicina Compendium*. Lugduni, 1609.

St. Jerome. *Breviarium in Psalmos*. [*PL* 026. 0821C–1278D].

—*The Chronicle of Eusebius of Caesarea*. Translated by Roger Pierce. https://tertullian.org/fathers/jerome_chronicle_03_part2.htm.

—*Commentarium in Isaiam Prophetam*. [*PL* 024. 0017–0705].

—*Commentarium in Joelem Prophetam*. [*PL* 025. 0947–0988D].

—*Divina Bibliotheca Pars Prima: Liber Numerorum*. [*PL* 028. 0337–0400D].

—*Liber Hebraicarum Questionum in Genesim*. [*PL* 023. 0935–1010].

Jocisco, Andrea. *Oratio De Ortu, Vita, et Obitu Ioannis Oporini Basiliensis*. Argentorati, 1569.

Jo(h)nstonus, Johannes. *Historiae Naturalis. De Insectis libri III, De Serpentibus et Draconibus libri II. Cum aeneis Figuris*. Francofurti ad Moenum, 1653.

—*Historiae Naturalis de Quadrupetibus Libri*. Francofurti ad Moenum, 1650. Amstelaedami, 1657.

—*Thaumatographia Naturalis, in Decem Classes Distincta, in quibus Admiranda*. Amstelodami, 1632, 1661, 1665.

Jordanus, Raymundus. *Contemplationes Idiotae*. Paris, 1519.

—*Divi Aurelii Augustini Hippon. Meditationes... . Meditationes Idiota Viri Docti De Amore Divino*. Coloniae Agrippinae, 1639.

Josselyn, John. *An Account of two Voyages to New-England*. London, 1674.

—*New-Englands Rarities Discovered in Birds, Beasts, Fishes, Serpents, and Plants of that Country*. London, 1672.

Jovius, Paulus. *Historiarum sui Temporis Tomus Primus*. Florentinae, 1550.

Julius Caesar. *De Bello Gallico Commentarius*. Edited by George Long. London: Whittaker, 1871.

Jung, Georg Sebastian. "Observatio CCL. Ovum Ovo Praegnans." In *Miscellanea Curiosa Medico-Physica Academiae Naturae Curiosorum*. Annus secundus. Jenae, 1672.348–49.

Jurin, James. "An Account of a remarkable Instance of the Infection of the Small Pox." *PT* 32.373 (October 31, 1722): 191–93.

—*An Account of the Success of Inoculating the Small Pox in Great Britain, for the Year 1724*. London, 1725.

—*An Account of the Success of Inoculating the Small-Pox in Great-Britain, for the Year 1726. With A Comparison between the Miscarriages in that Practice, and the Mortality of the Natural Small-Pox*. London, 1727.

—*An Account Of the Success of Inoculating the Small Pox in Great Britain. With A Comparison between the Miscarriages in that Practice, and the Mortality of the Natural Small-Pox*. London, 1724.

—"Advertisement" (December 11, 1723). *PT* 32.370 (March 31, 1722): n.p.

—*Invitatio ad Observationes Meteorologicas communi consilio institutendas*. *PT* 32.379 (October 31, 1723): 422–27.

—"James Jurin to Cotton Mather (May 20, 1725). In Rusnock, *Correspondence*. 300–01.

—"James Jurin to John Huxham" (August 3, 1727). In A. Rusnock, editor. *Correspondence*. 364–65.

—*A Letter to the Learned Caleb Cotesworth, M.D… . Containing, A Comparison Between the Mortality of the Natural Small Pox, And that Given by Inoculation. To which is Subjoined, An Account of the Success of Inoculation in New England …* . London, 1723.

—"A Letter to the learned Dr. Caleb Cotesworth, F.R.S. of the College of Physicians … containing a Comparison between the Danger of the Natural Smallpox, and that given by Inoculation." *PT* 32.374 (December 31, 1722). 213–27.

Justin Martyr. *Dialogue with Trypho, a Jew*. *ANF*. 1:194–270.

—*Exhortation to the Greeks*. *ANF*. 1:273–89.

—*The First Apology*. *ANF*. 1:159–87.

—*The Monarchy*. *ANF*. 1:290-93. [*PG* 006. 0311–0326.]

—*Second Apology*. *ANF*. 1:188–93.

(Pseudo-Justinus Martyr). *Quaestiones et Responsiones ad Orthodoxos*. Edited by J.C.T. Otto. *Corpus apologetarum Christianorum saeculi secundi*. Vol. 5 (3rd edition). Jena: Mauke, 1881.

Juvenal. *Satyra*. In *Juvenal and Persius*. Translated by George G. Ramsay. London: William Heinemann, 1928.

Keepe, Henry. *Monumenta Westmonasteriensia: or an Historical Account of the Original, Increase, and Present State of St. Peter's, or, The Abby Church of Westminster*. London, 1682.

Kennedy, Peter. *An Essay on External Remedies. Wherein is Considered, whether all Curable Distempers incident to Human Bodies, may not be cured by Outward Means*. London, 1715.

Kepler, Johannes. *Astronomia Nova*. N.P. 1609.

—*Epitome Astronomiae Copernicanae*. N.P 1618.

Kerckring, Theodor. *Opera Omnia Anatomica: Contenti Specilegium Anatomicum*. Editio Secundae. Lugduni Batavorum, 1717.

Kircher, Athanasius. *Arca Noë, in tres libros digesta*. Amstelodami, 1675.

—*Ars Magnesia*. Herbipoli, 1631.

—*Linguæ Ægyptiaca Restituta*. Amstelodami, 1643.

—*Magnes sive de Arte Magnetica*. Romae, 1641, 1654.

—*Mundus Subterraneus, in XII Libros digestus; Quo Divinum Subterrestris Mundi Opificium*. Amstelodami, 1665, 1678.

—*Oedipus Ægyptiacus*. Amstelodami, 1652–55.

—*Philosophia Magnetica, in qua Magnetis Natura penitus explicatur*. Romae, 1629.

—*Scrutinium Physico-Medicum Contagiosae Luis, quae dicitur Pestis*. Lipsiae, 1659.

Kirchius, G. "X. Brevis Descriptio clari cujusdam Luminis, Anno 1707, die 6 Martii." In *Miscellanea Berolinensia ad Incrementum Scientiarum*. Berolini, 1710. 1:135–37.

Kornmannus, Henricus. *De Miraculis Mortuorum, Opus Novum et Admirandum in decem partes Distributum*. Francofurti, 1610.

La Peyrère, Isaac. *Men Before Adam*. London, 1656.

—*Prae-adamitae*. Amsterdam, 1655.

—*Relation du Groenland*. Paris, 1647, 1662.

—*Relation d'Islande*. Paris, 1644.

—*A Theological Systeme Upon that Presupposition, That Men were before Adam*. London, 1655.

Lactantius. *Divine Institutes*. *ANF*. 7:9–255.

Laertius, Diogenes. *Lives of the Eminent Philosophers*. Translated by R.D. Hicks. Cambridge, MA: Harvard University Press, 1925.

Laetus, Julius Pomponius. *See* Camerarius, Philipp.

Lange, Johannes. *Epistolarum Medicinalium*. Francofurdi, 1589.

Langelott, Joël. "Observatio XIV: De Mirando Antipathiae Exemplo." *Miscellanea Curiosa Medico-Physica Academiae Naturae Curiosorum, sive Ephemeridum Medico-Physicarum Germanicarum*. Annus sextus et septimus. Francofurti & Lipsiae, 1677. 24–25.

Langwith, Benjamin. "Observations on the Figures of Snow… . In a letter to Dr. Jurin, R.S. Secr." *PT* 23.376 (January 1, 1723): 298–300.

Lanzoni, Joseph. "Observatio XLIX, De Pure in coxa tumore collecto, per alyum expurgare." In *Miscellanea Curiosa sive Ephemeridum. Decuriae II. Annus Nonun MDCXC*. Noribergae, 1691. 82.

Lapide, Cornelius à. *In Pentateuchum Mosis Commentaria. Editio ultima, acuta & recognita*. Lutetiae Parisiorum, 1626.

Lavater, Ludwig. *Of Ghostes and Spirits Walking by Nyght*. London, 1572.

—*De Spectris, Lemuribus et Magnis atque Insolitis Fragoribus*. Genevae, 1570.

Leguat, François. *New Voyage to the East-Indies*. London, 1708.

—*Voyage et Avantures de Francois Leguat, & de ses Compagnons, En Deux Isles Desertes Des Indes Orientales*. Londres, 1708.

Leibniz, Gottfried Wilhelm Freiherr von. *Dissertatio De Arte Combinatoria, In qua Ex Arithmeticæ fundamentis Complicationum ac Transpositionum Doctrina novis*. Lipsiae, 1666.

—"Sisteme Nouveau de la Nature et de la Communication des Substances, aussi bien que de L'Union qu'il y a entre L'Âme & le Corps. Par M.D. L[eibniz]." *LJS* XXV (Juin 27 1695): 294–306.

Leigh, Charles. *The Natural History of Lancaster, Cheshire and the Peak in Derbyshire*. Oxford, 1700.

Lentulus, Cyiracus. *Historia Admiranda De Prodigiosa Apolloniae Schreierae. Virginis in Agro Bernensi ... Tribus Narrationibus Comprehensa ... de ejusmodi prodigiosis inediis*. Bernae Helvetiorum, 1604.

Letter-Book. Manuscript letters from correspondents. Royal Society Archives.

Leunclavius, Johannes. *Annales Sultanorum Othomandiarum, A Turcis sua Lingua scripti*. Francofurti, 1588.

—*Historiae Musulmanae Turcorum*. Francofurti, 1591.

Licetus, Fortunius. *De His, Qui Diu Vivunt Sine Alimento Libri Quatuor*. Patavi, 1612.

—*De Monstris. Ex recensione Gerardi Blasii, M.D. & PP. Qui Monstra quædam nova & rariora ex recentiorum scriptis addidit. Editio Novissima*. Amstelodami, 1665.

—*De Monstrorum Caussis, Natura et Differentiis Libri Duo*. Patavi, 1616, 1634.

Lister, Martin. "An Account of a Stone Cut Out from under the Tongue of a Man; Lately Sent in a Letter of Mr. Listers to His Grace the Lord Arch-Bishop of York." *PT* 7.83 (May 20 1672): 4062–64.

—"An Extract of a Letter of July 28 1675, by Mr. Lister from York, containing … odd worms vomited by children." *PT* 10.117 (September 1675): 391–95.

—*Historiae Animalium Angliae Tres Tractatus*. Londini, 1678.

—"An Objection to the new Hypothesis of the Generation of Animals from Animalcula in Semine Masculino." *PT* 20.244 (September 30 1698): 337.

Livius, Titus. *Ab Urbe Condita. The History of Rome*. Translated by B.O. Foster. Cambridge, MA: Harvard University Press, 1919.

Locke, John. *Essay Concerning Humane Understanding in Four Books*. London, 1690.

Logie, Alexander. *Vox Corvi: Or, The Voice of a Raven. That Thrice spoke these Words distinctly. Look into Colossians the 3d and 15th*. London, 1694.

Lubieniecki, Stanisław. *Theatrum Cometicum, Duabus Partibus constance*. Amstelodami, 1668.

—*Theatri Cometici* (pars posterior) *Historia Universalis Omnium Cometarum, A Diluvio usque ad praesentem Annum vulgaris Epochae a Christo nato 1665. decurrentem, qui numero 415. ennarrantur* (second title page). Lugduni Batavorum, 1681.

Lucanus, Marcus Annaeus. *Lucan. Pharsalia*. Translated by J.D. Duff. Cambridge, MA: Harvard University Press, 1962.

—*Pharsalia*. Translated by Edward Ridley. London: Longmans, Green, and Co, 1896.

Lucian of Samosata. *Dialogues of the Gods*. Translated by H.W. Fowler et al. Oxford: The Clarendon Press, 1905.

Lucretius, Titus Carus. *De Rerum Natura. On the Nature of Things*. Translated by W.H.D. Rouse. Cambridge, MA: Harvard University Press, 1975.

Ludolphus, Hiob. *Ad suam Historiam Æthiopicam antehac editam Commentarius*. Francofurti ad Moenum, 1691.

—*Appendix Secunda ad Historiam Æthipicam Iobi Ludolfi continens Dissertationem De Locustis*. Francofurti ad Moenum, 1694.

—*Historia Æthiopica. Sive Brevis & Succincata Descriptio Regni Habessinorum*. Francofurti ad Moneum, 1681. Translated as *A New History of Ethiopia. Being A Full and Accurate Description of The Kingdom of Abessinia. In Four Books*. London, 1682.

Lupton, Thomas. *A Thousand Notable things, of sundry sorts*. London, 1579.

Lycophron. *Alexandria*. In *Callimachus Lycophron Aratus*. Translated by A.W. Mair et al. London: William Heinemann, 1921. 494–617.

Lycosthenes, Conradus. *The Doome warning all Men to the Iudgmente*. London, 1581.

—*Prodigiorum ac Ostentorum Chronicon*. Basileae, 1557.

Lyranus, Nicolaus. *Postilla Super Biblicam*. 1492.

Macrobius. *Saturnalia*. Translated by Gottfried Bass. Quedlinburg and Leipzig, 1852.

Magirus, Johannes. Editor. *Physiologiae Peripatetica Libri Sex*. Francofurti ad Moenium, 1597, 1616.

Magius, Hieronymus. *Variarum Lectionum, seu Miscellaneorum Libri IIII*. Venetiis, 1564.

Maitland, Charles. *Mr. [Charles] Maitland's Account of Inoculating the Small Pox Vindicated, from Dr. Wagstaffe's Misrepresentations of that Practice, with some Remarks on Mr. Massey's Sermon.* London, 1722.

Malebranche, Nicolas. *De La Recherche De la Vérité*. Paris, 1674–75. Translated as *Father Malebranche his Treatise Concerning the Search after Truth*. London, 1700.

Malpighi, Marcelli. *Anatome Plantarum. Cui subjungitur Appendix, Iterates & auctas ejusdem Authoriis de Ovo Oncubato Observationes continens*. Londini, 1675.

Manilius, Marcus. *Astronomica*. Translated and Introduced by G.P. Goold. Cambridge, MA: Harvard University Press, 1977.

Margravius, Georgius. *Historia Rerum Naturalium Brasiliae, Libri Octo*. Amstelodami, 1648. Bound with, Pisonis, Guilielmus.

Martialis, Marcus Valerius. *Epigrammaton Libri IV*. Translated by D.R. Shackleton Bailey. 3 vols. Cambridge, MA: Harvard University Press, 1993.

Massey, Edmund. An allusion to *A Sermon against the Dangerous and Sinful Practice of Inoculation. Preach'd at St. Andrew's Holborn, On Sunday, July the 8th, 1722*. The Second Edition. London, 1722.

Mather, Cotton. *An Account of the Method and Success of Inoculating the Small-Pox in Boston in New-England*. London, 1722.

—"Address to the Physicians." In "Curiosa" 6.2 *Variola Triumphata* (MS 134–41).

—*Agricola. Or, The Religious Husbandman: The Main Intentions of Religion*. Boston, 1727.

—"The Angel of Bethesda." Holograph Manuscript. 1693–1724. Worcester, Mass: American Antiquarian Society.

—*The Angel of Bethesda, Visiting the Invalids of a miserable World*. New London, 1722.

—*The Angel of Bethesda*. Edited by Gordon W. Jones, M.D. Barre, MA: American Antiquarian Society and Barre Publishers, 1972.

—"Another Brand Pluckt Out of the Burning." In Robert Calef. *More Wonders of the Invisible World*. London, 1700. 1–13.

—*Armour of Christianity. A Treatise, Detecting first, the Plots of the Devil against our Happiness. Declaring then, the Wiles by which those Plots are managed*. Boston, 1704.

—*Biblia Americana. America's First Bible Commentary. A Synoptic Commentary on the Old and New Testaments*. Edited by Reiner Smolinski, Jan Stievermann, et al. 10 vols. Tübingen: Mohr Siebeck, 2010–26.

—*Blessed Unions. An Union with the Sonof God by Faith*. Boston, 1692.

—*Boanerges. An Short Essay to preserve and strengthen the Good Impressions Produced by Earthquakes*. Boston, 1727.

—"Boanerges, or Work of the Day." c. 1716.
—*Bonifacius. An Essay upon the Good, that is to be Devised and Designed.* Boston, 1710.
—"A Brand Pluck'd out of the Burning." In Worthington C. Ford. "Mather-Calef Paper on Witchcraft." *PMHS* 47 (October 1913–June, 1914): 240–68.
—*The Boston Ephemeris. An Almanack for The (Dionysian) Year of the Christian Æra. MDC.LXXXIII. And of the Worlds Creation 5632.* Boston, 1683.
—*Brontologia Sacra: The Voice of the Glorious God in Thunder.* London, 1695.
—*Columbanus. Or, The Doves Flying to the Windows of their Savior.* Boston, 1722.
—*A Christian at his Calling.* Boston, 1701.
—*The Christian Philosopher: A Collection of the Best Discoveries in Nature, with Religious Improvements.* London, 1721.
—*The Christian Philosopher.* Edited, with an Introduction and Notes, by Winton U. Solberg. Urbana and Chicago: University of Illinois Press, 1994.
—*Christodolus. A Good Reward of A Good Servant.* Boston, 1725.
—*Compassions Called for. An Essay of Profitable Reflections on Miserable Spectacles. To which is added ... The Surprising Distresses and Deliverances of a Company lately Shipwreck'd, on a Desolate Rock, on the Coast of N.E.* Boston, 1711.
—"Cotton Mather to John Cotton" (c. November, 1678). In *MP* (383–84). *SLCM.* 7–8. *CJCJ.* 243–44.
—"Cotton Mather to John Winthrop" (15*d.* VIII*m.* [Oct.] 1716). In *MP* (419–21).
—"Cotton Mather to John Winthrop" (November 19, 1716). In *SLCM.* 224.
—"Cotton Mather to Jeremiah Dummer" (January 19, 1719). In *SLCM* (271–72).
—"Cotton Mather to John Winthrop" (9*d.* III*m.* [May] 1720). In *MP* (438–39); *SLCM* (305–06).
—"Cotton Mather to John Winthrop" (29*d.* VI*m.* [Aug.] 1720), *MP* (442–43).
—"Cotton Mather to John Winthrop" (26*d.* X*m.* [Dec.] 1720). In *MP* (444, 455).
—"Cotton Mather to John Winthrop" (12*d.* 1*m.* [March] 1722). In *MP* (455).
—"Cotton Mather to Richard Waller" (December 1, 1712). In Stearns, *Science.* 405–06.
—"Cotton Mather to Samuel Penhallow" (19*d.* 7*m.* [September] 1712). In *D* (2:175–76).
—"Cotton Mather to Sir William Ashurst" (18*d.* 8*m.* [Oct.] 1715). In *D* (2:330–32).
—"Cotton Mather to Zabdiel Boylston" (June 24, 1721). In *MM* 1 (1789): 778.
—*Decennium Luctuosum. An History of Remarkable Occurrences, in the long War, which New-England hath had with the Indian Salvages.* Boston, 1699.

—*The Diary of Cotton Mather* (*D* 1-2). Edited by Worthing C. For. Collections of the MHS. Seventh Series. vol. 7–8. Boston, 1911–12.
—*The Diary of Cotton Mather for the Year 1712* (*D* 3). Edited by William R. Manierre II. Charlottesville, VA: University Press of Virginia, 1964.
—*Diluvium Ignis. De Secundo ac Optando Jehovae-Jesus Adventu; Deque Secundo ac Tremendo Illum Comitante Diluvio*. Boston, 1726, 1730.
—*Elizabeth in her Holy Retirement*. Boston, 1710.
—*An Essay on Comets, Their Nature, the Laws of their Motions*. Boston, 1744.
—"An Extract of a Letter from the Rev. Cotton Mather, of Boston, to Dr. John Woodward, of London, F.R.S. Communicated by Joshua Coffin, A.M." *CNH* 3 (1832): 122–23.
—"IV. An Extract of several Letters from Cotton Mather, D.D. to John Woodward, M.D. and Richard Waller, Esq;" *PT* 29.339 (June 30, 1714): 62–71.
—"A Faithful Account of what has occur'd under the Late Experiments of the Small-Pox managed and governed in the Way of Inoculation." in *Boston Gazette*. No. 101 (October 30, 1721).
—"Fructuosus, or the Life of Mr. Samuel Mather." *MCA* (4.2.13).
—*Hatzar-Maveth. Comfortable Words; In a Short Essay on The Comforts Of One Living God, But Walking through the Valley of the Shadow of Death*. Boston, 1726.
—*Hor-Hagidgad. An Essay, upon An Happy Departure. Occasioned By the Decease of the Valuable Mr. William Waldron*. Boston, 1727.
—*India Christiana. A Discourse, Delivered unto the Commissioners, for the Propagating of the Gospel among the American Indians which is Accompanied with several Instruments relating to the Glorious Design of Propagating our Holy Religion, in the Eastern as well as the Western, Indies*. Boston, 1721.
—*Late Memorable Providences Relating to Witchcrafts and Possessions*. London, 1691.
—*A Letter, About a Good Management under the Distemper of the Measles*. Boston, 1713.
—*The Life and Death of Mr. John Eliot*. London, 1691.
—*Magnalia Christi Americana: Or, The Ecclesiastical-History of New-England*. London, 1702.
—*Malachi. Or, The Everlasting Gospel, Preached unto the Nations*. Boston, 1717.
—*Man Eating the Food of Angels*. Boston, 1710.
—*Manuductio ad Ministerium. Directions for a Candidate of the Ministry*. Boston, 1726.
—*The Mather Papers*. *CMHS*. Fourth Series. Vol. 8. Boston, 1868.

—"Mr. Mathers Letter from Boston in New England. March 10, 1721/22." In *Mr. [Charles] Maitland's Account of Inoculating the Small Pox Vindicated, from Dr. Wagstaffe's Misrepresentations of that Practice, with some Remarks on Mr. Massey's Sermon.* London, 1722. 58–61.

—*Memorable Providences, Relating to Witchcrafts and Possessions*. Boston, 1689.

—*The Negro Christianized. An Essay to Excite and Assist that Good Work, The Instruction of Negro-Servants in Christianity*. Boston, 1706.

—*A New Offer to the Lovers of Religion and Learning*. Boston, c. 1714.

—*Observable Things. The History of Ten Years Rolled away under the great Calamities of A War, with Indian-Salvages*. Boston, 1699.

— *Ornamental Piety*. Boston, 1707.

—*Paterna. The Autobiography of Cotton Mather*. Edited by Ronald A. Bosco. Delmar: Scholars' Facsimiles & Reprints, 1976.

—*Parentator; Memoirs of Remarkables in the Life and Death of the Ever-Memorable Dr. Increase Mather*. Boston, 1724.

—*Pietas in Patriam: The Life of his Excellency Sir William Phips, Knt. Late Captain General, and Governour in Chief of the Province of the Massachuset-Bay, New England.* London, 1697.

—"Problema Theologicum." In "Cotton Mather's 'Problema Theologicum': An Authoritative Edition." Edited by Jeffrey Mares. *PAAS* 104, Pt 2 (1994): 333–440.

—*Psalterium Americanum. The Book of Psalms, In a Translation Exactly conformed unto the Original; But all in Blank Verse*. Boston, 1718.

—*The Right Way to Shake off a Viper. An Essay Upon a Case Too commonly Calling for Consideration, What shall Good Men do, when they are Evil Spoken of.* London, 1711. Boston, 1720.

—*Seasonable Thoughts upon Mortality*. Boston, 1712.

—*Selected Letters of Cotton Mather.* Compiled with Commentary by Kenneth Silverman. Baton Rouge: Louisiana State University Press, 1971.

—*Sentiments on the Small Pox Inoculated.* Appended to Increase Mather's *Several Reasons*. 1–2.

—*The Serviceable Man*. Boston, 1690.

— (With Zabdiel Boylston). *Some Account Of what is said of Inoculating or Transplanting the Small-Pox. By the Learned Dr. Emanuel Timonius, and Jacobus Pylarinus. With some Remarks thereon*. Boston, 1721.

—*The Stone Cut out of the Mountain*. Boston, 1716.

—*Terribilia Dei: Remarkable Judgments of God.* Boston, 1697.

—*The Thankful Christian. A brief Essay, Upon those Returns of Thankfulness and Obedience to the Glorious God, in which there lies the most Acceptable and Ingenuous Thanksgiving to Him for His Benefits. Address'd unto All that have received the Favours of Heaven; But more especially unto*

them, who after the good Success of a Whaling-Season, would Express their Gratitude unto God their Savior. Boston, 1717.
—"Thaumatographia Pneumatica." In *MCA* (1702) 6.7. 66–83.
—*Things for a Distress'd People to Think upon*. Boston, 1699.
—*Thirty Important Cases, Resolved With Evidence of Scripture and Reason. [Mostly,] By several Pastors of Adjacent Churches, meeting in Cambridge, New-England*. Boston, 1699.
—*Thoughts for the Day of Rain. In Two Essay's*. Boston, 1712.
—*The Threefold Paradise of Cotton Mather. An Edition of 'Triparadisus.'* Edited by Reiner Smolinski. Athens and London: University of Georgia Press, 1995.
—*The Triumphs of Reformed Religion, in America*. Boston, 1691.
—*A Treacle Fetch'd Out of a Viper*. Boston, 1707.
—*Triumphs of Troubles*. Boston, 1701.
—*To the truly Honourable & Venerable ... Professors, of the Renowned University of Glasgow*. Boston, 1710.
—*Vanishing Things. An Essay, On Dying Man, Known in his Place no more, When Death hath once removed from it. A Sermon Preached on the Departure of the Shining, and yet Humble, Mr. Thomas Bernard, Late Pastor of a Church in Andover*. Boston, 1718.
—*Victorina. A Sermon preach'd on the Decease and at the Desire of Mrs. Katharin Mather*. Boston, 1717.
—*A Voice from Heaven. An Account of a Late Uncommon Appearance in the Heavens*. Boston, 1719.
—*Warnings from the Dead*. Boston, 1693.
—"The Way of Proceeding in the Small Pox Inoculated in New England. Communicated by Henry Newman, Esq; of the Middle Temple." *PT* 32.370 (March 31 1722): 33–35.
—*The Will of the Father submitted to*. Boston, 1713.
— *Winter-Meditations. Directions How to employ the Liesure of the Winter For the Glory of God*. Boston, 1693.
—*Winter Piety. A Very Brief Essay, on the Methods of Piety*. Boston, 1712.
—*The Wonderful Works of God Commemorated*. Boston, 1690.
—*Wonders of the Invisible World. Observations As well Historical as Theological, upon the Nature, the Number, and the Operations of the Devils*. Boston, 1693.
—*Work upon the Ark. Meditations upon the Ark as a Type of the Church*. Boston, 1689.
—*The World Alarm'd. A Surprizing Relation, Of a New Burning-Island Lately raised out of the Sea, near Tercera; with A Geographical and Theological Improvement of So astonishing an Occurrence*. Boston, 1721.

—*The Voice of God in a Tempest. A Sermon Preached in the Time of the Storm; Wherein many and heavy unknown Losses were Suffered at Boston, (and Parts Adjacent,) Feb. 24, 1722–3*. Boston, 1723.

Mather, Increase. *Angelographia, or A Discourse Concerning the Nature and Power of the Holy Angels*. Boston, 1696.

—*An Essay for the Recording of Illustrious Providences*. Boston, 1684.

—*Kometographia. Or A Discourse concerning Comets*. Boston, 1683.

—*The Life and Death of That Reverend Man of God, Mr. Richard Mather*. Cambridge, 1670.

—*Several Reasons Proving that Inoculating or Transplanting the Small Pox, is a Lawful Practice, and that it has been Blessed by God for the Saving of many a Live*. Boston, 1721.

—*Some further Account from London, of the Small-Pox Inoculated*. Second Edition. Boston, 1721.

—"To the Honourable William Stoughton, Esqr. Lieut. Governour of the Province of Massachusetts" (December 16 1698). In Samuel Sewall, *Diary*. 1:493–94.

Mather, Samuel. *The Life of the Very Reverend and Learned Cotton Mather D.D. & F.R.S. Late Pastor of the North Church in Boston. Who Died, Feb. 13. 1727, 8*. Boston, 1729.

—*Apology for the Liberties of the Churches in New England*. Boston, 1738.

Mattioli, Pietro Andrea Gregorio. *Opera quae extant omnia: Hoc est Commentarij in VI. libros Pedacij Dioscoridis Anazarbei de Medica materia*. Basiliae, 1598.

Medina, Michael. *Christianae Paraenesis, sive De Recta in Deum Fide libri septem*. Venetiis, 1564.

Mela, Pomponius. *De Chorographia Libri Tres*. (*De Situ Orbis*). Edited and translated by F.E. Romer. In *Pomponius Mela's Description of the World*. Ann Arbor: The University of Michigan Press, 1998.

—*De Situ Orbis, Libri Tres. Ex recensione Andreae Schotti*. Helmaestadii, 1635.

Mercuriale, Girolamo. *De Morbis Puerorum Tractatus Locupletissimi*. Venetiis, 1583.

Metcalfe, Francis. *Cortex Peruvianus*. Cambridge, 1663.

Mexía, Pedro. *The Forest or Collection of Historyes*. London, 1576.

—*Silva de Varia Lección*. Séville, 1540.

Meyerus, Jacobus Baliolanus, *Commentarii, Sive Annales rerum Flandricarum Libri Septendecim*. [Antverpiae, 1561] n *Annales, Sive Historiae Rerum Belgicarum, à Diversis Auctoribus ... ad Haec Nostra Usque Tempora conscripta deductaque*. Francofurti ad Moenum, 1580.

Minos, Claudius. *Omnia Andreae Alciati v.c. Emblemata: cum Commentariis*. Antverpiae, 1577.

Miqra'ot Gedolot. The Commentators' Bible. 5 vols. Edited, translated, and annotated by Michael Carasik. Philadelphia: Jewish Publication Society, 2005–15.

Miscellanea Curiosa sive Ephemeridum Medico-Physicarum Germanicarum Academiae Caesareo-Leopoldinae Naturae Curiosorum ... Annus. Lipsiae, Nurembergae, Jenae.

The Mishnah. Translated ... by Herbert Danby. 1933; Peabody: Hendrickson Publishers, 2011.

Moinichen, Henrik. *Observationes Medico-Chirurgicae*. Ferrariæ, 1688.

Molière, Jean-Baptiste Poquelin. *Le Malade Imaginaire* [1673]; i.e., *The Imaginary Invalid*.

Molyneux, Thomas. "An Essay concerning Giants. Occasioned by some further remarks on the large Humane Os Frontis, or Forehead-bone mentioned in the Philosophical Transactions of February 1684/84, Number 168." *PT* 22.261 (February 28, 1700): 487–508.

—"A Letter from Dr. Thomas Molyneux, Fellow of the Royal Society, to the Right Reverend St. George, Lord Bishop of Clogher; concerning Swarms of Insects, that of late Years have much infested some Parts of the Province of Connought in Ireland." *PT* 19.234 (November 30, 1697): 741–55.

—"Part of 2 Letters from Mr. Thomas Molyneux concerning a Prodigious Os Frontis in the Medicine School at Leyden." *PT* 15.168 (February 23, 1685): 880–81.

—"Remarks upon the aforesaid Letter and Teeth, by Thomas Molyneux, M.D. And R.S.S. Physician to the State of Ireland: address'd to his Grace the Lord Archbishop of Dublin." *PT* 29.346 (December 31, 1715): 370–83.

Monardes, Nicolás Bautista. *De Simplicibus Medicamentis ex Occidentali India Delatis, Quorum in Medicina Usus est*. Antverpiae, 1574.

Montaigne, Michel. *See* De Montaigne.

Monumenta Westmonasteriensia. *See* Henry Keepe.

Moodey, Joshua. "Joshua Moodey to Increase Mather" (25^{d}. 8^{m}. 1683). In *MP* (1868), pp. 361–62.

More, Henry. *An Antidote against Atheism, or An Appeal to the Naturall Faculties of the Minde of Man, whether there be not a GOD*. London, 1655.

—*The Immortality of the Soul, So farre forth as it is demonstrable from the Knowledge of Nature and the Light of Reason*. London, 1659.

—*The Theological Works of the most pious and learned Henry More. In Two Parts*. London, 1708.

Morgan, Joseph. *The History of the Kingdom of Basaruah*. Boston, 1715.

Morton, Charles. *An Essay toward the Probable Solution of this Question. Whence come the Stork and the Turtle, the Crane, and the Swallow, when they Know and Observe the appointed Time of their Coming*. London, 1703.

Mundinus de Liucci, Raymondo. *Anatomia Mundini: En Lector Libellum Mundini quem de Partibus Humani Corporis Inscripsit*. Lugduni, 1531.

N.A. "A Relation of the Effect of a Thunder Clap on the Compass of a Ship on the Coast of New England" and "A Letter of Sir R.S. concerning the former Relation (Lond. March 8th, 1682/82)." *PT* 14.157 (March 20, 1684): 520–21.

Nachmanides, Rabbi Moshen ben (Ramban). *Commentary on the Torah*. 5 vols. Translated and annotated by Rabbi C. Chavel. Brooklyn: Shilo Publishing House, 1999.

Needham, John Turberville. "A Summary of some late Observations upon the Generation, Composition and Decomposition of Animal and Vegetable Substances." *PT* 45.490 (January 1, 1748).

Nevile [Neville], Francis. "A Letter of Mr. Francis Nevile to the Right Reverend St. George Lord Bishop of Clogpher, R.S.S. Giving an Account of some large Teenth lately dugg up in the North of Ireland." *PT* 29.346 (December 31, 1715): 367–70.

Newton, Sir Isaac. *A New and most Accurate Theory of the Moon's Motion; Whereby all her Irregularities may be solved, ... Written by ... Mr. Isaac Newton, And Published in Latin by Mr. David Gregory in his Excellent Astronomy*. London, 1702.

—*Optics: or, A Treatise of the Reflections, Refractions, and Inflexions and Colours of Light*. London, 1704.

—*Philosophiae Naturalis Principia Mathematica*. Londini, 1687.

Nicene and Post-Nicene Fathers: A Selected Library of the Christian Church. (First Series). 14 vols. Edited by Philip Schaff. 1886. Peabody, MA: Hendrickson Publishers, Inc. 1999.

Nieuwentyt, Bernard. *The Religious Philosopher: Or, The Right Use of Contemplating the Conviction of Atheists and Infidels*. 3 vols. London, 1718–19.

Oethaeus, Jacob. *Gründtlicher Bericht / Lehr und Instruction / von rechtem und nutzlichem brauch der Artzney*. Dilingen, 1574.

Olaus Magnus (Olof Månsson). *Historia de Gentibus Septentrionalibus*. Romae, 1555; Antverpiae, 1558.

Oppian. *Halieutica, or Fishing*. Translated by Alexander W. Mair. Cambridge, MA: Harvard University Press, 1928.

O. R. "VIII. O.R. Descriptio-Luminis Borealis." *Miscellanea Berolinensis ad Incrementum Scientiarum ex scriptis Societati Regiae Scientiarum*. Berolini, 1710. 1:131–33.

Origen. *Origenis Hexaplorum Quae Supersunt*. Tomus I et Tomus 2. Oxonii: Typographeo Clarendoniano, 1855.

Orosius, Paulus. *Historiarum Contra Gentes Libri Septem*. [*PL* 031.663–1174*B*]

Orpheus. *Orphica: Lithica*. In *Orphei Hymni*. Edited by W. Quandt. Berlin: Weidmann, 1692.

Ovid, Publius Naso. *Amores*. In *Heroides. Amores*. 314–512.

—*Ex Ponto*. In *Tristia. Ex Ponto*. Translated by A.L. Wheeler. Cambridge, MA: Harvard University Press, 1924. 264–489.

—*Heroides. Amores*. Translated by Grant Showerman. Rev. G.P. Goold. Cambridge, MA: Harvard University Press, 1914.

—*Metamorphoses*. Translated by F.J. Miller. Second edition. 2 vols. Cambridge, MA: Harvard University Press, 1921.

Palaephatus of Paros. *De Incredibilibus. On Incredible Things. Palaephati ΠΕΡΙ ΑΠΙΣΤΩΝ. Mythographi Graeci*. Vol. 3. Fasc. II. Edidit Nicolaus Festa. Lipsiae: In Aedibus B.G. Teubneri, 1902.

Panciroli, Guido. *Rerum Memorabilium: iam olim Deperditarum & contrà Recens atque Ingeniose Inventarum: Libri Duo*. Ambergae, 1590; Francofurti, 1660. Latin translation by Henricum Salmuth. English translation: *The History of Many Memorable Things Lost, which were in Use among the Ancients*. London, 1638.

Paraeus, Ambrosius. *Ambroise Paré. On Monsters and Marvels*. Translated with an Introduction and Notes by Janis L. Pallister. Chicago: University of Chicago Press, 1982.

—"De Monstris et Prodigiis." In *Opera Chirurgica Ambrosii Paraei*. Francofurti ad Moenum, 1594.

—*Opera* [*Chirurgica*] *Ambrosii Parei Regis Primarii, et Parisinensis Chirurgi*. Parisiis, 1582.

—*The Workes of that famous Chirurgion Ambrose Parey*. Translated by Thomas Johnson. London, 1649.

Patrick, Simon. *A Commentary on the Fourth Book of Moses, called Numbers*. London, 1699.

—*A Commentary upon the Two Books of Kings*. London, 1705.

Patritius (Patricius), Franciscus. *De Legendae Scribendae quae Historiae Ratione, Dialogi Decem, ex Italico in Latinum Sermonem Conversi*. Basileae, 1570.

Paullinus, Christianus Franciscus. *Cynographia Curiosa seu Canis Descriptio*. Norimbergae, 1685.

Pausanias. *Description of Greece*. Translated by W.H.S. Jones et al. 4 vols. Cambridge, MA: Harvard University Press, 1918.

Penhallow, Samuel. *The History of the Wars of New-England With the Eastern Indians*. Boston, 1726.

Perkins, William. *A Discourse of the Damned Art of Witchcraft*. Cambridge, 1608.

Petiver, James. *Jacobi Petiveri Opera, Historiam Naturalem Spectantia;* or, *Gazophylacium. Containing Several 1000 Figures of Birds, Beasts, Reptiles, Insects, Fish, Beetles, Moths, Flies, Shells, Corals, Fossils, Minerals, Stones, Fungusses, Mosses, Herbs, Plants, &c. from all Nations, on 156 Copper-Plates, with Latin and English Names*. 2 vols. London, 1764.

—*Petiveriana III, seu Naturae Collectanea; Domi Forisque Auctori Communicata*. Londini, 1717. Rpt *Jacobi Petiveri Opera*. Vol. 2.

Petty, Sir William. *An Essay concerning the Multiplication of Mankind*. London, 1682.

Philo Judaeus. "*De Gigantibus*." *Works*. 152–57.

—"*De Vita Mosis*." *Works*. 459–517.

—*Philonis Alexandrini Opera qua supersunt*. Edited by L. Cohn et al. 6 vols. Berlin: Reimer, 1896–1915.

—"*Quaestiones et Solutiones in Genesin, I*." *Works*. 791–813.

—*The Works of Philo. Complete and Unabridged*. New Updated Edition. Peabody: Hendrickson, 1993.

Philostorgius, *Ecclesiastical History*. In *Epitome of the Ecclesiastical History of Philostorgius*. Compiled by Photius, Patriarch of Constantinople. Translated by Edward Walford. London: Henry G. Bohn, 1855.

Philostratus, Lucius Flavius. *Heroicus. Gymnasticus. Discourses 1 and 2*. Edited and translated by Jeffrey Rusten, Jason König. Cambridge, MA Harvard University Press, 2014.

— *Life of Apollonius of Tyana*. Edited and Translated by Christopher P. Jones. 2 vols. Cambridge, MA: Harvard University Press, 2005.

—*De Vita Sophorum. The Lives of the Sophists*. In *Philostratus and Eunapius*. Translated by Wilmer Cave Wright. London: William Heinemann, 1922. 2–315.

Phlegon, Publius Aelius. *De Rebus Mirabilibus* [*Peri Makrobion kai thaumasion*]. *Book of Marvels*. In *Opuscula De Rebus Mirabilibus De Longaevis*. Edidit A. Stramaglia. Berlin: De Gruyter, 2011.

Pictorius, Georg. *ΠΑΝΤΟΠΩΛΙΟΝ, continens omnium fermè quadrupedum, avium, piscium, Serpentum*. Basileae, 1563, 1568.

Pigafetta, Antonio Franceso. *Magellan's Voyage: A Narrative Account of the First Circumnavigation*. 1969; New York: Dover Publications, 1994.

—*Le Voyage et Navigation fait par les Espaignolz es Isles de Molucques*. Paris, 1525.

Pisonis, Guilielmi. *De Medicina Brasiliensi libri IV et Georgii Marcgravii de Liebstad, Historiae Rerum Naturalium Brasiliae libri octo, … cum Appendice de Tapuyis, et Chilensibus Joannes de Laet, Antwerpianus*. Amstelodami, 1648.

Pitcairn, Archibald. *Dissertationes Medicae*. Edinburgh, 1701, 1713.

—*The Method of Curing the Small-Pox: Written in the Years 1704. For the Use of the Noble and Honourable Family of March*. London, 1715.

—*Opuscula Medica*. Venetiis, 1716.

—*The Works of Dr. Archibald Pitcairn*. London, 1715.

Pitiscus, Batholomaeus. *Thesaurus Mathematicus Sive Canon Sinuum ad Radium*. Francofurti, 1613.

— *Trigonometriae. Sive De Dimensione Triangulas Libri Quinque*. Francofurti, 1595, 1600, 1612.

Platerus, Felix. *Observationum, In Hominis Affectibus plerisque, corpori & animo, functionum laesione, dolore, aliáve molestiá & vitio intommodantiubs, Libri Tres*. Basilae, 1614.

—*Praxeos Medica Tomi Tres*. Basileae, 1602.

—*Praxeos Se de Cognoscendis, Praedicendis, Praecauendis, Curandisque affectibus homini incommodantibus Tractatus Tertius et Ultimus De Vitiis, Libris duobus agens*. Tomus Tertius. Basileae, 1608.

Plato. *Complete Works*. Edited, with Introduction and Notes, by John M. Cooper. Indianapolis: Hackett Publishing Company, 1997.

—*Charmides*. In *Complete Works*. 639–63.

—*Phaedrus*. In *Complete Works*. 49–100.

—*Republic*. In *Complete Works*. 971–1223.

—*Symposium*. In *Complete Works*. 457–505.

—*Theaetetus*. In *Complete Works*. 157–234.

—*Timaeus*. In *Complete Works*. 1224–91.

Plautus, Titus Maccius. *Amphitryon, or Jupiter in Disguise*. Translated by Paul Nixon. Cambridge, Mass. Harvard University Press, 1916.

Pliny the Elder (Gaius Plinius Secundus). *Historiae Naturalis Libri XXXVIII*. Lugduni Batavorum, 1635.

—*Naturalis Historiae. Natural History*. Translated by H. Rackham et al. 10 vols. Rev. ed. 1938. Cambridge, MA: Harvard University Press, 1949.

Pliny the Younger. *Letters*. 2 vols. Translated by Betty Radice. Cambridge, MA: Harvard University Press, 1969.

Plot, Robert. "A Discourse concerning the most seasonable Time of Felling of Timber, Written by the advice of the Honorable Sam. Pepys Esq; Secretary of the Admiralty, and presented to his late Majesty. By Robert Plot LLD. And R.S. Soc." *PT* 17.192 (February 1, 1691): 455–61.

—*The Natural History of Oxford-shire*. Oxford, 1677.

—*The Natural History of Stafford-shire*. Oxford, 1686.

Plukenet, Leonard. *Almagesti Botanici Mantissa Plantarum Novissimè Detectarum ultra Millenarium Numerum Complectens*. Londini, 1700.

—*Almagestum Botanicum sive Phytographiae Pluc'netianae Onomasticon Methodo Syntheticâ digestum*. Londini, 1696.

Plutarch, Lucius Mestrius. *De sera numinis vindicta. On the Delays of the Divine Vengeance*. In *Moralia*. 7:180–300.

—*Moralia*. Edited with an English Translation by Frank Cole Babbit. 16 vols. Cambridge, MA: Harvard University Press, 1936.

—*Parallela Minora*. In *Moralia*. 4:252–317.

—*Sentiments Concerning Nature*. In *The Complete Works of Plutarch*. 6 vols. New York: Crowell, 1909. Vol. 3.

—*Symposiacs: Quaestiones Conviviales IV. In Moralia.* 8:283–371.
—*Vitae Parallelae: Cato the Younger*. In *Plutarch's Lives*. 8:236–414.
—*Vitae Parallelae: Pyrrhus*. In *Plutarch's Lives*. 9:345–462.
—*Vitae Parallelae: Romulus*. In *Plutarch's Lives*. 1:89–187.
—*Vitae Parallelae: Sertorius*. In *Plutarch's Lives*. 8:1–75.
—*Plutarch's Lives*. Translated by Bernadotte Perrin. 11 vols. Cambridge, MA: Harvard University Press. 1914–26.
Poiret, Pierre. *Cogitationes Rationalis (de Deo, Anima & Malo)*. Amsterdam, 1685.
Pomponazzi, Pietro. *De Naturalium Effectuum Causis, sive De Incantationibus. Opus Abstrusioris Philosophiae Plenum.* Basileae, 1556.
Poole, Matthew. *Synopsis Criticorum Aliorumque S. Scripturae Interpretum.* 5 vols. Londini, 1669–1676.
Pope, Alexander. *Dunciad*. London, 1728.
Postlethwayt, James, et al,. *A Collection of the Yearly Bills of Mortality, From 1657 to 1758 inclusive*. London, 1759.
Prince, Thomas. *An Account of a Strange Appearance in the Heavens on Tuesday-Night, March 6. 1716.* Boston, 1719.
—"Appendix Concerning the Operation of God in Earthquakes by Means of Electrical Substance." In *Earthquakes the Works of God*. Boston, 1755.
—"Mr. Prince's Account of the Northern Lights, when first seen in England, 1716." *CMHS* 2 (1793): 14–16.
Priscianus Caesariensis. *Periegesis*. In *Geographi Graeci Minores*. Edited by Karl Müller. 1882. 2:190–99.
Procopius of Caesarea. *The History of the Wars*. Translated by H.B. Dewing. 7 vols. Cambridge, MA: Harvard University Press, 1916.
Ptolemy, Claudius. *Geographia. Edidit Carolus Fridericus Augustus Nobbe.* Lipsiae, 1842–45.
Purchas, Samuel. *Hakluytus Posthumus or Purchase His Pilgrimes.* London, 1625; Glasgow: James MacLehose, 1905.
Pylarinus, Jacobus. "Nova & tuta Variolas excitandi per Transplantationum Methodus, nuper inventa & in usum tracta: Per Jacobum Pylarinum, Venetum, M.D. & Reipublicae Venetae apud Smyrnenses nuper Consulem." *PT* 29.347 (March 31, 1716): 393–99.
—*Nova et Tuta Variolas excitandi per Transplantationem Methodus; Nuper inventa & in usum tracta: Qua ritè peracta, immunia in posterum præservantur ab hujus mod contagio Corpora.* Venetiis, 1715.
Quercetanus, Josephus. *Liber De Priscorum Philosophorum Verae Medicinae Materia*. Gervasii, 1603.
Raleigh, Sir Walter. *The History of the World*. London, 1614.
Ramesey (Ramsay), William. *De Venenis, or, A Discourse of Poysons, their Names, Natures, & Vertues*. London, 1663.

Ramusio, Giovanni Battista. *Delle Navigatione et Vaggi*. Venetiis, 1559, 1583.

Randolph, Edward. "*A Letter of Sir* R.S. *concerning the former Relation. Lond. March* 8*th. 1684*." *PT* 14.157 (March 20 1684): 521–22.

RASHI. *See* Solomon Jarchi.

Rattray, Sylvester. *Aditus Novus ad Occultas Sympathiae et Antipathiae causas Inveniendas*. Glasgow, 1658.

—*Theatrum Sympatheticum Auctum exhibens Varios Authores, De Pulvere Sympathetico*. Noribergae, 1662.

Ray, John (and Francis Willughby). *Historia Piscium libri quatuor*. Oxonii, 1686.

—*Historia Plantarum Species hactenus editas aliasque insuper multas noviter inventas & descriptens Complectens*. 3 vols. London, 1686–1704.

—*Philosophical Letters Between the late Learned Mr. Ray And several of his Ingenious Correspondents, Natives and Foreigners*. London, 1718.

—*The Physico-Theological Discourses*. 3rd edition. London, 1713.

—*The Wisdom of God Manifested in the Works of the Creation*. London, 1691.

Redi, Franceso. "An Account of Three Books. I. Ezperienze Intorno alla Generatione Degl' Insetti, fatte da Francisco Redi … ." *PT* 5.57 (March 25 1670): 1175–76.

—*Esperienze Intorno alla Generazione Degl' Insetti*. Firenze, 1668.

—*Experimenta circa Generationem Insectorum*. Amstelodami, 1671.

Reineccius, Reinerus. *Origines Illustriss. Stirpis Brandeburgicae, seu Historiae Expositiones Geminae de Nobiliss. et Antiquiss. Welforum Prosapia*. Francofurti, 1581.

Report of the Record Commissioners of the City of Boston, Containing the Boston Records from 1700–1728, vol. 8. Boston: Rockwell and Churchill, City Printers, 1883.

Reynolds, John. *A Discourse Upon Prodigious Abstinence: Proving that without the Intervention of a Miraculous Power the Texture of Humane Bodies may be so altered, that Life may be long Continued without the usual supplies of Meat and Drink*. London, 1668.

—"*A Discourse Upon Prodigious Abstinence: Occasioned By the Twelve Moneths Fasting of Martha Taylor, the Famed Derbyshire Damosell. . . . Humbly offered to the Royall Society.*" London, 1669.

Rhodiginus, Lodovicus Caelius. *Lectionum Antiquarum Libri triginta*. Venetiis, 1516. Coloniae Allobrogum, 1620. Genevae, 1626.

—*Liber Quartus Naturalium De Actionibus et Passionibus Qualitatum Primarum*. Venetiis, n.d.

Rhodius, Ambrosius. *De Crepusculis*. Witebergae, 1611.

Rhodius, Johannes. *Observationum Medicinalium Centuriae Tres*. Patavi, 1657.

Riccioli, Giavanni Battista. *Almagestum Novum Astronomiam Veterem Novamque Complectens*. Bononiae, 1651.

Rivinus, Andreas (Bachmann). *Disputatio Philo-Physio-Logica; De Venilia & Salacia, nec non Malacia, Atque adeò sequiorum quoque seculorum Malina & Liduna*. Lipsiae, 1645.

Robie, Thomas. "An Account of a large Quantity of Alcalious Salt produced by burning rotten Wood. By Mr. Robie of Harvard College in Cambridge, New-England. Communicated by the Rev. Mr. Derham, Prebendary of Windsor, F.R.S." *PT* 31.366 (December 21, 1720): 121–24.

—*An Almanack of the Coelestial Motions, Aspects and Eclipses, for the Year of the Christian Aera 1717*. Boston, 1716.

—*A Letter to a Certain Gentleman desiring a particular Account may be given of a wonderful Meteor, that appeared in New-England, on Decemb. 11, 1719, in the Evening*. Boston, 1719.

—"Part of a Letter from Mr. Thomas Robie, Physician in New-England, to the Reverend Mr. Derham, F.R.S. Concerning the Effects of Inoculation; The Eclipse of the Sun in November 1722; And the Venom of Spiders." *PT* 33.382 (April 30, 1724): 67–70.

"The Records of the Cambridge Association." *PMHS* 17 (1879–1880): 254–81.

Rondelet, Guillaume. *L'Histoire entière de Poissons*. Lyon, 1558.

—*Libri de Piscibus Marinis, in quibus verae Piscium effigies express sunt*. Lugduni, 1554.

Rorarius, Girolamo Rorario. *Quod animalia bruta ratione utantur melius homine*. Parisiis, 1644.

Ross, Alexander. *The Alcoran of Mahomet. Translated out of the Arabique into French*. London, 1649.

Sala, Angelo. *Anatomia Vitrioli, in duos Tractatus divisa*. Aureliae Allobrogum, 1609.

—*De Variis tum Chymicorum, tum Gelenistarum Erroribus*. Francofurti, 1602.

Salonius Viennensis. *In Ecclesiasten Expositio Mystica*. [*PL* 53. 0993B–1012A.]

Saurin, Jacques. *Dissertations, Historical, Critical, Theological and Moral, On the most Memorable Events of the Old and New Testaments*. 2 vols. Translated by John Chamberlayne. 1720. Revised Edition. London, 1723.

—*Discours historiques, critiques, théologiques et moraux sur les événements les plus mémorables du Vieux et du Nouveau Testament*. Amsterdam, 1720.

Savonarola, Giovanni Michele. *Practica Major ... Omnia Morborum Genera a Capite usque ad Pedes Morta*. Venetiis, 1547.

Scaliger, Joseph Justus. *Elenchus Primae Orationis Chronologia*. Lugduni Batavorum, 1607.

—*Opus de Emendatione Temporum: Castigatus & Multis Partibus Auctius, ut novum videri posit*. Lugduni Batavorum, 1583.

—*Thesaurus Temporum Eusebiii Pamphili Caesarea Palaestinae Episci, Chronicorum Canonum*. Amstelodami, 1606.

Scaliger, Julius Caesar. *Exotericarum Exercitationum Liber Quintus Decimus, de Subtilitate, ad Hieronymum Cardanum*. Lutetiae, 1557. Francofurti, 1582.

Schellhammer, Günther Christoph. *De Auditu Liber Unus. Quo plerorumque omnium doctorum sententiae examinantur, & auditus ratio nova methodo, ex ipsus naturae legibus, explicatur*. Lugduni Batavorum, 1684.

Scheuchzer, John Gaspar. F.R.S. *An Account of the Success of Inoculating the Small-Pox in Great Britain, for the Years 1727 and 1728. A Comparison between the Mortality of the Natural Small-Pox, and the Miscarriages in that Practice*. London, 1729.

Schiller, Julius. *Coelum Stellatum Christianum*. Augustae Vindelicorum, 1627.

Scholzius (Schultzius), Simon. "Observatio XI: Alia singularis Idiosyncrasia." *Miscellanea Curiosa Medico-Physica Academiae Naturae Curiosorum sive Ephemeridum Medico-Physicarum Germanicacrum Curiosarum*. Annus Secundus. Jenae, 1671. 29–30.

—"Observatio CXX *De Usu Argenti Vivi in Vermibus*," *Miscellania Curiosa Sive Ephemeridum*. Annus Quartus et Quintus. Francofurti, 1676. 130.

—"Observatio CCXXIX: *De Lumbrico Lato*." *Miscellanea Curiosa Sive Ephemeridum*. Decuria 3, pars 2, Annus Tertius. Francofurti, 1673. 417–19.

Schönfelder, Philippus Jacobus. *Enarrationum et Curationum: Medicarum in certis locis*. Ratisbonae, 1681.

Schottus, Caspar. *Physica Curiosae, sive Mirabilia Naturae et Artis Libri XII Comprehensa* Herbipoli, 1662, 1667.

Schröder, Johann. *The Compleat Chymical Dispensatory, in Five Books*, Written in Latin, by Dr. John Schroder … and Englished, By William Rowland. London, 1669.

—*Pharmacopoeia Medico-Chymica, Sive Thesaurus Pharmacologicus*, Editio Secunda. Ulmae, 1644. Lugduni, 1665.

Scriverius, Petrus. *Respublica Hollandiae et Urbes*. Lugduni Batavorum, 1630.

Seger, Georg. *Miscellanea Curiosa Medico-Physica Academiae, Annus Tertius, Anni scilicet 1672*. Lipsiae & Francofurti, 1673.

Seidelius, Christopher Matthaeus. "IX. Observatio Luminis Borealis anni 1701." *Miscellanea Berolinensia ad Incrementum Scientiarum*. Berolini, 1710. 1:133–35.

Seneca (the Younger). *Controversiarum*. In *Seneca the Elder. Declamations*. Translated by Michael Winterbottom. 2 vols. Cambridge, MA: Harvard University Press, 1974.

—*Epistulae ad Lucilium* XV. In *Seneca Epistles*. 3 vols. Translated by Richard M. Gummere. Cambridge, MA: Harvard University Press, 1925. 1:96–97.

—*Medea*. Translated by Frank J. Miller. Cambridge, MA: Harvard University Press, 1917.
—*Natural Questions*. Translated by Thomas H. Corcoran. Cambridge, MA: Harvard University Press, 1922.
—*De Vita Beata*. Translated by John W. Basore. Cambridge, MA: Harvard University Press, 1932. 98–179.
Senguerdius, Wolferdus. *Philosophia Naturalis, Quatuor Partibus Primarias Corporum Species, Affections, Vicissitudines, & Differentias, Exhibens. Editio secunda, priore auctior.* Lugduni Batavorum, 1680, 1685.
Sennert, Daniel. *De Chymicorum cum Aristotelicis et Galenicis Consensus ac Dissensu*. Wittebergae, 1619.
—*Hypomnematum Phisicorum*. Wittebergae, 1636.
—*Institutionum Medicinae Libri V*. Lugduni, 1645.
—*Medicina Practica*. Lugduni, 1629; Parisiis, 1632.
—*Thirteen Books of Natural Philosophy*. London, 1660.
Sewall, Samuel. *The Diary of Samuel Sewall*. 3 vols. *CMHS*. Fifth Series. Vol. 5–7. Boston, 1878–82.
Silius, Tiberius Catius Asconius Italius. *Punica*. With an English translation by J.D. Duff. 2 vols. Cambridge, MA: Harvard University Press, 1961.
Sloane, Hans. "An Account of Elephants Teeth and Bones found under Ground." *PT* 35.403 (September 30, 1728): 457–71.
—"Of Fossile Teeth and Bones of Elephants. Part the Second." *PT* 35.404 (October 31, 1728): 497–514.
Socrates Scholasticus. *Ecclesiastical History*. *NPNFii*. 2:1–178.
Solinus, Caius Julius (Polyhistor). *Collectanea Rerum Memorabilium*. Ed. Thomas Mommsen. Berolini, 1895.
—*De Mirabilibus Mundi Libri LVII*. London, 1587.
Southwell, Robert. "III. Account of a Monstrous Calf with Two Heads." *PT* 20.238 (March 31, 1698): 79–80.
Spencer, John. *A Discourse concerning Prodigies: wherein The Vanity of Presages by them is reprehended, and their true and proper Ends asserted and vindicated.* Cambridge, 1663. Second edition. Corrected and Enlarged. London, 1665.
—*Discourse Concerning Vulgar Prophecies. Wherein the Vanity of receiving them as the Certain Indications of any future Event is discovered; And some Characters of Distinction between true and pretending Prophets are laid down.* Cambridge, 1665.
Spigelius, Adrianus. *De Humani Corporis Fabrica Libri Decem, Tabulis XCIIX. Aeri incises elegantissimis, nec antehac visis exornati*. Venetiis, 1627, 1632.
—*De Lumbrico Lato Liber. Cum ejusdem Lumbrici icone, & notis*. Patavii, 1618.

Spon, Jacob. *Voyage D'Italie, de Dalmatie, de Grèce et du Levant, fait aux années 1675 &1676*. Amsterdam, 1679.

Sprat, Thomas. *The History of the Royal Society of London, For the Improving of Natural Knowledge*. London, 1667.

Stow, John. *Abridgement of the English Chronicle, First collected by M. John Stow*. London, 1618.

— *Annals, or, A Generall Chronicle of England*. London, 1592, 1632.

Strabo. *Geographica*. Translated by Horace Leonard Jones. 8 vols. Cambridge, MA: Harvard University Press, 1917.

Stromer, Heinrich. *Decreta Medica et Senectute*. Lipsia, 1531.

—*De Morte Hominis Decreta aliquot Medica*. Lipsia, 1531.

Suetonius Tranquillus. *The Lives of the Twelve Caesars: Octavian*. Translated by J.C. Rolfe. 3 vols. Cambridge, MA: Harvard University Press, 1913–14. 2:123–287.

—*De Vita Caesarum: Divus Augustus*.

Suidas (Suda). *Suidae Lexicon, Graece & Latine. Tribus Voluminis. Textum Graecum cum Manuscriptis Codicibus collatum... . Versionem Latinam Æmelii Porti innumeris in locis correxit.* Ludolphus Kusterus. 3 vols. Cantabrigiae, 1705.

Sulpicius Severus. *Chronica. Historia Sacra. NPNFii* 11:71–122. [*PL* 20. 0095B–0160A].

Swammerdam, Jan. *Miraculum Naturae sive Uteri Muliebris Fabrica*. Lugduni Batavorum, 1729.

Swift, Jonathan. *A Modest Proposal For preventing the Children of the Poor People From being a Burthen to their Parents, or the Country*. Dublin, 1729.

Sydenham, Thomas. *Compleat Method of Curing Almost All Diseases*. 4th ed. London, 1710.

—*Dissertatio Epistolaris ad Spectatissimum Doctissimumq; Virum Gulielmum Cole, M.D. De Observationibus nuperis circa curationem Variolarum Confluentium Nec non de Affectione Hysterica*. Londini, 1682.

—"An Epistolary Discourse to the Learned Doctor William Cole, concerning some Observations on the confluent Small-pox, and of Hysterick Hiseases." In *The Whole Works of that Excellent Practical Physician Dr. Thomas Sydenham*. London, 1696. 404–78.

—*The Entire Works of Dr. Thomas Sydenham, newly made English from the Originals*. London, 1749.

—*Methodus Curandi Febres, Propriis Observationibus Superstructa.* Londini, 1666.

—*Observationes Medicae circa Morborum Acutorum Historiam et Curationem*. London, 1676.

—*The Whole Works of that Excellent Practical Dr. Thomas Sydenham* (1696). Second Edition. Translated by John Pechey. London, 1697. Fifth edition. Corrected. London, 1712.

Tachenius, Otto. *Antiquissimae Hippocraticae Medicinae Clavis*. Neapoli, 1697.

—*Hippocrates Chimicus*. Venetiis, 1666.

—*Otto Tachenius His Hippocrates Chymicus*. London, 1696.

Tacitus, Publius Cornelius. *Agricola*. Translated by M. Hutton et al. Cambridge, MA: Harvard University Press, 1914.

—*The Annals of Tacitus*. Translated by J. Jackson. Cambridge, MA: Harvard University Press, 1937.

—*De Origine et Situ Germanorum Liber. Germania*. In *The Germany and the Agricola of Tacitus*. Translated and introduced by Edward Brooks, Jr. Philadelphia: David McKay, Publishers, 1897.

Tagliacozzi, Gaspare. *De Curtorum Chirurgia per Insitionem Libri Duo*. Venetiis, 1597.

Talmud. *See The Babylonian Talmud.*

Temporarius, Johannes. *Chronologicarum Demonstrationum Libri Tres*. Francofurti, 1596.

Tertullian. *The Five Books Against Marcion*. In *ANF*. 3:269–475.

—*On the Soul*. In *ANF*. 3:181–235.

Thacher, Thomas. *Brief Rule To Guide the Common-People of New-England How to order themselves and theirs in the Small Pocks, or Measles*. Boston, 1677/78, 1702. Facsimile by Henry R. Viets, M.D. Baltimore: The Johns Hopkins Press, 1937.

Theocritus Syracusanus. *Idyllia*. In *Greek Bucolic Poets*. Translated by J.M. Edmonds. Cambridge, MA: Harvard University Press, 1912.

Theodoretus of Cyrus. *Commentary on the Psalms, Psalm 1–72*. Translated by Robert C. Hill. [The Fathers of the Church] Washington, D.C.: The Catholic University of America Press, 2000. Vol. 1.

—*Quaestiones et Responsiones ad Orthodoxos*. Edited by A. Papadopoulos-Kerameus. St. Petersburg: Kirschbaum, 1895. 1–150.

—*Quaestiones in Genesin CIII*. In. *Quaestiones in Octateuchum*. Edited by N. Fernández Marcos and A. Sáenz-Badillos. *Theodoreti Cyrensis quaestiones in Octateuchum*. [*Textos y Estudios Cardinal Cisneros* 17]. Madrid: Poliglota Matritense, 1979. 3–318.

Theophylact of Ochrida. *Expositio in Prophetam Jonam*. [*PG* 126. 0906–0969].

Theophylact Simocatta. *The History of Theophylact Simocatta*. Edited and translated by Michael and Mary Whitby. Oxford: Oxford University Press, 1986.

Theophrastus. *Historia Plantarum. Enquiry into Plants*. 2 vols. Translated by Arthur F. Hort. London: William Heinemann, 1916.

Thoresby, Ralph. "A Letter Communicated from Mr. Thoresby F.R.S. to John Evelyn Esquire, concerning the Cures done by Mr. Greatrix the Stroker." *PT* 21.256 (September 30, 1699): 332–34.

Thoreau, Henry David. *Walden; or, Life in the Woods*. Boston: Ticknor and Fields, 1854.

Tibulus, Albius. *Elegiae*. Translated by F.W. Cornish et al. Cambridge, MA: Harvard University Press, 1913.

Timonius, Emanuel. "V. An Account, or History, of the Procuring the Small Pox by Incision, or Inoculation; as it has for some time been practised at Constantinople. Being the extract of a Letter from Emanuel Timonius. Oxon and Patav. M.D. F.R.S. dated at Constantinople, December 1713." *PT* 29.339 (June 30, 1714): 72–82.

Tiraquelli, Andreae. *Semestria In Genialium Dierum Alexandri ab Alexandro*. Lugduni, 1586.

Topsel, Edward. *The History of Four-footed Beasts and Serpents… Collected out of the Writings of Conradus Gesner and other Authors*. London, 1607, 1658.

Torniello, Agostino. *Annales Sacri & Profani, Ab Orbe Condito, Ad Eundem Christi Passione Redemptum*. Colonia Agrippinae, 1622.

Trembley, Abraham. *Memoires pour servir A L'Histoire d'un Genre de Polypes D'Eau Douce, a Brase en Forme de Cornes*. Paris, 1744.

Tremellius, Immanuel & Franciscus Junius. *Testamenti Veteris Biblia Sacra, sive Libri Canonici Priscae Judaeorum Ecclesiae à Deo traditi*. Londoni, 1593.

Treubler, Johann Friedrich. "Observatio XXXVII: Miri Fulminis Effectus, Fulimine Percussorum Cura." *Miscellanea Curiosa sive Ephemeridum Medico-Physicarum* 1. Decuria 2. 1671. Jenae, 1671: 64–69.

Tuckney, Jonathan. "Jonathan Tuckney to Increase Mather" (February 25 1681/82). *MP*. P. 354.

Tulpius, Nicolaes. *Observationes Medicae*. Editio Nova. Amsterdami, 1652.

Turner, Daniel. *The Force of the Mother's Imagination upon her Foetus in Utero*. London, 1730.

—*De Morbis Cutaneis. A Treatise of Diseases Incident to the Skin*. London, 1714.

Turner, William. *A Compleat History of the most Remarkable Providences, both of Judgment and Mercy … being the Work set on Foot Thirty Years ago, by the Reverend Mr. Pool, Author of the Synopsis Criticorum*. London, 1697.

Tusser, Thomas. *Five Hundred Points of Good Husbandry*. London, 1573.

Twyne, John. *De Rebus Albionicis*. Londini, 1599.

Tylkowski, Adalbertus Wojciech. *Meteorologia Curiosa ad D. Boguslavm Leszcynski*. Cracoviae, 1669.

Tyson, Edward. *Lumbricus Hydropicus;* or An Essay to prove that Hydatides often met with in morbid Animal Bodies, are a Species of Worms, or Imperfect Animals." *PT* 17.194 (September 30, 1691): 506–10.

—"*Lumbricus Latus*, or a Discourse read before the Royal Society, of the Joynted Worm." *PT* 13.146 (April 10, 1683): 113–41.

—*Lumbricus Teres*, or some Anatomical Observations on the Round Worm bred in human Bodies." *PT* 13.147 (May 10, 1683): 154–59.

—"A Relation of two Monstrous Pigs, with the great resemblance of humane faces, and two young Turkeys joined by the Breast." *PT* 21.259 (December 31, 1699): 431–35.

—"*Vipera Caudi-Sona Americana*, Or the Anatomy of a Rattle-Snake, Dissected at the Repository of the Royal Society in January 1682/3." *PT* 13.144 (April 10, 1683): 25–58.

Ursinus, Johann Heinrich. *Arboreti Biblici Continuatio: Sivè Historiae Plantarum Biblicae, Libri Tres. Editio Nova.* Norimbergae, 1699.

Valentini, Michael Berhard. *Praxis Medicinae Infallibilis*. Francofurti, 1711.

Valsalva, Antonio Maria. *De Aure Humana Tractatus, In quo Integra Auris Fabrica, multis Novis Inventis, & Iconismis illustrate*. Bononiae, 1704.

Van Helmont, Jan Baptist. *De Magnetica Vulnerum Curatione*. Parisiis, 1621.

—*Ortus Medicinae, Id est Initia Physicae Inaudita. Progressus medicinae novus, in Morborum Ultinoem ad Vitam Longam.* Amstelodami, 1648, 1652. Lugduni, 1655.

Van Lansberge, Johann Philip. *Progymnasmatum Astronomiae Restitute de Motu Solis*. Middelburgi Zelandiae, 1628.

—*Tabulae Motuum Coelestium Perpetuae*. Middelburg Zelandiae, 1632.

—*Triangulorum Geometriae Libri Quatuor*. Amsterdami, 1631.

Van Leeuwenhoek, Antonie. "An Abstract of a Letter … about the Generation by an Animalcule of the Male seed. Animals in the seed of a Frog. Some Observations in the parts of a Frog." *PT* 13.152 (October 20, 1683): 347–55.

—*Arcana Naturae, Ope & Beneficio Exquisitissimorum Microscopiorum*. Lugduni Batavorum, 1696, 1708.

—"Microscopical observations on the Cortex Peruvianus." *PT* 25.312 (April 1, 1708): 2446–55.

—"Part of a Letter of Mr. Anthony van Leeuwenhoeck, giving his Observations on the late Storm." *PT* 24.289 (February 28, 1704): 1535–37.

Van Linschoten, John Huyghen. *His Discours of Voyages into the Easte & West Indies. Divided into Foure Bookes*. London, 1598.

Van Malderen. Johannes. *De Sancta Trinitate, Creatione in Genere, et Angelies.* Antverpiae, 1634.

Varenius, Bernardus. *Descriptio Regni Japoniae*. Amstelodami, 1649; Cambridge, 1673.

Varner, Gary R. *Creatures in the Mist: Little People, Wild Men and Spirit Beings around the World*. New York: Agora Publishing, 2007.

Varro, Marcus Terentius. *Antiquitates Rerum Humanarum et Divinarum*. Ed. Peter Hermann. Leipzig: Teubner, 1906.

—*De Re Rustica*. Translated by W.D. Hooper et al. Cambridge, MA: Harvard University Press, 1934.

Vartomannus (Varthema), Ludovicus. *Ludovicus Vartomani Patritii Romani Novum Itinerarium*. Milan, 1511.

—*Navigationis Aethipiae, Aegypti utriusque*. In Simon Grynaeus. Ed. *Novus Orbis Regionum ac insularum veteribus incognitarum*. Basileae, 1532. Fols 189–297.

—*The Travels of Ludovico Di Varthema in Egypt, Syria, Arabia Deserta and Arabia Felix, In Persia, India, and Ethiopia A.D. 1503–1508*. Translated by John Winter Jones. London: Hakluyt Society, 1863.

Velschius, Georgius Hieronymus. *Exercitatio de Vena Medinensi, Ad Mentem Ebusinae, Sive De Vermiculis Capillaribus Infantium*. Augustae Vindelicorum, 1674.

Vergilius, Publius Maro. *Aeneid*. In *Virgil, Books 7–12*. 2:2–365.

—*Eclogues*. In *Virgil*. 1:2–77.

—*Georgics*. In *Virgil*. 1:80–237.

—*Virgil. Eclogues, Georgics, Aeneid, Books 1–6*. Translated by H. Rushton Fairclough. 2 vols. Cambridge, MA: Harvard University Press, 1916.

Vermigli, Peter Martyr. *Common Places of the most famous and renowned Divine Doctor Peter Martyr, divided into foure principall Parts*. Translated by Anthonie Marten. London, 1597.

Verzascha, Bernard. "Observatio XXX *Convulsiones à lumbricis ortae*." *Observationum Medicarum Centuria Prima*. Basileae, 1677. 81.

Vincentius Bellovacensis. *Bibliotheca Mundi. Seu Speculum Maioris*. Tomus Secundus. Duaci, 1624.

Vitringa, Campegius. *Observationum Sacrarum Libri Sex... Tomus I*. Franequerae, 1712.

Vives, Juan Luis. *De Disciplinis libri XX*. Lugduni, 1551.

Volaterranus, Raphael. *Commentatoriorum Urbanorum libri XXXVIII*. Rome, 1506.

Von Auerbach, Heinrich. *Saluberrime adversus Pestilentiam Observationes*. Lipsia, 1516.

Von Kronland, Johann Marcus Marci. Πάν ἕν πάντων *seu, Philosophia Vetus Restituta*. Pragae, 1662.

Von Leibniz. *See* Leibniz.

Von Schonefeld, Stephan (Stephanus à Schonevelde). *Ichtyologia et Nomenclaturae Animalium Marinorum, Fluviatilium, Lacustrium, quae in*

Florentissimis Ducatibus Slesvici et Holsatiae & celebrerrimo Emporio Hamburgo occurrunt triviales. Hamburgi, 1624.

Vorstius, Guilielmus Henricus. "Animadversiones, in Pirke R. Elieser." Appended to פרקי רבי אליעזר *Capitula R. Elieser.* [*Chronologia Sacra-Profana Germen Davidis*]. Lugduni Batavorum, 1644.

Vossius, Gerard. *De Theologia Gentili et Physiologia Christiana, sive De Origine ac Progressu Idololatriae Libri IV*. Tomi 3. Amsterodami, 1641.

Waller, Richard. "A Catalogue of Simple and Mixt Colours, with a Specimen of each Colour prefixed to its Proper Name." *PT* 16.179 (February 28, 1686): 24–32.

—"Some Observations made on the Spawn of Frogs, and of the Production of Todpoles [*sic*] therein." *PT* 17.194 (September 30, 1691): 523–24.

—"Richard Waller to Cotton Mather" (July 22, 1713). RSL Holograph Manuscript (EL/W3/75–79).

Wallis, John. *Operum Mathematicorum*. Oxonii, 1657.

—"A Letter of Dr. Wallis to Dr. Sloane, concerning the Generation of Hail, and of Thunder and Lightning, and the Effects thereof." *PT* 19.231 (August 31, 1697): 653–58.

—"A Relation Of an Accident by Thunder and Lightning, at Oxford." *PT* 1.13 (March 12, 1666): 222–26.

Walter, Thomas. *Choice Dialogues* (1719) with his own (most likely coached by Mather)

—*Choice Dialogue Between John Faustus A Conjurer, and Jack Tory His Friend.* Boston, 1720.

Walton, Brian. Editor. *Biblia Sacra Polyglotta, Complectentia Textus Originales, Hebraicum, cum Pentateucho Samaritano, Chaldaicum, Graecum, Versionumque Antiquarum Samaritanae, Graecae LXXII Interp. Chaldaicae, Syriacae, Arabicae, Æthiopicae, Persicae, Vulg. Lat. Quincquid comparari poterat. Cum Textuum, & Versionum Orientalium Translationibus Latinis*. 6 vols. Londini, 1653–57.

Ward, Samuel. *Magnetis Reductorium Theologicorum Tropologicum*. Londini, 1637.

Ward, Seth. *Vindicia Academiarum*. Oxford, 1654.

Warren, Edward. *No Præexistence: Or a Brief Dissertation against the Hypothesis of the Humane Souls, Living in a State Antecedaneous to This*. London, 1667.

Watson, William. "XCVI. A Letter of Mr. Watson, F.R.S. in the Royal Society, concerning the electrical Experiments in England upon Thunder-Clouds." *PT* 47 (December 21, 1752): 567–70.

—"CI. Some Suggestions concerning the preventing Mischiefs, which happen to Ship and their Masts by Lightning." *PT* 52 (December 31, 1761): 629–35.

Webster, John. *The Displaying of supposed Witchcraft. Wherein is affirmed that there are many Sorts of Deceivers and Imposters. And Diverse person under a passive Delusion of Melancholy and Fancy. But that there is a Corporeal League made betwixt the Devil and the Witch, Or that he sucks on the Witches Body, has Carnal Copulation or that Witches are turned into Cats, Dogs, raise Tempests, or the like, is utterly denied and disproved.* London, 1677.

Weekly Memorials for the Ingenious: Or, an Account of Books Lately set forth in Several Languages. With Other Accounts relating to Arts and Sciences. London, 1683.

West, James. "XXI. Proposal of a Method of securing the Cathedral of St. Paul's from Damage by Lightning." *PT* 47 (March 9, 1769): 160–69.

Whiston, William. *Astronomical Lectures, Read in the Publick Schools at Cambridge ... Whereunto is added a Collection of Astronomical Tables.* London, 1715.

—*Athanasius Convicted of Forgery*. London, 1712.

—*The Calculation of Solar Eclipses without Parallaxes*. London, 1724.

—*A Collection of Ancient Monuments Relating to the Trinity and Incarnation*. London, 1713.

—*An Essay on the Revelation of St. John*. Cambridge, 1706.

—*An Essay Towards Restoring the True Text of the Old Testament*. London, 1722.

—*A New Theory of the Earth*. London, 1696.

—*Praelectiones Astronomicae Cantabrigiae in Scholis Publicis Habitae ... Quibus Accedunt Tabulae Plurimae Astronomicae Flamstedianae Correctae, Halleianae, Cassinianae, et Streetianae*. Cantabrigiae, 1707.

—*Primitive Christianity Revived*. 5 vols. London, 1711–12.

Whitbourne, Richard. *A Discourse and Discovery of New-found-land.* London, 1620.

White, John. *The Planters Plea. Or The Grounds of Plantations Examined, And visual Objections answered*. London, 1630.

Wilkins, John. *Discovery of a World in the Moone*. London, 1638.

—*Mathematical Magick, Or, The Wonders that may be performed by Mechanicall Geometry*. London, 1648.

—*Of the Principles and Duties of Natural Religion*. London, 1675.

Williams, John. *An Answer To a Late Pamphlet, Intitled, A Letter to a Friend in the Country*. Boston, 1722.

—*Several Arguments, Proving That Inoculating the Small Pox Is not contained in the Law of Physick, either Natural or Divine, and therefore Unlawful.* Boston, 1721.

Williams, Roger. *A Key into the Language of America: or, An help to the Language of the Natives in that part of America, called New-England*. London, 1646.

Willius, Johannes Valentinus. "LXXXIII. De Naevis Maternis." In *Thomae Bartholini Acta Medica et Philosophica Hafniensia Ann: 1674. 1675. 1676.* Hafniae, 1677. 152–53.

Willughby, Francis (and John Ray). *Historia Piscium libri quatuor*. Oxonii, 1686.

—*Ornithologiae Libri Test: Aves omens hactenus cognitae in methodum naturis suis*. Londini, 1676.

Winthrop, John. "Letter of John Winthrop to Cotton Mather" (September 12, 1717). *MP* 681–83. And *CMHS* 2 (1793): 12–13.

Winthrop, John (Governor). *A Short Story of the Rise, reign, and ruin of the Antinomians, Familists & Libertines, that infected the Churches of New-England*. London, 1644.

Witsius, Hermann. *De Oeconomia Foederum Dei cum Hominibus*. Leeuwarden, 1677.

—"*Oratio de Felici an Infelici Reipublicae Literariae, hoc seculo, statu*." *Miscellaneorum Sacrorum Tomus Alter. Continens XXII Exercitationes*. Amstelaedami & Ultrajecti, 1700. 2:937–51.

Woodman, Philip. *Medicus Novissimus; or, the Modern Physician: shewing The Principal Signs, Causes, and most Material Prognosticks ... according to the most Modern and Best Method of Practice now in Use*. London, 1712, 1722.

Woodward, John. *An Attempt Towards a Natural History of the Fossils of England in a Catalogue of the English Fossils in the Collection of J. Woodward, M.D.* London, 1729.

—*Brief Instructions For Making Observations in all Parts of the world: as also For Collecting, Preserving, and Sending over Natural Things*. London, 1696.

—*An Essay toward a Natural History of the Earth and Terrestrial Bodies, especially Minerals: As also of the Sea, Rivers, and Springs. With an Account of the Universal Deluge: And of the Effects that it had upon the Earth.* London, 1695, 1702, 1723, 1726.

—"Letter of John Woodward to John Winthrop" (April 3, 1721). *PMHS* 13 (1873): 110.

—"Letter of John Woodward to Cotton Mather" (April 3, 1721). *PMHS* 13 (1873): 110–11.

—*The Natural History of the Earth, Illustrated, and Inlarged: as also Defended, And the Objections against it, Particularly those lately publish'd by Dr. Camerarius, answered.* London, 1695, 1726.

—*Naturalis Historia Telluris Illustrata & Aucta Una cum Ejusdem Defensione; Praesertim contra Nuperas Objectiones D. El. Camerarii Med. Prof. Tubingensis*. Londini, 1714.
—*The State of Physick: And of Diseases; with an Inquiry into the Causes of the late Increase of them: But more particularly of the Small-Pox*. London, 1718.
Xenophon of Athens. *Memorabilia. Oeconomicus. Symposium. Apology*. Translated by E.C. Marchant et al. Cambridge, MA: Harvard University Press, 2013.
Yaakov ben Raash. *Perush Baal HaTurim al HaTorah al Ha Torah Chumash. The Torah: With the Baal HaTurim's Classic Commentary*. Translated by R. Eliyahu Touger. Edited, Elucidated and Annotated by R. Avi Gold. 5 vols. Brooklyn, NY: Mesorah Publications, 2004.
Yonge, James. "Concerning several Solid Bodies voided by Urine." *PT* 26.323 (October 1, 1709): 420–23.
—*Wounds of the Brain Proved Curable*. London 1682.
Zacchias, Paulus. *Quaestiones Medico-Legales*. Lugduni, 1621–35; Editio Nova. Lugduni, 1701.
Zacutus Lusitanus. *De Medicorum Principum Historia Libri Sex*. Lugduni, 1642, 1649.
—*Praxis Medica Admiranda*. Lugduni, 1637.
Zárate, Agustín de. *See* De Zárate.
Zuinger, Theodore the Elder. *Theatrum Vitae Humanae*. Basileae, 1565–1604.

Secondary Works

Albury, W.R. "Halley and the *Traité de la lumière* of Huygens: New Light on Halley's Relationship with Newton." *Isis* 62.4 (1971): 445–68.
Allen, Phyllis. "Medical Education in 17th-Century England." *JHMAS* 1 (1946): 115–43.
Allgemeine Deutsche Biographie. 56 vols. Leipzig: Duncker & Humblot, 1875–1912.
Almond, Philip C. *Heaven & Hell in Enlightenment England*. Cambridge: Cambridge University Press, 1994.
Anchor Bible Dictionary. Edited by David Noel Freedman. 6 vols. New York: Doubleday Anchor, 1992.
Andrews, William. *Famous Frosts and Frost Fairs in Great Britain*. London: George Redwall, 1887.
The American Heritage Medical Dictionary. Boston: Houghton Mifflin, 2007.
Arnold, Matthew. *Empedocles on Etna, and Other Poems*. London: B. Fellowes, 1852.

Asma, Stephen T. *On Monsters: An Unnatural History of our Worst Fears*. New York: Oxford University Press, 2011.

Babson, John J. *The History of the Town of Gloucester*. Gloucester, MA: Proctor Brothers, 1860.

Bailyn, Bernard, and Patricia L. Denault. Eds *Soundings in Atlantic History: Latent Structures and Intellectual Currents, 1500–1830*. Cambridge, Mass.: Harvard University Press, 2009.

Baker, David Weil. "Etymology, Antiquarianism, and Unchanging Languages in Johannes Goropius Becanus's *Origines Antwerpianae* and William Cambden's *Britannia*." *RQ* 72.4 (2019): 1326–61.

Baker, Emerson W. and John G. Reid. *The New England Knight: Sir William Phips, 1651–1695*. Toronto: University of Toronto Press, 1998.

Barnes, Daniel. R. "The Bosom Serpent: A Legend in American Literature and Culture." *JAF* 85.336 (1972): 111–22.

Barnett, Lydia. "Giant Bones and the Taunton Stone: American Antiquities, World History, and the Protestant International." In Paula Findlen (2019). 225–46.

Barry, Jonathan. "Educating physicians in seventeenth-century England." *SC* 32 (2019): 137–54.

—"John Beaumont: Science, Spirits and the Scale of Nature." *Witchcraft and Demonology in South-West England, 1640–1789*. New York: Palgrave Macmillan, 2012. 124–64.

Bates, A.W. *Emblematic Monsters: Unnatural Conceptions and Deformed Birth in Early Modern Europe*. Amsterdam: Editions Rodopi BV, 2005.

Bath, Jo and John Newton. "'Sensible Proof of Spirits': Ghost Belief during the Later Seventeenth Century." *Folklore* 117 (April 2006): 1–14.

Beall, Otho T. "Cotton Mather's Early 'Curiosa Americana' and the Boston Philosophical Society of 1683." *WMQ* 18.3 (1961): 360–72.

Beall, Otho T. and Richard H. Shryock. *Cotton Mather: The First Significant Figure in American Medicine*. 1954; New York: Arno Press, 1979.

Bell, James B. "A Handmaiden for Episcopacy: John Checkley of Boston." *A War of Religion: Dissenters, Anglicans, and the American Revolution*. New York: Palgrave Macmillan, 2008. 33–41.

Bell, Whitfield J., Jr. "The Reverend Mr. Joseph Morgan, an American Correspondent of the Royal Society, 1732–1739." *PAPS* 95.3 (1951): 254–61.

Bigotti, Fabrizio. "Corpuscularianism." *Encyclopedia of Early Modern Philosophy and the Sciences*. Edited by Dana Jalobeanu and Charles T. Wolfe. Cham: Springer International Publishing, 2020. 1–13.

Bird, Alexander. "Squaring the Circle: Hobbes on Philosophy and Geometry." *JHI* 57.2 (1996): 217–31.

Black, Deborah. "Rational Imagination: Avicenna on the Cogitative Power." In L.X. López-Farjeat and J.A. Tellkamp. Eds *Philosophical Psychology in Arabic Thought and the Latin Aristotelianism of the 13th Century*. Paris: Vrin, 2013. 59–81.

Blair, Ann M. *Too Much To Know. Managing Scholarly Information before the Modern Age*. New Haven: Yale University Press, 2010.

Blake, John B. "The Inoculation Controversy in Boston: 1721-1722." *NEQ* 25.4 (1952): 489–506.

—*Public Health in the Town of Boston, 1630-1822*. Cambridge, MA: Harvard University Press, 1959.

Bosco, Ronald A. Editor. *Paterna. The Autobiography of Cotton Mather*. Delmar, NY: Scholars' Facsimiles & Reprints, 1976.

Bowler, Peter J. "Preformation and Pre-Existence in the Seventeenth Century: A Brief Analysis." *JHB* 4.2 (Autumn 1971): 221–44.

Boylston, Arthur. "The Origins of Inoculation." *JRSM* 105 (2012): 309–13.

Braam, Maximilian. "The Tenderton Steeple and the Goodwin Sands." *Science* 40.1042 (December 18 1914): 893–94.

Brogan, Stephen. *The Royal Touch in Early Modern England: Politics, Medicine, and Sin*. London: Boydell & Brewer, 2015.

Brown, Thomas H. "The African Connection: Cotton Mather and the Boston Smallpox Epidemic of 1721–1722." *JAMA* 260.15 (1988): 42–49.

Buckingham, Joseph T. Editor. *Specimen of Newspaper Literature*. 2 vols. Boston: Charles C. Little and James Brown, 1850.

Burns, William E. *An Age of Wonders: Prodigies, Politics, and Providence in England, 1657–1727*. Manchester: Manchester University Press, 2002.

—"'Our Lot is Fallen Into an Age of Wonders': John Spencer and the Controversy Over Prodigies in the Early Restauration." *Albion* 27.2 (Summer 1995): 237–52.

Bush, Sargent, Jr. "Bosom Serpents before Hawthorne: The Origins of a Symbol." *AL* 43.2 (1971): 181–99.

Carrell, Jennifer L. *The Speckled Monster. A Historical Tale of Battling Smallpox*. New York: Dutton, 2003.

Catholic Encyclopedia: An International Work of Reference. Edited by Charles Herbermann. 16 vols. New York: Robert Appleton Company, 1907–14.

Chamberlain, Ava. "A Fish Tale: Jonathan Edwards and Cotton Mather on Jonah's Whale." *Jonathan Edwards and Scripture: Biblical Exegesis in British North America*. Edited by David P. Barshinger and Douglas A. Sweeney. New York: Oxford University Press, 2018. 144–62.

Chaplin, Joyce. *Subject Matter: Technology, the Body, and Science on the Anglo-American Frontier, 1500–1676*. Cambridge, MA: Harvard University Press, 2003.

Christianson, Eric Howard. "Makings Heads or Tails of It: Cotton Mather's 'Prodigious Worm.'" *PMHS*. Third Series. 105 (1993): 59–73.

Clemens, William M. Editor and compiler. *American Marriage Records Before 1699*. Pompton Lakes, NJ: Biblio Co., 1926.

Connell, Sophia. "Aristotle's Explanations of Monstrous Births and Deformities in 'Generation of Animals 4.4.'" *Aristotle's Generation of Animals. A Critical Guide*. Edited by Andrea Falcon and David Lefebvre. Cambridge: Cambridge Univeristy Press, 2017. 207–24.

Cook, David. "Comets and Meteors in the Islamic World." *Encyclopaedia of the History of Science, Technology, and Medicine in Non-Western Cultures*. Edited by Helaine Selin et al. Dordrecht: Springer Dordrecht, 2008. 632–34.

Cook, Sherburne F. "The Significance of Disease in the Extinction of the New England Indians." *HB* 45.3 (1973): 485–508.

Corneanu, Sorana. "Joseph Glanvill on Imagination, Method and the Art of Thinking." *Testimonies of Mind and States of the Body in the Early Modern Period*. Edited by Gideon Manning. Cham, Switzerland: Springer International Publishing, 2020. 139–68.

Coulton, Richard. "'What He Hath Gather'd Together Shall Not be Lost': Remembering James Petiver." *Notes and Records: The Royal Society Journal of the History of Science*. 74.2 (2020): 189–211.

Cowie, Leonard W. *Henry Newman. An American in London 1708–43*. London: SPCK, 1956.

Crawford, Julie. *Marvelous Protestantism: Monstrous Births in Post-Reformation England*. Baltimore: Johns Hopkins University Press, 2005.

Dasen, V. "Multiple Births in Ancient Medical Texts." *Gesnerus* 55.3-4 (1998): 183–204.

—"Twins in the Greco-Roman World: Medical Theories and Iconography." *Medicina Nei Secoli* 7.2 (1995): 301–21.

Davis, J.J. *Biblical Numerology. A Basic Study of the Use of Numbers in the Bible*. Grand Rapids, MI: Baker Books, 1968.

Delabarre, Edmund B. *Dighton Rock: A Study of the Written Rocks of New England*. New York: Walter Neale, 1928.

—"Early Interest in the Dighton Rock." *TCSM* 18 (1917): 234–99.

—"Middle Period of Dighton Rock History." *TCSM* 19 (1916): 46–149.

Delbourgo, James, and Nicholas Dew. Eds *Science and Empire in the Atlantic World*. New York: Routledge, 2007.

Delwiche, Theodore R. "Ezra Stiles and North America in the Early Modern Republic of Letters." *MIH* 20.1 (2023): 27–61.

Denny, Margaret. "The Royal Society and American Scholars." *The Scientific Monthly* 65.5 (1947): 415–27.

Devine, Arthur. "Idiota (Raymundus Jordanus)." In *CE*. Vol. 7.

Dewhurst, Kenneth. "Sydenham's Original Treatise on Smallpox." *MH* 3 (3 Jul. 1959): 278–302.

Dopffel, Michael. "Between Biblical Literalism and Scientific Inquiry: Cotton Mather's Commentary on Jeremiah 8:7." In *Cotton Mather and Biblia Americana – America's First Bible Commentary*. Edited by Reiner Smolinski and Jan Stievermann. Tübingen, Germany: Mohr Siebeck, 2010: 203–25.

Dow, Joseph. *History of the Town of Hampton, New Hampshire. From its Settlement in 1638, to the Autumn of 1892*. 2 vols. Salem, MA: Salem Press Publishing, 1893.

Drake, Samuel G. "Introductory." In *The History of King Philip's War, By the Rev, Increase Mather, D.D.* Boston: Printed for the Editor, 1862. v.–xxxii.

Duffy, John. *Epidemics in Colonial America*. Baton Rouge: Louisiana State University Press, 1953.

—"Smallpox and the Indians in the American Colonies." *BHM* 25.4 (1951): 324–41.

Elmer, Peter. *The Miraculous Conformist: Valentine Greatrakes, The Body Politic, and the Politics of Healing in Restoration Britain*. Oxford: Oxford University Press, 2013.

Ergil, Kevin V. *Ancient Healing: Unlocking the Mysteries of Health & Healing Through the Ages*. Morton Grove, IL: Publications International, 1997.

Esfandiary, Helen. "'A thankless enterprise': Lady Mary Wortley Montagu's campaign to establish medical unorthodoxy amongst her female network." *Notes and Records* 77 (2023): 235–57.

Farley, John. *Spontaneous Generation from Descartes to Oparin*. Baltimore: Johns Hopkins University Press, 1977.

Findlen, Paula. *Empires of Knowledge: Scientific Networks in Early Modern Europe*. London: Routledge, 2019.

—"Jokes of Nature and Jokes of Knowledge: The Playfulness of Scientific Discourse in Early Modern Europe." *RQ* 43.2 (1990): 292–331.

Finley-Crosswhite, Annette. "Henry IV and the Diseased Body Politic." In *Princes and Princely Culture: 1450–1650*. 2 vols. [Brill's Studies in Intellectual History. Vol 118/1.] Edited by Martin Gosman et al. Leiden-Boston: Brill, 2003. 131–47.

Fisher, Eric P. *Mighty Storms of New England: The Hurricanes, Tornadoes, Blizzards, and Floods that Shaped the Region*. Lanham, MD: Rowan & Littlefield, 2021.

Fischer-Homberger, Esther. "On the medical history of the doctrine of imagination." *PM* 9 (1979): 619–28.

Flynn, George J. (G. J. Flynn, "Micrometeoroid." *Encyclopedia Britannica*. https://www.britannica.com/science/micrometeoroid. (Accessed July 18 2023.)

Freemon, Frank R. "American Colonial Scientists Who Published in the 'Philosophical Transaction' of the Royal Society." *NRRS* 39.2 (1985): 191–206.

French, Roger, and Andrew Wear. Editors. *The Medical Revolution of the Seventeenth Century*. Cambridge, UK: Cambridge University Press, 1989.

Fughelli, Patrizia. "Antonio Maria Valsalva (1666–1723), The Smooth Revolution Made by a Gentleman." *CR* 124 (2019): 1704–06.

Gantet, Claire. *Der Traum in der Frühen Neuzeit. Ansätze zu einer kulturellen Wissenschaftsgeschichte*. Berlin: De Gruyter, 2010.

Gasking, Elizabeth. *Investigations into Generation, 1651–1828*. Baltimore: Johns Hopkins Press, 1966.

Glacken, Clarence J. *Traces on the Rhodian Shore: Nature and Culture in Western Thought from Ancient Times to the End of the Eighteenth Century*. Berkeley: University of California Press, 1967.

Glass, Bentley. "The Germination of the Idea of Biological Species." *Forerunners of Darwin, 1745–1859*. Baltimore: Johns Hopkins Press, 1968: 30–48.

Goulet, Benjamin E. Federico Roda, Robin Hopkins. "Hybridization in Plants: Old Ideas, New Techniques." *Plant Physiology* 173.1 (2017): 65–78.

Grafton, Anthony. *Cardano's Cosmos: The World and Works of a Renaissance Astrologer*. Cambridge, MA: Harvard University Press, 1999.

—*The Footnote: A Curious History*. Cambridge, MA: Harvard University Press, 1997.

—"Joseph Scaliger and Historical Chronology: The Rise and Fall of a Discipline." *History and Theory* 14.2 (1975): 156–85.

Grainger, Brett Malcom. "Vital Nature and Vital Piety: Johann Arndt and the Evangelical Vitalism in Cotton Mather." *CH* 81.4 (2012): 852–72.

Greene, Robert A. "Henry More and Robert Boyle on the Spirit of Nature." *JHI* 23.4 (1962): 451–74.

Greydanus, Donal E. and Musunkumuki Kadochi. "Reflections on the Medical History of the Kidney: From Alcmaeon of Croton to Richard Bright—Standing on the Shoulders of Giants." *JINA* 3.4 (2016): 101–08.

Griffin, J.P. "Venetian treacle and the foundation of medicines' regulation." *BJCP* 58.3 (2004): 317–25.

Gruesz, Kristina Silva. *Cotton Mather's Spanish Lessons: A Story of Language, Race, and Belonging in the Early Americas*. Cambridge, MA: Harvard University Press, 2022.

Hadass, Ofer. *Medicine, Religion, and Magic in early Stuart England: Richard Napier's Medical Practice*. University Park, PA: Pennsylvania State University Press, 2018.

Hagner, Michael. "Enlightened Monsters." In *The Sciences in Enlightened Europe*. Edited by William Clark et al. University of Chicago Press, 1999. 175–217.

Hall, David D. *Worlds of Wonder, Days of Judgment*. New York: Alfred A. Knopf, 1989.

Hawthorne, Nathaniel. "Egotism; Or, The Bosom Serpent." *United States Magazine and Democratic Review*. 12.55 (March 1843): 255–61.

Hedesan, Georgiana D. "Reproducing the Tree of Life: Radical Prolongation of Life and Biblical Interpretation in Seventeenth-Century Medical Alchemy. *Ambix* 60.4 (2013): 341–60.

Heffernan, Teresa, and Daniel O'Quinn. Eds *The Turkish Embassy Letters*. New York: Broadview Press, 2012.

Heyd, Michael. *Be Sober and Reasonable: The Critique of Enthusiasm in the Seventeenth and early Eighteenth Centuries*. Leiden: E.J. Brill, 1995.

Hindmarsh, D. Bruce. *The Spirit of Evangelicalism: True Religion in a Modern World*. Oxford, UK: Oxford University Press, 2018.

Hirai, Hiro. "Imagination, Maternal Desire and Embryology in Thomas Fienus." In Manning, G. et al., eds 239–53.

—*Medical Humanism and Natural Philosophy: Renaissance Debates on Matter, Life and the Soul*. Leiden: Brill, 2011.

Holifield, E. Brooks. "The Abridging of Cotton Mather." In R. Smolinski and J. Stievermann (2010): 83–109.

Holmberg, Megan E. *Anomalous Apparitions of Light in Colonial America: Visions of Comets, New Stars, The Aurora Borealis, and Rainbows*. Unpublished Ph.D. dissertation, Temple University, 2019.

Holmes, Sr. Oliver Wendell, M.D. *Medical Essays 1842–1882*. Boston and New York: Houghton, Mifflin, and Company. The Riverside Press, 1895.

Holmes, Thomas J. *Cotton Mather: A Bibliography*. 3 vols. Cambridge, MA: Harvard University Press, 1940.

Hornberger, Theodore. *Scientific thought in the American colleges, 1638–1800*. Austin, TX: University of Texas Press, 1945.

Hotson, Howard. *Johann Heinrich Alsted 1588–1638: Between Renaissance, Reformation, and Universal Reform*. Oxford: Oxford University Press, 2000.

Huet, Marie-Hélène. *Monstrous Imagination*. Cambridge, MA: Harvard University Press, 1993.

—"Monstrous Medicine." In *Monstrous Bodies / Political Monstrosities in Early Modern Europe*. Edited by Laura Lunger Knoppers and Joan B. Landes. Ithaca: Cornell University Press, 2004. 127–47.

Hughes, D.W. et al. "The History of Halley's Comet." *PT* A 323 (1987): 349–67.

Hunter, Michael. "John Webster, The Royal Society and *The Displaying of Supposed Witchcraft* (1677)." *Notes and Records* [*of the Royal Society*] 71 (2017): 7–19.

Hunter, William B. "The Seventeenth-Century Doctrine of Plastic Nature." *HTR* 43 (July 1950): 197–213.

Huth, Edward. "Quantitative evidence for judgments on the efficacy of inoculation for the prevention of smallpox: England and New England in the 1700s," *JRSM* 99 (2006): 262–66.

Inglehart, Ashley J. "Boyle, Malpighi, and the Problem of Plastic Powers." *Early Modern Medicine and Natural Philosophy*. Edited by Peter Distelzweig et al. Dordrecht: Springer, 2016. 295–321.

James, K.A. "'Humbly Dedicated': Petiver and the Audience for Natural History in Early Eighteenth-Century Britain." *Archives of Natural History* 31.2 (2004): 318–29.

Jesseph, Douglas M. *Squaring the Circle: The War Between Hobbes and Wallis*. Chicago: University of Chicago Press, 2000.

Jones, Gordon W. Jones. Editor. *The Angel of Bethesda.* Barre, MA: American Antiquarian Society and Barre Publishers, 1972.

—"Baptista Mantuanus – Amateur Physician." *BHM* 36.2 (1962): 148–62.

Jones, Matt B. "Biographical Notes on Thomas Walter's 'Grounds and Rules of Musick Explained.'" *PAAS* 1932: 235–36.

Jordan, Peter N. *Naturalism in the Christian Imagination: Providence and Causality in Early Modern England.* Cambridge, UK: Cambridge University Press, 2022.

Kaplan, Barbara Beigun. "Greatrakes the Stroke: The Interpretations of His Contemporaries." *Isis* 73.2 (1982): 178–85.

Kass, Amalie M. "Boston's Historic Smallpox Epidemic." *MHR* 14 (2012): 1–51.

Kennedy, Kathleen. "On Writing the History of So Much Grief: Cotton Mather's *Decennium Luctuosum* and the Trauma of History." *The Eighteenth Century* 58.2 (2017): 219–41.

Kennedy, Rick. *The First American Evangelical. A Short Life of Cotton Mather*. Grand Rapids, MI: William B. Eerdman's Publishing Company, 2015.

—"Thomas Brattle and the Scientific Provincialism of New England, 1680–1713." *NEQ* 63.4 (December 1990): 584–600.

Kilgour, Frederick. "Thomas Robie, Colonial Scientist and Physician." *Isis* 30 (1939): 473–90.

King, Helen. *Hippocrates Now. The 'Father of Medicine' in the Internet Age.* London: Bloomsbury Academic, 2020.

King, Lester S. *The Philosophy of Medicine: The Early Eighteenth Century.* Cambridge, MA: Harvard University Press, 1978.

—*The Road to Medical Enlightenment, 1650–1695*. London: Macdonald, 1970.

Kittredge, George L. "Cotton Mather's Election into the Royal Society." *PCSM* 14 (1912): 81–114.

—"Cotton Mather's Scientific Communications to the Royal Society." *PAAS* 26.1 (rpt 1916): 18–55.

—"Further Notes on Cotton Mather and the Royal Society." *PCSM* 14 (1912): 281–92.

—"Letters of Samuel Lee and Samuel Sewall Relating to New England and the Indians." *TCSM* 14 (February 1911–1913): 142–92.

—"Some Lost Works of Cotton Mather." *PMHS* (October 1911–June 1912): 418–79.

Knoppers, Laura L. and Joan B. Landes. Eds *Monstrous Bodies/Political Monstrosities in Early Modern Europe*. Ithaca and London: Cornell University Press, 2004.

Koo, Kathryn. "Strangers in the House of God: Cotton Mather, Onesimus, and an Experiment in Christian Slaveholding." *PAAS* 117 (2007): 143–75.

Kopperman, Paul E. and Jeanne Abrams. "Cotton Mather's Medicine, with particular reference to Measles." *JMB* 27.1 (2019): 30–37.

Kottek, Samuel S. "Nishmat Ḥayyim, The Breath of Life as seen by Cotton Mather the Puritan and Manasseh the Jew." *Koroth* 8.5–6 (Fall 1982): 146–56.

Koutroumpas, Dimitrios C., and Haroula Koetsi-Kounari. "Galen on Dental Anatomy and Physiology." *JHD* 60.1 (June 2012): 39–51.

Kraemer, Fabian. *A Centaur in London: Reading and Observation in Early Modern Science*. Baltimore: Johns Hopkins University Press, 2023.

Kunitz, Stephan J. "Mortality Change in America, 1620–1920." *HB* 56.3 (1984): 559–82.

Kyrkoudis, Theodoros. "Vaccination of the Ethnic Greeks (Rums) Against Smallpox in the Ottoman Empire: Emmanuel Timonis and Jacobus Pylarinos as Precursors of Edward Jenner." *Erciyes Medical Journal* 43.1 (2021): 100–06.

Landes, Margaret W. *The Philosophical Writings of Richard Burthogge*. London: The Open Court Publishing, 1921.

—"Richard Burthogge, His Life and Place in the History of Philosophy." *The Monist* 30.2 (April 1920): 253–66.

Laver, A.B. "Miracles No Wonder! The Mesmeric Phenomena and Organic Cures of Valentine Greatrakes." *JHMAS* 33.1 (1978): 35–46.

Lawn, Brian. *The Rise and Decline of the Scholastic 'Quaestio disputata' with Special Emphasis on Its Use in the Teaching of Medicine and Science*. Leiden, New York, Cologne: E.J. Brill, 1993.

—*Salernitan Questions. An Introduction to the History of Medieval and Renaissance Problem Literature*. Oxford: Clarendon Press, 1963.

Leeson, Peter T. and Jacob W. Russ. "Witchtrials." *The Economic Journal* 128 (2017): 2066–2105.

Lehoux, Daryn. *Creatures Born of Mud and Slime: The Wonder and Complexity of Spontaneous Generation*. Baltimore: John Hopkins University Press, 2017.

Leonard, David C. "Harvard's First Science Professor: A Sketch of Isaac Greenwood's Life and Work." *HULB* 29.2 (April 1981): 135–68.

Lepore, Jill. "An Horrid Snow." *The New Yorker* (February 1 2011). https://www.newyorker.com/news/news-desk/an-horrid-snow Accessed January 31 2024.

Levin, David. "Giants in the Earth: Science and the Occult in Cotton Mather's Letters to the Royal Society." *WMQ* 45.4 (1988): 751–70.

—*The Young Life of the Lord's Remembrancer, 1663–1703*. Cambridge, MA: Harvard University Press, 1978.

Lewis, Alonzo, and James R. Newall. *The History of Lynn, Essex County, Massachusetts*. Boston: George C. Herbert, 1865.

Lewis, Frederic T. "The Passenger Pigeon as observed by the Rev. Cotton Mather." *The Auk* 61 (October 1944): 587–92.

Lewis, Gillian. "The debt of John Ray and Martin Lister to Guillaume Rondelet of Montpellier." *Notes and Records of the Royal Society Journal of the History of Science* 66.4 (December 20 2012): 323–39.

Livingstone, David N. *Adam's Ancestors: Race, Religion, and the Politics of Human Origins*. Baltimore: John Hopkins University Press, 2008.

López-Salazar, Ana Isabel. "General Inquisitors and the Portuguese Crown in the Seventeenth Century: Between Political Service and the Defense of the Faith (1578–1705)." *MS* 21.2 (2013): 79–90.

Lounsberry, Alice. *Sir William Phips: Treasure Fisherman and Governor of the Massachusetts Bay Colony.* New York: Charles Scribner's Sons, 1941.

Ludlum, David McWilliams. *Early American Hurricanes 1492–1870*. Boston: American Meteorological Society, 1963.

Lutz, Cora E. "Ezra Stiles and the Bones of the Giant of Claverack." *Yale University Library Gazette*. 57.1/2 (October 1982): 18–25.

Lyons, H.G. "John Winthrop (Junior), F.R.S. Governor of Connecticut 1660–1676." *NRRS* 3.1 (1940): 110–15.

Maddux, Clark Harry. "Euhemerism and Ancient Theology in Cotton Mather's 'Biblia Americana.'" In *Cotton Mather and Biblia Americana – America's First Bible Commentary*. Edited by Reiner Smolinski and Jan Stievermann. Tübingen, Germany: Mohr Siebeck, 2010. 337–59.

Majno, Guido. *The Healing Hand: Man and Wound in the Ancient World.* Cambridge, MA: Harvard University Press, 1975.

Mallery, Garrick. *Picture-Writing of the American Indians: Extract from The Tenth Annual Report of the Bureau of Ethnology*. Washington, D.C.: Government Printing Office, 1894.

Manning, Gideon and Cynthia Klestinec. Eds *Professors, Physicians and Practices in the History of Medicine: Essays in Honor of Nancy Siraisi*. Heidelberg: Springer, 2017.

Martin, Debra L. and Alan H. Goodman. "Health Condition before Columbus: Paleopathology of Native North Americans." *WJM* 176.1 (2002): 65–68.

Mares, Jeffrey. Editor. "Problema Theologicum." In "Cotton Mather's 'Problema Theologicum': An Authoritative Edition." *PAAS* 104, pt 2 (1994): 333–440.

Marvin, Abijah P. *The History of the Town of Lancaster, Massachusetts*. 3 vols. Lancaster: Published by the Town, 1879.

Marvin, Gerald. "Zabdiel Boylston: Medical Pioneer of Colonial Boston." Unpublished dissertation, 1975, University of Illinois at Urbana-Champaign.

Matthews, Albert. Editor. "Massachusetts Royal Commissions 1681–1774." *PCSM* 2 (1913) 2:315–16.

Mayor, Adrienne. *The First Fossil Hunters: Paleontology in Greek and Roman Times*. Princeton: Princeton University Press, 2000.

—*Fossil Legends of the First Americans*. Princeton: Princeton University Press, 2005.

McDonald, Grantley. *Biblical Criticism in Early Modern Europe: Erasmus, the Johannine Comma and Trinitarian Debate*. Cambridge, UK: Cambridge University Press, 2016.

McGiuire, J.E. "Force, Active Principles, and Newton's Invisible Realm." *Ambix* 15 (1968): 154–208.

McHenry, Lawrence C., Jr. and Ronald Mac Keith. "Samuel Johnson's Childhood Illnesses and the King's Evil." *MH* 10.4 (1966): 386–99.

McIntyre, Sheila & Len Travers. Editors. *The Correspondence of John Cotton Junior*. Boston: The Colonial Society of Massachusetts, 2009.

McKie, Douglas. "The Hon. Robert Boyle's 'Essays of Effluviums' (1673)." *SP* 29.114 (October 1934): 253–65.

McLaughlin, Peter. "Spontaneous versus Equivocal Generation in Early Modern Science." *AHPB* 10 (2005): 79–88.

Mendillo, Michael. *Saints and Sinners in the Sky: Astronomy, Religion and Art in Western Culture*. Zurich: Springer, 2022.

Middlekauff, Robert. *The Mathers: Three Generations of Puritan Intellectuals, 1596–1728*. New York: Oxford University Press, 1971.

Miller-Keane Encyclopedia and Dictionary of Medicine, Nursing, and Allied Health. 7th ed. Edited by Marie T. O'Toole. New York: Saunders, 2003.

Minardi, Margot. "The Boston Inoculation Controversy of 1721–1722: An Incident in the History of Race." *WMQ* 61.1 (2004): 143–75.

Molina-Venegas, Rafael, and Rodrigo Verano. "The Quest for Homer's *moly*: exploring the potential of an early ethnobotanical complex." *JEE* 20.11 (2024): 1–12.

Morens, David M. "The Past is Never Dead – Measles Epidemic, Boston, Massachusetts, 1713." *EID* 21.7 (July 2015): 1257–60.

Morison, Samuel Eliot. *Harvard College in the Seventeenth Century*. 2 vols. Cambridge, MA: Harvard University Press, 1936.

Morris, Amy. "Geomythology on the Colonial Frontier: Edward Taylor, Cotton Mather, and the Claverack Giant." *WMQ* 70.4 (2013): 701–24.

Müller, Uwe. "Johann Laurentius Bausch und Philipp Jacob Sachs von Lewenhaimb. Foundation of the Academica Naturae." *Acta Historia Leopoldina* 49 (2008): 243–64.

Murdock, Kenneth B. Editor. *Cotton Mather. Magnalia Christi Americana.* Cambridge, MA: The Belknap Press of Harvard University Press, 1977.

—*Selections from Cotton Mather*. New York: Harcourt, Brace, and Company, 1926.

Murrin, John M. "'Things Fearful to Name': Bestiality in Colonial America." *PAMA* (Special Issue: Explorations in Early American Culture). 65 (1998): 8–43.

Needham, Joseph. *A History of Embryology*. 1959; New York: Arno Press, 1975.

Oberhelman, Stephen M. "Galen, *On Diagnosis from Dreams*." *JHMAS* 38 (1983): 36–47.

Ockenden, R.E. "Marco Antonio de Dominis and His Explanation of the Rainbow. *Isis* 26.1 (1936): 40–49.

Oudemans, Antoon, C. *The Great Sea-Serpent. An Historical and Critical Treatise*. London, 1892.

Parrish, Susan S. *American Curiosity: Cultures of Natural History in the Colonial British Atlantic World.* Chapel Hill: The University of North Carolina Press, 2006.

Pätzold, Jürgen. *Town Hall and Roland on the Market-Place of Bremen*. N.p.: CRC Press, 2021.

Park, Katharine and Lorraine J. Daston. "Unnatural Conceptions: The Study of Monsters in Sixteenth- and Seventeenth-Century France and England." *P&P* 92.1 (1981): 20–54.

—*Wonders and the Order of Nature 1150–1750*. New York: Zone Books, 1998, 2001.

Paulys Real-Encyclopedie der Classischen Altertumswissenschaft. 49 vols. Stuttgart: Metzlersche Verlagsbuchhandlung, 1894–1967.

Pedersen, Olaf. *Early Physics and Astronomy: A Historical Introduction*. Cambridge, UK: Cambridge University Press, 1974.

Perley, Sidney. *Historic Storms of New England.* Salem, MA: The Salem Press Publishing Company, 1891.

Peterson, Mark. *The City-State of Boston: The Rise and Fall of an Atlantic Power. 1630–1865*. Princeton: Princeton University Press, 2019.

—"Theopolis Americana. The City-State of Boston, the Republic of Letters, and the Protestant International, 1689–1739." In *Soundings in Atlantic History. Latent Structures and Intellectual Currents, 1500–1830*. Edited by Bernard Bailyn and Patricia L. Denault. Cambridge, MA: Harvard University Press, 2011.

Pinto-Correia, Clara. *The Ovary of Eve: Egg and Sperm Preformation*. Chicago: University of Chicago Press, 1997.

Pitt, Steven J. "Cotton Mather and Boston's Seafaring Tribe." *NEQ* 85.2 (2012): 222–52.

Plane, Ann Marie, and Leslie Tuttle. Eds *Dreams, Dreamers, and Visions: The Early Modern Atlantic World.* Philadelphia: University of Pennsylvania Press, 2013.

Platt, Peter G. *Wonders, Marvels, and Monsters in Early Modern Culture.* Newark: University of Delaware Press, 2000.

Pollini, John. "Burning Rome, Burning Christians." *The Cambridge Companion to the Age of Nero*. Edited by Shadi Bartsch et al. Cambridge, UK: Cambridge University Press, 2017.

Popkin, Richard H. *Isaac La Peyrère: His Life, Work, and Influence*. Leiden: E.J. Brill, 1987.

Prybot, Peter K. "Ebb & Flow: Revisiting Lanesville's unique 'moving rock," *Gloucester Daily Times*, August 27 2010. https://www.gloucestertimes.com/news/local_news/ebb-flow-revisiting-lanesvilles-unique-moving-rock/article_b10b1073-978b-5aa8-a1ad-8167c88630fe.html. (Accessed 31 January 2024).

Quincy, Josiah P. "The Supernormal in New England History." *PrMHS* 40 (1906–07), 439–53.

Rappaport, Rhoda. "Jokes of Nature and Jokes of Knowledge: The Playfulness of Scientific Discourse in Early Modern Europe." *RQ* 43.2 (1990): 292–331.

—*When Geologists Were Historians, 1665–1750*. Ithaca and London: Cornell University Press, 1997.

Rather, L. J. "Thomas Fienus' (1567–1631) Dialectical Investigation of the Imagination as Cause and cure of Bodily Disease." *BHM* 41.4 (1967): 349–67.

Records of the First Church in Boston, 1630–1868. Edited by Richard D. Pierce. *PCSM* 39. Boston: CSM and Anthoensen Press, 1961.

Reeves, Thomas C. "John Checkley and The Emergence of The Episcopal Church in New England." *Historical Magazine of the Protestant Episcopal Church* 34.4 (1965): 349–60.

Riches, Sir Eric. "The History of Lithotomy and Lithotrity." *ARCS* 43.4 (1968): 185–199.

Rivière, Janine. "Demons of Desire or Symptoms of Disease? Medical Theories and Popular Experiences of the 'Nightmare' in Premodern England." In Plane, Ann Marie et al. (2013): 49–71.

Robinson, John, and George F. Dow. *The Sailing Ships of New England, 1607–1907*. Salem, MA: Marine Research Society, 1922.

Roe, Shirley A. *Matter, Life, and Generation. 18th-Century Embryology and the Haller-Wolf Debate*. Cambridge, UK: Cambridge University Press, 1981.

Rogers, Pat. "'The Monster of Ragusa': Pope, Addison, and Button's Wits." *RES* 67.280 (2016): 496–522.

Rollins, Hyder E. "Notes on Some English Accounts of Miraculous Fasts." *JAF* 34.134 (October–December 1921): 357–76.

Rossi, Paolo. *The Dark Abyss of Time: The History of the Earth and the History of Nations from Hooke to Vico*. 1979; Chicago: University of Chicago Press, 1984.

Ruestow, Edward G. "Images and Ideas: Leeuwenhoek's Perception of the Spermatozoa." *JHB* 16.2 (1983): 185–224.

Rusnock, Andrea. Editor. *The Correspondence of James Jurin* (1684–1750). Amsterdam and Atlanta: Editions Rodopi, 1996.

—*Vital Accounts: Quantifying Health and Population in Eighteenth-Century England and France*. Cambridge, UK: Cambridge University Press, 2002.

Sanchez-Ramos, Juan R. "The rise and fall of tobacco as botanical medicine." *JHM* 22 (2020): 1–7.

Satterfield, Susan. "Livy, Cassius Dio, and Prodigies in the Early Roman Empire," *Klio. Beiträge zur Alten Geschichte* 104.1 (2022): 253–76.

Scarborough, John. "Galen's Investigations of the Kidney." *CM* 11 (1976): 171–77.

Scribner, Vaughn. "Mermaids and Mermen." *History Today* 68.5 (May 2018). https://www.historytoday.com/archive/feature/mermaids-and-mermen Accessed January 31 2024.

—*Merpeople: A Human History*. London: Reaction Books Ltd., 2020.

—"'Such Monsters Do Exist in Nature': Mermaids, Tritons, and the Science of Wonder in Eighteenth-Century Europe." *Itinerario* 41.3 (2017): 507–38.

Schleiner, Winfried. *Medical Ethics in the Renaissance*. Washington, D.C.: Georgetown University Press, 1995.

Schorger, Arlie W. "Unpublished Manuscripts by Cotton Mather on the Passenger Pigeon." *The Auk* 55 (July 1938): 471–77.

Se-Hwee Oh, Hamilton. "Organ aging signatures in the plasma proteome track health and disease." *Nature* 624 (December 6 2023): 164–72.

Shannon-Henderson, Kelly E. "Phlegon's Paradoxical Physiology: Centaurs in the Peri Thaumasion." *Medicine and Paradoxography in the Ancient World*. Ed. by George Kazantzidis. Berlin: De Gruyter, 2019. 141–62.

Silva, Cristobal. *Miraculous Plagues. An Epidemiology of Early New England Narrative*. New York: Oxford University Press, 2011.

Seeman, Erik. *Speaking with the Dead in Early America*. Philadelphia: University of Pennsylvania Press, 2019.

Semonin, Paul. *American Monster: How the Nation's First Prehistoric Creature Became a Symbol of National Identity*. New York: New York University Press, 2000.

Shapin, Steven, and Simon Schaffer. *Leviathan and the Air-Pump: Hobbes, Boyle, and the Experimental Life*. Princeton: Princeton University Press, 1985.

Shea, W.R.J. "Galileo's Claim to Fame: The Proof that the Earth Moves from the Evidence of the Tides." *BJHS* 5.2 (December 1970): 111–27.

Shroeder, John W. "Hawthorne's 'Egotism; or, The Bosom Serpent' and Its Source." *AL* 31 (May 1959): 150–62.

Sievers, Julie. "Drowned Pens and Shaking Hands: Sea Providence Narratives in Seventeenth-Century New England." *WMQ* 63.4 (October 2006): 743–76.

Silva, Cristobal. *Miraculous Plagues. An Epidemiology of Early New England Narrative*. New York: Oxford University Press, 2011.

Silverman. Kenneth. *The Life and Times of Cotton Mather*. New York: Harper & Row, Publishers, 1984.

—*Selected Letters of Cotton Mather*. Compiled with Commentary by Kenneth Silverman. Baton Rouge: Louisiana State University Press, 1971.

Siraisi, Nancy. *Taddeo Alderotti and his Pupils*. Princeton: Princeton University Press, 1981.

Siscoe, George L. "An historical footnote on the origin of 'Aurora Borealis." *TAGU* 58.12 (1978): 994–97.

Sivils, Matthew Wynn. "Dissecting the Pamphlet Literature of the Boston Smallpox Inoculation Controversy." *LM* 29.1 (2011): 39–57.

Slafter, Edmund F. "Memoir," in *John Checkley; or the Evolution of Religious Tolerance in Massachusetts* Bay. 2 vols. [Publications of the Prince Society]. Boston, 1897.

Smith, C.M. "Origin and Uses of *Primum Non Nocere – Above All, Do No Harm!*" *The Journal of Clinical Pharmacology* 45.4 (2005): 371–77.

Smith, Frank. *A History of Dedham, Massachusetts*. Dedham, MA: Transcript Press, 1936.

Smolinski, Reiner. Editor. *Cotton Mather. Biblia Americana. America's First Commentary on the Old and New Testaments*. Vol. 1: Genesis. Tübingen: Mohr Siebeck, 2010.

—*Cotton Mather. Biblia Americana. America's First Commentary on the Old and New Testaments*. Vol. 2: Genesis. Tübingen: Mohr Siebeck, 2019.

—*The Threefold Paradise of Cotton Mather: And Edition of 'Triparadisus'*. Athens and London: University of Georgia Press, 1995.

Smolinski, Reiner and Jan Stievermann. Eds *Cotton Mather and Biblia Americana: America's First Bible Commentary. Essays in Reappraisal*. Tübingen: Mohr Siebeck, 2010.

Smolinski, Reiner and Kenneth P. Minkema. Eds *A Cotton Mather Reader*. New Haven: Yale University Press, 2022.

Solberg, Winton U. Editor. *The Christian Philosopher*. Urbana and Chicago: University of Illinois Press, 1994.

Spiller, Robert E. "The Verdict of Sydney Smith." *AL* 1.1 (1929): 3–13.

Spinks, Jennifer. *Monstrous Births and Visual Culture in Sixteenth-Century Germany*. London and New York: Routledge, 2009.

Springborg, Patricia. "Hobbes and Cluverius." *HJ* 39.4 (1996): 1075–78.

Stanford, Donald E. "The Giant Bones of Claverack, New York, 1705." *New York History* 40.1 (January 1959): 47–61.

Stannard, David E. *American Holocaust: The Conquest of the New World*. New York: Oxford University Press, 1992.

Stearns, Raymond P. "Colonial Fellows of the Royal Society of London, 1661–1788." *NRRS*, 8.2 (April 1 1951): 178–246.

—"James Petiver. Promoter of Natural Science, c. 1663–1718." *PAAS* 62 (October 1952): 243–365.

—"Joseph Kellogg's Observations on Senex's Map of North America (1710)." *MVHR* 23.3 (1936): 345–54.

—"Remarks upon the Introduction of Inoculation for Smallpox in England." *BHM* 24.2 (1950): 103–22.

—*Science in the British Colonies of America*. Urbana: University of Illinois Press, 1970.

Stievermann, Jan. "The Genealogy of Races and the Problem of Slavery in Cotton Mather's 'Biblia Americana.'" In R. Smolinski and J. Stievermann. Eds (2010), pp. 515–76.

Stern, Tiffany. "'This Wide and Universal Theatre': The Theatre as Prop in Shakespeare's Metadrama." *Shakespeare's Theatres and the Effects of Performance*. Edited by Farah Karim-Cooper and Tiffany Stern. London: Bloomsbury Publishing Plc., 2014. 11–32.

Stothers, Richard. "Ancient Aurorae." *Isis* 70.1 (March 1979): 85–95.

Sweet, Timothy. *Extinction and the Human: Four American Encounters*. Philadelphia: University of Pennsylvania Press, 2021.

Thacher, James. *American Medical Biography: or Memoirs of Eminent Physicians who have Flourished in America*. Boston: Richardson & Lord and Cottons & Barnard, 1828. 1:185–92.

Tindol, Robert. "Getting the Pox off All Their Houses: Cotton Mather and the Rhetoric of Puritan Science." *EAL* 46.1 (2011): 1–23.

Tuttle, Julius H. "The Libraries of the Mathers." *PAAS*. New Series. 20 (1912): 269–356.

—"William Whiston and Cotton Mather." *TCSM* 13 (1912): 197–207.

Upton, Thomas V. "Aristotle on Monsters and the Generation of Kinds." *ACPQ* 77.1 (2003): 21–36.

Vail, R.W.G. "Random Notes on the History of the Early American Circus," *PAAS* 43.1 (April 1933): 119–20.

Vartanian, Aram. “Trembley’s Polyp, La Mattrie, and Eighteenth-Century French Materialism.” *JHI* 11.3 (1950): 259–86.

Veith, Ilza. “Michael Berhard von Valentini Non-Specialist in Medicine and Science: Physician of the Enlightenment.” *BHM* 52.1 (Spring 1978): 96–101.

Vermij, Rink. “Translating, Adapting, Mutilating: Or, How to Make an Enlightenment Classic.” *Isis* 109.2 (2018): 333–38.

Von Arnim, Johannes. Editor. *Stoicorum Veterum Fragmenta.* 2 vols. Stuttgart: Teubner, 1964.

Warner, Margaret H. “Vindicating the Minister’s Medical Role: Cotton Mather’s Concept of the Nishmath-Chajim and the Spiritualizing of Medicine.” *JHMAS* 36.3 (July 1981): 278–95.

Watson, Patricia A. *The Angelical Conjunction: The Preacher-Physician of Colonial New England.* Knoxville: University of Tennessee Press, 1991.

Weis, Frederick Lewis. *The Colonial Clergy and the Colonial Churches of New England.* Lancaster, MA: N.p, 1936.

Weishampel, David B. and Nadine M. White. Eds *The Dinosaur Papers 1676–1906.* Washington and London: Smithsonian Books, 2003.

Werrett, Simon. “Wonders Never Cease: Descartes’s “Météores” and the Rainbow Fountain.” *BJHS* 34.2 (2001): 129–47.

Weyburn, Samuel Fletcher. *Weyburn-Wyborn Genealogy: Being a History and Pedigree of Thomas Wyborn of Boston and Scituate, Massachusetts.* New York: Frank Allaben Genealogical Company, 1911.

Wickham, Parnel. “Conceptions of Idiocy in Colonial Massachusetts.” *JSH* 35.4 (2002): 935–54.

—“Idiocy and the Construction of Competence in Colonial Massachusetts.” In *Children in Colonial America.* Edited by. James A. Marten. New York: New York University Press, 2007. 141–54.

Wickman, Thomas. “The Great Snow of 1717.” *NN* 24 (2017), Special Issue 7. H81–H114.

Williams, Gareth. *The Angel of Death: The Story of Smallpox.* New York: Palgrave Macmillan, 2010.

Wilson, Philip K. “‘Out of Sight, Out of Mind?’ The Daniel Turner-James Blondel Dispute Over the Power of the Maternal Imagination.” *AS* 49.1 (1992): 63–85.

Winship, Michael P. “Prodigies, Puritanism, and the Perils of Natural Philosophy: The Example of Cotton Mather.” *WMQ* 51.1 (1994): 92–105.

—*Seers of God: Puritan Providentialism in the Restoration and Early Enlightenment* Baltimore: Johns Hopkins University Press, 1996.

Winslow, Ola E. *A Destroying Angel: The Conquest of the Smallpox in Colonial Boston.* Boston: Houghton Mifflin Company, 1974.

Winterer, Caroline. “Where is America in the Republic of Letters?” *MIH* 9.3 (November 2012): 597–623.

Wisecup, Kelly. "African Medical Knowledge, the Plain Style, and Satire in the 1721 Boston Inoculation Controversy." *EAL* 46.1 (2011): 25–50.

—"African Testimony, Dangerous Communications, and Colonial Medical Knowledge in the 1721 Boston Inoculation Controversy." *Medical Encounters: Knowledge and Identity in Early American Literatures*. Amherst: University of Massachusetts Press, 2013. 97–126.

—*Medical Encounters. Knowledge and Identity in Early American Literature*. Amherst: University of Massachusetts Press, 2013.

Wolf, Abraham. *A History of Science, Technology, and Philosophy in the 16th & 17th Centuries*. 2 vols. London: George Allen & Unwin, 1935.

—*A History of Science, Technology, and Philosophy in the 18th Century*. 2 vols. London: George Allen & Unwin, 1938.

Woods, Henry Ernest. Editor. *Vital Records of Newton, Massachusetts, to the Year 1850*. Boston: New England Historic Genealogical Society, 1905.

Woodward, Samuel B. "The Story of Smallpox in Massachusetts." *NEJM* 206.23 (1932): 1182–91.

Woodward, Walter W. *Prospero's America: John Winthrop, Jr., Alchemy, and the Creation of New England Culture, 1606–1676*. Chapel Hill: University of North Carolina Press, 2010.

Zirkle, Conway. *The Beginnings of Plant Hybridization*. Philadelphia: University of Pennsylvania, 1935.

—"More Records of Plant Hybridization before Koelreuter." *JH* 25 (1934): 3–18.

—"Plant Hybridization and Plant Breeding in Eighteenth-Century American Agriculture." *AH* 43.1 (1969): 25–38.

Index of Names

Index of Subjects